Australia
State of the Environment
1996

Australia
State of the
Environment
1996

An independent report

presented to the

Commonwealth Minister for the Environment

by the

State of the Environment Advisory Council

STATE OF THE
ENVIRONMENT
REPORTING

Australia: State of the Environment 1996
Cataloguing-in-Publication Data

Bibliography.
Includes index.
1. Environmental policy — Australia. 2. Natural resources —
Australia. 3. Man — Influence on nature — Australia.
I. Australia. Dept. of the Environment, Sport and Territories.
II. State of the Environment Advisory Council (Australia).

ISBN 0 643 05830 3

This book is printed on paper made in Australia from 100% waste
material. The fibre is waste paper and cotton. Both are renewable
resources and have been recovered from other industries.

The information in this report is current at the beginning of 1996.

Published by:

CSIRO Publishing
150 Oxford Street
Collingwood 3066
Australia

Managing Editor: Nicholas Alexander
Editor: Robin Taylor
Editorial assistance: Yvonne Roberts
Editorial consultant: Richard Eckersley
Layout and design: Tony Yap
Design concept: Melissa Spencer
Illustrations: Tibor Hegedis and Angela Halpin
Print management: Roy Osborne
Pre-press: Electronic Graphic Output Pty Ltd
Printer: A. E. Keating (Printing) Pty Ltd
Cover art: City Graphics
Cover photo: The Bookleaf Pea Flower (*Daviesia cordata*) is one of 100
    or so species found in Western Australia. It is common from
    Toodyay to the South Coast and grows to 1.5m, flowering between
    October and November.
Photo credits: (*front cover*) Lochman Transparencies
    (*back cover*) (Cornfield) FPG International Stock Photography
    (Ayers Rock/Uluru) The Photo Library of Australia
    (Gold Coast) Australian Picture Library
    (Fish/coral) Australian Picture Library
    (Red-eyed tree frog) The Image Bank
    (Jim Jim Creek, NT) DFAT Photo Library

This book is available from:
CSIRO Publishing, PO Box 1139, Collingwood 3066, Australia

Tel: (03) 9662 7666; Int +(613) 9662 7666
or 1800 645 051 toll free in Australia

Fax: (03) 9662 7555; Int +(613) 9662 7555
e-mail: sales@publish.csiro.au

Foreword

Over the last two decades, both in Australia and overseas, community awareness of the environment and expectations of a healthy environment have been growing. Through international agencies such as the United Nations Commission for Sustainable Development, the United Nations Environment Program and the Organisation for Economic Cooperation and Development the nations of the world are beginning to face up to the environmental challenges of the twenty-first century. Regular reporting on the state of the environment is being adopted across the globe as an important tool for governments and communities to better understand their environment. A community better informed about the pressures on and condition of its environment is a community better able to manage its environmental challenges successfully.

The National Strategy for Ecologically Sustainable Development adopted by all Australian Governments in 1992 calls for introduction of regular national state of the environment reporting to enhance the quality, accessibility and relevance of data relating to ecologically sustainable development. Following extensive public consultation a framework for state of the environment reporting for Australia was released in June 1994. This report is the first based on this reporting framework.

Australia has a great breadth and depth of internationally recognised environmental expertise. This document is a distillation of this expertise and represents an independent report to the Australian community by the State of the Environment Advisory Council, supported by seven expert reference groups. Approximately 170 of the nation's leading experts have worked under the broad direction of the Advisory Council on preparing and refereeing the main chapters of the report and the supporting technical papers. Many other people contributed by providing information and advice on particular subjects. The result is the best available assessment of the overall state of Australia's environment at the beginning of 1996. It carries important messages for governments and the Australian community.

Many Commonwealth and State/Territory government agencies, industry and community groups cooperated in providing information for this report. Drafts were widely circulated within Commonwealth and State/Territory agencies for review and many valuable comments were incorporated.

A feature of this report is its very innovative approach — it has been a pioneering enterprise in many respects. It is the first to cover all Australia and to use the internationally widespread approach of reporting on pressures on the environment, the condition of the environment and societal responses to these pressures and conditions. It has pioneered the recognition of natural and cultural heritage as part of our environment, and has taken a highly innovative approach to reporting on human settlements. It combines environmental, social and economic indicators into measures of 'livability' to reflect the condition of human society as a part of the environment. The importance of managing natural resources within the context of their ecosystems has come through as a major message of the report and the status of Australia's major ecosystems is presented.

As with all pioneering enterprises these first attempts are not perfect. In many cases national information is lacking or scientific knowledge is not advanced enough to do more than lay a foundation for future reports to build upon.

Contents

GRIFFITH UNIVERSITY

FACULTY OF SCIENCE AND TECHNOLOGY
Nathan Campus Kessels Road Nathan Brisbane
Please Contact
Telephone (07) 3875   Fax (07) 3875 7656
Reference

Minister for the Environment
Parliament House
Canberra

It is with pleasure that I present the first ever independent and comprehensive report on the state of Australia's environment. It has been prepared by the State of the Environment Advisory Council and seven expert reference groups. With the added assistance of referees and consultants, the report has drawn on the expertise of a broad section of the Australian scientific community. The whole task has taken more than two years.

The report shows that Australia has a beautiful, diverse and unique environment. It finds that some aspects of Australia's environment are in good condition by international standards. Indeed, our approach to environmental management has won international recognition in several areas.

Unfortunately, the report also shows that Australia has some serious environmental problems. These need to be tackled with determination if we are to achieve the goal of ecological sustainability, as is the agreed aim of the Council of Australian Governments.

I want to emphasise that no single government or sector is to blame for the problems identified in this report. They are the overall consequences of a range of factors: population growth and distribution, lifestyles, technologies and our demands on natural resources.

I also want to stress that the report is not a catalogue of problems. Overall, it is very positive. Most of the problems identified do have solutions. The report details many successful initiatives to improve the state of our environment.

It is the hope of the Advisory Council that the report will be of use to the general community, managers and policy-makers. It should help all of these people to be better informed about the state of our environment, the pressures we exert on it and the effectiveness of our responses. The report provides a guide to priorities for all those who make decisions affecting our environment.

I am pleased to commend the report to you and, through you, to the people of Australia.

Ian Lowe
Chair
State of the Environment Advisory Council

23 May 1996

Members of the State of the Environment Advisory Council

Ian Lowe	(Chair) Professor of Science, Technology and Society, School of Science, Griffith University
Don Blesing	Agribusiness adviser and farmer
Brian Boyd	Australian Council of Trade Unions Environment Committee and Victorian Trades Hall Council
Peter Cullen	Director, Co-operative Research Centre for Freshwater Ecology
Joan Domicelj	Principal, Domicelj Consultants
Roy Green *(until February 1996)*	Chief Executive, CSIRO
David James	Principal, Ecoservices Pty Ltd
Bill Jonas	Principal, Australian Institute of Aboriginal and Torres Strait Islander Studies
Michael Kennedy	Director, Humane Society International
Oleg Morozow	Manager, Environmental Affairs, Santos Limited
Graeme Pearman *(from February 1996)*	Acting Director, CSIRO Institute of Natural Resources and the Environment
John Towns	Chief Executive, ANI Krüger Pty Ltd
Virginia Walsh	Executive Director, Australian Library and Information Association
David Yencken	Elisabeth Murdoch Professor of Landscape Architecture and Environmental Planning, The University of Melbourne
Margaret Clarke* *(ex officio member)*	Assistant Secretary, Department of the Environment, Sport and Territories
Allan Haines *(ex officio member)*	Director, State of the Environment Reporting Unit Secretary to Council.

* replaced Nicola Pain in April 1995.

Acknowledgment

More than 50 experts in seven expert reference groups contributed to preparing the main sector chapters of the report. Their names are listed at the beginning of each of their chapters.

A large number of people also provided information, usually at very short notice. They included individuals in State and Territory Government departments, private industry, and voluntary organisations. Mention must also be made of the assistance provided by the CSIRO, the Environmental Resources Information Network, the Bureau of Resource Sciences and the Australian Bureau of Statistics. Their assistance is also gratefully acknowledged. Fifty external reviewers examined the report's main chapters in draft form. In addition, Commonwealth Government departments and members of the Commonwealth/State ANZECC State of the Environment Reporting Taskforce also helped identify errors of fact or omission.

State of the Environment Reporting Unit: (past) and present members

(Shaun Andrews), John Higgins, (Andrew James), (Marion Kelly), (Fern MacLachlan), Mathew Maliel, Ian Robertson, (Michelle Smith), Allan Spessa, (David Tait), (Karena Walls), Stephen Watson-Brown and (Charlie Zammit).

Public Affairs Branch:

Tracey Bell, Jennifer Bray, Patrick Fletcher, Bill Haniotis, Leona Jorgensen and Kylie Preece.

Executive Summary

Contents

Overall Message

- This is the first ever independent and comprehensive State of the Environment Report for Australia. It links land, water, air, plants and animals, human settlements and how we value them.

- An independent advisory council and seven expert groups prepared the report. It draws on the knowledge and skills of more than 200 eminent scientists and other experts.

- The report shows that Australia has a beautiful, diverse and often unique environment which is a priceless heritage and should be a source of pride to all Australians.

- Some aspects of the Australian environment are in relatively good condition by international standards. In some areas our approach to environmental management has won international recognition.

- In many other areas it is not possible to decide whether our environmental management is adequate. We urgently need better information and understanding, which will require data collection and research.

- The report also shows that Australia has some very serious environmental problems. If we are to achieve our goal of ecological sustainability, these problems need to be dealt with immediately. This will be no small task.

- The problems are the cumulative consequences of population growth and distribution, lifestyles, technologies and demands on natural resources over the last 200 years and more.

- No single government or sector is to blame for these problems. We are all responsible. Changes are needed in government policies and programs, corporate practices and personal behaviour.

- Australians are among the most environmentally aware people in the world. All sections of the community now recognise the need to do more to tackle environmental issues.

- Most of the problems identified in the report do have solutions. The report details many positive and successful initiatives.

- Our actions have been most effective where they have taken a comprehensive and systematic approach, integrating different aspects of the overall problem. By contrast, failures tend to be piecemeal efforts that treat symptoms rather than underlying causes.

- Australia has an international responsibility to protect its rich biological diversity and its unique environmental features such as the Great Barrier Reef and other World Heritage Areas. We also have a national responsibility towards future generations of Australians.

- Australia has a better opportunity than perhaps any other nation to protect its environment and use its natural and heritage resources sustainably. We need to do much more if we are not to lose this opportunity.

- Progress towards ecological sustainability requires recognition that human society is part of the ecological system and integration of ecological thinking into all social and economic planning.

Towards ecological sustainability

Sustainable development

Sustainable development is arguably the central issue of our time. Its basic aim is to meet the needs of the present without compromising the ability of future generations to meet their own needs. Support for this goal is now widespread. Following extensive consultation with all community sectors, Australian governments adopted in 1992 a National Strategy for Ecologically Sustainable Development (ESD).

The strategy defines ESD as a pattern of development that improves the total quality of life, both now and in the future, in a way that maintains the ecological processes on which life depends. One of the functions of this report is to assess progress towards the goal of ecological sustainability (see the box on page ES-6 for more details about the approach taken in the report, including its objectives).

Information needs

State of the Environment reporting is an important step in the essential process of refining the knowledge base on which decisions about the environment are made. That base is currently inadequate. For example, while we believe that more than 90 per cent of vertebrates and higher plants in Australia are identified and described, it is estimated that only about 50 per cent of the invertebrates and simpler plants are identified. We know even less about other species such as fungi and bacteria. With such limited knowledge, it is impossible to assess the impact of human activity on biodiversity — a critical aspect of ecosystem health and resilience.

Australia lacks the integrated national systems and databases to measure environmental quality, manage it and evaluate the effectiveness of that management. Until these deficiencies are rectified, we will remain unable to truly answer the question of whether our pattern of development is genuinely sustainable.

Our lack of knowledge and understanding of environmental issues emerges again and again in the report as a major obstacle to sound environmental management.

What we need to sustain

Sustainable development requires the maintenance of the following three key components of the environment:

- *biodiversity:* the variety of species, populations, habitats and ecosystems

- *ecological integrity:* the general health and resilience of natural life-support systems, including their ability to assimilate wastes and withstand stresses such as climate change and ozone depletion

- *natural capital:* the stock of productive soil, fresh water, forests, clean air, ocean, and other renewable resources that underpin the survival, health and prosperity of human communities

Global environmental context

We now recognise that we are part of a global ecosystem and an increasingly integrated global economy. This poses a new set of questions for sustainable development. Issues such as ozone depletion and climate change demand global responses. International action is succeeding in reducing the use of ozone-depleting substances. Halting and reversing the trends in greenhouse gas emissions will be much harder, requiring a fundamentally different approach to the use of energy.

Agenda 21 — a program of action for sustainable development worldwide — was adopted in 1992 by more than 178 governments, including Australia, at the UN Conference on Environment and Development. It highlights the pressures on the natural environment from population growth and associated poverty in the developing world, and unsustainable patterns of consumption in the industrialised world. Increasing global pressures to reduce resource consumption and waste production are a part of Australia's future.

There are various indications of global environmental stress. More than two-thirds of the world's bird species are in decline, vulnerable or threatened with extinction. Frog populations appear to be declining, although it is uncertain why this is happening. Coral reefs appear to be in trouble. Globally, key indicators of food production per head are falling, including production of beef and mutton, cereals and seafoods.

The Australian environment: constraints and driving forces

Australia's natural environment is a product of its geological history and the manifold impacts of human activity. Any analysis of the state of the environment has to begin from an appreciation of the unique geological, physical and biological characteristics of the country which constrain our actions. It has also to consider the driving forces which are altering the state of the natural world, including population size and distribution, and the patterns of human activity.

Australia is an old, weathered, eroded landscape, flat and generally dry, with a highly variable climate, especially rainfall. Its unique plants and animals reflect its long isolation from other land masses and their wildlife.

The population of Australia when Europeans arrived is estimated to have been between 300 000 and 1.5 million. Seventy years later, the European population had reached one million. By 1995, the total population stood at 18 million. Adding to the environmental pressures created by the increasing permanent population is the growing influx of tourists, who are often attracted to our most precious natural and cultural heritage areas. Other factors that influence our impact on the environment include the distribution of people

(Australia is one of the most urbanised of all nations, with most people living in a few large coastal cities), our lifestyles and our use of technology.

For at least 50 000 years, people have lived here and modified the environment, harvesting its biological resources, using fire to alter it, changing its shape and imposing their patterns of settlement. Since 1788, the human impact has increased dramatically, because of both the rapid growth in population and the introduction of farming, forestry, mining and large towns and cities.

Climate variability has been a constant problem for Australians because it causes natural fluctuations in vegetation and fauna. It is not this variability, or other features such as the poor soils, that puts pressure on our environment, but human activities that fail to take sufficient account of these features.

Two fundamental difficulties hinder the development of appropriate responses to environmental problems. First, we may not know the cause or causes, which can involve a complex interaction between a variety of factors; and second, there can be a long time lag between changes in human activities and any observable differences in natural systems.

Even where we are clear about the problems and what needs to be done, our institutional arrangements — such as those between and within governments — make it difficult to deliver coordinated responses.

Background to the report

State of the environment reporting

The principle of ecologically sustainable development is now widely accepted and supported in Australia. It acknowledges that our way of life depends critically on a range of natural assets — air, soil, water, forests and other biological systems — and that these assets must be safeguarded.

Sustainable development is impossible without adequate and accessible information about the environment. State of the Environment (SoE) reporting is a powerful tool for providing this information — to the public, industry, non-government organisations and all levels of government.

It allows regular reports on agreed sets of national indicators (once these are established) of changes and trends in environmental conditions, in much the same way as well-accepted economic indicators are used to report on the state of the economy. It describes the effects of human activities on the environment, and their implications for human health and economic well-being. It also provides an opportunity to monitor the performance of government policies against actual outcomes.

Thus SoE reporting can act as a report card on the condition of the environment and natural resource stocks.

Principles and objectives

Six principles, embracing rigour, objectivity, cooperation, openness, global vision, and ecological sustainability, guide SoE reporting. Key objectives include:

- To provide accurate, timely and accessible information about the condition and prospects of the Australian environment.

- To increase public understanding of these issues.

- To facilitate the development of an agreed set of national environmental indicators, and to review and report on these indicators.

- To provide an early warning of potential problems.

- To report on the effectiveness of policies and programs designed to respond to environmental change, including progress towards achieving environmental standards and targets.

Approach

The approach to SoE reporting in Australia is based on a modified version of the OECD's 'pressure-state-response' model. The model is based on the concept of causality: human activities exert pressures on the environment; these change its state or condition; society responds by developing or implementing policies that influence those human activities, and so change the pressures. Australia has modified this model to include cultural aspects of the environment, to recognize the inherent variability and lack of knowledge about the Australian environment and to allow for an interactive rather than a linear model.

Indicators

Developing a nationally agreed set of indicators for Australia is a high priority for SoE reporting. These are physical, chemical, biological or socio-economic measures that can be used to assess natural resources and environmental quality. Preparing this SoE report has involved selecting a 'first generation' of environmental indicators. These indicators provide a good foundation for future development.

Structure

Chapter 1 sets out the background to the report and its purpose and is summarised in this box.

Chapter 2 gives a general description of Australia's physical, biological, social and cultural landscape, and of how we have come to our present situation.

Chapters 3 to 9 discuss in detail the key facets of the contemporary environment, employing the pressure-state-response approach. The seven aspects of the environment are: human settlements, biodiversity, atmosphere, land resources, inland waters, estuaries and the sea, and natural and cultural heritage.

Chapter 10 draws together the major findings to give the State of the Environment Advisory Council's overall assessment of our progress towards ecological sustainability.

Key issues

The report identifies several issues that are critical to assessing and improving the state of Australia's environment:

A systems perspective

Our responses need to embrace a systems approach that reflects the complexity of the natural world and the cultural values associated with it. There is little likelihood of a coherent policy emerging from the traditional compartmentalised approach in which different departments or different levels of government each handle different, small parts of the problem.

Biodiversity

Biodiversity is the variety of all life forms: the different plants, animals and micro-organisms, their genes and the ecosystems of which they are a part. Australia has one of the widest ranges of biodiversity of any nation, with a large proportion of its species found nowhere else in the world. The major threats to biodiversity are: land clearing; loss of native forests; introduced species; the absence of some representative ecosystems in national parks and other reserves; and the lack of knowledge about our biodiversity.

Water

How we use the land affects the state and use of inland waterways, which in turn affect estuaries and the sea. A systems approach is essential: land management is more likely to be effective if it is based on biophysical regions, water management on integrated catchments, and oceans on an integrated approach to marine ecosystems. Water-related issues include scarcity, quality and nutrient loads, waste water disposal, rising groundwater and associated salinisation, and pollution of coastal waters.

Land degradation

There is widespread concern about the degradation of farm land. Pressures include clearing, over-stocking, cropping on marginal land, irrigation, and introduced species such as rabbits and goats. These pressures lead to a range of problems including erosion, salinisation, acidification, waterlogging and poor soil structure.

Global climate change

Global warming and other climatic changes which result from increased emissions of greenhouse gases pose a serious problem, both in terms of the direct impacts and the potential to aggravate other environmental problems such as biodiversity loss. While Australia's total emissions of greenhouse gases are small in global terms, our per capita emissions are among the highest in the world. We do not appear to be making much progress in stabilising, let alone reducing, these emissions.

The urban environment

While the environmental problems of our large cities may be less serious than those of many smaller settlements in terms of the impact per person, they are among our most important environmental concerns because of the scale of human activity these cities represent. Issues include: the dispersed, low-density character of our cities (and coastal development), and the resultant dependence on the private car for transport; urban waste, including vehicle emissions, stormwater and garbage; the destruction of remnant native vegetation and habitats; the continuing growth of urban settlements; and the protection of heritage in urban areas.

Social and cultural issues

The biophysical environment is imbued with cultural significance and encompasses the dwellings and settlements in which we live. The report identifies four key social and cultural issues: various forms of social stress such as emerging pockets of poverty and high unemployment; the loss of physical and cultural well-being of many indigenous Australians; the loss of cultural continuity under conditions of rapid change; and the inadequate protection of heritage.

These issues can affect the natural environment as well as being important in their own right. One criterion for sustainable development is social integrity; for this reason alone, social problems deserve attention. In the case of the first of the four key social and cultural issues a growing number of 'pockets of poverty' pose potential problems for the social environment, and a reduction in the size of households is contributing to urban sprawl. The other three issues expose inadequacies in the conservation of our natural and cultural heritage. Losing our past makes it more difficult to evaluate our future. Sustainable development will only be possible if it is firmly based on our natural and cultural heritage.

Overview of the state of the environment

The report aims to provide an objective picture of the state of Australia's environment. It contains both good news and bad about the state, and reveals some responses that are exemplary and others that are ineffective. Both positive and negative aspects are important in strengthening our resolve to achieve a development trajectory that is genuinely sustainable. One — the good news and the successes — offers hope; the other — the problems and the failures — encourages urgency.

The good news

In some areas Australia's environment is not threatened, and is in relatively good condition, especially compared with most other parts of the world.

- Unlike most industrial nations, we have no significant problem with sulfur dioxide and acid rain. Levels of some urban air pollutants, including lead, carbon monoxide, nitrogen oxides and large particulates, have declined in recent years.

- Urban drinking water quality is generally very good, as is the standard of our food, with low levels of chemical residues and metals.

- Oceans and estuaries, away from major cities and developed coastal areas, are in relatively good shape.

- Our urban housing is generally of good quality, and a well-established system exists for protecting significant places.

The bad news

Elsewhere, however, we are experiencing serious environmental problems.

- The loss of biological diversity is perhaps our most serious environmental problem. Whether we look at wetlands or saltmarshes, mangroves or bushland, inland creeks or estuaries, the same story emerges. In many cases, the destruction of habitat, the major cause of biodiversity loss, is continuing at an alarming rate.

- In cities, transport systems, stormwater and sewage and other waste disposal continue to have substantial adverse impacts on the environment, including biodiversity and water quality.

- Inland waters in southern Australia are in poor shape, largely because of poor management. Too much water is being taken from some systems, and nutrient and salt levels are of concern. Algal blooms may be becoming a more serious problem. In northern Australia, there is still time to prevent deterioration.

- The hole in the protective ozone layer over the Antarctic is growing larger and deeper, exposing humans and other species to increased levels of harmful ultra-violet radiation. Present indications suggest that the layer will slowly recover.

- Soil erosion from agricultural land remains a problem, especially given our poor, shallow soils and low rates of soil formation. Other forms of degradation, such as salinisation, are also not being adequately addressed. The continuing productivity of our agriculture may be at risk.

- Some aspects of the environment experienced by indigenous Australians remain poor. Indigenous people have not shared the improvements in health enjoyed by other Australians over the past 50 years. Their health remains a serious problem that is yet to be adequately addressed. The loss of traditional indigenous languages has been profound. Both these issues are important to our cultural heritage.

- Old growth forests continue to be logged.

The uncertainties

In a number of areas there is not enough information to assess the situation. Some areas warrant concern rather than alarm.

- The status of some marine species, including mammals, reptiles and some types of fish, is of concern.

- Some types of forest are threatened with disappearance and we cannot be certain that others are adequately protected to ensure their survival.

- Our system of reserves is patchy, with areas of poor biodiversity being better protected than areas of high biodiversity, because the poorer areas also have less economic value.

- For a range of issues, including aspects of urban air quality, water systems and natural and cultural heritage, the available data do not permit a clear description of the national picture.

Overview of responses

Australia's response to the state of its environment has been effective in some areas, in others questionable, and in yet other areas, either inadequate or counter-productive.

Commendable actions

Australia is doing well — in some cases setting an international example — for a range of issues.

- The listing of natural areas and cultural landscapes under the World Heritage Convention, and their subsequent protection, is a real success story, as is the increasing provision for other forms of reserve status, and the strengthening of State and Territory heritage legislation.

- Some of our structural solutions to complex management problems, such as the Great Barrier Reef Marine Park Authority, the Murray–Darling Basin Commission and the Board of Management of Uluru-KataTjuta, are recognised internationally as good models of response.

- The recent decision to limit further diversion of water from the rivers of the Murray–Darling Basin is a landmark in Australian water management. An interim cap is in place and a final cap will be made by 30 June 1997. However, in many cases water diversion remains above acceptable levels.

- New Fisheries Acts have recently been introduced throughout Australia with the aim of managing resources within the principles of ecologically sustainable development and economic efficiency in the face of increasing fishing pressure. This has resulted in much improved fisheries management, but it is too early to judge the effectiveness for those species which were over-fished.

- The Landcare program has been very successful in mobilising land-owners and communities to improve environmental conservation and encourage the adoption of sustainable land-use practices. The program needs to be broadened to embrace a more integrated, systems approach.

- Australia has taken prompt and purposeful action to phase out ozone-depleting substances such as CFCs. This action puts us ahead of international targets. We have also accelerated the transition to unleaded petrol.

- State governments, notably Queensland, Western Australia and South Australia, are acting to increase the use of renewable energy such as solar power.

- Recycling schemes have dramatically reduced the volume of waste in some areas; the Better Cities demonstration programs, water efficiency appliance schemes, the Commonwealth's National Pollutant Inventory and Cleaner Production are, like many other initiatives, helping to improve environmental standards.

Questionable actions

Another group of responses is questionable.

- The short-term economic objective of reducing the price of power presently drives reform of the electricity industry. With this goal, energy use — and so pollution and greenhouse gas emissions — is likely to increase.

- The move to regulate aircraft noise by taxation and legislation is a positive step, but only came after widespread public concern about the noise levels associated with the new runway at Sydney airport.

Poor responses

Finally, there are areas where our responses seem inadequate or counterproductive.

- The national ability to manage the environment is continually hamstrung by structural problems between different areas of government. Standards vary from State to State, and State and Commonwealth governments frequently battle over environmental issues. The recently established National Environment Protection Council will address some of these issues.

- Adequate measures are not yet in place to combat the threats to biodiversity.

- Concerns remain about whether the changes to fisheries management are enough to reverse the decline in some fish stocks.

- Despite the commitment to ecologically sustainable development, some government agencies still see their primary role as promoting economic development, with little regard to environmental costs.

- While land clearing is restricted in some States, in others it continues to be tolerated and even encouraged.

- Urban planning in general, and transport planning in particular, are still problems, with few effective attempts to contain urban sprawl or discourage the use of private cars. There is no concerted attempt to redirect our pattern of energy consumption in a sustainable direction.

- Australia is falling short of its greenhouse gas emission reduction targets. In recent years, energy-related carbon dioxide emissions have grown much faster than the OECD average.

Conclusions

Environmental awareness has increased dramatically in the past decade, penetrating all sections of the community. There is now ready acceptance of national and international environmental standards. Environmental management is growing rapidly, as a profession and a practice. However, we do not yet have an integrated, system-based approach to the management of natural resources. Until we do, environmental management will be characterised by *ad hoc* responses to urgent, emerging problems. Despite the adoption of national strategies for ecologically sustainable development and conservation of biological diversity, there is little evidence that this broader approach and commitment to sustainability has been fully integrated into decision-making.

Overall, economic planning appears to take little account of environmental impacts. It is assumed that the first priority should be a healthy economy, and that problems can always be solved using the wealth created. The economy is a subset of human society which, in turn, is part of the environment. Progress towards sustainability requires recognition of this fundamental truth, and a willingness to build environmental thinking into our economic planning.

Australia — the island continent

Australia is the smallest of the world's seven continents, but the sixth largest country and the only country to occupy an entire continent. It is the only permanently inhabited continent that lies wholly in the southern hemisphere.

The southern hemisphere is mainly water, and is markedly less affected by human activities than the northern. Australia lies in relatively unpolluted air and sea. Its comparative isolation means that transboundary pollution is not as significant an issue as it is for many northern hemisphere countries.

Continental Australia

Many of Australia's unusual features are important in assessing the state of its environment.

It is the world's lowest continent. Australia has the lowest average elevation and the lowest absolute relief (the difference between highest and lowest points). This is partly a result of its geological stability. Its few mountain ranges are very low by world standards, and most of the country is a broad flat platform.

It is geologically stable. Australia has some of the oldest land surface on Earth. Much of this surface has been weathering and eroding for long periods of geological time. It is the only continent that has no tectonic mountain building or active volcanoes.

It is the driest inhabited continent. Australia receives the lowest annual average rainfall, apart from Antarctica. Its latitude puts it in a low rainfall zone. The lack of mountains, which tend to increase rainfall, is also a factor. More than one-third of Australia is classified as arid (receiving an average of less than 250 mm of rain a year), and another third is semi-arid (receiving 250–500 mm).

Its climate, especially the rainfall, is highly variable. Averages can be very misleading: in some places much of the annual rainfall can fall in a week or so; in others, the annual rainfall can vary from a fraction of the annual average to several times its size. Changes in sea temperature and atmospheric pressure associated with a suite of events known as the El Niño–Southern Oscillation, or ENSO, are a major cause of this variability.

It has fewer sizeable rivers and less run-off than any other continent, except Antarctica. The low rainfall means that, for its size, Australia has few sizeable rivers, and accumulates little fresh water on its surface.

It has large, but diminishing, groundwater resources. Australia is well provided with (under)groundwater that can be tapped with bores and wells. The largest source, the Great Artesian Basin, is huge, underlying about 22 per cent of the continent. Groundwater is being used much faster than it is replenished.

Its soils are generally poor. Australian soils are among the most nutrient-poor and unproductive in the world. This is mainly because of the country's geological stability, with many soil components having been leached out during the land's long exposure to weathering. The dryness also mitigates against the formation of rich, deep soils. Less than 10 per cent of the country has reasonably productive soils that can sustain intensive agriculture or dense vegetation.

It is a land of natural hazards. These are part of Australia's natural environmental variability. Drought, flood, fire, earthquake, storm and tropical cyclone are normal features of the land. Large fires occur somewhere in Australia every month of the year. Much of the native vegetation is adapted to fire, and many species need it to reproduce. The frequency of fires is not just a result of the hot and dry climate, but also its variability. Wet periods allow vegetation to build up as fuel for fires in dry times. The flammability of eucalyptus oil and the dry leaves of many plants also contribute.

It is rich in minerals. Australia has more than 20 per cent of the world's stock of recoverable bauxite, iron ore, uranium, mineral sands and diamonds. Coal is another important mineral resource.

Australia's seas

Australia's seas are an important part of its environment, both in their own right and because of the influence they have on the terrestrial environment, especially its climate.

Mosaic of Australia taken by satellites at a height of 830 to 850 kilometres. The map is a mosaic of ten satellite pictures collected so as to minimise cloud cover over the land.

Although the satellite only has one visible band, the colour scheme used in the image has been selected so as to simulate natural colour. The visible channel has been used in the blue, a band which highlights vegetation has been coloured green, and the red is the 'thermal band' which highlights features which are dark in the visible.

The mosaic was prepared by GEOIMAGE Pty Ltd (in Brisbane) and published with permission.

Australia sits between three ocean basins — the Pacific, Indian and Southern. Its ocean territory, as defined by the 200-nautical-mile Exclusive Economic Zone (EEZ), covers about 12 million square kilometres, about one and half times its land area (7.6 million square kilometres). It is the third largest EEZ in the world.

The adoption of the EEZ in 1994 gives Australia rights to explore, exploit and manage the zone's natural resources. It also imposes responsibilities to conserve the zone's rich diversity of living organisms.

Australia has several important marine features.

It has warm ocean currents. Australia has a major ocean current passing down each side of the continent, bringing warm water from the tropics to both east and west coasts. This feature is unusual, as most other continents have a 'cold' current, from the polar regions, running along one side, and a 'warm' current, from the tropics, along the other.

Its seas are not highly productive by world standards. The lack of an upwelling of cold waters, rich in nutrients and oxygen, is one reason. Another is lack of run-off from the land and the low level of nutrients in the run-off. Despite the large size of our fishing zone, the total fish catch is low by world standards, ranking about 50th by production (although most of the catch has a high unit value).

It has the world's largest coral reef. The most famous feature of Australia's seas is the world's largest continuous coral reef complex, the Great Barrier Reef. The reef extends about 2500 kilometres along the north-east coast. Near its southern end lies the world's largest sand island, Fraser Island. Both the reef and the island are World Heritage Areas.

Australia's biology

Australia's unique physical characteristics have made its biology distinctive.

It has a rich biodiversity. Australia is one of the twelve most biologically diverse countries in the world, a status owed to its size, long isolation and many climatic zones.

Many of its plants and animals are unique. Australia has a very high degree of endemism — that is, many of its plants, animals and micro-organisms are not found elsewhere in the world. This is because they have evolved in isolation.

Its vegetation is dominated by two genera of trees — Eucalyptus and Acacia. Acacias are widespread in southern continents, although Australian species are endemic. Eucalypts are mostly confined to Australia and New Guinea, with a few species occurring in other islands to our north.

Most plants are sclerophylls. A particular quality of much of the Australian flora is the hard, dryish, leathery, spiny or small leaves. These plants are called sclerophylls. They are well adapted to dry conditions and low-nutrient soils. Their leaves lose less moisture and are tougher, so lasting longer and needing replacement less often.

Australia is the undisputed world centre for marsupials. Several families of marsupials (mammals whose young develop in a pouch) occur in South America and one species in North America; Australia has about 144 marsupial species.

Monotremes (mammals that lay eggs) are even more special. Only three species exist — the platypus and two echidnas (one is in New Guinea). There is also a large number of native placental mammals (in which the young develop in the womb), mainly bats, rodents and marine mammals.

Its tropical rainforests are small but rich. Australia's tropical rainforests are distinctive — again a consequence of the continent's long isolation. While covering only 0.1 per cent of the land area, they contain a large proportion of Australia's known plant and animal species. Many rainforest species have not been scientifically described, and there are probably many more, especially invertebrates, that remain undiscovered.

Introduced species have 'run wild'. In the past 200 years, introduced species of plants, animals and micro-organisms have caused dramatic and irreversible changes to the natural ecology. They range in size from the Asian buffalo to disease-causing viruses. Introduced animals such as the fox, cat, rabbit, goat and pig have been directly or indirectly responsible for many of Australia's native mammal extinctions. Exotic fish such as the European carp and trout have damaged freshwater environments. Introduced plants such as *Mimosa* and rubber vine are taking over large tracts of land and waterways. An introduced fungus, *Phytophthora cinnamomi,* is a major cause of dieback, a disease threatening whole ecosystems.

Australia's people

The first humans arrived in Australia at least 50 000 years ago. Prehistorians believe that they came from the north, through south-east Asia, probably at a time when sea levels were lower than now and sea voyages shorter. Waves of migration probably occurred over thousands of years. By about 20 000 years ago, people had spread to all parts of Australia.

The early Australians did not use agriculture. The persistence of hunting and gathering may well have been an adaptation to the low productive potential of most of the country. The people clearly knew a great deal about the land's resources and how to care for certain plants.

Like all people, they modified their environment to suit their needs — mainly through the use of fire. It is possible that continuous hunting, coupled with vegetation changes resulting from regular burning, drove some animals to extinction and changed the species composition of the fauna.

With the arrival of Europeans just over 200 years ago, the rate of human-induced environmental change accelerated dramatically. Sent in the belief that the land had boundless potential, the first European settlers found the environment harsher than anticipated, and both it and the original inhabitants soon came to be seen as generally hostile and alien. Introduced disease, especially smallpox, and violent conflict had a devastating impact on the indigenous people.

As the newcomers colonised more and more land — clearing it and introducing European agricultural practices and exotic plants and animals — the prevailing view changed from early dismay to unrealistic expectations. The environment could be 'tamed' to create great wealth. As the colony prospered, visions grew of a wealthy and powerful nation, rivalling the United States.

Only much later did people begin to realise that Australia's environment has many limitations, and that its 'development' has come at great cost.

Human settlements

Australia is a highly urbanised nation, with about 85 per cent of its population living in towns and cities of 10 000 or more people. Although these settlements occupy less than one per cent of the country's total land area, they have a pervasive influence on the natural environment.

The report examines Australian settlements in terms of the resources they consume, the wastes they produce and their 'livability'— a measure of social amenity, health and well-being. This 'extended metabolism' model allows settlements — whether capital cities or remote indigenous communities — to be assessed in terms of their human outcomes as well as their impact on the natural environment.

Australian settlements have high livability by international standards and, in general, it is improving. However, these patterns vary considerably between and within settlements.

Our settlements have higher metabolic flows — that is, they use more resources and produce more wastes — than those in other industrial nations. These flow levels have been increasing, both in total and per person, over the past few decades, as illustrated by the following examples.

- Domestic water consumption has risen significantly over the past 20 years because of the rising population and increasing use per head. Sydney's total consumption per head (domestic and industrial) rose 25 per cent between 1970 and 1990, from 144 to 180 tonnes.

- Australia's primary energy consumption per head increased by 37 per cent between 1970 and 1990, and its energy consumption per unit of GDP has fallen only marginally since 1970, while some other countries have reduced theirs by more than 30 per cent.

- Food consumption per head (measured by energy content) increased by more than 70 per cent between 1967 and 1992, not because we ate more, but probably because of more energy-intensive production and more wastage in processing.

- Australia produces more municipal solid wastes than other industrial nations — 681 kg per person per year, compared with an OECD average of 513 kg.

Cities

The big cities are generally more efficient in their metabolic flows than smaller cities and country towns. The large cities also tend to enjoy better livability. This suggests there is little to be gained environmentally by dispersing urban populations into other areas, especially the non-urban coastal zones, which are growing rapidly. Much can be done to reduce metabolic flows while further improving livability within cities.

On the other hand, the large cities, notably Sydney, are experiencing 'capacity' problems associated with photochemical smog, stormwater and waste water that demand changes if the cities are to continue to grow. Global constraints must also be faced, especially those arising from the greenhouse effect.

Within the large cities, the more compact core and inner areas consume fewer resources and produce less waste per head than outer and fringe areas, although there are some pressures on their

Human settlements: key threats to sustainability

Issue	Detail	Comment
Livability of remote indigenous communities	Critically low levels of social amenity and health are evident in indigenous communities.	A major problem. The situation is not improving.
Livability of inland towns	Many inland towns are declining in population with a consequent reduction in livability.	The situation is deteriorating.
Livability of coastal settlements	Despite the attraction of coastal locations and lifestyles, rapid growth of coastal areas is associated with a serious lack of economic opportunities and services.	Of some concern, with the situation steady.
Coastal development	Coastal lands are ecologically sensitive. The greatest concentrations of population and new developments are now on the coastline thus putting great pressure on it.	Coastal habitats are being destroyed and degraded by overdevelopment. The situation is deteriorating.
Metabolism of big cities	Australian cities are comparatively high in their consumption of resources (energy, water, land and building materials) and production of wastes (air, liquid and solid).	Although big cities have lower per capita flows than small cities, many resource usages and waste flows have been increasing at an unsustainable rate. Capacity problems are being experienced in the airsheds, streams and coastal waters of all big cities.
Livability of big cities	Sprawling car-based suburbanisation continues, though there is an increasing trend towards reurbanisation of older suburbs. Emerging 'pockets of poverty' stemming predominantly from unemployment are appearing across the cities.	Significant problems could emerge if negative trends continue.

infrastructure. Livability levels are similar across each large city, except the urban fringe, which suffers poorer social amenity (access to public transport, and health, educational, sporting and recreational facilities etc).

Livability is declining in emerging 'pockets of poverty' found right across the city. Unlike many cities in the United States, which have deteriorating cores, Australian cities are undergoing simultaneous processes of suburbanisation, with new suburbs being created at the fringe, and reurbanisation, with older areas being redeveloped. Suburbanisation is still dominant in terms of population numbers, but reurbanisation now accounts for more than 30 per cent of housing and 50 per cent of commercial development.

Towns

The urban fringe and coastal areas, particularly in northern New South Wales and Queensland, are expanding more rapidly — and less sustainably — than other settlements. This is evident from their metabolic flows, their pressure on sensitive environments and their livability.

Many inland towns are stagnating economically, many face significant 'capacity' problems with water and waste management, and are generally starved of adequate technological investment because of their limited growth potential.

The populations of the more remote provincial towns — apart from tourist centres — are declining, and this is reflected in their reduced livability. It suggests that population decline is not good for settlements that need investment and community commitment to address long-term environmental and livability problems. Diversifying the productive base is an economic, social and environmental priority.

Remote settlements

The number of remote indigenous communities is growing. They have low metabolic flows, but face some 'capacity' issues (for example, firewood). They also have extremely low livability on all indicators, particularly health. Powerful cultural forces have driven this 'return to the country', which is only now being helped by appropriate technology. The communities lack basic infrastructure, as well as social and economic development programs to improve livability.

Management

Australian settlements are falling well short of the environmental objectives expressed by governments. Increasing use of resources per head, growing population, and patchy implementation of the legislative, social and technological innovations necessary to ameliorate the environmental impact of settlements, all contribute to this situation.

The new technique of sustainable infrastructure planning, which evaluates the environmental impact of infrastructure and services, together with innovative Commonwealth government programs such as Better Cities and Cleaner Production, can reduce metabolic flows in all Australian settlements

and increase their livability. State of the environment reporting can assist this integrated process, particularly where local capacities need to be more carefully assessed and monitored.

Biodiversity

Biodiversity is the variety of all life forms, and comprises the different plants, animals and micro-organisms, their genes and the ecosystems of which they are a part.

Life in Australia has evolved in relative isolation for at least 50 million years. As a result, Australia possesses a rich diversity of unique and unusual plants, animals and micro-organisms.

About 85 per cent of flowering plants, 84 per cent of mammals, more than 45 per cent of birds, 89 per cent of reptiles, 93 per cent of frogs and 85 per cent of inshore, temperate-zone fish are endemic to Australia — that is, they are found nowhere else in the world. While a majority of these species have been studied, relatively little is known about the vast and less visible world of invertebrate animals and micro-organisms. In all, Australia is home to more than one million species, but less than 15 per cent have been described.

We depend on biodiversity for our survival and quality of life. The most significant impediment to the conservation and management of biodiversity is our lack of knowledge about it and the effects of human population and activities on it.

Preserving Australia's biodiversity is important for four reasons:

- *Ecosystem processes:* biodiversity underpins the processes that make life possible. Healthy ecosystems are necessary for maintaining and regulating: atmospheric quality, climate, fresh water, marine productivity, soil formation, cycling of nutrients and waste disposal.

- *Ethics:* no species — and no generation — has the right to sequester Earth's resources solely for its own benefit.

- *Aesthetics and culture:* biodiversity is intrinsic to values such as beauty and tranquillity. Many Australians place a high value on native plants and animals, which contribute to a sense of cultural identity, spiritual enrichment and recreation. Biodiversity is central to the cultures of Aboriginal and Torres Strait Islander peoples.

- *Economic:* Australian plants and animals attract tourists and provide food, medicines, energy and building materials. Our biodiversity is a reservoir of resources that remains relatively untapped.

Pressures

The greatest pressures on biodiversity come from the demands on natural resources by increasing populations of humans, their affluence and technology. Habitat modification — especially the removal of native vegetation for agriculture, urban development (including the depletion of coastal habitats) and forestry — has been, and remains, the most significant cause of loss of biodiversity.

Exotic organisms — including species introduced for production purposes and introduced diseases, and native species whose range and/or abundance has changed because of human activities — also exert pressure on biodiversity. Other pressures are harvesting of native species, pollution from industrial sites and urban areas, and fertilisers and other chemicals used in agriculture and forestry. Climate change, as yet difficult to assess, could also have a serious impact.

Impacts

Every ecosystem in and around Australia, with the possible exception of the deep ocean, has been modified in some way, with varying impacts on biodiversity. However, there is limited information on which species were present when Europeans arrived here, which are present now and which have been lost. It is difficult to respond to the pressures and the changes in the state of biodiversity without basic information on what that state is and how biodiversity is distributed.

For the land animals and plants about which we know enough to assess their current state, the trends are disturbing. Some 5 per cent of higher plants, 23 per cent of mammals, 9 per cent of birds, 7 per cent of reptiles, 16 per cent of amphibians and 9 per cent of fresh-water fish are extinct, endangered or vulnerable. Australia has the world's worst record of mammal extinctions. In the past 200 years, we have lost 10 of 144 species of marsupials and 8 of 53 species of native rodents.

Management

Since the 1980s, several major international agreements have addressed biodiversity as a key issue. It has become an important organising principle in a number of national policy statements,

Biodiversity: key threats to sustainability

Issue	Detail	Comment
Effects of human population and consumption	The overwhelming causes of the decline in Australia's biodiversity result from the human population, their lifestyles, technologies and demands on natural resources.	The situation continues to deteriorate as population and demands on natural resources increase.
Condition of ecosystems	Most terrestrial, freshwater and marine ecosystems are altered in structure and function to some extent.	Few ecosystems remain in a largely natural condition. The situation is deteriorating.
Distribution and abundance of species	Many species are undescribed or poorly studied; of those that are described, many are lost or threatened.	The loss of and decline in species continues and is cause for national concern.
Changes in genetic diversity	Little is known for most species although there is strong evidence of loss of genetic diversity for some.	While the degree of genetic diversity is unclear, it is almost certainly declining.
Land clearance and related activities	Land clearing destroys and modifies ecosystems, thus threatening biodiversity. The past extent and continuing rate vary greatly between States and Territories.	This is the single largest threat to biodiversity. The situation is deteriorating as threatening activities continue.
Impacts of introduced species	Most terrestrial, freshwater and marine ecosystems are affected or threatened, as are many native species.	Impacts have often been severe and the situation continues to deteriorate.
Harvesting native species	Some species have been and are being over-exploited. There are detrimental effects on habitat and non-target species.	Harvesting of native species is an important pressure on biodiversity in some areas. The situation is deteriorating.
Lack of knowledge of biodiversity	This affects ability to develop strategies for achieving sustainable production without further detrimental effects on biodiversity.	The knowledge base, while still inadequate, is slowly improving.
Effectiveness of conservation measures outside conservation parks and reserves	Most biodiversity will continue to rely on areas outside the system of conservation parks and reserves.	Better integration of management approaches in the local, regional and national spheres is required.
Adequacy of protected areas	The number and extent of protected areas is increasing but nature conservation is generally a residual land use in agricultural districts.	Some ecosystems and species are represented well, others poorly.
Adoption of integrated ecosystem-based management of natural resources.	This is necessary for achieving sustainable production without further detrimental changes in biodiversity.	Bioregional management requirements are partially recognised, but enormous efforts are still required to fully develop and implement them.

and is now influencing decision-making at national, State and regional levels. In particular, the Commonwealth, State and Territory governments have endorsed the National Strategy for the Conservation of Australia's Biological Diversity to guide the implementation of policies related to biodiversity.

However, governments are not providing enough resources to give effect to these policies. A lack of coordination between Commonwealth, State and local government is also a major impediment.

Australia is one of 12 nations in the world that contain major repositories of biological diversity. It is the only one that is industrially developed, has a relatively small human population, and occupies an entire continent. Thus we have a good opportunity, as well as the responsibility, to balance conservation, human population growth and demands, and economic development.

This can only be done by making substantial changes to the way we manage the land and ocean. Many current practices are not sustainable, and biodiversity-based industries such as agriculture, pastoralism, forestry, fisheries and tourism often erode the very resources upon which they depend.

Australia lacks major, co-ordinated programs for the discovery, monitoring, management and sustainable use of biodiversity. New strategies, particularly ecologically sustainable development, give us the opportunity to provide world leadership in the wise use of natural resources, including their conservation for future generations. Without this comprehensive approach, the future is bleak for much of Australia's unique flora and fauna.

The Atmosphere

The atmosphere makes life on Earth possible. Without it, or with a changed atmosphere, conditions on the planet would be very different. Not only is it essential for life, it is important to the quality of life: clean, fresh air adds to our sense of well-being.

The report defines the atmosphere as the air environment — from the envelope of gases surrounding the planet to the air inside our homes. It focuses on those aspects of the environment where human activities have a detectable effect. Pressures on the atmosphere are created mainly through the emission of gases and other substances into the air. The report considers these pressures and their impact on the state of the atmosphere, together with our response, at three different scales: global, regional and local.

Most of our atmospheric emissions result from burning fossil fuels for transport, power generation and industrial production. Many industrial, commercial and domestic processes also emit waste gases. Agriculture is a source of emissions, and affects the atmosphere indirectly — clearing vegetation, for example, increases carbon dioxide emissions and decreases its absorption.

The size and growth rate of the human population, the industrial and resource base, economic growth and lifestyles all influence the demand for energy and other resources, and so the level of emissions. For urban areas, the impact of these emissions is affected by the location of the major cities in coastal areas where a natural recirculation of polluted air sometimes occurs.

Global issues

Global issues result from long-lived emissions that become well-mixed through the global atmosphere — for example, the enhanced greenhouse effect and the loss of stratospheric ozone.

The enhanced greenhouse effect

Human activities are substantially increasing the atmospheric concentrations of a range of so-called greenhouse gases. These include carbon dioxide, methane, (tropospheric) ozone, nitrous oxide and chlorofluorocarbons (CFCs). Since pre-industrial times (since about 1750), the concentration of carbon dioxide has increased by more than 30 per cent, and methane by more than 145 per cent.

Greenhouse gases absorb infrared radiation emitted by Earth's surface and so keep the planet warmer than it would otherwise be. Climate models suggest that increases in atmospheric concentrations may lead to global warming because of an enhanced greenhouse effect. The models suggest an increase in the global surface temperature, relative to 1990, of about 2°C by 2100. Many uncertainties remain about the timing, magnitude and regional patterns of climate change. Nevertheless, the balance of evidence suggests that humans have a discernible influence on global climate.

Unless large reductions in emissions occur, the concentrations of most greenhouse gases will continue to rise well into the next century. Models indicate that carbon dioxide levels will more than double unless global emissions are reduced to well below 1990 levels. Australia produces between one and two per cent of global greenhouse emissions, which come mainly from fossil-fuel burning, land-clearing and agriculture.

Initiatives to reduce emissions under the National Greenhouse Response Strategy (NGRS) have achieved limited success. The Greenhouse 21C program includes cooperative agreements with industry and has the potential to bring about reductions in greenhouse gas emissions, though not of the magnitude required to achieve the goals of the NGRS. It is too soon to judge the success of this program.

Ozone depletion

CFCs, together with some other chemicals, have been implicated in another important global phenomenon — the depletion of ozone in parts of the stratosphere, which increases levels of damaging ultraviolet (UV) radiation at the earth's surface.

Losses of stratospheric ozone of between two and four per cent per decade have occurred in mid-latitudes since the 1950s, including over Australia.

More dramatic, has been the emergence in the 1980s of a 'hole' in the ozone layer over Antarctica each spring, in which more than 60 per cent of the total ozone is now destroyed over a region covering most of the continent.

Concentrations of CFCs are now levelling off, following the implementation of international controls. Australia has reduced production and use of ozone depleting substances well ahead of international obligations and plans are in place for further reduction. However, because of time lags, ozone depletion will continue to worsen for several years, and UV radiation will increase. Current trends suggest that the ozone layer should begin to recover at the beginning of the next century.

Regional air quality

Pollutants such as vehicle and industrial emissions last long enough to have effects far from their sources — for example, photochemical smog and brown haze.

Air quality is not monitored over 95 per cent of the continent. Certain important point sources of air pollution exist away from the major cities — mainly power stations and metal smelters. The main pollutants from these sources are sulfur dioxide, fluoride and lead. The largest single sources of sulfur dioxide occur in arid areas, where dry deposition of sulfate aerosols takes place downwind of the source.

Away from these major sources, air quality is probably good. However, it is difficult to assess trends because of the natural variability of the climate and because changes in demand for metals will affect smelting operations and hence emissions.

Urban areas

Motor vehicles are the single most important source of air pollution in cities. In general, the concentrations of common pollutants meet official health guidelines and are low by world standards.

Over the last decade, some aspects of air quality in cities like Sydney and Melbourne have improved. Atmospheric lead levels have declined significantly in recent years due to legislation, pricing measures and voluntary actions by industry. Important initiatives include the mandatory use of catalytic converters and unleaded petrol in motor vehicles manufactured or imported since 1986, the price differential between leaded and unleaded petrol, and a reduction in the lead content of leaded petrol.

Levels of some pollutants, especially carbon monoxide and lead, are likely to fall further. In the short term, a similar trend is expected in smog levels (from emitted nitrogen oxides and

Atmosphere: key threats to sustainability

Issue	Detail	Comment
Air quality data	Only 5% of Australia is routinely monitored for air quality. Little is known of the effects of air quality on human health. There are no national air quality standards and no national database.	Implementation of the Inter-Governmental Agreement on the Environment may lead to improvements.
Indoor air quality	Little is known about the long term effects of indoor air quality on human health. However, potential concerns include house dust mites in warm, coastal areas; tobacco smoke; Legionella and asbestos.	Little improvement is expected. Responsibilities for indoor air quality are not clearly assigned. Reductions in ventilation to save energy may worsen indoor air quality.
Stratospheric ozone loss	Stratospheric ozone loss, caused by anthropogenic emissions, has resulted in increased UV-radiation at the surface and poses a threat to human health and ecosystems.	The Antarctic ozone hole has increased in depth and extent since the early 1980s. Ozone levels have also decreased over Australia. Stratospheric ozone levels should begin to recover at the turn of the century as a result of current reductions in global emissions of ozone depleting substances.
Urban Air Quality	There are episodes in most years of extended periods with high levels of ozone and/or particulates in large cities. Major sources include vehicles, domestic burning, control burns and bushfires.	Measures have focused on ozone, generally with success. Current controls are unlikely to be adequate in the future. Lead emissions from vehicles have been reduced. Responses to particulate emissions have been inadequate, and little has been done on air toxics.
Enhanced greenhouse effect	Potential changes in regional climates pose a major threat to sustainability. Australia has a comparatively high level of greenhouse gas emissions per GDP. Australia produces only 1-2% of global emissions.	The balance of evidence suggests that there is a discernible human influence on global climate. Early initiatives to reduce emissions have achieved limited success.
Regional emissions	A small number of remote sources produce 60% of Australian sulfur dioxide emissions. Principal concern is the impact of acid deposition. Potential impacts on ecosystems are not yet comprehensively assessed.	Emissions from some major sources are being reduced through application of environmental protection policies.

hydrocarbons), but this may be countered by increasing vehicle use linked to population growth, urban design and possible deterioration of catalytic converters. In areas of rapid growth, such as south-east Queensland, Perth and western Sydney, vehicle emissions will remain a major concern.

Sulfur dioxide is not a major pollutant in urban areas because of the low sulfur content of Australian fuels and the location of most power stations outside the cities.

Major pollution problems in Australian cities are episodic rather than continuous, being influenced by seasonal and meteorological factors. Depending on local terrain and air flow patterns, pollutants may remain relatively undispersed as air recirculates. Australia's generally sunny climate promotes the formation of ozone and photochemical smog. As a result, air pollution levels in Sydney and Melbourne, particularly for ozone, can approach those in cities such as New York and Tokyo on some days. Perth's air quality is considered to be approaching the limits of its airshed.

High concentrations of fine particles in the air can reduce visibility, and endanger health. Sources include motor vehicles and woodfires, but pollen, sea salt and silica from soil also contribute to particulate levels. Levels in major Australian cities have generally declined over the past decade or so because of tighter emission controls and bans on backyard burning.

Air toxics emitted by motor vehicles — mainly volatile organic compounds and metallic compounds — are an emerging issue that may need further investigation.

Local air quality

Some pollutants are contained, or rapidly dispersed or inactivated. Odour and smoke arising from traffic, intensive agriculture, wood stoves, backyard incinerators, and even cooking, can be important local issues.

Indoor pollution

An important local air quality and human health issue is indoor pollution, which can include: house dust mites; tobacco smoke; toxic compounds including insecticides; legionella bacteria in airconditioning systems; and inadequate ventilation and volatile irritants from furnishing and equipment (creating 'sick' buildings).

Management

It is impossible to provide a comprehensive, quantitative assessment of Australia's air quality for national SoE reporting. Issues include:

- lack of a consistent national set of ambient air quality standards
- often inadequate monitoring in major cities
- lack of monitoring outside the major cities
- fragmented and inaccessible data

- inadequate understanding of Australian conditions
- lack of data on the effects of air quality on flora and fauna
- lack of a clear relationship between air quality and human health effects
- inadequate assessment of indoor air quality and health effects

Land resources

The lives of many Australians, particularly indigenous people and farming families, are closely tied to the land. For the 88 per cent of Australians who live in large towns and cities, and especially those who live in the capital cities, the land is a more distant entity. Nevertheless, responsibility for the stewardship of the land is widely accepted in the Australian community.

We make many different uses of land resources, and often multiple uses of the same patch of land. Australian soils are old and relatively infertile, and rainfall is low; only six per cent of our land is arable, compared with 20 per cent of the United States. The main vegetation is woodland and shrubland.

The predominant use is extensive grazing by introduced species, which occurs over 54 per cent of the country. While the area of intensively managed agricultural land (cropping and improved pasture) is proportionally small (six per cent), it is almost twice the total area of the United Kingdom and more than the total area of Japan.

Many pressures affect the state of land resources. General pressures include: population, the failure to allow sufficiently for the poor soil and climate variability, loss of biodiversity, and economic and social pressures. Other, indirect and direct pressures include: agriculture (land clearing, fertiliser use, tillage, changes to water flows, pollution from pesticides and herbicides), pastoralism, forestry, mining, human settlement (urban expansion, tourism and recreation, and transport and utility corridors), and changed management.

Land cover

Although clearing on the scale required to establish Australia's agricultural industries is no longer occurring (it peaked in the two decades after World War II), substantial areas are still being cleared. In some wheat-growing regions, less than 10 per cent of the native vegetation remains. Clearing is an important contributor to environmental problems, especially soil salinity, loss of biodiversity, and net greenhouse gas emissions.

Since the 1980s concern about land degradation, and the decline in native vegetation, has been widespread. As a result, the Commonwealth has established a number of programs such as One Billion Trees and Save the Bush. In addition, some State and Territory governments have established legislative or regulatory controls on clearing. Although the evidence suggests that legislation has been effective in South Australia and Western

Australia, there are not enough data to evaluate the effectiveness of controls in other States or Commonwealth programs there.

Rangelands

Rangelands cover about 75 per cent of the country. During the early years of pastoral development stocking rates were generally unsustainable and caused major changes to vegetation and soils. Although only a small part of the rangelands (two per cent) is regarded as severely degraded, a much larger area (15 per cent) is sufficiently affected to require destocking if it is to recover. The conservation value of rangelands varies greatly. Important sites, such as water holes, are the focus of domestic and feral animal activity and are seriously damaged. The Commonwealth, States and non-government organisations are developing a National Strategy for Rangeland Management.

Forests

Before the arrival of Europeans, nine per cent of Australia was covered with forest. Of this area, 40 per cent has been cleared, mostly for agriculture, and that the same area has been affected by logging. Logging can have severe impacts on biodiversity, soil and water. The National Forest Policy Statement, agreed by the Commonwealth and the States, is intended to address the issue of sustainable forest use. Regional Forest Agreements are being established to ensure a comprehensive and adequate reserve system for forests and a report is being prepared on the state of the forests. As an interim measure the Commonwealth has established deferred forest areas with the stated intention of protecting potentially important forests and reserves while comprehensive regional assessments are completed.

Pests and weeds

Vertebrate and invertebrate pests cost rural enterprises dearly. Losses occur both directly through consumption and damage to produce, and indirectly through degradation of the resource base. Since European settlement, more than 1900 new vascular plant species have become naturalised — that is, either deliberately introduced or accidentally released. Half of these are now regarded as weeds, and more than 220 have been declared noxious weeds. The major pests and weeds include some native species that have spread and thrive in ecosystems altered by European-style land uses.

Weeds are estimated to cost Australia about $3.3 billion annually. Insects cause annual losses to primary production of about $3.1 billion. A mouse plague in South Australia and Victoria in 1993 cost a total of $65 million, including $55 million in reduced yields. Rabbits are estimated to cost Australia at least $90 million (and possibly up to $60 million) a year in lost production and reduced land values.

Many Commonwealth government agencies are involved in pest control. The Bureau of Resource Sciences runs a vertebrate pest program, which deals with pest animals in agricultural production. The Australian Nature Conservation Agency has a feral pest program, which addresses the problems of pest management for conservation purposes. Other bodies, established by the Commonwealth, such as the Australian Plague Locust Commission, deal with specific pests.

Soils

Most areas of cropland and improved pasture in Australia are affected by soil degradation. However,

Land resources: key threats to sustainability

Issue	Detail	Comment
Land clearing	Land clearing has important implications for land degradation processes, especially salinity, biodiversity loss and greenhouse gas emissions.	Substantial areas are still being cleared for cropping, grazing and urban development. Programs to retain remnant vegetation and improve on-farm conservation are active, but the rate of replanting is much less than clearing. Deteriorating.
Agriculture	Many areas are subject to slow cumulative threats such as soil acidification, salinity and structural decline.	Substantial damage has been done. Practices are improving in some areas.
Rangelands	Only a small proportion of rangelands is in formal reserves. Some important conservation sites (e.g. waterholes) are threatened because they are the focus of domestic herbivore activity.	There are areas of severely degraded rangelands, but condition and trend vary between regions and are very sensitive to management practices.
Cropping lands	There is particular concern about rates of erosion in marginal cropping areas with shallow soils and variable climate.	Severe degradation will increase if current practices continue. The long term sustainability of cropping on sloping lands is threatened by soil erosion.
Forests	There is inadequate conservation of old growth forests and some other forest types. Logging competes with other uses in native forest areas.	Management methods are improving but monitoring is not yet adequate and primary data are limited.
Data	Compared with other developed nations, Australia has only rudimentary information on the condition and productive capacity of its land resources.	This lack of information limits our ability to use land in accordance with its capabilities and to monitor trends in its condition.

within these areas, its impact can vary enormously — even from farm to farm or paddock to paddock. It is impossible to give meaningful estimates of the cost of soil degradation, which occurs in several ways.

- *Soil structure decline:* the structure of many Australian soils is naturally poor or has been damaged by land uses such as tillage and over-grazing. Poor structure increases run-off and erosion and reduces productivity. It is costly to repair.

- *Water logging and salinity:* water tables have risen in parts of Australia as a result of land clearing and irrigation, causing water logging and salinisation. These effects reduce productivity and encourage erosion.

- *Water and wind erosion:* rates of soil erosion, even on the best managed land, may be ten times greater than rates of soil formation. On sloping lands, soil erosion poses a threat to the long-term sustainability of farming.

- *Soil nutrient balance:* Australian soils are infertile, making fertiliser use an essential part of most farming systems. The continental nutrient balance is positive — that is, more nutrients are being added than are being used or lost — but nutrients may be accumulating in fertilised pastures while declining in more naturally fertile soils.

- *Soil acidification:* acidification affects most agricultural land, leading to toxicity, poorer water and nutrient use and so lower yields. Causes include the use of fertilisers and legumes, and natural weathering. Applying lime is an effective remedy, but current rates of application are inadequate.

Food quality

Misuse of agricultural and veterinary chemicals is low in Australia, and the quality of our food compares favourably with other countries. Among the heavy metals, which can be toxic, cadmium exceeds permissible levels in a proportion of some foods including offal and some seafoods.

Sustainability

The question of the sustainability of land use in Australia remains difficult to answer.

Agriculture

Agricultural yields have increased significantly this century, but land degradation could undermine past gains. Inputs of fertilisers and energy into agricultural production have increased, as has our knowledge of better agricultural practices, such as trace element supplements and minimum tillage. Future sustainability will depend upon the efficient use of inputs and continuing to improve our understanding of the land's productive capacity. We also need to know more about the damage to other ecosystems before we can draw firm conclusions about the industry's sustainability.

Forests

The past 200 years have seen the widespread destruction of our forests, through clearing for agriculture and timber harvesting that, in the past, was often unsustainable. The national goal is not to allow our forest estate to be eroded further and to ensure that there is a comprehensive and adequate reserve system to protect at least 15 per cent of each of the pre-1750 forest types for conservation purposes. Stronger and more effective controls are being introduced to cover all uses and values of the remaining forests, such as timber, water catchment and recreation. However, old growth forests continue to be harvested despite several findings that this practice violates sustainability principles.

There is no clear answer to whether our use of the forests is sustainable. In one sense, the answer is that 'it can be'. However, simple and permanent solutions are unlikely to be found; sustainable use must be achieved in the context of a continuing debate by many elements of society, each with different goals and different values.

Management

Compared with other developed nations, Australia has only rudimentary information on the condition and productive capacity of its land resources, and the potential hazards associated with their use. This limits our ability to use land in accordance with its capabilities and to monitor trends in its condition.

There are several other major issues concerning the management of land resources:

- Our small population means we have only limited economic and technical resources to manage a huge land area.

- Northern Hemisphere solutions to management problems are often inappropriate, and even counter-productive.

- We have to contend with the impacts of many introduced animals and plants on a previously isolated continent.

The Landcare Program, which is enlisting widespread community support for the sustainable use of land resources, began very successfully and may provide a model of community involvement in environmental management. However, the approach must be extended and sustained, with adequate financial and technical support, if it is to have the necessary impact on land use practices.

Inland waters

Australia is the driest of the world's inhabited continents. Of all continents, it has the least river water, the lowest run-off, and the smallest area of permanent wetlands. One-third of the country produces almost no run-off, while two-thirds of the average annual run-off occurs in northern drainage systems.

Australia's rainfall and stream flow are the most variable in the world, and our inland streams are naturally turbid (muddy) and saline. Few permanent, freshwater lakes exist on the mainland; where lakes do occur, they are usually shallow, salt lakes that are dry more often than not.

The largest river system is the Murray–Darling, which drains about one-seventh of the continent. It ranks among the world's biggest in terms of river length and catchment area, but carries much less water than comparable river systems — in less than one day, the Amazon carries the Murray's annual flow.

Over large tracts of inland Australia, (under)groundwater is the only practical source of water for pastoral and mining industries and local communities. The Great Artesian Basin is among the largest groundwater systems in the world.

Since the early days of European settlement, the development of water resources to support the economy and human population has been a priority. Because of our dry and variable climate, Australia stores more water per head of population than any other country in the world. Sydney stores 932 kilolitres of drinking water per person, compared to New York's 250 and London's 18.2. We are now focusing more on the environmental impacts of this development, principally through an initiative of the Council of Australian Governments (COAG) known as the COAG Water Reform Framework.

Drinking water

Water quality in large cities is generally high, but in some rural and remote communities, it is less satisfactory. Problems are usually caused by a combination of micro-organisms, chlorination by-products, taste, odour, algal toxins, iron, manganese, turbidity, salt and 'hardness'. Drinking water in Australia is generally free of industrial pollutants.

Issues

The diversion of water for human purposes has seriously affected the natural environment — wetlands and rivers, in particular, need more water. The environmental stress caused by the over-allocation of water for human use and consumption is greatest along the eastern seaboard and in the Murray–Darling Basin. Some 80 per cent of the divertible or accessible water in the Murray–Darling Basin has now been developed for human use. Elsewhere, the environmental stress is far less severe (and absent in undeveloped areas).

Australia is effectively mining its groundwater, with reserves being used much faster than they are

Inland waters: key threats to sustainability

Issue	Detail	Comment
Dryland salinity	This is increasing in south-west Western Australia and eastern uplands; stream salinity is rising and will continue to worsen.	Much damage has already been done and the situation is deteriorating.
Wetlands	Deterioration of wetlands has been caused by drainage, changes to water regimes and increases in sediment run-off and nutrient inputs.	Wetlands continue to be under threat, and large numbers are already destroyed. The situation is very poor.
Over-allocation of water to consumption	'Droughtproofing' by damming has starved rivers of water, and drastically altered seasonal flow regimes in the most developed areas. Groundwater is being 'mined'	Particularly severe in the south-east of the country. Deteriorating.
Irrigation	The greatest use of water, and the cause of much over-allocation, causing waterlogging and salinisation, and nutrient and pesticide pollution.	A major pressure on inland waters. Infrastructure is ageing and will need replacement. Some land may need to be retired from existing uses. The situation is deteriorating.
Endangered species	Pollution, over-allocation of water, changed flow regimes and exotic and displaced species are all affecting native species.	Many species of aquatic animals are endangered, in decline or extinct. Deteriorating.
Nutrients	Catchment erosion and point-source discharges have contaminated many water bodies so they now produce blue-green algae.	Effects are greatest in the south-east of the country. Trends are unclear but the current situation is poor.
Water weeds	Several vigorous weed species are spreading, particularly *Mimosa pigra* and alligator weed.	Weeds affect the entire country and the situation is generally deteriorating.
Sediments	Although decreasing in some areas, sediments continue to have impacts on biota and water-treatment costs; trends differ between regions.	The outlook is improving in the south-east of the country, but is deteriorating elsewhere.
Monitoring	This is a low priority for most agencies and the technical base is weak. Co-ordination and key indicators, particularly biological, are lacking.	The position is poor and deteriorating.
Data	There are few national compilations from which to estimate state or trend; many basic data do not exist; little interpretation of existing data; poor archiving.	The situation is very poor nationwide. Deteriorating.

replenished. Groundwater is often very old, with some bores tapping water that entered the ground 1–2 million years ago.

Irrigation uses the most water, accounting for 70 per cent of all the 'developed' resource. Next come the major cities, with most of this use being domestic. Household water use has increased because of increasing populations and rising consumption per head. Industrial water use is not large overall.

A major environmental consequence of both irrigation and land clearing for dryland agriculture is salinisation — the result of rising water tables that bring dissolved salts to surface soils and waters. In some parts of the Murray–Darling Basin, the water table is rising by as much as 0.5 metres a year. In dryland catchments in south-east Australia, the water table has risen by up to 30 metres since the 1880s. About one-third of Victoria's irrigation area has been salinised.

The irrigated agricultural industry in Victoria and New South Wales will need major restructuring to deal with the economic and environmental problems salinity causes. Dryland agriculture has increased salt loads and concentrations in many rivers in south-east and south-west Australia. About one-quarter of the cleared land in south-west Western Australia will become salinised over the next few decades, and dryland salinity will also expand in the eastern uplands of Victoria, New South Wales and Queensland.

Sediment from erosion continues to foul rivers, increase the cost of water treatment and reduce the storage capacity of dams and reservoirs. Levels of nutrients, particularly phosphorus, from run-off, erosion and point sources such as sewerage outlets, remain unacceptably high in our rivers, lakes and reservoirs. Levels of pesticides can be expected to increase in both surface water and groundwater.

Consequences

All these problems — sediments, salt, nutrients, pesticides and, in a few cases, trace metals and organic wastes — threaten aquatic environments. High levels of phosphorus, in conjunction with reduced stream flows, have resulted in extensive blooms of toxic blue-green algae, whose frequency may be increasing. The biological impacts of pesticides are largely unknown, but in some cases they cause more damage to aquatic species such as frogs than to the target organisms in crops.

Drainage and the loss of wetlands also threaten aquatic environments. A large proportion of wetlands have been destroyed or seriously disturbed, and the banks of most rivers have been damaged. Introduced species of aquatic plants and animals are another pressure. Exotic fish have established wild populations, often at the expense of native fish and other species. Exotic water weeds have spread in many parts of the country. The most damaging exotic species are the European carp, trout and the giant sensitive plant, *Mimosa pigra*.

As a result of these changes to Australia's inland waters, many species of aquatic animals are endangered, in decline or already extinct. Native fish species have declined in abundance and diversity in most regions of Australia since European settlement. Some 32 species of frogs are reported to be in decline, with only limited data available on many others. The platypus is still found throughout its original range, but frequently in reduced populations.

Management

Australia lacks basic data on water quality and catchment characteristics. Where they do exist, figures are often not collated nationally or are unavailable because of issues of ownership. Information is often of poor quality, incomplete and not comparable between agencies, localities and over time.

Most environmental and water resource organisations give low priority to high-quality monitoring of phenomena other than flow, and the technical expertise behind many monitoring programs is poor. The Commonwealth-funded National River Health Program, dedicated to developing a nationally consistent methodology for assessing the health of rivers, is a major initiative in this area. It focuses on the use of biological indicators and includes research into environmental flow requirements of Australia's rivers.

Of the key responses that would improve the condition of Australia's inland waters, the most important is catchment and water management that acknowledges the high natural variability of rainfall and run-off. Catchment management is being implemented over many systems. It is becoming more effective with increasing knowledge and is spreading around the country. The Murray–Darling Basin Commission is the most obvious example of this approach. However, despite these gains, catchment management still lacks the technical support to meet stated goals for water quality.

Estuaries and the sea

As an island continent with a long coastline, Australia has many different marine and estuarine environments. These span a wide range of coastal types, climates, geological and biological regions. Most are far away from major population centres and are little affected by human activities. Large stretches of our coastline are among the least-polluted places on earth.

Australia's marine environments extend from its beaches, rocky shores and intertidal reefs to the boundary of its 200-nautical-mile Exclusive Economic Zone (EEZ). They include large areas of the seabed that are important for fishing, oil and gas production and possibly mining, and areas of water that, in places, are biologically highly productive. Generally, however, our marine waters are low in nutrients and therefore productivity. Our estuaries are ecologically important habitats, usually rich in nutrients and with high productivity and biological diversity.

On the whole, our marine and estuarine environments are in good condition. In areas of high population density or intense human activity, however, they are often degraded as a result of urban, agricultural and industrial development and tourist and recreational activities.

Nutrient enrichment and sedimentation

The most serious issue affecting Australia's marine and coastal environments is the decline in water quality caused by rising levels of nutrients and sediments. Soil erosion, fertiliser use, intensive animal production, and sewage and other urban and industrial discharges have increased the levels of nutrients (especially phosphorus and nitrogen) and sediments entering many bays and estuaries. Other pollutants include industrial chemicals and pesticides, heavy metals, pathogens (disease-causing micro-organisms) and litter.

Each year, Australia's sewerage systems discharge about 10 000 tonnes of phosphorus and 100 000 tonnes of nitrogen, much of which enters the sea. Yet most nutrients (possibly as much as 85 per cent) come from diffuse catchment sources. Scientists have estimated that, in Queensland, the amounts of sediment, nitrogen and phosphorus entering the sea each year have increased three to fivefold since European settlement. The rivers of Queensland's east coast catchments are estimated to deliver about 14 million tonnes of sediment to estuaries and coastal marine waters annually. Between Palm Beach and Cronulla in Sydney, 200 large stormwater outlets discharge water containing high levels of pollutants such as sediments, bacteria, nutrients, trace metals and organic chemicals.

High nutrient levels promote algal blooms — some toxic. These have increased in frequency, intensity and geographic distribution over the past 30 years. They can degrade ecosystems, reduce the recreational value of waterways, affect human health and destroy aquaculture production.

The Commonwealth is cooperating with State and Territory Governments on issues such as land-based sources of marine pollution, including nutrients and sediment.

Estuaries and the sea: key threats to sustainability

Issue	Detail	Comment
State of seagrasses	Seagrasses have declined in area, particularly the temperate species, and most species are slow to recover after disturbance.	Trends are unclear and will have to be watched closely.
Fisheries: status	Control of catches is needed so that pressure on marine species is kept at tolerable levels.	Stocks of several commercial species have been overfished and it is too early to tell whether recent changes will be effective.
Fisheries: effects	Trawling and dredging can severely affect benthic habitats and populations of non-target species.	Damage has been, and is being, done. It is not known whether changing fisheries regimes will improve the situation.
Integrated eco-system based management objectives	We lack coordination and integration of marine management based on ecologically sustainable multiple-use management objectives for large marine ecosystems.	Our record in this area is poor with no significant improvement.
Effects of nutrients	Many estuaries and bays are affected by excessive inputs of nutrients and/or sediment runoff from the land.	Effects have been serious in some places. While trends are not always clear, this is becoming a major problem.
Effects of coastal development	Increasing development on and near the coast is removing or degrading marine and estuarine habitats, and reducing water quality.	Lack of national and strategic planning is contributing to the continuing deterioration.
State of mangroves	The area of mangrove habitat is declining, particularly in the tropics.	Threats to mangroves continue through unplanned development.
Effects of introduced pests	Mariculture and natural ecosystems are threatened by introduced pests.	Some pests have had major impacts on mariculture. Trends are unclear but will probably get worse.
Coral reefs	Near shore reefs are being degraded by land run-off, tourism and recreation.	This is potentially a serious problem. The current condition is fair, but deterioration is accelerating.
Lack of representative marine protected areas	There is a lack of marine reserves to protect representative samples of marine ecosystems, habitats and species.	A national programme to build a representative system by 2000 is underway.
Lack of knowledge	This affects the ability to make objective decisions, predict the effects of economic development and propose adequate responses.	The knowledge base relating to most areas of the estuarine and marine environments is inadequate.

Exotic species

Introduced plants and animals are damaging marine and coastal environments. Some introduced species also threaten aquaculture, and pose risks to human health. At least 55 species of fish and invertebrates, plus several seaweeds, have been introduced into Australia, either intentionally for aquaculture, or accidentally in ships' ballast water or encrusted on their hulls. Population eruptions of certain native species — for example, the crown-of-thorns starfish — can also cause problems.

A number of measures have been, or are being, developed in response to these problems, most recently an Australian Ballast Water Management Strategy. A national Centre for Research on Introduced Marine Pests has also been established within CSIRO's Marine Laboratories in Hobart to research early warning tools, better prediction, improved control methods, and more effective assessment of risks and costs.

Fishing

Fishing, both commercial and recreational, imposes heavy pressure on marine species and their habitats. Most major Australian seafood species are now fully exploited. Some, such as the southern bluefin tuna and eastern gemfish, have been overexploited. With the exception of turtles and possibly dugongs, which remain at risk, reptiles (the saltwater crocodile) and mammals (whales and seals) that have been over-hunted in the past are now recovering.

All States and the Commonwealth are revising fisheries legislation to ensure it is directed towards the principles of ecologically sustainable development. Fisheries advisory committees comprising management and industry interests have been established to help develop management plans for individual fisheries.

Mining

Coastal and offshore mining activities can have an impact on marine and estuarine environments. Enterprises include sand and gravel mining, oil and gas exploration and production, coral (limestone) mining, and diamond dredging. Over the past 30 years, oil companies have drilled more than 1100 wells offshore and extracted 2800 million barrels of oil. Their environmental record has been very good, with only about 800 barrels of oil being spilt over this period.

Specific environments

Human activities have caused extensive localised losses of saltmarshes, mangroves and seagrass beds, particularly near major population centres. Australia has the third largest area of mangroves in the world, and the northern mangroves are among the world's most diverse. Our waters also have the greatest diversity of seagrasses, and some of the largest seagrass beds in the world.

Australia also has the largest area of coral reefs. By international standards, our reefs are still in good condition. However, they are now exposed to significant pressures, with those close to population centres showing the most signs of damage.

Marine protected areas are important for conserving specific environments. Although Australia has a very large number of these areas (more than 300 in 1992) most of those in the southern and eastern half of the continent are small and many environments or bioregions are not adequately represented. The Commonwealth is addressing this issue in cooperation with State and Territory governments by developing a national representative system of marine protected areas under the Ocean Rescue 2000 program.

Seafood quality

Environmental conditions affect seafood quality, with the main issue being the accumulation of contaminants in the flesh of seafood species. Our seafood is generally low in contaminants, but exceptions occur in species that accumulate heavy metals, biotoxins, microbes and chemical pollutants. Some survey information is not available for public scrutiny.

Management

An increasing awareness of the pressures on our estuaries and seas has resulted in a range of legislative and management responses to prevent, contain and reduce degradation. For example, licensing point sources of pollution, establishing marine protected areas, oil spill contingency planning, regulation of national and international navigation to reduce maritime accident hazards and developing fisheries management plans, can all be effective in reducing or managing pressures on the environment.

However, the management of our marine and coastal systems still lacks an integrated and coordinated framework that is built on a set of ecosystem-based goals and environmental performance indicators. Apart from the Great Barrier Reef Marine Park Authority, no agency is responsible for managing the marine environment on such a basis. We do not have the long-term research and monitoring to provide baseline information at a national level, and research and management are yet to be fully integrated as the basis for sustainable use of coastal and marine resources. Such programs as the Commonwealth's State of the Marine Environment Report, National Marine Information System, Coastal Atlas and the Commonwealth Coastal Policy are improving the knowledge base.

Natural and cultural heritage

Australia's natural and cultural heritage is an integral part of its environment. The state of our heritage is as important as the state of our atmosphere, land, water, seas, plants and animals. Its inclusion adds a new dimension to state of the environment reporting.

Australia's heritage comprises natural and cultural places and objects that have special significance and value to Australians because of the meanings we ascribe to them. They are imbued with human associations, stories, myths and traditions.

Heritage provides the cultural and physical links with the past. It is central to our cultural identity and a source of spiritual well-being. Natural landscapes — with their biological and physical diversity — and cultural landscapes — with their diversity of cultural records and layers of meaning, objects and stories — collectively give us our uniquely Australian 'sense of place'.

Heritage places

Many places remain unprotected for a number of reasons: their heritage values are not recognised; the community is not involved; social values are poorly considered in heritage studies; or resources to identify and conserve places are inadequate.

Tourism has significant effects — both positive and negative — on many natural and cultural places. It can lead to better management, renewed cultural activity and increased understanding, but it can also result in pollution, vandalism and cultural exploitation and debasement.

Urban development, especially in the capital cities, which have the most listed heritage places, creates pressures where, for example, places are demolished or re-used. Neglect is the main threat in areas of low or declining population.

There is no national overview of the physical condition of Australia's heritage places, and no national monitoring system. Historic places and indigenous sites of non-archaeological significance still have no legislative protection in Tasmania and such protection is weak in Western Australia and Queensland.

Heritage objects

Major pressures on heritage objects include poor national coordination, inadequate conservation facilities, loss of context (through removal from their original site, for example) and insufficient documentation. Over the past two decades, the level of technical and other support for biological collections in the major government-funded museums and herbariums has decreased significantly.

We don't know with any accuracy the physical condition of all objects in collections, but many are thought to be deteriorating. No national monitoring system is in place. Indigenous heritage objects in their natural locations receive legislative protection in all States.

Indigenous languages

Loss of traditional languages imposes particular pressure on the heritage of indigenous Australians

Natural and cultural heritage: key threats to sustainability

Issue	Detail	Comment
Knowledge about heritage places and objects	While heritage registers include considerable detail about heritage places, there are geographic and thematic gaps. Many collections of heritage objects are inadequately documented and there are major thematic gaps.	Knowledge is improving and the gaps are slowly being filled.
Physical condition of heritage places and objects	Few quantifiable data are available to assess the condition of heritage places and heritage objects. No national monitoring system is in place.	Loss and deterioration of heritage places continues. Deterioration of many collections of heritage objects is likely to continue. Heritage assistance programs are still inadequate, despite improvements.
State of traditional indigenous languages	Some traditional languages are declining, despite language maintenance programs. Few languages are used as primary means of communication.	Many languages have been lost ; others are declining.
Survival of heritage in areas of significant population change	Many places are under significant pressure from urban development, expansion and re-zoning in metropolitan areas and neglect in declining rural areas.	The situation is uncertain due to lack of data, but there are signs of improvement.
Laws to protect heritage places and objects	Indigenous archaeological places are protected in all States and Territories and historic places in all but one. Specific protection of heritage objects is lacking in some States.	Improving.
Community involvement	Cultural groups are not adequately involved in the identification and conservation of their heritage. In particular, indigenous Australians' involvement in policy, decision-making, administration and management of their heritage is still low.	The involvement of indigenous Australians in heritage issues is increasing. Trends in the involvement of other elements of the community in heritage issues are unclear.
Impact of tourism	Increased tourism affects heritage places, heritage objects and cultural values in positive and negative ways, including how they are understood and valued. There are inadequate data to assess impacts of tourism despite major government tourism policies.	The situation is unclear and may be a concern.

because these languages provide the most appropriate way of transmitting and maintaining cultural knowledge and the traditions relating to places.

Heritage registers and collections

The size of heritage registers and collections has grown considerably over the last 20 years.

Eleven Australian sites have been inscribed on the World Heritage List, an international register of places of outstanding universal value. The Register of the National Estate (at June 1994) lists 10 772 places, of which seven per cent are Aboriginal and Torres Strait Islander, 16 per cent are natural and the remainder historic.

Although our knowledge about the state of Australia's heritage places has improved substantially (since the early 1980s), major gaps remain for some geographic and subject areas, including places of social value — places which often seem ordinary but which are so highly valued by the community that threats of their destruction provoke strong protests.

Millions of heritage objects are located in major collecting institutions. Collections of natural objects far outnumber those of cultural items. Some types of objects are poorly represented in both natural and cultural collections. For example, museums contain relatively few objects representing the experience of migrants, working class people and women.

Management

There have been many responses to the state of Australia's heritage and the pressures on it. Community groups, heritage professionals and some corporations and industry groups have actively supported heritage identification and conservation activities. Governments have developed specific policies and implemented major programs to help identify and conserve heritage places and objects, although some programs appear to lack adequate resources. For example, in 1994–95, the National Estate Grants Program, established to assist in identifying, conserving and presenting the National Estate, received 866 applications totalling $28 million for the $4.7 million available.

Legislative protection of heritage places has improved significantly, with governments passing ten major acts covering cultural places since 1987. Programs have been initiated for the national co-ordination of museum and other collections. All State and national museums now employ conservators, and museum policies are changing to redress imbalances in their collections.

Despite numerous positive responses to assist the identification, documentation and conservation of Australia's heritage, it is too early to assess the effectiveness of many recent initiatives. Before we can evaluate these initiatives, we need to agree nationally on which indicators are the most useful to measure the state of our heritage resources.

Major issues

The Australian community realises it has an obligation to protect significant objects and places for future generations. We lack quantifiable national data on the physical state of our heritage, the pressures affecting it, and society's response to these. Without this information, ensuring the sustainable use of our heritage resources is difficult.

The strong links between places, objects and the meanings people give to them are not reflected in current policies and institutional approaches, and no national heritage strategy exists to integrate these elements. Also, heritage considerations do not yet effectively integrate natural and cultural values. The concept of cultural landscapes — parts of the environment that, in being significantly modified by humans, express their attitudes, values and interactions with the environment — provides a powerful mechanism to assist this.

Australia lacks the necessary coordination between and within the three levels of government to ensure effective heritage identification and conservation and the integration of heritage values in early stages of policy and program development.

While we have made progress in many areas of indigenous heritage, indigenous communities still do not have enough say and involvement in identifying, protecting and interpreting their heritage, and in cultural tourism and language maintenance.

All groups in the Australian community need to be involved in heritage matters to ensure their heritage is acknowledged, and to assist in conserving and transmitting it to future generations in good condition.

Acknowledgment

The State of the Environment Advisory Council acknowledges the assistance of Richard Eckersley, Senior Strategic Analyst, CSIRO, in the preparation of this summary.

Photo credit

Page ES-1: The Photo Library of Australia

Chapter 1

Introduction

Prepared by

Ian Lowe (Chair), State of the Environment Advisory Council

Contents

Why produce a state of the environment report?

In Australia the principle of ecologically sustainable development is now broadly supported in the community, by governments of the Commonwealth, States and Territories and by local government. This support arises from the recognition that all lifestyles depend critically on a range of natural assets: air, soils, water, mineral resources, forests and other biological systems.

The National Strategy for Ecologically Sustainable Development called for the introduction of regular national state of the environment reporting to enhance the quality, accessibility and relevance of data relating to ecologically sustainable development. In Australia, two parliamentary inquiries have called for state of the environment reporting at the national level.

In the past two decades the governments of many countries, including Australia, Bahrain, Canada, Hungary, Italy, Japan, Kuwait, the Philippines, the Scandinavian nations, Turkey and the United States, have published reports on national environmental conditions. All OECD members have agreed to produce regular state of the environment reports; most have done so. In several countries these reports are thoroughly integrated into national economic policy formulation. For instance, in the Netherlands an independent research institute, the RIVM (Rijksinstituut voor Volksgezondheid en Milieuhygiene) produces comprehensive environmental outlook reports that assess current conditions and trends against criteria for ecological sustainability. These public audits are presented to the Dutch Parliament. They monitor and influence the workings of the Dutch National Environment Policy Plan, which guides Dutch economic and environmental policy towards ecological sustainability by the year 2010.
The Canadian Government published state of the environment reports in 1986 and 1991; it has also published a preliminary set of environmental indicators for such reporting.

A number of non-government organisations, including the World Resources Institute (Washington) and the World Conservation Monitoring Centre (Cambridge), also produce state of the environment reports, as do international agencies like the United Nations Environment Programme and the OECD. The 1991 OECD report on the state of the environment had the following aims:

• to assist member countries in the definition, implementation and evaluation of environmental policies

• to help those countries incorporate environmental concerns in economic decision-making in order to progress towards ecologically sustainable development

• to provide environmental information to the public

In addition, the OECD reviews the environmental performance of individual member countries in meeting their domestic policy goals and international commitments. Australian state of the environment reports will be an important source of information for these international reviews.

Establishing a pattern of sustainable development is not possible without adequate and accessible information. There is widespread and understandable concern about some aspects of environmental quality, such as air pollution, degradation of waterways, loss of biological diversity and erosion of agricultural lands. Decision-makers need reliable data on these and other key indicators of the state of the environment. They also need to know how the environment is changing. Without adequate, accessible information, it is possible to make two sorts of errors: we may inadvertently do irreparable damage to the natural systems on which all life depends; and we may forgo opportunities for desirable developments through lack of detailed understanding of the potential impacts.

State of the environment reporting is one of the most powerful tools for informing the public about their environment. It describes the effects of human activities on the condition of the environment, as well as the implications of this for human health and economic well-being. It also provides an opportunity to monitor actively, directly and accountably the performance of government policies against actual environmental outcomes, which makes it, effectively, a 'report card' on the condition of our environment and natural resources. This allows discussions about future economic and social development, and consequent policy, to be based on accurate and commonly agreed perceptions of environmental conditions and trends. If these conditions and trends are identified as they develop, decision-makers in industry and government would be in a position to avoid policies that might be environmentally unsustainable. Such policies could otherwise be socially and economically inequitable and costly.

There is growing community concern about the environment. 'Friends of Merri Creek' replant the banks of the creek with native grasses.

What constitutes the environment?

Twenty years ago the Commonwealth Government defined the environment as including '... all aspects of the surroundings of human beings, whether affecting human beings as individuals or in social groupings' (section 3 of the *Environment Protection (Impact of Proposals) Act 1974*). More recent definitions of the environment have taken a broader perspective. For example, the *Local Government (Planning and Environment) Act 1990* (Qld) defined environment in section 1.4 (1) as including:

(a) ecosystems and their constituent parts including people and communities

(b) all natural and physical resources

(c) those qualities and characteristics of locations, places and areas, however large or small, which contribute to their biological diversity and integrity, intrinsic or attributed scientific value or interest, amenity, harmony, and sense of community

(d) the social, economic, aesthetic and cultural conditions which affect the matters referred to in paragraphs (a), (b) and (c) or which are affected by those matters.

The New Zealand *Resource Management Act 1991* adopts a similar definition.

The broad range of environmental issues facing Australia today, and the need to assess progress towards ecological sustainability, has necessitated this report covering more than just people's surroundings. A more comprehensive scope was required. The report therefore covers terrestrial, atmospheric, marine, inland aquatic, and human environments, and how we value them.

It is a major innovation in this report to include natural and cultural heritage aspects of our environment. The chapter on human settlements has adopted a similarly innovative approach. It combines environmental, social and economic data into measures of 'livability' to reflect the condition of human society as part of the Australian environment.

Reporting framework

Purpose

To meet the need for improved environmental information and to satisfy its international commitments, the Commonwealth Government has established a system for regular reporting. The State of the Environment Reporting Unit of the Department of the Environment, Sport and Territories has responsibility for this task. A framework for state of the environment reporting for Australia was articulated in 1994 (DEST,1994). This report relies on the guiding principles and aims of that framework.

Guiding principles

This first report has been guided by the following principles:

- rigour — to use the best available scientific information, methods and advice, and to present accurate data and information in a balanced and accessible way

- objectivity — to present data and information without bias or modification

- cooperation — to prepare the report cooperatively, and to use a range of data sources including those of Commonwealth, State and Territory Government agencies as well as a number of non-government organisations

- openness — to ensure open access to information about Australia's environment

- global perspective — wherever possible, to present information in a comparative manner, seeking to place local and regional information in national and international contexts

- ecological sustainability — to assess environmental information and issues against the principles of ecologically sustainable development

Aims

The report has the following broad aims:

- to provide the Australian public, its governments, and decision-makers with accurate, timely and accessible information about the condition of and prospects for the Australian environment

- to increase public understanding of the Australian environment, its condition and prospects

- to facilitate the development of an agreed set of national environmental indicators

- to provide an early warning of potential problems

- to report on the effectiveness of policies and programs designed to respond to environmental change, including progress towards achieving environmental standards and targets

- to contribute to the assessment of Australia's progress in protecting biological diversity and maintaining ecological processes and systems

- to create a mechanism for integrating environmental information with social and economic information, thus providing a basis for incorporating environmental considerations in the development of long-term, ecologically sustainable economic and social policies

- to identify gaps in our knowledge of Australia's environmental conditions and trends and recommend strategies for research and monitoring to fill these gaps

- to contribute to Australia's international environmental reporting obligations (see the box on page 1-11)

- to help decision-makers to make informed judgments about the broad environmental consequences of social, economic and environmental policies and plans

Users and products

It is now widely accepted in other countries, and confirmed by public consultations in Australia, that a wide range of people use state of the environment information. The main user groups include: the general public and specific community interest groups and sectors; government decision-makers and policy analysts at Commonwealth, State, Territory and local levels; cultural and natural resource planners and managers; scientists; primary, secondary and tertiary educational institutions; industry groups; the print and electronic media; and international agencies.

Each group has different needs. For example, while scientists and environmental planners require very detailed information, the general public, secondary school students and community groups want broader assessments of the state of the environment. This hierarchy of information needs (see Fig. 1.1) applies at all scales from local to global.

All groups expect the information, in whatever form it is provided, to be up-to-date, accurate and accessible. Meeting these expectations was a fundamental goal of the present report.

Approach to state of the environment reporting

The approach adopted for state of the environment reporting for Australia has two major elements:

- the conceptual structure — essentially the OECD's 'pressure–state–response' model, which is broad in scope, but with significant local modifications that are now attracting international attention

- relevant techniques — development of environmental indicators, establishment of baseline information, development of a predictive capacity, selection of issues, information management and evaluation of the reporting system and its products.

Figure 1.1 The hierarchy of information needs

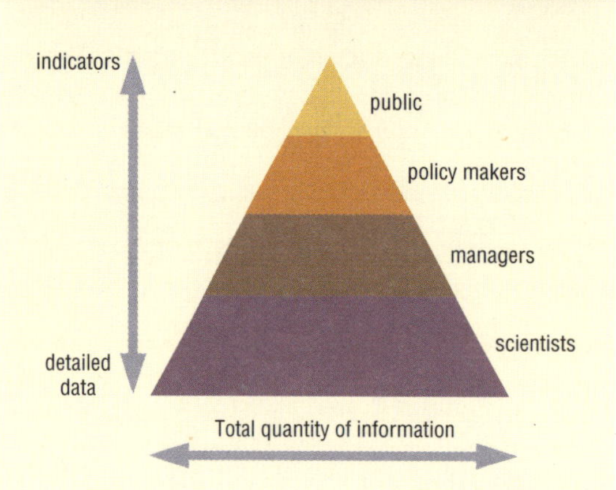

Structure and scope

In recent years, groups around the world have made considerable progress in developing a consistent conceptual structure for reporting on the state of the environment. Australia has adapted the OECD's pressure–state–response model for its reporting system (see the box opposite). The model is based on the concept of causality: human activities exert pressures on the environment and change its state or condition. Society responds to this changed state by developing and implementing policies, which complete the cycle and influence those human activities that exert pressure on the environment.

The OECD model is not the only approach. Any structure will change as community values change and our understanding of environmental problems increases. Furthermore, the OECD model implies simple relationships in the interaction between human activity and the environment: this should obscure neither the complexity of ecological relationships nor the difficulties in taking into account the natural variability of ecological systems. The Australian refinement of the OECD model includes some additional interactions.

Responses are sometimes directed to changing the state rather than relieving pressures: instead of restricting vehicle use and economic activities that cause urban air pollution, for example, we tend to look for ways to clean up the air-shed. The state of the environment can itself affect the pressures, as when depletion of a fishery reduces the level of fishing or the littering of a beach by tourists reduces its appeal. Finally, the responses we develop are significantly shaped by our perception of the pressures causing the problem. This Australian refinement of the OECD model has been adopted by the international group set up by the United Nations Environment Programme (UNEP) to develop a report on the Global Environmental Outlook. Thus the more sophisticated model developed for this report has influenced environmental reporting at the global level.

State of the environment reporting is guided by the principles of ecologically sustainable development, as outlined in the National Strategy for Ecologically Sustainable Development endorsed by the Council of Australian Governments in 1992.

The present report has a comprehensive coverage. It reviews significant environmental pressures and changes caused by human activities and broadly links socio-economic forces with changes in the environment.

The report covers all Australian States and mainland Territories, the continental shelf, the external territories (with the exception of the Australian Antarctic Territory and the Territory of Heard Island and McDonald Islands) and areas within the 200-nautical-mile Exclusive Economic Zone.

Where possible, the report uses biophysical and ecological rather than administrative boundaries to present information. This practice is now widespread among governments that report on the state of the environment. It is consistent with current approaches to natural resource management and integrated management of human settlements.

Approaches to regionalisation are complex. The authors have determined criteria for identifying regions and their spatial extent according to the purpose of the analysis concerned. Different regionalisations were found to be appropriate for different issues. Among the spatial units used are major groundwater basins, major airsheds, urban settlements and statistical and administrative regions used for national surveys and population censuses. The report also includes the distribution of selected plant and animal species and communities.

For each issue, the state of information relating to Australian environments is reviewed. In 1990 the Australian Science and Technology Council found that Australia lacks the following:

- an integrated national system for measurement of environmental quality

- a national data set of sufficient calibre to assess and manage environmental quality

- appropriate national baseline data to evaluate the effectiveness of strategies

It is not the function of the state of the environment reporting system to maintain national data of this kind. It is, however, part of its function to assess the adequacy of current environmental monitoring and data-management systems, in order to identify gaps in monitoring effort, coverage and knowledge of suitable indicators and their use. Chapter 10 of this report identifies needs for improving the knowledge base necessary for effective environmental assessment and national state of the environment reporting. This point cannot be over-emphasised. In many important areas, Australia just does not have the data, the analytical tools or the scientific understanding that would allow us to say whether current patterns of change to the natural environment are sustainable. We are effectively driving a car without an up-to-date map, so we cannot be sure where we are. Improving our view of the road ahead by enhancing the environmental data base is a very high priority. Our intended destination is a sustainable pattern of development, but it is not always clear which direction we need to take to get there.

Environmental indicators

These are physical, chemical, biological or socio-economic measures that can be used to assess natural resources and environmental quality. The OECD notes two particular functions for environmental indicators.

- They reduce the number of measures that would normally be required to give an 'exact' representation of a situation.

Australian refinements to the OECD pressure–state–response reporting model

The authors of this report have adopted the following refinements and clarifications:

- pressures are defined as human-induced

- natural conditions are primary states (for example, soil salinity, climate variability, soil nutrients, topography and natural hazards)

- inappropriate human actions, including responses to such natural conditions as droughts, are pressures

- states reflect pressure and the effectiveness of responses

- responses can be aimed at both pressures and states

- appropriate responses reduce pressures

- lack of action can be a pressure

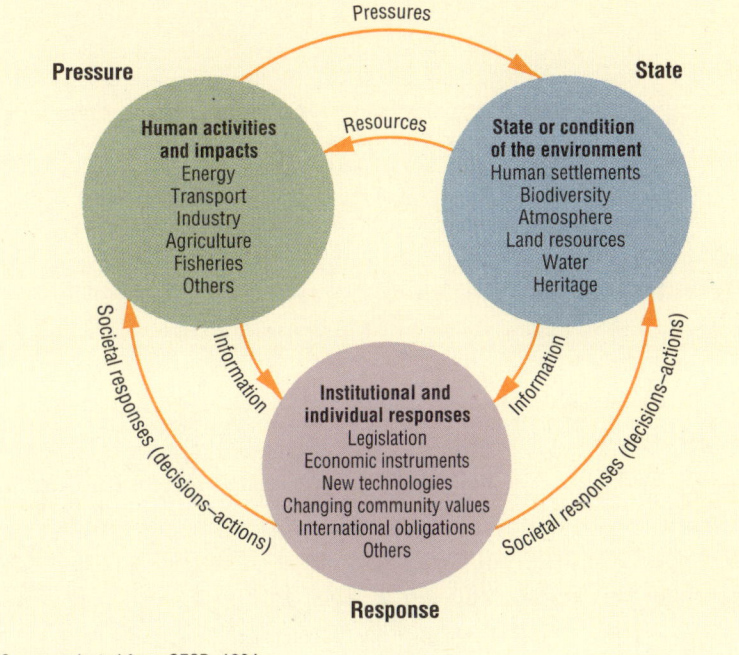

Source: adapted from OECD, 1994.

- They simplify the communication process by which information about the results of measurement is provided to the user.

Environmental indicators are usually developed for a specific purpose. They differ from other measures in providing meaning that extends beyond the attributes directly associated with them. In a well-developed system, each matter of environmental concern will have spawned its own specific indicator or indicators. Access to widely accepted, easily measured indicators of environmental quality is essential for informed decision-making, since these express the best available knowledge.

For Australian state of the environment reporting, indicators are being developed to specifically measure the pressures on the environment, the state of the Australian environment and the societal responses taken in the light of the pressures and state.

Indicators of environmental pressures describe the extent of human activities as they affect the environment. Indicators of environmental conditions — the state of the environment — describe the quality of the environment. Measurement of those conditions can be extremely difficult and expensive, and the OECD notes that measurement of environmental pressures is often used as a substitute for measurement of the conditions.

Indicators of response show the extent and effectiveness of society's responses to environmental changes and concerns. Responses include individual and collective actions aimed at mitigating, adapting to or reversing negative impacts on the environment and reversing environmental damage already done. They also include actions to improve the preservation and conservation of the environment. Policy performance indicators are also relevant.

Monitoring environmental indicators over time can provide an effective early-warning system. Using them in monitoring programs to report on the condition of the environment can therefore serve a range of objectives, such as identifying where present social behaviour and economic policies could lead to future environmental degradation and associated economic and social costs.

Procedures for choosing environmental indicators are discussed in many state of the environment reports and in the literature on ecologically sustainable development. The box on the left summarises the criteria that are considered most useful for selecting indicators.

Development of a nationally agreed set of environmental indicators for Australia is a high priority for state of the environment reporting. It is, however, a complex task that will take a number of years to complete. Major interested parties, including key community and industry sectors, Commonwealth, State, Territory and local governments, scientific communities and other research groups will need to be involved. It will be necessary to progressively identify a scientifically credible set of environmental indicators and associated monitoring requirements for state of the environment reporting, and to reach national agreement on a set of indicators.

The role of this first report under the new reporting framework has been to identify the key environmental issues at the national level, using the best available data sets, and thus it has not been based on a formal set of approved environmental indicators. However, it is an essential first step in the development process because it has led to an initial identification/characterisation of the 'first generation' of national environmental indicators. Only now is it possible to start identification and technical specification of environmental indicators appropriate to Australian conditions.

Some data sets used in this report suggest themselves as indicators, but they will need to be refined by an iterative checking and testing process. Other indicators will need to be developed from scratch because existing data sets that are directly or indirectly relevant were collected for other purposes and are inadequate to serve as the basis for specific environmental indicators.

Not all of the environmental indicators used overseas, for example by the OECD, are directly relevant to Australia. Many will have to be developed specifically for Australian conditions and issues. In addition, data sets may not exist for emerging environmental issues. The box above summarises part of the core set of environmental indicators developed by the OECD.

Selection criteria for national environmental indicators

A number of criteria guide the choice of indicators for state of the environment reporting. An indicator should:

- serve as a robust indicator of environmental change

- be sensitive to environmental change

- reflect a fundamental or highly valued aspect of the environment

- be either national in scope or applicable to regional environmental issues of national significance

- provide an early warning of potential problems

- be capable of being monitored to provide statistically verifiable and reproducible data that show trends over time and, preferably, apply to a broad range of environmental regions

- be scientifically credible

- be easy to understand

- be monitored regularly with relative ease

- be cost-effective

- be as aggregative as possible (that is, amenable to combination with other indicators to produce more general information about environmental conditions)

- have relevance to policy and management needs

- contribute to monitoring of progress towards implementing commitments in nationally significant environmental policies

- where possible and appropriate, facilitate community involvement

- contribute to the fulfilment of reporting obligations under international agreements

- where possible and appropriate, use existing commercial and managerial indicators

- where possible and appropriate, be consistent and comparable with other countries and State and Territory indicators

Source: Department of the Environment, Sport and Territories, 1994

Need for baseline information

Evaluation of environmental change depends on the presence of a baseline to measure against. In general, baselines are established so we can measure significant change in a selected attribute. In that sense, baselines form part of the set of environmental indicators. They can reflect change over time or they can reflect the difference between spatial areas at a particular time. Such evaluation is hampered by the natural variability of Australian conditions and the limited availability of data from scientifically rigorous monitoring of environmental change.

The OECD's core environmental indicators

The OECD has identified the following list of issues to reflect current environmental challenges: climate change; ozone layer depletion; eutrophication; acidification; toxic contamination; urban environmental quality; biodiversity; landscape; waste; water resources; forest resources; fish resources; and soil degradation. The first nine issues can be considered 'sink-oriented', dealing with environmental quality, whereas the remaining four are 'source-oriented', focusing on the quantity aspect of natural resources.

Not all indicators can be directly associated with a specific environmental issue such as population growth or economy-wide environmental expenditure, so a category of general indicators has been added to their core set.

Using the pressure–state–response reporting framework, the OECD identified a number of specific indicators for these identified environmental issues. Below is an extract from its core set for three of its listed issues.

Issue	Pressure	State	Response
	Indicators of environmental pressure	Indicators of environmental condition	Indicators of societal response
Climate change	• Index of greenhouse gas emissions (M) • Carbon dioxide emissions (S)	• Atmospheric concentrations of greenhouse gases (S) • Global mean temperature (S)	• Energy efficiency (M) • Energy intensity (S) • Economic and fiscal measures (M)
Ozone layer depletion	• Index of apparent consumption of ozone-depleting substances (S) • Apparent consumption of CFCs and halons (S/M)	• Atmospheric concentrations of ozone depleting substances (M) • Ground level UV-B radiation (M)	• CFC recovery rate (M)
Eutrophication	• Emissions of nitrogen and phosphorus in water and soil (–> nutrient balance) (L) • Nitrogen from fertiliser use and from livestock (S) • Phosphorus from fertiliser use and from livestock (S)	• Biological oxygen demand/dissolved oxygen (S/M) • Concentration of nitrogen and phosphorus in inland and marine waters (M/L)	• Percentage of population serviced by biological and/or chemical sewage treatment plants (M/L) • Percentage of population connected to waste-water treatment plants (S) • User charges for waste-water treatment (M) • Market share of phosphate-free detergents (S/M)

Note: each indicator is followed by a character specifying its availability: S = data available in the short term; M = data expected to be available in the medium term; and L = data expected to be available in the long term.

Not all OECD indicators — for example, those dealing with acid rain — are relevant to Australian conditions. The OECD report acknowledges that the set of environmental indicators is dynamic and may change as knowledge and perception of environmental problems evolve.

Source: OECD, 1994.

Much of the significant environmental change to Australia — such as clearing of native vegetation, erosion of topsoil and pollution of waterways with heavy metals — occurred during the nineteenth and early twentieth centuries. The environment continues to change today in response to pressures from human activities but, to represent accurately the importance of current trends and conditions, state of the environment reporting must place impacts in a historical context whenever possible.

Long-term monitoring is required in order to screen out natural variability and establish reliable baselines. State of the environment reporting aims to use baselines that reflect the full extent of impacts of human uses of the environment.

The reporting process

State of the Environment Advisory Council

The State of the Environment Advisory Council, representing a broad range of expert and community interests, has shaped and overseen the production of this report. Membership of the Council is primarily drawn from outside government and includes eminent persons from the conservation movement, industry, the scientific community, academia, and Aboriginal and Torres Strait Islander communities. Members were appointed by the Minister for the Environment, Sport and Territories on the basis of merit and standing in community and professional circles.

The Council's role is as follows:

• to provide advice on national state of the environment policy and planning

• to assist in the identification of environmental information needs

• to evaluate national state of the environment reporting

• to review drafts of state of the environment publications to ensure their objectivity and credibility

• to assist in enhancing public awareness of the findings of reports

Reference groups

The credibility of this report depends on its accuracy, relevance and impartiality. Expert reference groups provided the mechanism for identifying important issues and the kinds of information needed to report on them. Chapters 3–9 were each prepared by an expert reference group, chosen to bring together a comprehensive knowledge base in those specific areas. Group members were drawn from the academic and research community, as well as government and non-government scientific, technical and professional groups. More than 50 experts were involved. The members of the relevant reference groups are listed in the chapters. The report could not have been produced without their professional contribution, which went well beyond the call of duty. Every reference group included a member from the State of the Environment Unit, acting as the point of contact between the reference group, the Advisory Council and the Department, as well as facilitating the group's operation.

State of the Environment Reporting Unit

The Unit is part of the Environmental Strategies Directorate of the Department of the Environment, Sport and Territories. It has three main areas of responsibility.

• It manages the production of regular state of the environment reports and associated products.

• It develops, in cooperation with State and Territory governments and non-government agencies, a national state of the environment reporting system. The development of nationally agreed environmental indicators is a high priority in bringing greater cohesion to the national reporting system.

• It deals with Australia's international responsibilities with respect to state of the environment reporting.

It also serves as secretariat to the Advisory Council, has contracted the reference groups and handled the administrative aspects of the report production. It has joint responsibility with the Public Relations and Education Unit of the Department for implementing the communication strategy for the report.

Referees

As part of the process of developing this report, each of Chapters 2–9 was reviewed by a panel of expert referees. The aim of this review process was to ensure the report's scientific accuracy and independence. About 40 independent experts were involved. The final report was improved considerably by the valuable comments of these referees, who are listed in the chapters concerned.

A draft of the report was also sent to all State and Territory governments through ANZECC SoE Reporting Taskforce members, to all Commonwealth departments and to the Australian Local Government Association for comment on its factual accuracy. Their comments were also valuable, and much appreciated.

How to use this report

The best way to get the 'big picture' of the state of the Australian environment is to read the Executive Summary. The entire report should be seen as a reference work rather than a narrative. Chapter 2 sets out a general description of the nature of Australia as a physical, biological, social and cultural landscape, describing how we have arrived at the present situation. Each of Chapters 3 to 9 describes in detail a particular facet of the contemporary environment. Each concludes with a summary table, listing the main points of the analysis. Chapter 10 draws together the key findings of the report to give an overall picture of our progress toward a pattern of development that would be ecologically sustainable.

While the Advisory Council has tried to give the complete report a style and approach that is as uniform as possible, the differences in the subject matter dictate different ways of reporting. In that sense, Chapters 3 to 9 resemble a collected set of reference works, each giving an expert view of one aspect of the contemporary environment. A number of technical reports, commissioned for the reference groups, are also being published. These provide more detailed information on aspects of Chapters 2 to 9 and are referred to in those chapters.

Please note that the report is only as comprehensive as is possible within the constraints of this volume and the limited time available to prepare it. Every one of the reference groups could easily have used two or three times as much space to give a more detailed picture. Finally, the report presents as accurately as possible a picture of the state of the environment in 1995, taking into account limitations of available data when the typescript was handed to the publishers. As it points out, the most obvious characteristic of the environment is change. Its findings will not remain an up-to-date picture for the complete period between the publication of this report and the next one in this series. The longer the period since its publication, the more one needs to take care in establishing what changes have occurred.

Australia's international environmental reporting commitments

The following are examples of Australia's international reporting commitments.

International conventions

The Convention of the World Meteorological Organization, ratified by Australia in 1948, commits Australia to international cooperation in monitoring, research and data exchange in respect of the atmosphere, oceans and inland waters.

The Framework Convention on Climate Change, ratified by Australia in 1992, focuses on research, data collection and monitoring. It commits countries to periodic reporting on their greenhouse gas emissions and sinks.

The Vienna Convention and subsequent Montreal Protocol for the Protection of the Ozone Layer, ratified by Australia in 1990, focus on research, data collection, monitoring and periodic reporting in their commitment to control ozone-depleting emissions of chlorofluorocarbons, halons, carbon tetrachloride and methyl chloroform.

The Basel Convention on the Control of Transboundary Movement of Hazardous Wastes and their Disposal was ratified by Australia in 1992. Under this convention, Australia is obliged to transmit information to the Secretariat of, among other things, all transboundary movements of hazardous wastes in which Australia has been involved.

The Convention on Biological Diversity, ratified by Australia in 1993, requires parties to provide periodic reporting on measures taken to implement the provisions and meet the objectives of the Convention, which aims to promote the conservation and sustainability of biological diversity and the equitable sharing of benefits from genetic resources.

The Convention Concerning the Protection of the World Cultural and Natural Heritage, ratified by Australia in 1974, requires annual monitoring reports on the management of Australia's World Heritage areas.

The Protocol on Environmental Protection to the Antarctic Treaty (Madrid Protocol) designates Antarctica as '... a natural reserve devoted to peace and science' and places an indefinite ban on mineral resource activity in Antarctica. When in force, it will entail environmental reporting obligations for Australia.

The London Convention on the Prevention of Marine Pollution by Dumping of Wastes and Other Matter, ratified by Australia in 1985, requires annual reports of approvals to dump or incinerate wastes and other matter at sea.

The International Convention for the Prevention of Pollution from Ships, ratified by Australia in 1987, requires annual reporting of incidents involving pollution from ships.

Australia signed the Convention to Combat Diversification in 1994 and is currently considering the question of ratification. The Convention will require reporting on the strategies established to combat diversification and mitigate the effects of drought. Development of and hence reporting on implementation of a national action program would be voluntary for Australia if the convention is ratified.

The Convention on the Means of Prohibiting and Preventing the Illicit Import, Export and Transfer of Ownership of Cultural Property, ratified by Australia in 1989, requires periodic reporting to UNESCO of actions taken under the national protection provisions.

The Convention on Conservation of Migratory Species of Wild Animals (Bonn Convention), the Convention on International Trade in Endangered Species of Wild Fauna and Flora (CITES Convention) and the Convention on Wetlands of International Importance especially as Waterfowl Habitat (Ramsar Convention) also have reporting requirements.

Reporting requirements of international organisations

Australia's membership of the OECD, the United Nations Environment Programme (UNEP), the United Nations Economic and Social Commission for Asia and the Pacific (ESCAP) and the World Meteorological Organization (WMO) brings with it reporting obligations for various aspects of the condition of the Australian environment.

Agenda 21

The 1992 United Nations Conference on Environment and Development (UNCED) created a high-level Commission on Sustainable Development, which considers national reports on implementation of Agenda 21, the principal program arising from the Conference, and investigates ways to improve the collection and dissemination of environmental data at the global level.

Improving the reporting process

This document is a first attempt at an ambitious task; it will certainly need to be refined in the light of experience. The Commonwealth Government, the State of the Environment Advisory Council and the State of the Environment Reporting Unit are very keen to involve the whole community in refining the reporting framework. After all, it is our environment that is at stake. Any comments you may have, on the reporting task and on its approach, will be most welcome.

To contact the Council, write to:

> The Secretary (State of the Environment Advisory Council)
> State of the Environment Reporting Unit
> Department of the Environment, Sport and Territories
> GPO Box 787
> Canberra ACT 2601

For more information about the report, please telephone the Community Information Unit of the Department of the Environment, Sport and Territories on 1800 803 772.

References

Commonwealth Environment Protection Agency (1992). 'Development of a National State of the Environment Reporting System - Discussion Paper.' (CEPA: Canberra.)

Council of Australian Governments (1992). 'National Strategy for Ecologically Sustainable Development.' (AGPS: Canberra.)

Department of the Environment, Sport and Territories (1994). 'State of the Environment Reporting: framework for Australia.' (DEST: Canberra.)

Organisation for Economic Co-operation and Development (1991). 'State of the Environment.' (OECD: Paris.)

Organisation for Economic Co-operation and Development (1994). 'Environmental Indicators — OECD Core Set.' (OECD: Paris.)

Acknowledgments

Ian Robertson (Department of the Environment, Sport and Territories) assisted in the preparation of this chapter.

Photo credits

Page 1-1: The Photo Library of Australia
Page 1-4: Ray Radford, 'Friends of Merri Creek'.

Chapter 2

Portrait of Australia

Detail from *'Container Train in Landscape'*
by Jeffrey Smart, donated to the Victorian
Arts Centre by Eva and Marc Besen, 1984.

Prepared by

Roger Beckmann
under the direction of the
State of the Environment Advisory Council.

Contents

An unusual continent

Earth's southern hemisphere consists mainly of water. Only two continents, and some sizeable islands, lie entirely within it and of the two exclusively southern continents, only one — Australia — is permanently inhabited. It is the world's smallest continent, with an area of about 7.6 million sq km. But the nation of Australia, which occupies the entire continent and many outlying islands, is the world's sixth-largest country, and the only one to have responsibility for a whole continent.

Australia is a federal nation divided into six States and two main Territories, each with its own internal government and capital city. The country has 770 local government divisions. It also has seven external territories: Norfolk Island; Christmas Island; Cocos (Keeling) Islands; the Coral Sea Islands Territory; the Territory of Ashmore and Cartier Islands; the Australian Antarctic Territory; and Heard Island and McDonald Islands (see Fig. 2.1). Norfolk Island and Cocos (Keeling) Islands are locally self-governing. The federation of all the States and Territories forms the Commonwealth of Australia, of which Canberra, in the Australian Capital Territory, is the capital. Most environmental responsibilities fall under the shared jurisdictions of local governments, the individual States and Territories, and the Commonwealth Government. The Commonwealth has primary responsibility for international relations, for the sea and seabed outside three nautical miles and for the

management of certain areas of national or international significance.

Because Australia has no land boundaries, we do not suffer any great problem of pollution coming from other countries via the air or rivers. Of course, some general air- or water-borne contaminants from around the globe inevitably enter Australia and emissions of greenhouse gases and ozone-depleting substances have an impact. Our nearest neighbour is Papua New Guinea, about 130 km from the tip of the mainland, but only a few kilometres from the northernmost islands of the Torres Strait. The issue of transboundary pollution is starting to arise in terms of the effect of the discharge of rivers carrying waste from mining operations within Papua New Guinea into the Torres Strait. In theory, types of air pollution from Indonesia could also reach our northern shores. A major quarantine concern is the potential for organisms to be introduced from overseas.

Although we have no land boundaries, Australia does have marine jurisdictional boundaries with five neighbours — Indonesia, Papua New Guinea, New Zealand, New Caledonia and Solomon Islands — and the international waters beyond us are shared with and used by many countries.

The southern hemisphere remains, in general, markedly less affected by human activities than the northern. Australia lies in relatively unpolluted air and sea. But its position ensures that it is affected

This Land

Give me harsh land to wring music from,
brown hills, and dust, with dead grass
straw to my bricks.

Give me words that are cutting-harsh
as wattle-bird notes in dusty gums
crying at noon.

Give me a harsh land, a land that
swings, like heart and blood,
from heat to mist.

Give me a land that like my heart
scorches its flowers of spring,
then floods upon its summer ardour.

Give me a land where rain is rain
that would beat high heads low.
Where wind howls at the windows

and patters dust on tin roofs
while it hides the summer sun
in a mud-red shirt.

Give my words sun and rain,
desert and heat and mist,
spring flowers, and dead grass,
blue sea and dusty sky,

song-birds and harsh cries,
strength and austerity
that this land has.

By Ian Mudie (1911–1976) In *The New Oxford Book of Australian Verse.* (Chosen by Les A Murray. Oxford University Press, Melbourne 1987).

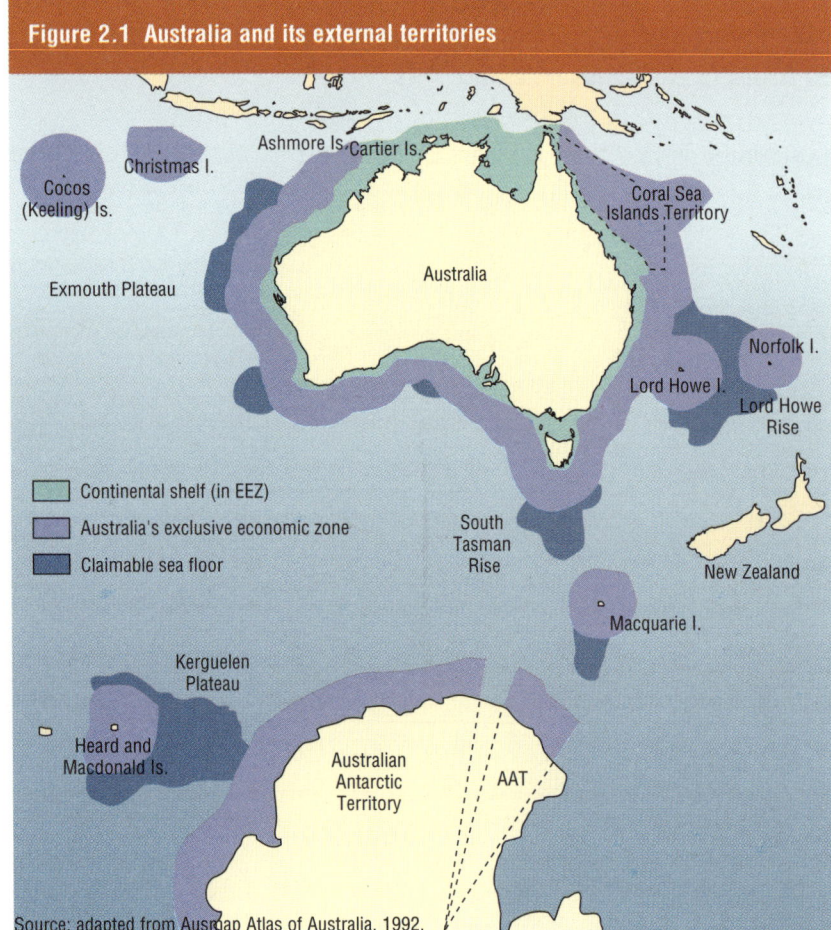

Figure 2.1 Australia and its external territories

Christmas I.
Cocos (Keeling) Is.
Ashmore Is. Cartier Is.
Coral Sea Islands Territory
Exmouth Plateau
Australia
Norfolk I.
Lord Howe I.
Lord Howe Rise
South Tasman Rise
New Zealand
Macquarie I.
Kerguelen Plateau
Heard and Macdonald Is.
Australian Antarctic Territory
AAT

☐ Continental shelf (in EEZ)
☐ Australia's exclusive economic zone
☐ Claimable sea floor

Source: adapted from Ausmap Atlas of Australia, 1992.

by certain natural features of the southern hemisphere ocean and atmosphere that cause great climate variability. In particular, Australia lies near the centre of action of the so-called El Niño–Southern Oscillation (ENSO), described in the box on page 2-8.

The imprint of the ENSO phenomenon can be detected in the continuous records of nearly all climatic variables in Australia, but it has its greatest impact on rainfall and air temperature. The most pronounced variability in the Australian region occurs over the eastern two-thirds of the continent, where ENSO accounts for 30–40 per cent of rainfall variability, which is such a feature of our continent.

As well as the ENSO, changes in sea-surface temperature in the Indian Ocean contribute to climatic variability through their effect on the passage of north-west rain-bearing cloud bands over the continent. In some seasons, such cloud bands can bring increased winter rainfall to southern and western parts.

The land

Mainland Australia (see Fig. 2.2) extends some 3180 km from the tip of Cape York in the tropical north to South Point on Wilson's Promontory in temperate Victoria. The southern tip of Tasmania is a further 500 km south. From east to west Australia stretches about 4000 km. In area, the country is about 32 times larger than the United Kingdom, more than 20 times larger than Japan, two-and-a-half times larger than Indonesia, and nearly as large as the United States without Alaska. However, in terms of population, it is much smaller than any of these countries.

Along the continent's eastern margin lies the Great Escarpment. In part coincident with this, but often up to 300 km further inland, is the Great Dividing Range. Both these formations are 'great' in length rather than height. Their highest points rarely exceed 1600 metres, and their features are interrupted in places by river valleys, such as those of the Fitzroy River near Rockhampton and the Hunter River near Newcastle. The highest parts of the uplands occur in the continent's south-east, where a small truly alpine area, snow-covered for more than half the year, includes Australia's highest mountain, Mt Kosciusko (2228 m above sea level). Australia's lowest temperature of –23° C was recorded at Charlotte Pass, New South Wales, in this alpine region on 29 June 1994.

West of the eastern uplands are the lowlands of the Great Artesian Basin and further west, the Western Plateau with emergent ranges, such as the MacDonnell, Hamersley, Stirling and Flinders Ranges. The large, flat, limestone Nullarbor Plain (the name means no trees) is a distinctive feature. This arid area was once the bottom of a shallow sea during the Tertiary period (65 to 20 million years ago).

The Murray–Darling system forms the largest drainage basin, but it has a small volume of flow (0.15 ML/sec on average) in comparison with drainage basins in other parts of the world.

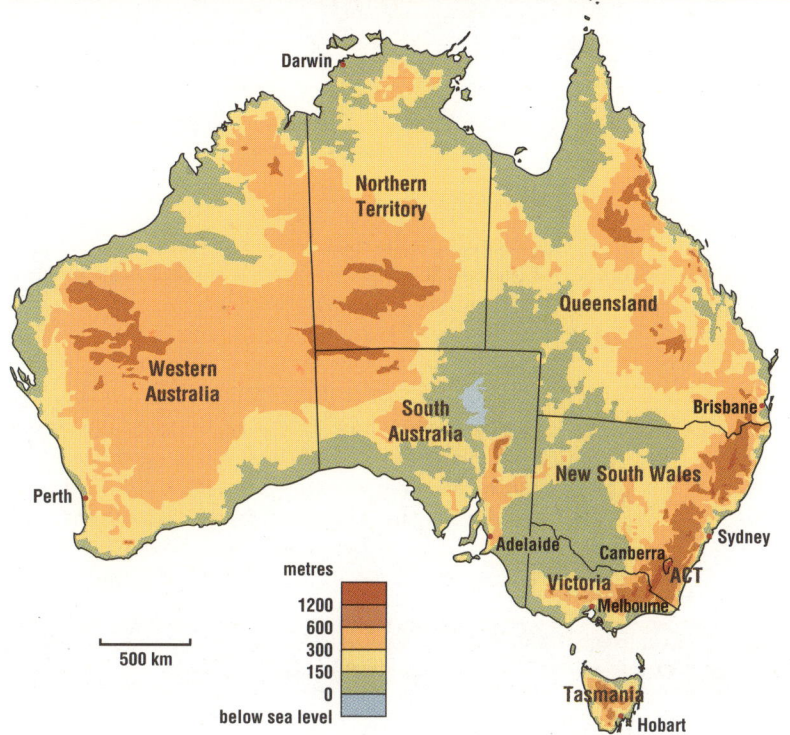

Figure 2.2 Australia's continental elevation, capital cities and States

Darwin

Northern
Territory

Queensland

Western
Australia

Brisbane

South
Australia

New South Wales

Perth

Sydney

Adelaide

Canberra
ACT

Victoria

Melbourne

metres
1200
600
300
150
0
below sea level

500 km

Tasmania

Hobart

For example, the Mississippi River has an average annual flow of 17 ML/sec and the Amazon River has 180 ML/sec. The system drains rainfall from part of Queensland, a large part of New South Wales, the Australian Capital Territory and much of Victoria. It flows into the sea on the eastern side of South Australia. The Murray is about 2500 km long, while the Darling and Upper Darling Rivers together total about 2000 km.

World's lowest continent

Australia has the lowest average elevation and also the lowest absolute relief (the difference between highest and lowest points) of any continent, partly as a result of its long period of geological stability (see Fig. 2.3).

Figure 2.3 Average elevation and rainfall of the world's inhabited continents

Average elevation (metres)	elevation	Average yearly rainfall (millimetres)
1200	rainfall	1800
1000		1500
800		1200
600		900
400		600
200		300
0		0

Eurasia   North America   Africa   South America   Australia

Source: AAS, *Environmental Science*, 1994 and ABS, *Yearbook Australia*, 1995.

Its few mountain ranges are exceedingly low by world standards, and most of the country is a broad flat platform, broken by low hills and basins in a few places. The eastern uplands provide an important feature, ensuring a fairly reliable rainfall to the eastern seaboard. Partly because it has no high mountains, Australia is also the only continent with no permanent year-round snow.

Stable geology

Australia is the only continent without a geologically active volcano, although both western Victoria and northern Queensland have experienced major volcanic events in the last 20 000 years. The country does not have great rift systems or large permanent lakes. Large areas have been stable for very long periods of time and in these places the surface features have changed relatively little. Much of western and north-western Australia and large parts of the eastern areas have been stable dry land for 400 million years, and most of the continent has been extensively weathered for at least 60 million years, with surface features and rocks greatly eroded. While we have some of the oldest rocks yet recorded on the planet, the great antiquity of our landscapes is not so well appreciated.

Lowest rainfall of any inhabited continent

Australia's latitude ranges from about 10° S to about 43° S. This puts much of it in a zone of high atmospheric pressure and therefore low rainfall (see Fig. 2.4), which is exacerbated by the absence of high mountains. Indeed, it is the driest inhabited continent (Antarctica is the absolute driest, partly because it sits in the polar high-pressure region).

More than one-third of Australia is classified as arid, which means it receives an average annual rainfall of less than 250 millimetres. Another third is semi-arid, with an annual average between 250 and 500 mm (see Fig. 2.5). By comparison, the country's cities receive much higher rainfalls and thus are not representative of most of the continent (see Table 2.1). However, definitions of aridity based on rainfall take little account of differences in evaporation rate. Perhaps a better indicator of Australia's dryness is the minimum rainfall necessary for successful crop-growing without irrigation in the different latitudes. In the far south of Australia, rain-fed crops need about 250 mm of rain per year; in most of New South Wales they need at least 375 mm a year; this rises to 500 mm in northern New South Wales and south-eastern Queensland, and to 750 mm a year in the far north of the country. The important point is that areas in the north can receive much higher rainfalls than those in the south and still be arid.

Despite its aridity, Australia has few areas of true desert — defined either as areas of no perennial plant growth or as places receiving less than 100 mm annual average rainfall. (The region around Lake Eyre in South Australia receives the lowest rainfall in the country with an annual average of about 100 mm.) Environmental conditions in the arid zone are surprisingly variable, partly because rain, when it does come, can be very localised and also because water may accumulate in natural features, which can serve as relatively lush refuges during dry times.

Diverse climate

Australia's geographic location and its size mean that it experiences a wide range of climate zones: temperate climates in coastal Tasmania and the southern parts of the mainland; Mediterranean in the south-west and south-central areas; tropical climates in the north; subtropical along the warm east coast; and a small region of alpine climate occurring in the south-east of the mainland and in central Tasmania (see Fig. 2.5).

Much of the continent lies in the earth's desert zones, which are between about 20° and 35° north and south of the equator. This area experiences high atmospheric pressure, formed from descending air that has arisen in the tropics. High-pressure cells travel across the country from west to east. In the

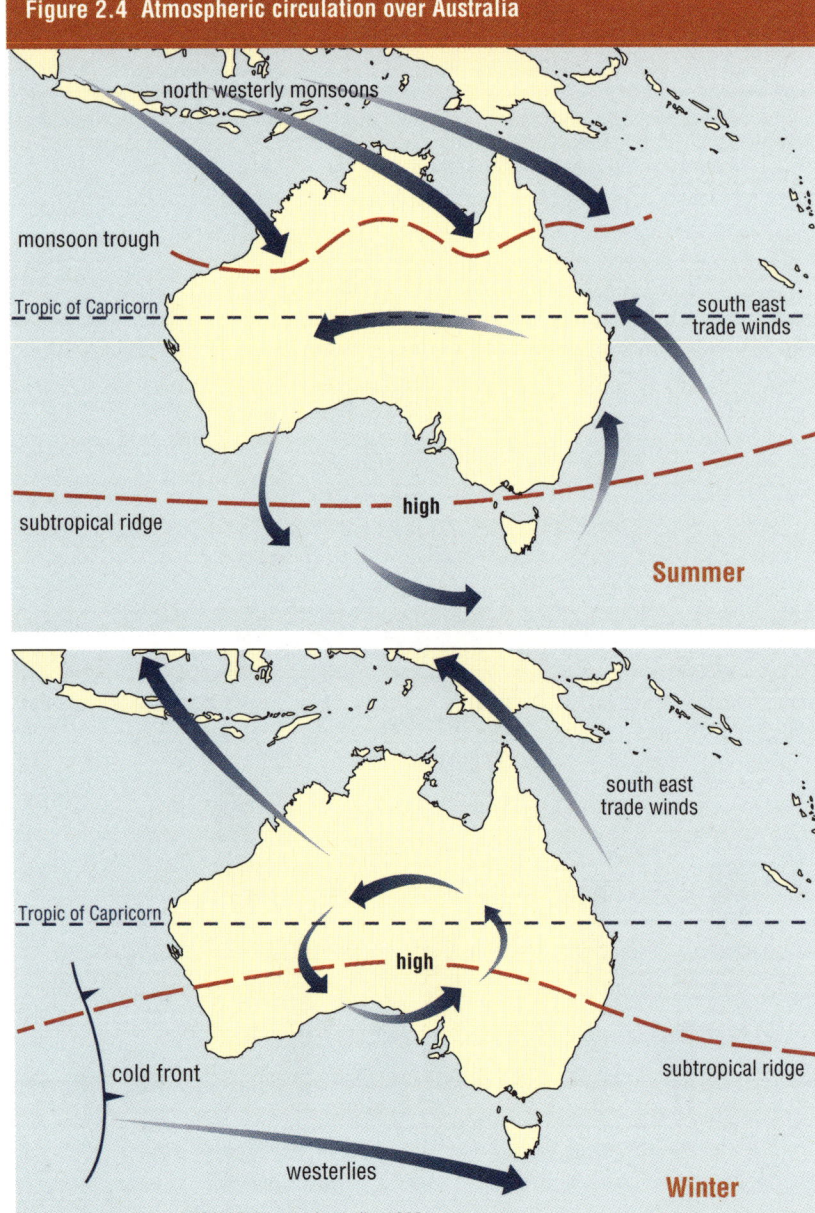

Figure 2.4 Atmospheric circulation over Australia

north westerly monsoons

monsoon trough

Tropic of Capricorn

south east trade winds

subtropical ridge

high

Summer

south east trade winds

Tropic of Capricorn

high

cold front

subtropical ridge

westerlies

Winter

Source: adapted from AUSMAP Atlas of Australia, 1992.

cooler half of the year (May to October) the high-pressure systems pass right across the continent — often remaining stationary for several days. Northern Australia falls under the influence of mild, dry south-east trade winds, while southern areas experience cooler, moist westerly flows. In the warmer half of the year (November to April), the highs become centred well to the south of the continent. Easterly winds predominate and most of southern Australia experiences fine warm weather; but great heat — with daytime temperatures often greater than 40° C — can build up throughout much of the inland.

More than one-third of Australia lies in the tropics (see Fig. 2.4) where cyclones are a feature of summer. These bring heavy rains and strong winds and their effects can reach much further south. They are unpredictable in their frequency and likely trajectory, so areas where rainfall is mainly cyclone-derived — such as much of the inland north — have an inherently unpredictable climate pattern. Cyclones also cause natural disturbance in the coastal zones, where they are at their strongest. The northern Queensland rainforest, for example, relies on the passage of cyclones to uproot trees and provide gaps in the dense canopy for saplings to grow. The human and economic cost of cyclones is very real, the worst in Australia's recorded history being the devastation of Darwin by Cyclone Tracy over Christmas 1974.

Most annual field crops require a three- to six-month growing season — that is, a time when rainfall, temperature and light are all within the tolerable range for the particular species being grown. In many parts of Australia the growing season is determined by lack of rainfall, or limiting high temperatures; but in southern upland and inland Australia, winter frosts limit the choice of crops and may check the growth and development of adapted crops.

The extremes of climate over much of the continent limit human settlement and cause social, economic and planning problems.

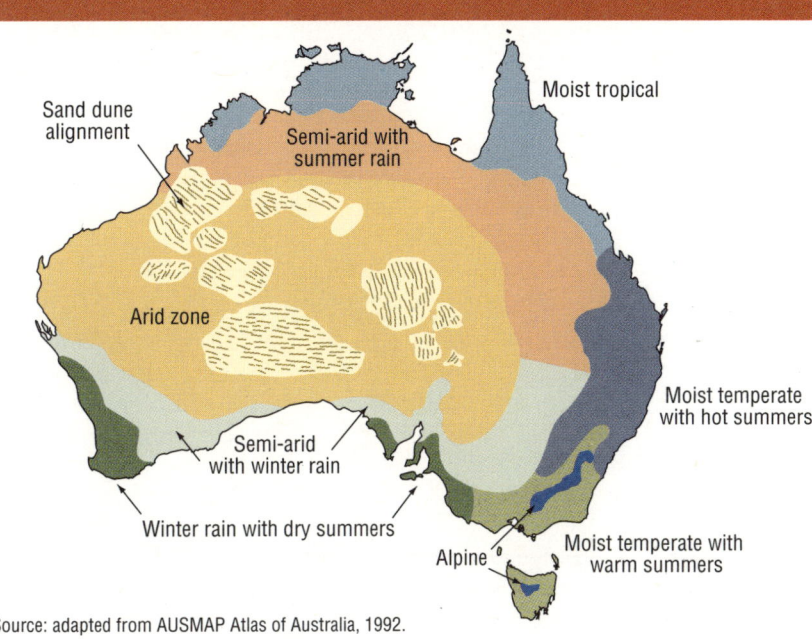

Figure 2.5 Australia's climatic zones

Source: adapted from AUSMAP Atlas of Australia, 1992.

Variable climate and rainfall

Combining the range of annual temperature and the degree of variability in annual rainfall gives an index of climatic variability for different parts of the country. The result is shown in Fig. 2.6. Clearly the central zone varies the most, while the whole eastern coast, including Tasmania, and the south-west tip of the continent show reasonably stable climate patterns. Not surprisingly, these areas have been most heavily settled by European and other recent immigrants in the last 200 years. They are also the areas with the most intensive agriculture and, increasingly, the greatest conflicts over land use.

Figure 2.6 Australia's climatic variability

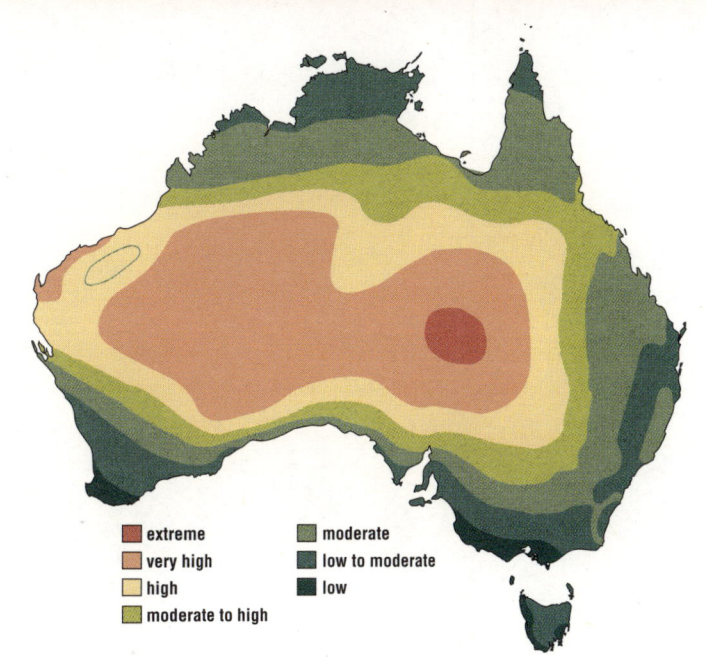

extreme / very high / high / moderate to high / moderate / low to moderate / low

Source: ABS, *Yearbook Australia*, 1988.

Table 2.1 Rainfall in our major cities

	Average annual rainfall (mm)	Number of rain days/yr
Sydney	1214	148
Melbourne	655	143
Brisbane	1151	123
Adelaide	578	119
Perth	873	119
Hobart	628	160
Darwin	1661	108
Canberra	626	109
Alice Springs	285	44
Cairns	2032	155

Source: Bureau of Meteorology, Climatic Averages of Australia, 1988.

El Niño–Southern Oscillation

South American fishermen gave the name El Niño to the sudden change in sea-water temperature associated with the disappearance of anchovies that occurred irregularly every few years off the Pacific Coast of South America near Christmas time. The Southern Oscillation, which occurs irregularly about every 2–7 years, is a meteorological term that refers to a large disturbance in the atmospheric circulation over Australia and the eastern Pacific. It is a measurable see-sawing change in average atmospheric pressure between the mid Pacific and northern Australia (measured at Tahiti and Darwin respectively).

For most of the time, pressure remains high over the mid-Pacific, and lower over Australia and the Indian Ocean. However, every few years the surface water in the central and eastern Pacific undergoes a remarkable warming, which leads to substantial changes in the atmospheric circulation throughout the entire Asia–Pacific region. The atmospheric pressure situation reverses, with persistent below-average pressure in the mid-Pacific and higher pressure over eastern Australia. Much of Australia — particularly the eastern part — then experiences a drought. The generic term El Niño–Southern Oscillation (ENSO) refers to the whole suite of events. Some ENSO impacts are also felt in the northern hemisphere.

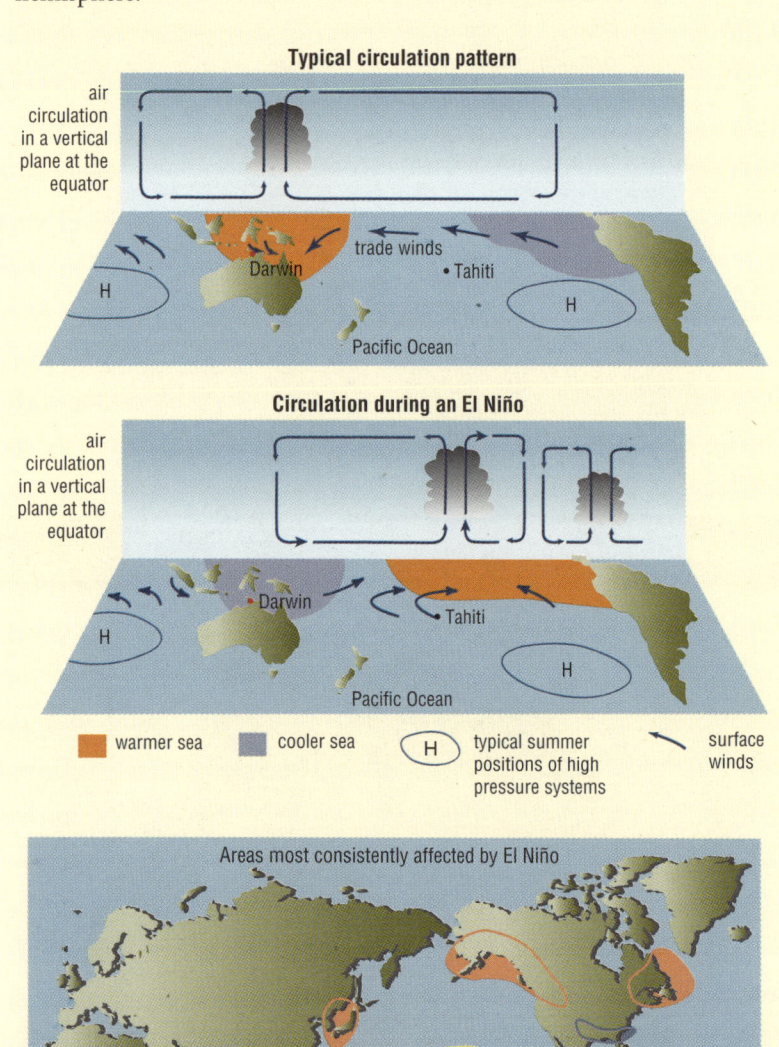

Typical circulation pattern

air circulation in a vertical plane at the equator

trade winds

Darwin • Tahiti

H   H

Pacific Ocean

Circulation during an El Niño

air circulation in a vertical plane at the equator

• Darwin

Tahiti

H   H

Pacific Ocean

■ warmer sea   ■ cooler sea   (H) typical summer positions of high pressure systems   → surface winds

Areas most consistently affected by El Niño

■ wet   ■ dry   ■ warm

Tremendous fluctuations in annual rainfall occur commonly throughout the country. Sequences of years with high rainfall, and consequent lush plant growth, can give a false picture. The variability is associated with the El Niño–Southern Oscillation, especially for the eastern half of the continent. It occurs across all time scales — within a year, between years and across decades and centuries. In some places, where total rainfall is cyclone-dependent, a large proportion of the annual average can fall within days, with little or none for the rest of the year. Hence, averages are misleading. The town of Alice Springs in the centre of the continent has an annual average rainfall of 270 mm, but the actual quantity that falls in a year has varied from 60 mm to 900 mm since records began.

Other areas of the country, such as south-west Western Australia, have strongly seasonal rainfall patterns, with long, hot and dry summers and mild wet winters. Variability, seasonality and evaporation rates influence agricultural production and the location of various agricultural enterprises. Many cities require large water storages to last during the inevitable, but unpredictable, dry spells.

In eastern Australia, the climatic fluctuations between years are partly accounted for by the El Niño effect.

The natural variability of Australia's climate affects many aspects of the country's environment — from the fauna and flora to the patterns of settlement and the water supply. It also tends to exacerbate many of the human impacts on the natural environment.

Generally unproductive soils

Australian soils are among the most nutrient-poor in the world — mainly because many components and minerals have leached out during the land's long exposure to weathering. The continent has been geologically stable for a long time; with little glaciation, mountain-building or seismic activity and scarce formation of new soil during the past few million years.

Much of the interior is covered with sandy soils that cannot hold water for long, even when it does rain. Other soils are naturally salty, low in nutrients and organic matter and, in some places, extremely stony. Unstable cracking clay soils, which swell when wet and shrink when dry, cover about 10 per cent of the country, creating problems for agriculture and building.

Some soils lack trace elements such as cobalt or copper. Some plants, especially native ones, may be able to grow in these soils, but grasses grown to feed sheep or cattle will become deficient in the element and die. Over many parts of the country plant growth was always sparse, but native plants helped enrich and retain the soil. The reduction in land vegetation cover has undoubtedly increased soil erosion in many areas.

The little volcanic activity that has occurred during the last 10 million years has been responsible for some areas of good soil. However, less than 10 per cent of the country is thought to have reasonably productive soils — that is, soils that can sustain intensive agriculture or dense vegetation. Many areas of otherwise favourable soil occur in regions that are too dry for non-irrigated agricultural use.

World's most fire-prone continent

Large fires can occur somewhere in Australia at any time of the year (see Fig. 2.7). This is not just because of the generally hot and dry climate. It is also because of the climate variability: wet periods allow vegetation to build up as fuel for a fire that can take hold after a dry period. The characteristics of much of the natural vegetation are also important. The flammability of eucalyptus oil and the dry leaves of many plants encourage the spread of fire. In some natural communities, fire is important as a way of recycling nutrients, a task performed by soil organisms in wetter environments. Much of the native vegetation is adapted to fire — indeed many species, including trees such as the mountain ash (*Eucalyptus regnans*), that live in normally 'wet' forests, need fire before seeds will set or germinate. This suggests that fire has long been an important feature of much of the country. As well, the fire-lighting practices of the Aboriginal people may have created more grassy areas, which burn more easily.

Few rivers, low run-off

After Antarctica, Australia has the fewest rivers and the least run-off of any continent. Its low rainfall means that, for a land area of its size, Australia has few sizeable rivers, and few accumulations of fresh water. Many of the rivers we do have are neither fast-flowing (because most of the land is flat) nor regular (because of the rainfall variability). In most parts of the world, run-off usually drains into creeks and rivers that eventually enter the sea. But in Australia, many rivers drain inland, where all the water eventually evaporates.

Most inland lakes are dry for long periods and many are brackish or salty. Only a minute proportion of the continent is permanently covered with fresh water, and recent human activity is

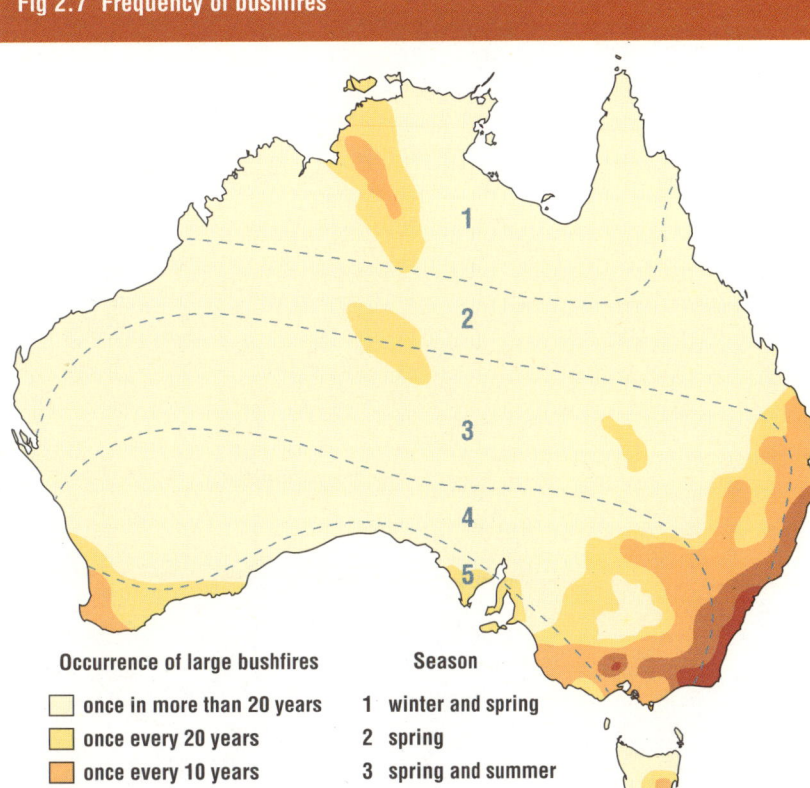

Fig 2.7 Frequency of bushfires

Occurrence of large bushfires

- ☐ once in more than 20 years
- ☐ once every 20 years
- ☐ once every 10 years
- ☐ once every 5 years
- ☐ once every 3 years

Season

1 winter and spring
2 spring
3 spring and summer
4 summer
5 summer and autumn

Source: adapted from AUSMAP Atlas of Australia, 1992.

increasing salinity in major waterways such as the Murray–Darling system.

Australia has a very low total run-off per unit area of catchment compared with other large landmasses, and a much more variable one. Nearly one-third of the country, the western plateau, has no significant run-off at all. By contrast, Tasmania, which accounts for just 0.8 per cent of the land area, provides 13 per cent of the run-off because of its high rainfall and low rates of evaporation. So, the fresh water that is available is not evenly distributed.

◄ Burnt forest in southern Australia; large fires occur in different parts of Australia throughout the year.

Figure 2.8 The Great Artesian Basin

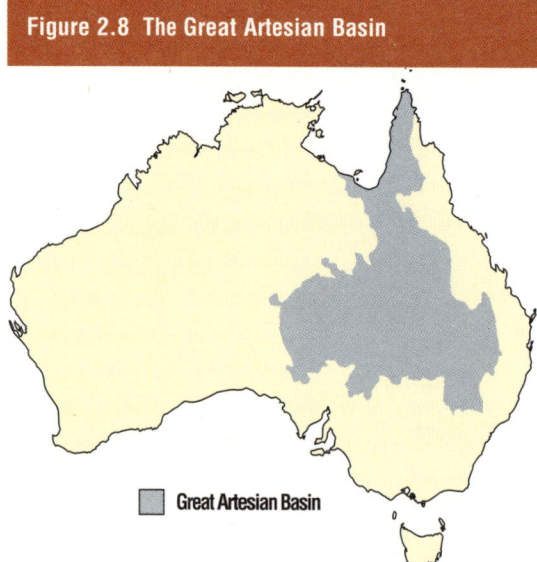

■ Great Artesian Basin

Source: Australian Water Resources Council.

Urban areas greatly affect the pattern of run-off. Impermeable roads and buildings stop water from soaking into the ground, so it must be channelled into drains and then discharged — often in a polluted form.

Large, but diminishing, groundwater resources

The arid and semi-arid zones are well provided with water that has accumulated beneath the ground and can be used to supply wells or bores. The Great Artesian Basin, our biggest single source of this groundwater, underlies about 22 per cent of the continent (see Fig. 2.8). More than 4000 flowing

Figure 2.9 Australia's mineral deposits

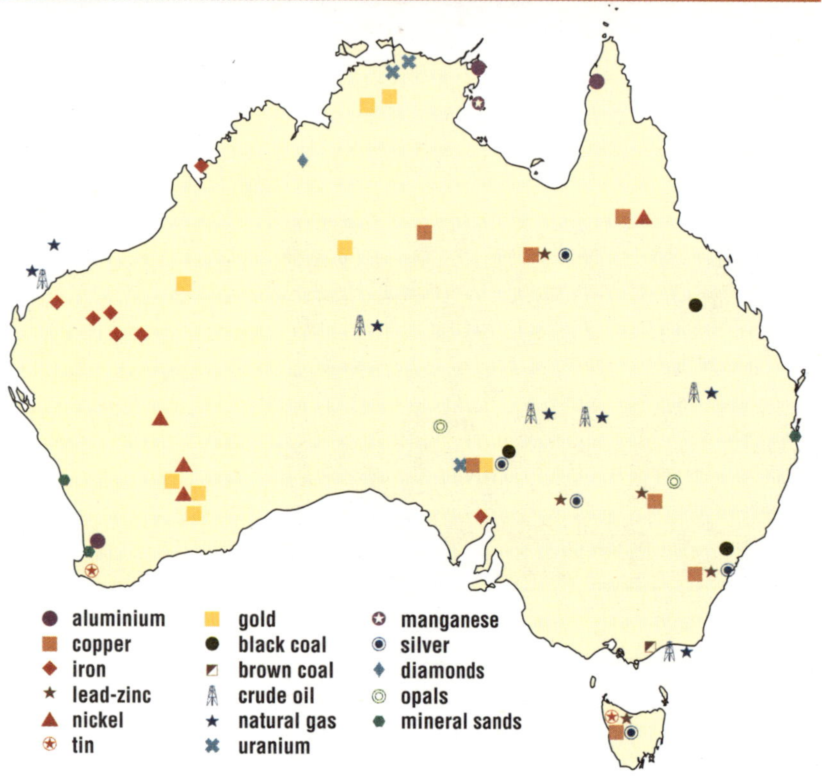

● aluminium   ■ gold   ✪ manganese
■ copper   ● black coal   ◉ silver
◆ iron   ◨ brown coal   ◆ diamonds
★ lead-zinc   ⚒ crude oil   ◎ opals
▲ nickel   ★ natural gas   ● mineral sands
✳ tin   ✖ uranium

Source: adapted from AUSMAP Atlas of Australia, 1992.

wells have been sunk into this basin, but by 1990, just over 1000 bores had stopped flowing.

About 60 per cent of the country by area is totally dependent on groundwater, and a further 20 per cent uses appreciable amounts of it. Its quality varies; in some places it contains many dissolved minerals and may not even be suitable for irrigation. In others, it is useable for agriculture and as drinking water for humans. This groundwater resource has underlain much of the development of inland towns like Alice Springs, and of the pastoral properties in the arid zone. However, the rate of extraction has been increasing. In some aquifers, it is now much greater than the rate of recharge and so, inevitably, the resource will run out. Groundwater can also be — and in some places already is — subject to pollution, which can remain unnoticed for a long time (see Chapter 7).

Rich in minerals

Minerals of economic significance occur throughout Australia (see Fig. 2.9). The abundance of mineral resources per head of population is much greater than in most other countries. We have more than 20 per cent of the world's stock of recoverable bauxite, iron ore, uranium, mineral sands and diamonds, as well as other important mineral deposits, including large quantities of coal (most of which is low in sulfur). Extractable oil exists on land in the Amadeus, Bowen/Surat, Cooper/Eromanga and Perth basins and also under Bass Strait and off the coast of north-western Australia. However, the country is not self-sufficient in oil, with about 25 per cent of requirements being imported.

Hazards — *a natural part of life*

Drought, fire, cyclones, floods, tornadoes, venomous creatures and even earthquakes are all part of Australia's natural environment. Any picture of our environment must include these phenomena, and so too must our planning. Risk and uncertainty pervade every aspect of life here.

The sea

Our marine environment provides us with resources in the form of minerals, finned fish and shellfish, and important assets for recreation and tourism. Much of our trade is carried by sea, which provides a natural barrier to invasions of exotic species.

Australia's coastline (including that of its 12 000 or so islands) extends for about 70 000 km and comprises a range of conditions. In international law, Australia has primary management responsibility for waters extending for 200 nautical miles (about 300 km) from the coast of the continent and islands that are part of its territory — unless another country's zone intervenes (see Fig. 2.1). These waters, known as the exclusive economic zone (EEZ), cover a greater area than the country's land. In some cases, Australia has internationally recognised rights and obligations concerning the seabed of the continental shelf beyond the 200 nautical mile EEZ.

Hazards such as flood (above), fire and drought are a natural part of Australia's environment. Cyclone Tracy (left) devastated the city of Darwin on Christmas Day, 1974.

The continent lies between three ocean basins: the Pacific, the Indian and the Southern. The sea around the country spans four world ocean temperature zones, from tropical (sea-surface temperature of 25–31°C) to subpolar (5–10°C). Polar waters (-2–5°C) surround Macquarie Island and the Australian Antarctic Territory.

Because the sea moderates the climate, coastal areas generally experience much lower temperature extremes than the inland. The ocean takes much longer to warm up and cool down than the air. The sea also influences rainfall and patterns of air movement in coastal areas. Cyclones are generated over tropical seas.

Australia has a major ocean current flowing southwards on each side of the continent (see Fig. 2.10). Coming as they do from the tropics, these currents bring warm water to both western and eastern coasts. This feature of two warm currents is unusual, as most other continents have one 'cold' current (coming from polar regions) down one side, and a warm one, from the tropics, moving along the other side.

The lack of upwelling of nutrient rich cold water is one reason why Australia's marine areas are not as highly productive as marine zones in many other countries. Another reason is that run-off into the sea on the shelf is low in nutrients. The total fish catch in our waters comes to only a small proportion of the world harvest, despite the large fishing zone.

Australia's most famous marine feature is the Great Barrier Reef, the world's largest continuous coral reef complex, which is about 2500 km long and contains thousands of separate reefs. We have many other reef systems, including coral growing in areas far from the tropics, such as Lord Howe Island. In fact, we have the largest area of coral reefs of any nation (see Fig. 8.10).

Our reefs face many pressures, including: sediments and nutrients from land disturbance, run-off of applied fertiliser and discharges of sewage; tourism and fishing in certain areas; and oil spills. Other potential threats are outbreaks of coral predators, such as the crown-of-thorns starfish in the outer central and northern Great Barrier Reef and snails (*Drupella* species) in Ningaloo Reef, Western Australia. The conservation and tourism values of Australia's reefs are becoming more important as the condition of other coral reefs in the world deteriorates. Coral reefs are now relatively well represented in marine protected areas, by far the largest of which is the Great Barrier Reef Marine Park, with an area of about 344 000 sq km. The management of the Great Barrier Reef, in particular, is a major environmental undertaking for the country because of the competing needs for: preservation of the natural area; increasing recreational tourism and coastal and island development; fishing (traditional, recreational and industrial); and the passage through the reef waters of large transport vessels.

Figure 2.10 Ocean currents surrounding Australia

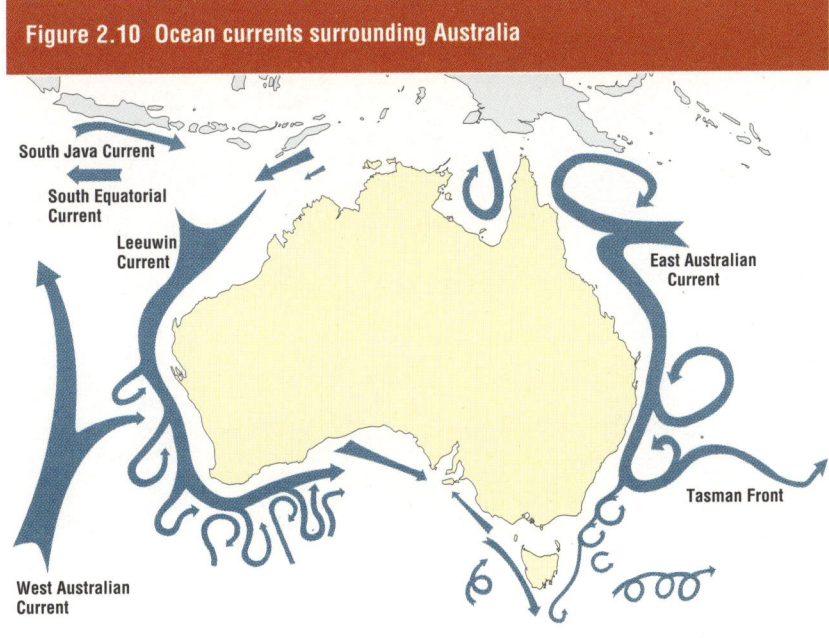

South Java Current

South Equatorial Current

Leeuwin Current

East Australian Current

West Australian Current

Tasman Front

Source: CSIRO Division of Oceanography.

◀ Fraser Island; part of the World Heritage Great Sandy Region contains many unique features such as perched lakes.

Off the southern end of the Great Barrier Reef lies the world's largest sand island — Fraser Island, near Maryborough in Queensland. It is part of a 'sandy region', and contains many unique features, such as perched lakes. Both the Barrier Reef and Fraser Island are World Heritage Areas.

Australia's biological characteristics

Biodiversity

Australia is one of the twelve most biologically diverse countries in the world. Unlike the other 11, it has the development and wealth to research and protect its diversity. However, it has a small population and a large land area, and the number of ecosystems and species to research is also large. Consequently, the continent's biodiversity is still far from fully known. Many aspects of this diversity,

such as our richness in lichens, ants and fungi, are not apparent to the casual eye.

Endemism

Australia and New Guinea share a common biological inheritance, now greatly modified by environmental differences. Australia has considerable variety in its living organisms and accommodates many groups of endemic organisms (those that exist nowhere else – see Tables 2.2 and 2.3). Its biological uniqueness has come about partly because it has been so isolated from the world's other landmasses for so long.

Two genera dominate the continent's trees

Like other continents in the southern hemisphere, Australia has mostly evergreen trees, but the

---

Table 2.2 Endemism in Australian mammals

	Total no. of species[1]	Endemic species	% endemism
Monotremes[2]	2	1	50
Marsupials[3]	141	131	93
Bats[4]	69	40	58
Rodents[5]	65	57	88
Seals etc[6]	4	1	25
Dugong[3]	1	0	0
TOTAL	282	231	82

Notes:
1. Excludes oceanic islands and Antarctica.
2. Strahan, 1988.
3. R Strahan 1993, pers comm.
4. G Richards 1993, 1994, pers comm. This figure includes only described species; a further 18 nominal species await description, and 16 of these are thought to be endemic.
5. R Strahan 1994, pers comm.
6. Strahan 1988. This figures includes only those species which breed or used to breed on the Australian mainland or Tasmania (including King Island). A further four species are casual or accidental visitors.

Source: adapted from DEST, 1994.

Table 2.3 Numbers of primitive angiosperms

Plant family	World species	Australian species
Order magnoliales		
Annonaceae	574	29
Austrobaileyaceae*	2	2
Eupomatiaceae	2	2
Himantandraceae	2	1
Winteraceae	60	10
Myristicaceae	120	1
Order Laurales		
Atherospermataceae	8	8
Gyrocarpaceae	7	1
Hernandiaceae	21	3
Idiospermaceae*	1	1
Lauraceae	1260	62
Monimiaceae	95	19
Trimeniaceae	7	1

*endemic families

Source: adapted from DEST, 1994.

majority are not conifers. Instead, eucalypts and acacias define much of the landscape. Few other land areas in the world are so completely dominated by just two genera of trees. Acacias as a group are not only found here. Their distribution reflects the existence of the super-continent Gondwana to which Australia once belonged. But, apart from a few species in New Guinea and other nearby islands, more than 500 species of eucalypt are uniquely Australian.

Sclerophylly

One particular quality of much of our flora is the presence of hard, dryish, leathery, spiny or small leaves. Plants with these features are called sclerophylls. Such leaves lose less moisture and are also tougher, so they last longer and need replacement less often. The leaves are also less attractive to grazing animals and may be harder to digest. As a result, sclerophyllous vegetation is well adapted to dry conditions and low-nutrient soils — a great advantage across much of present-day Australia. This may explain why sclerophylls are so prevalent among our native plants. The low soil nutrient level in many areas is not the problem for them that it is for many of the introduced species used in modern agriculture.

Marsupials and monotremes

Australia holds the undisputed title as the world's centre for marsupial mammals. These animals, born in a tiny, immature state, complete much of their development in a pouch. Although several marsupial groups thrive in South America, which was once connected to this continent through Antarctica, and one species of opossum is found in North America, Australia harbours the widest range of marsupial families. We have 144 marsupial species, including koalas, wombats, kangaroos and wallabies. They have been extremely successful.

Monotremes are even more special to Australia. These break some of the familiar 'rules of mammals' by laying eggs, like reptiles and birds, but producing milk to nurture their young, and by having less effective body temperature regulation. There are only three species of monotreme in the world: the platypus and two species of echidna. One echidna species occurs in New Guinea, but the platypus is uniquely Australian.

Other fauna

Despite our legacy of unique marsupials, most native mammals are not marsupials. Most of them are placentals (the young remain in the womb for much longer than marsupials and there is no pouch). These include: bats, such as flying foxes; rodents, such as the water rat; and marine mammals like seals, dugongs and dolphins. Dingoes are not native but came here with people at least 3000 years ago.

Much of Australia's faunal richness lies in its birds, reptiles, amphibians and invertebrates. About 93 per cent of the frogs and 89 per cent of the reptiles are found nowhere else. Our arid zones are particularly rich in lizards compared with similar

Two genera, the eucalypts (left) and the acacias (below), dominate the woodlands and drier forests of Australia.

areas on other continents (see Table 2.4). We also have a great variety of ants. The arid zone supports a rich and diverse array of termites that do the work of decomposing plant matter, a job carried out by other soil fauna and fungi in wetter areas. Termites and their distinctive mounds are extremely numerous — it has been calculated that the biomass (total weight) of termites per hectare in many parts of the semi-arid zone is greater than that of the cattle that graze these regions. Indeed, despite the prevalence of kangaroos, there is no doubt that the continent's major herbivores are ants and termites.

Table 2.4 Number of species of lizards in 10 hectare desert study sites on three continents

Lizard family	North America	Africa (Kalahari)	Australia
Agamidae		1	2–8
Chameleontidae		1	
Gekkonidae	1	4–7	5–9
Helodermatidae	1		
Iguanidae	3–8		
Lacertidae		3–5	
Pygopodidae			1–2
Scincidae		3–5	6–18
Teiidae	1		
Varanidae			1–5
Xantusidae	1		
Total	**4–11**	**12–18**	**18–42**

Source: adapted from Pianka, 1986 as cited by DEST, 1994.

People brought the dingo to Australia at least 3000 years ago (top).

The koala, one of Australia's 144 marsupial species (centre).

Introduced species, such as the rabbit, have been responsible for many of the extinctions of native animals (above).

Introduced species that 'all ran wild'

Australia's flora and fauna have changed dramatically, and irreversibly, over the last 200 years. Species ranging in size from the Asian water buffalo to the AIDS virus have been introduced. The result has been a complete and rapid change to much of the country's natural ecology. Introduced species, including domesticated stock, have been responsible in part, for many of the extinctions of native mammals in the central zone of the continent — the region with the least human habitation!

Australia's new human arrivals didn't just bring in their northern hemisphere species, they also modified the environment in ways that helped many of the introduced species to spread. Roads, railways and other corridors in the bush often fragmented native populations and gave access to many introduced species — especially plants. The new weeds colonised roadside verges. Four-wheel drive vehicles and even bushwalkers can still, unwittingly, spread plants and soil fungi into new areas.

Introduced species — whether rabbits, cane toads, bitou bush or *Mimosa pigra* (see Fig. 2.11) — occasionally enter our awareness, but the effects of the many introductions go much further. Australians today depend totally on introduced species. Our agriculture relies on introductions that

have changed the fabric of the land — its vegetation cover, sand dunes, water holes and creeks. New ecological associations are forming between native and introduced species. For example, new soil fungi interact with the roots of native trees. Introduced bees pollinate native plants and native birds eat seed from introduced grasses. Despite efforts at eradication, it seems that most introduced species are here to stay — even if we succeed, in some cases, in keeping their numbers in check. That they will continue to affect the natural environment is obvious, but exactly how remains to be seen.

Tropical rainforest — rare but rich

As might be expected, true tropical rainforest is rare in Australia. Patches exist in northern Queensland (see Fig. 2.12). In addition, small stands of monsoon rainforest occur in parts of the Northern Territory, Western Australia and elsewhere in Queensland. The real tropical rainforest receives about ten times the annual rainfall of the continent's arid zone and, most importantly, its rainfall is predictable.

The tropical rainforests of northern Queensland occur from sea level to more than 1600 metres altitude, and include a number of different types of vine forests and communities, often in small pockets interspersed with open forests, woodlands, swamps and mangroves. They are of special value both nationally and globally, and provide some of the world's most important ancestral links in the history of plant evolution. They are, for example, the centre of diversity of the cycads, and contain the world's greatest concentration of primitive flowering plants: of the 19 known families of primitive flowering plants, 13 are found in north-east Australia and two of these are found nowhere else. Some 1161 species of plants have been recorded in the area. One-third of them belong to genera of which they are the sole representatives. An extraordinary number of flowering plants

Figure 2.11 Distribution of the pest weed *Mimosa pigra*

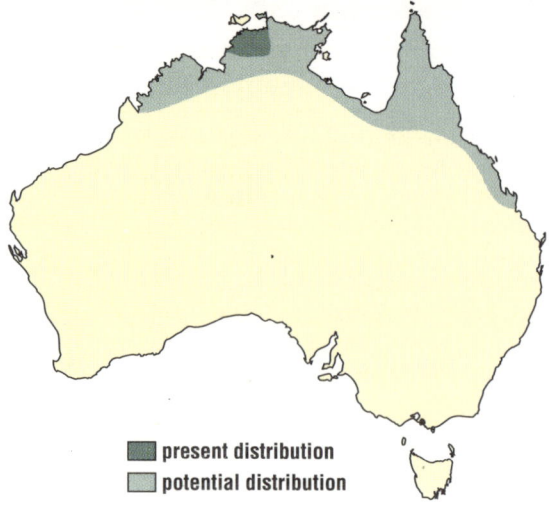

■ present distribution
■ potential distribution

Source: compiled by ANCA.

Figure 2.12 Distribution of tropical rainforest

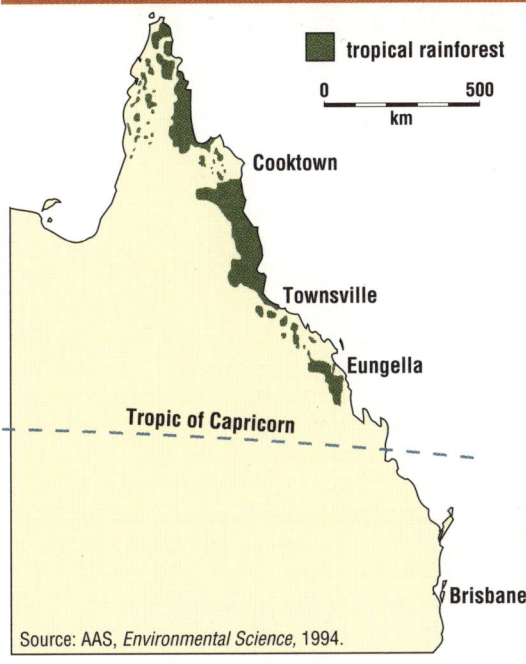

tropical rainforest

0　　　　500
km

Cooktown

Townsville

Eungella

Tropic of Capricorn

Brisbane

Source: AAS, *Environmental Science*, 1994.

belong to these monotypic genera: 43 have been recorded, of which 37 are endemic to Australia and 28 to the area. Many rainforest plants are known from only a few populations.

Faunal diversity in the wet tropics is similarly rich, with 47 species of frogs (23 per cent of the Australian total) found there — 19 of them endemic to it. Some 128 species of birds enrich the area, and 89 mammal species, including some 60 per cent of Australia's bat species. Twenty three per cent of our reptiles also occur there. The number of invertebrate species is likely to be enormous, but remains largely unknown. A survey at Bellenden Ker found more than 4385 species of invertebrates. Of the 1514 beetle species identified, 86.5 per cent were unnamed. The mountain tops within the tropical rainforests are like geological islands, and host many species that appear to be of great antiquity, with some perhaps unchanged from Gondwanan times.

Birth of a continent

The instantly recognisable outline of Australia today has not always existed — and the biological and physical features of our continent have not always been the way they are now. If we view things on long-enough time scales, Earth's surface is a constantly moving jigsaw of landmasses and oceans: nothing is static. What seems so permanent now is merely one frame in a long film full of movement and change. About 1000 million years ago (a time span less than one-quarter of Earth's age), the planet would have been unrecognisable. At that time, no life existed on land, and only the simplest forms in the sea. There would have been no soil; and the atmosphere was unbreathable, as living cells in the sea had not yet released enough oxygen. The land and its features would bear no relation to any of today's maps. Yet part of what was to become

Australia was there. Much of the western two-thirds of today's continent (the most ancient part) existed within a larger landmass.

Starting about 450 million years ago life appeared on land. Much later (about 160 million years ago), through the slow movement of the vast geological plates riding on Earth's mantle, the super-continent Pangaea, containing the forerunners of all today's major landmasses, split into two. A northern super-continent, comprising most of North America, Greenland and Eurasia was completely separated from a southern super-continent called Gondwana. As well as much of present-day Australia and New Guinea, Gondwana included what is now Antarctica, India, South America, Africa, Madagascar and Arabia, and a mass that later became New Zealand. (At various times, different parts of the Gondwanan landmass were below the sea.)

Eventually, Gondwana started to break up (see Fig. 2.13). Land destined to become India, Africa and Arabia separated off at different times and started to fan out northwards. New Zealand broke away about 80 million years ago. The area of land that is now Antarctica lay between Australia and South America, joining all three together. (Antarctica had no permanent icecap then, because of its different position, and it was forested.) As continents drifted, they moved into different climate zones. The changed position of the continents could affect the climate of a region, because ocean and atmospheric currents are all influenced by the presence of landmasses. On the slowly moving continents, geographical features — such as the sizes and directions of river systems, and rates of erosion — also changed. So too, of course, did the types of living organisms.

Australia's tropical rainforests are rich in endemic species of plants. Many of them provide important ancestral links in the history of plant evolution.

The development and spread of Australia's main groups of higher plants occurred while Gondwana was still intact, but undoubtedly flora and fauna varied across the massive super-continent. Those parts more distant from what is now Australia probably had less-closely related organisms before the break-up. That is why we share many biological groups with South America and some with New Zealand, and have fewer in common with Africa and India.

Very little mountain-building has occurred in the land that is now Australia — for about 100 million years or more — in marked contrast to the creation of the Himalayas, the European Alps and the American Rockies. These are all quite recent in

geological terms. New Guinea has been, and continues to be, the site of mountain-building and volcanic activity as the Australian landmass pushes into Asia. The absence of new mountains here (apart for some uplift in eastern Australia that continued through the Tertiary period) has affected the present-day nature of our soils and water flows. Many of our rivers have occupied their present positions for far longer than those in most other countries.

About 65 million years ago, a mass extinction occurred across the world. Many groups of organisms quickly disappeared from the fossil record, including the great variety of dinosaurs. As a result, many new organisms evolved over the succeeding millions of years, and the previously inconspicuous mammals diversified into a range of different forms. Fifty million years ago, Gondwana (comprising the land for only the three future continents) was probably a place of damp forests where tree ferns, conifers and some flowering plants thrived in a fairly constant cool to warm and wet climate. Mammals — both placental and marsupial — did well after the removal of the dinosaurs and related reptiles, and both types probably existed in Gondwana. Later, marsupials declined in South America and placentals gained the ascendancy there. Marsupials became extinct in the cold of Antarctica, but they thrived in Australia, where new forms continued to evolve. Recent evidence suggests that early types of placental mammals, which died out, may have existed here. Mammals of the third type — the monotremes — are now only found in Australia, although they once occurred in South America.

Australia started to separate from Antarctica about 53 million years ago. At that time its southern boundary was near the Antarctic Circle, and daylight would have been almost non-existent during the winter. Despite this, plants seem to have thrived, probably because the whole world was warmer and wetter then.

The last point of connection to the Antarctic landmass was through land that is now Tasmania, but this broke about 38 million years ago, and a channel of deep water formed between the two continents. Australia became a separate continent — and started to develop the unique characteristics found today. Eucalypt pollen first appeared in the fossil record at about that time. Since then — and until very recently in geological time — Australia and Antarctica have been quite isolated from all other landmasses (unlike South America, India and Africa, which all eventually came into contact with northern landmasses). The separation caused circular currents in the ocean and the atmosphere — the so-called circumpolar currents — to develop around Antarctica. These restricted the transport of heat from tropical latitudes. Antarctica became colder and, although marine life and birds survived, eventually all higher terrestrial life across the continent disappeared, so removing many of the living organisms most closely related to those in Australia. The isolation of Australia allowed its flora and fauna to develop for about 30 million years

Figure 2.13 The break up of Gondwana

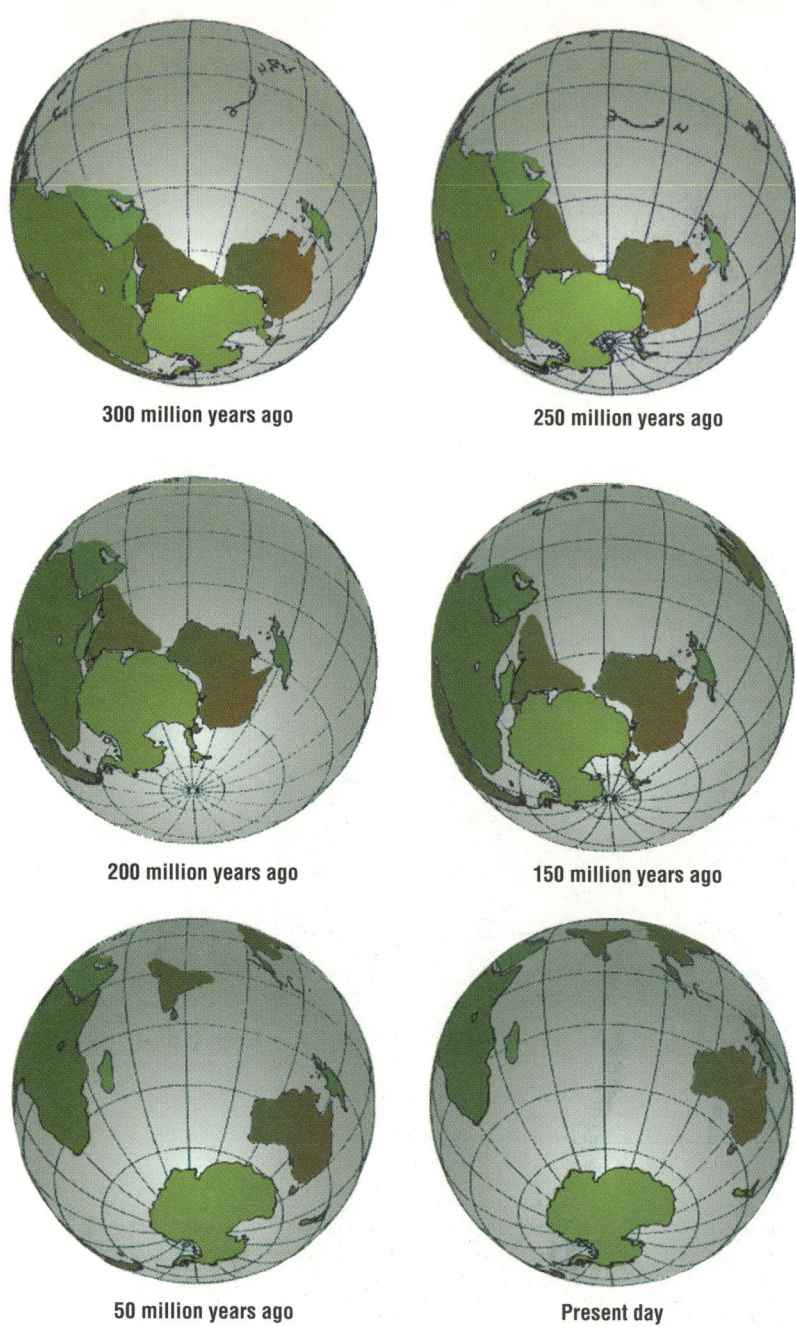

300 million years ago

250 million years ago

200 million years ago

150 million years ago

50 million years ago

Present day

Source: Bureau of Mineral Resources, Geology and Geophysics, as cited in ABS *Yearbook Australia*, 1988.

without any significant input from elsewhere. Many Gondwanan species survived relatively unchanged.

Australia drifted further northwards. By about 20 million years ago it started to become drier, especially in the southern inland. Sclerophyllous plants were at an advantage in these less stable conditions and so they spread. Eucalypts, banksias and hakeas developed into many forms. Grasslands replaced areas of lush forest. New marsupial species evolved. Acacia pollen first appeared in abundance about 25 million years ago, probably as a result of the increased aridity. (It is likely that acacias had been present from long before, as their distribution today — in Australia, Africa, India and South America — suggests that they were Gondwanan plants.)

In general, Australia's flora and fauna are a mixture of the original Gondwanan stock, modified over the millions of years of separation, and the more recent arrivals from Asia.

During the last few million years, the main influences on our climate have been alternating glacial and interglacial phases. For most of this time Earth has been in the grip of so-called Ice Ages, or glacial periods, when the planet-wide average temperature is lower and the atmosphere is drier. During peak glacials of a few thousand years, water becomes frozen at the poles and in glaciers, so sea levels are lower. The briefer peak interglacials are characterised by warmer, moister conditions when sea levels rise and glaciers retreat. Phases of cooler, wetter weather and warm, dry phases intersperse the tens of thousands of years between these extremes. During the peak of the last glacial period, about 18 000 years ago, the sea level was some 120 metres lower than it is today. Several parts of the world, now separated by sea of that depth, were therefore joined. As Fig. 2.14 shows, the islands of New Guinea and Tasmania were joined to mainland Australia and the Gulf of Carpentaria was land, although a large lake existed in the middle of it. For much of the past 1.8 million years, Australia and New Guinea formed one land mass. Unlike much of the northern hemisphere, mainland Australia was too warm to suffer from widespread glaciation, but the climate became cool, dry and windy during the glacial periods. At these times dunes covered more of the country, and rivers and wind carried away a lot of dust from the soil. Elsewhere, an icecap covered the central highlands of Tasmania. Large glaciers originating from it carved deep features in the landscape.

Earth is now in an interglacial period, which has lasted for about 10 000 years. It was during an earlier glacial period that the first humans would have reached a much larger Australia across smaller stretches of water from Asia, so starting a series of new changes to the continent.

Australia's slow northwards drift, at a rate of about 100 km every million years (or 10 cm a year) continues today. Now we are closer to Asia, but our geological past explains why our flora and fauna resemble those found in far-away South America and are, in general, so different from those of most of our near neighbours.

The Nullarbor Plain is a distinctive feature of the Australian landscape. It was once the bottom of a shallow sea.

Figure 2.14 Sea level changes over the last 140 000 years

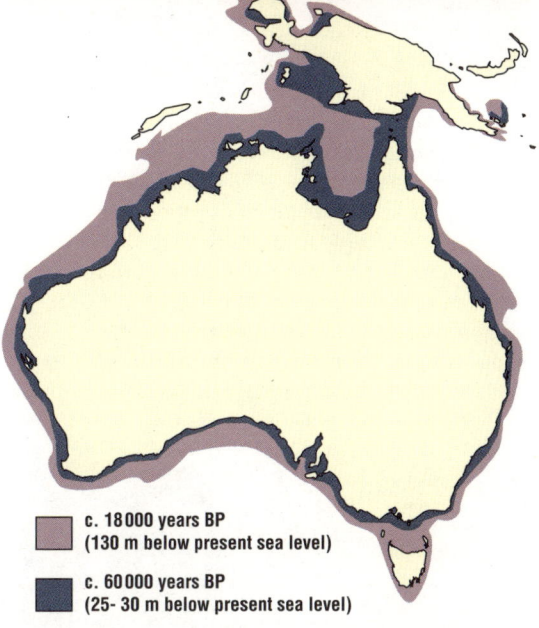

■ c. 18 000 years BP (130 m below present sea level)
■ c. 60 000 years BP (25–30 m below present sea level)

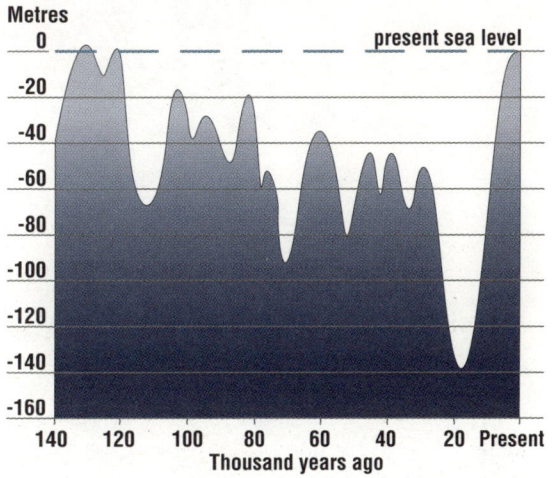

Source: AAS, *Environmental Science*, 1994.

Humans in the environment

It may never be known when the first humans arrived in Australia. Many Aboriginal people hold the traditional belief that their ancestors have always dwelt here. Archaeological evidence shows that human settlement goes back at least 50 000 years, and some archaeologists believe it goes back even further.

The consensus of opinion among prehistorians is that the first people came from the north, through south-east Asia. They came at a time when sea levels were lower than today's. Their journeys across water would have been shorter but still perilous. Different groups could have come at different times over thousands of years. By at least 20 000 years ago people had spread to all parts of the greater land mass of Australia, including Tasmania and New Guinea. Rising sea levels later cut off the land bridge to Tasmania about 10 000 years ago, effectively isolating the people there.

Despite the use of agriculture by the inhabitants of New Guinea and the Torres Strait Islands, and the likelihood that people in northern Cape York knew of it, it seems that the Aboriginal peoples did not develop formal agriculture on the Australian mainland. The persistence of hunting and gathering might well have resulted from the low productive potential of most of the country compared with the lush fertility of most of New Guinea. However, it is clear that Australian Aboriginal peoples knew a great deal about the resources of the land, and how to care for certain plants. Like all humans, they modified the environment to improve their own well-being — mainly through the use of fire. It is also possible that continuous hunting brought many larger animal species to extinction, although extinctions could have resulted from climate change, or use of fire, or a combination of all three. This area is still subject to scientific debate.

▶
Philip Langley, from the Aboriginal and Islander Dance Theatre breathes life into an Aboriginal tradition going back many thousands of years.

Lalai (Dreamtime)

Dreamtime,
The first ones lived, those of long ago.
They were the Wandjinas—
Like this one here, Namaaraalee.
The first ones, those days,
shifted from place to place,
In dreamtime before the floods came.
  Bird Wandjinas, crab Wandjinas
Carried the big rocks.
They threw them into the deep water
They piled them on the land.
Other Wandjinas—
all kinds—
She the rock python,
He the kangaroo,
They changed it.
They struggled with the rocks,
They dug the rivers.
These were the Wandjinas. They talk with us
at some places they have marked.
Where the sun climbs, over the hill and the river
they came
And they are with us in the land.
We remember how they fought each other
at those places they marked—
It is dreamtime there.

Some Wandjinas went under the land,
They came to stay in the caves
And there we can see them.
Grown men listen to their Wandjinas.
  Long ago, at another time,
these Wandjinas changed the bad ones
into the rocks
And the springs we always drink from.
These places hold our spirits,
These Wunger places of the Wandjinas.

Recounted by Sam Woolagoodjah, elder of the Worora people, north-west Australia.
Extract from translation by Andrew Huntley from the prose version by Michael Silverstein. In *The New Oxford Book of Australian Verse.* (Chosen by Les A Murray. Oxford University Press, Melbourne 1987.)

Coupled with the vegetation changes following regular burning, the variety of animal species changed.

People spread across the whole of the continent and adapted to the various climates and resources available. Their relative isolation from the rest of humanity, particularly in southern Australia, lasted for tens of thousands of years, but some interchange with the peoples of the island of New Guinea and nearby Indonesian islands continued.

Various European and Asian navigators were aware of part of the coastline of the continent by the 1600s. The Englishman Captain Cook mapped much of the east coast in 1770. More than a decade later, the British Government decided to start colonising the country. In particular, the Government wanted to use the opportunity to deal with overcrowded jails — and a large population of what it deemed criminals — by sending the convicts away, although strategic and trade considerations also played an important part in the decision to start colonisation. Amid optimistic ideas of boundless plenty in the new land, England dispatched a collection of ships (the First Fleet), which landed in Botany Bay (now part of the greater Sydney metropolis) in 1788 to start a penal settlement.

The settlers had no understanding of their new environment. They could not know of the great variability in climate, the low nutrient status of the land and the importance of fire as a land management tool. Their main imperative was survival and the prevention of famine. The environment turned out to be more difficult than expected and both it and its original human inhabitants soon came to be viewed as generally alien and hostile.

In 1789 Arthur Phillip, the Lieutenant Governor of the new colony, wrote:

> '... in the whole world there is not a worse country. All that is contiguous to us is so very barren and forbidding.... here nature is reversed [and] nearly worn out.'

Of course, not all the early settlers felt the country was hostile. Gradually, attitudes to the environment swung to a different extreme, and an over-optimism about its potential came to prevail.

Conflict between the settlers and the original inhabitants soon began and, despite the efforts of various peaceful individuals on both sides, tragically continued in violent form into this century. As was the case wherever Europeans colonised, diseases, especially smallpox, preceded the advance of settlement and made colonisation easier by devastating the Aboriginal peoples. Despite resistance in some parts of the country, colonists killed Aboriginal peoples outright (by gunshot or poison) or indirectly (by taking their lands). In Tasmania the tribes were almost wiped out. But the north of the country was much more difficult for the European settlers because they were not used to the harsh conditions. Many early settlements there failed.

The Aborigines, having lived on this continent for so long, were much better adapted to its ways than the newcomers. They had a detailed knowledge of their land and a deep emotional and spiritual attachment to it. However, their land was wrested from them and the way the natural environment was treated changed greatly as a result.

The newcomers tried to impose their agriculture, using species adapted to humid, temperate and predictable climates, onto a country that did not have the soil, the water or the climatic stability for it. They did not understand the use of fire as a land management tool, or the possibility of using the well-adapted native plants and animals for food. But, above all, they failed to appreciate the constant inevitability of both flood and drought.

As the decades passed and the new settlements became self-sufficient, European settlers spread further into the continent, with great difficulty,

▲

'The prince at the Rabbit Warren, Barwon Park' from the *Illustrated Australian News*, 20 December, 1867. European settlers brought with them an alien culture to their new land.

always trying to introduce agriculture and make money from the land. Every so often they came across areas of good plant growth, and were deluded by the apparent lush aftermath of good rains. This, of course, was not the norm; but the settlers felt that the dry spells were just a run of bad luck and that 'normal' conditions of good rain would quickly be reinstated. Even today, many people consider droughts freak 'acts of God' for which compensation must be paid, rather than as naturally recurring events that are a normal part of the Australian environment.

European settlement of Australia started at the margins — the coastline — which was all that was accessible to the colonists. Today's pattern of urbanisation reflects this — all the major cities are coastal and are sited on fresh-water supplies and near harbours. Each major settlement became a separate colony in its own right; each colony later became a State. Agricultural areas surrounded the early cities, and eventually they prospered because they were located on trade routes and in well-watered areas. After making some adaptations to Australian conditions, agriculture succeeded, but continued to rely, as it still does, on introduced species. Throughout, the new arrivals had to learn to come to terms with the land they had acquired — especially those parts of it that they could not change.

The new settlers colonised more and more land, and opinions changed from the early dismay into unrealistic expectation. In some respects the environment was still hostile but, the pioneers felt, it could be 'tamed' and transformed into a land of great wealth. Societies were set up to bring in many species from Europe to increase the feeling of being 'at home' in the new land. Other alien species established themselves when they escaped from farms, were let loose or came in by accident. As a result, thousands of plant and animal species never before known here now exist in the wild in Australia. These have changed the natural environment utterly and irrevocably.

Figure 2.15 Changes in vegetation since 1788

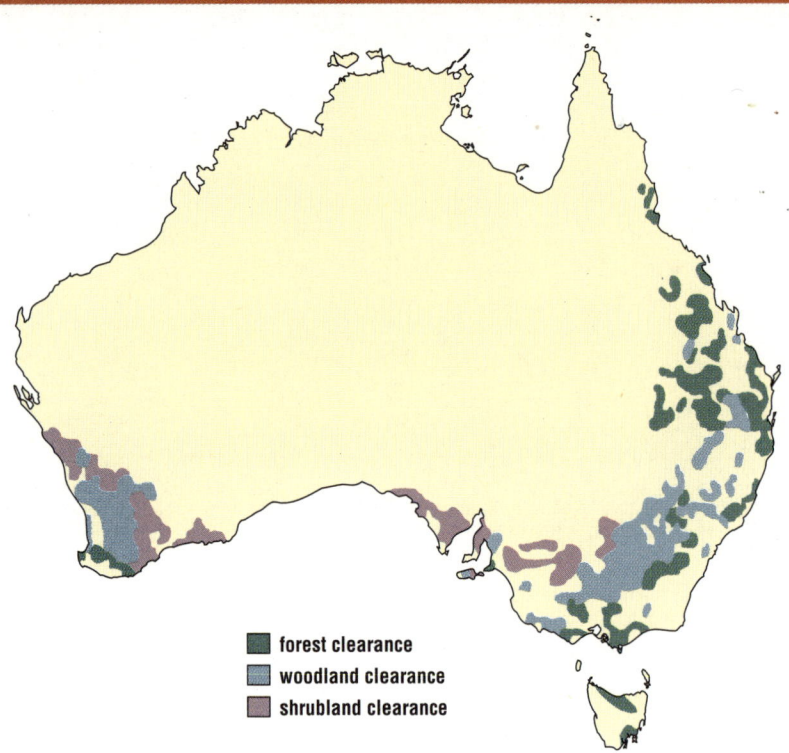

forest clearance
woodland clearance
shrubland clearance

Source: AUSMAP Atlas of Australia, 1992.

between the States over resources, trade and defence provided some of the early imperatives for the new national government.

The rapid build-up of people with new cultures and ways of treating the environment imposed many pressures on the country. The introduced species — especially the rabbit — became a curse. They disturbed the agriculture that was so important to the nation, as well as devastating native ecosystems. Feral cats and foxes preyed on native animals, and Australia now has the world's worst known rate of mammalian extinction. High populations of grazing animals during good times became a liability during droughts, and then overgrazed the remaining vegetation, leading to soil erosion. The clearance of deep-rooted native trees for agriculture in many areas caused the water table to rise, bringing salts close to the surface and killing crops.

Only this century did people begin to appreciate the many limitations of Australia's environment. The idea of a huge, hostile land that would slowly be tamed and brought under control, then to offer limitless opportunity, died hard. For example, in 1847 it was claimed that the Big Scrub of northern New South Wales would need five or six centuries of toil before it could be cleared and the land used by the settlers. It must have seemed almost limitless. But by 1900 it was mostly gone. Technology had greatly speeded up the rate of change. In another case, the thylacine, a marsupial predator in Tasmania, was seen as a threat by farmers who thought it would kill their sheep. The government paid bounties for its destruction. Three decades of hunting and trapping reduced the thylacine to the verge of extinction. Bounty payments ceased, but too late to save the thylacine.

Population

We now realise that Australia has a small area of arable land compared with other continents, and that its relative lack of water puts restrictions on dense settlement in many areas. The question of how many people the place can support — at a reasonable lifestyle and in a sustainable fashion — is now the subject of national debate.

As the colony started to export its agricultural and, later, mined produce, it prospered. Visions grew of a wealthy and powerful nation, rivalling the United States of America. Politicians and journalists in the late nineteenth and early twentieth centuries spoke of Australia as a land of boundless plenty with a potential population of 80–100 million of British stock in the new millennium. The sense of future glory spurred people on and, coupled with developing national pride, started the movement to join up, or federate, all the separate colonies into one nation. In 1901 the nation we know today — the Commonwealth of Australia — came into being with the act of federation and the creation of a federal government. Dealing with conflicts

Table 2.5 Population growth rates

	Average annual rate of growth (%) 1986–91
Australia	1.6
Canada	1.3
China	1.5
Indonesia	2.3
Japan	0.4
Korea	1.0
New Zealand	0.8
Papua New Guinea	2.1
United Kingdom	0.2
United States of America	1.0

Source: Australian Demographic Statistics, 1995 and United Nations Demographic Yearbooks.

Table 2.6 Australia's capital cities

City	Population 1993 ('000)
Sydney	3 719.0
Melbourne	3 187.5
Brisbane	1 421.7
Adelaide	1 070.2
Perth	1 221.3
Hobart	193.3
Darwin	77.4
Canberra	298.6

Source: Australian Demographic Statistics, 1995 and unpublished ABS data.

Estimates of the Aboriginal and Torres Strait Islander population of Australia when Europeans arrived in 1788 range between 300 000 and 1 500 000. Whatever the figure, it is clear that following European settlement the population of Aboriginal people declined rapidly. After 70 years, the European population had reached one million, but the Aboriginal population was not recorded. In March 1995, the Australian population stood at 18 million. The latest one million Australians were added in just five years. With a birth rate of about 15 per 1000 people and a death rate of about seven per 1000, the country's population, in theory, should be growing by natural increase at the rate of about eight per 1000 per year, or 0.8 per cent. This is fairly low by global standards, but one of the highest rates of any developed country.

Added to this is the growth due to immigration. During the 1980s the total population growth rate was about 1.6% annually. This had declined to 1.1% in 1993/94. (The precise figure fluctuates with changing annual immigration quotas.) This decline in the rate of population growth is due to a reduction in the level of migration to Australia; net inward migration declined from around 0.8% during the 1980s to 0.2% per annum in 1993/94 (see Table 2.5 and Fig. 2.16).

Even if net migration were reduced to zero, Australia's population growth would continue by natural increase until about 2026, when it would peak at about 20 million. This is despite the fact that the fertility of the population is below replacement level (defined as each woman of reproductive age having one surviving daughter). It is often not appreciated that a population can continue to grow for a time even when fertility is below the replacement value. Without migration or a change in the fertility rate, the Australian population would be expected to start slowly declining after 2026. Figure 2.17 shows projections of population to 2086 with different annual net migration figures.

Where we live

The aspirations of many Australians are for a detached (preferably single-storey) house on a block of shaded land, near the beach, but well-provided with amenities. Although not all live this way, it has been called the 'Australian dream'.

While Australia has a very low population density by world standards, the average figure for the whole country is entirely misleading, as the distribution of population is very uneven, with most of the people concentrated in a small area. More than 80 per cent of the entire population live on just one per cent of Australia's land surface, most of them in the eight capital cities. Of these, the largest two — Sydney and Melbourne — together account for 39 per cent of the country's population. Each is more than double the size of the next largest city, Brisbane (see Table 2.6).

In terms of numbers of people, the capital cities tend to dominate their respective States or Territories. Only Tasmania, Queensland and the Northern Territory have capitals containing fewer than half their populations.

Sydney Harbour; most Australians live in the eight capital cities, with Sydney and Melbourne accounting for 39 per cent of the country's population.

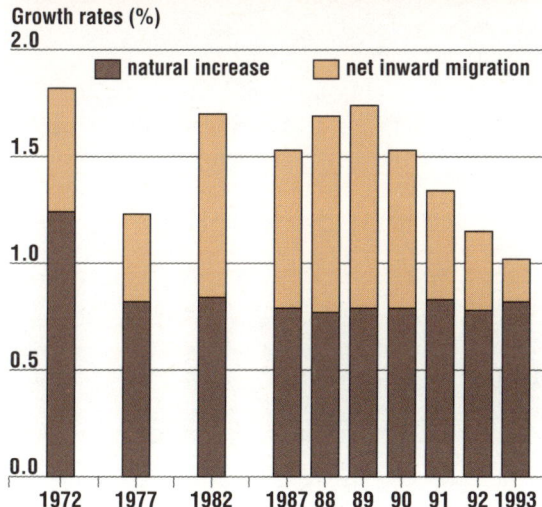

Figure 2.16 Population growth

Source: ABS, *Yearbook Australia*, 1994.

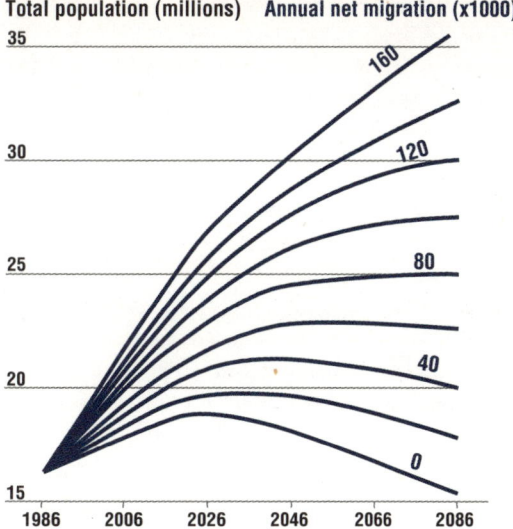

Figure 2.17 Projections of population growth

Source: AAS, *Environmental Science*, 1994.

▲
Outer suburb, Canberra; urban sprawl has resulted in the creation of new suburbs away from the city centre.

Patterns of settlement

Most settlement lies in two widely separated coastal regions — the south-east crescent (which stretches along the coast from Bundaberg to Ceduna) and the south-west, stretching southwards from Perth. The former is by far the larger region. With the exception of a few small towns, the interior of the continent is not densely settled. However, it is far from being the 'dead heart' that has sometimes been portrayed.

The high level of urbanisation has major implications for the environment, because it means that many pressures are concentrated in a small area. Even though most people live in a very small part of the country, other human impacts have extended to every part of the continent, including the largely uninhabited arid areas. Indeed, the arid zone has suffered more from extinctions than the zones of denser human habitation.

Cities

All the major Australian cities are coastal, and occupy areas with reasonable and reliable rainfall.

Unlike the older ones of Europe, for example, Australian cities have experienced most of their development since the advent of the railway and motor vehicle, and this has determined their sprawling character. They are not compact, as were cities when people moved on foot or by horse. This early pattern has been exaggerated by rapid population growth, and particularly the desire of migrants to settle in the existing metropolises. As a result, cities have expanded outwards, encroaching on well-watered, often fertile land, as well as stretching along the coastlines and transport links.

The population densities of our cities remains low by world standards (see Table 3.5) and many urban Australians enjoy living in one- or two-storey homes with a garden. But countering this is the phenomenon of urban sprawl, where cities encroach on natural areas that serve as their water catchment and for recreational opportunities in natural settings. As well, extended cities mean longer journeys to and from work, which has generally meant an increase in private car ownership and use, and hence in air pollution. In addition, a population spread out over a greater area can consume more materials per capita in building, transport and infrastructure than a more compact one.

Urban sprawl has also resulted in the rapid creation of new suburbs, far from the city centre, often with few amenities in their early stages. This can cause either social isolation or car-dependency and congested roads. However there are some positive aspects of the new suburbs. Land use conflicts can be avoided and there is more open space. The suburban backyard is the location of many recreational activities (usually undervalued or un-valued in economic terms). These new suburbs have allowed many Australians to realise their goal of owning their own home which they could not otherwise afford.

How we live

Most Australians live in suburbs, own a private vehicle and own, or are buying, one of the 5.8 million occupied dwellings in the country. Of the labour force of about 8.6 million, most are employed in the service, finance and retail trades, while about 16 per cent work in manufacturing and industry, and a mere six per cent farm vast expanses of land. About eight per cent of the workforce was unemployed in 1995 (as the nation recovered from the economic downturn of the early 1990s). For those adults in full-time work the average wage was $692 a week for males and $550 a week for females.

In comparison with most of the world, the health of the population as a whole is good, and the death rates are low. Most babies born survive to adulthood: Australians can expect to live to be 74 years old, if male, and to 80 if female. The major killers could be described as 'diseases of affluence' (cardiovascular conditions and late-onset cancer, which is not apparent in societies where adults die young).

About three million young Australians attend school, nearly three-quarters of them going to government schools. Older students can attend one of the country's 36 universities, or receive technical and vocational training at colleges of Technical and Further Education.

Most Australians speak English, the national language; about 2.4 million speak another language at home, the main ones being Italian, Chinese and Greek. Nearly all Australians are literate. A great variety of languages are represented here; however, before the arrival of Europeans, there were even more, because of the large number of quite distinct Aboriginal languages — many of which have now been lost.

In their spare moments (when not working, travelling or shopping), Australians spend most time watching television. The next most popular activities (by time spent) are reading and organised sport. Fishing is a particularly popular outdoor pastime. It is estimated that about 4.5 million Australians go fishing at least once a year.

The average figures for Australian wealth, health and life expectancy — comparable with the best in the world — hide the fact that a proportion of the population show the characteristics of a much poorer and less developed country. The Aboriginal and Torres Strait Islander people — the original inhabitants — have lower earnings, higher infant mortality rates, a greater incidence of preventable disease, lower life expectancies, a greater rate of population increase and lower literacy.

What the nation does

Australia is rich in natural resources and its early economy was built exclusively around them. Nowadays, although primary products and commodities still earn much of the country's foreign exchange, a far smaller proportion of the workforce is engaged in working to extract them. Instead, the service industries have increased greatly in importance (see Table 2.7 and Fig. 2.18).

Much employment and economic activity in Australia takes place in what are classified as small to medium business enterprises. These are particularly important in the areas of construction, community services, retail trade, transport and storage, property and business services and recreation and personal services.

Services and manufacturing

The three main areas of private-sector industry in Australia, in terms of numbers of people employed, are wholesale and retail trade, manufacturing and the joint category of finance, property and business services. These were also the three top contributors to the gross domestic product. Over the last few decades, the broad category of service industries has taken over the dominant role in the economy. Agriculture's contribution has declined, and so has that of manufacturing, although the latter still accounts for about 15 per cent of GDP. However, finance, property and business services have nearly doubled their share in the 30 years to 1993. Tourism has become increasingly important — both for employment and as a source of national revenue. International visitors increased nearly threefold (to three million) from 1981 to 1993. As well as being a major earner of foreign currency, tourism clearly has the potential to generate great environmental pressures — especially in certain 'hot spots'.

About 166 000 Australians work in the food, beverage and tobacco industries, making this the largest part of our manufacturing base. Other important areas are machinery and equipment, paper and its products and publishing. New South Wales and Victoria dominate manufacturing, together accounting for two-thirds of manufacturing turnover and employment.

The wholesale industry is dominated by businesses dealing in machinery and equipment, and minerals, metals and chemicals. Computer wholesaling has grown greatly and is now an important part of the machinery and equipment category. The retail sector is dominated by supermarkets, grocery stores, department stores and specialised food-retailing.

Table 2.7 Australian exports of goods and services at current prices

	($ million) 1987–88	1992–93
Rural Exports (f.o.b.)		
Meat and meat preparations	2 557	3 752
Cereal grains and cereal preparations	2 298	2 953
Sugar, sugar preparations and honey	701	1 073
Wool and sheepskins	5 806	3 365
Other rural	3 979	5 936
Total rural	**15 341**	**17 079**
Non-rural exports (f.o.b.)		
Metal ores and minerals	5 480	7 942
Coal, coke and briquettes	4 866	7 620
Other mineral fuels	1 786	3 913
Gold	3 107	4 583
Other metals	3 863	5 211
Machinery	1 836	4 347
Transport equipment	1 022	2 020
Other manufactures	3 006	6 017
Other non-rural	1 208	1 283
Total non-rural	**26 174**	**42 936**
Total exports of goods (f.o.b.)	**41 515**	**60 015**
Exports of services	**9 565**	**14 863**
Total exports of goods and services	**51 080**	**74 878**

Source: compiled from ABS, *Yearbook Australia*, 1995 (Table 26.4).

Figure 2.18 Australian labour force in paid employment, by occupation

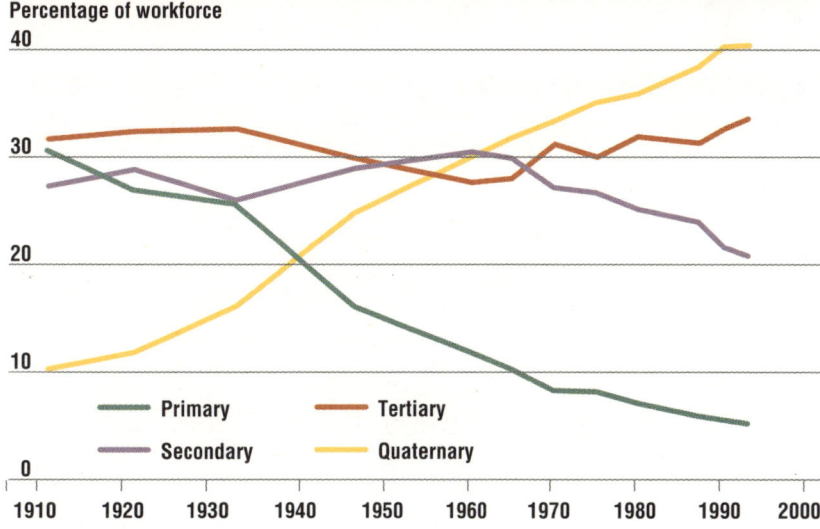

Percentage of workforce

Legend: Primary, Tertiary, Secondary, Quaternary

Note: Primary sector includes agriculture, forestry, fishing, mining, quarrying and oil extraction. Secondary sector includes manufacturing, building and construction. Tertiary sector covers tangible economic services involving the processing or transfer of matter and/or energy. Quaternary sector covers intangible services such as the processing of information.

Source: Censuses of the population of Australia as cited by B. Jones (1995).

The wholesale and retail trade sector employs about 1.6 million people. Motor-vehicle selling and services are also important, as might be expected in a country that, at the last count, contained 567 private motor vehicles for every 1000 people. Australia is second only to the United States in its rate of car ownership.

Agriculture

Despite the high levels of employment and turnover in the retail, finance and other aspects of the service sector, the primary industries — agriculture and mining — remain important in terms of generating employment in the other sectors of the economy, of export income and of their great potential to affect the natural environment over large areas. Agriculture may not be a big employer any more, but it is certainly our major form of land use, taking up about 60 per cent of the country's area (see Fig. 2.19). Grazing — of cattle, mainly for beef, and of sheep for wool — is the main form of agriculture in terms of dedicated land area. By comparison, conservation reserves take up a little over five per cent of the land area, and forestry about 4.8 per cent. Mining operations take up less than one per cent.

In 1994, Australia supported about 134 million sheep, 25 million cattle and 69 million chickens. Australian agriculture not only feeds the country's human population of 18 million, but also provides sustenance for millions of people in other countries. A rough calculation suggests that our exports feed about 55 million people at Australian levels of consumption. We also produce about one-third of the world's wool. Although its importance to the nation has been declining since earlier this century, wool still contributes about five per cent of our merchandise exports.

Australian agriculture tends to be extensive rather than intensive, characterised by the raising of domesticated animals on outdoor pasture. Stocking rates are low compared with other countries because the quality and quantity of feed tends to be poor as a result of low rainfall and poor soils. Many of our sheep and cattle are grazed in the better-watered mixed-farming zones, with a much smaller proportion in the arid and semi-arid rangelands. Wool and beef are the main products; dairying is now concentrated in the south-east and in irrigated areas in the Murray Valley.

Arable farming is confined to a far smaller area than pastoralism (see Fig. 2.19). Winter cereals (wheat, barley and oats) are the major crops. The wheat belt stretches in an inland ribbon from central Queensland, through New South Wales and Victoria and into part of South Australia. A separate belt occurs in Western Australia in the south-west of the State. Just over 9.5 million hectares of land are devoted to wheat-growing. The other main food crops are sugar cane (confined to the tropics) and rice. In addition, the country produces a wide range of fruit and vegetables. Cotton is the principal non-food crop.

Soil nutrient deficiencies are common and fertilisers, particularly phosphate, are widely used. However, their use is limited over much of the country, where low rainfall limits potential yield and thus the return on fertiliser expenditure. High rates of fertiliser use do occur with irrigated crops and those in humid areas where nutrient run-off may have a potentially serious impact on waterways and coastal zones.

About two million hectares are irrigated. Most water for irrigation comes either from the main river system, the Murray–Darling, or from groundwater aquifers.

The success of the nation's agriculture has come at an environmental price. Land degradation (soil erosion, soil structural decline and salinity) is one of the most serious problems affecting the country's terrestrial environment. Much of the reduction in the biological diversity of the continent has also been driven by the destruction or disturbance of natural habitat and the introduction of non-native species of plants and animals for use in agriculture.

Minerals

About three-quarters of Australia's mineral production is exported, especially to Japan. The mining industry contributed 4.3 per cent of Australia's GDP in 1992–93. In terms of exports, mining now has more importance to the economy than agriculture.

Australia is the world's largest single producer of bauxite (the ore from which aluminium is refined), accounting for some 38 per cent of the global supply. The country is rich in rare mineral sands,

Figure 2.19 Land use in Australia

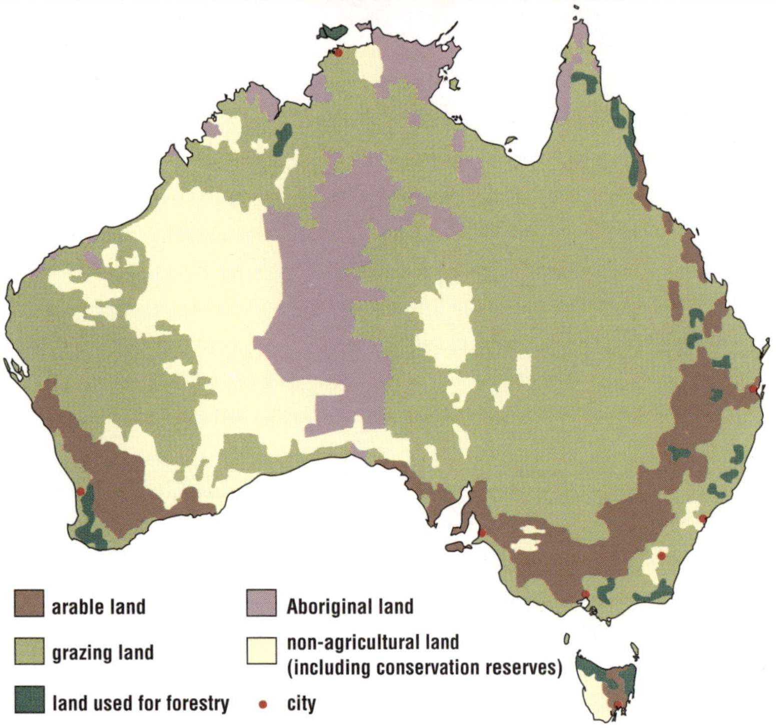

arable land

grazing land

land used for forestry

Aboriginal land

non-agricultural land (including conservation reserves)

city

Source: adapted from AUSMAP Atlas of Australia, 1992.

and again is the world's largest supplier, providing about half of all ilmenite and rutile, and nearly 40 per cent of zircon. In terms of sheer quantity, the major mined products are black coal, iron ore, brown coal and bauxite. In terms of value, coal, gold and iron ore are the most important. Some coal is consumed domestically, as most of Australia's electrical power comes from coal-burning power stations. Gold is now our biggest export earner, having overtaken wool in 1991. Gold-mining is the second largest employer in the mining sector after coal.

Forestry

Apart from treeless Antarctica, Australia in its natural state is the world's least-forested continent. Since European occupation started 200 years ago, the continent's already small stock of forest has been reduced by about 40 per cent. Much of the clearance has occurred to provide land for agriculture and settlement. The total area of native forests remaining is about 41 million hectares. Some 16 per cent of this area stands in national parks or heritage areas, and 27 per cent on private land. The rest is either managed by State forest agencies, or lies on vacant or leased Crown land. The area of plantations is growing and now covers some one million hectares. Most of these plantations comprise exotic species of pine, although the area of native hardwood and softwood plantations is also increasing.

Forests are managed for a variety of sometimes competing uses. The timber industry is an important employer in certain parts of the country. Australia produces nearly 70 per cent of its sawn-timber needs, and some of its own paper products. It exports woodchips, paper and paperboard products. But, in terms of value, it imports more forest products — mainly as paper and sawn timber — than it exports.

Fisheries

Fishing has developed rapidly since the 1950s, with a dramatic rise in the tonnage caught. Large quantities of crustaceans and molluscs (such as lobsters, prawns, abalone and scallops) are taken, mainly for export. While most finfish are caught for domestic consumption, increasingly more of the catch is exported. The main finfish caught are tuna species for export markets and some domestic consumption. The south-east trawler fleet catches species such as whiting, gemfish, morwong and orange roughy, a deep-dwelling bottom fish. Aquaculture is a growing industry.

From 1984 to 1990, Australian consumption of seafood rose from 6.3 kg to 7.9 kg per person per year (about 60 per cent of this was imported). The fishing industry is Australia's fifth-largest primary industry exporter and over the period 1987–88 to 1989–90 seafoods comprised four per cent of the value of primary product exports. In 1992 the Industry Commission found that most Commonwealth fisheries were over-fished and over-capitalised.

Tandala, Aboriginal Culture Centre, Adelaide; tens of thousands of years of continuous human settlement have left a rich legacy of culture and history.

A rich human heritage

On top of its unique natural heritage, tens of thousands of years of continuous human settlement have left a rich legacy of culture and history in Australia. But the early European settlers (with their own religious and social beliefs) found Aboriginal cultures and languages quite alien, difficult to understand and irrelevant to them and so tended to ignore or deliberately destroy them. As a result, much of Australia's original cultures — especially those based on oral tradition — have been lost. But some remains, and is being built on, continuing in strong contemporary forms today. As well, traditions brought to the continent by other peoples over the last two centuries have taken root. At their best, the various cultures and traditions together make a total sum of distinctive human and cultural heritage that is as important and as recognisable a part of Australia's environment as kangaroos or gum trees.

When the European colonisation of Australia started in 1788, the country was inhabited by an unknown number of Aboriginal peoples organised into social groups that covered the whole continent and the islands of the Torres Strait. Literally hundreds of complex, intricate languages, as distinct as the different languages of Europe, were spoken and traditional legal systems operated. Exchange networks were long established across the continent — they linked distant societies in alliances that were social and ceremonial as well as serving as trade routes. The exchange networks of eastern central Australia are amongst the most extensive known from small-scale, non-urban, non-industrialised societies and attest to the cultural achievements and sophistication of Australia's prehistoric societies (see Fig. 2.20). Most of the linguistic and cultural diversity has now been lost, especially in the southern part of Australia, as the Aboriginal peoples died in large numbers (often from introduced diseases to which they had no resistance), moved or were removed from their land or were integrated — forcibly or otherwise — into European culture.

The great majority of original Australian languages have become extinct, or are becoming so — existing with only a few elder speakers. The death of a language is the loss of something unique and

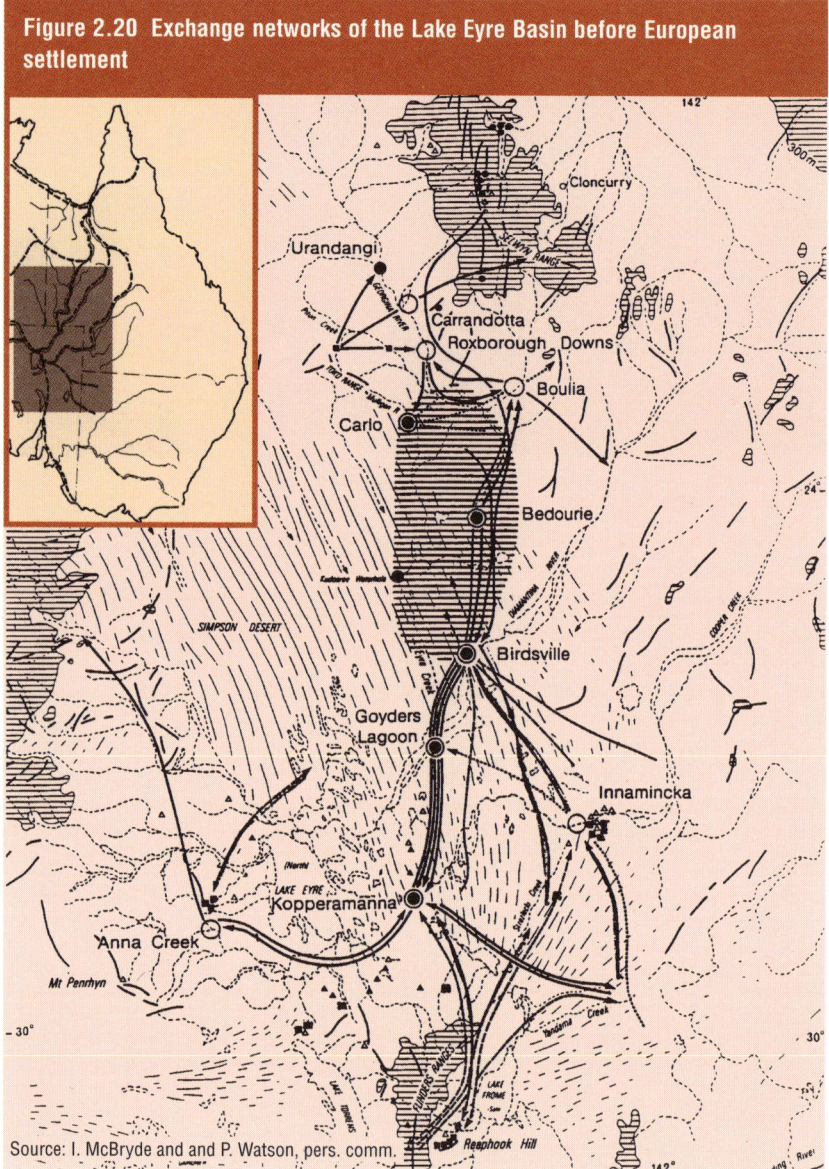

Figure 2.20 Exchange networks of the Lake Eyre Basin before European settlement

Source: I. McBryde and and P. Watson, pers. comm.

Managing our environment

Human impacts

Humans have interacted with the natural environment in Australia for tens of thousands of years and have brought about changes as a result. But in the last 200 years the pace of change has accelerated dramatically. Now much of the natural environment — not to mention the original cultural environment — is irreversibly altered.

The environment that so amazed and then dismayed the first European explorers and settlers in Australia — and which in the recent past had been thought of as 'pristine' — was clearly affected by an earlier human presence. However, the extent and seriousness of the impact is the subject of debate. As we do not know the total population of the continent when Europeans first came, or the population densities that possibly occurred in certain highly settled areas, it is hard to determine the scale of pressure on the environment at that time.

Aboriginal hunting may have affected the numbers of certain animals. Early explorers' reports indicate that some species of edible wildlife (such as kangaroos, wallabies, koalas, possums, bandicoots. as well as brolgas, swans and ducks) were nowhere as abundant as they became subsequently. Some animals, such as the red kangaroo, benefited from land clearing, planting of crops and the provision of water for stock in the outback by European settlers. The original inhabitants also affected the vegetation, probably reducing forest cover in places.

The first Australians are thought to have possessed raft- or canoe-building skills in order to reach the country, and a detailed knowledge of fire and its effects. Aboriginal people at the time of European arrival used fire to manage their land. It had the dual purpose of stimulating seed-setting and germination, and of driving out game animals or attracting large marsupial grazers to areas of new shoots. Continuous, regular burning must have modified vegetation patterns and species compositions in many parts of the country. Areas that now support forest and woodland were grassy open woodland with widely spaced trees 200 years ago (according to the descriptions and paintings of the time). These changes have occurred almost certainly as the result of the decline in regular burning. Aboriginal peoples built fish traps in rivers and on the coast, which could also have had environmental effects.

Whether any extinctions can be attributed to the first Australians is a matter of scientific debate — the evidence remains circumstantial. About 40 species of marsupials (mainly very large ones) became extinct after the arrival of people, and by 20 000 years ago few mammals survived that were larger in weight than the by-then-widespread human predators. The later arrival (about 3000 years ago) of the dingo, possibly brought by voyagers from South-East Asia, must have had effects too. Some would argue that the competition from the dingo led to the extinction of the larger marsupial carnivores, the thylacine and the Tasmanian devil on the mainland. Dingoes did not

irreplaceable, and is also an indicator of the loss of cultural heritage. But some elements of the original Australian culture, adapted to modern life, remain in certain areas still inhabited by the traditional owners, and some have been adopted, albeit in token or modified form, into the mainstream identity of Australia. Since the Bicentennial in 1988, the cultural heritage of Aboriginal and Torres Strait Islander peoples has been recognised more widely. Some Aboriginal languages are still spoken, and a few — such as Pitjantjatjarra — have even increased their area of coverage, becoming a 'lingua franca' for central Australian Aboriginal peoples. Aboriginal languages have also given us many Australian place names and unique words in Australian English.

The great post-war immigration wave has continued, bringing many languages, along with cultural traditions from many parts of the world. In 1947, more than 90 per cent of Australians had been born here. Today, nearly one-quarter of the population was born overseas. The range of countries from which migrants come has been increasing in line with changes to immigration policy. Australia is now truly a multicultural society.

Australia 1970

Die wild country, like the eaglehawk,
dangerous till the last breath's gone,
clawing and striking. Die
cursing your captor through a raging eye.

Die like the tigersnake
that hisses such pure hatred from its pain
as fills the killer's dreams
with fear like suicide's invading stain.

Suffer, wild country, like the ironwood
that gaps the dozer-blade.
I see your living soil ebb with the tree
to naked poverty.

Die like the soldier-ant
mindless and faithful to your million years.
Though we corrupt you with our torturing
mind,
stay obstinate; stay blind.

For we are conquerors and self-poisoners
more than scorpion or snake
and dying of the venoms that we make
even while you die of us.

I praise the scoring drought, the flying dust,
the drying creek, the furious animal,
that they oppose us still;
that we are ruined by the thing we kill.

By Judith Wright (1915–)
In *The Collins Book of Australian Poetry.*
(Chosen by Rodney Hall. Collins, Sydney,1981).
Reproduced with permission of Harper Collins
Publishers.

Patch burning in Uluru National Park as a management technique follows traditional Aboriginal practice (top).

The giant herbivore, *Diprotodon optatum* (right) was one of about 40 species of marsupials that became extinct after people arrived in Australia.

reach Tasmania because by then it was separated from the mainland by a wide sea barrier — Bass Strait. The thylacine survived in Tasmania until the 1930s.

The original inhabitants of Australia altered the biophysical environment in more subtle ways than we do today. Whatever their impacts, their continued survival showed that they must have lived sustainably within the resources of the continent for tens of thousand of years.

Of far greater import were the effects of the settlers from 1788 onwards, who were responsible for:

- introducing and spreading agriculture
- introducing exotic plants and animals
- widespread change of land cover (by deforestation/clearing and ringbarking)
- changing fire regimes
- introducing land use practices suited to Europe
- commercial hunting for hides, skins, plumes and blubber oil
- large-scale mining
- urbanisation
- industrialisation

The broad categories of effect from these actions are:

- land degradation
- loss of biodiversity
- depletion and/or modification of water resources
- air and water pollution and soil contamination
- changes to the coastal and marine environment

In addition, humans now impose a pressure on the various environments that they have created for themselves.

Environmental management

Environmental management in Australia is guided by the National Strategy for Ecologically Sustainable Development (ESD). The strategy defines ESD as a pattern of development that improves the total quality of life, both now and in the future, in a way that maintains the ecological processes on which life depends.

The environment itself does not respect the borderlines that humans draw — whether these be national, administrative or intellectual boundaries. But the fragmentation, and sometimes duplication, of society's decision-making processes contradict the nature of environmental issues themselves. Thus the structure, organisation and functioning of Australian society and government are themselves key factors, and sometimes pressures, affecting the environment.

Although government in various forms is responsible for much of what is generally called management, many other groups in society — and especially the everyday decisions of millions of individuals — play important roles when it comes to affecting the environment.

Individuals and the community

In recent years, opinion polls have consistently shown that environmental issues are high on the list of concerns for most Australians. The Australian Bureau of Statistics has found that between 1986 and 1994 the number of people expressing concern for the environment had increased by 20 per cent to 70 per cent. Community interest in global environmental issues, and in the response of governments at all levels to those issues, is also high.

Across the country, many environment and conservation organisations exist, with hundreds of thousands of members collectively. Most of their resources come from public donations although a number receive some funds from the Commonwealth Government.

And, when multiplied millions of times, the behaviour of individuals — choosing to use phosphate-free detergent, conserving water in the garden, recycling household paper, or reducing energy consumption — can be significant. Most Australians make these everyday, small 'environmental management' decisions. Little coordination of this takes place, beyond single-issue campaigns conducted by government and other organisations, and the dissemination of factual information, of which this report forms a part.

Major environmental issues in Australia's offshore territories

Territory	Main Environmental Issues	Comments
Ashmore and Cartier Islands	Protection of Ashmore Reef National Nature Reserve	The Australian Nature Conservation Agency maintains a presence during the fishing season to protect wildlife and monitor Indonesian traditional fishing activities as well as the activities of any other visitors.
	Oil/gas developments in Bonaparte Basin	Aerial surveillance flights monitor the potential threat of pollution drifting from the adjacent fields to the Territory. No such pollution has as yet escaped from the fields.
Norfolk Island	Conservation of the Kingston/Arthur's Vale Historic Area	The Commonwealth and the local government provide ongoing financial support for the conservation and restoration of this Historic Area, which contains the finest collection of colonial Georgian architecture in Australia and is listed on the Register of the National Estate.
	Protection of the Island's National Park and endangered species	The Australian Nature Conservation Agency administers the National Park which is being extended to include Phillip Island. The Agency also manages 3 endangered species recovery programs for the Island's green parrot, morepork owl and a composite program for some 40 threatened plant species.
Cocos (Keeling) Islands	Protection of the Cocos Atoll including North Keeling Island	North Keeling Island is listed on the Register of the National Estate. The Island is an important major breeding ground for sea birds that are protected under various international treaties. Because of this and the fact that it includes the last significant remnants of Cocos Islands' original vegetation, the island is presently in the process of being declared a national park.
Coral Sea Islands Territory	Protection of Lihou Reef and Coringa-Herald National Nature Reserves	The Australian Nature Conservation Agency patrols the reserves with assistance from the Royal Australian Navy and Coastwatch and conducts wildlife management programs. Permits are required for camping, for conducting scientific research and for any commercial activity including fishing or diving charters.
Christmas Island	Protection of Christmas Island National Park and endangered species	Christmas Island's isolation has resulted in a unique ecological environment attracting its own bird species such as the endangered Abbot's booby. Much of the island's original rainforest remains intact which enhances the survival of the species and other rare plant and animal life.

Note:
Australian Antarctic Territory and the Territory of Heard Island and McDonald Islands are not included.

Source: DEST, Office of Territories, unpublished data.

Industry and commerce

Australia's manufacturing and industrial base plays a crucial role in the life of the nation. The private sector provides a high proportion of jobs, as well as goods and services. Increasingly, businesses are becoming aware of the importance of environmental matters, and their accounting includes expenses related to responsible environmental behaviour.

Companies consider environmental management for a range of reasons: they may be pragmatic — not wishing to fall foul of the relatively new environmental laws and regulations ranging from control of emissions and toxic-waste disposal to rehabilitation of mined areas; they may see that environmental auditing and management are good for business public relations; or it may make good economic sense. This has been stimulated by the recent rise in green consumerism in Australia and throughout much of the Western world. Many businesses realise that adopting cleaner production through better use of resources and waste minimisation enables them to save money. Others may quite simply be motivated by an environmental concern that goes beyond the economic rationalism of the short-term 'bottom-line'.

Many sectors of business have highly developed environmental divisions, which have existed for decades, but other areas still remain relatively unconcerned about the environment. However, in general the business sector's awareness of environmental matters and of its responsibilities has seen a rapid upturn.

The increasing importance of environmental management, monitoring and pollution control present considerable business opportunities. Australia already exports both expertise and hardware in these fields.

Government

Australia is a democratic, politically stable country headed by a constitutional monarch, with several spheres of government. These are the Commonwealth, State, Territory, and local governments. However, in certain areas, the traditional law- and decision-making processes of Aboriginal and Torres Strait Islander peoples continue — and the significance of this has recently been strengthened by the High Court's decision on the Mabo case concerning the traditional ownership of land. Beyond government in Australia lies the international community, in which Australia participates through its membership of many world bodies, through being a party to agreements, treaties and protocols, and by being accountable through the mechanism of the World Court.

The nation known as the Commonwealth of Australia came into existence in 1901 following the federation of six separate British colonies (the forerunners of the present States). An important issue in the debate that led up to Federation was the extent of conflicts between the colonies — notably trade and defence, although some related to

Parliament House Canberra; the Commonwealth has substantial powers to enact laws affecting the environment and sustainable development.

resource-use issues, in particular the Murray–Darling river system. The Constitution formulated at Federation defined the powers of the Commonwealth Government, which are designed to encompass matters relating to the country as a whole. These include defence, external affairs, banking, immigration, trade and commerce, taxation, posts and telecommunications and social welfare. Since Federation, various powers, in particular those set out in section 51 of the Constitution, have become particularly relevant to implementing the Commonwealth's environmental responsibilities, especially since 1970, including the following:

- the conduct of external affairs (in particular the signing of certain international treaties and agreements that require actions within Australia)

- taxation and the power to grant financial assistance (with conditions) to State and Territory Governments

- the 'corporations power' whereby the Commonwealth can impose various conditions on corporations concerning the management of their land

- the management of fisheries throughout Australia's exclusive economic zone

- the issuing of export licences for major resource developments

- quarantine across the whole country

- Aboriginal and Torres Strait Islander affairs

Thus, the Commonwealth Government relies on a wide variety of powers to give effect to its policies.

Australia has about 770 local governments, which are responsible for the provision of local services such as environmental health regulation, road building and maintenance, municipal waste management, land use planning and development control, pollution control and monitoring, traffic management, parks and open space, recreation facilities and community services.

One of the most 'high-profile' of the environmentally relevant international treaty agreements that Australia has entered into is the UNESCO World Heritage Convention of 1972, which was ratified in Australia in 1974 and came into force in 1975. The aim of the Convention is to protect, anywhere in the world, natural and/or cultural heritage of such outstanding universal value that its conservation is of concern to all people. Places on the World Heritage List should be conserved for all time.

Member countries must ensure that they identify, protect, conserve and present their listed World Heritage properties. The 11 World Heritage properties in Australia, as of June 1995, are listed in Chapter 9.

The Australian Constitution does not specifically deal with environmental powers and they are not the sole province of any one sphere of government. Most environmental legislative responsibilities rest with the State and Territory Governments although the Commonwealth does have substantial powers to enact laws affecting the environment and sustainable development. In some cases, the High Court has tested and confirmed these powers. However, many day-to-day government decisions that affect the environment occur at the level of local government. The interconnectedness of the environment means that very few aspects can be managed solely by one sphere of government, or one agency, in isolation from other spheres and agencies; and no aspect of the environment can be managed in isolation from the community. And plenty of other environmentally relevant decisions neither involve nor emanate from governments at all, but rather are taken by industry, various special-interest groups and individuals acting alone or collectively.

Various means of overcoming the existing fragmentation in the way we view and respond to the environment are in place, both nationally and internationally. These are briefly touched upon below.

National environmental management and policy-making

The national processes to help provide a more holistic form of environmental management can be classified into the following eight categories.

- *The Council of Australian Governments (COAG)*, composed of the Prime Minister and the leaders of the States and Territories and local government, is the peak body overseeing closer cooperation between Australian Governments on issues concerning clarification of roles and responsibilities in areas of shared responsibility, micro-economic reform, regulation and the environment. Among COAG's first tasks when it was established in 1992 was overseeing the finalisation of the InterGovernmental Agreement on the Environment (IGAE) and the National Strategy for Ecologically Sustainable Development.

- *Ministerial councils and their advisory groups* involve Commonwealth, State and Territory ministers responsible for various common matters meeting together regularly. There are many such councils relevant to the environment.

- *National inquiries* are initiated at the political level by Parliament, by governments, by individual ministers, or by ministerial councils. At their best, inquiries can be multidisciplinary in scope, and can pull together information from different jurisdictions and sections of the community. In this way they can arrive at a point of view dictated by the region or phenomenon under investigation rather than emanating from particular sectors.

- *The InterGovernmental Agreement on the Environment* was agreed to in 1992. It was designed to avoid damaging disputes about environmental matters between the different levels of government in the country. It establishes conditions under which Australia's various governments will interact and includes a set of principles to guide the development of environment policies.

- *National strategies* include the National Strategies for Ecologically Sustainable Development, Greenhouse Response and Biodiversity as well as various sectoral strategies. Their development involves consideration by all spheres of government and consultation with industry and the community. When adopted, all governments, at least in theory, use national strategies to guide their decision making.

- *Strategic arrangements for management on a drainage basin or ecosystem basis* are exemplified by the Murray–Darling Basin Commission. The river catchment that it manages straddles four States and one Territory. It has been estimated that 33 different government departments have some responsibility for the river system, and 268 local governments have a stake in it. Countless land-owners and farmers use the water or have an effect on it. The Great Barrier Reef Marine Park Authority and the Wet Tropics Management Authority are other good examples.

- *Commonwealth legislation* is the only way some aspects of environmental management can be consistently governed throughout Australia; for example the *Wildlife Protection (Regulation of Exports and Imports) Act 1982*, the *Antarctic Marine Living Resources Conservation Act 1981*, the *Hazardous Waste (Regulation of Exports and Imports) Act 1989* and the *World Heritage Properties Conservation Act 1983*.

- *Complementary legislation* is legislation mirrored in each jurisdiction to ensure a national approach; for example the legislation establishing the National Environmental Protection Council.

Other functions of government

The work of modern government is too complex and intricate to detail here. Suffice it to say that the Commonwealth, State and Territory Governments maintain various departments that deal with a range of matters related to the environment. The names and precise divisions of responsibility between departments frequently change in response to political agendas.

International perspectives

Certain aspects of the natural environment, such as migratory species and water or atmospheric pollutants, although they may originate in a particular country, cannot be contained by any human boundaries. Other features, like the open ocean or the atmosphere, are 'common goods' that do not belong, in the legal sense, to any single nation. International agreements regulate certain human activities, such as air and sea transport, high-seas fishing and human uses of the Antarctic region. As a result of the interconnectedness of so much of the natural environment and international human activity, many environmental problems — for example, the enhanced greenhouse effect or over-exploitation of oceanic resources — require global action if we are to have any hope of solving them.

Since the 1940s, the pace and breadth of human activities affecting the global environment have increased greatly; and so too have awareness of and concern for our environment. Over the last few decades it has become clear that there is a need for international agreements to preserve many places (such as Antarctica) or regulate activities (such as whaling) for the benefit of all nations or for the sake of other species or unique natural environments.

The world has responded to the increase in human pressures and to the growing realisation of the 'wholeness' of nature by creating many international organisations, treaties, conventions and agreements in an effort to deal with issues that transcend national boundaries.

Australia participates in the international community and contributes to the global effort to manage the planet. It is a party to one of the world's earliest environmental treaties, the Antarctic Treaty of 1959, which designates Antarctica as an area to be used for peaceful purposes only. A number of international agreements now involve Australia in global environmental issues, such as the enhanced greenhouse effect, ozone depletion in the stratosphere and the trade in endangered species. Appendix 1 includes a full list.

Between 1990 and 1992 the United Nations developed a global action plan for sustainable development. The United Nations Conference on Environment and Development, also called the 'Earth Summit', in Rio de Janeiro in June 1992 adopted this plan — Agenda 21. The Conference also saw the signing of two new global Conventions — on Climate Change and on Biological Diversity — the signing of a statement of principles for forest management and the adoption of a declaration on the principles of sustainable development ('The Rio Declaration'). Agenda 21 sets out actions that nations, communities and international organisations can all take to help achieve the goal of global sustainability in the 21st century. The Conference also led to the establishment of a new United Nations organisation, the Commission on Sustainable Development, which meets annually to review progress in the implementation of Agenda 21.

The environment is now firmly on the international political agenda and is the subject of negotiation and discussion in many forums. The aim of all these activities is to protect the environment and conserve natural resources thereby enhancing human well-being and development and the security of nations. The institutions we design to monitor and protect the environment will themselves be complex and imperfect. This is because the environment includes everything and everyone, the impacts of human activity can be so far-reaching, and the world is divided into nations and regions with different aspirations.

Conclusion

Australia has a distinctive environment and heritage. Its biological diversity is special in global terms. Its climate and rainfall constrain people's activities. Its cultural landscapes are of great antiquity. The broad spatial distribution of its population has energy usage implications.

How well is Australia managing its environment as we near the end of the 20th century? How can community, industry and government be empowered to make decisions that protect the future environment and heritage of Australians? State of the environment reporting is a tool to provide a broad picture of the Australian situation. Such reporting, including monitoring of environmental indicators, provides the facts needed now by all decision-makers in the move to a more ecologically sustainable society.

Further reading

Australian Bureau of Statistics (1988 and 1995). 'Yearbook Australia', Cat. No. 1301.0. (ABS: Canberra.)

Australian Bureau of Statistics (1992). 'Striking a Balance: Australia's Development and Conservation.' (ABS: Canberra.)

Australian Bureau of Statistics (1995). Australian Demographic Statistics, March Quarter 1995, Cat. No. 3101.0.

Beckmann, R. (1994). 'Environmental Science.' (Australian Academy of Science: Canberra.)

Department of the Environment, Sport and Territories (1994). 'Australia's Biodiversity — an Overview of Selected Components.' (DEST: Canberra,)

Johnson, K. (1992). 'The AUSMAP Atlas of Australia', Australian Surveying and Land Information Group. (Cambridge University Press: Melbourne.)

White, M. (1994). 'The Greening of Gondwana.' (Reed Books: Sydney.)

White, M. (1994). 'After the Greening: The Browning of Australia.' (Kangaroo Press: Sydney.)

Acknowledgments

The chapter was reviewed by:

Professor Graeme Davison (Australian National University)

Professor Roger Kitching (Griffith University)

Professor Henry Nix (Australian National University)

The chapter has benefited from contributions by:

Professor Isabel McBryde (Australian National University)

Unisearch Ltd (University of New South Wales)

Stephen Watson-Brown and other staff of the Department of the Environment, Sport and Territories assisted in the compilation of this chapter.

Photo credits

2-1: Artist — Jeffrey Smart (Westpac Gallery, Victorian Arts Centre)

2-9: CSIRO

2-11: (*from left*) DEST; Bill Van Aken (CSIRO Division of Water Resources)

2-12: Nick Alexander (Oryx Films)

2-13: Malcolm Paterson (CSIRO)

2-14: (*from top*) Nick Alexander (Oryx Films); DFAT; CSIRO

2-15: Mike Prociv (Wetropics)

2-17: ERIN

2-18: DFAT

2-19: Reproduction by courtesy of the Latrobe Library, Victoria

2-21: DFAT

2-22: DEST

2-25: DFAT

2-27: (*from top*) ANCA;
    Artist — Peter Murray (Reed Books)

Chapter 3

Human Settlements

Perth's central business district.

Prepared by

Peter Newman (Chair), Murdoch University

Bob Birrell, Monash University

Doug Holmes, Coopers and Lybrand

Colin Mathers, Australian Institute of Health and Welfare

Peter Newton, CSIRO Division of Building, Construction and Engineering

Graeme Oakley, Australian Bureau of Statistics

Alice O'Connor, Ministry for Planning, Western Australia

Bruce Walker, Centre for Appropriate Technology

*Allan Spessa (State of the Environment Reporting Unit member),
    Department of the Environment, Sport and Territories (Facilitator)*

*David Tait (Former State of the Environment Reporting Unit member),
    Department of the Environment, Sport and Territories (Former
    facilitator)*

Contents

Introduction

Australia's population is highly urbanised. In 1991, 85 per cent of Australians lived in settlements with populations of 10 000 or more. The remaining 15 per cent of the population lived in small country towns, on farms or in remote settlements (see Table 3.1). Although Australia's largest settlements occupy less than one per cent of the nation's land area, they have a considerable influence on the natural environment of their hinterlands.

This chapter examines the influence of demographic, economic, social and technological pressures on Australian settlements. The condition of our settlements is assessed with regard to livability, social amenity, health and resource inputs and waste outputs. The country's urban, rural and remote settlements are considered as three different kinds of interaction with the environment and various pressures, conditions and responses in these three types of settlement are examined. Finally, the chapter reviews government responses to identified problems.

Table 3.1 Proportion of Australians living in urban, rural and remote settlements, 1991

	Number	Percentage
Urban		
Big Cities[1] (above 1 million)	10 062 003	59.7
Other Cities (80 000 to 1 million)	2 025 803	12.0
Rural		
Large Rural Towns (25 000 to 80 000)	962 041	5.7
Small Rural Towns (10 000 to 25 000)	853 051	5.1
Rural Other[2] (less than 10 000)	2 452 264	14.6
Remote[3]		
Remote Towns[4] (above 5 000)	203 137	1.2
Indigenous Settlements[5]	73 297	0.4
Remote Other[6]	209 973	1.2
Total[7]	16 850 540	100.0

Notes:
1. Includes Sydney, Melbourne, Brisbane, Perth and Adelaide.
2. Includes people on farms and in small towns in the agricultural region.
3. Outside the cleared agricultural areas, i.e. areas used mostly for pastoral, mining, tourist and Indigenous purposes consistent with Holmes (1988).
4. Includes a substantial group of Indigenous communities whose members are residents of country towns mixed in with a predominantly non Indigenous population.
5. Data on the number and location of discrete Indigenous settlements and population of those settlements are drawn from ATSIC (1993). Settlements in this table are those discrete indigenous townships, outstations or groups of indigenous people living in an identifiable location often located on indigenous land and likely to be responsible for their own municipal services or which are without such services altogether.
6. Mostly pastoral and mining settlements, including communities between 100 and 5000 that have sizeable indigenous representation associated with them.
7. Includes offshore and migratory population (8971). Census data do not match the estimated resident population data published in some of the following tables.

Source: ABS, 1991.

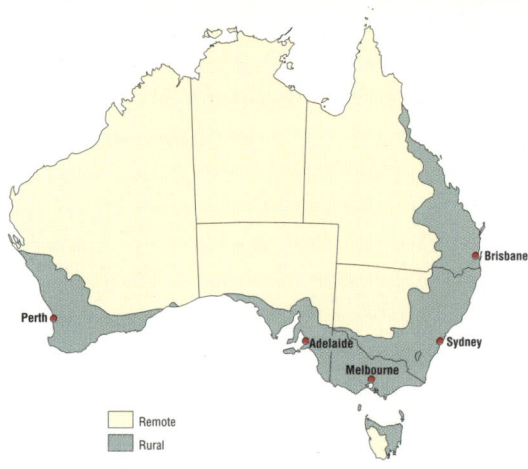

Figure 3.1 Australia's five big cities, rural and remote areas

Remote
Rural

Note:

In this report the definition of urban, rural and remote is based on a functional approach, i.e. urban includes those surrounding small settlements which are functionally now part of the urban area although they may have once been quite separate and still look somewhat rural. The ABS urban classification does not include as many of these peripheral areas and hence they report only 76% of people live in cities above 10 000 (ABS, 1994g). Rural is defined as including cleared agricultural areas, and remote as beginning in the pastoral regions beyond the rural zone.

Metabolism of settlements

Extended metabolism model

Metabolism, the flow of resources into and waste outputs from settlements, provides a model for assessing the environmental impacts of settlements (Wolman, 1965; Boyden *et al.*, 1981; Girardet, 1992). This chapter extends the basic metabolism model to include the dynamics of settlements and livability (see Fig. 3.2). Livability is defined as the human requirement for social amenity, health and wellbeing and includes notions of individual and community wellbeing in both the human and wider environment.

Settlements are places where communities evolve economically and socially in an attempt to improve their quality of life. The products of economic and industrial developments have value for livability as well as environmental impact. This chapter looks not only at how settlements in Australia affect the environment by using resources and producing wastes, but also at the way quality of life within them can be measured using various socio-economic and health indicators. The management of the built environment and the natural environment need not necessarily be in conflict. This is the challenge of the National Strategy for Ecologically Sustainable Development (ESD) as adopted by Australian governments in 1992.

Settlements manage their resource inputs and waste outputs by engineering systems at city, industry and household scales. Such systems have various levels of efficiency and impact depending on their size and the priority given to infrastructure and technological innovation. State of the environment

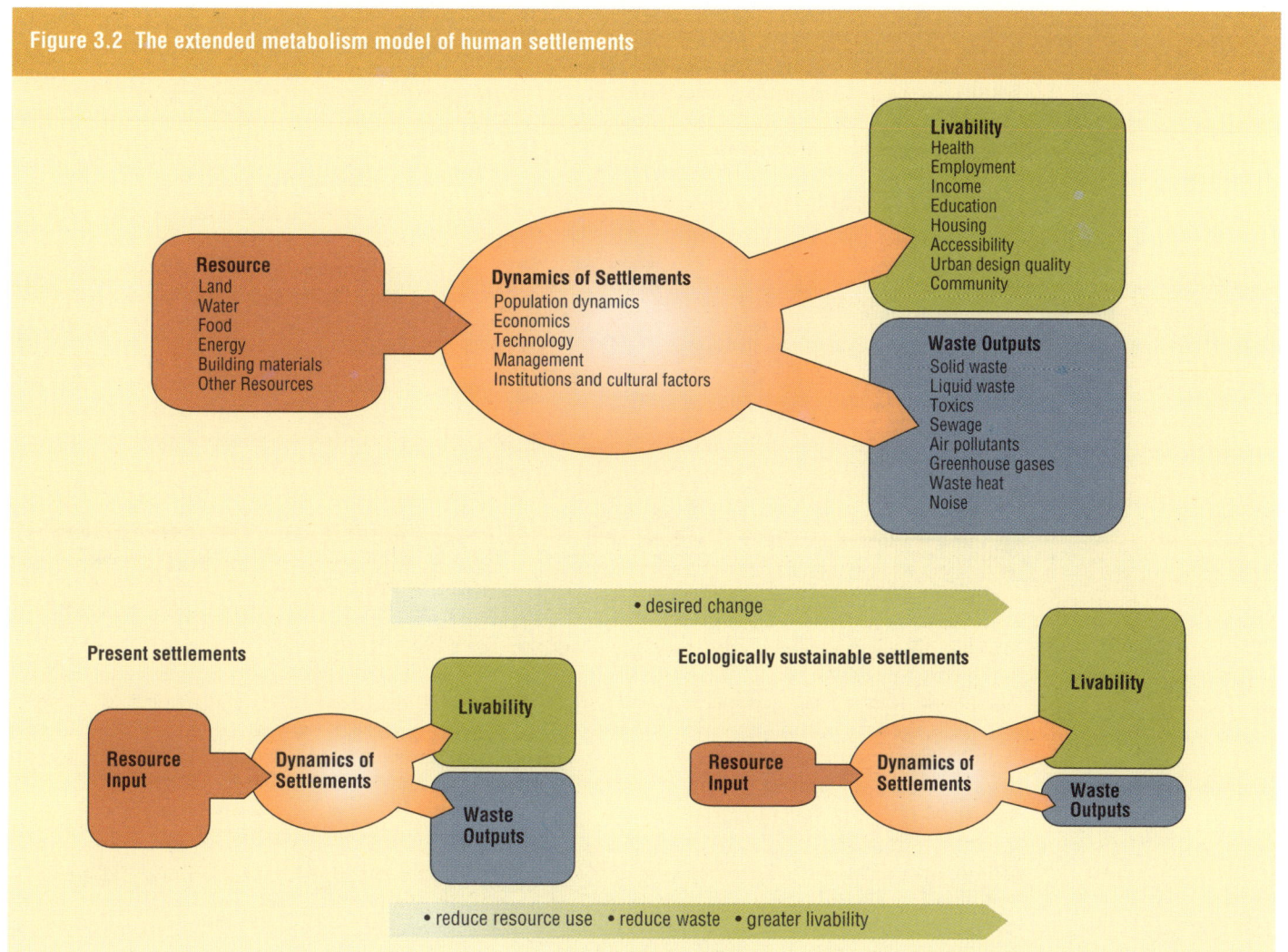

Figure 3.2 The extended metabolism model of human settlements

reporting is one way that we can assess both our natural resource capacities and quality of life.

Other models

A city's pressure on the environment extends beyond its own geographic limits, as it draws on resources such as oil, minerals, timber and food, not only from its own region, but from other parts of Australia and the world. Wastes also have effects beyond the local area, especially greenhouse gases, ozone-depleting substances and hazardous materials. Researchers are developing techniques for estimating the extended environmental impact of settlements due to their resource consumption. One such method is William Rees' 'ecological footprint' (see the box on the right).

The 'ecological footprint' model and other systems that focus on resource flows in settlements help us understand their extended impact on the environment. But they do not help to assess whether or not we can reduce requirements for resources while maintaining or improving livability. For this reason the extended metabolism model is used in preference to the ecological footprint model in this chapter.

Carrying capacity and the ecological footprint

The term 'carrying capacity' is a concept usually used to determine the population of a particular species that an ecosystem can sustain without damaging its essential functions. Determining carrying capacity for human settlements is complex because not only do their consumption requirements reach well beyond settlement boundaries, so too do the effects of their waste.

Canadian William Rees has taken the concept of carrying capacity and applied it to human resource use. He defines human carrying capacity as 'the maximum rate of resource consumption and waste discharge that can be sustained indefinitely without progressively impairing the functional integrity and productivity of relevant ecosystems' (Rees, 1992).

Rees refers to the total area of land and water required to sustain a settlement as its 'ecological footprint', and has developed a method of measuring the area required to provide its resources and absorb its wastes. Because settlements are remote from much of the land producing the resources they consume, each one's 'ecological footprint' represents its 'appropriated carrying capacity'. With the increasing intensity of world trade in resources, this concept is an attempt to measure the true impact of settlement consumption patterns.

In calculations for metropolitan Vancouver, Rees showed that its population requires 20 times more land than the region represents for food production, forestry products and energy requirements alone (Wackernagel *et al.*, 1993). The bulk of the 'footprint' land is consumed for energy, because the calculation estimates greenhouse gas emissions in terms of the forest needed to take up the CO_2 or to produce the energy from biomass-derived ethanol. This is purely illustrative of the possible extent of the city's impact, and should not be seen as a direct impact.

Pressure

The major forces shaping the recent evolution of Australian settlements are those related to technological and economic development and associated population dynamics.

Technological and economic forces

Globalisation

The process of globalisation is affecting all human settlements. National and local economies are increasingly being opened to the competition of international markets and other global trends. They are being exposed to the global flows of information, capital, people, goods and services. The effects of such forces are amplified within Australia as the Commonwealth Government continues to dismantle barriers to competition by, for example, reducing tariffs and deregulating financial markets. Globalisation is changing the nature of manufacturing in Australia (reduced blue-collar production-line employment, fewer smokestack industries), reducing the need for warehousing (just-in-time production), increasing office employment (world-wide electronic communication and computerisation) and providing increased employment in information-oriented service industries.

Between 1981 and 1991, Australian employment in urban manufacturing declined by 10 per cent, resulting in the net loss of some 104 000 jobs in this sector. During the same period producer services (finance, real estate, information and communications) increased by 51 per cent, personal services grew by 36 per cent and social services increased by 30 per cent (see Table 3.2).

A later section will suggest how these changes are reflected in current patterns of urban development: in the spatial distribution of income and unemployment; patterns of housing affordability; and in processes of suburbanisation and re-urbanisation.

Globalisation of the economy can be a powerful positive or negative force for sustainability. Globalisation is also a powerful force in terms of ideas and pressures on governments. International agreements to reduce metabolic flows and improve livability are setting the global agenda. Australian settlements can adopt appropriate innovations in urban management and technology and use them to help promote wealth, livability and sustainability.

Transport/information technology

People cluster together in cities because they can do more as a community than as individuals. The technologies for people to move around and between cities and to share information are critical to how they manage their metabolic flows and create livability.

Research shows that people in settlements tend not to spend more than about half an hour on average travelling to destinations (Manning, 1978; Pederson, 1980). This has been a pattern throughout history.

The walking city (before 1860) seldom exceeded a five km radius and so was densely populated (100–200 people per ha) with narrow streets. Many cities in developing countries retain this walking-based urban form.

The transit city (1860–1940), which could spread 10 to 20 km, tended to be linear and focused on railway stations or along tram lines, with medium-density houses and work locations (50–100 people per ha) and a strong emphasis on the central business district (CBD). Many European cities retain this transit-based urban form.

The automobile city (from 1940 on) could spread 20–40 km wherever roads were built; the density was subsequently lower (10–20 people per ha) and a much more spacious city became possible, although it used much space for roads and parking. The CBD became mainly an office centre with most other work dispersing to the suburbs. Many United States and Australian cities reflect this automobile-based urban form.

In recent decades a fourth type of city has emerged.

The multi-nodal/information city is emerging in large cities where the distance of travel, even by automobile or fast train, from the periphery to the CBD is now well beyond the half hour limit. Partly as a response, the multi-nodal/information city comprises a range of smaller subcentres with global information processing and networking capabilities equivalent to the CBD, as well as other urban services. Although linked to the rest of the city, these nodal/information subcentres can generate a large degree of self-sufficiency in their immediate urban region. The layout of these cities varies considerably: European versions are more densely settled and transit-oriented, with their subcentres strung out like pearls along a string (Cervero, 1995); the North American multi-nodal/information centres, which are more dispersed and car-oriented, are sometimes called 'edge cities' (Garreau, 1991).

All four types are represented in Australian cities to varying degrees. Remnants of high-density walking cities, like The Rocks, Fremantle and some CBD areas, are often pedestrianised. Many pre-1940

Table 3.2 Industry and employment changes in urban Australia				
Employment category	**1981**	**1991**	**Absolute growth**	**Percentage growth**
Extractive industries	50 925	49 377	-1 548	-3.0
Transformative industries	1 017 906	914 015	-103 891	-10.2
Distributive industries	927 646	1 115 941	188 295	20.3
Producer services	369 942	560 046	190 104	51.4
Social services	743 078	969 171	226 093	30.4
Personal services	176 978	276 362	99 384	36.0
Total	**3 286 475**	**3 884 912**	**598 437**	**18.2**

Source: Newton, *et al.*, in press.

transit-based, medium-density suburbs, often called the 'inner city', are extensive in Sydney and Melbourne, but also occur in Brisbane, Perth, Adelaide, Newcastle and Hobart. Large expanses of low density car-based suburbs exist in every city, providing the main source of accomodation for the 10 million people added to the population since 1945. Emerging nodal/information subcentres are found in all major cities.

The older core and inner areas have much lower dwelling occupancies due to the age of their occupants (for example, children leave home and go to new suburbs while parents remain) and their greater housing diversity — including many more units that are occupied by smaller households, often single people (see Table 3.3).

In the 1940s, Australia's big cities had an occupancy rate of 3.9 people per dwelling. This declined to 2.6 in 1991. About one-third of the development pressure for housing in Australian cities in the post-war period came from the decline in household size which was caused by a combination of factors, including reduced birthrates, higher divorce rates, children leaving home earlier and lifestyle preferences.

This chapter often refers to two processes — suburbanisation and re-urbanisation — that are simultaneously helping to shape the new multi-nodal/information city.

Suburbanisation of work and services to outer areas is occurring along with new housing. Most new development in outer areas and the vast majority of fringe areas in major Australian cities are now beyond the 40-km limit that allows comfortable access by car (or train) to the city core in half an hour. These more dispersed locations have to function largely without relating to the historical core of the city. The resultant multi-nodal/information subcentres in these suburbanising regions include: Parramatta, Blacktown (Sydney); Dandenong, Knox, Frankston (Melbourne); Joondalup (Perth); Elizabeth (Adelaide); and Ipswich (Brisbane).

Re-urbanisation of core, inner and middle areas means that more households can live within easy access of major employment centres and services. In these re-urbanising areas there is a revitalising of older subcentres that are now becoming nodes as described above. Examples include: Chatswood (Sydney); South Melbourne, Port Melbourne, Box Hill (Melbourne); Fremantle (Perth); Tea Tree Gully (Adelaide); and Toowong (Brisbane).

Changing transport technology and globalisation processes also help to explain the settlement patterns of rural and remote areas. Many older provincial cities on the coast acted as ports for rural hinterlands and were quite contained. The spread of dispersed coastal settlements now reflects the greater availability and use of the car for transport. Many remote settlements now depend on air transport, with 'fly-in/fly-out' forming the basis of most new remote mining towns. Aeroplanes and an increasing number of four-wheel drive vehicles also provide transport for remote indigenous

▲
"The Rocks", Sydney.
The shape of cities past and present.

settlements. Many of these settlements depend on new small-scale, renewable technologies for power, water and communications.

Developments in transport technology that allow easier access to remote and undeveloped regions have boosted the demand for ecotourism. This nature- or culture-based experience and activity, along with the 'return to country' of indigenous Australians, has become a source of growth in remote settlements.

Other forces shaping urban settlements

Nineteenth century Australian settlers found one resource in abundance — space. They could realistically aspire to owning a detached house of their own or a farm. Such aspirations were encouraged by the anti-city, pro-rural ideas dominant in Great Britain at the time and romanticised in Australia by writers such as Adam Lindsay Gordon and Banjo Patterson. The relatively high income of Australian workers and the proliferation of building societies that provided

Table 3.3 Population and dwellings in each zone in Australia's big cities and their adjoining fringes

	Population	No. of Dwellings	Occupancy Ratio
Core	841 588	384 660	2.19
Inner	1 111 990	476 208	2.34
Middle	3 676 856	1 380 300	2.66
Outer	4 431 569	1 542 198	2.87
Fringe	260 509	103 552	2.52
Total	**10 322 512**	**3 886 918**	**2.66**

Note:
In the case of Melbourne and Sydney, the core incorporates the heart of the city up to 6 kilometres from the centre and the inner zone of pre-1940s housing 6-10 kilometres further out. For Brisbane, Adelaide and Perth the boundaries do not extend quite so far. The middle area includes established suburbs outside the inner area where the first wave of post 1940s suburbanisation is largely complete. The outer area of each city includes the suburbs where most population growth is now occurring and the fringe is the area beyond the formal metropolitan boundaries where scattered development is occurring and at least 25% of those employed travel to jobs located within the metropolis.

Source: ABS, 1991.

Re-urbanisation is occurring both in inner city areas (Melbourne, above) and in older suburban sub-centres (Fremantle, right).

another urban tradition of residents wanting easy access to urban services, work and the kind of inner-urban community life available in areas like Paddington, Balmain, Carlton, North Adelaide and Fremantle. Many people have left these inner areas for new suburban locations, but at the same time many have moved in from outer suburbs. Preferred locations for living in Australian cities continue to be a combination of these two cultural traditions (ABS, 1981).

After World War II, immigration provided the major boost to growth in Australian metropolitan settlements (other than Brisbane), with Melbourne receiving the highest numbers of migrants until the 1970s. Since then, Sydney has received more migrants than Melbourne. Most of the growth in both cities since the mid 1980s has resulted from immigration. Perth, too, has grown rapidly mainly through overseas migration. In 1994, the proportion of the population, aged 15+, born overseas was 35 per cent in Sydney and Melbourne and 37 per cent in Perth. This is considered high by international standards (Birrell, 1994).

accessible finance helped make the home-owning dream a reality. In Melbourne, for example, by the 1880s some 40 per cent of households owned or were purchasing their homes, probably the highest level for any city in the world at that time (Davison, 1978).

The Victorian colonial governments fuelled the suburbanisation process by their massive investment in suburban railways. Prospective home-owners seized the opportunities these transport networks created, in part because of anxiety to leave the pollution of the inner city. In the case of Melbourne, poor sewerage and drainage meant that, by the late nineteenth century, 'Melbourne stood ankle deep in its own wastes'. As such, Melbourne was a prototype of the industrial transit city, but with a higher-than-average spread of housing. By the 1890s, Melbourne's density averaged 54 people per hectare — only about one-third as crowded as the major British cities of the time (Dingle and Rasmussen, 1991). Sydney always had a higher density (about 100 per hectare) but cities like Perth, Adelaide and Brisbane were even less crowded than Melbourne.

People embraced the wide spaces available in the New World cities of America and Australia. Twentieth century town planning and reformist movements designed to create more 'healthy' and 'morally upright' urban residents imposed measures such as minimum density standards and segregation of land uses (King, 1978; Boyer, 1983; Newman, 1992). However, Australia also has

Population dynamics and patterns of Australian settlements

Urban settlements

The two main distinguishing characteristics of Australia's settlement pattern are the spread of urbanisation along the coastline and the concentration of Australia's population in five large cities. No inland urban centre other than Canberra is growing at an appreciable rate.

Australia's settlement pattern is dominated by the seven major capital cities: Sydney, Melbourne, Brisbane, Perth, Adelaide, Canberra and Hobart. With the possible exception of Canberra, each has traditionally held a relatively secure role as the unchallenged site of administrative, commercial and welfare services for its respective State or regional population and as the centre of industrial activities. The increasing internationalisation of the Australian economy is now challenging this dominance at a number of levels.

The outcome of increased competition between cities depends on their ability to provide goods and services for the national or international marketplace. Success can be measured by the number of basic or export-oriented firms located in a city, the extent to which it is interlinked with national and global telecommunication networks (as distinct from local or regional ones), the numbers of people with key skills (human capital), and the number of multinational corporation offices. The 1995 Business Review Weekly top 200 companies in Australia were located in Sydney (83), Melbourne (63), Perth (25), Brisbane (13) and Adelaide (eight). Some 46 per cent of Australia's outgoing overseas business telephone calls are made from Sydney, 26% from Melbourne, seven per cent from Brisbane, eight per cent from Perth and four per cent from Adelaide (Newton, 1995). Other indicators are the number of international firms based in the cities and the relative success of each city in attracting capital investment. These show the same general patterns (Stimson, 1995).

On these criteria, Sydney is rapidly increasing its dominance in the nation's urban hierarchy. It is now Australia's pre-eminent international city in terms of the location of global corporations and communications networks. Melbourne is also an international city to a lesser extent. Both differ from other major Australian cities in terms of their concentration of export-oriented industries (other than raw or lightly processed rural and mineral commodities), innovative human capital and infrastructure supporting manufacturing and producer service industries (Newton, 1995; O'Connor and Stimson, 1994 and in press). Brisbane, Perth and Adelaide remain largely regional cities but ones whose global orientations are growing — particularly in the past decade.

In this emerging new economic order, a city's population growth is less related to its economic growth than in the past. Indeed, population shifts such as the move of population from Sydney to north-east coastal New South Wales and the Queensland coast (outlined below) appear to reflect lifestyle and housing-cost factors more than the attraction of employment.

Intra metropolitan restructuring

During the past 20 years many of the manufacturing firms once located in Australia's inner metropolitan areas have either disappeared or have moved to outer metropolitan areas (Newton *et al.*, in press). At the same time, service industries have also been moving. Service industries fall into two principal classes. The first consists of a range of specialist professional and managerial activities ('producer' services) required by those organisations that successfully compete in the national and international marketplace. The second comprises what could be termed 'people services', provided by both public and private sectors, and relating to the full spectrum of social (for example, health, welfare, education etc.) and personal (for example, restaurant, hairdressing, recreational etc.) services.

The major outcome of intra-metropolitan restructuring in Australian cities has been the increased concentration of producer services in core and inner metropolitan areas, while outer metropolitan areas (particularly where nodal subcentres have emerged) are increasingly becoming the workplaces for manufacturers and the providers of people services. Reflecting the concentration of different economic activities within the major cities, people with higher incomes, with professional and managerial

Table 3.4 Population change in Australia's big cities 1986–1991 by urban sector		
	Core/Inner/Middle	Outer
	1986-91	1986-91
Perth	+ 32 697	+106 065
Adelaide	- 3 087	+ 56 446
Melbourne	+ 6 100	+ 182 950
Sydney	+17 300	+ 184 050
Brisbane	+ 29 449	+ 111 200

Source: ABS,1993a, b, c, d, e, and 1994a, b.

occupations, tend now to be increasingly concentrated in the core and inner suburbs (see Fig. 3.3). It is also notable that the concentration of such households in inner areas is greater in Sydney than elsewhere, in line with Sydney's location at the top of the urban hierarchy. House prices reflect this trend. Conversely, more blue-collar and routine white-collar workers are moving to the lower-cost residential areas in the middle and outer suburbs, thus feeding the suburbanisation process.

The doughnut effect and pockets of poverty

In the 1970s, Australian cities rapidly lost population in their old inner suburbs as settlement dispersed outwards, in a pattern similar to that in American cities — the so-called 'doughnut effect'. Since then a simultaneous process of suburbanisation and re-urbanisation has occurred.

Suburbanisation is still a dominant process in Australia's major cities. Although the population of the core, inner and middle sectors of major cities has remained fairly stable over the 1986–91 period, almost all metropolitan population growth has occurred in outer suburbs (see Table 3.4).

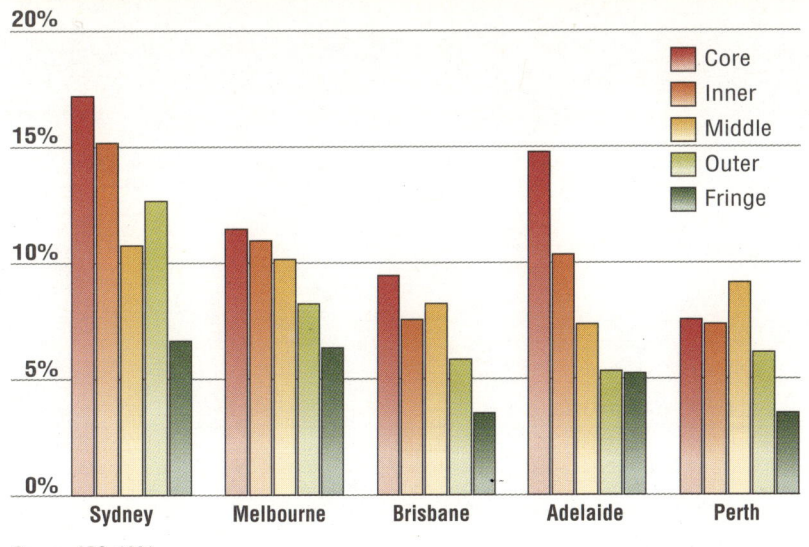

Figure 3.3 Proportion of households earning $70 000 or more per annum by urban sector in Australia's major cities, 1991

Source: ABS, 1991.

The inclusion of households moving to the extra-metropolitan fringe would highlight the continued suburbanisation pattern even further.

Although suburbanisation continues, re-urbanisation in the older areas of Australian cities is also evident, in the form of higher-density dwelling construction, the replacement or renovation of older homes and the addition of extra units on blocks (dual occupancy).

Between 20 and 25 per cent of new dwellings built in Melbourne and Sydney since 1985 can be classified as re-urbanisation. That is, construction occurred in built-up urban areas that required existing housing to be replaced or former office or industrial sites re-used for housing purposes. In Melbourne, most of this re-urbanisation consisted of detached housing, but in Sydney, higher-density development predominated. Indeed by 1993–94, 48 per cent of all residential development in Sydney consisted of medium- and higher-density dwellings. In Perth and Brisbane too, the re-urbanisation of established urban areas significantly increased in the early 1990s, mainly as higher-density units. About one-third of all housing constructed in Brisbane in the 1990s comprised medium- and high-density units.

The situation in Australian cities therefore does not resemble that of the United States, where the 'doughnut' phenomenon has left a largely poor, black and immigrant community to languish in deteriorating inner city areas. The re-urbanisation process described above suggests that a substantial proportion of households wish to relocate to more central urban areas of Australian cities, although small household sizes mean the trend is not yet significantly increasing population in these areas. As well as dwellings, the core, inner and middle areas are receiving about 50 per cent of all non-residential development, consistent with the location of new employment and global-oriented firms in these areas.

The issues of suburbanisation and re-urbanisation are highly controversial. Proponents of suburbanisation claim it provides privacy for families and space for firms, while its critics stress the environmental impacts resulting from 'sprawl' and the associated dependence on cars. They also criticise the capital costs of new infrastructure, especially when infrastructure in inner areas is sometimes underutilised. In response to these concerns, political leaders and urban planners alike have espoused policies to help consolidate Australian cities through re-urbanisation. Concerns are now arising in Australia about the rapid gentrification of inner urban areas and the loss of access to centres of employment for those on low incomes.

At a broader level comes evidence of a new and disturbing trend towards 'two Australias' emerging in urban environments. Gregory and Hunter (1995) describe an increasing disparity between richer and poorer neighbourhoods in Australian cities over the past 15 years. They show that the poorer areas now have significantly lower-income populations and are dominated by high levels of unemployment. These 'pockets of poverty' in our urban areas are scattered through traditional inner-city working class areas, in newly settled suburban areas, and in non-metropolitan areas on the fringe and coast. They are associated with the rationalisation of some previously protected manufacturing industries and reduced public sector employment. They cluster around low-cost housing.

Density still low

Urban density remains low in Australian cities compared with that in other international cities (see Table 3.5). Sydney is the only Australian city to show densities in its older suburbs comparable with those found overseas.

The data indicate that, in a global context, we have scope for re-urbanisation, although as discussed later, design is critical to achieving this goal.

The tentacles of growth

A growing pattern of low-density settlement is spilling beyond the formal boundaries of each of

Table 3.5 Urban density in people per hectare by city region and the variations within the cities by inner and outer area, 1980

	Inner Density	Outer Density	Total Density
Asian cities	464	115	160
European cities	91	43	54
US cities	45	11	14
Toronto	57	17	25
Sydney	39	16	18
Other Australian cities	21	12	13

Notes:
Inner is defined as the pre-war urban area and essentially corresponds to the definition used here.
Urban density takes into consideration only that land which has been developed. All cities incorporate their region into the outer area and total city.

Source: Newman and Kenworthy, 1989; Kenworthy and Newman, 1994.

Table 3.6 Population growth in big cities and nearby urban centres, 1986–1993*

	Population growth 1986–1993	Total Population 1993
Perth	188 823	1 261 524
Adelaide	78 500	1 128 953
Melbourne	253 400	3 480 993
Sydney	313 650	4 485 850
Brisbane	358 965	1 930 321
Total	**1 209 465**	**12 287 638**
Share of Australia	72.4%	69.6%

*Note: Includes areas defined as fringe plus nearby large urban areas.

Source: ABS 1993a,b,c,d,e and 1994a,b,c,d,e,f.

the capitals. Most of these ex-urban settlements are linked to the metropolis by people who commute to jobs within the city boundaries. For example, about half the work-force of Ballan, Romsey, Kilmore and Bacchus Marsh (all rapidly growing areas outside the official boundaries of Melbourne) commute to employment in the city.

A second pattern connects previously independent provincial cities — like Geelong , Wollongong and Newcastle; and the Gold Coast and Sunshine Coast — with their nearby metropolitan capitals via extended freeway and other transport links. Infill settlement connects these previously disparate cities into one mass that shares complementary economic opportunities and resources.

By 1993, 70 per cent of Australia's population lived in these urban agglomerations (see Table 3.6). They accommodated nearly three-quarters of Australia's population growth between 1986 and 1993. Suburban development consumes about 1300 sq m of land per person — so Australia's five largest cities consumed 160 000 ha of rural land (or 100 000 football fields) as a consequence of the tentacles of urban growth in those seven years.

Migration patterns

Although Sydney's net rate of population growth has slowed, the city continues to be the major destination for overseas migrants. Between 1986 and 1991, 91 per cent of its population growth was due to international migration, compared with 55 per cent for Melbourne and 68 per cent for Perth (Newton and Bell, in press). Some of the migrants moving to Sydney are attracted by its success as a centre for globally competitive

enterprises, but other migrant flows reflect the prior establishment of substantial ethnic communities.

Interstate migration is having a reverse impact. Since the 1970s a shift in population has occurred, mainly from Sydney and Melbourne, to the north-east coast of New South Wales and to Queensland. Sixty thousand people who were living in Sydney in 1986 and 30 000 from Melbourne had moved to south-east Queensland by 1991. However, since 1991, a higher proportion have moved from Melbourne. Perth also attracted significant net inflow of interstate movers in the 1980s (about 10 000 per year), but in the 1990s the net flow has been much less (about 4000 per year).

An Australian sunbelt

In the United States, the State of Florida has trebled in population since the early 1950s. Some observers believe Queensland will experience similar growth (Holmes, 1994). Between 1986 and 1991, 28 per cent of Australia's population growth occurred in mainland non-metropolitan coastal regions (see Table 3.7). This increased to 35 per cent between 1991 and 1993. Yet in 1991 only 18 per cent of Australia's population lived in these regions.

It is only the warmer coastal zones that are attracting significant numbers of people (see the box on page 3-12). The interstate migration data indicate that most of the sunbelt locations including the Gold Coast and the Sunshine Coast are drawing new residents primarily from the southern States. A tiny proportion of Gold Coast and Sunshine Coast residents commute to work

Table 3.7 Population growth in coastal non-metropolitan areas

Coastal Location	Population 1991	Share of Australian population 1991(%)	Population growth 1986–91	Population growth 1991–93	Share of Australian growth 1986–91(%)	Share of Australian growth 1991–93(%)
Qld Gold Coast	248 768	1.4	57 914	18 637	4.7	4.9
Qld Sunshine Coast	164 936	1.0	45 563	19 075	3.7	5.0
Other Qld coast	669 304	3.9	65 950	36 823	5.3	9.6
NSW N-E Coast	356 670	2.1	59 300	20 070	4.8	5.2
Newcastle Area	464 000	2.7	29 850	11 400	2.4	3.0
Wollongong Area	244 930	1.4	11 910	5 150	1.0	1.3
NSW S Coast	123 810	0.7	21 830	7 190	1.8	1.9
Victoria Gippsland	75 670	0.4	6 470	1 430	0.5	0.4
Geelong-Bellarine	200 350	1.2	14 350	2 070	1.2	0.5
Other Vic Coast	72 560	0.4	2 420	500	0.2	0.1
SA Metro Fringe	5 853	0.03	1 325	757	0.1	0.2
Other SA Coast	157 434	0.9	-1 596	-530	-0.1	-0.1
WA Metro Fringe	29 223	0.2	9 782	4 971	0.8	1.3
Other WA Coast	250 449	1.4	22 482	5 186	1.8	1.4
Total	3 063 957	17.7	347 505	132 729	28.1	34.7

Source: ABS 1993a,b,c,d,e,and 1994a,b,c,d,e,f.

within the Brisbane Statistical Division, but in no locality does this exceed 15 per cent, even in nearby local areas (ABS, 1991).

Older people and retirees form an important component of migration to south-east Queensland but, contrary to popular perception, they are not the dominant source. For the period 1986–91, only 15 per cent of interstate movers to the Gold Coast and 18 per cent of those moving to the Sunshine Coast were aged 60+, although this migration has contributed to the creation of relatively old communities. By 1991, 20 per cent of the Gold Coast and 22 per cent of the Sunshine Coast populations were aged 60 or more (Barker, 1993). This compares with 12 per cent for Australia's total population.

Most of those moving to north-coast locations are people of working age, many of whom were displaced by the recession of the early '90s and the restructuring of older industries in the south. They often find difficulty gaining work.

As a result, many people living in coastal areas now depend on government benefits. The proportion of the population aged 15+ living on the Gold and Sunshine Coasts who depended on a Commonwealth pension or benefit reached 34 per cent and 37 per cent respectively in 1994. This compares with 29 per cent for all Australian residents. For the Gold Coast, 19 per cent of the 15+ population were receiving an age or veteran's pension and 7.5 per cent unemployment benefits. Comparable figures for the whole country in 1994 were 15.3 per cent and six per cent respectively (Birrell, *et al.*, 1995). The booming coastal area of Mandurah, south of Perth, has 23 per cent unemployment. It appears that government benefits and private superannuation are helping to fuel a relocation process quite independent of the productive base of these growing coastal communities. The process is self-reinforcing for a time, as the provision of housing and social and physical infrastructure (often with substantial government subsidy) creates additional job opportunities, thereby attracting more job-seekers to these locations. The pressure on the coastal environment from this population growth is a significant focus of this chapter.

Other cities are growing too

Many north coast provincial cities and towns, which cater for a population seeking a recreation-oriented lifestyle, are booming. In Queensland the major growth points, other than the Gold and

Settlement in the sunbelt — the south-east Queensland experience

Recent decades have witnessed a trend of population dispersal along the coast. Like other areas of coastal settlement in Australia, south-east Queensland has followed the pattern of low-density ribbon development stretching along the coastline. Few available sites now remain. Urban development is now focusing on the nearby hinterland, including large areas of low-density rural residential or 'acreage' developments, as they are known in Queensland. The historical fragmentation of land-ownership in the area, plus extensive zonings for 'acreage' purposes, especially in Albert Shire, means that the impact of human settlement is diffusing widely into the coastal hinter-land and along the estuarine streams that drain the region.

Coastal urbanisation has already destroyed much of the original local ecosystem. Between 1974 and 1989, 33 per cent of the coastal bushland along the south-east Queensland coast was lost (Catterall and Kingston, 1993). Some 20 percent of the mangrove fringes in the Moreton Bay area of Brisbane have also been cleared (see Chapter 8). These losses are seriously disrupting ecosystem health and native habitats along the coastal strip which has a notable biodiversity. Further low-density development has increasingly fragmented the remaining vegetation leading to more 'edge' effects that also threaten biodiversity. These effects include the creation of numerous niches for invasion by opportunistic weeds and feral animals; and changes to the physical environment along the borders of vegetation remnants.

The diffusion of urban development has also increased the nutrient and sediment loads deposited in local marine waters. All but two of the sewage plants located in the region discharge effluent — which has not been treated for nutrient removal — into local estuaries (the exceptions are the Gold

Canal development at Surfers Paradise.

Coast City and Caloundra plants, which discharge into ocean outfalls). As well, few residential developments can hold and treat stormwater flushes before these add their load of nutrients, sediments and other pollutants to the river systems. The pollutants threaten marine ecosystems in the poorly flushed areas of Moreton Bay.

In Queensland, local authorities, who control land use zonings, tend to compete for development projects. In addition, Queensland law allows foreshore areas and riverine edges to be alienated to private land-owners, thus facilitating destruction of the natural vegetation. However, it is now widely recognised in south-east Queensland that further urban growth along past lines could destroy many of the coastal values that attracted people to the area in the first place.

Hopetoun — a story of country town decline

Hopetoun's population fell by 19 per cent between 1976 and 1991. It now has just 703 people. This downturn is the result of a drastic decline in the income of the district's farmers.

When the cereal- and sheep-farmers don't have money to spend it's not long before the small businesses begin to struggle. Hopetoun has lost many of the services which made it the hub of the Shire of Karkarooc. Gone are the former State Rivers and Water Supply Office, the solicitor, the Westpac Bank, the court house, Elders office, the Massey Ferguson dealership and the weekly visit from the dentist. The doctor lives in the town only during the week. If you need urgent care from a doctor on the weekend it is a one-hour drive to Birchip in a car or ambulance. A solicitor visits the town occasionally.

The schools are declining; teachers leave and are not replaced. The only four apprentices in the town include the butcher, the baker and a mechanic. Most of the school graduates go to larger centres like Melbourne or Bendigo to work, or go to university. Some drop out and remain unemployed.

The nature of work has changed noticeably as many people try to supplement their income by doing extra jobs. Members of farming families work in town or elsewhere on a part-time basis. Shearers work at the silos and the swimming pool over summer. The hats and tasks have changed because of the reduced income from farms, businesses and services.

Despite its decline, Hopetoun is still a lively community. It has a wide range of sporting facilities and many active groups. But its ageing population contains more than 80 widows and widowers living alone. This trend, along with a continuing loss of young people and low farm incomes, places the town in a serious position of decline.

— *Kerry Conway, Hopetoun resident and farmer*

Sunshine Coasts, are Hervey Bay (next to Fraser Island) and Cairns. Cairns is emerging as an international city servicing foreign tourists through its gateway airport to the the Great Barrier Reef. The State has other coastal cities with an industrial or port base, like Bundaberg, Gladstone, Rockhampton and Mackay, which are growing but at a considerably slower rate.

Elsewhere in Australia, provincial cities other than those servicing coastal recreation have generally maintained their population growth. Most are benefiting from the relocation of residents from smaller country towns and rural hinterlands. These cities play important roles as suppliers of community services and as centres for wholesaling and retailing, often at the expense of nearby smaller towns.

Some provincial towns also benefit from increased industrial growth due to their relatively low land and labour costs or their proximity to primary resources and producers (Beer *et al.*,1994).

Rural settlements

Rural Australia (defined as cropped or cultivated zones) covers a wide range of climatic and land use areas, from northern Queensland to Tasmania and across to south-west Western Australia, and as far inland as receives sufficient rainfall to support agricultural activities. Generally, rural areas — particularly those with links to metropolitan or provincial cities — are maintaining their populations. However, in the drier wheat/sheep belt, population numbers are declining across all States, with losses generally between zero and five per cent of population between 1986 and 1991. The Conargo, Jerilderie and Bland areas in the western Riverina of New South Wales, for example, lost 13 per cent, seven per cent and six per cent of their populations respectively over the period (McKenzie, 1994).

Population decline, both on the land and in the small rural towns servicing farming communities, undermines the economic viability and livability of these towns. The Hopetoun story (see the box above) illustrates this process. Although not subject to the pressures of rapid growth like the coasts, inland settlements with declining populations sometimes struggle to cope with environmental problems. Farmers, communities, businesses and shires do not have the necessary money to invest in rehabilitating degraded land in their areas or to improve urban services like sewage treatment and recycling.

Remote settlements

Although only a small proportion of Australia's population lives in remote settlements, they service vast areas of the continent and are a significant part of what defines our country. Remote-area land use patterns are diverse, ranging from pastoralism, tourism, mining and indigenous communities. Remote settlements demonstrate a high degree of functional diversity relative to their size, acting as

Fossil Downs — a case study of pastoral settlement

Although it shares many of the characteristics of pastoral communities in remote areas, Fossil Downs station is unique in other respects. These days, most pastoral stations are run by managers acting on behalf of absentee lessees. Fossil Downs is one of the few in north-western Australia to have been consistently owner-operated over the last century. In recent years, indigenous corporations have purchased a number of pastoral station leases (25 in the Kimberley region).

Fossil Downs station covers slightly less than 400 000 ha. It has a good supply of water and the homestead is surrounded by an oasis of lawns, trees and shrubs, which are watered from a bore. It has been subjected to the same economic pressures and climatic vagaries as other stations.

In response to these pressures, the number of Aboriginal stockmen has fallen from 30 in the 1960s to three today. There are only three or four other employees.

Among the ramifications of this widespread decrease in station populations, every individual — owner and employee alike — now does the work previously undertaken by two or three people. Family members often work in nearby towns during the day, as well.

Fitzroy Crossing, the nearest settlement, has a dentist and a small hospital staffed by two doctors. A health sister visits an indigenous community near Fossil Downs once a fortnight and services the station at the same time. Primary and high schools are located in Fitzroy Crossing. Children on pastoral stations generally remain home doing either School of the Air or correspondence lessons until year seven, when they leave home to attend boarding schools in either Perth or Brisbane. The area has limited sporting and recreational events and facilities.

Supplies and stores for Fossil Downs are sea-freighted from Perth to Derby, where the pastoralists pick them up in their own truck — a journey of 280 km each way. Floods during the wet season often make the roads impassable, so the family place the annual store order for staple items like salt, rice, sugar and flour in October or November, just before the 'wet'. The station used to produce its own fruit, vegetables, bread and ice-cream, but because of reduced staff these items are now delivered from Perth each fortnight — roads permitting.

Annette Henwood, owner of Fossil Downs.

foci for their service regions and as links to services outside their region (Holmes, 1988).

Remote area land use is being rationalised and restructured by continuing downward trends in commodity prices and through efforts to protect fragile lands (through programs such as Landcare).

Pastoral areas

The effects of climatic variation, deterioration in pasture quality, increases in soil erosion, onset of cattle tuberculosis and its subsequent eradication, and shrinking global markets have all contributed to a decline in the number of cattle and sheep in the pastoral industry. Lower stock numbers tend to make the whole industry less financially viable, which affects the economic and social infrastructure of remote areas. For example, all abattoirs in the Kimberley region of Western Australia and a number in the Northern Territory have now closed.

Pastoralists on less viable stations cannot afford to keep pace with new production techniques and management methods, or to maintain the land resource itself. Some pastoralists are diversifying into tourism.

Tourist areas

Remote settlements receive a much lower proportion of tourists than do the large urban settlements, but their relatively small size makes the impact of tourism more significant.

The rapid growth of tourism — especially in remote areas — places both ecological carrying capacity and recreational carrying capacity under pressure, and has caught many natural-resource managers and nearby remote settlements by surprise (Dowling, 1993). Many tourist centres in remote locations lack the necessary resources to monitor the impacts of tourism on the environment and culture of their settlements.

Ecotourism is receiving much attention as a way of reducing the environmental impacts of tourism. Although usually consisting of activities that are environmentally friendly, it can still affect fragile areas unless limited to 'the maximum number of people who can use a site without an unacceptable decline in the quality of the experience gained by the visitors' (Mathieson and Wall, 1982).

Mining settlements

Resource extraction has given rise to the establishment of many settlements within Australia since the first phase of European settlement. The gold rushes of the mid 19th century spawned mining towns such as Ballarat and Bendigo that still exist today. Other towns have declined and died as their resource base was exhausted.

Australia has pursued four different mining settlement strategies.

• *New single-industry towns*
Since 1960, more than 25 new towns and ports have been constructed adjacent to mine sites (Robinson and Newton, 1988).

The Argyle diamond mine, an example of a fly-in/fly-out community.

• *Expansion of existing communities*

Additional growth created by resource projects is attached to an established community within daily commuting range of the resource site(s). In many instances the existing community has an economic base other than mining (for example, Capella — agriculture; Port Hedland — transport and services; Leeman — fishing).

• *Combine town*

Several mining companies operating in the same general area have built a new 'combine' town to house and service the workforce at a central location within daily commuting range (for example, Moranbah in the Bowen Basin of central Queensland or Jabiru in the Northern Territory).

• *Fly-in/fly-out*

In more recent times companies have used the 'oil rig' philosophy to fly workers on a rotational basis into a resource site from a distant established community (usually a major urban centre). Workers are accommodated in a hostel at the site. They return to their home community upon completion of each work shift.

Remote mining ventures have increasingly opted for fly-in/fly-out, and thus direct more wealth and development to the big coastal cities instead of to remote settlements, although this may result in less impact on ecologically sensitive remote areas. However, the redirection in population creates lower levels of service and reduced livability in remote regions.

Remote indigenous communities

In the early 1970s, many indigenous people began to return from mission or government settlements to their traditional lands. The outstation or homelands movement became a milestone in the development of indigenous communities and emphasised a commitment to their traditional lifestyle and culture.

These remote communities have undergone a 'locational trade-off', which involves reduced access to an already limited labour market as well as education, training and other services in remote areas. However, many of them have increased public funding and a growing spirit of self-determination and cultural revitalisation. The process has been made feasible by small-scale technical innovations including solar technology.

Population changes in remote areas have been made feasible by small-scale, solar-based innovations in technology. These give the communities access to power, water and communications in areas that sometimes are 500–1000 km from the nearest powerline, water pipe or electricity grid.

In a 1992 survey, the Aboriginal and Torres Strait Islander Commission (ATSIC) identified 1385 indigenous communities throughout Australia — 819 of which are in remote regions (see Table 3.8).

But despite this 'return to country', many indigenous Australians have dispersed to big cities. In 1961, only five per cent of the total indigenous population were located in major cities compared with 33 per cent 30 years later, while 67 per cent now live in rural and remote areas (see Table 3.9). In the 1940s, almost all Torres Strait Islanders lived in the Torres Strait, whereas today only one-fifth of the total population reside there.

An indigenous community

This small Aboriginal community of 300–400 people is in an arid desert region several hundred kilometres from the nearest service centre. It is not named to protect the privacy of its residents. The people have returned to their country after a period of 25 years living in a government settlement. They have strong affiliations with their country and have established a number of outstations within a radius of about 100 km around their small settlement.

Natural resources

The climate is hot and dry in summer and cold in winter; temperatures range from sub-zero in winter through to the high forties in summer. The dramatic temperature changes bring intense wind gusts. Rain is usually torrential and cuts access on the unsealed road.

Three bores provide groundwater to the settlement. Although water falls within the accepted guidelines for drinking, its relative hardness creates problems with deposits on plumbing. Only 45 per cent of hot-water-supply systems are functioning. The community uses about 30 per cent more water per head than cities 'down south', but nearby mining towns use four times as much. Significant leakage occurs through pipes and taps.

Diesel generators, which provide the power, are regularly shut down for routine maintenance. Each 50-litre electric hot-water heater costs $2500 per year to run, not including significant maintenance costs. The people prefer wood as a fuel, but they now have to travel up to 50 km from the community to find it.

The economy

The settlement's economy depends largely on public-money transfers and the store. Most people in the community draw social security benefits. Although household incomes appear high ($400 per week), occupancy rates are around 12–15 per house. The Community Development Employment Project (commonly known as 'work for the dole') provides some work. Of their total budget (about $1.2 million per year from all sources) up to $700 000 comes from the community-controlled health program. The housing budget of $100 000 to $300 000 per year, depending on grant allocations, provides enough to build up to three houses.

Some members of the community are well-respected Aboriginal artists and derive an income from commissions. Others would like to work in a nearby mining town or establish an ecotourism venture, but often lack the specific skills or financial backing for such work, and little useful training is available to them.

The store is the community's largest source of economic activity and its only retail outlet. Supplies arrive by truck each week and the store cashes cheques. The extensive packaging required for rugged transit creates problems of waste disposal. The community owns and operates the store and regularly divides its profits among the various family groups. In most situations people use this money to buy motor cars so that they can move around their country and to town, and usually share in buying older cars. They have a limited choice of models that they can maintain to survive in the bush. However, the roads ensure that these second-hand vehicles rarely last more than six months.

Land and housing

People live on reserve land that is set aside for Aboriginal people. They have no freehold title to it and they cannot lease the land for enterprises. The community has responsibility for all housing, and imports all building materials.

Most people spend 80 per cent of their time living outside in the area around the house. Over a one-year period, many move between a number of houses.

Society and culture

The community has a very active social and cultural life involving traditional business, hunting and sport. People move widely across their own and surrounding country, often in large groups of up to several hundred people — men, women and children — maintaining their links to land and social connections. For successful hunting and gathering the people need to travel to favourite sites in four-wheel drive vehicles. They still move on foot across country to maintain their links with the land. They keep a large number of dogs.

Intense sporting activity takes place at some times of the year. The men form football teams and regularly travel

	Number of discrete indigenous centres[1]	Population of discrete indigenous centres	Average size of discrete indigenous centres
Table 3.8 Indigenous communities in remote Australia by States and Territory			
NSW	37	4 203	114
SA	88	3 861	44
QLD	82	16 672	203
WA	182	18 602	102
NT	430	29 959	70
Total	**819**	**73 297**	**106**

Note: 1. Only includes those settlements comprised of predominantly indigenous people.
Source: ATSIC, 1993.

distances of up to 500 km or more for a game. In addition to a strong emphasis on music and painting, people talk about wanting to improve gardens and grow trees. A satellite dish and microwave communications open their links to the world, and some local radio and television broadcasting takes place through the Broadcasting for Remote Aboriginal Community System (BRACS).

Health

Since it began operating the health service, the community's health profile has improved slightly. The health service employs a part-time doctor and several nurses, as well as nine Aboriginal health workers and traditional healers to deliver primary health care.

Last year, only six births were recorded and four people died. Over 90 per cent of the children under five years have needed to be evacuated from the community and admitted to hospital for serious illness. Half the schoolchildren have active trachoma and 26 per cent suffer from conditions that predispose them to permanent kidney damage. Diabetes and obesity are chronic in the older population.

The community has a very strong anti-alcohol stand and uses Avgas for cars (a non-toluene-containing fuel) to deter petrol sniffing.

There are a large number of dogs within the community, many of them carrying diseases.

Education

Attendance rates at the primary school vary. Young people are encouraged to leave and go to high school in large urban centres, but while there, they learn skills that don't necessarily suit their small remote settlement. This gives them a feeling of disorientation and creates difficulties for the community about how to employ them.

Up to 70 different broker groups or agencies come into the community to talk about issues ranging from basic finance, welfare payments, training programs, and various government initiatives, to sporting and cultural programs, and art and craft sales.

The Community Council, which is the legally constituted body to govern the community, is unpaid, and generally not well educated in terms of management or finance control. Most of the operations are left to the work of an adviser or town clerk, and an accountant who generally lives off the settlement.

The people are concerned about land title issues and health and employment prospects for their young people, but are totally united in their fierce desire to stay in their own country rather than live on the fringe of settlements with different cultural and lifestyle perspectives.

Table 3.9 Distribution of indigenous Australians in relation to total population

Location	Total Australian population (No.)	Total Australian population (%)	Total indigenous population (No.)	Total indigenous population (%)	Indigenous population (% of total population)
Five big cities	10 062 003	59.7	62 544	23.6	0.6
Other cities	2 025 803	12.0	26 037	9.8	1.3
Rural	4 267 356	25.3	88 578	33.4	2.1
Remote	486 407	2.9	88 142	33.2	18.1
Total[1]	16 850 540	100	265 378	100	1.6

Note: 1. Includes offshore and migratory components.
Source: ABS, 1991.

State

Livability — the human dimension

Human settlements in Australia vary in terms not just of their physical features but also of their human qualities and the wider social amenity they offer. People in different settlements enjoy greater or lesser measures of health and happiness, have access to housing, employment and community services that are more equitable or less so and live

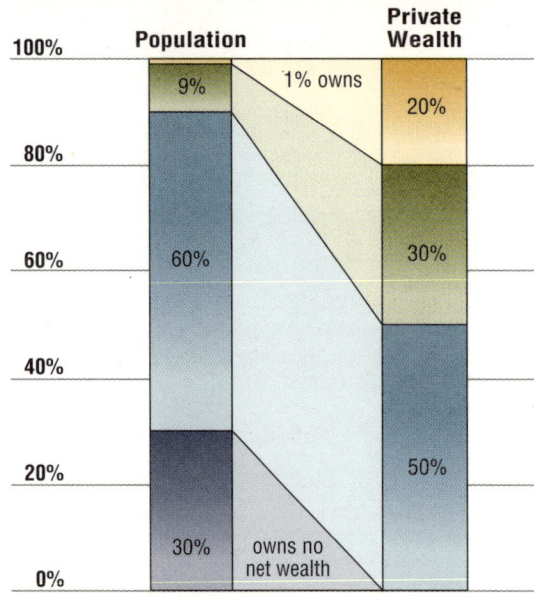

Figure 3.4 Estimated distribution of wealth in Australia

Source: Travers and Richardson, 1993.

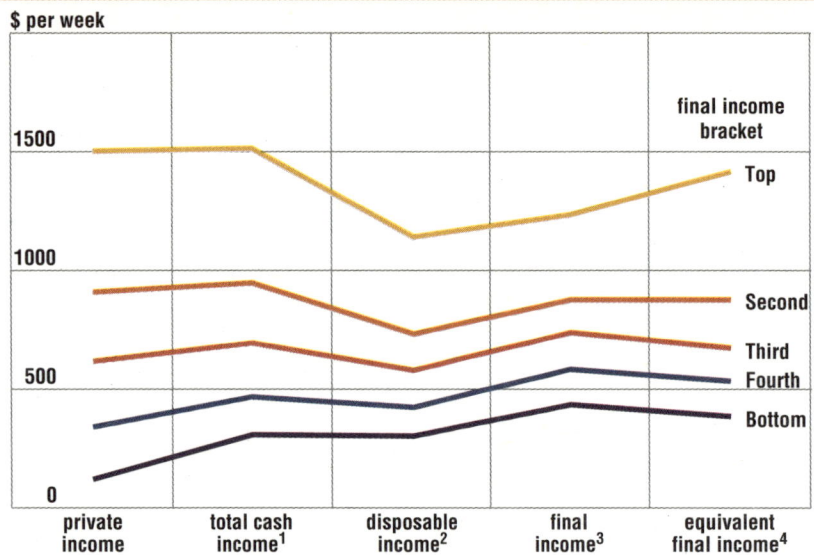

Figure 3.5 The influence of taxes, government benefits and the social wage on the final income of different households

Notes:
1. Total cash income equals private income plus cash transfers.
2. Disposable income equals total cash income less income tax.
3. Final income equals disposable cash income plus non-cash benefits including the value of public health care, education and housing services — the "social wage".
4. "Equivalent final income" uses Henderson equivalence scales to allow for household size.

Source: NATSEM, 1995.

in settlements that are more or less safe and well designed. Settlements and communities rich in these qualities are also rich in livability.

Most Australian environmental legislation defines the environment broadly — incorporating health, social and economic factors. 'Ecological footprints' and the metabolic processes that underpin them help to describe the ecological effects of human settlements. However, the structures created by those settlements (housing, urban infrastructure and services, transport systems, industrial plants and commercial facilities, together with their associated urban ecological processes) are an environment in their own right.

Social and economic priorities also affect the natural environment. It is now widely understood that economic weakness, social inequality and poor health standards are important forces in determining this impact.

Equality and the environment are linked in two broad ways. Firstly, poor communities or the poor in communities lack the necessary resources to manage their environments adequately. Without sufficient resources, the poor are often forced to exploit natural resources more heavily and to push natural systems beyond their sustainable level. While this point may appear to have more relevance to developing countries, it is also true in countries like Australia. For example, high rates of unemployment in country towns can influence forest-policy decisions, declining terms of trade for farm products can fuel the clearing of remnant vegetation and communities in economic decline cannot upgrade their technology and infrastructure for water, waste and transport systems.

Secondly, inequality may generate alienation and indifference towards the public realm. Large disparities in health and social amenity may undermine concern for the public realm or the 'commons', whether built or natural, whether inside cities or outside them. Even though inequality takes its heaviest toll on the poor and unemployed, economic decline and material insecurity can easily infect the mood of an entire society. Intergenerational equity in human settlements depends on equity within a generation.

This section focuses on social amenity and health issues, with particular emphasis on whether significant patterns of inequality are developing in different parts of Australian settlements.

Social amenity issues

Wealth

In the financial year ending 30 June 1994, Australia's private wealth grew in real terms by 10 per cent. Following a slump in the early 1990s, the rate of growth returned to its historical average of the past two decades, so the national private asset base now stands at about $1 531 billion (Commonwealth Treasury of Australia, 1995). This places Australia in the higher-than-average per capita wealth category among OECD countries. A recent estimate of wealth distribution in Australia is shown in Fig. 3.4.

Income

The real incomes enjoyed by Australians have generally increased throughout the post-war period and until the late 1970s income distribution was also becoming more equal. However, since then, this trend has become uncertain (Saunders, 1994). Reviewing a large number of income distribution studies, the Economic Planning and Advisory Council has concluded that 'from the 1970s and into the 1980s, the distribution of income appears to have become less equal' due both to domestic social changes and Australia's changing place in a globalising economic order (EPAC, 1995). A subsequent study conducted by the National Institute of Economic and Industry Research, however, suggests the opposite. Over the 1981–94 period, the distribution of income has, despite new challenges, become more equal (Johnson, *et al.*, 1995). These differences appear to be largely a result of the fact that the latter study attempts to incorporate a wide range of government non-cash benefits or social wage goods (including education, health, child care, public housing subsidies and other government concessions) into its assessment of changing income distribution. In announcing the completion of the study, the Prime Minister stated that the 'social wage' had grown by 41 per cent over the period and that it had played an important role in the redistributive process.

It will be some time before differences between the various studies are resolved but, in the meantime, it is worth commenting briefly on the positions common to most recent income distribution studies. First, it is widely agreed that the overall effect of government intervention (encompassing income taxes, transfer payments and social wage goods) is to moderate the income inequalities likely to result from the operation of market forces alone (see Fig. 3.5). Second, it is also agreed that processes of economic globalisation represent an important new force which tends to tip the balance in favour of market determined outcomes and greater inequality. This seems to be confirmed by a wider international trend towards greater income inequality. Despite Australia's standing as one of the world's 12 most equal societies, globalisation is likely to involve processes which continue to challenge our notions of equality and arrangements for social protection (Travers and Richardson, 1993). Some income inequalities in Australia are of longer standing. While income differences based on gender are closing, they are still significant (see Fig. 3.6). Income inequalities between indigenous Australians and other Australians remain stark (see Figs 3.7 and 3.15).

From the point of view of human settlements and state of the environment reporting, however, the spatial distribution of income and other resources — encompassing variations between regions rather than individual households — may be more important. This issue is explored briefly below in two ways: on a broader regional basis (using the ABS index of economic resources) and on a more closely focused neighbourhood (census collector's district) scale.

Regional income inequality

- Core and inner areas contain significantly greater proportions of high-income earning households (17.1 per cent in the case of Sydney's core) than do outer and fringe areas (see Fig. 3.9). This points to the emerging bi-modal character of inner cities, as urban manufacturing and blue-collar workers give way to a growing service economy and an expanding professional workforce. However, poorer households still retain a strong presence in inner and core areas.

- Outer areas are characterised by households with sufficient incomes to purchase a house, but often not to acquire significant wealth beyond that. The urban fringe and some of the new coastal areas are notable regions of low income in Australian cities.

Neighbourhood income inequality

In the period between 1976 and 1991:

- Household incomes in higher-status neighbourhoods (top 5%) increased by $12 500 (23%) (see Fig. 3.10).

- Household incomes in lower-status neighbourhoods (bottom 5%) declined by $7400 (23%) (see Fig. 3.10).

- Additional weekly income needed by households enjoying median incomes to catch up to areas in the top percentile doubled, from $442 to $885.

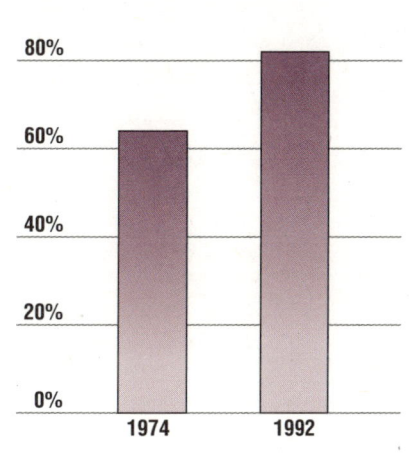

Figure 3.6 Income inequality and gender

Female earnings as a percentage of male earnings (full-time employment)

Source: ABS, 1992a and 1994g.

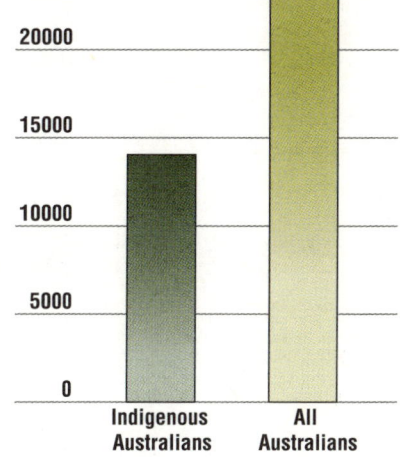

Figure 3.7 Income inequality and race

Mean annual income ($ per head) for people aged over 13 years

Source: 1994g and 1995a,b,and c.

Distribution of economic resources across Australia

- Larger and more economically diverse urban settlements have a clear relative economic advantage over smaller towns and centres.

- Remote indigenous settlements are areas of extreme disadvantage.

- A relatively uniform pattern occurs across large cities, apart from fringe areas where households command significantly fewer economic resources.

- Significantly greater variations occur in smaller urban regions.

- The coastal areas of New South Wales and Queensland are growing rapidly but are characterised by high levels of unemployment and a large number of welfare benefit holders.

Distribution of resources across regions

The recently developed ABS index of economic resources (ABS, 1994h) provides a way of exploring the pattern of economic advantage/disadvantage across urban, rural and remote regions of Australia (see Fig. 3.8). Although the detailed construction of this index could be improved, its attempt to incorporate employment and housing as well as income circumstances of housholds has moved in the direction of a full income accounting approach.

Distribution of resources across neighbourhoods — pockets of poverty

The spatial inequalities and variations shown above reveal that, apart from indigenous communities, the greatest differences occur between poor and affluent suburbs of big cities. In United States and British cities, these disparities have led to ghettos of poverty which, in turn, have had seriously adverse impacts on the human environment. In these cities, poverty, which tends to be concentrated in inner suburbs, has been a major cause of suburbanising and exurbanising processes. Australian cities do not have this kind of inner city problem (see Fig. 3.9). But the differences between affluent and poor areas located in both inner and outer areas of Australian cities point to disparities which may be more localised (see Fig. 3.8, -inside urban Australia and -inside Sydney region).

Gregory and Hunter (1995) have focused on neighbourhoods (defined as ABS collectors' districts and typically including 200–300 people) across urban Australia and demonstrated that disparities between them measured in terms of gross income (private and public) have increased dramatically in the period between 1976 and 1991 (see Fig. 3.10). Even though these census-based

Figure 3.8 Distribution of economic resources across Australia

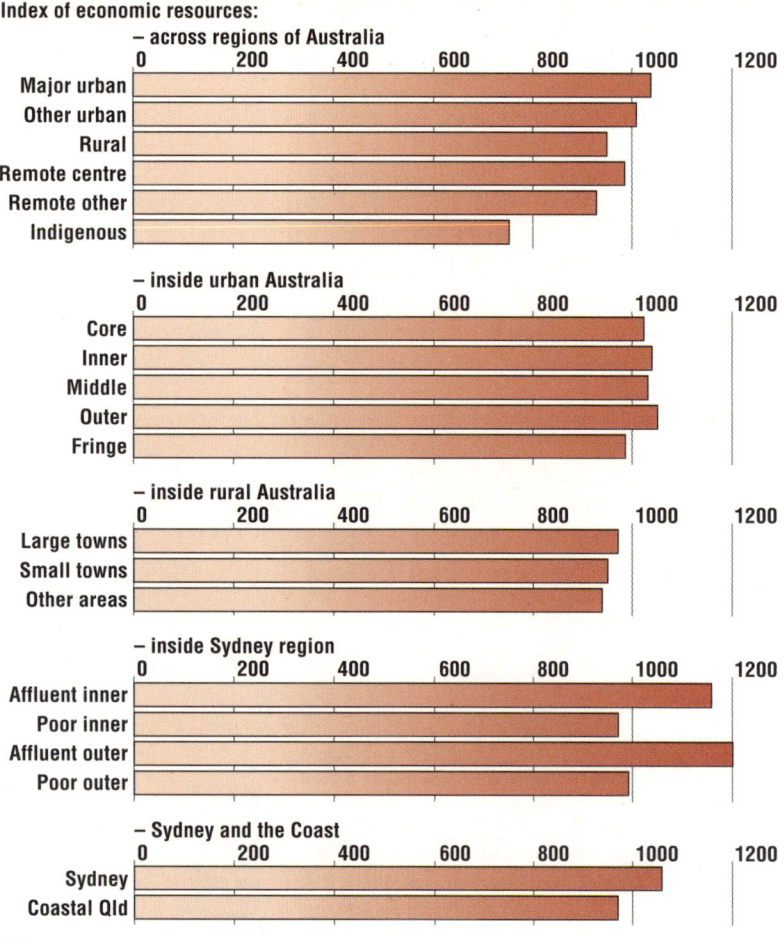

Index of economic resources:

Notes:
1. The index measures deviation from a national average score of 1000 on a standardised scale. The graphs indicate broad relativities only.
2. Because the index of economic resources assigns equal weight to home owning and home purchasing – owned houses and newly mortgaged houses can be very different assets – it may well overstate the advantage of outer areas where purchasers prevail over owners – see glossary.
3. The affluent and poor refer to selected local government areas which were at the high and low ends respectively of income distribution.

Source: Derived from ABS, 1991 and 1994h.

Figure 3.9 Households in urban Australia earning more than $70 000 or less than $35 000 per year by zone

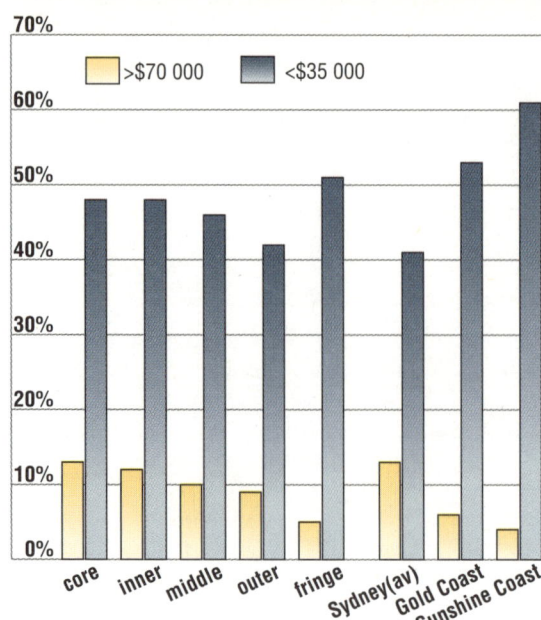

Source: Derived from ABS,1991.

income data exclude the impact of taxation and social wage goods, the changes being explored are quite distinct from changes in income distribution measured at the level of individual households. The results tell us that the most disadvantaged are congregating in particular quarters of cities to a much greater extent than previously — and perhaps forming embryonic ghettos with the range of problems these may bring.

The explanation which the researchers give for the changes in neighbourhood income that they have observed provides a further reason for concern. Widening disparities in the distribution of income across neighbourhoods and the increasing tendency for the least well-off in our cities to congregate is largely due to the availability of employment — and only to a lesser extent to a widening of salary and wage dispersion. If 'two Australias' are emerging, then the difference between them is that households in well-off Australia can find at least one job and often two. Households in poor Australia, by contrast, are experiencing increasing difficulties in finding even one. A generous social wage, including adequate cash benefits, is important for cushioning the effects of unemployment. However, for those who want jobs — because of the meaning and dignity which work brings — income maintenance can never be a satisfactory substitute. These emerging pockets of poverty have implications for regional development and ecological sustainability.

Employment

The early 1970s were a watershed in the condition of working Australia. They were preceded by a quarter of a century of high growth, during which unemployment rarely exceeded 1.5 per cent. Slow employment growth from the mid 1970s resulted in unemployment rising to more than six per cent by the end of the decade.

The recession of the early 1980s increased it even further, to more than 10 per cent. While the rate fell with subsequent strong employment growth in the late 1980s, to around six per cent, it rose again to peak at 11.2 per cent in late 1992 (Langmore and Quiggin, 1994). In recent years, strong jobs growth has caused the unemployment rate to fall to 8.3 per cent in April 1995 (DEET, 1995).

While the Commonwealth Government is aiming to significantly reduce the unemployment rate by the year 2000, economic opinion remains divided about the prospect of futher significant falls in coming years and some commentators now refer to a new 'natural' unemployment rate of eight per cent (Mitchell, 1993). They argue that progress beyond this may involve risks of inflation or calls for more drastic wage-cutting and restraint — bringing with it America's problems of a 'working poor'.

Persisting high levels of unemployment and the apparent difficulty for governments in finding remedies, again reflect Australia's new openness to global economic forces. The changing and increasingly flexible labour market suggests the

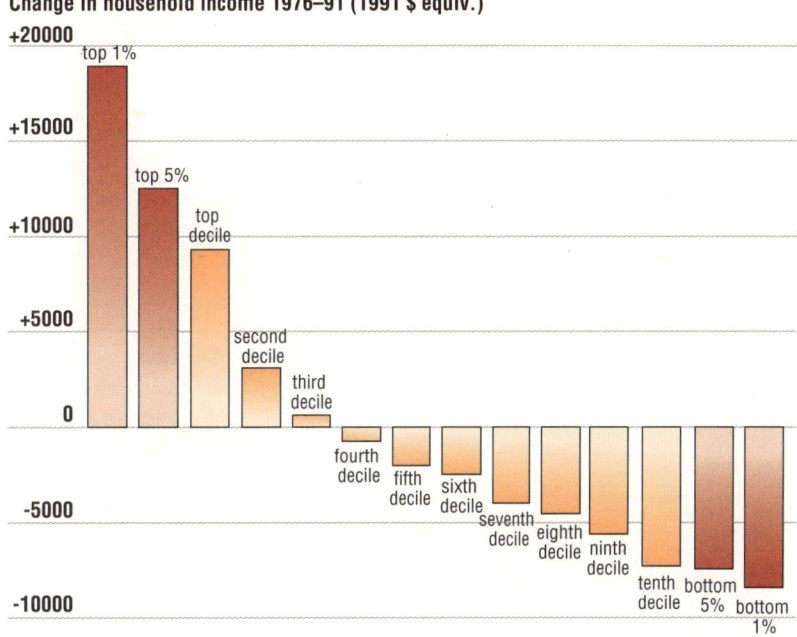

Figure 3.10 Changing income share of urban neighbourhoods by socio-economic status group, 1976 –1991

Change in household income 1976–91 (1991 $ equiv.)

Notes:
1. Socio economic status groups defined using ABS Urban and Rural Indexes of Relative Advantage.
2. Neighbourhood analysis involve the presentation of data as group averages from Collectors Districts – the smallest geographic areas for which Census data are available (typically 200-300 dwellings).
3. Income includes gross (untaxed) monetary income from all sources including pensions and benefits.

Source: Gregory and Hunter, 1995.

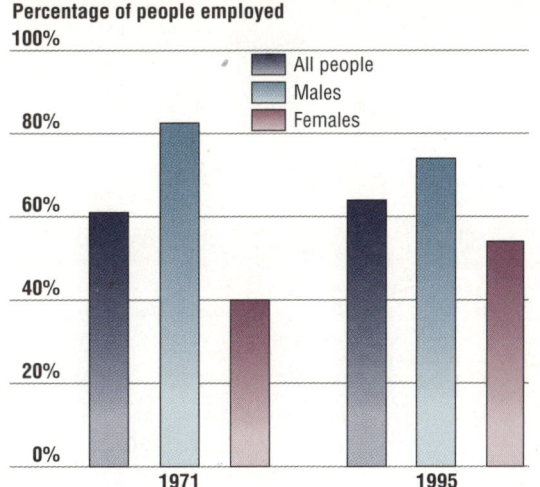

Figure 3.11 Trends in labour market participation

Percentage of people employed

Source: ABS, 1992c and DEET 1995.

Overworked women

Even though women are joining the paid workforce in greater numbers (see Figs 3.11, 3.12 and 3.13) research indicates that men are not shouldering a corresponding share of necessary unpaid domestic duties (ABS, 1995d).

same openness in different ways. Against this background, there are changing patterns of employment in terms of gender, hours worked, and the incidence of long-term unemployment (see Figs 3.11, 3.12, 3.13 and 3.14).

Apart from those people living in lower socio-economic neighbourhoods, a number of groups are exceptionally vulnerable to high rates of unemployment. These include people of non-English-speaking backgrounds, those with disabilities, older people, youth and indigenous Australians. Young people have unemployment rates more than three times as high as the general rate, and indigenous people are even worse off.

Indigenous Australians are at a serious disadvantage in the labour market. Their rate of unemployment is four times higher than average, and long-term and youth unemployment are much more severe problems for them (see Fig. 3.15). The figures

would be even higher if not for the Community Development Employment Project (CDEP), which accounts for 26 per cent of all indigenous employment and which redirects social security benefits into employment-generating community projects. In the absence of CDEP, indigenous unemployment would rise to 57 per cent (ABS, 1995a).

Regional distribution of unemployment

Unemployment data indicates that larger and more diverse settlements tend to be more economically robust (see Fig. 3.16). While the evidence for remote settlements suggests otherwise, their lower rate of unemployment is due to their often highly dedicated economic character and the mobility of their populations. Many people move to remote settlements in order to work — in some cases even commuting by air — and leave when their employment ceases.

The relatively high incidence of unemployment in the urban core (see Fig. 3.16) points again to its bi-modal character. However, the low level of unemployment on the urban fringe suggests that the fringe is a more uniformly low-income area.

Education

One in three Australians is now enrolled in educational or training courses of some kind. The new emphasis on education again points clearly to the dynamics of a globalising economy, increasing competition, rapid technological change and demand for a flexible and highly skilled workforce (see Fig. 3.17). Although significantly higher in absolute terms, public expenditure on education as a proportion of GDP is now slightly lower than it was in the 1970s — down from a post-war high of

Figure 3.12 Part-time employment

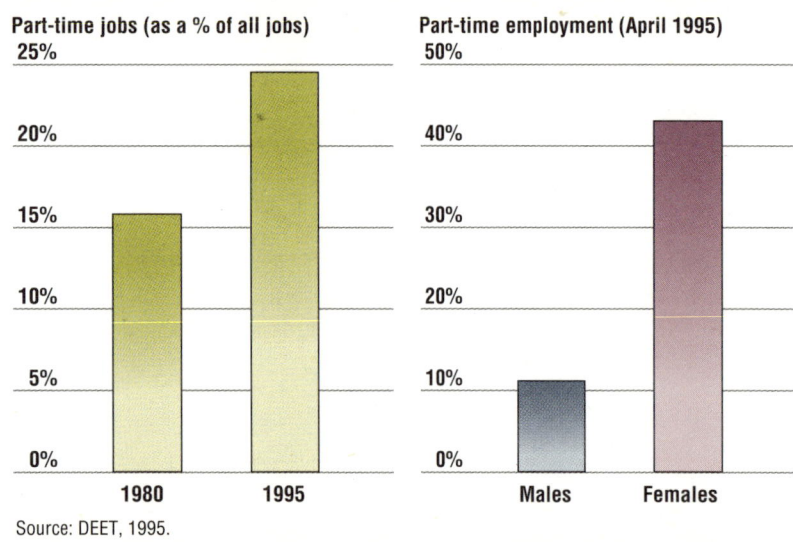

Source: DEET, 1995.

Figure 3.13 More work for the working

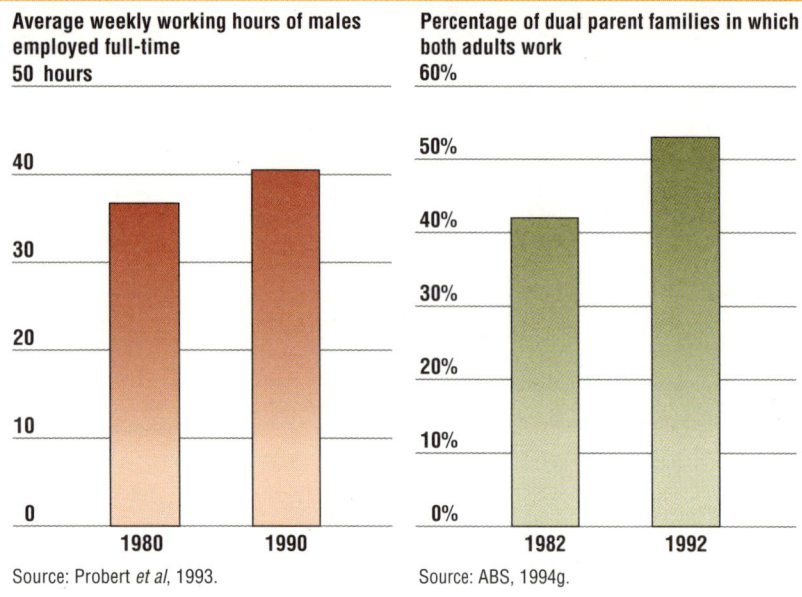

Source: Probert *et al*, 1993.   Source: ABS, 1994g.

Figure 3.14 Unemployed males and females who are long-term unemployed

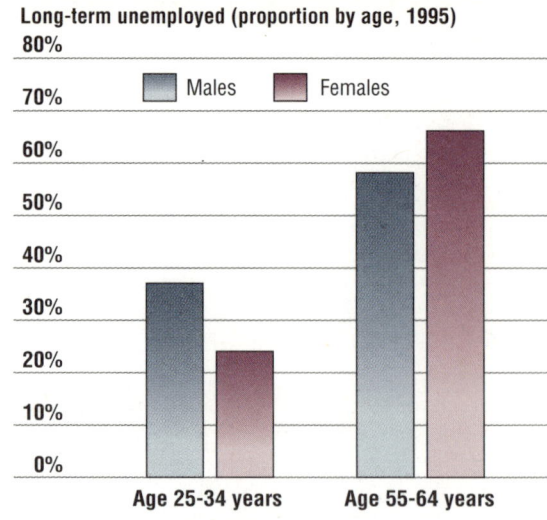

Once unemployed, older people clearly experience greater difficulties in finding alternative jobs. Between 1983 and 1993, median period of unemployment almost doubled: from 15 to 29 weeks (ABS, 1994g). Reflecting the general improvement noted above, however, the number of long-term unemployed fell by 25% between March 1994 and 1995. (DEET, 1995).
Source: DEET, 1995.

5.7 per cent of GDP in 1975 to 5.5 per cent in 1993–94 (Marginson, 1993; DEET, 1995).

People living in rural and remote areas have less access to higher education and training facilities (see Fig. 3.18). Continuing attempts by government to redress the problem through initiatives like tertiary Open Learning courses, external education and the relocation of post-secondary education facilities in regional centres have made a difference. In many fields jobs that require specialised training depend on enterprises that enjoy large markets and economies of scale. This remains a severe constraint on the development of specialised educational and training facilities in smaller settlements.

Housing

Between the 1940s and '80s, households grew smaller and houses larger. The average household size has fallen from 3.9 to 2.6 people, and by 1986 some 50 per cent of all households consisted of one or two people. Over the same period, houses increased in size: the percentage of houses containing less than five rooms, for example, has fallen from 37 per cent to 22 per cent (ABS, 1992a). For much of the post-war period housing has become more affordable: homes owned or being purchased rose from 54 per cent of all dwellings in the 1940s to a post-war high of 71 per cent in 1966 (ABS, 1992a). In 1992, this figure stood at 69 per cent (ABS, 1994g). Since the mid 1970s, however, the proportion of houses being purchased has declined markedly — from 35 per cent in 1976 to 28 per cent in 1992 (ABS, 1992a and 1994g). If this trend continues, a growing

number of young Australians will join the rental market and thus forgo the financial and non-income-related advantages of home ownership.

Almost one in five Australian households experience housing stress: six per cent are inadequately housed and twice as many experience difficulty in paying for their accommodation (see Fig. 3.19). Major cities experience less after-housing poverty than small towns or rural areas.

Figure 3.16 The distribution of unemployment across urban, rural and remote Australia and within urban Australia, 1991

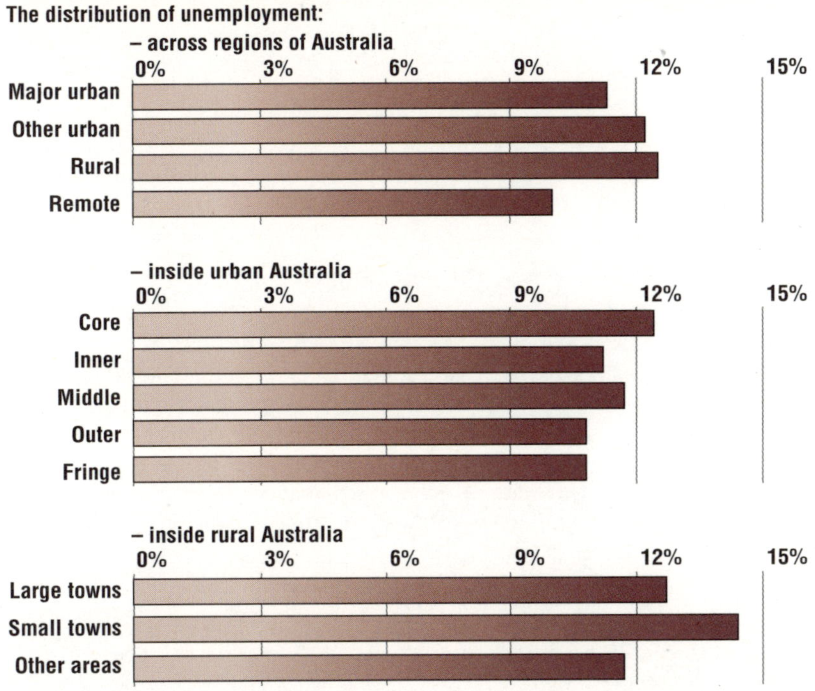

Source: derived from ABS, 1991.

Note: Unemployment among indigenous people living in rural and remote Australia was 30% in 1994 (Jones, 1994).

Figure 3.15 Comparative labour market profile for indigenous Australians, 1994

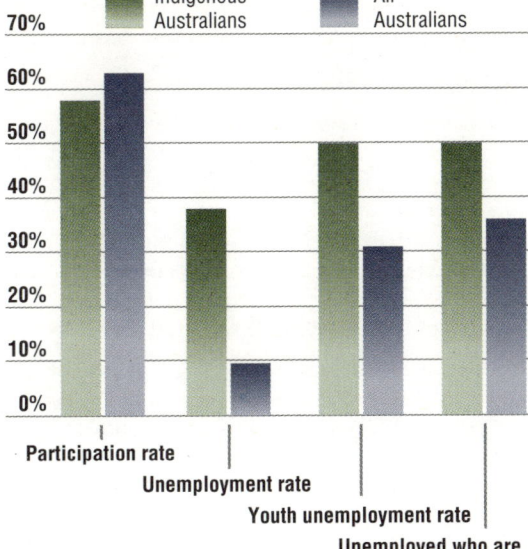

Source: ABS,1995a; DEET, 1995.

Figure 3.17 Key educational commitment indicators

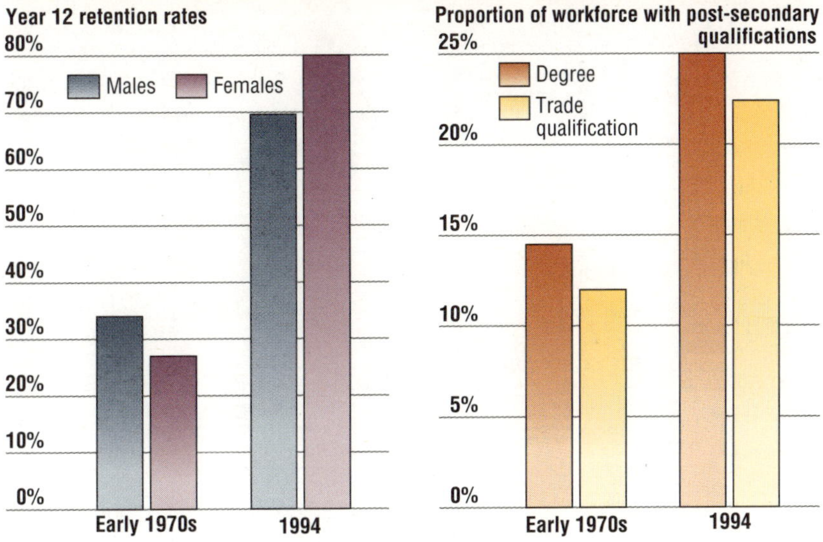

Sources: McLelland, 1994; Margison, 1993 and DEET, 1995.

Indigenous Australians are far worse off in terms of housing wherever they live (see Fig. 3.20). Indigenous families are 20 times more likely to be homeless than their non-indigenous counterparts (Jones, 1994).

Public housing authorities are failing to keep up with demands for assistance (see Fig. 3.22).

Housing affordability in major cities

As discussed in the section on population dynamics (see page 3-8), the impacts of globalisation are leading to increasing employment concentrations in core and inner city areas. While this is fuelling the process of reurbanisation in central areas, it may also be reducing the stock of affordable housing in these areas. Data on the cities of Melbourne (Maher, 1992 and 1994) and Perth (REIWA, 1994) suggest this is likely. The phenomenon may well be common to all cities. Government regulations plus demand for more accessible housing are generating pressures for

consolidation throughout the city, including outer areas. Although core and inner areas already contain the highest proportions of high- and medium-density housing, a majority of dwellings in the inner and middle area remain detached (see Fig. 3.21).

Accessibility and locational disadvantage
The Australian suburb

The typical image of Australian urban life depicts a family living in a detached house with generous front and back gardens in a suburb zoned exclusively for housing. The family owns at least one car — often two or three — which provides for most of its transport needs. This popular image expresses important truths, but it is also a little too simple and overlooks important changes.

While detached dwellings are still the dominant housing form in Australia, trends towards higher-density housing in established areas are slowly beginning to change this pattern.

Australian households do rely very heavily on cars. As low-density suburbs mushroomed in the post-war period so did rates of car ownership: from 144 motor vehicles per thousand people in 1948 to 572 per thousand in 1990 (ABS, 1955 and 1992b). Again, however, indications suggest that the rate of growth of car dependence may be slowing (see page 3-37).

The proportion of households that include children belong to a declining minority — 43 per cent in 1992 compared with 47 per cent in 1982 (ABS, 1994g). It is true, however, that the outer suburbs do contain both the youngest households (ABS, 1991) and the highest proportion of those with children — 55 per cent of outer suburban households had children compared with 30 per cent in inner areas (NHS, 1992a).

In reality, the typical suburb is not as comfortable as the image suggests. A number of recent studies (for example, the Social Justice Research Program into Locational Disadvantage, 1991–95 and the National Housing Strategy, 1992a and b) have raised questions about lack of social amenity in the outer suburbs of major Australian cities. These studies have documented low levels of service provision, particularly in lower-income areas.

Conflicting interpretations

There is general agreement about the relative lack of services in outer suburban areas but considerable disagreement over its causes and wider significance. Not surprisingly, those who favour continuing low density suburban development ('suburbanisers') are less inclined to see serious or intractable problems than those who favour alternative 'reurbanising' strategies and more compact urban forms.

Suburbanisers point to the temporary nature of 'lagging' suburban services — needlessly delayed, they sometimes argue, by short-sighted fiscal restraint (Stretton, 1994; Troy, 1992). They draw attention to the attractions of suburban living and to the strong preferences urban residents continue to show for it — as reflected in market behaviour and relevant consumer surveys (NHS, 1992a;

ABS index of education and occupation

The ABS index of education and occupation indicates the distribution of educational and training qualifications across different types of settlement and within major cities. Once again it highlights the relative advantage of larger and more diverse settlements and the disadvantage of remote indigenous ones. Educational status and distance from the city centre have a strong correlation — confirming again the two-sided nature of the inner and core areas and their status as areas in social and economic transition. It also adds a further dimension to the picture of disadvantage characterising the urban fringe.

Figure 3.18 Distribution of educational and training qualifications across urban, rural and remote Australia

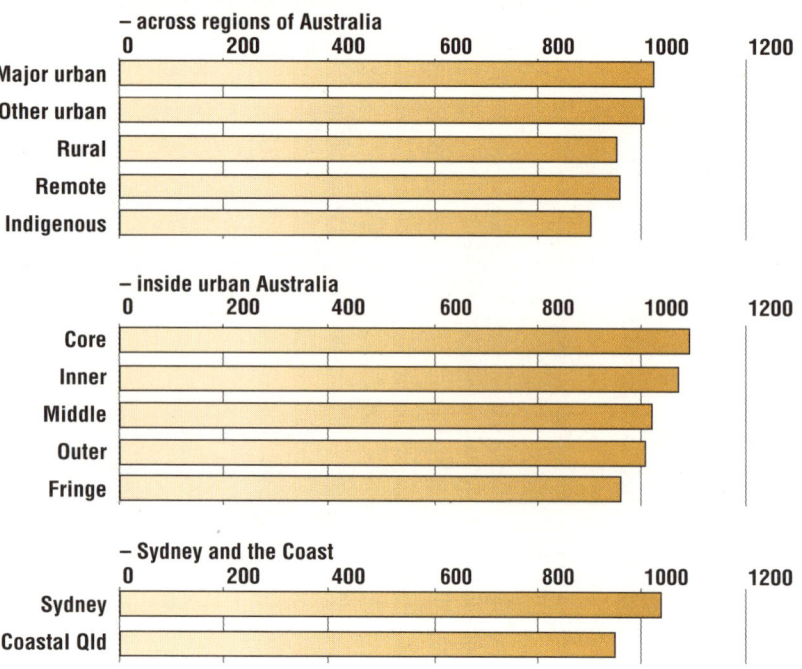

Index of Education and Occupation:

Note: The index measures deviation from a national average score of 1000 on a standardised scale.
Source: Derived from ABS, 1991 and 1994h.

McDonald and Moyle, in press). Suburban 'poverty' is judged to be a temporary problem, most often the passing experience of young outer suburban households yet to enjoy rising incomes and increasing equity in their homes. More serious urban poverty, this argument also sometimes suggests, is to be found in inner urban locations (Maher *et al.*, 1992; Wulff *et al.*, 1993). Suburbanisers argue that the private resources generally posessed by outer suburban households — especially cars — enable them to overcome the problems of distance and poor access to services (Maher, 1995). And now, also, these problems are said to be diminishing as a result of urban employment dispersion (Brotchie, 1992).

Re-urbanisers disagree. They point to examples of more enduring outer suburban deprivation that result not just from short-sighted fiscal restraint but also from the inherent inefficiency of low-density forms unable to generate sufficient economies of scale (Newman *et al.*, 1992; AURDR, 1995a). Housing markets, they argue, tend to be driven by supply rather than demand, offering, until very recently, only a narrow range of choices and few examples of attractive higher density living (Newman and Kenworthy, 1992; Sarkissian and Marcus, 1986; AURDR, 1995b). They also point to evidence of longer-term outer-suburban poverty and discontent (Richards, 1994; McDonald, 1995; Wynhausen, 1995), and to the vulnerability of outer-suburban residents who have only limited access to cars (Tranter, 1994). Finally, they draw attention to the adverse impacts of increasing car use on the social, built and natural

Figure 3.19 Australians in housing stress

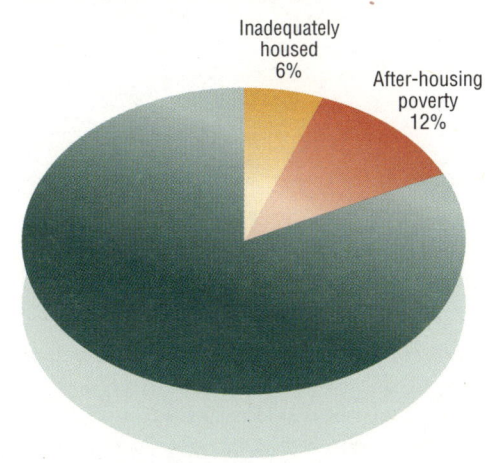

Percentage of households under stress

Source: Jones, 1994.

Figure 3.20 After-housing poverty

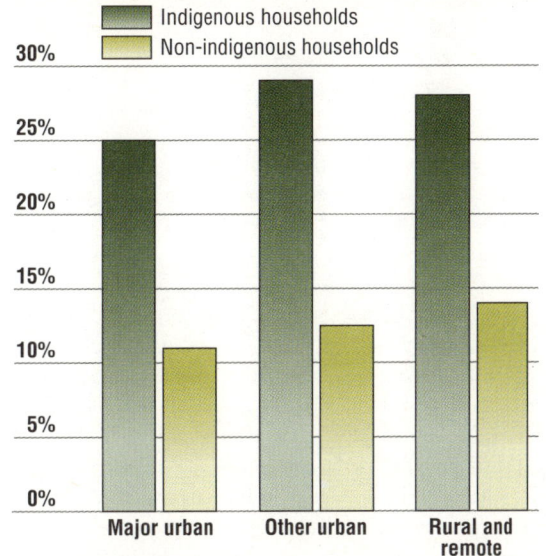

Source: Jones, 1994.

Figure 3.21 High-, medium- and low-density housing in urban Australia

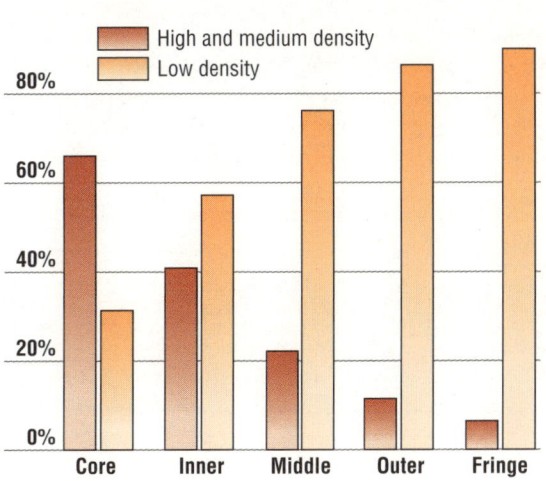

Note: Some houses are not stated in the above categories.
Source: derived from ABS, 1991.

Figure 3.22 Demand for public housing outstrips supply

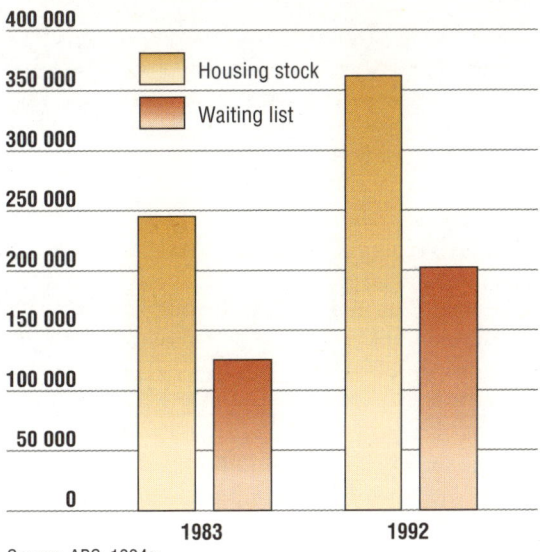

Source: ABS, 1994g.

The street is a very significant part of any community.

environments of our cities and dispute the claim that these are being relieved by continuing urban deconcentration (Prime Minister's Urban Design Taskforce, 1994; Engwicht, 1992; Newman *et al.*, 1993).

While it is true that inner urban areas contain significant areas of social disadvantage, two other facts need to be noted in this context. First, low income households in inner areas report below average levels of access difficulties to urban services — while high income households in outer areas report above average difficulties (Newman *et al.*, 1992). Second, affordable inner city housing options may well be shrinking — see the discussion of housing affordability above. This suggests a trend likely to force outward migration not just of younger aspiring home owners making a start, put of poorer renting households as well — especially where public housing policy favours cheaper outer urban development locations. Alternatively, people in these circumstances may move further afield to fringe and rural locations, thus sacrificing the access advantages of inner areas (Flood *et al.*, 1991).

Although planners continue to argue about the relative merits of suburbanising and re-urbanising strategies, people are clearly choosing to move to both inner and outer areas of Australian cities. One survey, taken before the recent globalisation-related trends, showed that those wanting to move outwards were roughly matched by those wanting to move in the opposite direction (ABS, 1981). The development resulting from these demands needs to be managed more carefully and should more effectively incorporate the principles of ESD, community development and quality urban design.

Community

We can look at the idea of community from many angles, shaped as it is by a multitude of factors, among them being education, mass media, cultural and religious diversity, the role of the arts and the importance assigned to community participation in

political life (Sarkissian and Walsh, 1995). However, the focus here is on the role of urban design which was highlighted in the recent report of the Prime Minister's urban design task force. Other urban planners argue that the unsustainability of settlements in terms of their metabolic flows is closely linked to their loss of community vitality — and that both are related to the way cities have been designed (Engwicht, 1992; Hayward and McGlynn, 1993).

Urban design task force

The recent report of the Prime Minister's Urban Design Task Force (1994) commented on the importance of public spaces and places:

'Australians devote great care to their private places. Yet many Australian cities struggle with a neglected stock of public spaces because of the premium placed on individual choice and because of inappropriate government and industry structures... The state of Australian streets tells the story starkly. Throughout the ages urban street networks have provided cities' essential civic communication and movement channels: in the modern city, because of the primacy of the motor car...streets have become almost exclusively conduits for cars. Yet we all know that streets have other roles... In our cities, streets must retain their function as the backbone of our society's public domain, and be made attractive for pedestrians, for children's play, for meeting other people, for resting and eating. Like other parts of the cities, streets must be designed to serve these purposes well.'

The task force outlines problems in urban design that emphasise the loss of diversity in the environment in new areas relative to older mixed-land-use areas with a greater range of housing types. It is critical of the car dependence in Australian cities and finds the coastal sprawl particularly damaging, not only to the natural environment but to community values as well.

Community and urban design

For most people, a city is much more than just a place to live and work. It is a place to belong to and be proud of, a place in which to make and maintain connections and one to enjoy in common. It is a shared and public space, which not only surrounds and connects many sites of private endeavour but also supports a diverse public and cultural life (Gehl, 1992). It is also, as Jacobs (1961) has argued, a source of vital learning, amusement and adventure for growing children and young adults.

The neglect of such spaces inside and beyond urban Australia noted by the Prime Minister's task force is also a neglect of our settlements as crucibles of community. This neglect, of course, is not universal, and there are also encouraging signs of change — particularly in some older towns and urban centres that retain richer and denser fabrics and pedestrian-friendly forms. The move towards nodal/information subcentres is partly due to the desire to create community centres. These are forming, therefore, in older parts of the city as well as in new outer areas. In these subcentres people are trying to recreate — or to create anew — convivial public spaces in which the adverse effects of cars are controlled and more human contact is possible.

City dwellers are concerned about the importance of community and its connection with urban design. The Western Australian government recently conducted a survey and consultative process in Perth, which highlighted the fact that people felt strongly that community was disappearing from their suburbs. They were searching for a village concept in urban design (Community and Family Commission, 1992). They recognised that the development of a greater sense of community meant sharing in the management of their neighbourhoods and having closer access to each other and their local services. They saw that this was an important element in fighting the growth of crime or the fear of crime in their suburbs.

We do not have well-developed indicators of community or good urban design for Australian settlements — apart from some recent work in Perth, Melbourne and Sydney CBDs (Gehl, 1994a and 1994b; SCC, 1993). However, the Prime Minister's urban design task force suggests a growing role for these indicators in state of the environment reporting.

Health in Australian settlements

The international context

Australia is one of the healthiest countries in the world and our health continues to improve (see Fig. 3.23). Expenditure on health has been stable at about eight per cent of GDP for the last 15 years. Nevertheless, some population groups suffer wide disparities in health, with substantial room for improvement.

Our average life expectancy has risen continuously during the 20th century, apart from a period during the 1960s when deaths from cardiovascular disease increased, particularly for men (see Table 3.10). The increases that occurred in the first half of the century were due to rapid declines in infant and maternal mortality, particularly the lessening impacts of the infectious diseases associated with childhood and early adulthood. Access to better housing, sanitation and education, a trend to smaller families, growing incomes, the introduction of public health measures such as immunisation against infectious diseases and the development of antibiotics in the 1940s further contributed to these improvements.

Since the 1960s gains in life expectancy have been concentrated among the middle aged and older population. Some causes of death have declined dramatically. Between 1968 and 1992, for example, age-adjusted death rates from cardiovascular disease declined by 56 per cent for men and 55 per cent for women (AIHW, 1994).

The decline in deaths due to infectious diseases was accompanied by increases in both the death rates and the proportion of deaths attributed to diseases of the circulatory system and to cancers. In 1921, these diseases accounted for about 22 per cent of deaths, whereas by 1994 they were responsible for nearly 72 per cent.

In contrast to this good record, Australia's indigenous people have life expectancies and patterns of health more comparable to those of developing countries (see Table 3.11).

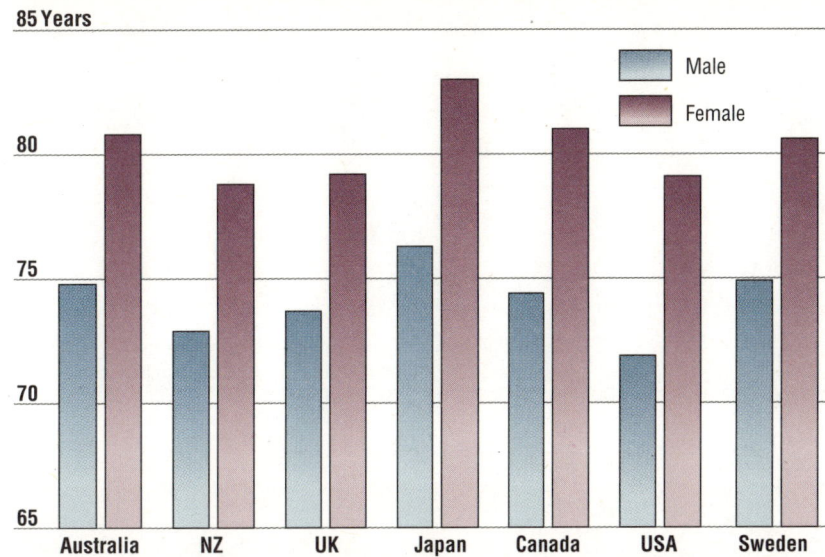

Figure 3.23 Life expectancy at birth for a number of OECD countries

Source: WHO, 1994.

Table 3.10 Life expectancy at birth and at age 65, 1905–1993, Australia

	1905	1921	1947	1966	1993
At birth					
males	55.2	59.2	66.1	67.6	75.0
females	58.8	63.3	70.6	74.2	80.9
At age 65					
males	11.3	12.0	12.3	12.2	15.7
females	12.9	13.6	14.4	15.7	19.5

Source: Australian Life Tables, ABS, 1994i.

Table 3.11 Life expectancy at birth for indigenous people and the total Australian population, in selected states by sex, 1990–1992

	Indigenous WA	Indigenous SA	Indigenous NT	All Australia
Males	56.3	57.8	56.8	74.5
Females	64.2	63.7	60.6	80.4

Source: AIHW, 1994.

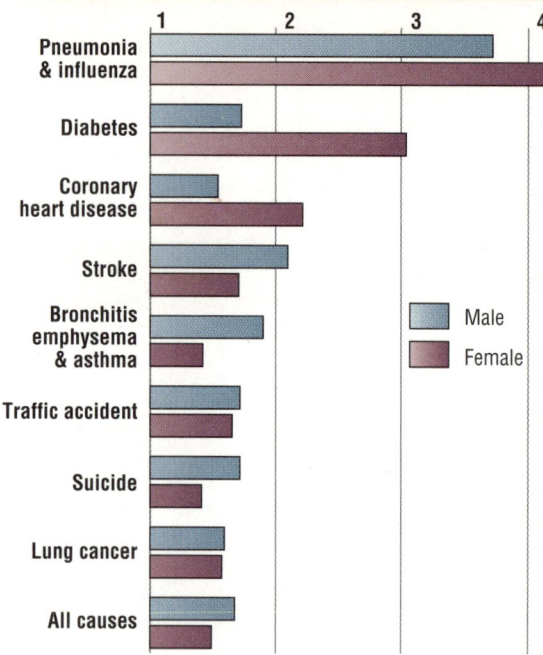

Figure 3.24 Ratios of death rates for Australians living in the most over the least disadvantaged areas

Note: The above rate ratios compare the standardised death rates of the 20% of Australians aged 25-64 years living in the most disadvantaged areas with those of the 20% of the Australians living in the least disadvantaged areas.

Source: Mathers, 1994a.

Table 3.12 Selected health indicators by broad region, 1990–1992

Indicator	Metropolitan	Other cities	Rural	Remote
Life expectancy at birth (years)	77.7	77.3	76.6	71.7
Infant mortality rate (per 1000 live births)	7.11	6.76	7.63	13.12
Age-standardised mortality rate (per 100 000 population)	679	692	734	985

Source: AIHW, 1995.

Table 3.13 Cardiovascular diseases: standardised mortality ratios[1] by settlement type, 1990–1992

Cause of death[2]	Metropolitan	Other cities	Rural	Remote
Cardiovascular diseases (ICD9 390–459)	0.97	1.00	1.05	1.31
Rheumatic heart disease (390-398)	0.97	0.97	0.98	2.79
Ischaemic heart disease (410-414)	0.97	1.02	1.06	1.29
Cerebrovascular disease (430-438)	1.00	0.97	1.01	1.20

Notes:
1. Standardised mortality ratios - see Glossary
2. Causes of death are classified according to the International Classification of Diseases (9th Revision) or ICD9.

Source: AIHW, 1995.

Social amenity and health

Elements of the social environment seen as important determinants of health include family income and wealth, individual education level, occupation and the working environment, marital status, social networks and social support, the living environment and culture (Mathers, 1994 a and b).

Regardless of the measure used — income, education level, occupation or other areas of socio-economic disadvantage — Australians from less-advantaged backgrounds have higher death rates and report worse health and higher levels of illness than their better-off counterparts (see Fig. 3.24).

It is difficult to make links between specific environmental hazards and particular human diseases because of the number of intervening factors, the quality and availability of data and the time-lag between exposure and the onset of disease. Many factors — ranging from genetic through to individual lifestyle and the social and physical environment — interact to cause major health problems.

The following comparisons of patterns of health across the various categories of human settlements in Australia identify differences in death rates for particular diseases that are partly caused by the social and/or physical environment.

Death rates are based on place of usual residence at the time of death and so relate to populations living in each of the settlement types at a particular

The health of Indigenous Australians

On almost every measure, indigenous Australians suffer poorer health than other Australians.

* Death rates are between two and four times those of the total Australian population.

* Mortality is much higher for young and middle-aged adults — males in the 35- to 44-year age group die at a rate more than eight times that for non-indigenous males.

* Life expectancy at birth is between 16 and 18 years less.

* Infant mortality rates are between two to three times as great.

* Indigenous babies weigh, on average, 200 grams less than non-indigenous babies at birth.

* Indigenous Australians are admitted to hospital about twice as often as non-indigenous Australians.

* Preventable communicable diseases continue to contribute disproportionately to high mortality and hospitalisation.

* The growing impact of non-communicable diseases — particularly cardiovascular disease and diabetes — without much decline in infectious-disease mortality is a phenomenon peculiar to indigenous Australians.

time. They also reflect the effect of the migration of older people associated with retirement or increased dependency.

Urban, rural and remote variations in health

The health of Australians as measured by broad indicators shows little variation between metropolitan and non-metropolitan settlements (see Table 3.12), with the exception of remote settlements, where the substantially worse health status of indigenous people results in significantly worse figures for the region. For some specific diseases and for some age groups, significant health differences exist between human settlements in Australia.

Cardiovascular diseases

Despite a dramatic decline in mortality related to it over the past 25 years, cardiovascular disease remains Australia's biggest health problem and is responsible for nearly half of all deaths in Australia each year. Coronary heart disease, which is the most common form, accounts for 25 per cent of all deaths, followed by cerebrovascular disease (stroke), which causes 10 per cent. The major causes of these diseases in the Australian population are thought to be a number of dietary factors (particularly those related to high blood pressure, high cholesterol and obesity), tobacco-smoking and physical inactivity.

Rural and remote settlements have higher mortality rates for coronary heart disease than metropolitan ones (see Table 3.13). Stroke death rates are also higher in remote, but not in rural, settlements. These higher ratios are partly attributable to the higher levels of cardiovascular disease among indigenous people.

The standardised mortality ratio of 2.8 for rheumatic heart disease in remote communities reflects its alarmingly high rate in indigenous people (up to 20 times more common in some remote communities than in Australian communities, generally).

Smoking and lung cancer

Most lung cancers are due to smoking and could be avoided. Age-standardised death rates for the disease increased seven-fold for men between 1945 and 1982, reflecting the very high prevalence of smoking following World War II (up to 70 per cent). After peaking at 68 deaths per 100 000 men in 1982, the rate has declined slowly (to 58 per 100 000 in 1992).

Since 1945, the female death rate due to lung cancer has increased enormously, with a six-fold increase to 17.5 deaths per 100 000 in 1992, as a result of increased cigarette consumption by women since the 1940s. The death rates for women have not yet started to decline, although in the last few years, women's smoking rates have fallen — albeit at a slower rate than men's.

Cancer

Death rates for cancers have changed little over the last 30 years, with small increases in the age-standardised death rates from 2.0 per 1000 males and 1.3 per 1000 females in 1965, to 2.3 and 1.4 respectively in 1993. Despite this relatively small absolute increase, the decrease in total mortality rates has meant that the proportion attributed to the disease actually increased from 15 per cent in 1965 to 27 per cent in 1992, making it the second leading cause of death after cardiovascular diseases. Death rates for most cancers are higher in remote settlements (particularly lung and cervical cancer), but there are fewer variations in cancer mortality among other settlement types (see Table 3.14).

Lung cancer is by far the major form causing deaths, accounting for 20 per cent. The mortality rate for lung cancer is rising in women, but falling in men. Breast cancer is the most common form of cancer in women. Death rates for skin, prostate and liver cancers are rising in men.

Much cancer mortality could be prevented. Researchers estimate that approximately one-third of cancer deaths may be attributed to tobacco smoke, another one-third to diet and about five per

Australians have the highest rate of skin cancer in the world.

▼

Table 3.14 Cancer standardised mortality ratios by settlement type, 1990–1992

Cause of death	Metropolitan	Other cities	Rural	Remote
All Cancers (ICD9 140-239)	1.00	1.00	1.00	1.08
Digestive organs (150-159)	0.99	0.98	1.02	1.11
Lung (162)	1.01	0.98	0.96	1.19
Skin (172 & 173)	0.99	1.04	1.00	1.16
Breast (174)	1.00	1.02	1.02	0.74
Cervix (180)	0.97	1.08	0.92	2.96
Prostate (185)	0.95	1.03	1.11	0.96
Lymph, leukemia etc (200-208)	1.02	1.01	0.96	0.72

Source: AIHW, 1995.

cent to a range of physical environmental factors, including carcinogenic chemicals and materials, ionising radiation and electric and magnetic fields (Giles, *et al.*, 1987).

Sunlight exposure is of particular concern. Australians have the highest rate of skin cancer in the world. Cases of melanoma are estimated to have quadrupled in the last two decades, with the highest incidence in Queensland.

Injuries

Injury is a leading cause of death in Australia, accounting for about six per cent of deaths, more than 10 per cent of hospital episodes and about 25 per cent of all handicap cases. While degenerative diseases such as heart disease and cancer occur primarily in older people, injury disproportionately affects the young and is the leading cause of death of both males and females between the ages of one and 44 years. Injury death rates are substantially higher in rural and remote settlements than in cities (see Table 3.15).

In 1993, the major causes of deaths due to injury were: motor vehicle crashes; suicide; falls (particularly in older people); homicide and drowning (particularly among toddlers). Suicide now causes more deaths in Australia than motor-vehicle crashes. The age-standardised suicide

mortality is increasing by two per cent per year for men and falling by one per cent per year for women. Among young males, suicide has been rising since the 1950s. For women it peaked in the 1960s.

Mental health problems are of increasing concern. The Commonwealth, State and Territory governments have adopted a national policy and are planning a national survey to identify the prevalence of such problems in the population.

Road crashes in Australia account for more years of working-life lost than do all forms of heart disease, and more than half the loss through all cancers (Ginpil *et al.*, 1992). When we consider that these affect a much younger age group than heart disease and cancer, their social impact is particularly disturbing. In the next decade, one in every ten Australian families will be directly affected by a road death or serious injury (Federal Office of Road Safety, 1992). Death rates due to motor vehicle accidents are considerably higher for people living in rural and remote areas.

Chronic diseases caused by occupational exposures to toxic and carcinogenic substances and to radiation cause considerable controversy. A large number of occupationally caused cancers are well documented — for example, the primary cause of mesothelioma occurring in Australia and other industrialised countries is recognised as occupational exposure to asbestos fibres. However, experts disagree considerably about the extent to which modern industrial and agricultural development have increased the levels of cancer and congenital abnormalities (Doll, 1992; Landrigan, 1992).

The workplace

Most Australians spend a great part of their life working. Their workplaces are many and varied, but a central element (indeed a Commonwealth and State responsibility) of work is the need for a healthy and safe working environment (see the box opposite). Occupational health and safety not only affects the workers' ability to make a living, but can also have an impact on their employer's productivity levels.

Respiratory diseases

About eight per cent of deaths in Australia are due to respiratory diseases, chiefly pneumonia, influenza, bronchitis, emphysema and asthma. However, the number of deaths due to influenza is generally low, about 60 in each of the last two non-epidemic years (1990 and 1991).

The extent of asthma in the community varies with age. Recent studies have estimated the prevalence of asthma to be up to 30 per cent in children, between 15 and 20 per cent in adolescents and about seven per cent in adults. During the 1980s, asthma mortality rose significantly in Australia, and since then it has not changed for males but has increased by two per cent per year for females.

Prevalence of the condition in children appears to have doubled in the last decade (Peat *et al.*, 1994).

Table 3.15 Injury and poisoning: standardised mortality ratios by settlement type, 1990–92

Cause of death	Metropolitan	Other cities	Rural	Remote
Injury and poisoning (ICD9 E800-E999)	0.90	1.02	1.17	2.06
Motor vehicle accidents (E810 E819)	0.84	0.98	1.35	2.15
Accidental drowning (E910)	0.81	1.41	1.21	2.26
Suicide (E950-E959)	0.96	1.06	1.07	1.29
Homicide (E960-E969)	0.92	0.88	0.85	4.53
Production injuries[1]	0.73	0.95	1.65	2.38

Note:
1. A number of causes of death (being struck by a falling object; accidents involving machinery; falls from ladders, scaffolds; being caught or crushed; and deaths involving an electric current) are known to be mostly work related. This combination of causes has been used as an indicator of production related death.

Source: AIHW, 1995.

Table 3.16 Respiratory diseases: standardised mortality ratios by settlement type, 1990–1992

Cause of death	Metropolitan	Other cities	Rural	Remote
Respiratory diseases (460-519)	0.97	0.96	1.04	2.26
Pneumonia, influenza and bronchitis (466, 494-496, 480-492)	0.95	0.98	1.05	2.45
Asthma (493)	0.96	0.85	1.16	1.05

Source: AIHW, 1995.

It is now the most common chronic illness and the main cause of hospital admission in Australian children. Asthma mortality is highest in rural settlements (see Table 3.16). This may be related to higher levels of exposure to airborne allergens (such as pollens) among rural communities.

Environmental factors implicated in triggering asthma attacks include outdoor air pollutants (particularly ozone, nitrogen oxides and automotive emissions), indoor air pollutants (nitrogen dioxide and tobacco smoke), grass pollens, certain foods and preservatives, household pets and house dust-mites.

Health differences within and between urban settlements

People living in the core — and to some extent in the inner urban areas — have higher infant and total mortality rates. This reflects higher rates for several causes of death (see Table 3.17). For example, the core metropolitan area has a mortality

Work-related injury

Workplace injury and disease have a far greater impact than many people realise. Each year, about 500 Australians are killed in workplace or work-related accidents. If mortality from chronic diseases and cancers related to occupational exposures (which may have occurred many years earlier) are also taken into account, it has been estimated that up to 2700 Australians may die from work-related health problems each year and up to 650 000 suffer work-related injuries and illnesses.

At any time, work-related injury and ill-health can mean: up to 115 000 workers cannot continue at full capacity; about 200 000 workers have to permanently reduce their work-hours or change their jobs; and almost 200 000 people are prevented from working at all — disturbingly, the vast majority of these have not worked for more than a year. Work-related health problems also affect retired people — figures indicate that up to 285 000 people over the age of 65 are suffering from them.

Work-related fatal injuries are substantially higher in rural and remote settlements than in cities.

Table 3.17 Standardised mortality ratios for selected causes of death, by metropolitan settlement type, 1990–92

Cause of death	Core	Inner	Middle	Outer	Other cities
Infectious and parasitic diseases (1-139)	1.79	1.17	0.96	0.98	0.78
Cancers (140-239)	1.12	1.05	0.97	0.99	1.00
Digestive organs (150-159)	1.08	1.04	0.97	0.98	0.98
Lung (162)	1.21	1.02	0.98	1.02	0.98
Skin (172 & 173)	1.29	0.95	0.95	0.97	1.04
Breast (174)	0.99	1.08	0.98	0.98	1.02
Cervix (180)	1.03	1.02	0.92	0.99	1.08
Prostate (185)	0.90	0.98	0.92	0.98	1.03
Lymph, leukemia etc (200-208)	1.17	1.09	1.01	0.99	1.01
Diabetes mellitus (250)	1.01	1.02	0.96	0.91	0.82
Mental disorders (290–319)	1.62	1.10	0.94	0.87	1.04
Diseases of the nervous system & sense organs (320–389)	1.12	1.17	0.93	0.89	0.98
Circulatory system (390–459)	1.08	1.01	0.95	0.95	1.00
Acute rheumatic fever (390-398)	0.96	0.94	0.98	0.92	0.97
Ischaemic heart disease (410-414)	1.07	0.99`	0.95	0.95	1.02
Cerebrovascular disease (430-438)	1.14	1.06	0.97	0.97	0.97
Respiratory system diseases (460-519)	1.10	0.97	0.93	0.97	0.96
Pneumonia, influenza & bronchitis (466, 494-496, 480-492)	1.11	0.96	0.92	0.94	0.98
Asthma (493)	1.10	0.92	0.90	1.01	0.85
Digestive system diseases (520–579)	1.34	1.06	0.90	0.90	0.99
Genitourinary system diseases (580–629)	1.12	0.98	0.98	0.95	0.89
Congenital anomalies (740-759)	1.44	0.93	1.09	0.86	1.00
Perinatal conditions (760-779)	1.39	1.03	1.03	0.93	0.85
Sudden infant death syndrome (798.0)	0.93	0.77	0.92	1.01	0.99
Injury and poisoning (E800-E999)	1.23	0.94	0.84	0.85	1.02
Motor vehicle accidents (E810-E819)	0.78	0.69	0.79	0.88	0.98
Accidental drowning (E910)	1.05	0.72	0.74	0.80	1.41
Suicide (E950-E959)	1.39	1.06	0.90	0.87	1.06
Homicide (E960-E969)	1.93	1.00	0.86	0.78	0.88
Production injuries*	0.70	0.65	0.61	0.81	0.95
All causes	1.14	1.02	0.94	0.95	0.99

*Note: A number of causes of death (being struck by a falling object; accidents involving machinery; falls from ladders, scaffolds; being caught or crushed; and deaths involving an electric current) are known to be mostly work related. This combination of causes has been used as an indicator of production related death.

Source: AIHW, 1995.

Backyard swimming pools — a health hazard?

Drowning is the most common cause of death (16 per cent) among one- to four-year-olds in Australia. It is estimated that for every child who drowns, between four and ten children are admitted to hospital for near-drowning, and between five and 10 per cent of these will suffer some neurologic damage.

The drowning death rate for children aged one to four increased about 60 per cent between 1965 and 1972 at a time when, generally, child death rates were falling. This increase was associated with the construction of large numbers of above-ground and in-ground home swimming pools. By the early 1980s, child drowning rates varied more than 30-fold between different metropolitan settlements, linked strongly to the different legal requirements for fencing around pools.

Over the last decade, the introduction of legislation requiring fencing that isolates children from pools has achieved marked success. Accidental drowning rates among one- to four-year-olds have declined by nearly 50 per cent since the mid 1970s.

Table 3.18 Standardised mortality ratios for selected causes of death, by rural/remote settlement type, 1990–92

Cause of death	Rural			Remote		
	Large	Small	Other	Centres	Other	Indigenous
Infectious & parasitic diseases (1-139)	0.86	0.83	0.72	2.14	2.38	18.40
Cancers (140-239)	1.04	1.02	0.97	1.12	1.00	1.47
Digestive organs (150-159)	1.07	1.03	0.99	1.28	0.97	1.26
Lung (162)	0.95	1.06	0.92	1.12	1.17	1.76
Skin (172 & 173)	1.10	0.96	0.98	0.98	1.32	0.94
Breast (174)	1.04	1.04	1.00	0.69	0.76	0.84
Cervix (180)	0.92	0.89	0.93	2.22	2.47	9.98
Prostate (185)	1.09	1.07	1.13	1.06	0.97	0.19
Lymph, leukemia etc (200-208)	0.99	0.99	0.93	0.69	0.75	0.69
Diabetes mellitus (250)	0.97	1.11	1.13	1.88	2.25	10.63
Mental disorders (290–319)	1.04	0.95	0.93	1.69	1.15	5.01
Diseases of the nervous system & sense organs (320–389)	1.22	1.06	1.02	1.25	1.18	2.55
Circulatory system (390–459)	1.08	1.08	1.03	1.25	1.24	2.46
Acute rheumatic fever (390-398)	0.84	1.00	1.05	1.85	2.33	12.28
Ischaemic heart disease (410-414)	1.11	1.08	1.04	1.25	1.23	2.14
Cerebrovascular disease (430-438)	1.04	1.08	0.96	1.16	1.17	1.81
Respiratory system diseases (460-519)	1.01	1.10	1.02	1.79	1.89	8.89
Pneumonia, influenza & bronchitis (466, 494-496, 480-492)	1.02	1.12	1.03	2.00	1.96	10.27
Asthma (493)	1.13	1.16	1.18	0.86	1.09	1.78
Digestive system diseases (520–579)	1.05	1.04	1.06	1.55	1.69	2.70
Genitourinary system diseases (580–629)	1.08	1.06	0.98	1.94	1.76	12.14
Congenital anomalies (740-759)	1.10	1.14	0.96	0.95	1.06	1.78
Perinatal conditions (760-779)	1.03	0.90	0.94	1.36	1.23	4.48
Sudden infant death syndrome (798.0)	0.96	0.93	1.01	1.72	1.55	2.76
Injury and poisoning (E800-E999)	1.09	1.03	1.25	1.53	2.15	3.77
Motor vehicle accidents (E810-E819)	1.10	1.11	1.56	1.36	2.39	4.08
Accidental drowning (E910)	0.86	1.03	2.21	1.49	2.52	3.82
Suicide (E950-E959)	0.95	0.96	1.39	1.04	1.47	1.31
Homicide (E960-E969)	1.13	1.03	1.09	3.28	3.81	13.34
Production injuries*	0.84	1.09	0.77	1.84	2.75	2.62
All causes	1.06	1.06	1.02	1.33	1.36	3.34

*Note: A number of causes of death (being struck by a falling object; accidents involving machinery; falls from ladders, scaffolds; being caught or crushed; and deaths involving an electric current) are known to be mostly work related. This combination of causes has been used as an indicator of production related death.

Source: AIHW, 1995.

rate for infectious and parasitic diseases 79 per cent higher than the national average.

Although this urban variation is not as great as that between cities and remote areas, the very high standardised core-area mortality ratios for mental disorders (1.62), suicide (1.39) and homicide (1.93) could be linked to the pockets of disadvantaged people still resident in core areas, including the homeless who tend to congregate there.

The data on variations in mortality within rural and remote settlements (see Table 3.18) show that remote indigenous settlements have much higher mortality ratios for a range of causes of death.

Some types of settlement have standardised mortality ratios that differ significantly from the general pattern. In remote settlements, for example, skin cancer is 32 per cent higher than the national average, while in indigenous settlements it is six per cent less (as would be expected). However, the data highlight much higher general mortality rates for remote indigenous communities (apart from some cancers and suicide). Their extremely high levels of infectious and parasitic diseases for example, indicate the lack of basic and appropriate sanitation in many of them.

Key results of the 1992 ATSIC survey of national housing and community infrastructure needs indicate that 2760 people do not have a water supply that is maintained to an acceptable standard, 137 communities are without a sewerage system and 57 communities have sewerage systems not working satisfactorily (ATSIC, 1993).

Poor living conditions, inappropriate housing, poor nutrition, overcrowding and poor hygiene and a lack of basic services such as clean water and sewerage, contribute to high rates of infectious disease, rheumatic heart disease, respiratory disease, genito urinary diseases and cervical cancer.

Although many remote communities have maintained much of their social and cultural integrity, in some places the disruption of traditional indigenous society and chronic unemployment have led to high rates of alcohol abuse and petrol-sniffing, cigarette-smoking, accidents and violence.

Livability indicators

A set of social amenity and health indicators in any ongoing state of the environment reporting process would significantly assist in monitoring the state of livability in Australian settlements. The parameters and indicators set out in Table 3.19 could be used.

Table 3.19 Suggested indicators for social amenity and health

Parameter	Indicator
Wealth inequality	• percentage of private wealth owned by the richest 10 per cent of the population
Income inequality	• trends in (full) income inequality (see Johnson *et al.*, 1995) • female income as a proportion of male income • income of indigenous Australians as a proportion of national average
Unemployment	• total rate • youth rate • indigenous rate • median period
Education and training	• year-12 retention rates • proportion of workforce with post-secondary qualifications • public expenditure on education as a proportion of GDP
Housing	• percentage inadequately housed (as defined by Jones, 1994) • percentage in housing-related financial stress • percentage of indigenous households inadequately housed (as defined by Jones, 1994) • percentage of indigenous households in housing-related financial stress • national public housing waiting list

Accessibility and urban design	• modal split (%) for journey to work — or for all journeys:
	  – car
	  – public transport
	  – walking and cycling
	• local employment availability
	• local housing availability
	• design of new developments
	• percentages of medium and high density developments
	• percentage with public transport within 500 m
	• mix of office/retail/residential
	• parking space provision (sq m)
	• quality urban design (see Gehl, 1994a,b; PM's Urban Design Task Force, 1994)

Health	• life expectancy • infant mortality • cause-specific mortality rates • disability-adjusted life years lost (burden of disease)

Health effects of lead exposure

In recent years, the exposure of children to lead has emerged as a particular health problem. Australian and overseas research indicates a strong correlation between concentrations of lead in the blood of young children and neurological malfunction, learning disability and retarded mental development. The major source of airborne lead in most Australian urban areas is leaded fuel used in motor vehicles. (Exceptions occur in residential areas in close proximity to lead smelters in Port Pirie, Boolaroo and Broken Hill.) Other sources of lead exposure are soil, water, dust and deteriorating lead based paint. The introduction of unleaded petrol in 1985 and the reduction of the lead content in standard petrol in 1993 as part of the lead abatement strategy have considerably reduced the amount of lead being discharged to the atmosphere by motor vehicles.

The Australian Institute of Health and Welfare has recently performed the first national survey of blood lead levels in young children on behalf of the Commonwealth Environment Protection Agency. Although the result (93 per cent had blood lead levels below 10 micrograms per decilitre (μg/dL)) is within the NH&MRC goal for 90% of children to have blood lead levels below 10 μg/dL by 1998, it is still a matter of concern that seven per cent of children exceed the target level.

Figure 3.25 Sydney's resource inputs and waste outputs, 1990

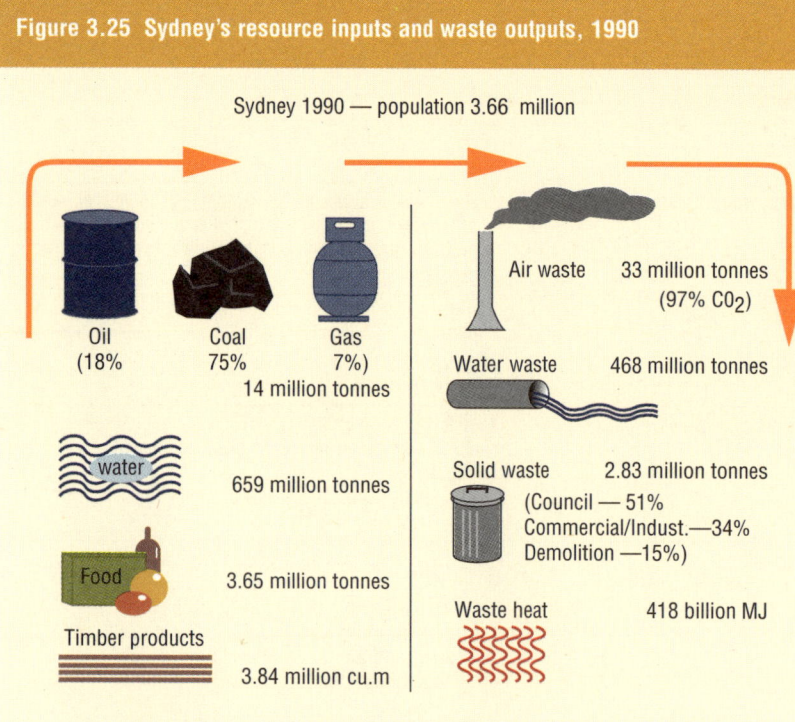

Sydney 1990 — population 3.66 million

Oil (18%) Coal 75% Gas 7%)
14 million tonnes

water
659 million tonnes

Food
3.65 million tonnes

Timber products
3.84 million cu.m

Air waste   33 million tonnes
(97% CO_2)

Water waste   468 million tonnes

Solid waste   2.83 million tonnes
(Council — 51%
Commercial/Indust.—34%
Demolition —15%)

Waste heat   418 billion MJ

Note: 1. Waste water data do not include stormwater and waste water outside sewerage system.
2. Timber products and food data derived from national per capita data.
3. 1991 water consumption and disposal data used.

Table 3.20 Trends in resource flows, Sydney, 1970 and 1990

	Sydney 1970		Sydney 1990	
Population	2 790 000		3 656 500	
Resource inputs per head				
Energy (MJ)	88 589		115 377	
Domestic		10%		9%
Commercial		11%		6%
Industrial		44%		47%
Transport		35%		38%
Food intake[a] (tonnes)	0.52		1.0	
Water (tonnes)	144		180	
Domestic		36%		44%
Commercial		5%		9%
Industrial		20%		13%
Agricultural/gardens		24%		16%
Miscellaneous		15%		18%
Waste outputs per head				
Solid waste (tonnes)	0.59		0.77	
Sewage (tonnes)	108.0[b]		128.0[c]	
Hazardous waste (tonnes)	n/a		0.04	
Air waste (tonnes)	7.6		9.3	
Carbon dioxide (kg)	7210.0		9050.0	
Carbon monoxide (kg)	204.9		177.8	
Sulfur oxides (kg)	20.5		4.5	
Nitrous oxides (kg)	19.8		18.1	
Hydrocarbons (kg)	63.1		42.3	
Particulates (kg)	30.6		4.7	

Notes:
(a). Derived from food sales data, not consumption data. It reflects an increased use of primary foodstuffs (eg grains) in the production of meat and processed foods.
(b). Includes stormwater,
(c). Waste water within sewerage systems only

Source: NSW Office of Energy, 1995; ABARE, 1991; 1993; ABS, 1993f; 1995e; EPA NSW, 1993; Nix, 1973; Butlin, 1976; SWB 1991a, b.

Metabolism in Australian settlements

The metabolism model of human settlements (as described at the beginning of this chapter), uses an integrated approach to assess both the state of the environment in those settlements and their effects on the wider environment. This section illustrates the performance and trends in Australian settlements with respect to these metabolic processes.

No comprehensive input and output data are available for most Australian settlements. These material flows could become standard indicators on urban environments for future State of the Environment reporting. Such indicators provide a start in reducing metabolic flows at the household, industrial and city-wide levels. No Australian settlement presently collects such data in a standardised way, with data on rural and remote settlements particularly hard to obtain. This makes it impossible to present a composite picture of the metabolic flows in Australian cities. An application of the metabolism model to a large settlement is presented below. It illustrates the resource and waste flows in Sydney for the period 1970 to 1990 (see Fig 3.25 and Table 3.20). Even for Sydney, data had to be collected from a variety of sources and in some cases had to be estimated.

Metabolism and scale

A look at Sydney's resource flows between 1970 and 1990 shows that not only did the city's total population grow by 31 per cent over the 20-year period but consumption and waste-generation per head increased as well (see Table 3.20).

Since 1970, levels of air pollutants per head have fallen — particularly emissions of sulfur dioxide and particulates. This is probably due to a combination of stricter pollution control on emission stacks (the major component) and improved technology, combined with a gradual decline in industrial activity within the city. Although the level of gases which form photochemical smog (nitrogen oxides and hydrocarbons) decreased per head, nitrogen oxides increased by 20 per cent in total. The level of hydrocarbons fell by 33 per cent on a per head basis but only 12 per cent in total. These emissions are primarily from motor vehicles (see Chapter 5).

Some gains are only made at the cost of something else; for example, carbon monoxide levels have been reduced by better and more efficient motors converting carbon monoxide into more carbon dioxide — although this is a definite health and amenity improvement in cities it still contributes to greenhouse gas emissions.

Between 1970 and 1990, Australia's average consumption of primary energy per head increased by 37 per cent to 156 567 MJ, compared with 30 per cent for Sydney over the same period. This difference might be due to urban economies of scale and the high level of mineral processing outside Sydney. By 1990, the city's annual emissions of carbon dioxide had risen from 7.21 to

9.05 tonnes per head — but this is about half the New South Wales level of 18 tonnes.

As settlement size increases, resource-use efficiency also increases. This is probably due to a combination of factors, including the availability of waste-recycling facilities, bigger markets for recycled products, greater access to globally innovative technology, easier access to more-efficient forms of energy generation and greater population size and density allowing for greater economies of scale, better public transport and more efficient use of land (see Table 3.21).

The table highlights some interesting differences between the two remote settlements, which reflect their differences in both cultural attitudes and economic base. The mining settlement of Yandicoogina is considerably more energy-, water- and waste-intensive than the indigenous community described earlier (see pages 3-16 and 3-17), because it is trying to provide all the comforts of the city for a work force in a remote and arid location, without the benefits of scale to do so efficiently.

Resource inputs and their indicators

Water

Australia is the driest inhabited continent, yet we have one of the highest total water consumption levels per head by international standards (OECD, 1995). So it is important for Australian human settlements to minimise their water use. After irrigation, the next biggest water use occurs in the urban areas of large cities, where both consumption and storage are very high (due to Australia's low and erratic rainfall, and the prevalence of lawns and gardens in our urban areas) (AURDR, 1995a). For example, Sydney has to store 930 cubic metres per head compared with 250 in New York and 18.2 in London (Munro, 1974).

While urban water supplies must cater for industrial and commercial needs, the major demand for water is domestic. Levels of consumption vary significantly across the major cities according to rainfall, mean temperatures and humidity, availability of water, water pricing and education. Within the city, data on water use suggest that the size of the block of land is the most important factor affecting consumption because of garden- and lawn-watering. Thus, core and inner areas can consume two or three times less water per head than outer suburbs (Mouritz and Newman, 1995).

As the figures on Sydney illustrate (see Table 3.20), domestic water consumption has increased significantly over the past 20 years because of rises in both population and individual consumption rates, which can be attributed to an expectation of higher living standards and the increased proportion of houses in outer areas. Adelaide suffers real constraints to water supply and increasing water abstraction is an environmental issue for both Perth and Sydney.

An industry's water use can be monitored on the basis of its consumption per unit of GDP output. However, most industries do not have such data readily available.

Energy

Resource capacity

Australia is well endowed with a wide range of energy resources and is climatically well placed to exploit renewable energy opportunities (see Table 3.22).

Our energy consumption per head is a little higher than the average for OECD nations. In terms of energy per unit of GDP, Australia's position has only marginally improved since 1970, while Canada, the United States and the United Kingdom have all improved by more than 30 per cent. From the mid 1980s until the early 1990s,

Table 3.21 Selected per capita resource flows in four different-sized settlements, 1990

	Sydney pop. 3 656 500	Warrnambool pop. 24 720	Indigenous community pop. 300–400	Yandicoogina pop. 79–159[a]
Resource inputs (per head)				
Water (tonnes)	180	182	241	946
Food (tonnes)	1 [b]	1 [b]	1.07 [c]	0.74 [c]
Energy (MJ)	115 377	102 997 [d]	29 000	177 630
Waste outputs (per head)				
Solid waste (tonnes)	0.77	0.94	0.2	1.58
Sewage (tonnes)	128	104	n/a	n/a

Notes:
(a) 1992 figure (first full year of operation).
(b) Based on average Australian food consumption/head.
(c) Based on settlement stores inventory.
(d) Includes electricity use figures for 1991.

Source: NSW Office of Energy, 1995; ABS, 1993f and 1995e; City of Warrnambool, 1995; SWB, 1991 a and b; Newman *et al.*, 1994; EPA NSW, 1993; Vic DEM, 1995; CAT, 1995; BHP Pty Ltd, 1995.

Table 3.22 Australian energy reserves, production and use, 1991–92

	Demonstrated economic resources (PJ)	Production 1991–92 (PJ)	Total domestic use (PJ)
Black Coal	1 400 000	4 647	1 176
Brown Coal	410 000	497	497
Crude Oil	14 000	1 158	1 441
Natural Gas	37 000	931	677
Uranium	263 000	2 223	–
Renewables[1]	(not estimated)	230	230

Note: 1. Comprises wood, bagasse from sugar cane, hydro-electricity and domestic solar hot water heaters only. Official statistics do not currently count other uses of solar energy, for example, commercial hot water, building and swimming pool heating, photovoltaic power for communications, salt drying and clothes drying.

Source: ABARE, 1993.

Table 3.23 Energy consumption in Australia by sector, 1991–92

	Energy consumed (PJ)	Proportion of total energy consumption(%)
Agriculture	58.5	1.5
Mining	172.1	4.3
Manufacturing	1068.2	26.7
Electricity Generation	1114.0	27.3
Construction	38.8	1.0
Transport	1016.8	25.4
Commercial	152.3	3.9
Residential	334.4	8.4
Other	48.3	1.6
Total	**4003.2**	**100**

Source: ABARE, 1993.

Table 3.24 Comparison of domestic energy consumption in Australian cities with other national data for mid 1970s

Domestic energy use per head (GJ)	
Canada	62.13
United States	55.85
Netherlands	40.87
United Kingdom	27.87
Hobart	26.20
France	23.89
Canberra	23.10
Melbourne	**20.00**
Italy	15.84
Adelaide	**14.82**
Japan	13.15
Sydney	**12.65**
Brisbane	**8.99**
Perth	**8.74**
Hong Kong	4.06

Source: Newman, 1982.

Table 3.25 Trend in domestic energy use by fuel in Sydney, 1976 and 1993–94

Fuel Source	1976 energy use (GJ/head)	1993–94 energy use (GJ/head)
Electricity	7.29	9.28
Gas	3.27	2.44
Oil	1.76	1.35
Coal	0.29	0.12
Wood	0.05	3.01
Total (incl wood)	**12.65**	**16.2**
Total (excl wood)	**12.60**	**13.19**

Source: Newman, 1982; NSW Office of Energy, 1995.

Australian energy efficiency actually declined by about four per cent (AURDR, 1995a; ABARE, 1993). Thus we have considerable potential to improve our energy efficiency, and still improve our quality of life.

Table 3.23 shows that manufacturing, electricity generation and transport account for 79 per cent of energy use in Australia. Electricity is not constrained in terms of the base resources — either fossil fuels or renewables — but its contribution to air quality and greenhouse gas emissions is significant.

Domestic energy

Domestic energy use varies mostly with climate and to a lesser extent, household income, and household size (Newman, 1982; ABARE, 1993) (see Table 3.24). It is dominated by water- and space-heating energy services, followed by refrigeration and cooking. In 1991–92, the residential sector accounted for 12 per cent of total energy production in Australia. Owing to transmission wastage, this translates to eight per cent of energy end-use (see Table 3.23).

A range of fuels supply domestic energy (see Table 3.25). Sydney increased its domestic energy

Figure 3.26 Australian domestic fuel use, 1991–92

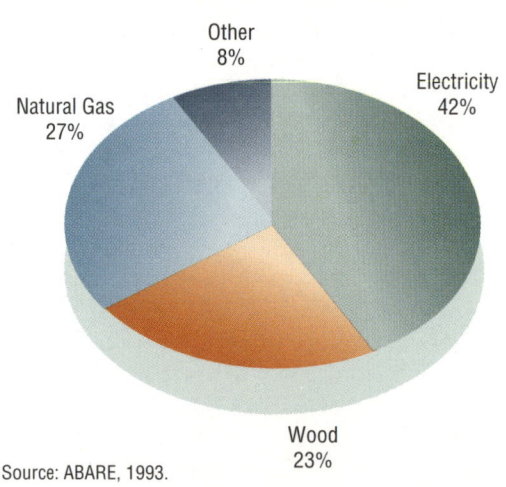

Source: ABARE, 1993.

Table 3.26 Per head fuel consumption by urban region in Melbourne

	inner areas	middle areas	outer areas
Distance from CBD (km)	5.5	15.5	38.8
Urban density (people per ha)	31.9	20	9.6
Jobs to population ratio	0.8	0.3	0.2
Car trips to work (%)	57.7	74.5	82.1
Gasoline use per person (MJ)	13 244	20 303	26 881

Source: Newman and Kenworthy, 1990.

consumption by 27 per cent per head from 1976 to 1993–94 largely due to an increase in the use of wood for heating and electricity. By comparison, New South Wales increased its energy consumption by 80 per cent over the same period (NSW Office of Energy, 1995).

Transport energy

Road transport accounted for one-quarter of the energy consumed in Australia in 1990–91 (ABARE, 1993). Of this, urban areas used 63 per cent directly and much of the rest was consumed in satisfying the requirements of urban areas or travelling between them (AURDR, 1995a). Australia has 20 per cent higher fuel use per head than the OECD urban average (OECD, 1994). Our vehicles have a poor average fuel economy of 11.8 L/100 km, which did not improve between 1971 and 1991. By contrast, the vehicle fleet efficiency in the United States improved by 29 per cent, to 10.8 L/100 km, between 1980 and 1988 (AURDR, 1995a).

Transport fuel use varies in Australian cities by around 15 per cent, with Perth at the high extreme and Melbourne and Sydney at the other (see Table 3.27). Larger cities generally have lower fuel use per head (Naess, 1993) because of better transit and higher urban densities.

Rates of fossil fuel consumption per head are twice as high in the outer regions of cities, where most people live (see Table 3.26). However, people feel the impact of these levels of car dependence everywhere in the city.

Table 3.28 shows how jobs in Australian cities have grown in the city core, but much less than in the rest of the city which is often often difficult to reach by public transport. This has implications for sustainable urban development. However, the trend may be offset by a greater proportion of local jobs

and by the development of nodal/information centres in the suburbs which are more public-transport-oriented.

Over the last three decades, the rate of growth of car use in Australian cities has declined compared with American trends although car usage is still high compared with that in European and Asian

Table 3.27 Transport fuel use per head in Australian cities compared with selected international cities

City	Motor Vehicles		Public Transport		Total
	Petrol (GJ)	Diesel (GJ)	Diesel (GJ)	Electric (GJ)	(GJ)
Houston	74.51	9.40	0.29	0	84.20
Los Angeles	58.47	6.40	0.65	0	65.50
New York	44.03	6.17	0.55	1.34	52.09
Perth	32.61	6.82	0.82	0	40.25
Brisbane	30.65	5.80	0.80	0.06	37.30
Adelaide	28.79	5.40	1.01	-	35.21
Melbourne	29.10	4.54	0.25	0.35	34.24
Sydney	27.99	5.69	0.66	0.32	34.66
Toronto (Metro)	22.67	7.93	0.98	0.51	32.09
Hamburg	16.67	5.94	0.15	0.31	23.07
London	12.43	5.17	0.64	0.48	18.72
Amsterdam	6.84	2.94	0.59	0.10	10.47
Tokyo	8.49	4.58	0.25	0.72	15.35
Singapore	5.96	3.50	1.56	0	11.04
Hong Kong	1.99	2.56	0.79	0.07	5.41

Source: Newman and Kenworthy, 1989; Newman, 1994; plus new data on Toronto from Kenworthy and Newman, 1994.

Table 3.28 Changing work place destinations in Australia's major cities

	Worked at home	City core	Local area[1]	Other metro area	Total
1981 (% of all jobs)	3.93	31.17	29.12	35.28	100
1991 (% of all jobs)	3.89	27.12	30.28	38.66	100
Absolute growth 1981–91 (jobs)	21 804	14 966	219 580	342 087	

Note: 1. Equivalent to a Statistical Local Area
Source: Gipps *et al.*, 1994.

Table 3.29 Trends in use of cars and transit in global cities

		US cities	Australian cities	Canadian cities	European cities	Asian cities
Car Use (vehicle kilometres travelled per head)	1970	7 334	4 628	na	2 750	470
	1980	9 168	5 850	4 807	3 798	531
	1990	11 559	6 589	5 680	4754	1178
Public Transport (trips per head)	1970	48	118	154	249	418
	1980	57	93	202	290	430
	1990	64	91	210	359	535

Source: Newman and Kenworthy, 1989; Kenworthy and Newman, 1993; 1994.

Australian cities are high transport energy consumers due to low efficiency cars, and heavy car dependence.

Urban remnant vegetation is much valued. Perth is particularly fortunate in having a large amount of native vegetation — Kings Park and Botanic Gardens — in its core area.

cities. Our level of public transport use has stabilised, and remains higher than in the United States, although it is much less than the levels in European and Asian cities, which have grown sharply (see Table 3.29). However, if the pattern of dispersed fringe and coastal development continues, along with extensive freeway development, it may well be that Australia will follow the American pattern of rapidly increasing car dependence.

Constraints on transport fuel depend on global oil availability, which is far more constrained as a resource than electricity. Analysts predict that a global peak in production will occur in the early part of next century (Fleay, 1994). Australia's oil production peaked in 1994–95 and is expected to decline rapidly thereafter (AIP, 1995).

Food

The resource flow to produce food sees nutrients and energy being taken from predominantly non-urban areas for predominantly urban consumers. The remaining nutrient load is rarely returned to the area of production, but is usually discharged into the ocean or local urban environments.

Australia's food production feeds more than just Australians, although the proportion produced for export has fallen from 68 per cent in 1967 to 52 per cent in 1992. During this period, the amount of food produced increased from 214×10^9J in 1967 to 440×10^9J in 1992. At the same time, Australia's population also rose, by 5.7 million, and per capita food consumption increased by more than 70 per cent (Newman *et al.*, 1994). However, actual calorific intake has remained stable throughout this period (ABS, 1993f). One factor in this surprising equation is the increase in feedlotting; another may be food wastage occurring in the processing of food.

Raw materials and forest products

Generally, Australian settlements are unconstrained by the availability of basic materials like steel, aluminium and bricks. However, our forests provide only 92 per cent of the national demand for wood and wood products. This figure is predicted to decline further unless large-scale plantation-growing occurs (Newman *et al.*, 1994).

Increasing efficiency, more-efficient product design and the use of recycled and alternative low-energy materials are some strategies that may minimise the use of scarce resources while still improving amenity. Few data are presently available to enable the development of indicators of resource input. The most obvious constraint on settlements is the availability of firewood for remote indigenous communities, where many people must now travel more than 50 km to get wood for domestic purposes.

Table 3.31 Remnant vegetation of rural areas of Western Australia

	Area of remnant vegetation (ha)	Proportion of total area (%)
Coastal shires	852 721	8
Inland shires	1 328 934	7

Source: WA Ministry for Planning, 1995

Table 3.30 Open space per head in Melbourne

Melbourne city region	Total open space per head (sq m)
Core	61
Inner	45
Middle	62
Outer	105
Fringe	225

Note: For every new person added to the city, about eight per cent of land lost is turned into public open space.

Source: Kenworthy and Newmann, 1991

Table 3.32 Remnant vegetation in regions of Perth

City Region	Area of remnant vegetation (ha)	Proportion of total area (%)
Core	385	24
Inner	735	5
Middle	2512	9
Outer	258 416	53
Total	**262 048**	**49**

Source: WA Ministry for Planning, 1995

Land

The loss of land to urban development often indicates a loss of productive agricultural land or of important regional bushland. Monitoring the rate at which land is converted to urban use thus provides an important environmental indicator. Between 1961 and 1971, Australian cities consumed 1042 sq m per person for each unit of population increase. The loss increased to 1207 sq m between 1971 and 1981 (Newman *et al.*, 1992). This is very high by world standards, reflecting Australian cities' low density. It is not surprising therefore to find substantial loss of biodiversity near large Australian cities. In Sydney only a few per cent of the original forest remains, and about 400 of the 900 native plant species in western Sydney are endangered (Benson and Howell, 1990; Metro Strategy, 1992). Around Brisbane, only 600 ha of the original 6000 ha of rainforest and only 450 ha of the original 13 000 ha of melaleuca woodland remain (ESS, 1988).

In coastal areas, the loss is serious as the land's nearness to the water's edge makes it generally more ecologically sensitive. Apart from the land resources consumed in the fragmented spread of Australian settlement along extensive sections of coastline, nearby development can seriously affect not only foreshore and littoral environments but intertidal and nearby marine coastal waters as well. South-east Queensland, for example, lost 33 per cent of its coastal bushland to development between 1974 and 1989 (Catterall and Kingston, 1993). Around Perth, 70 per cent of the original wetlands have been cleared; and drainage, filling or mining have affected the remainder (GOWA, 1992).

These patterns of coastal development have serious implications for coastal biodiversity. CSIRO estimates, for example, that 60 per cent of Queensland's rare, threatened or endangered plants lie in the urban growth areas of the State's south east (Hamilton and Cocks, 1994). Even so called 'sensitive' development poses risks to the integrity of remaining natural ecosystems, with the potential introduction of feral animals and invasive weeds, possible changes to drainage patterns and soil structure from building and road construction and altered nutrient levels from run-off and septic tanks. Development close to undisturbed areas can create 'edge effects' that, over time, can seriously affect the integrity of natural ecosystems by providing an opportunity for some of these destructive elements.

Retaining land for open space in a city can be important for managing biodiversity as well as providing a valuable source of human amenity. The area of open space per head is best monitored on a regional basis (see Table 3.30). Although small in area, many significant tracts of bush still occur in Australian cities. Table 3.31 shows the amount of remnant vegetation in each region of Perth and the proportion of the total land area.

Perth is particularly fortunate in having a large amount of native vegetation. The core area has 24 per cent, due to Kings Park and Botanic Gardens,

Table 3.33 Suggested indicators for resource inputs to human settlements

Resource inputs	Environmental indicators for urban, rural and remote settlements
Water	• Per head water consumption - domestic, commercial and industrial • Industrial water consumption per unit of GDP output • Water quality - domestic, recreational and industrial
Energy	• Per head energy consumption • Per head transport fuel consumption • Per head consumption of imported oil • Industrial energy consumption per unit GDP output
Food	• Per head food consumption (Kcals and gms protein)
Raw Materials	• Per head consumption of non-forest building materials (bricks, steel, aluminium)
Forest Products	• Per head consumption of forest products (including firewood) • Per head consumption of imported forest products
Land	• Per head consumption of urban land • Per head availability of open space • % remnant vegetation • Area of contaminated land

which was set aside in the early part of this century as a visionary project to preserve native bush next to the city centre. Perth's inner and middle suburbs have some also and the outer suburbs still have a lot of remnant vegetation.

Most of the shires in the south west of Western Australia have been almost totally cleared for agriculture, and often do not have as high a proportion of remnant vegetation as the metropolitan area (see Table 3.32). This highlights the importance of managing remnant bush in urban areas as well as the critical state of remnants in country ones.

Land contaminated by industry or hazardous waste is a major resource that rarely gets accounted for properly. Programs to bring some of this land back into use — in particular the Better Cities program — have been successful at pioneering techniques in the rehabilitation of contaminated land.

Waste outputs and their indicators

Industrial waste

Pollution from industrial waste can be measured either by quantifying and categorising emitted wastes according to type and toxicity, or by measuring the cumulative levels and impacts of discharged waste in air, water and soil. Limited data are available for major industries and particular factories, and from pollution monitoring sites in major centres (ESD Working Group, 1991). Recently, a number of studies have attempted to quantify and categorise waste generated and disposed of by industry. Such studies have relied on records kept by State government environmental protection and waste management authorities, and on surveys of industry. All these studies identified large data gaps, highlighting the need for more detailed record-keeping.

Figure 3.27 Proportion of industrial waste by weight

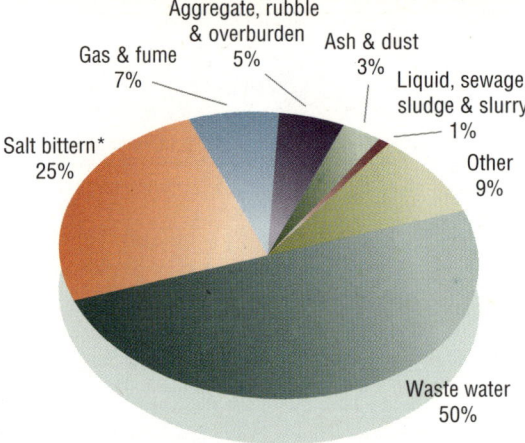

- Aggregate, rubble & overburden 5%
- Ash & dust 3%
- Gas & fume 7%
- Liquid, sewage, sludge & slurry 1%
- Salt bittern* 25%
- Other 9%
- Waste water 50%

Note: These are approximations only.
*The quantity of salt bittern (a calcium carbonate, gypsum, potassium and magnesium-salt rich by-product of rock and table salt) recorded by the CSIRO survey came almost entirely from two sources, both salt production plants. These wastes are disposed of to the ocean, and contain potentially valuble products such as potash and magnesium.

Source: Beretka and Whitfield, 1993

Table 3.34 Heavy metal surveys of mussels in Corio Bay, Victoria, 1976–78 and 1987–88

	Amount of heavy metal (µg/g dry weight)	
	1976–78	**1987–88**
Cadmium	17.9	7.8
Chromium	Not available	1.0
Copper	5.8	5.3
Iron	374	267.7
Manganese	12.0	8.9
Nickel	Not available	2.3
Lead	5.9	< 7*
Zinc	219.1	180.5

*Note: Lead samples in 1987–88 were only recorded as 'less than 7 µg/g'

Source: Nicholson, *et al.*, 1992.

Table 3.35 Liquid hazardous wastes in five Australian cities

	Tonnes/year	Rate (kg/person)
Sydney	62 000	18
Melbourne	90 000	30
Brisbane	17 000	14
Adelaide	40 000	40
Perth	26 000	24
TOTAL	235 000	av. 24

Note: Much of New South Wales' process industry now occurs outside the Sydney area.

Source: Samuel, 1989.

Table 3.36 Hazardous waste generation by sector in Sydney, 1990

Sector	Contribution to Hazardous waste (%)
Manufacturing	76.7
Electricity gas and water	1.9
Construction	0.6
Wholesale & Retail Trade	4.6
Transport and Storage	7.1
Public administration & defence	5.0
Community services	2.8
Other	1.3

Source: EPA NSW, 1993.

Of 1000 Australian companies with 10 or more employees surveyed by CSIRO in 1993, 354 responded. These businesses generated more than 25 000 000 tonnes of waste per year. Waste water accounts for about half of all industrial waste (see Fig. 3.27). Impacts on the environment are being reduced by cleaner production techniques which reduce inputs and outputs from industry.

The Victorian EPA has indicated that discharge conditions for certain heavy metals in Victorian waterways have been tightened in recent years. The gradual decline in the levels of heavy metals in mussel tissue collected from Corio Bay, for example, reflects the effectiveness of these controls (see Table 3.34). It also illustrates how discharges licence conditions can yield valuable information regarding contaminant levels of water discharged by industries.

A number of changes in Australian settlements are affecting waste streams. These include: restructuring; moving of old industries to offshore locations and partly to new rural locations; the start of new value-added manufacturing with associated wastes; and the move to cleaner production, which is reducing waste streams generally.

In a similar way to measuring resource-use efficiency, we can measure the intensity of waste generation in terms of the quantity of waste generated per unit of product manufactured. Such indicators should be developed for state of the environment monitoring of Australian settlements.

Historically, most manufacturing industries have been concentrated in or near urban centres. The recent trend is to relocate these industries to regional and rural centres (Beer *et al.*, 1994), which has numerous implications for their future environmental management. Pressures resulting from these industries may have quite different impacts in regional (often inland) centres from those in metropolitan (coastal) centres. For example, unless best-practice technology is used, trade waste disposed to sewerage systems discharging to inland waters is likely to have a greater impact on the environment than that discharged to ocean outfalls, where greater dilution occurs.

There is considerable variability in the level of liquid hazardous waste generated between the State capital cities, reflecting their different levels of industrialisation. Likewise, a number of different types of hazardous wastes occur. In 1986 the Commonwealth Government introduced national guidelines for their classification, but again, accurate centralised records of the quantities of waste generated are not kept. Manufacturing is the greatest source of industrial hazardous wastes in Sydney (see Table 3.36). Those wastes that cannot safely be recycled, or disposed of directly to landfill or via sewerage, require on-site storage and transporting to specialist treatment, disposal or storage facilities.

Solid waste

Municipal solid waste comprises wastes produced by households, commercial and industrial premises, building and demolition operations, and in the provision of services and maintenance of streets, public spaces and utilities.

Australia has a much higher production of municipal solid waste per head than the OECD average — 681 compared with 513 kg per year (AURDR, 1995a), and is second only to the United States in its per capita production of domestic solid waste (see Table 3.37).

A CSIRO study of the flow of construction materials in Melbourne has found that 69 per cent is already being re-used, reprocessed or reduced (broken down) for use in building and non-building activities (for example, road base) (see Fig. 3.28). For total municipal solid wastes, estimates suggest that only between six and seven per cent was being recycled in 1989 (AURDR, 1995a). However, more recent data show some local authorities have achieved a higher recycling rate than this, which indicates considerable potential still exists. Sydney households reduced their weekly solid waste from 21.6 kg in 1993 to 18.6 kg in 1995; and increased their weekly amount of recyclables from 2.8 kg to 3.8 kg (LRRA, 1995). In 1994, large cities recycled 50 per cent of newspaper (up from 44 per cent in 1990). The main potential is garden waste and food scraps which make up 50 per cent of kerbside waste. Many councils see home composting as the preferred option for this material.

Waste in urban, rural and remote areas

Even though there are robust data on solid waste generation and disposal at the regional level, these data have yet to be harmonised for Australia. A common classification system has been agreed but has yet to be implemented. The Industry Commission (1990) estimated that in 1989, the average amount of collected household waste was lower in the State capital cities (336 kg per head per year) than in the rest of Australia (427 kg per head per year). This provides supporting evidence that larger cities have a more efficient metabolism than smaller settlements.

Waste in landfill emits greenhouse gases and leachate that can contaminate groundwater and damage waterways. Methane from landfills is

◄ Australian cities have high solid waste generation rates. Although recycling is growing there is much more to do.

Table 3.37 Selected national waste indicators in the late 1980s

Indicator	Aust	Canada	USA	OECD	Aust/OECD ratio
Municipal solid waste (kg/per head/yr)	681	632	864	513	1.327
Industrial waste per unit of GDP (tonnes/$USmillion)	146	155	186	146	1.0

Source: AURDR, 1995a.

Figure 3.28 Construction waste flow from Melbourne central business district

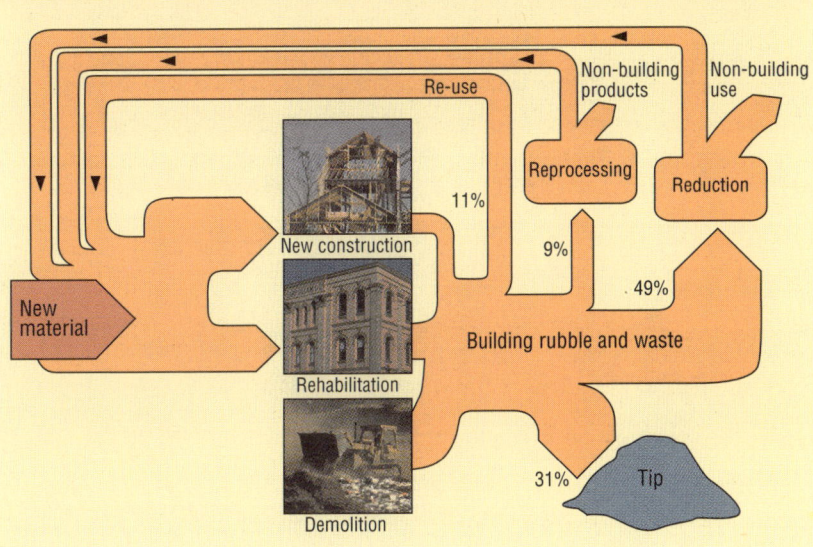

Source: Solomonsson and MacSporan, 1994.

Most coastal settlements discharge sewage without tertiary treatment. The near ocean environment absorbs 10 000 tonnes of phosphorus and 100 000 tonnes of nitrogen per year.

estimated to contribute 5.1 per cent (in carbon dioxide equivalents) of Australia's total greenhouse gas emissions (NGGIC, 1994). Increasing numbers of landfill sites are being used to tap methane for energy.

Remote communities' problems associated with waste differ from those in other settlements (CAT, 1991). In many of them, the costs of landfills and waste-disposal services are prohibitive, and so above-ground dumping or burning are common. Virtually all solid waste comes from domestic sources and the maintenance of structures and living areas. The waste stream consists largely of packaging and other post-consumer waste such as fuel drums, appliances, tyres and vehicles. Often the inadequacy of maintenance services results in 'repairable' appliances and vehicles being disposed of or dumped.

The solid wastes generated by remote communities affect visual and social amenity and, in some instances, human health. The most common method of waste disposal in these communities is trench landfill, but many settlements do not have operational bulldozers to backfill and cover tipped waste.

It is uneconomical to pick up re-usable and recyclable items in most such places, so virtually all materials transported into them eventually become waste. This problem presents a challenge to the providers of products as well as those trying to minimise the effects of waste in the communities.

Sewage

Sewerage systems are designed to convey water contaminated by urban activities to places where it can be disposed of without jeopardising human health. Consequently, human settlements can have a marked effect on the environmental quality of aquatic systems both within and beyond settlements. Although large urban centres have the highest proportion of population connected to deep sewerage, this varies from about 75 per cent in Perth to 98 per cent in Sydney. All urban centres in Australia with populations of more than 500 000 people are located on the coastal fringe of the continent and discharge most of their effluent to the ocean or tidal estuaries (DPIE, 1991). They generally provide primary treatment for effluent discharged via long outfalls, and secondary treatment prior to discharge via short outfalls. This means that each year around 10 000 tonnes of phosphorus and 100 000 tonnes of nitrogen are discharged to the near-ocean environment. Public opposition to such outfalls is growing.

Sewage-treatment facilities serving smaller settlements generally achieve higher levels of treatment than those serving larger populations (see Table 3.38). This is due, in part, to difficulties in treating large volumes of sewage, but also to the availability close to large urban centres of the ocean which can disperse large volumes of effluent (see Fig. 3.29). Smaller regional centres and rural towns more frequently dispose of waste to inland waterways or land, where health and pollution considerations demand higher levels of treatment prior to discharge.

Table 3.38 Level of sewage treatment and size of settlement

Population size served by sewerage	Proportion of population (%)	Proportion of population receiving particular level of sewage treatment (%)			
		Untreated	Primary	Secondary	Tertiary
5 000–20 000	3.2	1.1	4.0	82.3	12.6
20 000–100 000	10.2	–	8.0	74.5	17.6
100 000–500 000	23.0	–	25.9	62.9	11.2
>500 000	63.6	–	49.2	50.8	–
National average	–	<0.1	38.2	57.0	4.8

Source: Extrapolated from DPIE, 1991. This study surveyed the method and level of sewage treatment for different sewage treatment facilities across Australia.

Figure 3.29 Level of treatment for different methods of sewage disposal

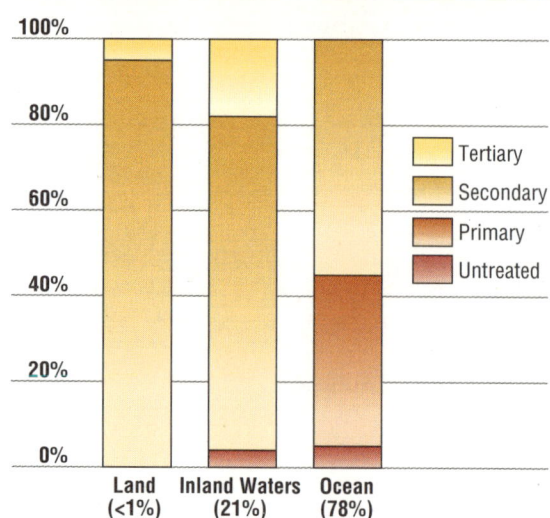

Legend: Tertiary, Secondary, Primary, Untreated

Land (<1%) Inland Waters (21%) Ocean (78%)

Source: DPIE, 1991.

Small inland towns contribute significantly to eutrophication of rivers and need to adopt tertiary treatment. This can be a significant constraint on future growth.

Many small coastal settlements (populations less than 5000), discharge untreated sewage into the ocean (see Table 3.38), usually into high-energy waters. The cumulative impact of this on fast-growing coastal areas is of considerable concern (Zann, 1995).

Sewage overflows are a major source of pollution. Sydney is estimated to have between 6000 and 10 000 sewerage overflow points where overflows may occur between 50 and 100 times per year (O'Loughlin *et al.*,1992). Population growth and consequent increased sewerage connections, and/or flooding due to rainfall, can exceed the capacity of facilities and result in the discharge of untreated sewage to waterways.

Major pollutants in sewage

The discharge to sewers of heavy metals and toxic materials by domestic and commercial/industrial premises can damage aquatic and marine environments and reduce the potential for recycling sewage waters and sludges for irrigation and compost.

Government agencies regulate the types and quantities of heavy metals and toxic wastes discharged to sewer by industry to forms and concentrations considered low enough to avoid damage to the environment. Wastes often need to be pre-treated. Most discharges from industrial sewerage consist of wash-waters and non-toxic suspended solids, which are readily treated by sewage-treatment facilities and pose little environmental threat. However, in many (particularly urban) sewerage systems, the cumulative effects of heavy metal and toxic pollutants remain at levels that may reduce the potential for recycling.

Domestic disposal of unwanted household hazardous waste — such as pesticides, paints, solvents, detergents, acids, alkalis and oils — also contributes to this problem (see Table 3.39). Of particular concern is the discharge of old stocks of banned hazardous chemicals such as organochlorides (for example, DDT) and arsenate pesticides. No one knows accurately the scale of this problem, but a typical household is thought to generate between 1.5 and 2.0 kg of hazardous waste per year (Melbourne Water, 1992).

In many cases it is prohibitively expensive to collect, transport and store these wastes. Currently, no waste-minimisation strategies exist to reduce the quantities of hazardous waste that households generate.

Nutrients such as phosphorus, which are introduced into the waste stream via household cleansers and detergents, have a major impact on Australia's coastal and marine ecosystems, as these are adapted to low-nutrient conditions (EPA Vic., 1992).

Table 3.39 Household hazardous waste recovered and destroyed between October 1990 and September 1992 in Melbourne

	Quantities collected (kg)	Destroyed by Sept 1992 (kg)
Arsenical compounds	617	0
Heavy metals	1 567	901
Poisons	1 494	849
Organochlorines	4 773	111
Other pesticides	6 340	0
Oils, paints and solvents	51 886	37 162
Acid and alkalis	3 366	2 914
Other	19 533	16 735
Total	**89 576**	**58 672**

Source: Melbourne Water, 1992.

Stormwater

Human settlements dramatically increase stormwater run-off from land. Streets, roofs and other cleared and sealed surfaces channel surface run-off, increasing the susceptibility to flooding and creating other stormwater-management problems down-stream. Traditionally, drains and other engineering works have been constructed to rapidly drain and dispose of stormwater. This approach exerts pressures on the natural and human environments by causing:

• downstream flooding

• erosion and turbidity

• contaminantion of waterways

• loss of recharge and alteration of groundwater hydrology

• overloading of sewerage systems

• loss of the fresh-water resource

This report deals with the effects of stormwater on the local environment and receiving waterways in detail in Chapter 8.

Stormwater is a major cause of water pollution. Diversion through artificial wetlands creates local amenity and habitat for wildlife.

▼

Most waste-water management systems are designed to cope with peak flows rather than provide accurate flow measurement. Given the pollution potential of waste water (in particular, stormwater), the scale of the problem needs to be accurately determined to enable adequate management. The crisis facing many settlement water supplies has prompted a number of demonstration projects attempting to turn stormwater from a pollution problem into a water resource (DEST, 1992).

The re-urbanisation of cities can increase the amount of impervious land surface and thus cause increased stormwater problems. Cities are recognising the need to retain if not increase the proportion of porous surfaces as well as establishing water-sensitive design through techniques like stormwater swales, holding basins and treatment areas such as artificial wetlands. The rehabilitation of urban creeks, which in the past have mostly been seen only as drains, is now taking place in our settlements.

Air waste

The main air wastes from Australian settlements are carbon dioxide, carbon monoxide, nitrogen oxides and volatile organic compounds (see Chapter 5). All, except carbon dioxide, are potentially toxic to humans and other living organisms. Motor vehicles are the dominant source of air waste in most Australian cities. In the case of Melbourne in 1990, winter week-day emissions from motor vehicles were responsible for 73 per cent of nitrogen oxide, 70 per cent of carbon monoxide, 40 per cent of volatile organic compounds, 11 per cent of sulfur dioxide and 11 per cent of particulates emitted (Carnovale *et al.*, 1991). The energy sector is the largest contributor to greenhouse gas emissions in Australia. Within this sector, motor vehicles contribute 18 per cent, which is close to twice that of industry, making them the single largest end-use source of greenhouse gases. (Electricity generation is by far the highest primary source) (NGGIC, 1994). No one has yet undertaken a full assessment of greenhouse gas emissions resulting from vehicle manufacture and road construction.

Photochemical smog (produced by nitrogen oxides and volatile organic compounds reacting in the presence of sunlight) has become a major problem in cities. It is related to particular climatic conditions (see the box on page 5-25). Although new vehicle technology is helping to reduce levels of smog, projections are that it will increase unless the growth in motor vehicle use is curtailed. The airsheds of Sydney, Melbourne, Brisbane and Perth are showing constraints to any further growth in photochemical smog — principally toxic, tropospheric ozone.

Carbon dioxide emissions from the production and use of energy in Australia are estimated to have increased by nearly eight per cent between 1987–88 and 1990–91. Most of this increase occurred in electricity generation, which is predominantly sourced from coal and accounts for 48 per cent of energy-sector carbon dioxide emissions. Petroleum products account for 34 per cent and natural gas 13 per cent (ABARE, 1993). In 1990, carbon dioxide accounted for 73 per cent, methane 23 per cent, nitrous oxide three per cent and other emissions one per cent of total greenhouse gases (NGGIC, 1994). (Note, the National Greenhouse Gas Inventory (NGGI) does not include chlorofluorocarbons (CFCs) which are strong greenhouse gases. These are better known as substances that deplete beneficial, stratospheric

---

Table 3.40 Estimated quantities, percentage composition and proportion of industrial emissions contributing to total air waste in the Sydney airshed, 1990

Air Waste	Industrial emissions (Kilotonnes)	Proportion of all manufacturing emissions of air waste (%)	Contribution of manufacturing to total waste load (%)
Carbon dioxide	32 300.0	99.5	37.0
Volatile organic compounds	81.0	0.3	52.6
Carbon monoxide	65.0	0.2	10.0
Nitrogen oxides	11.0	<0.1	16.7
Sulfur dioxide	12.6	<0.1	76.7
Particulate matter	7.3	<0.1	42.9
TOTAL	32 476.9	100.0	-

Note: There were approximately 10 800 manufacturing establishments in the Sydney area at this time.

Source: derived from EPA NSW, 1993.

Table 3.41 Suggested indicators for resource outputs from human settlements

Waste outputs	Environmental indicators for urban, rural and remote settlements
Solid Waste	• Per head domestic solid waste produced • Proportion of industrial solid waste recycled • Proportion of domestic solid waste recycled • Proportion of landfill gas being tapped for energy • Proportion of domestic hazardous waste collected
Sewage	• Proportion of settlement connected to sewage • Proportion of sewage treated at least to secondary level • Number of days per year beaches or rivers above WHO levels for sewage pathogens
Industrial Waste	• Level of industrial waste recycling • Industrial waste per unit of product • Proportion of industrial waste treated before disposal • Proportion of hazardous waste treated
Stormwater	• Proportion of stormwater recycled (not just discharged) • Proportion of stormwater with solids collected
Air Waste	• Per head and total output of greenhouse gases • Per head and total output of ozone-depleting substances • Per head and total output of volatile organic compounds • Per head and total output of nitrogen oxides • Per head and total output of carbon monoxide • Per head and total output of sulfur dioxide • Per head and total output of lead • Per head and total output of suspended particles
Waste Heat and Noise	• Heat generated per hectare of urban space • Number of people affected by noise above WHO recommended limits

The airsheds of Sydney, Melbourne and Perth are near capacity for motor vehicle emissions

ozone. They are separately controlled under the Montreal Protocol) (see Chapter 5). Methane emissions associated with coal-mining in particular are estimated to have increased substantially in the three-year period, due to a 22 per cent increase in black coal production (ABARE, 1993).

Australia's high dependence on fossil fuels to generate electricity means that electricity generation is the major source of greenhouse gas emissions. Activities inside urban areas are the chief users of that electricity, although electricity generation occurs mainly outside or on the fringe of major urban areas. Of the greenhouse gas emissions from energy production, about two-thirds are from activities in urban areas. In total, urban areas account for about 40 per cent of Australia's greenhouse gas emissions. Chapter 5 deals with the principles determining greenhouse effects in detail.

For the Sydney airshed, EPA NSW (1993) lists manufacturing industry as an important source of carbon dioxide and the main source of volatile organic compounds and sulfur dioxide. It also contributes significantly to the quantity of suspended particulates in the Sydney airshed (see Table 3.40).

Domestic air waste comes mainly from home heating (wood, gas, oil and briquettes) and incineration of waste. Compared with those of transport, industry and electricity generation, its contribution to global and regional atmospheric waste is minor, although it may result in localised concentrations of particulates and other respiratory irritants. Most large city local authorities in Australia now ban domestic waste incineration in backyards and are moving to regulate solid fuel heating.

Waste heat and noise

The end products of all energy use are dissipated in the form of chemical emissions and in the mechanical processes of waste heat and noise. Some industrial processes and all power plants produce waste heat that is generally cooled in nearby water bodies and often affects aquatic systems.

Waste heat from all energy-using processes in a city can cause the 'urban heat island effect' which traps air waste like a dome over a city. The most obvious way to alleviate this is by minimising energy inputs, although it is also possible to reduce waste heat by using it for co-generation of electricity and by using heat pumps.

Noise is becoming an increasingly important issue in the management of Australian settlements. The many sources of civic noise disturbance include aircraft, road and rail traffic, industry, entertainment venues, intruder alarms, domestic animals and, of course, human social behaviour itself. A recent report (AURDR, 1995a) describes the undesirable effects of noise:

- amenity effects — sleep interference, annoyance
- health effects — hearing impairment, tension, headaches and fatigue may contribute to cardiovascular and digestive system ailments
- communication effects — interference with business and social communication, reduction in enjoyment of activities

A number of Australian studies point to the size of the problem, as the following examples show.

- In a national survey, 40 per cent of respondents claimed that noise interfered with listening or sleep, with traffic noise being the main source (Hede *et al.*, 1986).
- More than nine per cent of Australians are subject to 'excessively high' levels (>68dB(A)) of noise, with 39 per cent subject to 'undesirable' levels (>58dB(A)) (Brown, 1993).
- Of people exposed to environmental noise in New South Wales, 73 per cent were affected by road traffic, 17 per cent were affected by aircraft, six per cent were affected by rail and four per cent were affected by industry (EPA NSW, 1993).

Noise can be a political issue, as the conflict over the third runway at Sydney Airport demonstrates.

Table 3.42 Summary

Element of the environment/Issue	State	Adequate Info.	Response	Effectiveness of response
Big Cities Rapid growth and sprawl of suburbs.	Sprawl continuing but there is an an increasing trend towards re-urbanisation.	✓✓✓	Urban consolidation programs, 'Better Cities', user pays on new infrastructure, PM's Urban Design Task Force.	Too early to assess. Models for re-urbanisation not yet mainstream.
Dominance of car-based planning in new suburbs.	Airshed limitation and urban runoff difficulties in all big cities, together with growing congestion and noise from transport.	✓✓	'Better Cities' and PM's Urban Design Task Force.	Good demonstrations, good policy options outlined, but little integration to transport providers and planners in mainstream. No national approach to public transport.
Improving livability in an equitable way.	Comparatively high levels of livability. Poverty tends to be local and may be increasing and moving further outwards.	✓✓✓	Innovation programs, transfer payments and the social wage for broad equity goals; employment schemes and local government development program for local variations.	Generally effective but globalisation of the economy is causing increased differences in livability (pockets of poverty) which are not yet targeted.
Reducing resource inputs.	High consumption of energy, water and land with much lower levels in older inner suburbs. No data on building materials.	✓	Energy Management Program, Energy Card, ethanol subsidy, fleet fuel efficiency goals, water demand management and appliance efficiency.	Building energy and vehicle energy programs not mandatory and not effective yet. Water efficiency programs beginning. Energy, water and land use planning not sufficiently integrated. No building material programs.
Reducing waste outputs.	Comparatively high levels of waste.	✓	'Cleaner production', ISO 14000, Motor Vehicle Emissions Program, Greenhouse 21C, strategies on water quality, waste minimisation and recycling, pollution inventory.	Cleaner production occurring, vehicle emissions improving but insufficient in itself without travel demand management, Greenhouse 21C strategy just beginning, strategies on waste are presently voluntary and not universally applied.
Coastal Areas Rapid growth exceeding provision of services.	Insufficient social, economic and environmental services.	✓✓	Coastal development studies, coastal management program and State of Marine Environment Report.	A few demonstrations of integrated planning but mostly little change yet.
Unsustainable land use planning.	Degrading coastal environment., loss of biodiversity and social amenity.	✓	Regional development plans and organisations, local government development program, coastal planning.	Unsustainable patterns are still accelerating.
Livability and metabolism issues.	Reduced livability (especially high unemployment and benefit holders) and worse metabolism in comparison to cities.	✓	Regional development plans and organisations, local government development program.	Unsustainable patterns are still accelerating.
Small inland towns Declining population (and economic viability) associated with major regional environmental problems.	Reduced capacity for environmental rehabilitation.	✓✓✓	Regional development plans and organisation, local government development program, land care programs and employment programs.	Effectiveness of programs diminished by economic decline.
Livability and metabolism issues.	Reduced livability and higher metabolic flows than in cities due to lower scale efficiencies, services and markets.	✓✓	Regional development plans and organisation, local government development program, land care programs and employment programs.	Effectiveness of programs diminished by economic decline.
Remote settlements Fly-in/fly-out for mining towns (and pastoral areas).	Population decline and lack of service provision in remote areas by mining companies limits human impacts but minimises settlement development.	✓	Encouragement from government.	Growing ecotourism pressures may require greater need for government services in remote settlements.
Rapid increase in indigenous communities and outstations raises land management issues.	Indigenous claims on land increasingly being recognised and necessity for joint management of adjacent public land.	✓	Native title legislation and proposed social justice package; regional agreements and joint ventures.	Just beginning—Examples of successful joint management exist.
Livability and metabolism issues.	Low livability in all remote settlements especially in indigenous communities with severe health problems. Some related metabolism problems exist, eg, firewood shortage, and lack of water/waste systems.	✓✓✓	National policies on health, housing and education target indigenous needs with special measures.	Long term effectiveness depends on continuity of special measures.

Response

Improving sustainability in Australian settlements implies reducing resource inputs and waste outputs while improving livability (see Fig. 3.2). Such change occurs as a result of education, regulation, innovation in socio-technical systems and the use of appropriate economic incentives. This section examines existing and potential responses.

The most positive changes in attitude from the point of view of future movement towards a sustainable pattern of Australian human settlements, are:

- the widespread acknowledgment among Commonwealth, State and local government bodies of the importance of ESD goals for settlements

- the recognition in government and private sector reports that Australian communities need to implement major reforms in the design, location and resource flow aspects of their settlements

- acceptance at all levels — from international agreements through to industry and community perspectives — that developmental and environmental goals need not necessarily conflict

- a tentative but growing appreciation by industry in general that they need to include environmental accounting in their corporate and business plans, recognising both market forces and social responsibilities

- the high degree of positive support in Australian settlements for environmental goals, as measured by public opinion polls and by the evidence of people's participation in environmental issues (see Chapter 10)

Despite this, the state of Australia's human settlements falls well short of the environmental goals described for them and of our international commitments. Increasing per capita use of resources, increasing population numbers and patchy implementation of the economic, legislative, social and technological innovations necessary to ameliorate the environmental impact of the settlements all contribute to this state.

Program and policy responses and their effectiveness

Table 3.42 summarises a number of current programs and policy responses, along with some comments on their effectiveness. The focus is on the Commonwealth Government and its initiatives, but in most cases other levels of government, industry and the community jointly share the responsibility. Much is happening and much is still required. The programs are a mixture of positive, inadequate and uncertain responses.

Positive programs

Three positive programs appear to be the most significant:

- 'Better Cities' is a $1000 million program that is demonstrating integrated solutions to Australian

The planned Homebush Olympic Village is designed to reduce metabolic flows, and have high human livability.

settlements by re-urbanising old industrial sites, focusing suburbanisation in new subcentres and revitalising regional centres. It involves innovative transit, new water management, toxic land remediation and improved community-oriented design.

- Australian urban and regional development review — a three-year project — provides a comprehensive opportunity for Australian settlements to be reassessed and prepared for the next century.

- Cleaner production — a major initiative of the Commonwealth EPA — provides a best-practice clearing house and demonstrations in 10 top companies on how to reduce metabolic flows and create more wealth.

Inadequate programs

In three other areas, this chapter suggests the level of response is not effective.

- Coastal development requires comprehensive, integrated strategies for preventing the excesses of recent developments. Authorities have not yet implemented recommended strategies (RAC, 1993; Zann,1995).

- Fuel efficiency regulations lag behind other countries. Australia's inability to keep up with fuel efficiency trends overseas is due to a desire not to lose the local motor vehicle manufacturing industry with its medium- to heavy-vehicle niche. If we do not impose stricter regulations on new vehicles, then gains can only come from stronger regulations on older vehicles plus more effort to reduce the need for travel or to use other transport modes.

- Public transport and integrated planning are inadequate. The critical role of transport in shaping settlements has been highlighted. A greater role for the Commonwealth in public transport (particularly in demonstrating new technologies such as light rail and demand-responsive systems) has been suggested (AURDR, 1994). We also need integrated planning of settlements. Most of these initiatives have come through 'Better Cities' and are not yet part of the Commonwealth approach to transport funding which remains single-purpose road funding.

Uncertain programs

Many other programs have uncertain potential to improve settlements' livability while reducing their metabolic flows. Three examples are listed here.

• According to the Hilmer report (ICICPA, 1993) lower prices for electricity, gas and water are likely, if its recommendations are implemented. However, this will probably lead to increased consumption of these resources, unless full cost accounting is used in pricing, and/or measures are taken on more efficient and innovative delivery of end-use services. Transport (road funding) is yet to be treated under this framework.

• Many national strategies have been developed, such as waste minimisation and recycling, water quality, transport planning, ESD, biodiversity, greenhouse, landcare, tourism, rangeland management, housing and energy management. Virtually all are based on educational and voluntary processes. Such programs would be more effective if supported by regulation and economic incentives. Such measures can be used to achieve both environmental and livability goals. The new hypothecated tax on airport noise remediation shows that there is popular support for measures that can demonstrate improvements in both environmental and livability indicators.

• The Prime Minister's Urban Design Task Force (1994) picks up many of the themes discussed here. It is as yet uncertain how governments will deal with these issues.

Two questions are critical to all these responses in Australian settlements: where geographically should governments be focusing their attention? And, how can infrastructure help to ensure these different settlement types (and areas within them) become more sustainable?

Development constraints and opportunities

Large settlements have more efficient metabolic flows than small ones because of better infrastructure and services made possible by.economies of scale, as well as the wealth and expertise of big cities. They also have higher standards of living. Thus there is a net increase in metabolic flows and loss of livability if people move from big cities to small towns.

This, however, should be balanced against some of the real constraints on large Australian cities since in some areas they can reach capacity constraints before small settlements. An example is Sydney's scale problems associated with photochemical smog, sewerage and waste water. The effect of the

latter on the Hawkesbury–Nepean catchment is illustrated in the box on page 3-50.

A city like Sydney has substantial problems coping with any extra growth, and so, if it is to grow in a sustainable way, planners need to carefully assess these capacity constraints. However, the recent dispersal of Sydney's growth to other parts of Australia may lead to an overall loss in ESD benefits. Indeed, some areas — particularly the new coastal developments — suffer obvious deprivations.

Large cities have relatively low metabolic flows in their compact areas, where land use is more efficient and less car-dependent. Levels of livability appear to vary little between these areas and dispersed ones, although this is less certain in regard to social amenity issues than to health. Livability varies far more between particular privileged and underprivileged parts of the city, within both compact and dispersed areas. Targeting these 'pockets of poverty' for development would be justified on both equity and ESD grounds.

Both suburbanisation and re-urbanisation can occur in ways that help to minimise metabolic flows and improve livability. Where the emphasis has been on the development of nodal/information subcentres with compact, mixed-use activities, then such centres have improved the local economy. They have also created a less resource-intensive urban form. The redevelopment of Fremantle provides an example of re-urbanisation with sensitivity to the city's heritage qualities. The subcentre is also a growing local economy and has one-third of the transport fuel use of new subcentres (Campbell and Newman, 1989). Similar nodal developments can occur throughout our cities.

Dispersed coastal development in Australia appears to involve high metabolic flows with more obvious environmental damage due to the sensitivity of the area and the speed of growth outstripping the capacity of poor infrastructure and services. The standards of livability of the coastal areas also appear to be reduced. However, individuals trade off this reduced livability (fewer services and opportunities) for the cheaper land and personal environmental amenity — which in turn is threatened by the scale of the development process. Restricting dispersed development along Australia's coast has been highlighted as a priority in many Commonwealth and State reports.

Inland settlements are mostly declining, unless they are tourist centres or have absorbed the functions of surrounding smaller towns. Many are caught in the spiral of declining investment, and often have inadequate technologies for water and waste management. Strategies for improving the local economy as well as the environment are therefore closely linked.

One important option for inland towns within a 100-km radius or so of a large city is to link them to the city by fast rail. This has been attempted in Melbourne, with links to Ballarat. If a journey to the city takes less than one hour, the town becomes a viable centre to help absorb growth that exceeds the city's capacities. It is not certain whether this 'exurbanisation' strategy leads to net environmental gains. State of the environment reporting could assist with this evaluation.

Remote indigenous communities have low metabolic flows but suffer significant livability problems. Some are exceeding their settlements' capacity to supply, for example, firewood and basic sanitation. These problems are due to a complex range of social and economic factors, including the rapid growth of the settlements since the 'return to country' movement began, frequently allied with indigenous peoples' lack of any real title to the land. The result has been a lack of basic infrastructure.

Sustainable infrastructure plans and indicators

Provision of infrastructure and services helps shape the pattern of settlements — both the internal structure of the city and the extent and type of regional development. Governments are responsible for regulating infrastructure and services and often provide much of the funding.

The Commonwealth Government has announced that it will require sustainable infrastructure plans on all new development requiring an input of Commonwealth funds. Such plans may become standard practice in urban development. The National Housing Strategy Background Paper 15 provides a checklist for achieving ESD goals in urban development (Newman *et al.*, 1992).

State of environment reporting can inform the application of sustainable infrastructure plans. Some suggestions follow.

Large growing cities

Most new investment in infrastructure and services takes place in the large cities. As described earlier, three simultaneous processes seem to be occurring: re-urbanisation, suburbanisation and exurbanisation. There is much discussion in Australia on the merits of these processes and how they relate to the environment. State of the environment indicators can inform sustainable infrastructure plans which provide a way of enhancing the processes in a more sustainable way.

Re-urbanisation

Re-urbanisation can improve the settlement environment and amenity by using under-utilised urban infrastructure and providing more people with a low-energy, low-water-use lifestyle. However, it can also mean development that overstretches infrastructure and can be insensitive in design and impact on the local community.

The Commonwealth Government program 'Better Cities' is designed to demonstrate how the re-urbanisation process can produce more efficient, socially just and ecologically sustainable urban development. Australian cities are moving rapidly towards a post-industrial form based on the

▲ The new 'Sprinter' trains connect Ballarat to Melbourne enabling the city to be reached in less than one and a half hours.

processing of information and provision of services. Government demonstration projects outlined in Better Cities are designed to assist in this transition. Many demonstrations are dense housing developments on abandoned industrial land in inner areas, and most are designed to bring more people to use the many existing services of these areas. Analysis of one such project in East Perth, for example, showed some $121 million in savings on infrastructure and transport in just 1000 dwellings and 100 workplaces, as well as considerable savings in fuel and greenhouse gases (Kenworthy and Newman, 1992).

State of the environment reporting can be used to monitor such demonstrations. Where benefits are revealed, this can provide the basis to redirect public subsidies to developments of this type.

Suburbanisation

This will probably remain as the dominant urban development process over the next 20 years unless significant disruptions (such as a world oil crisis) occur. The 'Better Cities' development of integrated urban villages around the extended Gold Coast railway is designed to show how suburbanisation can become more sustainable. State of the environment reporting might encourage more developments of this type: for example, to prevent development in sensitive areas; to ensure adequate densities are integrated to the infrastructure; to provide more water-sensitive design; and to facilitate development of the nodal/information city with the development of subcentres.

Exurbanisation

This chapter has highlighted the unsustainability of development occurring on the urban fringe and along the coast of non-metropolitan areas. The strategy outlined above — of providing inland country towns within 100 km or so of major capital cities with a fast rail service — may do little but feed the continued growth of dispersed land uses that cause much larger metabolic flows and reduced livability due to inadequate planning and services. However, a rail service can lead to land development that is more compact and integrated to good infrastructure and services, particularly environmental technology. State of the environment reporting can help assess whether these types of developments lead to a more sustainable use of infrastructure.

Transport/information infrastructure planning

Most governments recognise the critical role that transport infrastructure plays in shaping the city. However, the emphasis has been on providing high-capacity roads to fringe and coastal areas — a major and continuing force behind the development of these areas. That kind of infrastructure planning needs to be monitored, perhaps using state of the environment indicators.

As part of its Greenhouse 21C statement, the Commonwealth Government committed itself to requiring transport impact statements for commercial, industrial and residential type infrastructure projects with Commonwealth funding. This will ensure transport and land use planning are fully coordinated in relevant projects by assessing all options and impacts at the planning stage. Thus, planning can be guided by state of the environment indicators such as those developed here.

A more significant role for the Commonwealth in new public transport infrastructure in Australian cities has been suggested (AURDR, 1994) for two reasons.

Firstly, public transport is critical in the shaping of urban form. Urban designers like those on the Prime Minister's task force are calling for a more transit-oriented form. Road-based development has led not to compact subcentres but to the kind of dispersed land use that is now under question.

Secondly, disadvantaged areas in particular need new public transport infrastructure. The new 'pockets of poverty' in Australian cities are sometimes well away from the traditional good transit services like rail lines. Such areas are very poorly provided with public transport and yet residents are often without cars (Social Justice Research Program into Locational Disadvantage, 1994).

Despite this growing awareness at the Commonwealth level, a major program of expansion is occurring in Australian cities, mostly by the States, in the provision of new high-capacity roads rather than new public transport. It ignores the already-heavy car dependence shown in this report. The consequent investment of many billions of dollars of public money has not been evaluated in terms of environmental indicators. For example, it will be extremely difficult to explain how such investment can meet international commitments on reducing greenhouse gas emissions or most of the other indicators suggested in this report.

Although most governments are aware of the effect of transport infrastructure on cities, the same cannot be said about information infrastructure. The need for Australian cities to fully participate in the global information system (for economic, social and environmental reasons) points to the need for information superhighway infrastructure and training in its use. This can be part of the solution to the 'pockets of poverty'. Environmental indicators can help monitor the ecological sustainability of new transport and information infrastructure.

Affordable housing

This report has highlighted the growing poverty of some pockets of Australian cities. It has found strong ESD and economic arguments as well as social justice reasons, for focusing development in these areas. Critical to that revitalisation is the provision of diverse and affordable housing near to shops and services such as good public transport infrastructure. 'Better Cities' has demonstrated that it can be done with housing that is compact, integrated with urban services and without the stigma of 1960s public housing.

Rouse Hill development area — integrated water management

The Hawkesbury–Nepean River is Sydney's major river system. Because of its mixture of agricultural, urban and recreational uses, it has come under pressure from increases in nutrient flow, turbidity, pollutants and algal growth. Water quality in much of the Hawkesbury–Nepean exceeds New South Wales EPA goals for nutrients, algal growth, suspended solids and faecal coliforms. At present, the rivers receive treated sewage effluent from about 10 per cent of Sydney's population — a pressure that is likely to grow with the Hawkesbury–Nepean catchment being the focus for much of Sydney's urban development over the next 10 to 15 years.

One place earmarked for future urban development is the Rouse Hill area, which is planned ultimately to accommodate about 235 000 people. To avoid further stressing the Hawkesbury–Nepean system, the Sydney Water Corporation has developed an integrated water management strategy that seeks to combat urban development 'side effects' caused by stormwater run-off, sewage effluent and untreated waste water.

The strategy aims to minimise impacts on the Hawkesbury–Nepean through:

- integrated management of water supply, waste water and drainage
- management of run-off via constructed wetlands, litter and silt traps, detention basins and grassed stormwater floodways
- tertiary sewage treatment, including nutrient stripping and disinfection
- recycling of tertiary-treated water for gardens and toilets
- mandatory installation of water-efficient appliances

The Rouse Hill strategy is a move towards a more sustainable infrastructure plan for water management but there is little to suggest a more sustainable transport infrastructure.

Social and economic infrastructure

Alleviating poverty has been shown in this report to be an integral part of providing sustainable settlements with good human and physical environments. Thus, state of the environment reporting appropriately monitors the impact of education, community services and employment programs aimed at reducing the fundamental problem of structural unemployment, as well as physical infrastructure. These social infrastructure items can also be part of sustainable infrastructure plans.

Provincial towns

While many older industries have been phased out of the big cities, a simultaneous move of process industries to smaller provincial towns has been occurring (Beer, *et al.*, 1994). These moves are receiving increasing help from governments in terms of financial incentives and regulatory procedures. The process enables many industries to improve their technology as well as reducing their land and operating costs, although it could also be a way that unsustainable industries receive subsidies and reduce environmental assessment.

State of the environment reporting can help monitor the ecological sustainability of industrial processing in regional towns.

Coastal areas

This chapter has drawn attention to the rapid growth pressures being experienced by coastal areas. The data suggest that a large part of this growth is not sustainable economically or environmentally. Managing further coastal development is not as easy as drawing new lines on maps. We need to understand the processes driving it.

In Sydney, the expansion of the urban area through suburbanisation is now limited by the land that is available. New suburbs, like Rouse Hill, are constrained to carry out expensive environmental mitigation (see the box opposite), which pushes up the price of housing in standard subdivisions to levels beyond the reach of average families — creating an extra pressure to re-urbanise. Global economic pressures are encouraging high-income groups to locate in core and inner areas. If re-urbanisation does not proceed quickly enough to provide for that housing demand, then the price of houses and land in core and inner areas will also grow well beyond the reach of average families. This pincer movement in suburbanising and re-urbanising areas means that many families are fleeing the high prices of Sydney.

The preferred location for such families appears to be in small towns along the coast in New South Wales and Queensland, thus fanning the major environmental concerns outlined above. Not only are the metabolic resource flows high, but the coastal land being subdivided and the associated water areas are some of the most ecologically sensitive in Australia. The need to reduce this pressure on the coast raises obvious responses,

▲

Demonstration model of a remote area community ablution block with low-water-use toilet, wood and solar water heating, shower, hand-pumped washing machine and photovoltaics for electricity, if required.

which include reduced population growth, greater re-urbanisation, stronger planning controls and investment in better infrastructure and services to mitigate the direct ecological problems.

Each State (coordinated at the Commonwealth level) can scrutinise coastal areas more closely by monitoring sustainable infrastructure plans. Such monitoring can help determine where further coastal developments result in net economic and environmental gains.

Declining rural areas

The economic processes that have led to the decline of rural industries are well documented. Governments have provided financial assistance in many forms, but the economic forces are not likely to change. So, continuing decline is likely in many of those areas based on traditional products unless industries diversify. The Commonwealth Government has an active regional development program designed to halt rural decline.

Unfortunately, a vicious circle is operating here, in that the smaller the population becomes, the greater the need for cross-subsidies by governments to provide postal, education and transport services. The winding down of these services has cut into the range of job opportunities essential to the maintenance of the small town communities that once flourished in declining rural areas (Holmes, 1994).

New infrastructure is often seen as out of the question for such declining towns but, as with the urban 'pockets of poverty', it will be important not to neglect such areas for environmental as well as social and economic reasons.

Maintaining and expanding regional infrastructure and services is an essential part of reversing further rural decline. To help areas diversify economically, governments can use state of the environment reporting and sustainable infrastructure plans to guide regional economic development.

Remote areas

All remote settlements have heavy subsidies for infrastructure and services. The special allocations for establishing indigenous communities as they 'return to country' is part of a long tradition of assisting remote areas. Previous subsidies to remote areas for other development were given for a range of social and political reasons such as defence and long-term economic development goals.

The increasing costs to governments and the private sector of providing infrastructure and services is the main reason why mining settlements are increasingly fly-in/fly-out.

This report has highlighted the plight of many indigenous communities. It has shown that their levels of social amenity and health are well below those regarded as normal by the rest of Australia. If government reduces the investment in infrastructure in remote areas, conditions are unlikely to improve, rather they will probably deteriorate further. At the same time, the need for development of a greater economic base for remote indigenous communities is also apparent. We need to encourage the participation of these communities in remote-area growth industries such as mining, tourism, land management, farming of native species and the application of innovative technologies. State of the environment reporting can be used to monitor environmental flows and livability, including basic health standards, in these communities.

Settlement capacity

Ehrlich's formula for assessing carrying capacity is based on population, per capita consumption (lifestyle) and technology (efficiency) (Ehrlich *et al.*, 1977). Human settlements need to fit within a global and local ecological carrying capacity as they seek to manage their metabolic flows of resources and wastes. Global capacity is increasingly being defined by United Nations conventions for wastes like greenhouse gases or hazardous wastes, as well as by the global marketplace for resources like oil or timber. Local carrying capacity refers to resources like water or land and wastes like sewage, stormwater or photochemical smog.

Capacity for each settlement needs to be assessed. For many global resources and wastes, national and international policies that constrain the ability of settlements to expand in their usage will be increasingly significant — for example, the United Nations Framework Convention on Climate Change which Australia ratified in 1992. For local resources, the state of the environment reporting process can help settlements to monitor indicators and institute policies for living within constraints.

The policy responses outlined above tend to be those dealing directly with government agencies that manage resources such as land, water, waste and transport. But many other arenas can also significantly affect settlements. These include:

- population growth
- lifestyle choices that affect per capita consumption and are influenced by media, education and other forces
- the extent to which Australia can develop technological innovation that is both more sustainable and appropriate for our settlements

This report has highlighted capacity problems for each Australian settlement type. State of the environment indicators can help provide the necessary data to evaluate these capacity issues.

- Large cities — airshed capacities for photochemical smog, watershed capacities for stormwater and treated sewage as well as land development and water extraction impacts.
- Coastal areas — capacities to absorb land development on the sensitive border between land and water and also those of near-shore estuarine and marine environments to absorb waste.
- Smaller inland towns — watershed capacities and the effects on limited inland river flow.
- Indigenous communities — capacities for firewood, water and sewage and the basic infrastructure that can enable livability to improve.

Conclusion

The Commonwealth Government (in many cases in cooperation with the other spheres of government) has established a large number of initiatives covering many important environmental issues in human settlements. These have made progress in some areas. However, Australian settlements continue to grow in their metabolic flows and while livability is generally high, it is not shared equally. Potential ghettos are emerging in the big cities, although nothing like those in the United States. New coastal settlements appear to be growing at an unsustainable rate. Some inland towns, by contrast, are in sharp decline. Indigenous settlements clearly have the severest livability inequities.

This response section has gathered suggestions on how Australian settlements can reduce their metabolic flows while improving their livability. A significant portion of these initiatives require a holistic approach, and therefore a wide range of government agencies and other organisations need to adapt or consider their principles and implications. This will particularly apply to state of the environment reporting for human settlements as it crosses so many different areas of responsibility. Nationwide state of the environment approaches and guidelines, and more cooperative processes between the different levels of government, will be necessary to remove some of the impediments to addressing environmental issues in Australian settlements.

References

ABARE (1991). 'Australian Forest Resources.'(Australian Bureau of Agricultural and Resource Economics: Canberra.)

ABARE (1993). 'Energy Demand and Supply Projections Australia 1992–93 to 2004–05.' (Australian Bureau of Agricultural and Resource Economics: Canberra.)

ABS (1955). 'Commonwealth Year Book.' (Australian Bureau of Statistics: Canberra.)

ABS (1981). 'Housing survey 1978, Sydney, Newcastle and Wollongong- Part 3, Anticipated Residential Movement and Satisfaction with Current Housing Conditions.' (cat. no. 8713.1) (Australian Bureau of Statistics NSW Office: Sydney.)

ABS (1991). 'CDATA91' (cat.no. 2721.0) (Australian Bureau of Statistics: Canberra.)

ABS (1992a). 'Housing Australia: A statistical overview.' (cat.no. 1320.0), (Australian Bureau of Statistics: Canberra.)

ABS (1992b). 'Commonwealth Year Book.' (Australian Bureau of Statistics: Canberra.)

ABS (1992c). 'Social Indicators Australia 1992 No5.' (cat.no. 4101.0) (AGPS: Canberra.)

ABS (1993a). 'Estimated Resident Population and Components of Change in Population in NSW 1986 to 1991.' (cat.no. 3208.1) (ABS NSW Office: Sydney.)

ABS (1993b). 'Estimated Resident Population in Statistical Local Areas, Victoria.' (cat. no. 3203.2). (ABS Vic. Office: Melbourne.)

ABS (1993c). '1986 to 1991 Estimated Resident Population Qld.' (cat. no. 3212.3). (ABS Qld Office: Brisbane.)

ABS (1993d). 'Estimated Resident Population in Statistical Local Areas, SA 1991.' (cat. no. 3202.4). (ABS SA Office: Adelaide.)

ABS (1993e). 'Estimated Resident Population Age and Sex in Statistical Local Areas, WA 1991.' (cat. no. 3203.5) (ABS WA Office: Perth.)

ABS (1993f). 'Apparent Consumption of Foodstuffs and Nutrition, Australia, 1992.' (cat. no. 4306.0) (Australian Bureau of Statistics: Canberra.)

ABS (1994a). 'Estimated Resident Population by Age and Sex in Statistical Local Areas, SA 1993.' (cat. no. 3204.4) (ABS SA Office: Adelaide.)

ABS (1994b). 'Estimated Resident Population in Statistical Local Areas, WA 1993.' (cat. no. 3204.5) (ABS WA Office: Perth.)

ABS (1994c). 'Estimated Resident Population in Statistical Local Areas, NSW 1993.' (cat. no. 3210.1) (ABS NSW Office: Sydney.)

ABS (1994d). 'Estimated Resident Population in Statistical Local Areas, Vic. 1993.' (cat. no. 3203.2) (ABS: Vic. Office, Melbourne.)

ABS (1994e). 'Age and Sex Distribution of Estimated Resident Population, Qld 1993.' (cat. no. 3224.3) (ABS Qld Office: Brisbane.)

ABS (1994f).'Australian Demographic Statistics 1993.' (cat. no. 3101.0) (AGPS: Canberra.)

ABS (1994g). 'Australian Social Trends 1994.' (cat.no. 4102.0) (Australian Bureau of Statistics: Canberra.)

ABS (1994h). 'Socio-Economic Indexes for Areas.' (cat. no. 2912.0) (Australian Bureau of Statistics: Canberra.)

ABS (1994i). 'Deaths Australia 1993.' (cat. no. 3302) (Australian Bureau of Statistics: Canberra.)

ABS (1995a). 'National Aboriginal and Torres Strait Islander Survey 1994: Detailed Findings.' (cat.no.4190.0) (Australian Bureau of Statistics: Canberra.)

ABS (1995b). 'ABS Household Expenditure Survey 1993-94, Australia, States and Territories.' (cat. no. 6533.0) (Australian Bureau of Statistics: Canberra.)

ABS (1995c). 'Estimated Resident Population by Sex and Age: States and Territories of Australia June 1993 and Preliminary 1994.' (cat. no. 3201.0) (Australian Bureau of Statistics: Canberra.)

ABS (1995d). 'Occasional Paper: Recent Changes in Unpaid Work' (cat. no. 4154.0) (Australian Bureau of Statistics: Canberra.)

ABS (1995e). Data provided in correspondence to Human Settlements Reference Group, 1996 SoE Project.

AIHW (1994). 'Australia's Health 1994: the Fourth Biennial Report of the Australian Institute of Health and Welfare.' (Australian Institute of Health and Welfare: Canberra.)

AIHW (1995). Data provided in correspondence to Human Settlements Reference Group, 1996 SoE Project.

AIP (1995). 'Oil and Australia Forecasts 1995-2004.' *Supplement to Petroleum Gazette* 1995(1). (Australian Institute of Petroleum: Melbourne.)

ATSIC (1993). '1992 National Housing and Infrastructure Needs Survey'. *Final Report Stage.* (Aboriginal and Torres Strait Islander Commission: Canberra.)

AURDR (1994). 'Urban Public Transport Futures' *Workshop Papers* No4. (Australian Urban & Regional Development Review:Melbourne.)

AURDR (1995a). 'Green Cities' *Strategy Paper* No 3. (Department of Housing and Regional Development: Canberra.)

AURDR (1995b). 'Places for Everyone'. *Research Report* No1. (Department of Housing and Regional Development: Canberra.)

Barker, R. (1993). Interstate migration to south east Queensland: an analysis. *People and Place.* 1(4), pp. 30–6.

Beer, A., Bolam, A., and Maude, A. (1994). 'Beyond the Capitals: Urban Growth in Regional Australia.' (AGPS: Canberra.)

Benson, D.H., and Howell, J. (1990). Sydney's vegetation 1788–1988- utilization, degradation and rehabilitation. In 'Australian Ecosystems: 200 years of Utilization, Degradation and Reconstruction.' Eds D.A. Saunders, A.J.M. Hopkins and R.A. How. *Proceedings. of a Symposium in Geraldton WA, September 1988.*' (Surrey Beatty and Sons: Chipping Norton, NSW.)

Beretka, J. and Whitfield, D. (1993). 'Survey of Industrial Process Wastes and By-Products Generated in Australia.' (CSIRO: Melbourne.)

BHP Pty Ltd (1995). Data provided in correspondence to Human Settlements Reference Group, 1996 SoE Project.

Birrell, R. (1994). The overseas born component of Australia's major cities. *People and Place.* **2**(2), pp. 53.

Birrell R, Newman P and Newton P (1995). Sunbelt-rustbelt revisited: The case of south-east Queensland. *People and Place.* **3**(4), pp. 53–61.

Boyden, S., Millar, S., Newcombe, K., and O'Neill, B. (1981). 'The Ecology of a City and its People.' (ANU Press: Canberra.)

Boyer, M.C. (1983). 'Dreaming the Rational City: The Myth of American City Planning.' (Massachusetts Institute of Technology Press: Cambridge, USA).

Brotchie, J. (1992). The changing structure of cities *Urban Futures Special Issue* **5**, pp. 13–34.

Brown, A. (1993). 'Exposure of the Australian Population to Road Traffic Noise,' *Report prepared for ANZEC.* (AGPS: Canberra.)

Butlin, N. (1976). 'Sydney's Environmental Amenity 1970–1975.' *Botany Bay Project Report* No1. (ANU Press: Canberra.)

Campbell, R.H., and Newman, P.W.G. (1989). Local Government and Transport Energy Conservation. *Planning and Administration* **89**(1), pp. 68–75.

Carnovale, F., Alviano, P., Carvalo, C., Deitch, G., Jiang, S., Macauley, D., and Summers, M. (1991). 'Air Emssions Inventory for the Port Phillip Bay Control Region: Planning for the Future.' (EPA Victoria: Melbourne.)

Catterall, C.P., and Kingston, M. (1993). 'Remnant Bushland of South East Queensland in the 1990s: Its Distribution, Loss, Ecological Consequences and Future Prospects.' (Institute of Applied Environmental Research, Griffith University and Brisbane City Council: Brisbane.)

CAT (1991). 'Remote Controlled Waste.' (Centre for Appropriate Technology: Alice Springs.)

CAT (1995). Data provided in correspondence to Human Settlements Reference Group, 1996 SoE Project.

Cervero, R. (1995). Sustainable new towns: Stockholm's rail-served satellites. *Cities* **12**(1), pp. 41–51.

City of Warrnambool (1995). Data provided in correspondence to Human Settlements Reference Group, 1996 SoE Project.

Commonwealth Treasury of Australia (1995). 'Economic Roundup.' *Summer Issue.* (AGPS: Canberra.)

Community and Family Commission (1992). Speaking out, Taking part, *Final Report of the Community and Family Commission* . (WA Government: Perth).

Davison, G. (1978). 'The Rise and Fall of Marvellous Melbourne.' (Melbourne University Press: Melbourne.)

DEST (1992). 'Urban Stormwater: A Resource Too Valuable to Waste.' (AGPS: Canberra.)

DEET (1995). Data provided in correspondence to Human Settlements Reference Group, 1996 SoE Project.

DPIE (1991). 'Review of Effluent Disposal Practices.' *Australian Water Resources Council Water Management Series,* No. 20. (AGPS: Canberra.)

Dingle, T. and Rasmussen, C. (1991). 'Vital Connections, Melbourne and its Board of Works.' (McPhee Gribble: Melbourne.)

Doll, R. (1992). Health and the Environment in the 1990s. *American Journal of Public Health* **82**, pp. 933–40.

Dowling, R. (1993). An environmentally based approach to tourism planning. *PhD thesis, School of Biological and Environmental Sciences, Murdoch University, WA.*

ESS (1988). 'Brisbane Bushland Strategy'. *Consultancy Report to the Brisbane City Council.* (Environmental Science and Services: Brisbane.)

ESD Working Group (1991). Manufacturing. (AGPS: Canberra.)

EPA NSW (1993). NSW State of the Environment 1993. (EPA: Sydney.)

EPA Vic. (1992). 'Assessment of Coastal Discharges for their Potential Environmental Impact.' (EPA Vic.: Melbourne.)

EPAC (1995). 'Income Distribution in Australia. Commonwealth of Australia.' (Economic Planning and Advisory Council: Canberra.)

Engwicht, D. (1992). 'Towards an eco-city: Calming the traffic.' (Enviro Books: Sydney.)

Ehrlich, P., Ehrlich, A.H., and Holdren, J.P. (1977). 'Ecoscience: Population, Resources and Environment.' (Freeman: San Fransisco.)

Federal Office of Road Safety (1992). 'National Road Safety Strategy.' (Federal Office of Road Safety: Canberra.)

Fleay, B.J. (1994). Liquid petroleum is peaking: Decline of the oil age. *Occasional Paper* No 3/94. Institute for Science and Technology Policy, Murdoch University.

Flood, J., Maher, C., and Newton, P. (1991). The Determinants of Internal Migration in Australia. *Report to Indicative Planning Council for the Housing Industry, Dept of Industry, Science and Technology.* (AGPS: Canberra.)

Garreau, J. (1991). 'Edge City: Life on the New Frontier.' (Doubleday: New York.)

Girardet, H. (1992). 'The Gaia Atlas of Cities.' (Gaia Books: London.)

Gehl, J. (1992). The challenge of making a human quality in the city. In 'Perth Beyond 2000: A challenge for a city'. *Proceedings of the City Challenge Conference: Perth.*

Gehl, J. (1994a). 'Places for People.' (City of Melbourne: Melbourne.)

Gehl, J. (1994b). 'Public Life and Public Spaces in Perth.' (Department of Planning and Urban Development: Western Australia.)

Giles, G.G., Armstrong, B.K., Smith, L.R. (1987). 'Cancer in Australia 1982.' (Australasian Association of Cancer Registries and Australian Institute of Health: Canberra.)

Ginpil, S., Schneider, R., Stone, S. (1992). 'Years of Potential Life Lost through Road Crashes: 1990.' (Federal Office of Road Safety: Canberra.)

Gipps, P.G., Brotchie, J.F., O'Connor, K., and Hensher, D. (1995). 'The Journey to Work, Employment and the Structure of Australian Cities: an Empirical Study.' Report prepared by CSIRO Division of Building, Construction and Engineering for the Urban Futures Research Program (Department of Housing and Regional Development: Canberra.)

Government of Western Australia (1992). 'State of the Environment Report.' (Government of WA: Perth.)

Gregory, R., and Hunter, B. (1995). 'The Macro Economy and the Growth of Ghettos and Urban Poverty in Australia.' *Paper presented to National Press Club 26/4/1995* .

Hamilton, N., and Cocks, D. (1994). 'Australia's Coastal Population: Environmental Impacts Today and Tomorrow.' *Australian Population Association Workshop on Mobility in Australia, November 1994.*

Hayward, R., and McGlynn, S. (1993). 'Making Better Places: Urban Design Now.' (Butterworth Architecture: Oxford.)

Hede, A., Meagher, D., and Watkins, D. (1986). National noise survey - 1986. *Acoustics Australia* **15**(2), pp. 39-42.

Holmes, J. (1988). New challenges within sparselands: The Australian experience. In 'Land, Water and People: Geographical Essays in Australian Resource Management, eds R.L. Heathcote and J.A. Mabbutt. (Allen and Unwin: Sydney.)

Holmes, J. (1994). Coast versus inland: Two different Queenslands? *Australian Geographical Studies*, **32** (2), pp. 167–82.

ICICPA (1993). 'National Competition Policy.' (AGPS: Canberra.)

Industry Commission (1990). 'Waste Management and Recycling: Survey of Local Government Practices.' (Paragon Printers: Canberra.)

Jacobs, J. (1961). 'The Death and Life of Great American Cities.' (Random House: USA.)

Johnson, D., Manning, I., and Hellwig, O. (1995). 'Trends in the Distribution of Cash Income and Non-cash Benefits: An Overview.' (AGPS: Canberra.)

Jones, R. (1994). The housing needs of indigenous Australians 1991. *Australian National University Research Monograph* No. 8. (Centre for Aboriginal Economic Policy Research: ANU.)

Kenworthy, J.R., and Newman, P.W.G. (1991). 'Moving Melbourne a Public Transport Strategy for Inner Melbourne.' (Inner Melbourne Regional Association: Melbourne.)

Kenworthy, J.R., and Newman, P.W.G. (1992). 'The Economic and Wider Community Benefits of the Proposed East Perth Redevelopment.' *A commissioned report to the East Perth Redevelopment Authority.* (Institute for Science and Technology Policy, Murdoch University: Perth.)

Kenworthy, J.R., and Newman, P.W.G. (1993). 'Automobile Dependence- The Irresistable Force?' (Institute for Science and Technology Policy: Murdoch University.)

Kenworthy, J.R., and Newman, P. (1994). Toronto - Paradigm regained. *Australian Planner.* **31**(3), pp. 137–47.

King, A. D. (1978). Exporting planning: The colonial and neocolonial experience. *Urbanism, Past and Present.* 5, pp. 12–22.

Landrigan, P.J. (1992). Environmental disease - A preventable epidemic. *American Journal of Public Health* **82**, pp. 941–3.

Langmore, J. and Quiggin, J. (1994) 'Work for All.' (Melbourne University Press: Melbourne.)

LRRA (1995). '1995 Garbage Bin Analysis and Recycling Container Audit: Sydney.' (Litter and Recycling Research Association: Sydney.)

Maher, C. (1992). Residential Property Markets in late 1980s: Price and Affordability Issues in Australian Cities. *Institute of British Geographers Conference, Swansea, Britain.*

Maher, C. (1994). Housing prices and geographic scale: Australian cities in the 1980s. *Urban Studies.* **31**(1), pp. 5–27.

Maher, C. (1995). 'Locational Disadvantage and Concentrations: a Review and Evaluation of the Findings. ' Paper prepared for the Seminar on *Spatial Aspects of Inequality.* 20th Oct. 1995. (Department of Housing and Regional Development: Canberra.)

Maher, C., Whitelaw, J., McAllister, A., Francis, R., Palmer, J., Chee, E., Taylor, P. (1992). 'Mobility and Locational Disadvantage within Australian Cities: Social Justice and Implications of Household Relocation.' *Report of the Social Justice Research Program into Locational Disadvantage.* (Department of Prime Minister and Cabinet: Canberra.)

Manning, I. (1978). 'The Journey to Work.' (George Allen & Unwin: Hornsby, NSW.)

Marginson, S. (1993). 'Education and Public Policy in Australia.' (Cambridge University Press: Cambridge.)

Mathers, C.D. (1994a). Health differentials among adult Australians aged 25-64 years. *Australian Institute of Health and Welfare Health Monitoring series* No1.

Mathers, C.D. (1994b). Health differentials among older Australians. *Australian Institute of Health and Welfare Health Monitoring series* No2.

Mathieson, A., and Wall, G. (1982). 'Tourism: Economic, Physical and Social Impacts.' (Longman: Essex.)

McDonald, K. (1995). Morals is all you've got. *Arena Magazine* No 20. 18–23.

McDonald, P., and Moyle, H. (in press). Household reference zones in cities. In 'Population Shift', eds P.J. Newton and M. Bell. (AGPS: Canberra.)

McKenzie, F. (1994). 'Regional Population Decline in Australia: Impacts and policy implications.' (AGPS: Canberra.)

McLelland, A. (1994). Impacts of change on women and children. In 'Social Security: Issues and Options', eds J. Disney and L. Briggs (AGPS: Canberra.)

Melbourne Water (1992). 'Environmental Performance Report 1992.' (Melbourne Water: Melbourne.)

Metro Strategy, Environment and Health Working Group (1992). *Draft report to the task force.* (NSW Department of Planning: Sydney.)

Mitchell, M. (1993). 'The Future of Work.' (Australian Council of Social Services: NSW.)

Munro, C.H. (1974). 'Australian Water Resources and their Development.' (Angus and Robertson: Sydney.)

Mouritz, M., and Newman, P. (1995). 'Sustainable Urban Water Systems: Issues and Opportunities.' (Urban Water Research Association of Australia: Melbourne.)

Naess, P. (1993). Energy Use for Transport in 22 Nordic Towns. *NIBR Report* No 2. (Norwegian Institute for Urban and Regional Research: Oslo.)

National Housing Strategy (1992a). Housing Location and Access to Services *Issue Paper* 5. (AGPS: NSW.)

National Housing Strategy (1992b). 'Australian Housing: The Demographic, Economic and Social Environment.' (Commonwealth of Australia: Canberra.)

NATSEM (1995). Income distribution Report Issue 2. (National Centre for Social and Economic Modelling: University of Canberra.)

NGGIC (1994). 'Australian National Greenhouse Gas Inventory 1988 and 1990.' (AGPS: Canberra.)

Newman, P.W.G. (1982). Domestic energy use in Australian cities. *Urban Ecology*, 7: pp.19–38.

Newman, P. (1992). Planning and healthy cities: the renaissance of a movement to cope with automobile dependence. *National Healthy Cities Conference, Brisbane, May 1992.*

Newman, P. (1994). Energy conservation in transport and urban settlements. In 'Sustainable Energy Systems', ed. S. Dovers. (Cambridge University Press: Cambridge.)

Newman, P.W.G., and Kenworthy, J.R. (1989). 'Cities and Automobile Dependence: An International Sourcebook.' (Gower: Aldershot.)

Newman, P.W.G., and Kenworthy, J.R. (1990). 'Transport Energy Conservation Policies for Australian Cities: Strategies for Reducing Automobile Dependence.' (Institute for Science and Technology Policy: Perth.)

Newman, P.W.G., and Kenworthy, J.R. (1992). Transit orientated urban villages: design model for the 90s. *Urban Futures.* 2(1), pp. 50–8.

Newman, P., Kenworthy, J., and Robinson, L. (1992). 'Winning back the Cities' (Pluto Press: Sydney.)

Newman, P.W.G., Kenworthy, J.R., and Vintila, P. (1992). Housing transport and urban form. *Background Paper 15 for National Housing Strategy, October 1992.*

Newman, P., Kenworthy, J., and Vintila, P. (1993). Can we build better cities? *Urban Futures* 3 (2), pp. 17–24.

Newman, P., Bathgate, C., Bell, K., Dawson, M., Hudson, K., Lehman, J., Mason, S., Myhill, P., McEnaney, L., McCreddin, C., Pidgeon, C., Pratt, C., Ringrose, C., Ringvall, K., and Sargent, M. (1994). 'Australia's Population Carrying Capacity: An analysis of eight natural resources.' (Institute for Science and Technology Policy: Perth.)

Newton, P.J. (1995). Changing places? Households, firms and urban hierarchies in the information age. In 'Cities in Competition: Productive and Sustainable Cities for the 21st Century', eds J.F. Brotchie, M. Batty, E. Blakely, P. Hall and P.W. Newton (Longmans: Melbourne.)

Newton, P.J., and Bell, M. (in press). 'Population Shift.' (AGPS: Canberra.)

Newton, P.J., Brotchie, J.F., and Gipps, P.G. (in press). Cities in transition: changing economic and technological processes and Australia's settlement system. *Report prepared for Human Settlements Reference Group, DEST.*

Nicholson, G.J., Fabris, J.G., and Gibbs, C.F. (1992). 'Heavy metals in the mussels of Corio Bay.' (EPA Vic.: Melbourne.)

Nix, H.A. (1973). The City as a Life System? *Proceedings Ecological Society Australia Symposium, University of NSW, 12–13 August 1972.* (Ecological Society of Australia: Canberra.)

NSW Office of Energy (1995). Data provided in correspondence to Human Settlements Reference Group, 1996 SoE Project.

O'Connor, K., and Stimson, R.J. (1994). Economic change and the fortunes of Australian cities. *Urban Futures*, 4(2 & 3), pp. 1–12.

O'Connor, K., and Stimson, R.J. (in press). Australian regions in growth and decline: convergence and divergence of demographic and economic trends. In 'Population Shift', eds P.J. Newton and M. Bell. (AGPS: Canberra.)

O'Loughlin, E.M., Young, W.J., Molloy, J.D. (1992). 'Urban Stormwater: Impacts on the Environment.' *CSIRO Division of Water Resources, Consultants Report 92/29.*

OECD (1993). 'OECD Environmental Data Compendium 1993.' (OECD: Paris.)

OECD (1994). 'Climate Change Initiatives.' *1994 Update, Vol 1* (OECD: Paris.)

OECD (1995). 'OECD Environmental Data Compendium 1995.' (OECD: Paris.)

Peat, J.K., van den Berg, R.H., Green, W.F., Mellis, C.M., Leeder, S.R., Woolcock, A.J. (1994). Changing prevalence of asthma in Australian children. *British Medical Journal* **308**, pp. 1591–6.

Pederson, E. O. (1980). 'Transportation in Cities.' (Pergamon: New York.)

Prime Minister's Urban Design Task Force (1994). 'Urban Design in Australia.' (Commonwealth of Australia: Canberra.)

Probert, B., Freeland, J., and Gregory, B. (1993). The future of work. *Impact.* **23**(1), pp. 11-17, 19-20.

Rees, W.E. (1992). Ecological footprints and appropriated carrying capacity: What urban economics leaves out. *Environment and Urbanization* **4**, pp.121-130.

REIWA (1994). 'Suburban House Price Survey: Ten Year Wrap Up (1983–93) WA.' (Real Estate Institute of Western Australia: Perth.)

Resources Assessment Commission (1993). 'Coastal Zone Inquiry.' *Final Report.* (AGPS: Canberra.)

Richards, L. (1994). Suburbia: domestic dreaming. In 'Suburban Dreaming' ed. L. Johnson. (Deakin University Press: Geelong.)

Robinson, I., and Newton, P.W. (1988). Settlement options for non-renewable resource development in Canada and Australia: a comparative evaluation and decision framework. In 'Resource Communities: Settlement and Workforce Issues', eds T.B. Brealey, C.C. Neil and P.W. Newton. (CSIRO: Melbourne.)

Samuel, E.S. (1989). Hazardous waste management-Australian review. *Australian Water and Wastewater Association 13th Federal Convention.*

Sarkissian, W., and Marcus, M. (1986). 'Housing as if People Mattered.' (University of California: USA.)

Sarkissian, W., and Walsh, K. (1995). 'Community Participation in Practice.' (Institute for Science and Technology Policy: Murdoch University.)

Sarkissian, W., Walsh, K., Linstad, A. and Roberts, S. (1995). 'Urban Land Authority Children's Impact Assessment Preliminary Draft Report.' (Urban Land Authority: Melbourne.)

Saunders, P. (1994). Poverty, inequality and social security. In 'Social Security Policy: Issues and Options', eds J. Disney and L. Briggs (AGPS: Canberra.)

SCC (1993). 'Living city: Sydney City Council's Blueprint for Sydney.' (Sydney City Council: Sydney.)

Social Justice Research Program into Locational Disadvantage (1991–95). (Department of Prime Minister and Cabinet: Canberra.)

Solomonsson, G.D., and MacSporan, C. (1994). Recycling of materials in building construction. In 'Buildings and the Environment.' *Proceedings of the 1st International Conference, CIB Task Group* Number 8, (Building Research Establishments: Watford, UK.)

Stimson, R.J. (1995). Processes of globalization, economic restructuring and the emergence of a new space economy of cities and regions. In 'Cities in Competition. Productive and Sustainable Cities for the 21st Century', eds. J.F. Brotchie, M. Batty, E. Blakely, P. Hall, and P.W. Newton. (Longmans: Melbourne.)

Stretton, H. (1994). Transport and the structure of Australian cities. *Australian Planner,* **31** (3), pp. 131–6.

SWB (1991a). 'Drinking Water Basics.' (Sydney Water Board: Sydney.)

SWB (1991b). 'Sydney Water Board Clean Waterways Programme - Stategic Plan for Wastewater.' (Sydney Water Board: Sydney.)

Tranter, P. (1994). Behind closed doors: Women, girls and mobility. *On the Move* Conference, Adelaide.

Travers, P., and Richardson, S. (1993). 'Living Decently.' (Melbourne University Press: Melbourne.)

Troy, P. (1992). The new feudalism. *Urban Futures.* **2**, pp. 36-44.

Victorian Department of Energy and Minerals (DEM) (1995). Data provided in correspondence to Human Settlements Reference Group, 1996 SoE Project.

Wackernagel, M., McIntosh, J., Rees, W.E. and Woolard, W. (1993). 'How Big is our Ecological Footprint? A Handbook for Estimating a Community's Appropriated Carrying Capacity.' (Canadian Task Force on Planning Healthy and Sustainable Communities: Vancouver.)

WA Ministry of Planning (1995). Data provided in correspondence to Human Settlements Reference Group, 1996 SoE Project.

WHO (1994). 'World Health Statistics Annual, 1993.' (World Health Organisation: Geneva.)

Wolman, A. (1965). The Metabolism of the City. *Scientific American.* **213**, pp. 179.

Wulff, M., Flood, J., and Newton, P. (1993). 'Population Movements and Social Justice.' *Social Justice Research Program into Locational Disadvantage Report* No.11 (Department of Prime Minister and Cabinet: Canberra.)

Wynhausen, E. (1995). Struggle street. *The Weekend Australian (Review Section) Dec 16-17 1995.*

Zann L (1995). 'Our Sea, Our Future: Major Findings of the State of the Marine Environment Report for Australia.' (Great Barrier Marine Park Authority for DEST, Ocean Rescue 2000 Program: Townsville.)

Acknowledgments

Research Team:

Tony Burgon, CSIRO

Neil Blake, Murdoch University

Bill Grant, Monash University

Virginia Rapson, Monash University

Kurt Seemann, Centre for Appropriate Technology

Peter Vintila, Murdoch University

The following people reviewed the chapter in draft form and provided constructive comments:

Professor Mike Berry (Australian Housing and Urban Research Institute)

Professor Stephen Boyden (Australian National University)

Dr Kevin O'Connor (Monash University)

Dr Tom Spurling (CSIRO Division of Chemicals and Polymers)

Professor Hedley Peach (University of Melbourne)

In addition, Commonwealth Government departments and members of the Commonwealth/State ANZECC State of the Environment Reporting Taskforce also helped identify errors of fact or omission.

Photo credits

3-1: Dept Foreign Affairs and Trade

3-07: Nick Pitsas (Oryx Films)

3-08: (*from top*) Nick Alexander (Oryx Films); Peter Newman

3-12: Gold Coast Tourism Bureau

3-13: Margie McKenzie

3-14: Dione Davidson (*The West Australian*)

3-15: Argyle Diamond Mines Pty Ltd

3-17: Bruce Walker

3-26: Nick Pitsas (Oryx Films)

3-29: Dept Foreign Affairs and Trade

3-33: Bill Van Aken (CSIRO Division of Water Resources)

3-37: Peter Newman

3-38: Kings Park and Botanic Gardens

3-41: Peter Newman

3-42: Dept Environment, Sport and Territories

3-43: Geoff Diver (ISTP)

3-45: *Sydney Morning Herald*

3-47: Greenpeace Australia

3-49: Public Transport Corporation (Victoria)

3-51: Peter Newman

Chapter 4

Biodiversity

'Still Flying' from the painting of a Wandering Albatross by Richard Weatherly.

Prepared by

Denis Saunders (Chair), CSIRO Division of Wildlife and Ecology

Andrew Beattie, Centre for Biodiversity and Bioresources, School of Biological Sciences, Macquarie University

Susannah Eliott (Research Assistant/Science Writer), Centre for Science Communication, University of Technology, Sydney

Marilyn Fox, School of Geography, University of New South Wales

Burke Hill, CSIRO Division of Fisheries

Bob Pressey, New South Wales National Parks and Wildlife Service

Duncan Veal, Centre for Biodiversity and Bioresources, School of Biological Sciences, Macquarie University

Jackie Venning, State of Environment Reporting, South Australian Department of Environment and Natural Resources

Mathew Maliel (State of the Environment Reporting Unit member), Department of the Environment, Sport and Territories (Facilitator)

Charlie Zammit (former State of the Environment Reporting Unit member), Department of the Environment, Sport and Territories (former Facilitator)

Contents

Boxes

Introduction

Australia has an immense number of unique and unusual plants, animals and micro-organisms. More than one million species (including micro-organisms) are thought to live in Australia, but less than 15 per cent have been formally described. Not only is this living resource part of our cultural identity, we depend on it for our survival and quality of life. Lack of knowledge about the diversity of life on this continent and the effect of our activities on this fundamental resource pose the most significant impediment to its conservation and management.

Biodiversity is the variety of all life forms — the different plants, animals and micro-organisms, the genes they contain and the ecosystems of which they form a part. Consequently, biodiversity is considered at three levels: ecosystem diversity, species diversity and genetic diversity.

Genetic diversity: These starfish belong to a single species (*Patiriella calcar*), but various colour patterns result from different genes.

▼

▶

Species diversity: A small area of reef can contain a wide range of species — fish, corals, feather stars and algae.

Ecosystem diversity: Coastal coral reefs, sandy shores, rainforest and grassland provide different ecosystems that support different plants and animals.

▼

The species in a given area interact with each other, and with their environment, to form complex networks known as ecosystems. These differ from place to place, thus creating ecosystem diversity. Each ecosystem differs from all others because it contains a unique combination of species (and therefore genes) and because these species interact with each other and with each environment in distinctive ways.

Species diversity is the number of species and their relative abundance in a defined area.

Genetic diversity is the variety of genes contained in all the species in a given area. There are so many genes and different possible combinations of genes that, for most types of organisms, every individual, population and species is genetically distinct.

As a party to the 1992 international Convention on Biological Diversity, Australia is committed to the conservation of biodiversity, to the sustainable use of ecosystems, species and genetic resources and to the equitable sharing of any benefits arising from the utilisation of its genetic resources. The Australian Commonwealth, State, and Territory governments are responsible for managing human activities that threaten biodiversity. It is of real concern that species, including many that are as yet unrecorded, could become extinct due to human activities. For this reason, lack of scientific certainty regarding the impacts of human activities on biodiversity cannot be used to justify postponing measures to prevent environmental degradation.

Australia has evolved in relative isolation for at least the past 50 million years. This has resulted in a rich diversity of unique life forms. For example, 85 per cent of our flowering plants, 84 per cent of our mammals, 45 per cent of the birds, 89 per cent of the reptiles and 93 per cent of our frogs are found nowhere else (that is, they are endemic). Of the 600 inshore species of finfish in the southern temperate zone, about 85 per cent are found only in Australian waters.

While most of the large organisms, such as flowering plants and vertebrates, have been described and in some cases studied in detail, relatively little is known about the vast and less-visible world of invertebrates and micro-organisms. Australia has an invertebrate fauna estimated at around 225 000 species, of which only half have been described. Where limited data are available, however, some groups are known to be endemic. For example, 90 per cent of Australian springtails (Collembola) and 80 per cent of tiger beetles (Cincindelinae) are found nowhere else. The high level of endemism found in our plants and animals is also likely to be reflected among the micro-organisms. However, even less is known about micro-organisms, their ecology and their importance in ecosystems and the vast majority of them have yet to be identified.

Areas like the Great Barrier Reef, the species-rich rainforests of northern Queensland and the Southwest Botanical Province of Western Australia (with over one-third of Australia's plant species, of which 70 per cent are endemic) are internationally recognised major centres of biodiversity. Australia is the second-driest continent in the world (after Antarctica) and its inner arid core accounts for more than 70 per cent of the total land mass. Contrary to popular belief, this arid zone comprises many different kinds of communities and habitats, some with their own unique species and many that are highly sensitive to disturbance.

The four main reasons for preserving biodiversity relate to ecosystem processes, ethics, aesthetics and culture, and economics.

Biodiversity provides the critical processes that make life possible and that are often taken for granted. Healthy, functioning ecosystems are necessary to maintain the quality of the atmosphere, and to maintain and regulate the climate, fresh water, soil formation, cycling of nutrients and disposal of wastes (often referred to as ecosystem services). Biodiversity is essential for controlling pest plants, animals and diseases, for pollinating crops and for providing food, clothing and many kinds of raw materials.

Ethical values reflect the view that all species have an inherent right to exist. Biodiversity belongs to the future as well as the present and no species or generation has the right to sequester it for their exclusive use.

Biodiversity has intrinsic values, such as beauty, tranquility and isolation. Many Australians place high values on native plants and animals. This regard has contributed to our sense of cultural identity and is important for spiritual enrichment and recreation. Retaining biodiversity is also critical for maintaining the culture of Aboriginal and Torres Strait Islander peoples.

Some elements of biodiversity have economic value and can be used to create wealth. Australia's plants and animals attract tourists and provide food, medicines and other pharmaceutical products, energy and building materials. The commercial fishing industry, for example, produced catch worth 1.7 billion dollars in 1994–95 (ABARE, 1995). Tourists to six protected areas in 1991–92 spent more than two billion dollars (Driml, 1994). Ecotourism, which is a growing environmental business in Australia, has the potential to enhance the effect of other conservation activities, such as education and donations.

The Pilbara, Western Australia; Australia's arid zone contains unique ecosystems, plants and animals.

Another environmental business attracting attention is biodiversity prospecting. Sponges, for example, appear to be particularly rich in anti-cancer compounds. Other potentially valuable compounds include a sunscreen derived from a chemical that shallow-water corals use to protect themselves from ultraviolet rays.

Genetic resources from the wild

Natural compounds

Many chemical compounds produced in nature are used in medicine and technology. While an unknown number remain to be discovered, high levels of investment by the pharmaceutical industry suggest major future commercial prospects. Genetic resources are particularly abundant in species-rich habitats, such as rainforests and the ocean, but useful genes can be found in any type of environment.

Wild crop relatives

Crops usually have less genetic diversity than the wild populations from which they were derived because they have been intensively selected for many generations. Wild relatives often harbour extremely useful genes, like those that protect against disease or drought. Plant breeders can utilise the genetic diversity of wild crop relatives by hybridising domestic and wild varieties and choosing offspring with improved characteristics.

The soybean (*Glycine max*), for example, has a very narrow genetic base (Brown *et al.*, 1985). Plants cultivated throughout the world are descended from a few founders from a single geographic area. However, the genus *Glycine* contains about 16 species of wild soybeans, many of which are native to tropical northern Australia. These wild species are an important resource for the improvement of domesticated varieties. Research is underway to transfer the genes for rust and mould resistance found in native Australian species into agricultural varieties.

Like soybean, cotton has native Australian relatives including Sturt's desert rose (*Gossypium sturtianum*), a popular garden plant. Wild Australian cotton species in the genus *Gossypium* are being studied for their potential for crop improvement. Cottonseed is used as a stock food, but must be processed to remove the naturally occurring insecticide that the plant manufactures for its own defence. Australian *Gossypium* spp. have low levels of this compound in their seeds but normal levels in the rest of the plant. Experiments are being conducted to transfer this genetic trait into cultivated cotton varieties.

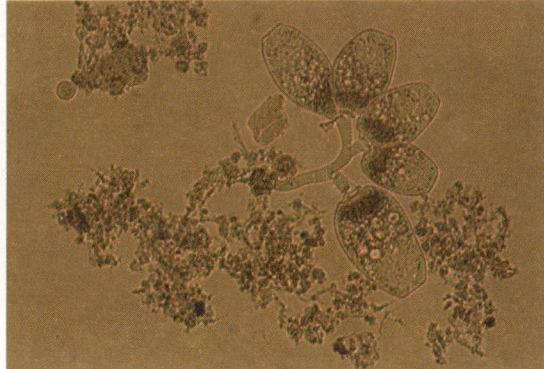

Micro-organisms in activated sewage sludge. Microbial biodiversity is essential for the many ecosystem services such as the breakdown of waste material and cycling of nutrients.

Elabana Falls, Lamington National Park; biodiversity values include beauty, tranquility and also vital ecosystem services: natural vegetation acts as an effective filter for sediment in catchments. When the vegetation is removed, the run-off becomes discoloured with sediment and creeks silt up.

Biodiversity provides a vast library of genetic material for use now and in the future, for a variety of industries including agriculture, medicine and gene technology. A loss in genetic diversity would result in a loss of potential for these industries. Thus, the challenge for Australians is to balance the exploitation and preservation of biological resources to ensure that exploitation does not compromise the options for the future.

The rate at which development has occurred in Australia is almost unprecedented. For example, in less than 200 years, land use in the wheat and sheep zones of Australia has changed from one of predominantly hunter–gatherer to one of intensive and extensive harvesting, compared with more than 10 000 years for the same evolution in the Middle East (Hobbs and Hopkins, 1990). Since European settlement, Australia has lost an estimated 75 per cent of its rainforests (Winter *et al.*, 1987) and about 40 per cent of its total forest area (AUSLIG, 1990). Nearly 70 per cent of all native vegetation has been removed or significantly modified by human activity since 1788 (see pages 6-8 and 6-9). The rate of land clearance has accelerated over time, with as much cleared during the last 50 years as in the 150 years before 1945.

While primary ecological processes are well understood, we know little about the ecological role of individual species. The maintenance of soil structure and fertility, for example, depends largely on the activity of groups of poorly understood organisms that constitute soil biodiversity. Loss of these organisms results in the disruption of processes essential to agriculture, such as water intake, nitrogen fixation and other types of nutrient cycling. Thus, by failing to take appropriate action to conserve biodiversity, Australia could be losing species vital to the sustainability of its rural industries.

State of the Environment reporting provides a framework to monitor and report on aspects of biodiversity essential to Australia's future, and to report on responses to changes in its condition. This chapter examines the pressures on biodiversity, as well as its current condition or state, and reports on responses to these pressures and states.

The vast majority of the organisms comprising soil biodiversity are small and generally unknown

Soil biodiversity

The highly diverse biota of the soil performs a wide range of important ecosystem services. They maintain soil health and fertility by cycling carbon, nitrogen, sulfur and phosphorus and by maintaining soil structure.

Human activities — such as cultivation, irrigation and the application of fertilisers and pesticides — alter both chemical and physical characteristics of the soil. Information on the effects of these impacts on soil biodiversity and key ecosystem services is fragmentary. However, it has become apparent that increasing acidification of soil in many parts of Australia is adversely affecting the nitrogen-fixing bacteria so important to the growth of many crop and native plants. Loss of the soil's biodiversity, as a consequence of cultivation, is associated with loss of its structure, making it subject to erosion. The destruction of its mycorrhizal fungi, which form a symbiotic association with many plants, can adversely affect the prospects of recolonisation of those plants.

While most soil biodiversity is yet to be discovered, it is clear that species of soil biota are tightly linked functionally to above-ground biota. Thus, preservation of familiar plant and animal biodiversity is dependent on a diverse soil biota.

Pressure

The ecosystem context

The major pressures on biodiversity today result directly and indirectly from the increasing human population and our lifestyles and expectations: our needs and desires for food, water, housing, energy, transportation, recreation and many other aspects of modern living.

The pressures develop because all human activities are carried out within, and thus form part of, some kind of ecosystem. This is most obvious in the country, where farms are established in grassland or woodland ecosystems, remnants of which survive between cultivated fields or in the corners of paddocks. Fishing is clearly an activity where species are removed from aquatic ecosystems. Towns, suburbs, industrial areas and cities also form part of ecosystems that existed long before development took place. Fragments of these original ecosystems persist even when the environment appears to be entirely one of human construction; one such remnant, for example, is the urban wildlife that colonises inner city suburbs or abandoned industrial sites.

Ecosystems are the result of long-term interactions between the physical environment and living species that produce characteristic processes and structural features. Well-established processes include, for example, food chains and soil formation, the development of particular groups of species called communities and, within each species, gene pools distributed among populations of different sizes. Evolution continuously modifies ecosystems, communities and species, making change a natural part of each of these entities. So too is disturbance; many species are unable to gain a foothold in established communities but flourish when dominant species are disturbed. The scale of disturbance may be as small as soil turnover at a wombat burrow or as large as a bushfire, but both can cause changes in the species present.

Human activities almost invariably affect the pace and alter the direction of change and the extent of disturbance, challenging the ability of ecosystems, communities and species to respond (Saunders *et al.*, 1990). Responses are often far-reaching because of the interactions between species and between different species and their environment, and so human pressures cause a cascade of effects.

Cascading effects

The best-known examples result from the removal of vegetation for agriculture, forestry or urban development. Clearing is often so rapid and extensive that natural systems cannot recover. The removal of plant species results in the loss of food for herbivores and, consequently, carnivores further up the food chain. Removal of plant cover leads to the loss of soils through erosion, or of soil nutrients through leaching. Both processes reduce the vast complexes of minute species that comprise soil biodiversity. Urbanisation and pastoral and agricultural programs that suppress the regeneration of native vegetation make these

▲

The 'living dead'; many long-lived organisms such as some trees in agricultural landscapes do not regenerate due to changes resulting from human activities.

changes and losses long-term, perhaps permanent. Loss of vegetation such as mangroves or seagrasses at coastal sites may have far-reaching effects because the juvenile nursery grounds of fish, crabs and prawns are destroyed.

Clearing rarely involves entire landscapes, and much effort has been made to understand the biodiversity of remnant vegetation and other communities isolated, like islands, in 'oceans' of rural or urban development. This process, known as fragmentation, often conceals cascading effects that are subtle and hard to detect, but nevertheless highly threatening. For example, trees in vegetation remnants in city bushland or rural paddocks are often very long-lived and give the appearance of healthy greenery. However, many of these remnants have no seedlings or saplings. It could be that the insects that once pollinated the flowers have become locally extinct, so the trees do not produce seed, or that ecological conditions have changed so the species can no longer regenerate.

Table 4.1 Pressures on plant biodiversity: major causes of extinction and past and present threats to endangered plant species

Threat/cause	Number of species presumed extinct	Endangered species Past threat	Endangered species Present and future threat
Agriculture	44	112	50
Grazing	34	51	55
Low numbers	-	10	85
Weed competition	4	12	57
Roadworks	1	8	57
Industrial and urban development	3	20	21
Fire frequency	-	10	17
Forestry	-	10	10
Collecting	-	6	17
Mining	1	3	11

Note: Many species are affected by more than one threat. In some cases the past threat may have ceased and new ones arisen. Other threats include recreation, dieback, clearing, railway maintenance, salinity, insect attack, quarrying, trampling by pigs and buffalo, drainage and flooding.

Source: Leigh and Briggs, 1994.

Table 4.2 Sources of current threats to Australian birds, marsupials, rodents, reptiles and freshwater fish

Threatening process	Birds C	Birds S	Marsupials C	Marsupials S	Rodents C	Rodents S	Reptiles S	Freshwater fish C	Freshwater fish S
Habitat clearance and/or fragmentation	32	4	13	3	3	4	35		
Altered fire regimes	16	35	1	16		2	10		
Grazing and/or trampling	10	35	5		1	6	21		
Fishing		3							
Disease		3		1					
Pollution		7					7		
Erosion	1	1							
Environmental weeds	2	9					5		
Forestry operations	3	14	2	1		1	6		
Changed hydrological regimes	1	3						5	4
Climatic variations	2	7					5		
Shortage of nest hollows	3	20	1						
Predation	8	29	9	13	1	4	14		
Competition	3	20	1	11	1				
Direct exploitation	10	33	2					3	1
Cropping							21		
Urban development	4	3					14		
Pasture improvement							12		
Soil degradation							9		
Visitor disturbance							8		
Mining	2	4					6		
Rabbit grazing				11	1	2	6		
Habitat drainage							4		
Rock removal							4		
Geomorphic alteration								12	6
Water quality								4	1
Introduced exotic species			9		1	10		5	10
Introduced native species				14					3
Loss of genetic diversity			1	1					2
Road kills			1						
Unknown				4		3			

Note: **C** — confirmed **S** — speculative

The figures in each column are the number of species affected by each process. However, a species may be affected by more than one process.

Source: Garnett, 1992 (a & b); Cogger *et al.*, 1993; Wager and Jackson, 1993; Lee, 1995; and Kennedy, 1992.

Many micro-organisms are crucial for the decomposition of organic material. Fallen branches and logs in forests are broken down by the action of fungi whose reproductive structures are mushrooms, toadstools and bracket fungi.

Another reason for the absence of seeds can be loss of genetic diversity. As a population shrinks, so does its gene pool. Genes that help species to adapt to changing environments, to win competitive interactions with other species or to combat disease, disappear along with the individuals that carry them. These genes are then unavailable to the survivors. In small populations, the level of inbreeding increases, which further decreases the ability to adapt to current and future challenges. Many species become highly vulnerable when populations decline because they have strong mechanisms to prevent inbreeding and thus individuals have few suitable mates.

Biologists believe that many small populations enter an 'extinction vortex', which occurs when small numbers of individuals harbour such low levels of genetic variation that few offspring survive. This further reduces their genetic resources and the chances of finding a suitable mate. The process continues until a final generation has so few individuals and such reduced genetic diversity that reproduction is no longer possible. This may happen quickly in small, short-lived species such as butterflies, but may take a century or more for bigger, long-lived species such as large trees.

Cascading effects commonly follow the introduction of exotic plants, animals or micro-organisms. Introduced weeds have effects that start at the base of the food chain. They displace native species and even entire communities of native plants. The effects flow on to animals, such as insects and birds, that depend on them for food and shelter. Higher up the food chain, introduced predators feed on native animals, both diminishing their numbers and competing with native predators. Marine animals from other parts of the world, introduced into coastal waters from ships' ballast, enter and change native food chains and, like weeds, can dominate local communities. Introduced micro-organisms, such as dieback fungus (*Phytophthora cinnamomi*), invade plant communities, killing selected species, and disrupting ecosystem processes and food webs.

Ecosystem health

Assessing the health of ecosystems — that is, determining whether or not their structure and processes are intact and they contain the expected variety of species — has become an important research objective. In forestry, for example, the removal of trees for sawlogs and for chipping disrupts forest structure, function and floristics. Forest-dwelling species rely on the structure or architecture provided by trees, shrubs and other plants for food, nesting places and concealment. The vegetation also plays a major part in regulating nutrients in the forest. When it is removed, branches and fallen trees that normally rot on the forest floor and recycle nutrients back to the soil can no longer do so, while rainfall, which is normally scattered by the tree canopy, falls directly to earth, flushing soil nutrients and organic matter out of the ecosystem.

Human populations

Our increasing human population with its affluence and technology, has resulted in increasing demands for natural resources and, in turn, increasing pressures on biodiversity (see Fig. 4.1).

Since 1788 when Europeans colonised Australia, about eight generations have lived on the continent and wrought vast changes to the natural environment. Habitat modification, particularly removal of vegetation for agriculture, urban development and forestry, has been and still is the most significant cause of loss of biodiversity (see Tables 4.1 and 4.2). Tourism, altered fire regimes and the effects of introduced plants and animals also adversely affect Australia's biodiversity.

Urban development

Australia has a relatively small, highly urbanised population, with about 86 per cent of people living on the coastal fringe (RAC, 1993). About 66 per cent live in coastal towns and cities of more than 100 000 people, located on harbours and estuaries with considerable biodiversity and habitat richness (NPC, 1992). The concentration of people in these areas generates a range of pressures on biodiversity throughout the continent, caused by destruction of natural habitat, harvesting of plants and animals, the spread of exotic species and pollution (see pages 4-21 and 8-8).

Huge parts of the continent are used for primary production to feed and clothe Australia's population as well as tens of millions of people living in other countries. The export of goods is a major factor in maintaining the high standard of living that most Australians enjoy. Some of the most intense pressures on biodiversity are therefore in agricultural and pastoral districts that are relatively sparsely populated.

Tourism and recreation

Little research has been carried out on the effects of tourism on biodiversity. The rapid growth of the tourist industry in Australia is a potential threat that requires continual monitoring and regulation. Since tourism focuses on areas of particular appeal that often have high biodiversity, the pressures it exerts can be disproportionately high and often result in the destruction of the qualities that initially attracted tourists to the area.

Recreational activities frequently affect biodiversity in ways that are not immediately obvious. For example, some popular beaches on the New South Wales coast have been protected from sharks since the late 1930s and some in Queensland since the early 1960s. In Queensland, more than 30 000 large sharks had been captured in shark nets or on lines by 1988 (Paterson, 1990). Catch rates for large sharks declined by 75 per cent over 25 years, suggesting a substantial fall in the shark population. Similar declines have been reported for whaler and white pointer sharks off New South Wales (Reid and Krogh, 1992). Large mesh shark nets also catch other large animals that are generally slow-growing with a low replacement

Figure 4.1 Major pressures on biodiversity

Human population
(size, affluence and technology)

Population patterns
Harvesting
Land use
Introduced species
Pollution
Mining
Climate change

Biodiversity
Ecosystem—Species—Genetic

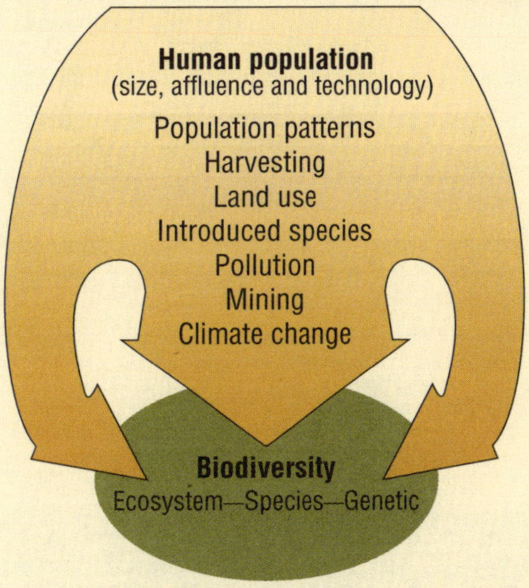

Increasing population size with its attendant demands for natural resources places increasing pressure on biodiversity

rate. Records from shark nets off Newcastle show mean annual catches of 111 rays, seven dolphins, four turtles, 25 jewfish and four tuna.

Like other types of urban growth, tourist developments — such as resort hotels and golf courses in wetland areas — can have a large impact on biodiversity in the surrounding region. The developments often take place in sensitive ecosystems, such as at the edge of national parks, on floodplains or close to beaches, accentuating the effect. An example is the increase in numbers of silver gulls (*Larus novaehollandiae*) and other scavenging birds, due to the availability of food from urban waste dumps. Populations of silver gulls are increasing at a rate of 10 to 13 per cent per year (Blaber *et al.*, in press). Waste dumps at tourist resorts on islands encourage gulls, which also prey on eggs of other seabirds, increasing

Changes in urban birds

The County of Cumberland (4273 sq km) surrounds and includes Sydney. It is the site of the first European settlement in Australia and represents the largest concentration of the human population, with 27.5 per cent of Australians living there (Cocks, 1992). Development of this area has had a major impact on the biota, as illustrated by changes to the bird community. At the time of European settlement 283 species of birds were believed to have occurred there (excluding seabirds and rare vagrants) (Hoskin *et al.*, 1991). Of these, 11 species (four per cent) are now locally extinct, 76 (27 per cent) have decreased in range and/or abundance and only 39 (14 per cent) have increased in range and/or abundance. As well, five Australian species have invaded the area because the changes imposed on the landscape suited them and 20 exotic species were deliberately released and have established viable populations.

pressure on other species. On Rottnest Island, for example, predation by silver gulls poses a serious threat to fairy terns (*Sterna nereis*), which breed there.

Pressures on biodiversity are synergistic, the impact of several activities compounding to produce a much greater impact than any one alone. The mallee woodland in Western Australia, for example, has already been severely reduced by clearing for agriculture. Demand in the tourist industry for didgeridoos made from mallee is an added pressure that on its own would be insignificant. Other impacts from tourism that can create cumulative effects are soil erosion caused by four-wheel drive vehicles, increases in the frequency of wildfires and localised nutrient-enrichment of the soil and water from food scraps and other wastes.

▶ Seabed on the north-west shelf of Australia; communities of sponges and large fish are found in several areas around Australia.

▲ Prawn trawler bycatch: in addition to prawns, the trawl collects a variety of animals. Although these will be dumped back into the sea, most (except for hardy species such as bivalves and crabs) will not survive. Dolphins, seabirds and sharks feed on dead animals at the surface while fish and crabs eat them on the seabed (Hill and Wassenberg, 1990). They make up an important part of the diet of these scavengers and appear to have contributed to increases in populations of some, such as crested terns (*Sterna bergii*) in the Gulf of Carpentaria (Blaber *et al.*, in press).

Harvesting resources and land use

A number of Australian industries are based on the exploitation of natural resources. They include fisheries, forestry and industries based on the use of wildlife and wildflowers. While harvesting natural resources is not necessarily detrimental to biodiversity, many harvesting practices are not sustainable. These can cause habitat loss, fragmentation and/or modification and reduced populations through overexploitation, or a combination of these pressures.

Fisheries

Australia exploits its marine biodiversity — with commercial harvesting of four species of crabs, four of lobster, 12 of prawn, three each of abalone and scallops and about 300 species of finfish. Under good management, the number of fish that commercial fishers take can be sustainable, especially if they use species that are present in large numbers and that have a high reproductive rate. However, high fishing pressure, especially on long-lived species, usually leads to over-exploitation. For example, catches of southern bluefin tuna (*Thunnus maccoyi*), which can live to over 20 years, peaked at 80 000 tonnes in 1961 but have since declined steadily (see Fig. 8.13). The breeding stock of this species is now estimated to be less than 10 per cent of the original (CSIRO Division of Fisheries, unpublished data). While the commercial catch of many species has been regulated and quantified, the recreational catch has been largely uncontrolled and for some popular angling species, such as snapper (*Pagrus auratus*), exceeds the commercial catch.

Heavy fishing pressure can be severe when populations are low due to adverse environmental factors. The gemfish (*Rexea solandri*) was once an abundant species off New South Wales but appears to have been depleted by a combination of adverse oceanographic conditions and heavy fishing (see Fig. 8.14).

Some forms of harvesting — trawling and dredging — are non-selective and kill large numbers of non-target species. The unwanted catch (bycatch) can be very large, as in the case of the prawn fishery in northern Australian waters, which catches and discards a total of about 38 000 tonnes of other species. This is four to six times the weight of prawns caught (Pender *et al.*, 1992). Fisheries research has concentrated on the impact of fishing on commercially valuable species and the effects of bycatches on other species are poorly understood.

Trawling can cause long-lasting damage to the seabed, resulting in a loss of biodiversity in trawled regions. In some areas, the seabed has a complex community of sponges, soft and hard corals and other bottom-dwelling organisms as well as species of large edible fish. In the 1970s, foreign trawlers exploited the fish resource on the north-western coast of Australia and, in the process, damaged and removed some of the seabed structure. This resulted in loss of habitat and a dramatic decrease in the numbers of commercially valuable fish. Large areas were subsequently closed to trawling and alternative methods of fishing encouraged. The fishery and seabed fauna are now recovering (Sainsbury, 1988).

A minimum size limit — usually set at the size at which the animals first mature — protects many commercially and recreationally important species of fish and marine invertebrates. If the stocks are subjected to heavy fishing pressure, few individuals larger than the minimum legal size survive to spawn and so the successful spawners are those individuals that do so at a smaller size. This process acts as a selection pressure favouring individuals with genes associated with spawning at a smaller size. For example, it appears to be responsible for a downward shift in size at maturity of King George whiting (*Sillaginodes punctata*) in South Australia and could cause a reduction in genetic diversity with the loss from the population of genes for maturity at a larger or older stage (Hill, 1992).

Forestry

Forests cover a small fraction of the continent (see pages 6-8 and 6-9 and Fig. 4.2). Logging operations, including the building of access roads, result in major structural changes in the vegetation and soil from which forests and their fauna may take a long time to recover. Recovery of biodiversity depends upon many factors, including the ages and species of the trees remaining, the extent of soil erosion and loss of nutrients and the areas left as sources for recolonisation by the forest flora, fauna and micro-organisms. Researchers have studied the recovery of biodiversity from previous logging operations in a range of Australian forest environments, but as yet they have identified few generic principles.

The fragmentation of previously continuous areas of forest can have destructive effects on the forest flora and fauna. Some species will become locally extinct when confined to small 'islands' or remnants of habitat that do not support viable populations (see page 4-13).

Other uses can compound the pressures associated with forestry operations. Grazing, for example, often destroys or alters the composition of the understorey and the species that depend upon it, and may result in soil erosion. Tourism and recreational activities such as hiking, camping, off-road driving, horse-riding, hunting and fishing all have some impact on biodiversity. Individually the effects may be localised, but together they can exert major pressures on forest biodiversity.

Figure 4.2 Australian forest area by tenure, 1990

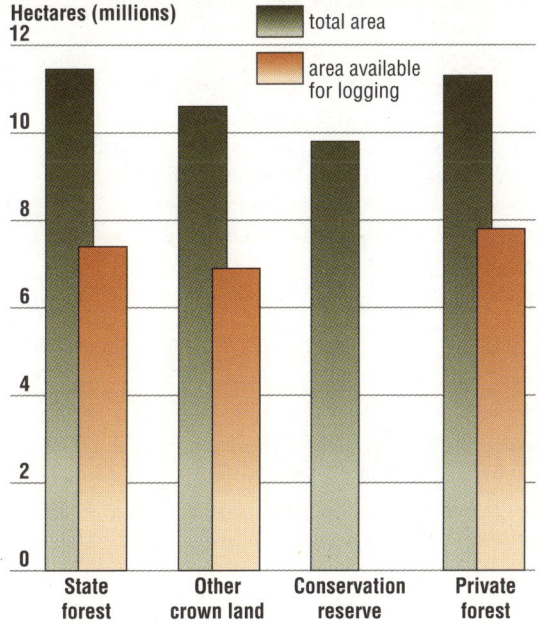

Source: Resource Assesment Commission, 1992.

Figure 4.3 The effects of different logging methods on forest ecosystem structure

Clearfelling is the localised removal of most or all trees followed by burning of debris. The soil tends to be exposed to erosion, and regrowth is often less diverse, both in terms of species and the age of trees.

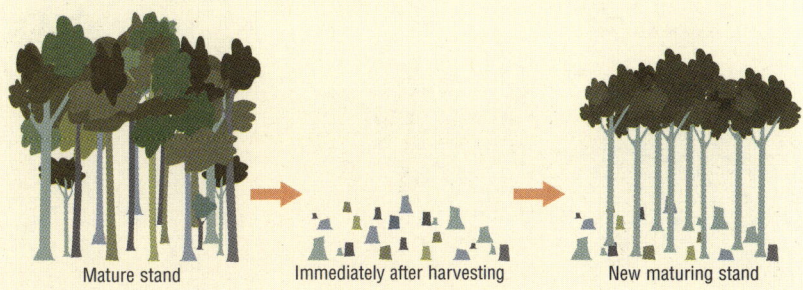

Mature stand — Immediately after harvesting — New maturing stand

Modified clearfelling keeps some trees for conservation purposes such as mammal habitat and to allow further growth of immature trees. The resulting forest retains a greater diversity of species and age classes.

Mature stand — Immediately after harvesting — New maturing stand

Shelterwood logging in highland forests minimises snow or frost damage to seedlings. Shelter trees are retained and then felled once some regrowth is established.

Mature stand — Regeneration after harvesting

Selective logging removes individuals or patches of trees at relatively short intervals. There are many variations including focus on a particular species or thinning of small trees. The general aim is to retain diversity of species, sizes and ages.

Single scattered trees are harvested — Trees above a specified girth are harvested — Selected groups of trees are harvested

Source: Resource Assessment Commission, 1992.

▲
A fence-line separating grazed from ungrazed land illustrates how soil erosion and loss of vegetation can result from overgrazing.

Pastoralism

The pastoral industry covers an area of 5.4 million sq km, or about 70 per cent of the continent. In the arid zone, despite lower stock densities, the impact of grazing on biodiversity can be greater than it is in high rainfall zones because low productivity limits forage and stock compete with native animals for limited resources. Introduced livestock trample vegetation and degrade soil structure, leading to changes in native vegetation cover. In turn, this results in erosion and loss of species associated with native vegetation. The provision of water through bore holes, earth tanks and dams, where water was formerly limiting, together with introduced predators, feral herbivores and altered fire regimes, contribute to changes in arid zone biodiversity.

Overgrazing of native pastures has three main effects on the diversity of native animals. Firstly, as stock remove long grass, native animals' shelter sites are depleted and the animals become exposed to predation and weather. Secondly, stock compact the soil and alter the soil texture, destroying burrows and making burrowing difficult. Finally,

stock compete with native herbivores for food and water, a situation which becomes critical in arid regions where such resources are scarce during drought. In some cases, the extra watering points associated with pastoralism have resulted in increases in the abundance of native species able to exploit the situation, like kangaroos (*Macropus* spp.) and emus (*Dromaius novaehollandiae*). This also increases grazing pressure on native vegetation.

Grazing in arid and semi-arid regions is thought to be partly responsible for the extinction of 34 plant species (41 per cent of the total number of plant species lost from Australia since European settlement) and continues to threaten a further 55 species in the rangelands or 24 per cent of plant species currently listed as endangered (Leigh and Briggs, 1992) (see Table 4.1).

Although the arid zone appears unchanged, about one-third of mammal species in the sandy and stony desert ecosystems are known to be regionally extinct (Burbidge *et al.*, 1988; Burbidge and McKenzie, 1989). This is the highest regional extinction rate on the Australian mainland (Kennedy, 1992). Many birds are also in decline. For example, eight per cent of arid zone birds are classified as rare and threatened nationally and a further five per cent are uncommon species that have declined or are at risk in two or more arid regions (Reid and Fleming, 1992).

Research agencies and pastoralists have introduced, and continue to introduce, many exotic grasses in an attempt to make rangelands more profitable. Since 1947, 463 exotic plant species have been introduced as pasture. Only five per cent of these have proved useful as fodder, yet 13 per cent have become problem weeds. These include mission grass (*Pennisetum polystachion*), which invades native bushland, outcompeting native grasses and changing fire regimes, and para grass (*Brachiaria mutica*), which has spread into Kakadu National Park, reducing habitats for waterbirds. As exotic pasture species are still being introduced, pastoralism poses a continuing threat to biodiversity in rangeland environments.

Agriculture

Agricultural practices adversely affect biodiversity in many ways. Following the initial destruction of native vegetation, soil tillage and burning of stubble lower the microbial biomass in the soil, affecting its structure and ecological processes such as decomposition rates and nutrient cycling. Soil micro-organisms help prevent erosion by binding soil particles together. Thus tillage can have a long-term impact on soil health, with adverse consequences for other soil organisms (see Chapter 6).

The long-term use of pesticides, herbicides and fertilisers has had direct and indirect effects on biodiversity. Pesticide residues enter the soil and aquatic ecosystems and may be ingested by organisms, with the harmful effects magnified up the food chain. Pesticides have been linked with the death of aquatic organisms in many areas such as the Namoi Valley in New South Wales and Maroochy River in Queensland. Their use is also

Figure 4.4 Percentage of Statistical Local Areas used for intensive production

▶
Intensive production refers to sown pastures and all crops for which the Australian Bureau of Statistics collects data. Cropping, the main activity, is dominated by wheat production, but includes other products as diverse as sugarcane, cotton, bananas and soybeans. Much of the eastern and south-western parts of the continent are used to some extent for intensive production, but not the arid interior or the monsoonal north, which are widely used for grazing on native rangelands.

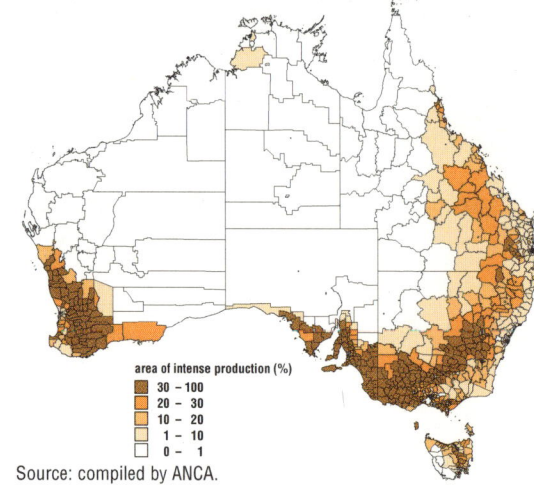

area of intense production (%)
- 30 – 100
- 20 – 30
- 10 – 20
- 1 – 10
- 0 – 1

Source: compiled by ANCA.

Vegetation clearance and fragmentation

Following European settlement, governments actively encouraged the clearing of native vegetation to make way for pastoralism and agriculture in the higher-rainfall areas. While it is difficult to get accurate estimates of changes in vegetation cover over the past 200 years, native vegetation (including regrowth) is still being cleared at a rate of over 600 000 hectares per year — about half the rate of clearing in the Brazilian Amazon in 1990–91. Most of this is taking place in Queensland and New South Wales (see page 6-40). Vegetation clearance and fragmentation are the major causes of loss of biodiversity (see Tables 4.1 and 4.2).

The situation in Kellerberrin, a shire in the central wheat belt of Western Australia about 200 km east of Perth, illustrates the effects on biological diversity of vegetation clearance and fragmentation (Hobbs and Saunders, 1993).

Clearing of native vegetation was not carried out randomly. The heavier soils on valley floors were cleared first and more extensively, as they were regarded as having most potential to support agriculture. In the landscape illustrated above, only five per cent of the valley floors (green and blue) are still covered with native vegetation compared with 63 per cent of the highest parts of the landscape (red).

In common with other intensively cleared regions in Australia, much of the remnant native vegetation is privately owned (more than 75 per cent in the diagram below) and is not regarded as part of the Crown conservation estate. Significant numbers of endangered or restricted species live on these privately owned remnants and so sympathetic management is required to ensure their survival.

Changes in ecosystem processes

Clearing of native vegetation leads to changes in rainfall interception and evapotranspiration, with more water flowing across the landscape and infiltrating the water table. In this case, little or no run-off or groundwater recharge occurred before clearing. After crops and pasture replaced native vegetation, rainfall run-off measured 25 per cent and recharge of groundwater was seven per cent. This has caused saline groundwaters to rise by more than 20 metres in some areas since clearing began. Consequently, soil salinity has increased, with various effects on remnant vegetation depending on its position in the landscape. On valley floors, for example, which have an average of 13 500 tonnes of total soluble salts per hectare stored in the soils, vegetation is highly susceptible and fresh-water ecosystems are under threat.

Increases in soil erosion cause topsoil and nutrients to move down slopes and into watercourses. Clearing also results in

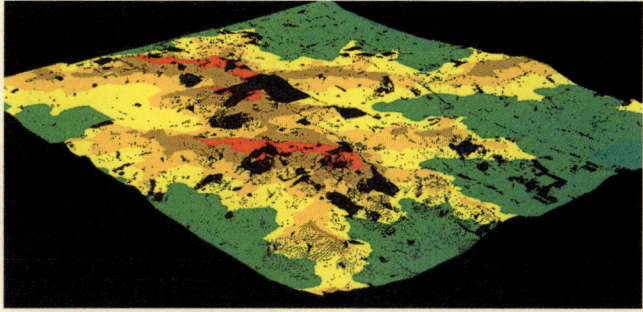

Digital elevation model of the Kellerberrin area; remnant vegetation is shown in black. Only five per cent of the valley floors (green and blue) are still covered with native vegetation compared with 63 per cent of the higher parts of the landscape (red). Source: CSIRO.

changes in the microclimate; radiation at ground level increases and higher soil temperatures occur. Cleared areas are subject to a much greater range of extreme temperatures. Wind speed is higher closer to the ground in cleared areas, leading to more rapid loss of soil moisture.

Changes in species distribution and abundance

Clearing and fragmentation of native vegetation result in loss of species dependent on it and increases in species dependent on the cleared agricultural matrix. For example, of 131 bird species recorded in Kellerberrin, 38 have declined or have become locally extinct since it was cleared and 18 have increased. These include 11 species that were introduced or invaded the area when suitable habitat was created in the agricultural land. Only nine of the 15 recorded species of native mammal (excluding bats) are still found there.

Isolation of species dependent on remnant vegetation

When a remnant is isolated from other native vegetation, it will be carrying more species than it can carry over time and so some will be lost. This loss will be most rapid for: species that depend entirely on native vegetation, for example, the yellow-plumed honeyeater (*Lichenostomus ornatus*), which is dependent on eucalypt woodlands, was the most common honeyeater in the Kellerberrin area and is now extinct there; those that require large territories, for example, the mallee fowl (*Leipoa ocellata*); and those that exist at low densities, such as the broad-faced potoroo (*Potorous platyops*). Loss will be slowest for eucalypts and other species with long generation times, like the salmon gums (*Eucalyptus salmonophloia*). These still occur along road verges, in agricultural land and in some remnant patches, but few seedlings are growing.

East Yorkrakine Nature Reserve (see left) was isolated from surrounding native vegetation in the late 1920s. Since 1974, four species of bird dependent on undisturbed native vegetation have disappeared from the reserve and others will be lost in future because the populations are too small and too isolated (they will not cross open agricultural land) to persist over time. Remnant vegetation is degrading and species are being lost because of: the creation of edges and an increase in edge specialists; the invasion of weeds; nutrient enrichment from agricultural practices (fertilisers or livestock excrement); grazing by domestic livestock, which is eliminating understorey and litter layers; and changed disturbance regimes such as fire frequencies.

 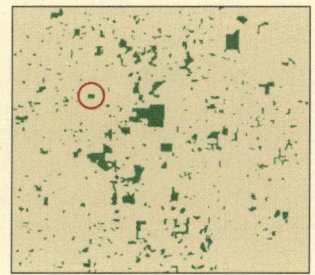

Distribution of native vegetation in the Kellerberrin area pre-1920 and in 1984. East Yorkrakine Nature Reserve is circled. Source: CSIRO.

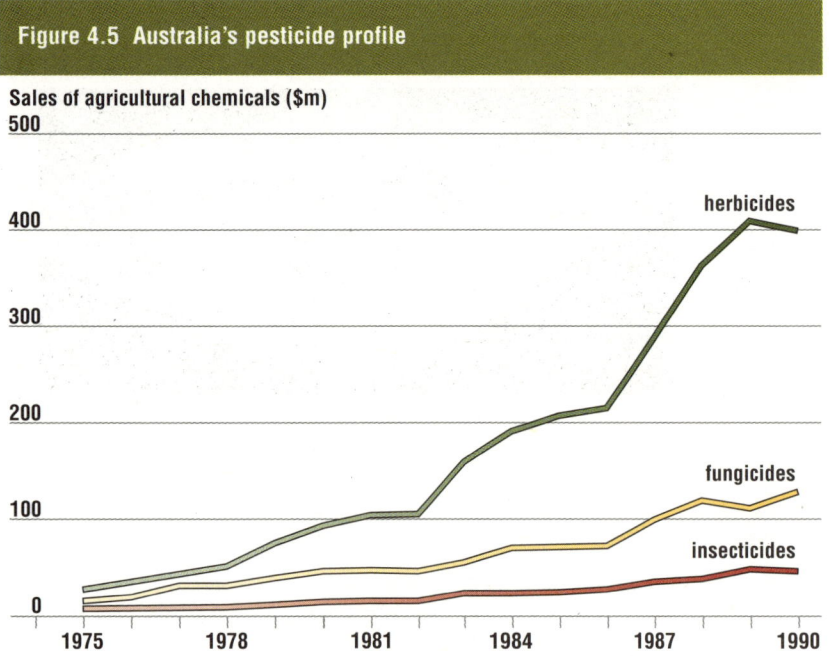

Figure 4.5 Australia's pesticide profile

Sales of agricultural chemicals ($m)

herbicides

fungicides

insecticides

Pesticides and herbicides are still widely used in Australia, as shown by the increasing sales trends. These chemicals kill native species as well as pests and so adversely impact on biodiversity. Source: AVCA, cited in Short, 1994.

associated with eggshell thinning in several species of Australian birds, particularly birds of prey and pelicans (*Pelecanus conspicillatus*) (Olsen *et al.*, 1993). Herbicides used to kill unwanted weed species also kill native plants.

Australia has large areas of low-nutrient soils so, to grow crops and improve pasture, farmers have made widespread and heavy use of fertiliser. The use of fertilisers over long periods has disturbed many soil and aquatic ecosystems by increasing their nutrient levels. Native vegetation is adapted to nutrient-poor soils and is resistant to invasion by exotic species unless soils are disturbed or enriched by nutrients. As this occurs, so does invasion by introduced weeds that are better-adapted to higher nutrient levels and to exploiting disturbed sites. Increasing nutrient levels in watercourses has led to algal blooms and an increase in other micro-organisms (see the box on page 4-15 and Fig. 4.6).

Vegetation clearance has caused major changes in the hydrological balance because native plants use more water by transpiration than most of the annual species that replaced them, and so more water flows across the landscape and enters the water table.

Water harvesting

Agricultural and urban development in Australia has dramatically altered the flow of many rivers and streams. These changes have been brought about because of our demand for water and a desire to mitigate floods. Many rivers, particularly in the south-east of the continent, have been severely affected by human demand for water. For example, 10 megalitres (one ML is a million litres) of an annual volume of 12.2 ML of water entering the Condamine–Ballone–Culgoa–Darling–Murray watercourse from the Murray–Darling basin is diverted for human use, thus significantly reducing the actual volume discharged (McComb and Lake, 1990). Many streams in the Mount Lofty Ranges of South Australia, which were permanent 50 years ago, no longer flow in summer. In fact, humans have significantly modified the flow of most rivers in south-eastern Australia.

Such changes have significant effects on biodiversity, which are most obvious when large fish such as Macquarie perch (*Macquaria ambigua*) or trout cod (*Maccullochella macquariensis*) become endangered. However, fish like these are near the top of complex food chains and it is important to realise that changes to the vertebrate fauna often reflect less obvious alterations in biodiversity. For example, the River Murray crayfish (*Euastacus armatus*) has declined in range and abundance and 13 of 14 native snails have disappeared from the banks of the Murray due to artificial changes in water level.

The threatening process responsible for loss of biodiversity in fresh-water ecosystems is habitat modification, which is caused by factors such as pollution, salinity, changes in water temperature or low oxygen. These conditions are often exacerbated by reduced flow rates connected with water harvesting.

Changes in the flow of a river may also affect the terrestrial biota surrounding a watercourse. For example, changing water levels in the Murray River due to flood control have been associated with the death of large numbers of trees and consequent loss of habitats. Water impoundments capture and store water that is subsequently redirected to other uses, such as irrigation schemes. These reduce the amount of channel submerged in water downstream resulting in a loss of aquatic habitat. Above the impoundment, they will have changed the environment from flowing water associated with periodically flooded wetlands to still water associated with a permanently flooded environment. Declines in wetland, riverine habitat and water quality are primary causes for the decline of several species of frog, aquatic tortoise and lizard. Some 32 species of frog have been recorded as being in decline, and only limited data are available for many other species. Such changes will have a significant impact on both the aquatic and surrounding terrestrial communities.

Changes in wetland habitats through alteration in water regimes, water quality and physical disturbance may all favour the spread of introduced or translocated aquatic plants. Infestations of many aquatic plants indicate that significant degradation has occurred or is occurring in infested rivers and wetlands.

In uncleared areas of arid and semi-arid regions of Australia, native flora and fauna utilise practically all water from rainfall or it evaporates and thus only rarely finds its way into the groundwater. This balance can easily be disrupted by clearing native vegetation (see the box on page 4-13) or by excessive irrigation. Irrigation involves drenching the soil with water, some of which will be discharged into the groundwater. Water deep in underground aquifers is usually saline (about 30 000 ppm) and thus a rise in water levels can result in salinisation of the soil and nearby rivers.

Pressures on aquatic biodiversity from land use

Estuaries and the sea

On many parts of the east coast of Australia, estuaries and floodplains of coastal rivers have been drained for a variety of agricultural and pastoral enterprises or for the establishment of townships and tourist facilities. This often results in the exposure of sediments containing iron pyrite. Oxidation of this mineral leads to the formation of highly acidic soils (acid sulfate soils) with highly acidic run-off, which can kill fish when flushed into rivers and estuaries (Callinan *et al.*, 1993). The fish may die from high acidity and other chemical properties of the water such as high levels of dissolved aluminium. Natural exposure of the acid-producing sediment also causes fish deaths.

Flood mitigation structures have exacerbated these problems by exposing more sediment in drainage works and changing drainage patterns. Floodgates prevent the normal tidal flushes that would neutralise or dilute the run-off water.

In southern Queensland and northern New South Wales, observers have noted seasonal fish kills where fish show symptoms of a skin disease identified by characteristic red spots that become lethal ulcers. The organism responsible, an exotic fungus (*Aphanomyces* sp.), cannot invade the intact skin of healthy fish. However, skin damage resulting from exposure to run-off from acid sulfate soils may allow the fungus to invade. Scientists have established strong links in the temporal and spatial patterns of rainfall, drainage from acid sulfate soil areas and outbreaks of 'red spot' disease.

Fish deaths are the most obvious symptom of acid run-off, but less mobile organisms such as crabs, worms, molluscs and seagrass beds are also severely affected. All of these changes disrupt estuarine food chains and have economic impacts on fisheries.

Rivers

Periodic outbreaks of blue-green algal 'blooms' in inland rivers are becoming predictable spring/summer events (see page 7-48). The timing of the blooms coincides with periods of low water levels and high temperatures. However, the main causal agents are the enrichment of river water with nutrients from farming land and effluents from rural cities and towns. The area sown to cotton has increased sharply over the past two decades (see the diagram above and Fig. 7.8). The nutrient load for the Darling Basin for an average year is 440 tonnes of phosphorus and 1890 tonnes of nitrogen.

A variety of organisms use the nutrients in the water, particularly fresh-water algae that, when they decay, use up the dissolved oxygen and create conditions unsuited to other aquatic organisms. Fish kills are one indication of deteriorating water conditions. As blue-green algae continue to grow, their toxic by-products increase in concentration and the water becomes poisonous to land animals. For example, waterfowl and stock may die from drinking the river water and towns can no longer rely on it for domestic purposes.

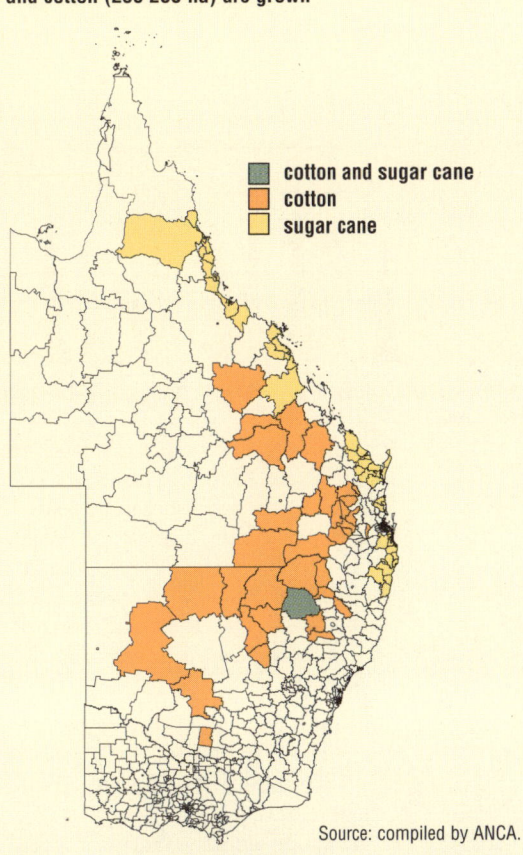

Statistical Local Areas in which sugarcane (162 500 ha) and cotton (235 200 ha) are grown

- cotton and sugar cane
- cotton
- sugar cane

Source: compiled by ANCA.

This green water taken from the Darling River indicates the presence of an algal bloom. High nutrient levels can cause algal blooms, sometimes resulting in fish kills.

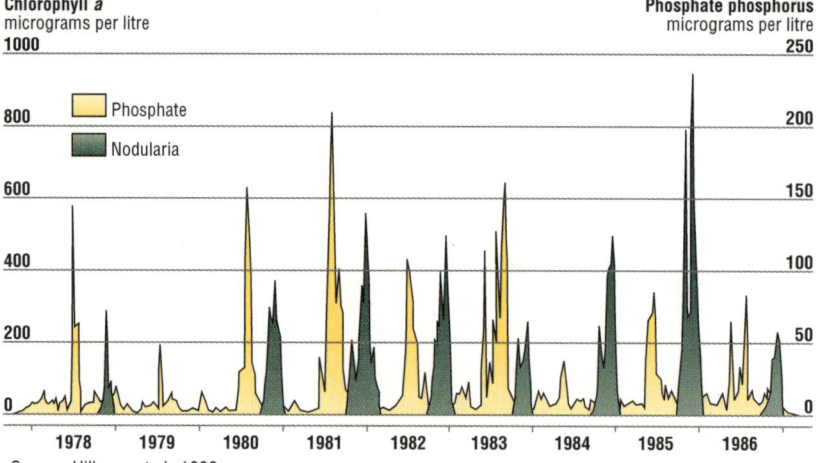

Figure 4.6 Long-term pattern of cyanobacterial blooms (*Nodularia spumigena*) in the Peel–Harvey estuary, Western Australia and its relationship to phosphate discharge from agricultural fertilisers in the previous winter period

Chlorophyll *a*
micrograms per litre

Phosphate phosphorus
micrograms per litre

☐ Phosphate
☐ Nodularia

Source: Hillman *et al.*, 1990.

In many regions, water tables are rising and, in some cases, bringing subsurface salt to the topsoil. In parts of the wheat belt of south-western Australia, saline groundwater tables have been rising by between 0.1 and 1.0 metres per year since removal of native vegetation began. When these saline waters reach the root zone of plants they cause widespread death of remnant vegetation (see page 4-13).

Fragmentation of native vegetation disrupts activity patterns of mobile species, particularly those that will not move over open agricultural land. Some honeyeaters, for example, may now be restricted to patches of remnant vegetation (Saunders and de Rebeira, 1991) and this may prevent them from carrying out their role in important plant pollination processes.

Introduced species

Many species, for a number of reasons, now occur in areas they formerly did not. Introduced species can include exotic organisms, both those introduced for production purposes and introduced diseases, as well as genetically modified organisms and native species whose range and/or abundance have changed because of human activities. Examples occur in terrestrial and aquatic ecosystems.

Introduced species exert a major pressure on biodiversity. They consume native fauna and flora and compete with native species for habitat, often to the detriment of those species.

Vertebrates

At least 18 exotic mammals have established feral populations in Australia, including cats, dogs, foxes, pigs, water buffalo, donkeys, goats and horses. Many species, such as sparrows (*Passer* spp.), trout (*Trutta* spp.) and salmon (*Salmo trutta*), were introduced by acclimatisation societies in the 1800s in an attempt to make Australia more like Europe; others, like the black rat (*Rattus rattus*), brown rat (*R. norvegicus*) and house mouse (*Mus musculus*), were introduced accidentally. Many domesticated animals were introduced for production purposes (Fox and Adamson, 1986).

Cats and foxes prey on a wide range of native animals and have been implicated in the decline, if not the extinction, of a number of species. The role of foxes in the decline of native species is well established. However, much less information exists on the role of cats. Studies of the red-tailed black cockatoo (*Calyptohypchus magnificus*) in Western Australia showed that feral cats climbing into tree hollows and preying on nestlings, caused the failure of up to 17 per cent of nests (Saunders, 1991).

Early European settlers introduced rabbits, which reached plague proportions over much of Australia, affecting native vegetation cover and competing with native fauna for scarce resources. Rabbits also take over burrows from native mammals such as bandicoots and bilbies, seriously reducing the breeding of these species.

All States and Territories have populations of exotic fish. Most widespread of these are trout, mosquito fish (*Gambusia affinis*), goldfish (*Carassius auratus*), European carp (*Cyprinus carpio*), redfin perch (*Perca fluviatilis*), and tench (*Tenca tenca*). All have expanded their ranges since first introduction, most with human assistance. Tilapia (*Oreochromis mossambicus*), one of the most recent fish introductions, is now found in many coastal creeks and estuaries in Queensland. Its spread is expected to continue, since the species can survive in fresh and sea water, and is extremely adaptable and a prolific breeder. Trout have been stocked intensively since the late 1800s, although climatic conditions have largely limited their spread since the mid 1900s. Goldfish are the most widespread of the exotic fish species in Australia, being found in every major drainage system from the Fitzroy River in Queensland to south-western Australia and Tasmania.

Exotic fish have been implicated in the decline of nine endangered, eight vulnerable and five rare or common native fish species. Trout alone are assumed wholly or partially responsible for declines in the abundance and range of nine native fish species as well as for changes in species composition and abundance of stream invertebrates.

Endangerment categories

Presumed extinct

A species is presumed extinct at a particular time if it has not been located in nature during the preceding 50 years, or during the preceding 10 years if thorough searching has been undertaken during the period.

Endangered

Species in danger of extinction include those whose survival is unlikely if the causal factors (threats) continue operating. This also applies to species whose numbers have been reduced to such a critical level, or whose habitats have been so drastically reduced, that they are in immediate danger of extinction.

Vulnerable

These are species believed likely to move into the 'endangered' category in the next 25 years if the threats continue operating.

Figure 4.7 Occurrence of six species of introduced mammals

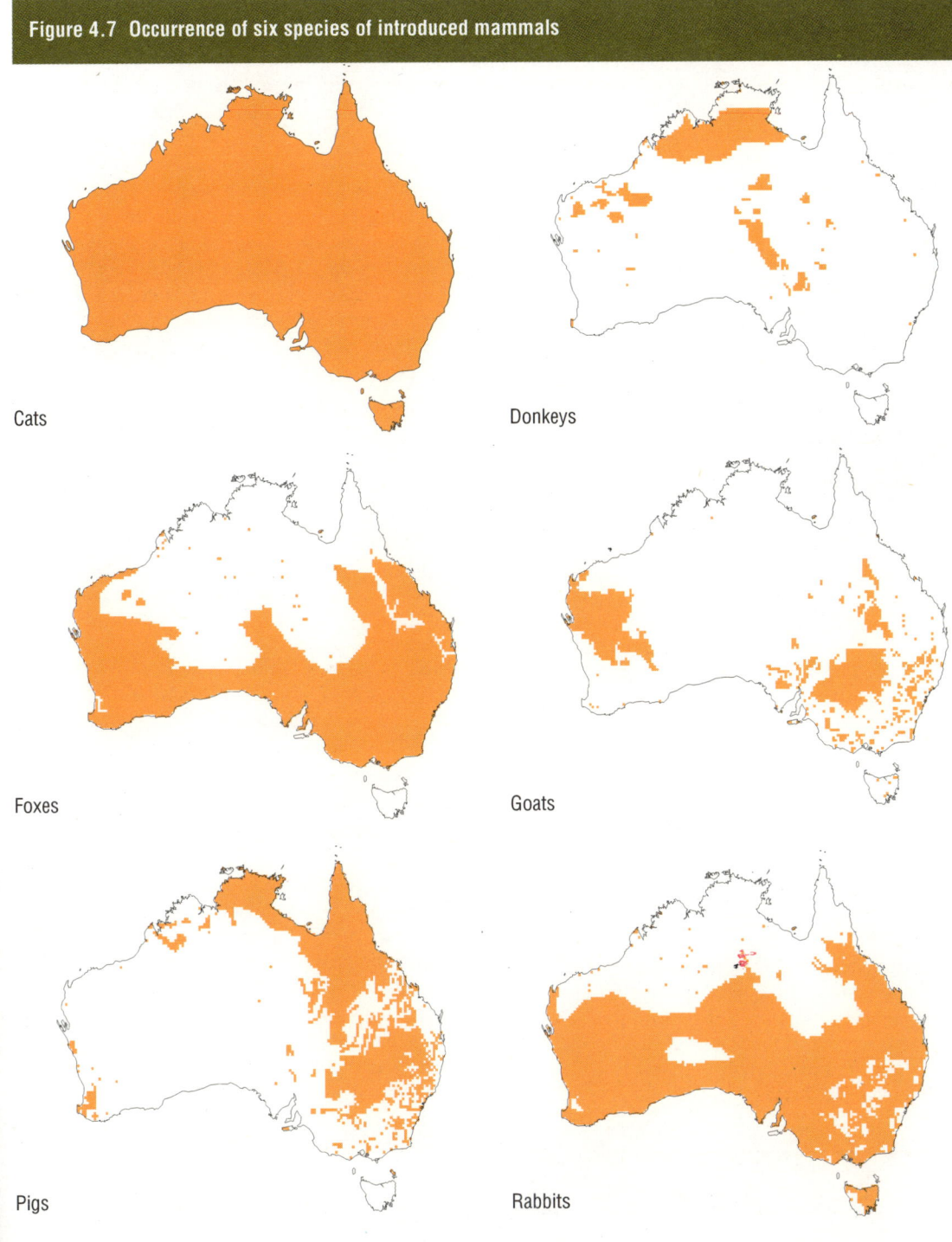

Cats

Donkeys

Foxes

Goats

Pigs

Rabbits

The entire continent is occupied by at least one of these feral species. Cats occur virtually everywhere, rabbits and foxes occur mainly in the south and not in the far north, pigs are found in the east, north and south-west, goats in the east and west and donkeys mainly in the north-west.

Grid cells (one degree of latitude by one degree of longitude).

Source: Compiled by ANCA.

Invertebrates

Many invertebrate species have been introduced from overseas, but most have attracted little attention because they are small and little is known about their effects on native species and ecosystems. However, some introductions are both obvious and destructive.

The European wasp (*Vespula germanica*) appears to have arrived in timber shipments. The nests were first seen in the Sydney area in 1978 and since then the wasp has not only spread widely, but its biology has changed significantly. In its native Northern Hemisphere, colonies are generally annual, the onset of winter killing all individuals

except the fertilised queens, which have to start colonies from scratch the following year. In Australia, however, the relatively warm winters do not kill colonies, which consequently become perennial and grow large with enormous numbers of workers. They prey on native insects, including commercially important pollinators, and attack soft fruits.

Pacific oysters (*Crassostrea gigas*) were introduced into Australia between 1947 and 1970 to establish an oyster industry in the southern States. Native Sydney rock oysters (*Saccostrea commercialis*) are not suitable for culture in the cool waters of Tasmania and Victoria. Despite attempts to

The Gouldian finch *(Erythura gouldiae)* has been affected by an introduced mite and has disappeared from large parts of its former range.

prevent the introduced oysters from invading New South Wales, they gradually spread up the coast where, because of their long breeding season, prolific production of spat and rapid growth, they are replacing the indigenous Sydney rock oysters by settling on and smothering them along with other rock fauna (Holliday and Nell, 1987).

The Gouldian finch (*Erythura gouldiae*), which now has a very patchy distribution in tropical northern Australia, has undergone a serious population decline over the past 20–30 years. Such has been the extent of its decline that large flocks are rarely encountered and the birds have disappeared altogether from parts of their former range. A study of possible causal factors (Tidemann *et al.,* 1992) demonstrated that finches in the wild suffered high incidences of infestation with an air-sac mite (*Sternostoma tracheacolum*). Aviary populations are susceptible to infection by the mite, which interferes with breathing. Affected birds may live for many months or die quickly depending on the severity of the infection. The mite was first described in aviary birds in South Africa and to date has not been widely reported in wild bird populations in the Southern Hemisphere.

Other introduced terrestrial invertebrates that have hit headlines are tiny springtails that displace native species (important in the decomposition of leaves), garden snails, the so-called Argentine ants, West Indian termites, a millipede and several varieties of cockroaches.

Introduced species also affect marine ecosystems. About 70 species are known to have been introduced into Australian waters. Many have come in as fouling organisms on the hulls of ships. Others have been introduced in ballast-water discharge. A total of 121 million tonnes of ballast water from overseas are released into our waters annually from international shipping and a further 34 million tonnes are translocated between Australian ports. This water carries many exotic marine organisms, including plants, animals and micro-organisms and at least 20 exotic species are believed to have been introduced into Australian waters this way. Many of these may have little effect, but several will impact on marine biodiversity.

Northern Pacific starfish (*Asterias amurensis*) were probably introduced into Tasmanian waters in ballast water from Japan. They can live in water temperatures of up to about 24°C and are thus capable of spreading to much of southern Australia. These are large starfish — adults can reach up to 40 cm across — and are prolific breeders, each individual producing up to 19 million eggs at a spawning. They can grow rapidly — reaching a size of 12 cm in less than a year. Tens of thousands have been removed from the Derwent estuary in Hobart, where they have reached densities of up to five per sq m. Pacific starfish have been found to eat many species that occur on the seabed, and thus the large numbers of starfish may have a devastating effect on populations of seabed organisms (see the box on page 8-17).

Plants

Introduced plants are an acute and insufficiently appreciated ecological problem. On a national scale, populations of the most invasive species are expanding. Plant species not native to Australia now account for about 15 per cent of our total flora. About half of them invade native vegetation and about one-quarter are regarded as serious environmental weeds or have the potential to be serious weeds (see Table 4.3). The largest proportion of environmental weeds are horticultural species that have escaped from cultivation. Almost all of Australia's native vegetation has been, or is likely to be, invaded by exotic species that could result in changes to the structure, species composition, fire frequency and abundance of native communities. Those species of greatest concern include rubber vine (*Cryptostegia grandiflora*), blue thunbergia (*Thunbergia grandiflora*), the semi-aquatic grasses hymenachne (*Hymenachne amplexicaulis*) and aleman grass (*Echinochloa polystachia*), para grass (*Brachiaria mutica*), giant sensitive plant (*Mimosa pigra*) and athel pine (*Tamarix aphylla*).

The rubber vine, which entangles trees and other vegetation and eventually smothers them, is spreading at an alarming rate through the river systems of southern Cape York and the Queensland part of the Gulf of Carpentaria and along the coast as far south as the Burnett River near Bundaberg, destroying the riverside vegetation in these regions.

The numbers refer to those species officially listed as noxious by the States and Territories and indicate the many problems with introduced plants considered to have serious adverse effects on biodiversity and production. The map does not necessarily indicate the extent to which ecosystems or native species are affected by weeds; some serious environmental weeds occur both in areas of high and low weed diversity.

Grid cells (one degree of latitude by one degree of longitude).

The boundaries indicate Interim Biogeographic Regions for Australia.

Figure 4.8 Number of weed species by area

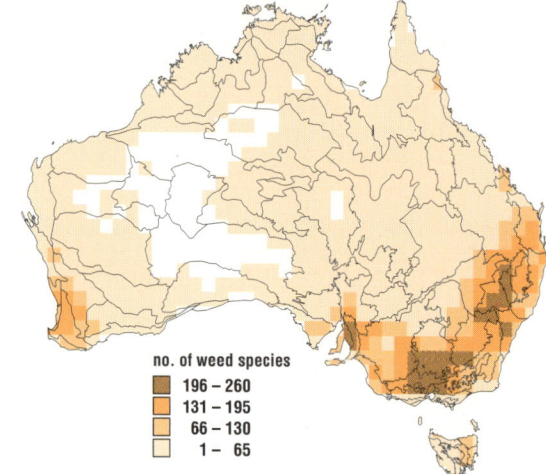

no. of weed species
- 196 – 260
- 131 – 195
- 66 – 130
- 1 – 65

Source: compiled by ANCA

Table 4.3 Australia's worst environmental weeds

Species	Key community/habitat affected	Nature of impact/threat
Prickly acacia *(Acacia nilotica)* small tree/shrub	Mitchell grasslands	Replaces perennial *Astrebla* spp. with annuals or bare soil and is a long-term threat to the Mitchell grass biome; converts grassland to shrubland
Para grass *(Brachiaria mutica)* semi-aquatic	Wetlands and streams in the wet-dry and wet tropics and sub tropics	Planted for ponded pasture but spreading into non-target areas destroying waterbird breeding habitats and choking tropical streams; replaces native vegetation
Buffel grass *(Cenchrus ciliaris)* groundcover	Moist 'refuges' and river banks in the arid zone	Threatening keystone habitats by displacing native vegetation and altering the fire regime; likely to reduce fauna resources
Bitou bush *(Chrysanthemoides monilifera rotundata)* **Boneseed** *(C.m. monilifera)* shrub	Range of coastal systems: foredune, heath, littoral rainforest; range of coastal and sub-coastal systems	Displaces native vegetation with unknown effects on fauna
Rubber vine *(Cryptostegia grandiflora)* shrub/vine	Gallery and other riparian communities in the wet-dry tropics; dry rainforest (vine thickets)	Smothers trees and shrubs and shades out the ground layer; destroys riparian vegetation including gallery forests threatening associated fauna; forms impenetrable thickets in Queensland's Gulf river systems
Water hyacinth *(Eichhornia crassipes)* aquatic	Standing surface waters especially where nutrient levels are high; occurs in all mainland States but particularly tropics and sub-tropics	Aggressively invades open water with potential for very rapid growth; still spreading in Australia despite extensive control measures; alters aquatic ecosystems
Aleman grass *(Echinochloa polystachia)* semi-aquatic	Wetlands in the wet-dry and wet tropics; grows in water up to 2 m	As for para grass; recent introduction and not yet widespread but larger than para grass with greater potential for damage
Reed sweetgrass *(Glyceria maxima)* semi-aquatic	Margins of creeks, rivers and ponded areas up to 1 m deep; temperate species eastern States	Used as a pasture or ornamental plant but is spreading to non-target areas; chokes the habitat
Hymenachne *(Hymenachne amplexicaulis)* semi-aquatic	As for para grass but can grow in water up to 2 m	Recently introduced as ponded pasture species, so not yet widespread, but has potential to modify tropical wetlands totally if not controlled
Giant sensitive plant *(Mimosa pigra)* small tree/shrub, semi-aquatic	Disturbed areas especially flood plains in the wet-dry tropics	Totally displaces native species leaving bare mud if removed; spreads by floods
Bridal creeper *(Myrsiphyllum asparagoides)* creeper	Spreading through wide range of habitats in southern Australia	Smothers ground and shrub layers
Parkinsonia *(Parkinsonia aculeata)* small tree/shrub	Ephemeral wetlands and riparian communities in the wet-dry tropics	Invades mesic habitats and seasonal wetlands threatening waterbird habitats of continental significance
Mission grass *(Pennisetum polystachion)* groundcover	Dry forests and woodlands of the wet-dry tropics	Displaces native sorghum changing the fire regime, which potentially reduces recruitment potential of woodland species of high conservation significance
Mesquite *(Prosopis spp.)* small tree/shrub	Semi-arid and arid riparian and other communities; Mitchell grasslands	Similar to prickly acacia but has a wider range of soil tolerances
Salvinia *(Salvinia molesta)* aquatic	Stationary and slow-moving water bodies, especially where nutrient levels are high; all mainland States and Territories	Aggressively invades open water with potential for very rapid growth; still spreading in Australia despite extensive control measures; alters aquatic ecosystems
Athel pine *(Tamarix aphylla)* small tree	Dryland river systems; currently small infestations	Displaces native trees; salinises soil; changes hydrology and geomorphology; reduces fauna resources
Blue thunbergia *(Thunbergia grandiflora)* vine	Tropical lowland rainforest in far north Queensland, especially along watercourses	Vigorous vine rapidly spreading and smothering native vegetation to the canopy; infestation in early stages
Japanese kelp *(Undaria pinnatifida)* marine kelp	Near-shore habitats along east coast of Tasmania	Spreading at rate of 10 km per year with potential to spread along southern coastline

Source: Humphries *et al.*, 1991

This kelp *(Undaria)* was probably introduced from Japan in ballast water and is a conspicuous underwater plant on the Tasmanian coast.

Another species spreading along the river systems is the athel pine, which displaces native vegetation — usually river red gum (*Eucalyptus camaldulensis*) — and can change river flow and sedimentation regimes. It is considered a threat to all the watercourses of arid Australia. So far it has become established along several hundred kilometres of the Finke River, the largest river system in the arid zone.

Some 65 species of aquatic plants have become weeds in Australian inland waters. About 15 of these are significant pests and 13 have the potential to become so. These species have been introduced either intentionally, for economic, aesthetic or management reasons, or accidentally through the aquarium and agricultural industries. One example of an introduced aquatic plant is the giant sensitive plant, which infests wetlands of the Northern Territory. It rapidly reduces the number of species of native plants and animals, and alters seasonal wetlands from predominantly native grassland to shrubland dominated by a single species.

Japanese kelp (*Undaria pinnatifida*) was probably introduced via ballast water discharge and is spreading widely in southern waters. This species is known to displace native seaweeds, changing seabed habitats and altering species composition.

Micro-organisms

Introduced species sometimes carry diseases that can infect native species. The introduced fungus *Phytophthora cinnamomi*, causes a disease, known as dieback, that is devastating many native communities in southern Australia and is now the most important threat to the biodiversity of Stirling Range National Park in Western Australia. By killing key species in those systems, the pathogen is threatening entire communities and ecosystems. In highly susceptible genera (such as *Banksia, Grevillea* and *Dryandra*), 80 to 100 per cent of infected individuals may die, exposing ground that is then invaded by weeds. The numbers of birds in affected areas in Two Peoples Bay Nature Reserve in Western Australia have been reduced significantly because of the effects of *Phytophthora* infection on the vegetation (Hart, in press).

The impact of other introduced micro-organisms on native animals is largely unknown, although an introduced fish virus (epizootic haematopoietic necrosis or EHN virus) may have been responsible for the decline of the Macquarie perch in Lake Eildon.

Native species out of place

Outside their natural range, native species may be as serious a threat to biodiversity as exotic ones. Many are spreading beyond their usual habitat or increasing in abundance due to human activities such as clearing, cropping, pastoralism, tree-planting on farms and in gardens and landscaping of roads and railways.

Much of the pastoral land in the central-western areas of New South Wales and Queensland has become densely infested with shrubs. These so-called woody weeds (including *Eremophilia mitchellii, E. sturtii* and *Cassia nemophila*) occur naturally in the pastoral regions but have spread at the expense of other native species due to overgrazing and changed fire regimes. Before European settlement, regular fires checked their spread. However, overgrazing, which removes the perennial grass cover, reduces the capacity of the land to carry fire. Longer periods between fires have allowed young plants to develop beyond their vulnerable stage, and once established they are able to survive fire. Dense thickets of these shrubs prevent pasture growth and impede mustering of stock.

The galah (*Cacatua roseicapilla*) was formerly associated with the river systems of the arid zone. However, development has created vast areas of suitable habitat (grasses, cereal crops and abundant water), encouraging the bird to colonise much of Australia. This expansion has brought the galah into contact with other species formerly outside its range. One of these is Carnaby's cockatoo (*Calyptorhynchus latirostris*), a black cockatoo of south-western Australia, which has disappeared from more than one-third of its range in the last 25 years due to loss of food and habitat. This cockatoo also suffers from competition with galahs over nest sites. In addition, galahs remove bark from the trunks of some of their nest trees, which in severe cases leads to the death of the trees and has become a significant cause of mortality in extensively cleared semi-arid woodlands with poor recruitment of trees.

Not only can introduced species reduce the numbers of indigenous organisms, they may hybridise with closely related endemic species, changing the genetic composition of the population. In some cases, hybrids resulting from interbreeding between endemic and closely related non-endemic species are more vigorous than the original stock. This can lead to reduction or eventual elimination of the indigenous species. The escape of rosemary grevillea (*Grevillea rosmarinifolia*) from gardens, for example, has resulted in hybridisation with a rarer endemic species, *G. glabella*. The resulting hybrids are more vigorous and may replace the parental *G. glabella*.

Two Peoples Bay, Western Australia; red, purple, orange and light red represent dieback locations. The other colours represent dieback-free areas. Source: Behn and Campbell, 1992.

Galahs remove bark from some of their nest trees. In extreme cases this may lead to the death of the tree.

Pollution

The release of pollutants into the environment is an actual and potential threat to biodiversity, particularly in regions close to industrial sites and highly urbanised areas. Urban stormwater may contain high levels of contaminants such as faecal bacteria, nutrients, chromium, cadmium, lead, nickel, hydrocarbons and chlorinated hydrocarbons. In rural areas, irrigation run-off often contains insecticides, fertilisers and herbicides that have been applied to crops and may affect fresh-water and marine ecosystems. Nutrients from urban effluent and agricultural chemicals, for example, are polluting inshore reefs of the Great Barrier Reef, killing coral and encouraging the growth of sessile algae.

Pollutants can act in a synergistic way to cause uncertain long-term impacts. These impacts are compounded by the cyclic nature of ecosystem processes, which disperse pollutants widely from their sources and may affect biodiversity at considerable distances from the original source.

Potential effects of pollutants on ecosystems include changes in the abundance of species, interruption of energy and nutrient flows, modification of habitats, reductions in soil, water and air quality, and changes to the stability and resilience of ecosystems. As an example, industrial discharges into Cockburn Sound in Western Australia have been associated with massive loss of seagrasses and substantial levels of contamination of sediments and fish (see page 8-25).

Natural enrichment in the sea occurs mainly in upwelling areas where water rich in nutrients rises to the surface and is used by algae, chiefly diatoms. A variety of small animals, which form part of the extremely diverse marine plankton and which include larval stages of many fish and crustaceans, feed on these diatoms. More than 5000 species of diatoms are found in Australian waters. Effluents from sewage and from agricultural run-off are high in nitrogen and phosphorus, but are lacking in silica. Thus diatoms — which require silica — cannot make use of these nutrients and they are used by other species of algae — especially dinoflagellates — that do not require silica. This results in a major shift in species composition,

Fire and Australia's biodiversity

The January 1994 wildfires in eastern New South Wales provided a vivid reminder of how destructive some of the natural forces affecting our landscape can be. Fire is a natural disturbance and the Australian biota has evolved under its influence. However, its contemporary pattern differs from the one that occurred before the arrival of Europeans.

The pattern comprises the frequency (expected return-time) of fire. This in turn affects its intensity, since litter accumulates between outbreaks, and the longer the interval the greater the amount of litter or fuel that can accumulate. The third important part of the pattern is the season in which fire usually occurs. In much of Australia, some or all of these components of the fire regime have changed. This in turn affects the way that communities of plants, animals and micro-organisms respond to fire.

Most often, people, by their actions, bring about the changes in fire frequency (and intensity) or seasonality. Some areas are protected. Thus, fires may be suppressed in urban bushland close to houses, schools and commercial buildings to prevent destruction of property. Such practices will cause biological changes. More commonly, bushland areas are burnt more frequently than they were prior to European settlement. For some species of plants and animals, the time between fires may be insufficient to allow their natural reproduction. For example, in shrubland around Sydney the heath banksia (*Banksia ericifolia*) first flowers in the seventh year after a fire, and the fruit are not mature until at least the eighth year. This means that if bushland such as Royal National Park is burnt more frequently than every eight years, that species of banksia will become locally extinct.

Flowering times for selected plants from sandstone shrublands near Sydney

Species	Time since fire (years)
	2 3 4 5 6 7 8
Mitre weed (*Mitrasacme polymorpha*)	▇▇ (2–4)
Flannel flower (*Actinotus helianthi*)	▇▇▇ (2–5)
Woollsia (*Woollsia pungens*)	▇▇▇▇ (3–7)
Sydney boronia (*Boronia ledifolia*)	▇▇▇▇ (3–7)
Pink wax-flower (*Eriostemon australasius*)	▇▇▇ (4–7)
Guinea flower (*Hibbertia monogyna*)	▇▇▇ (5–8)
Sweet-scented wattle (*Acacia suaveolens*)	▇▇▇ (5–8)
Wedge-pea (*Gompholobium grandiflorum*)	▇▇ (6–8)
Heath banksia (*Banksia ericifolia*)	▇ (7–8)
Dagger hakea (*Hakea teretifolia*)	▇ (7–8)

Source: Benson, 1985.

Mine sites such as this one at Queenstown, Tasmania can have a severe impact on biodiversity.

Table 4.4 Ten vertebrate species most threatened by climate change

Species	Vertebrate order	Loss of core climate area (%)
Kowari (*Dasyuroides byrnei*)	Mammal	99.6–100
Red-tailed phascogale (*Phascogale calura*)	Mammal	99.4–100
Central rock-rat (*Zyzomys pedunculatus*)	Mammal	94.0–100
Forty-spotted pardalote (*Pardalotus quadragintus*)	Bird	86.0–100
Swan galaxias (*Galaxias fontanus*)	Fish	81.8–100
Dusky hopping-mouse (*Notomys fuscus*)	Mammal	78.9–100
Heath rat (*Pseudomys shortridgei*)	Mammal	71.7–100
Broad-headed snake (*Hoplocephalus bungaroides*)	Reptile	65.1–98.9
Northern hairy-nosed wombat (*Lasiorhinus krefftii*)	Mammal	59.4–100
Carpentaria grass wren (*Amytornis dorotheae*)	Bird	51.6– 100

Climate change due to increasing greenhouse gases is a potential threatening process to Australia's biodiversity. Some vertebrate species may be driven to extinction due to loss of core habitat.

Source: Dexter *et al.*, 1995

from a plankton community dominated by diatoms to one dominated by dinoflagellate algae. Dinoflagellate blooms cause a range of serious problems from massive fish kills associated with red tides to paralytic shellfish poisoning in humans (see Chapter 8).

Mining

Mining affects the environment and associated biota through the removal of vegetation and topsoil, the displacement of fauna, the release of pollutants into the air and water and the production of mine overburden. When pyrite (iron disulfide) is brought to the surface during the mining of coal and metal ores it is oxidised to sulfuric acid, which in turn mobilises heavy metals. This acid mine waste can severely pollute rivers, causing loss of biodiversity. The Rum Jungle Mine in the Northern Territory, for example, released 130 tonnes of copper, 100 tonnes of manganese, 40 tonnes of zinc and 13 000 tonnes of sulfate into the Finnis River in one year. In submarine mining operations, equivalent disruptions occur in the immediate environs of the mine or drilling site. Whether terrestrial or marine, mining sites are numerous but generally of relatively small total area (see Fig. 6.3). Extensive mining operations, such as open-cut extraction of coal, bauxite and manganese, as well as sand mining in coastal heathlands, have caused long-term changes to biodiversity despite recent attempts at rehabilitation (Fox, 1990).

Climate change

The impact of climate change on biodiversity is difficult to assess since it depends largely on the rate of change and the compounding effects of other pressures such as habitat loss and fragmentation. While most scientists agree that the climate is changing, it is not yet possible to distinguish human-induced change from natural climatic variations.

As the global climate warms, the preferred climatic conditions for a species will shift to higher altitudes and latitudes. Survival will depend on its ability to relocate quickly enough and the availability of alternative habitats. Species most at risk are those with small population sizes that have slow growth rates with poor dispersal abilities and recruitment. The disruption of migration paths by human activities will also influence the ability of organisms to adjust geographically to climate change. Regions of urban development, agriculture and pastoralism will act as barriers, preventing the movement of many species from one remnant habitat to another.

A recent study into the impact of global warming on the distribution of vertebrates (Dexter *et al.*, 1995) found that the habitats of many of Australia's endangered vertebrates are likely to contract significantly in the advent of climate change (see Table 4.4). Under one scenario, 46 out of 57 endangered species examined would contract in range.

State

The state of ecosystem diversity

Biogeographic regionalisations for Australia

Ecosystems can be defined at many scales because there are many ways of representing them, depending on the purpose and the particular ecosystem components and processes that need to be emphasised. The main reason for developing the terrestrial regionalisations discussed here was to provide a framework for identifying gaps in the national system of protected areas and to assist in allocating priorities for funding to projects concerned with conservation of biodiversity (Thackway and Cresswell, 1995). The rationale for the marine regionalisation was similar. Quite different classifications of ecosystems could be appropriate for different purposes.

The terrestrial ecosystems discussed here are those outlined in the Interim Biogeographic Regionalisation for Australia (IBRA) (see Fig. 4.9), which divides the continent into 80 biogeographic regions representing major environmental units (Thackway and Cresswell, 1995). It is the only continent-wide regionalisation agreed to by all States and Territories and was initially developed in February 1994 by combining State and Territory land classifications at a scale appropriate for the whole continent. The system is still evolving as methods are developed to refine boundaries, make levels of classification consistent between parts of the country and reconcile differences in approach between jurisdictions.

The biogeographic regions vary in size from 2372 sq km (Furneaux, in Bass Strait) to 423 751 sq km (Great Victoria Desert). The smaller regions occur within 300 km of the coastline, have relatively high rainfall in the growing season and, in many cases, are mountainous. Of the largest regions, most are in arid or semi-arid areas with broad climatic gradients and little topographic relief. The regions can be progressively subdivided into smaller units based on, for example, major vegetation structural types, vegetation communities, local topographic variations in communities and, at extremely fine scales, water-filled tree hollows and small sections of the soil surface (see the box below). Any of these scales of ecosystems can be characterised in terms of distinctive biological and physical patterns and processes.

In 1986, the Australian Committee for IUCN proposed the classification used here for major marine regions (see Fig. 4.9), which is an interim one. The Department of the Environment, Sport and Territories (DEST) is reviewing the inshore classification with the States and the Northern Territory for waters under their jurisdiction — the marine areas within three nautical miles of the coast. DEST has also commissioned CSIRO to develop a new offshore regional classification that will integrate information on depth, type of substratum, water column characteristics, such as temperature, salinity, currents and eddies, and the distribution of more than 4000 species of fish (the best known group of marine organisms). Of the interim marine regions, 12 are predominantly coastal or near-coastal, defined arbitrarily within the 200-m-depth contour, which is relatively close to the coast in many areas but extends well offshore in Bass Strait and the Great Australian Bight and along the Great Barrier Reef and much of the northern coastline.

Biogeographic regions — a closer look

The New South Wales north coast (A) is one of the major biogeographic regions in eastern Australia (see Fig. 4.9). The region has been subdivided into six major vegetation structural units (B) — rainforest, moist open forest, dry open forest, woodland, and sclerophyll complexes on the coast and tablelands. Each of these units can be considered an ecosystem at a finer scale than the region as a whole. At a finer scale still, forest types (C) can be distinguished within the major vegetation types. So rainforest in the Dorrigo area can be subdivided into dry, warm temperate, cool temperate and subtropical and the open forest into many more types. Each of these forest types can be characterised by particular environmental variables and physical processes and a distinctive flora and fauna. At another level of detail, the ground surface of any one forest type can be mapped as a mosaic of surfaces (D), including the bases of trees, accumulated bark nearby, bare rock, disturbed bare soil, litter of leaves and small branches and fallen logs. The different surfaces also have distinctive physical and biological characteristics and large populations of macro-invertebrates and micro-organisms.

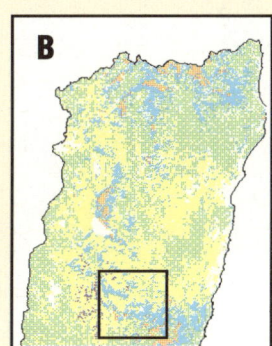
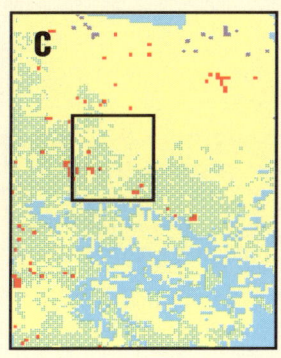
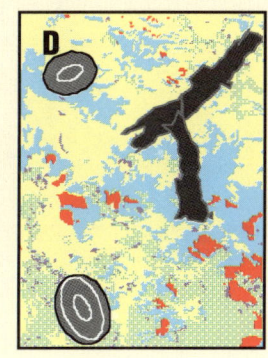

Source: compiled by NSW Parks and Wildlife Service.

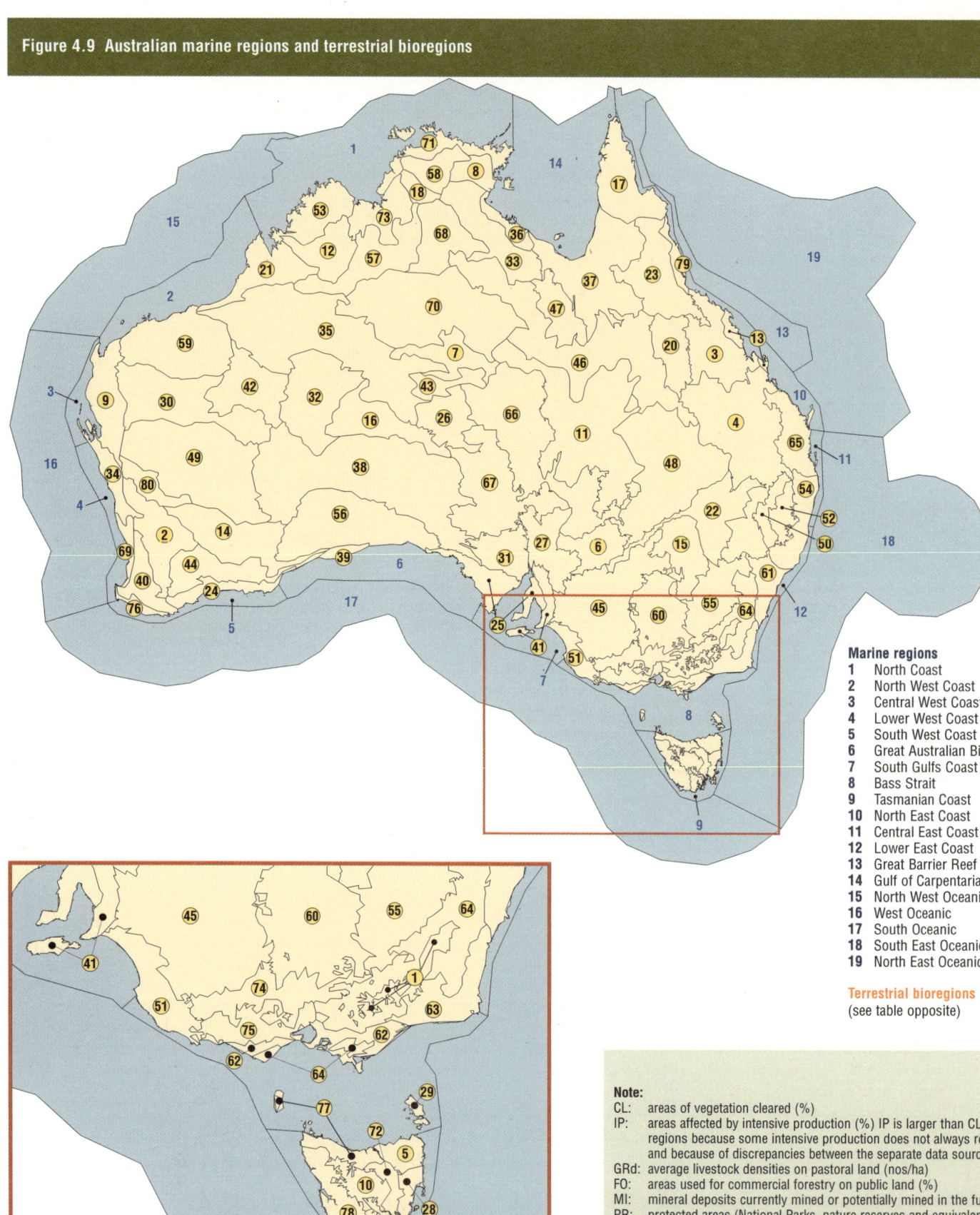

Figure 4.9 Australian marine regions and terrestrial bioregions

Marine regions

1 North Coast
2 North West Coast
3 Central West Coast
4 Lower West Coast
5 South West Coast
6 Great Australian Bight
7 South Gulfs Coast
8 Bass Strait
9 Tasmanian Coast
10 North East Coast
11 Central East Coast
12 Lower East Coast
13 Great Barrier Reef
14 Gulf of Carpentaria
15 North West Oceanic
16 West Oceanic
17 South Oceanic
18 South East Oceanic
19 North East Oceanic

Terrestrial bioregions
(see table opposite)

Note:
CL: areas of vegetation cleared (%)
IP: areas affected by intensive production (%) IP is larger than CL in so
  regions because some intensive production does not always require
  and because of discrepancies between the separate data sources.
GRd: average livestock densities on pastoral land (nos/ha)
FO: areas used for commercial forestry on public land (%)
MI: mineral deposits currently mined or potentially mined in the future (
PR: protected areas (National Parks, nature reserves and equivalent tenu
Some small or irregularly shaped IBRA regions were merged for the analys
intensive production and grazing because the data for these enterprises ar
available only for Statistical Local Areas (SLAs). Merging produced the foll
12 combinations of regions: Australian Alps and South Eastern Highlands;
and eastern Tasmania: Ben Lomond, Central Highlands, D'Entrecasteaux, F
Furneaux, Tasmanian Midlands and Woolnorth; Burt Plain and MacDonnell
Central Arnhem, Daly Basin, and Pine Creek–Arnhem; Esperance Plains an
Gulf Fall and Uplands and Gulf Coastal; Hampton and Nullarbor; Jarrah For
Swan Coastal Plain and Warren; Murchison and Yalgoo; Nandewar and Ne
England Tableland; Naracoorte Coastal Plain, Victorian Midlands, and Victo
Vocanic Plain; South East Coastal Plain and South East Corner.

Data compiled by ANCA.

Table 4.5 State of terrestrial biogeographic regions indicated by extent of major threatening processes and coverage by conservation reserve

		CL	IP	GRd	FO	MI	PR
1	Australian Alps	38	35	2.08	27	3	18.9
2	Avon Wheatbelt	88	81	0.96	0	26	0.5
3	Brigalow Belt North	60	17	0.76	2	29	1.2
4	Brigalow Belt South	64	16	1.01	9	28	2.0
5	Ben Lomond	37	19		44	29	9.3
6	Broken Hill Complex	0	0	0.15		8	1.4
7	Burt Plain	0	0	0.09		5	2.3
8	Central Arnhem	2	1	0.14		2	7.6
9	Carnarvon	0	7	0.08		3	6.9
10	Central Highlands	37	19		18	4	9.3
11	Channel Country	0	0	0.15	0	3	6.6
12	Central Kimberley	0	0	0.15		8	0.0
13	Central Mackay Coast	34	63	1.12	9	2	6.6
14	Coolgardie	3	13	0.08	1	325	7.6
15	Cobar Peneplain	33	25	0.50	1	14	0.9
16	Central Ranges	0	0	0.06		3	0.0
17	Cape York Peninsula	0	0	0.10	1	4	11.7
18	Daly Basin	2	1	0.14		6	7.6
19	D'Entrecasteaux	37	19		42		9.3
20	Desert Uplands	15	19	0.44		4	1.6
21	Dampierland	0	0	0.17		3	0.3
22	Darling Riverine Plains	51	28	0.85	0		0.5
23	Einasleigh Uplands	7	4	0.34	1	72	0.8
24	Esperance Plains	43	47	1.41	0	17	19.3
25	Eyre and Yorke Blocks	76	48	0.59	0	6	6.6
26	Finke	0	0	0.6		2	0.0
27	Flinders and Olary Ranges	7	4	0.18		25	10.1
28	Freycinet	37	19		28		9.3
29	Furneaux	37	19				9.3
30	Gascoyne	0	0	0.04		31	1.9
31	Gawler	0	11	0.06		5	9.9
32	Gibson Desert	0	0	0.0			12.0
33	Gulf Fall and Uplands	0	0	0.07		15	0.9
34	Geraldton Sandplains	52	56	0.73		4	14.0
35	Great Sandy Desert	0	0	0.10		12	1.9
36	Gulf Coastal	0	1	0.07		12	0.0
37	Gulf Plains	0	1	0.25			0.0
38	Great Victoria Desert	0	2	0.04		8	17.1
39	Hampton	0	1	0.04			28.6
40	Jarrah Forest	44	55	2.82	34	25	8.1
41	Lofty Block	81	64	1.58	1	6	4.9
42	Little Sandy Desert	0	0	0.03		2	4.9
43	MacDonnell Ranges	0	0	0.09		1	2.3
44	Mallee	43	47	1.41		32	19.3
45	Murray–Darling Depression	39	33	0.55	2	7	12.8
46	Mitchell Grass Downs	3	2	0.28		9	0.3
47	Mount Isa Inlier	0	0	0.21	0	52	2.3
48	Mulga Lands	8	5	0.30			1.6
49	Murchison	2	4	0.05	0	393	1.1
50	Nandewar	74	45	2.41	1	2	2.1
51	Naracoorte Coastal Plain	78	73	2.52	6		4.7
52	New England Tablelands	74	45	2.41	6	9	2.1
53	Northern Kimberley	0	0	0.15		1	12.0
54	NSW North Coast	37	21	1.78	21	64	8.0
55	NSW South Western Slopes	80	55	2.16	1	30	1.2
56	Nullarbor	0	1	0.04			28.6
57	Ord-Victoria Plains	0	1	0.18		7	5.4
58	Pine-Creek Arnhem	2	1	0.14		47	7.6
59	Pilbara	0	0	0.08		224	5.6
60	Riverina	72	38	1.05	2	4	0.2
61	Sydney Basin	26	22	1.22	6	33	32.4
62	South East Coastal Plain	41	37	0.97	4		16.6
63	South East Corner	41	37	0.97	53	4	16.6
64	South Eastern Highlands	38	35	2.08	25	46	18.9
65	South Eastern Queensland	60	23	1.33	12	26	3.0
66	Simpson-Strzelecki Dunefields	0	0	0.09		1	26.7
67	Stony Plains	0	0	0.05		5	4.9
68	Sturt Plateau	0	0	0.15			0.0
69	Swan Coastal Plain	44	55	2.82	5	33	8.1
70	Tanami	0	0	0.11		33	0.0
71	Top End Coastal	1	1	0.19		4	9.8
72	Tasmanian Midlands	37	19		4		9.3
73	Victorian Bonaparte	1	1	0.17		4	11.2
74	Victorian Midlands	78	73	2.52	9	44	4.7
75	Victoria Volcanic Plain	78	73	2.52	2		4.7
76	Warren	44	55	2.82	36	14	8.1
77	Woolnorth	37	19		29	12	9.3
78	West and South West	5	7	2.95	18	19	49.0
79	Wet Tropics	23	22	0.93	35		16.1
80	Yalgoo	2	4	0.05	0	13	1.1
		CL	IP	GRd	FO	MI	PR

Figure 4.10 Percentages of terrestrial biogeographic regions used for commercial forestry on public land

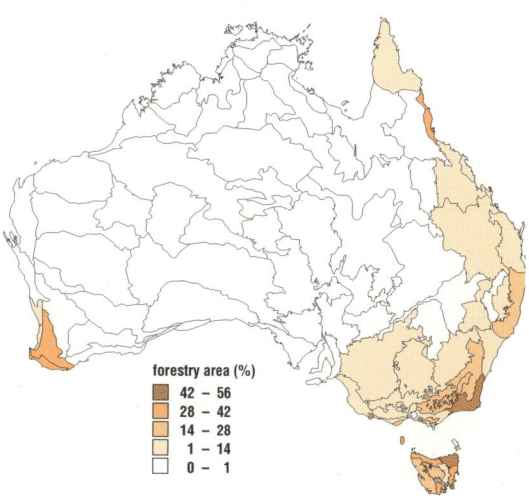

forestry area (%)
- 42 – 56
- 28 – 42
- 14 – 28
- 1 – 14
- 0 – 1

Source: compiled by ANCA.

Commercial forestry affects far fewer regions than clearing or grazing and is strongly concentrated in the south-east and south-west. Its overall effects on biodiversity can, however, still be substantial because forests are richer biologically than other terrestrial habitats.

Figure 4.11 Percentage areas of vegetation cleared in terrestrial biogeographic regions

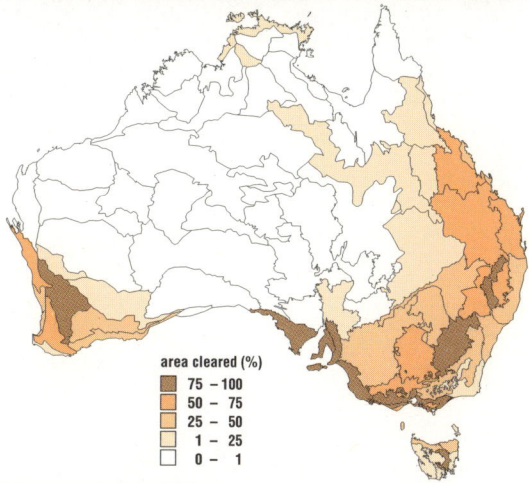

area cleared (%)
- 75 – 100
- 50 – 75
- 25 – 50
- 1 – 25
- 0 – 1

Source: compiled by ANCA.

The most extensive clearing has been in regions that are both topographically and climatically suitable for large-scale cropping or improved grazing. These form a band from east to west in the southern half of the continent and include Nandewar, New England Tablelands, New South Wales South-Western Slopes, Riverina, Victoria Midlands, Victoria Volcanic Plain, Naracoorte Coastal Plain, Lofty Block, Eyre and Yorke Blocks and the Avon Wheatbelt.
Source: estimates of cleared areas are derived from two sources. The major one is the Australian Heritage Commission wilderness inventory which is a compilation of information supplied by the States and Territories, except Western Australia. For south-western Australia, clearing data come from interpretation of satellite imagery by the Environmental Resources Information Network.

Marine regions, like terrestrial ones, can be defined at many scales. For example, the Gulf of Carpentaria is a large marine region with much internal variation in physical and biological characteristics. The eastern side contains large estuaries with extensive mangroves, fine sediments, high turbidity and lowered salinity in the monsoon season. On the western side, much of the coastline is rocky, sediments are mainly coarse, the water is relatively clear and extensive beds of seagrasses grow in sheltered areas. Further subdivision of these large areas is also possible using biological and physical data.

Ecosystem diversity

Researchers have used satellite imagery to examine changes in land cover on the Australian continent (Graetz *et al.*, 1995). They found that 39 per cent of the continent was in the intensive-land-use zone (characterised by clearing for cropping, pasture, plantation and urban development) and 61 per cent in the extensive-land-use zone (the central core of the continent characterised by pastoral activities on native vegetation and unallocated land).

The extent of clearing of native vegetation in the intensive-land-use zone ranged from 30 to 99 per cent with a mean value of 47 per cent. Within this zone, the remaining land cover was highly fragmented. In the extensive-land-use zone, 26 per cent of the area had significant disturbance to land cover and 15 per cent had been substantially disturbed. Land cover over 48 per cent of the continent is either significantly or substantially disturbed and no vegetation types remain completely unaffected by human activities.

Processes threatening major terrestrial ecosystems

In contrast to species, ecosystems have no nationally agreed classification system and, more importantly for this report, no agreed listing of the extent to which they are threatened. It is also difficult to specify when an ecosystem is extinct. How many of the components have to be lost and what degree of alteration of processes is necessary

to make such a proclamation? This report uses the extent of various threatening processes in each biogeographic region to summarise the state of our ecosystems. In this way, it is possible to make some national generalisations about the patterns of impacts on major Australian ecosystems and to augment these with some examples of altered ecosystems at finer scales than the national classification.

Figures 4.10 and 4.11 illustrate the continental patterns of two threatening processes in terrestrial environments — forestry and clearing/intensive production. The overall extent of both is summarised in Table 4.5. Clearing combines the impacts of human settlements and transport systems, cropping and the mechanical removal of woody vegetation for increased carrying capacity of stock. Of these, clearing for human settlements and associated activities is relatively minor in extent and strongly concentrated on the south-eastern seaboard, with smaller areas in the south-west and north-east. Human settlements, therefore, affect mainly the biogeographic regions in coastal parts of the continent and cover small percentages of their total areas. Nevertheless, urbanisation can have severe impacts on ecosystems defined at finer scales than IBRA regions. For example, some of the vegetation types in the Sydney region have been largely eliminated by the expansion of the metropolitan area while others are still extensive (Benson and Howell, 1990).

By far the largest proportion of clearing has been for cropping and grazing, either on improved or unimproved pastures. The extent of these impacts is demonstrated by the fact that 49 of the 80 biogeographic regions are affected to some extent (see Table 4.5). Incentives for clearing are largely determined by rainfall, soil type and available stream flow for irrigation. Clearing is therefore concentrated in the south-west, the east and the monsoonal north (see Figs 4.11 and 6.6). Large variations in the percentage of area cleared exist between biogeographic regions. The highest percentages are in regions that are both topographically and climatically suitable for large-scale cropping or improved grazing. Overall, most of the southern half of the continent is highly disturbed by clearing.

Percentages of biogeographic regions cleared give a national picture but hide variation in land use at finer scales. In the Avon Wheatbelt of Western Australia, clearing has avoided the more rugged outcrops (see the box on page 4-13). Similarly, in the New South Wales North Coast biogeographic region, overall clearing is about 37 per cent but this has been highly selective (see the image on page 4-27). Some ecosystems such as the 'Big Scrub', formerly the largest tract of subtropical rainforest in Australia, are almost completely cleared because they possess rich soils and gentle topography. Other large parts of the region are still trackless wilderness because they are topographically unsuitable for intensive land uses.

The national patterns of clearing are complemented by accounts of clearing in other

A snapshot of change to some of Australia's ecosystems: 1788–1995

- Seagrass beds in temperate areas have declined significantly.
- About 43 per cent of forests have been cleared.
- More than 60 per cent of coastal wetlands in southern and eastern Australia have been lost.
- Nearly 90 per cent of temperate woodlands and mallee have been cleared.
- More than 99 per cent of temperate lowland grasslands in south-eastern Australia have been lost.
- About 75 per cent of rainforests have been cleared.

ecosystems. Forests, for example, are among the least-extensive ecosystems in Australia, but are rich biologically. Of the original nine per cent of the continent that was once forested, only five per cent remains (AUSLIG, 1990) (see Chapter 6). Rainforests are thought to have once covered more than 80 000 sq km, but are now reduced to 20 000 sq km. Much of the remaining rainforest is in steep and inaccessible areas, on infertile soils, whereas most of that cleared was in lowland and tableland areas with more fertile soils (Winter *et al.*, 1987). Given the often specific requirements and limited distribution of rainforest organisms, there is a real possibility that clearances have already resulted in massive loss of biological diversity.

Until recently, the importance of grasslands to the conservation of biodiversity has been neglected. For a long time, graziers have regarded many grasslands as natural pastures. Only about 10 000 ha of the original grassland vegetation of south-eastern Australia is still intact, which represents only 0.5 per cent of the original area (Kirkpatrick *et al.*, 1995). Less than 0.2 per cent of Victoria's original grasslands now remain in fragmented remnants, mostly restricted to roadsides, railway lines, cemeteries and lightly grazed unimproved pastures. Some 21 species of vertebrates originally found in grasslands in Victoria are no longer present in that State and six are extinct (Victorian Department of Conservation and Natural Resources, 1992). More than 83 per cent of the original lowland grasslands and grassy woodlands in Tasmania have been destroyed or greatly modified.

Only about five per cent of the more than six million hectares of brigalow (*Acacia harpophylla*) in Queensland still remain (Biodiversity Unit, DEST, 1995). Other woodland communities have been entirely eliminated. For example, no natural stands remain of *Eucalyptus crenulata*, a Victorian endemic. Others are close to elimination: grassy woodlands of white box (*E. albens*) on the central and western slopes of New South Wales have only 0.01 per cent of their original area in approximately natural condition — these survive in four small sites on roadsides and in cemeteries (Prober and Thiele, 1993).

Grazing by domestic stock is the most widespread of the threatening processes listed in the Table 4.5. It affects most of the continent to some extent, occurring in almost all of the 80 biogeographic regions. While grazing often involves clearing in wetter regions, the most extensive grazing in the country is on native vegetation in the arid and semi-arid zones. Low densities of livestock in arid areas can be at least as damaging to soil, vegetation and fauna as higher densities in wetter regions. This is because the arid rangelands are less productive and so have a much lower carrying capacity for stock. It is also possible that stock grazing in the arid zone has serious impacts on some restricted ecosystems such as the isolated, relatively moist refugia on which many native as well as introduced animals rely during droughts (Morton, 1990; Morton *et al.*, 1995).

Landsat image of part of the New South Wales North Coast biogeographic region; altogether, about 37 per cent of the region is cleared, but, as the image shows, clearing has been highly selective. Some environments are still completely forested while others have almost no natural vegetation remaining.

Commercial forestry affects relatively few biogeographic regions and relatively small areas of the continent (see Fig. 4.10 and Table 4.5), although the high biological diversity of forests calls for careful management of logging and associated activities. Forestry alters hydrology (increasing erosion and sedimentation), removes the major structural components of the ecosystem such as foliage, canopy and root systems and reduces the availability of suitable habitat for many animals including forest-dependent endangered species, particularly arboreal mammals.

No national figures are available on the extent of mining, but the overall distribution of this threatening process is indicated by the number of major mines and potentially exploitable mineral deposits (see Table 4.5 and Fig. 6.3). Although some mines are extensive, the total area of land directly affected by mining in Australia is small relative to the other threatening processes shown in the table. Much progress has been made in recent decades with techniques for rehabilitation of mined sites, although serious impacts from some mined areas persist. These include long-term changes to soil profiles and topography after sand-mining in coastal dunes. Sand-mining on the coast of New South Wales and Queensland has also destroyed extensive areas of coastal forest, heathland and swamps. Processing of minerals has led to pollution with heavy metals in areas such as estuaries in Tasmania, with unknown effects on marine organisms. On land, smelter gases have devastated vegetation in places such as Queenstown in Tasmania (see photograph on page 4-22).

Other threatening processes discussed earlier in this chapter and not covered in Table 4.5 are pollution, introduced species and climate change. These pressures do not lend themselves to geographic analyses in the same way as those in the table and, in the case of climate change, the potential effects at the ecosystem level are still uncertain.

The important threatening processes in each terrestrial biogeographic region are summarised in Fig. 4.12 (Thackway and Cresswell, 1995). The results support the analyses in Table 4.5 regarding

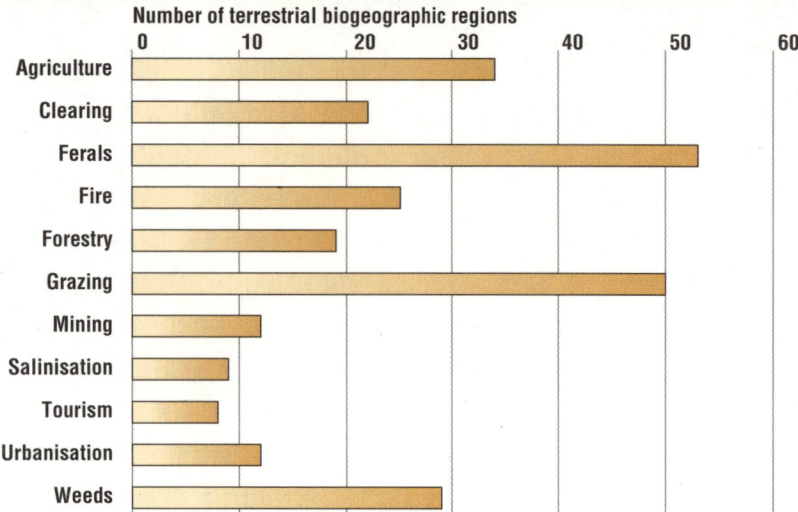

Figure 4.12 Number of terrestrial biogeographic regions in which each of 11 threatening processes are considered important by State and Territory conservation agencies

Number of terrestrial biogeographic regions

Source: Thackway and Cresswell, 1995.

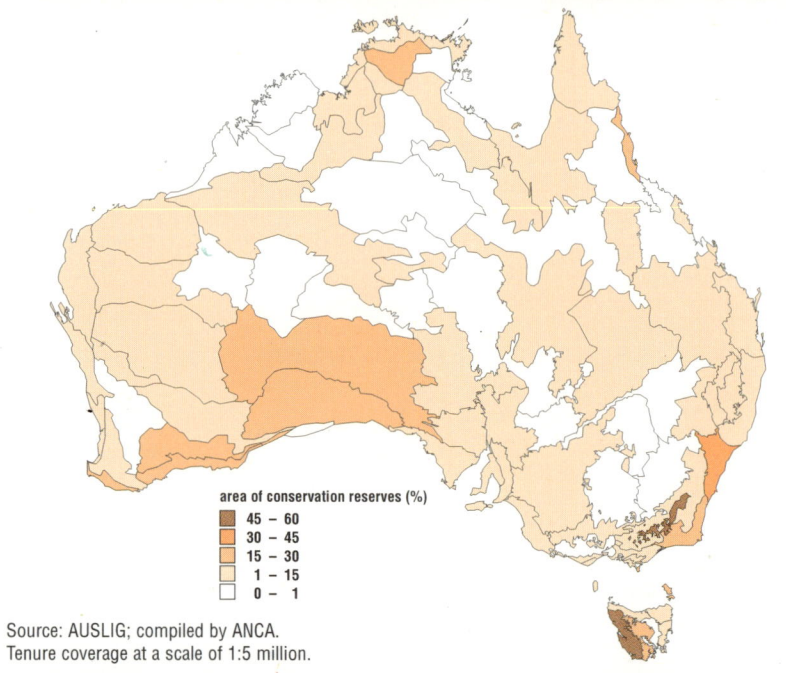

Figure 4.13 Percentage of biogeographic regions covered by nature conservation reserves

area of conservation reserves (%)
- 45 – 60
- 30 – 45
- 15 – 30
- 1 – 15
- 0 – 1

Source: AUSLIG; compiled by ANCA.
Tenure coverage at a scale of 1:5 million.

Table 4.6 Assessments of the condition of terrestrial biogeographic regions

Assessment by State and Territory conservation agencies	No. of regions
Natural ecosystems dominant with no known risk	5
Natural ecosystems dominant with no widespread degrading land uses but processes of disturbance present	19
Natural ecosystems present but coexisting with pastoral or timber industries	40
Modified ecosystems dominant with natural ecosystems occupying a very small proportion of the regions	16

Source: Derived from Thackway and Cresswell, 1995.

the relative extent of clearing, grazing, forestry and mining. They also emphasise the widespread problem of feral animals and weeds for nature conservation at the regional level and indicate that land managers in many regions perceive altered fire regimes as an important issue.

The information indicates that all biogeographic regions have been affected to some extent by one or more threatening processes. Some have been largely altered from their pre-European condition and others severely changed over small areas. While it is not possible to create a reliable index of ecosystem condition from these data, several pieces of information help to illustrate the condition of terrestrial ecosystems in Australia.

A general recognition that some ecosystems are more prone to serious impacts than others has led to a process for proposing and listing endangered ecological communities under the *Endangered Species Protection Act 1992*. In New South Wales, Benson (1991) has rated major vegetation groups according to reservation status and degree of threat in a similar way to previous ratings for plant species (Briggs and Leigh, 1988). Of 432 vegetation groups, he considered 27 (six per cent) to be endangered and 84 (19 per cent) to be vulnerable. A national view comes from assessments by State and Territory conservation agencies of the overall condition of each biogeographic region (Thackway and Cresswell, 1995) (see Table 4.6). Only five regions (six per cent) are considered to be largely natural with no significant impacts. Alteration is widespread in 56 (70 per cent) and almost complete in 16 (20 per cent).

Analysis of major threatening processes gives no indication of the extent of impacts on fresh-water habitats. Large areas of Australian wetlands have been seriously altered by factors such as drainage, filling, alteration of stream flows, changed burning regimes and invasion by weeds and introduced animals (Pressey and Adam, 1995). Overall, the greatest loss and degradation of wetlands and impacts on flowing waters have occurred in agricultural districts (see Chapter 7).

Table 4.7 summarises the condition of marine ecosystems for each interim marine region. For more detail on marine and estuarine habitats, see Chapter 8.

Protected areas

The condition of ecosystems can sometimes also be inferred from the extent to which they are represented in protected areas. Like threatening processes, the patterns of protected-area coverage vary widely between biogeographic regions. There is no clear geographical pattern (see Table 4.5 and Fig. 4.13), but many biogeographic regions are below the 10 per cent level of representation recommended by the IUCN. This assessment says something about the need for more protected areas and the degree to which regions are separated from threatening processes. However, the figures need to be qualified carefully. Extent is not equivalent to effectiveness for protecting biodiversity for a number of reasons.

Table 4.7 Summary of condition of marine regions

North Coast
Estuaries and coast generally in good condition. Impact of prawn trawling not known but large amounts of bycatch are discarded.

North West Coast
Coast and estuaries generally in good condition. Some wetlands destroyed for salt production. Significant traffic in bulk carriers for gas and iron ore. Impact of prawn trawling not known. Offshore, the seabed is recovering from damage from fish trawling.

Central West Coast
Coast generally in good condition. Impact of prawn trawling and lobster fishing not known.

Lower West Coast
Estuaries in the south are eutrophic. Inshore waters generally in good condition, especially in the north. The impacts of a large commercial rock lobster fishery and extensive recreational fisheries are not known. A natural predator *(Drupella)* is damaging corals.

South West Coast
Localised pollution, severe in some places, has been reduced in recent years. Impacts of fisheries for shark and abalone as well as a purse seining fishery are not known. Inshore waters in good condition.

Great Australian Bight
Coastal waters generally in pristine condition.

South Gulfs Coast
Gulf St Vincent and northern parts of Spencer Gulf are polluted by nutrients and heavy metals. Major loss of seagrasses associated with sewage discharges. Heavy recreational and commercial fishing pressure in finfish, abalone and lobster. This pressure is exacerbated by large-scale illegal fishing for high value shellfish. Impacts of the prawn trawl fishery are unknown.

Bass Strait
Bass Strait has exceptionally high biodiversity of seabed organisms which, in the case of Port Phillip Bay, appear to be crucial in preventing eutrophication from the high input of nutrients. Extensive seagrass loss has occurred in Westernport Bay, possibly through siltation or pollution. Other impacts include introduced marine pests and scallop dredging. Effects of the large recreational and commercial fisheries on finfish and abalone are not known but populations of rock lobsters have declined and poaching of shellfish is increasing the pressure of harvesting.

Tasmanian Coast
Some inshore areas such as the Derwent estuary are badly polluted by heavy metals. Introduced species such as starfish, macroalgae and dinoflagellates are posing threats to a large aquaculture industry. The high wave energy coasts are in good condition. Impacts of large abalone and lobster fisheries are not known.

North East Coast
Much of the coast is adversely affected by run-off of nutrients and silt from coastal catchments. Extensive urban development in the south and central areas have affected coastal habitats. There has been a major loss of wetlands through ponding of coastal flats for pastures. Impacts of the large trawl fishery for prawns and scallops are not known.

Lower East Coast
Extensive impacts from urban and tourist developments, agriculture, and commercial and recreational fishing on the estuaries which are subject to pollution. The coast is mainly in good condition except near centres of population. Impacts of fish and prawn trawling are not known.

Central East Coast
Extreme impacts from urban and tourist developments, agriculture,and commercial and recreational fishing on the estuaries which are subject to pollution. Acid sulfate soils are a problem for water quality caused by drainage of the coastal floodplains. Impacts of prawn trawling not known.

Great Barrier Reef
Subject to an overall management plan with zoning for conservation, tourism, fishing and research. Generally in good condition but there has been loss of coastal coral reefs, apparently from nutrient enrichment of coastal waters due to nearby land use. Impacts of prawn trawling and recreational and commercial line fishing are not known.

Gulf of Carpentaria
Generally pristine but localised impacts from dredging for port access. Impacts from prawn trawling in coastal waters are not known but large amounts of bycatch are discarded.

North West Oceanic
Generally pristine. Adverse impacts could come from the offshore extraction of gas and petroleum

West Oceanic
Apparently pristine

South Oceanic
Longline fishery in the south affecting seabirds. Impacts of finfish trawl fishery in the Bight are not known.

South East Oceanic
Deepwater seabed in the northern section is generally pristine. Deepwater trawl fisheries (up to 1000 m depth) in the south on seamounts could have adverse impacts on seabed fauna. Longline fishery around the edge of the continental shelf adversely affecting seabirds.

North East Oceanic
Pristine.

Major ecosystems are heterogeneous in terms of physical and biological characteristics. High overall percentages in reserves can mask the fact that much internal variation is unprotected. In north-eastern New South Wales, for example, about seven per cent of the region is covered by national parks and nature reserves but some environments are protected poorly or not at all (Pressey, 1995). High levels of bias in the distribution of reserves within ecosystems are common in Australia (Thackway and Cresswell, 1995). Within many ecosystems, protected areas are concentrated in environments least prone to disturbance from intensive land uses while the most vulnerable environments are missed (see the box on page 4-52).

The effectiveness of protected areas also depends on the types of threatening processes operating, funding for management and the size and boundaries of the areas. Strict reserves are generally good at preventing the effects of clearing, grazing by domestic stock, forestry and mining. For other threatening processes, they are less useful. Reserves do not protect against introduced species without intensive management to control weeds and feral animals. They can be subject to run-off and sedimentation from surrounding land uses if their boundaries are not aligned with catchments. No matter how well managed the reserve itself, species can be lost if their food sources in one season are outside the reserve and lost to clearing. Species can also be lost if reserves are too small to maintain populations of sufficient size to escape the problems of chance population changes that lead to extinction. In fragmented landscapes, only relatively small areas of natural habitats remain. Although this limits the viability of some species, many others will persist. Fragments offer the only chance of protecting parts of ecosystems that are heavily cleared and are vitally important for conserving biodiversity.

The state of species diversity

Of the 12 nations in the world that contain major repositories of species diversity, Australia is the only developed country. Others include Indonesia with its wealth of islands and different habitats, Zaire in equatorial Africa and Brazil with its expanses of rich tropical rainforests, rivers and mountains. The state of species diversity in Australia — particularly the large proportions of some groups that are threatened by the pressures outlined earlier — is cause for national concern.

The species is the basic unit of biological classification. It is defined as a group of interbreeding (or potentially interbreeding) natural populations that are reproductively isolated from other such groups. Certainly for organisms like larger plants and animals this is a useful definition. For micro-organisms, however, it poses considerable problems with identification. The species concept is difficult to apply to organisms with different modes of reproduction from the 'higher' organisms. Even some larger life forms such as algae and lichens are under-represented in censuses of biodiversity because of problems of identification.

Table 4.8 Estimated extent of Australia's species diversity and of those formally described

Major group	Estimated no. of species	Percentage described
Micro-organisms		
Protozoans	65 000	40
Fungi	160 000	5
Bacteria	40 000	0.1
Invertebrates		
Arthropods		
Coleoptera (beetles)	30 000	67
Lepidoptera (moths and butterflies)	20 000	53
Hymenoptera (ants, wasps and bees)	23 000	33
Diptera (flies and mosquitoes)	11 000	75
Other insects	15 000	20
Arachnids (e.g. spiders and mites)	39 000	14
Crustaceans (e.g. crabs and prawns)	18 000	5
Springtails	2 500	14
Other arthropods	?	?
Molluscs (e.g. snails, oysters, squid)	19 000	?
Sponges	1 400	28
Nematodes	150 000	1
Other invertebrates	?	?
Vertebrates	5588	90+
Plants		
Higher plants	20 000	90+
Algae	22 000	50

Estimates were taken from several published sources, each of which is not comprehensive. For some groups the level of knowledge is so poor that estimates are unavailable and are indicated by '?'. Sources are Kennedy, 1990; Williams, 1990; Castles, 1992 and Biodiversity Unit, DEST, 1994. Updated information was obtained from CSIRO Division of Entomology, Australian Museum, Queensland Museum and ABRS.

Species inventory

Despite these difficulties, we still need to estimate the extent of biodiversity. To do this, we usually count published, recognised species. For groups that are poorly collected and described, we use estimates of species richness based on the rate at which new species are being discovered and described (see Fig. 4.14). This rate is high and not diminishing. Table 4.8 illustrates the strong bias in knowledge towards large, conspicuous life forms and shows that most biodiversity is either invertebrate or microbial. Australian flowering plants are reasonably well known, as are the vertebrate animals, but many of our invertebrate groups are poorly known — both poorly collected and not yet adequately described. However, even for conspicuous life forms, new species are still being discovered (see the box on the Wollemi pine).

The outlook is not good for those groups of organisms where scientists estimate only a small fraction have been described. Many of these species may become extinct before they are collected and described. This lack of information (highlighted on page 4-51) about something as basic as inventory points out one of the most important issues relating to biodiversity — our ignorance. It is extremely difficult to respond to changes in the state of biodiversity if we do not possess the basic information about what it is and how it is distributed. The best approach may be to select groups for study that have known biological significance rather than trying to catalogue parts of all groups.

Australia is one of the world's six floristic realms — regions of the world supporting a characteristic flora. Most of our plant species are found naturally only here. However, the special nature of our flora is not limited to its global uniqueness. We have some of the world's most primitive plants, now found only in the rainforests of northern Australia.

Much of our terrestrial vegetation is dominated by eucalypts and acacias. Most forests and woodlands feature several species of eucalypts, while acacias dominate the extensive shrublands of the inland. Each genus has more than 700 species, almost all of which occur naturally only here. Neither group is distributed evenly across the continent. The acacias are particularly abundant in the semi-arid region of southern Western Australia while the eucalypts are richest in the south-eastern region. The recent assessment of the Blue Mountains in New South Wales for World Heritage Listing showed that parts of this area contain the greatest variety of eucalypt species anywhere.

The animal kingdom provides similar striking examples. Australia has two of the only three species of monotremes (egg-laying mammals), it has a large proportion of the living marsupials (pouched mammals) and it is the only nation to have the large arid-adapted kangaroos.

The numbers of taxa listed in different families of plants depend on the extent to which they have been studied. Recent research on eucalypts for example, has doubled the known number of species. Our knowledge of the distribution of species is also incomplete; for some groups, a map at one degree resolution is unreliable. Even for well-known groups, the data become patchy as we move to finer scales. In one of the best surveyed parts of Australia, the south-east forests of New South Wales, all the survey plots add up to only 0.1 per cent of the landscape.

Number and distribution of species

Of the four groups of vertebrate animals for which data are available, bird species are most numerous along the east coast and in the south-eastern region; their numbers diminish in the arid interior (see Fig. 4.15). In contrast, reptile species-richness is high in the arid zone and the tropics. The pattern for mammals is different again, with high numbers along the eastern margin of the continent and in the far north, the south-west and part of the arid zone. Lastly, amphibian species-richness is highest for parts of the eastern region and across the north, with some pockets in the south-west.

Our seas also support many forms of life unique to this region. For example, Australia has an extremely diverse flora of macroalgae (seaweed). Temperate waters alone contain nearly 1200 species, of which 62 per cent are endemic. Bass Strait has more than 500 species. The introduction of Japanese kelp to Tasmania, probably in ballast water, appears to be threatening indigenous macroalgae because it grows rapidly and smothers other species. This invader is spreading along the Tasmanian coast at a rate of between four and 40 km a year.

Australia has about 30 species of seagrasses (see page 8-24 and Table 8.8). They form rich beds in shallow water and are important as nursery grounds for the juvenile stages of many species of crustaceans and fish. Seagrasses have suffered severe reductions in recent years from a variety of causes, including pollution and siltation as well as natural causes such as floods and cyclones. The most recent major loss resulted from the construction of the third runway at Sydney Airport, which destroyed an important seagrass bed in New South Wales by filling part of Botany Bay.

Although a high proportion of our temperate marine plants and animals are endemic, most tropical marine fauna are Indo-West Pacific. This means that many species are also found in the Indian and Western Pacific Oceans. For example, the blue swimmer crab (*Portunus pelagicus*), also known as the manna or sand crab, occurs in the Pacific and Indian Oceans and the Red Sea and has even extended into the eastern Mediterranean by migrating through the Suez Canal. Such wide distributions have, however, not lessened the threat to many species. Massive clearance of mangroves in South-East Asia for aquaculture as well as destructive fishing practices, including the use of explosives on coral reefs, have severely depleted many tropical marine habitats in the Indo-Pacific.

Australia now has the largest intact coral reef in the world and some of the best-preserved mangrove forests. Some species, such as turtles and dugongs

The Wollemi pine — a relic from the age of dinosaurs

The discovery in December 1994 of a new kind of tree near Sydney caused great excitement, as Australia's flora is considered to be well known. The Wollemi pines (*Wollemia nobilis*) reach about 35 m in height with a main trunk up to one metre in diameter. The discovery of a new species of tree, especially one that grows to such an impressive height, is extremely unusual. The habitat of the trees — protected steep-sided canyons north-west of Sydney, which acted as refuges from fires that frequently burn the adjacent plateaus — contributed to their continued existence. The tree, given the common name of 'pine', is a conifer but is closer to the Norfolk Island pine than to the true pines. The discovery is a dramatic demonstration that parts of our biological heritage remain unknown.

Figure 4.14 The number of new species described each year world-wide for five groups

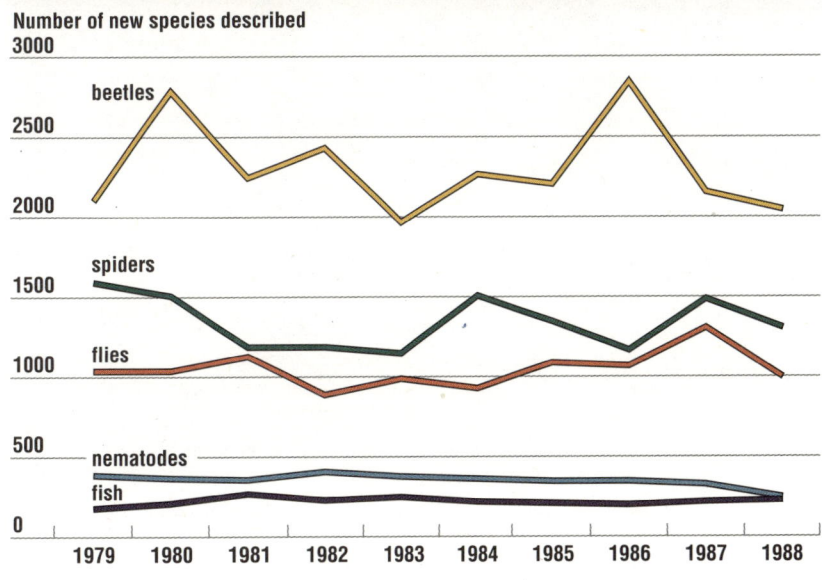

Source: derived from World Conservation Monitoring Centre, 1992.

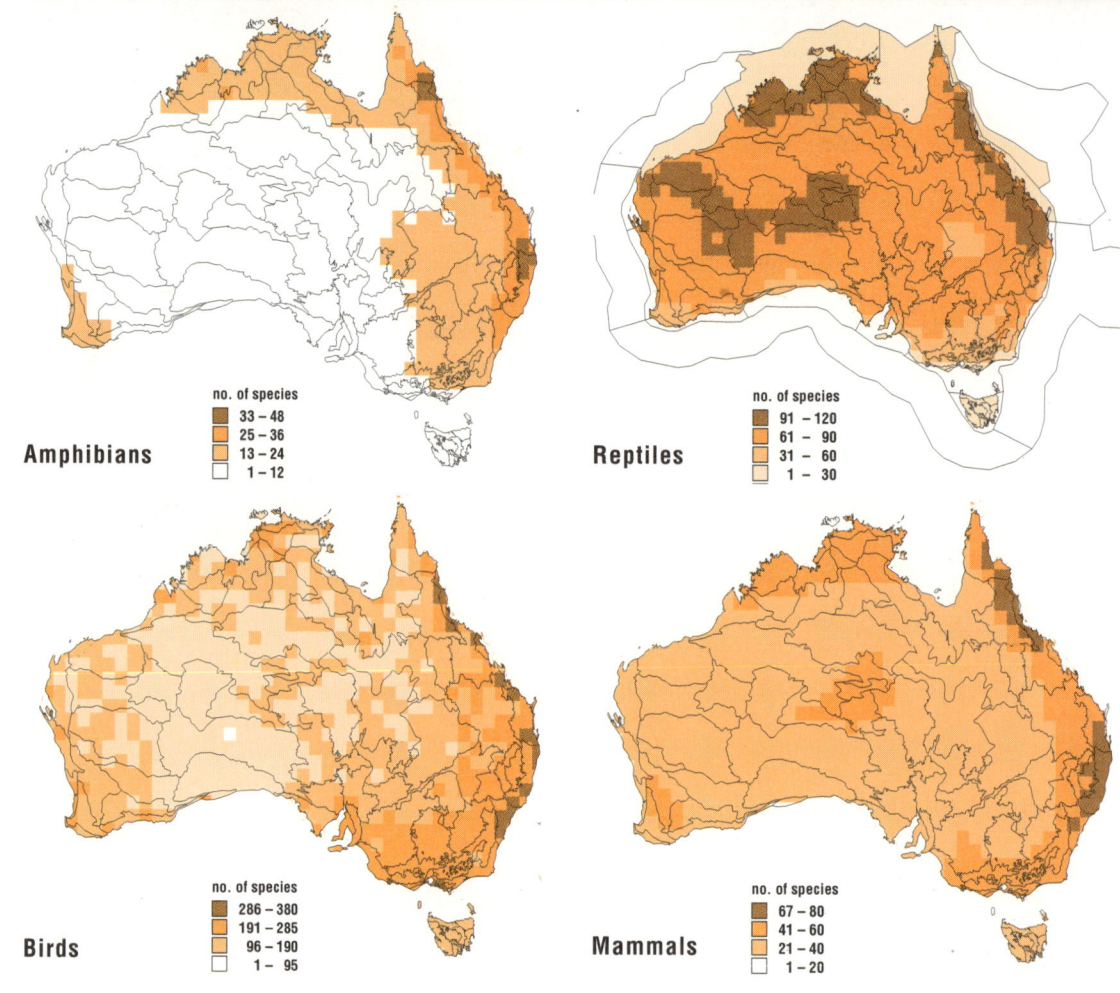

Figure 4.15 The distributions of species-richness for four vertebrate groups

Grid cells (one degree of latitude by one degree of longitude). The boundaries indicate Interim Biogeographic Regions for Australia.

Source: compiled by ANCA

Amphibians

no. of species
33 – 48
25 – 36
13 – 24
1 – 12

Reptiles

no. of species
91 – 120
61 – 90
31 – 60
1 – 30

Birds

no. of species
286 – 380
191 – 285
96 – 190
1 – 95

Mammals

no. of species
67 – 80
41 – 60
21 – 40
1 – 20

Figure 4.16 Areas of endemism based on distribution patterns of birds

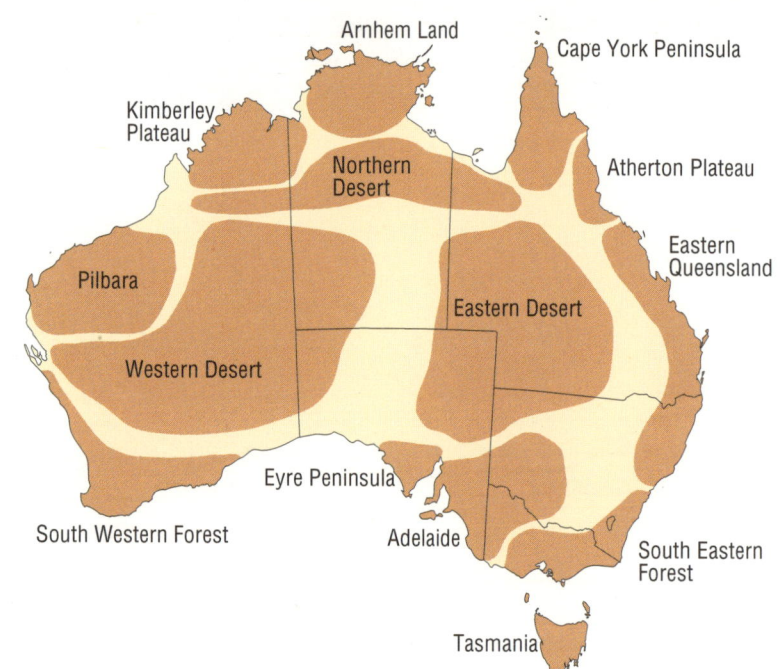

Each of the 14 recognised areas has a unique assemblage of bird species.
Source: Cracraft, 1991.

(*Dugong dugon*), may eventually survive only in Australian waters. Hence, the conservation of Australia's tropical marine flora and fauna is of global significance.

Some of the sea's most-threatened species are edible ones that, unlike most marine species, are (or were) abundant. Modern fishing techniques are extremely efficient and overseas experience shows it is possible to fish formerly abundant species to near extinction. Many exploited marine species are now subject to management plans that seek to exploit them at sustainable levels or try to rebuild depleted stocks. Nevertheless, several formerly abundant animals have been reduced to seriously low levels. They include eastern gemfish and eastern rock lobster off New South Wales, school shark and southern rock lobster off Victoria, King George whiting off South Australia and southern bluefin tuna in the Southern Ocean.

Status of species

All groups of higher plants and vertebrates have species that are highly threatened. The real situation is worse than indicated in Table 4.9 because many groups include species in decline. The cumulative effect on birds of the threatening processes will be accelerated loss of bird species paralleling the loss of mammal species (Recher and

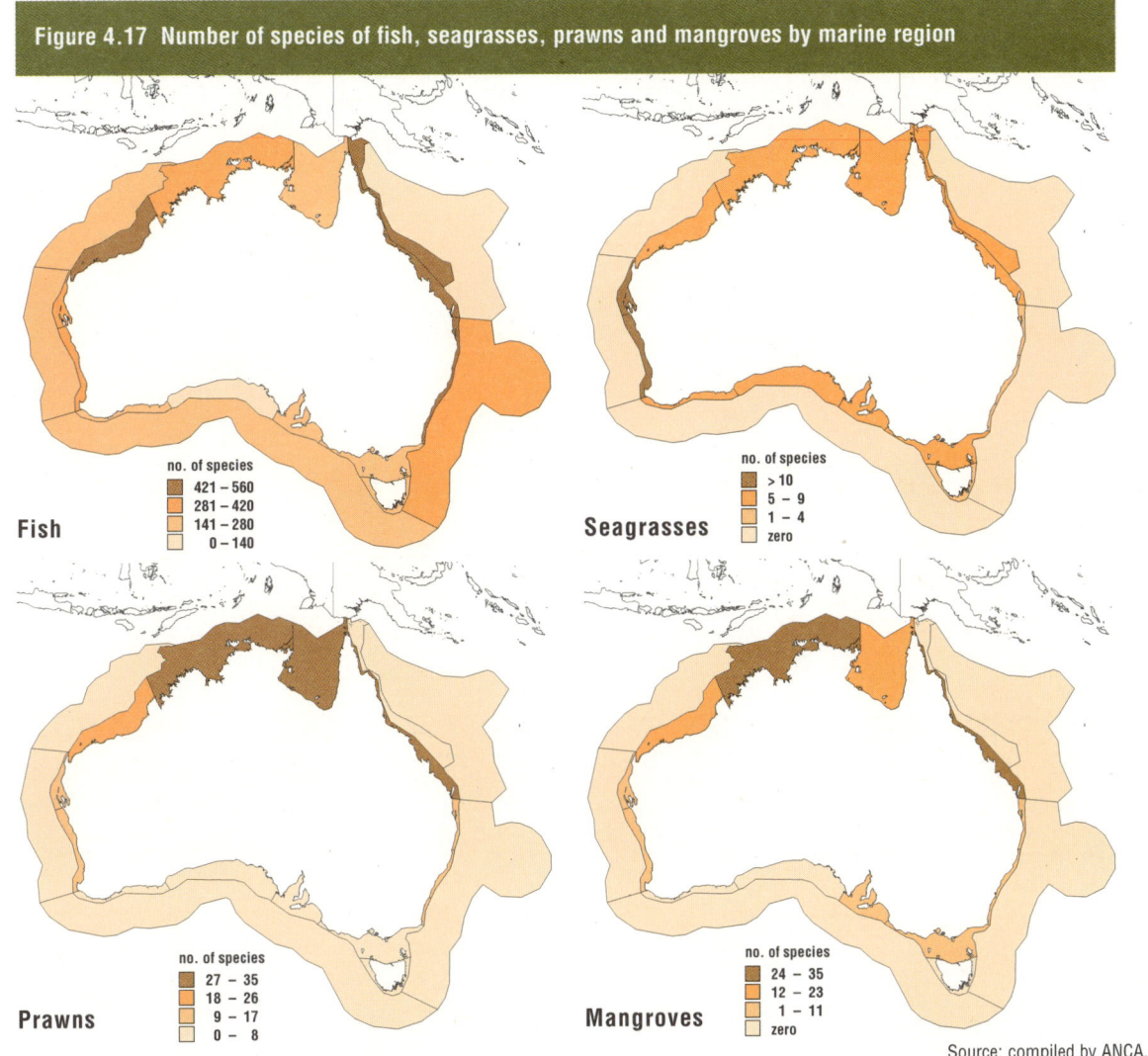

Figure 4.17 Number of species of fish, seagrasses, prawns and mangroves by marine region

Fish

no. of species
421 – 560
281 – 420
141 – 280
0 – 140

Seagrasses

no. of species
> 10
5 – 9
1 – 4
zero

Prawns

no. of species
27 – 35
18 – 26
9 – 17
0 – 8

Mangroves

no. of species
24 – 35
12 – 23
1 – 11
zero

Source: compiled by ANCA.

Lim, 1990). Many species of frogs are declining in part of their range and several, including the unique gastric brooding frog (*Rheobatrachus silus*), have disappeared in recent years (Tyler, 1994). Tables 4.9 to 4.18 and Fig. 4.18 indicate the current conservation status of various groups of Australian plants and animals.

Australia's record of mammal species extinctions is the worst for any country. In the past two centuries, the country has lost ten species of the original marsupial fauna of 144 species and eight of the 53 species of native rodents (see Table 4.17). More than one hundred mammal species are considered endangered, vulnerable or potentially vulnerable. This number includes marine mammals such as dugong. Some marine species, like whales and seals, which were hunted in Australian waters until recently, now show signs of recovery. The sightings of whales close to the coast, even within Sydney Harbour, herald a positive response to embargos imposed on harvesting.

Many species were illustrated or even photographed before becoming extinct. For example: the Tasmanian tiger (*Thylacinus cynocephalus*) — last specimen died in 1936; the pig-footed bandicoot (*Chaeropus ecaudatus*) — last recorded sighting 1907; the desert rat kangaroo —

(*Caloprymnus campestris*) — 1935; lesser bilby (*Macrotis leucura*) — 1931; long-tailed hopping-mouse (*Notomys longicaudatus*) — 1901; lesser stick-nest rat (*Leporillus apicalis*) — 1933; Alice Springs mouse (*Pseudomys fieldi*) — 1895; Darling Downs hopping-mouse (*Notomys mordax*) — 1840s.

Not surprisingly, the main regions containing threatened species coincide with those of high species-richness. However, there are some

The lesser bilby *(Macrotis leucura)* was last recorded in 1931.
▼

Table 4.9 The status of flowering plants and vertebrate animals

Major group	Estimated number of species	Percentage of endemic species	Number presumed extinct (since 1788)	Number of endangered species	Number of vulnerable species	Number of naturalised (introduced) species
Flowering plants	20 000	85	76	301	708	1500– 2000
Fish						
freshwater	195	90	0	9	8	21
marine	4000	13 (tropical inshore) 85 (temperate inshore)	0	0	-	8
Amphibians	203	93	3	10	19	1
Reptiles	770	89	0	11	40	2
Birds	777 (1074 species and subspecies)	45	20[1]	25	25	32
Mammals (terrestrial)	268	84	19	25	18	25

1. Nineteen once existed on Australian territorial islands, including Lord Howe and Norfolk Islands; only one is extinct on the mainland.

Table 4.10 The conservation status of Australia's freshwater fish

Conservation status	Number of species	Percentage of estimated total
Presumed extinct	0	0
Endangered	9	4.1
Vulnerable	8	3.7
Poorly known	19	8.8
Rare	36	16.6

Approximately one-third of the species are either threatened or rare. Source: derived from Wager and Jackson, 1993.

Table 4.12 The conservation status of Australia's amphibians

Conservation status	Number of species	Percentage of estimated total
Presumed extinct	3	1.5
Endangered	10	4.9
Vulnerable	19	9.4
Indeterminate	2	1.0

A total of 203 species of Australian amphibians has been identified from a world total of approximately 4000. It is estimated that a further 25–30 species in Australia await discovery.

Source: derived from Tyler, 1994.

Table 4.11 The conservation status of Australia's reptiles

Conservation status	Number of species	Percentage of estimated total
Presumed extinct	0	0
Endangered	11	1.4
Vulnerable	40	5.2
Rare or insufficiently known	148	19.2

A total of 770 species of Australian reptiles has been identified from a world total of approximately 6500 species.

Source: derived from Cogger *et al.*, 1993; H.G. Cogger (pers. comm.).

The meaning of rarity and threat

Rarity, while it can predispose some species to extinction, can be a natural, inherent characteristic that has nothing to do with the pressures discussed earlier in this chapter. In some cases species are rare because of human pressures, particularly habitat destruction. However, rarity may be the result of a natural condition such as specialised habitat requirements. Most marine species, for example, are rare. A large proportion of the flora of south-western Western Australia is endemic to the region and many of these species are rare. Threat, on the other hand, is explicitly the result of pressures. The conjunction of large numbers of species (including many rare ones) and many threatening processes — especially vegetation clearance for intensive agriculture — leads inevitably to large numbers of threatened species.

Table 4.13 The conservation status of Australia's non-marine invertebrates

Conservation status	Number of species	Percentage of estimated total
Presumed extinct	3	?
Endangered	40	?
Vulnerable	78	?
Indeterminate or insufficiently known	291	?

The number of non-marine invertebrates is unknown, hence no percentage of total can be calculated.

Table 4.14 The preliminary conservation status of Australia's algal, lichen and bryophyte species

Conservation status	Lichens	Bryophytes	Algae
Presumed extinct*	2	12	-
Endangered	94	83	-
Vulnerable	74	43	1
Potentially vulnerable	31	12	6

*Not found during recent surveys.

Information on the conservation status of these groups is poor and there is insufficient information to nominate any species or communities as endangered or rare.

Source: 'Overview of the Conservation of Non-Marine Lichens, Bryophytes, Algae and Fungi', report for the Endangered Species Program, ANCA, 1994.

Table 4.15 The conservation status of Australia's higher plants

Conservation status	Number of species*	Percentage of estimated total
Presumed extinct	76	0.4
Endangered	301	1.5
Vulnerable	708	3.5
Rare	1570	7.9
Poorly known	2376	11.9

*The numbers of species of vascular plants listed in the 1995 CSIRO Rare and Threatened Plants (ROTAP) publication.

Table 4.16 The conservation status of Australian native rodents

Conservation status	Number of species	Percentage of estimated total
Presumed extinct	8	15.0
Endangered	6	11.3
Vulnerable	5	9.4
Rare	4	7.5
Insufficiently known	7	13.2

Rodents, along with marsupials, have taken the brunt of mammal extinctions in Australia since European settlement. Source: derived from Lee, 1995.

Table 4.17 The conservation classification of Australian terrestrial native mammals recorded since 1778

Major group	No. of described extant species	Extinct	Endangered	Endangered as a percentage of total extant
Monotremes	2	0	0	0
Marsupials	144	10	19	13.2
Placentals				
Dingo	1	0	0	0
Rodents	53	8	6	11.3
Bats	68	1	0	0

Principal sources: derived from Kennedy, 1990 and 1992; Williams, 1990; Castles, 1992; Endangered Species Protection Act, 1992, schedules 1, 2 & 3, July 1994; Biodiversity Unit, 1994 and Census of Australian Vertebrate Species, 1994.

Table 4.18 The conservation status of Australian monotremes and marsupials

Major group	Total	Presumed extinct	Endangered	Vulnerable	Insufficiently known
Monotremes	2	-	-	1	-
Dasyurids	51	1	6	12	1
Bandicoots and bilbies	11	3	2	4	-
Marsupial mole	1	-	-	1	-
Wombats, possums koala and kangaroos	81	6	11	23	1

Australian marsupials are facing many threats. The bandicoots and bilbies are particularly vulnerable.
Source: derived from Kennedy, 1992.

Figure 4.18 The proportion of Australia's birds (by sub-species) that are extinct, endangered, vulnerable, rare or insufficiently known

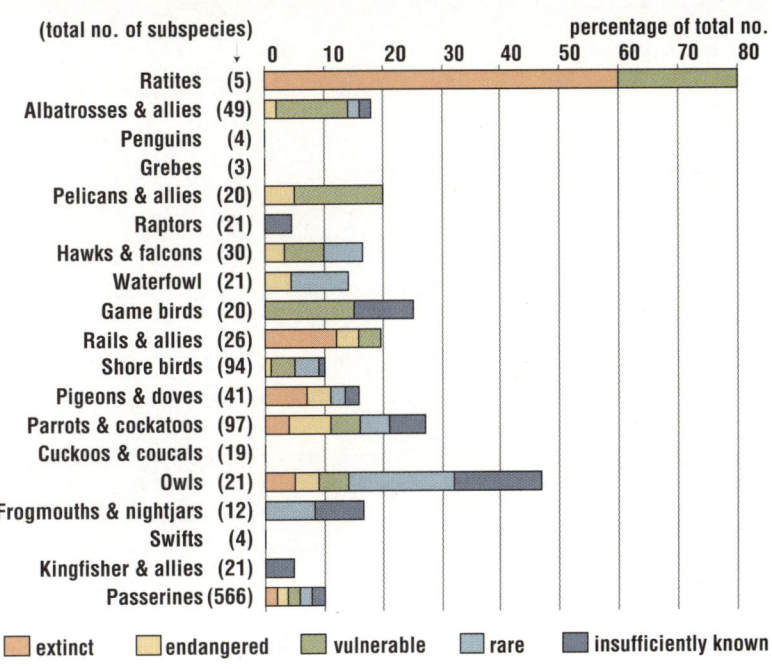

Source: Garnett, 1992b.

Figure 4.19 Australia's threatened vertebrate species

▶
Numbers of threatened species from each of the four groups for which data are available are shown. The various threat categories are combined. Grid cells (one degree of latitude by one degree of longitude). The boundaries indicate Interim Biogeographic Regions for Australia.

Source: compiled by ANCA.

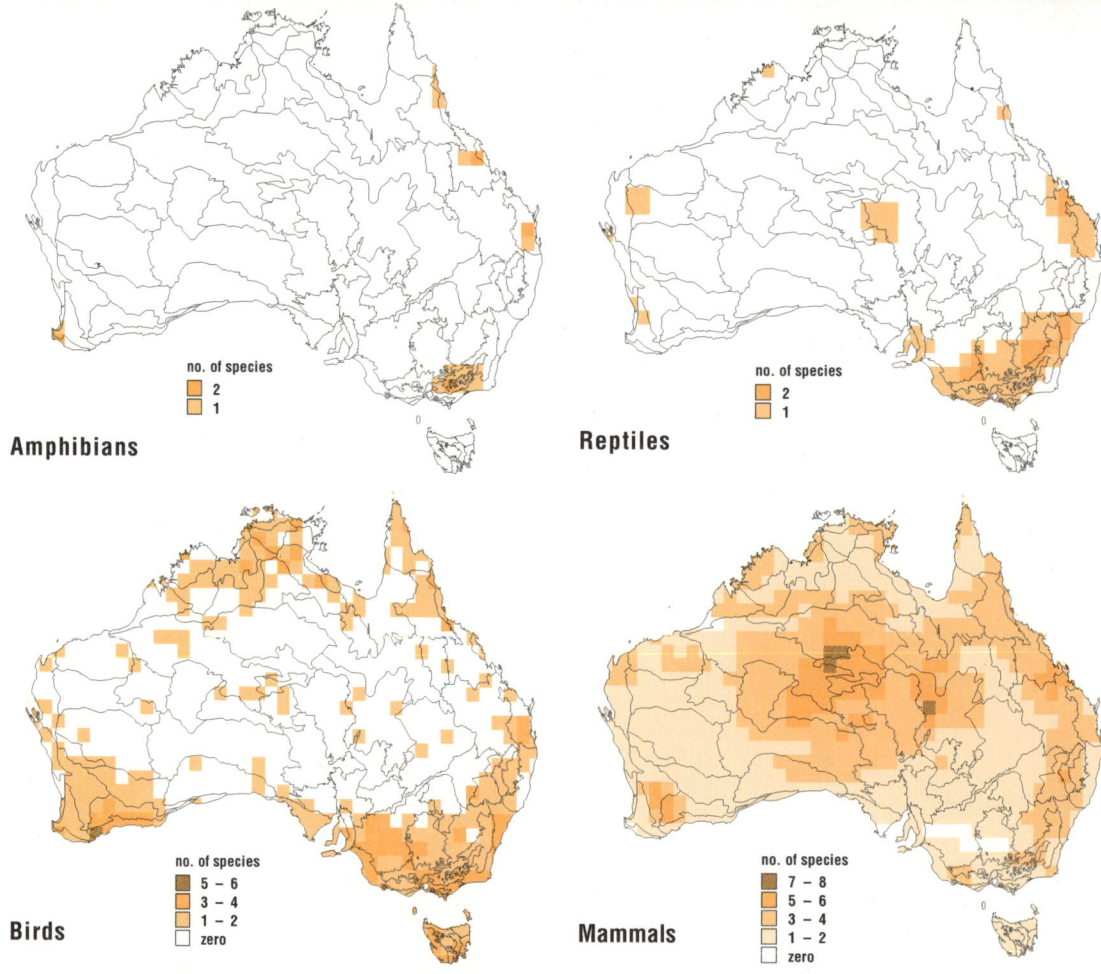

Amphibians

no. of species
- 2
- 1

Reptiles

no. of species
- 2
- 1

Birds

no. of species
- 5 – 6
- 3 – 4
- 1 – 2
- zero

Mammals

no. of species
- 7 – 8
- 5 – 6
- 3 – 4
- 1 – 2
- zero

Figure 4.20 Numbers of threatened plant species

▶
Grid cells (one degree of latitude by one degree of longitude). The boundaries indicate Interim Biogeographic Regions for Australia.

Source: compiled by ANCA.

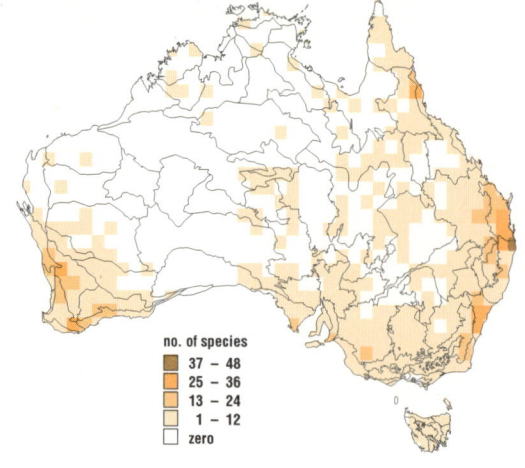

no. of species
- 37 – 48
- 25 – 36
- 13 – 24
- 1 – 12
- zero

exceptions. For birds, threatened species are mainly concentrated in the south-east, the north and the far south-west. Reptile species are richest in the arid interior, although threatened species are most numerous in the high-rainfall forested south-east of the continent. A strong pattern emerges for mammal species, with arid Central Australia clearly the zone of pressure. The amphibian pattern is a cluster of discrete regions along the eastern margin.

Figure 4.18 shows the proportions of Australian birds (by Order) that are presumed extinct, endangered, vulnerable, rare or insufficiently known. A similar situation exists for Australia's mammals (see Table 4.17). Alarmingly high proportions of each Order are in one of the threat categories. Similarly, 17 species of freshwater fish (see Table 4.10) and 29 species of amphibians (see Table 4.12) are either endangered or vulnerable. Many and varied pressures or threats have led to this situation. One study has been made of the current and former threats affecting Australian birds (see Table 4.2). The main differences between the perceived current threats and those that operated earlier are: altered fire regime, grazing, forestry operations and shortage of nest hollows.

Looking at the continental distribution of threatened plant species (see Fig. 4.20), the south-western region and parts of eastern Australia are clearly regions for concern but action is needed over most of the continent to offset threatening processes. Of Australia's presumed extinct and endangered plant species the overwhelming number are from woodland habitats (Leigh and Briggs, 1994). This reflects both the pressures on this habitat and its vast geographic extent. Combining woodland and scrub reveals 97 instances of endangered plant species — almost half the national total for all habitat types.

Genetic diversity

One measure of the degree of genetic diversity within a species is the recognition of identifiable subspecies. For example, the dingo (*Canis familiaris dingo*) is a subspecies of the domestic dog (*C. familiaris*). Although most inventories are made at the species level, the full diversity may be better expressed at the subspecies level. For example, worldwide the number of butterfly species is thought to total about 17 500; however, the number of currently recognised subspecies is closer to 100 000. The genus *Calyptorhynchus* contains four species of black cockatoo in Australia, but it has nine recognised subspecies.

Good scientific evidence now exists that, as populations decline, the amount of genetic diversity that remains in them is crucial to their survival. Very often, small populations lose genetic diversity and, consequently, suffer lowered fertility and lowered ability to adapt to changing environments. Because populations of many Australian species have declined through habitat loss or fragmentation, genetic issues are crucial to the management of biodiversity for conservation. Although there are relatively few Australian examples, some studies do provide clear illustrations of the genetic effects of habitat destruction and the consequent declines in population size.

Habitat loss and degradation

When habitats shrink, populations decline and lose genetic diversity. This reduces their ability to compete, fight disease or adapt to changing conditions.

Northern hairy-nosed wombat

Unlike the more abundant southern hairy-nosed wombat (*Lasiorhinus latifrons*), which is distributed in forests and grasslands in southern Australia, the northern hairy-nosed wombat (*L. krefftii*) is an example of a rare Australian animal that has experienced severe loss and degradation of habitat and consequent loss of genetic diversity. It is one of Australia's rarest mammals, existing as a single colony of about 65 individuals in Epping Forest, central Queensland.

By using a genetic technique known as microsatellite technology, scientists can detect losses in genetic diversity with astonishing accuracy using a tiny amount of DNA — the amount contained in a single wombat hair is enough. Every individual or population has a distinct complement of microsatellite labels and scientists have strong evidence that the northern species has lost significant amounts of the genetic diversity it once possessed (Taylor *et al.*, 1994). The genetic diversity of the northern species is less than half that of the southern hairy-nosed wombat. Because the two species are closely related and fill similar ecological niches, it is reasonable to expect that they should have similar measures of genetic diversity, especially heterozygosity. The fact that they don't is most likely a direct result of the steep decline in the number of animals, together with a process known as genetic drift, which occurs when the breeding population is so small that too few offspring are born in each generation to successfully carry all of the genetic variability in the parent population.

The population at Epping Forest cannot regain the lost genetic variability except by the long-term process of random mutation, but it is hoped that the problems associated with inbreeding and genetic drift can be overcome through careful management.

Koala

The koala (*Phascolarctos cinereus*) in south-eastern Australia has suffered severe population declines since European settlement due to loss of habitat. An extensive but *ad hoc* program of restocking has led to recolonisation of many regions throughout its original range. Many of the animals have come from the island populations of Westernport Bay in Victoria. These colonies were themselves founded artificially with low numbers and represent a narrow genetic base.

Comparisons of mainland populations derived from island colonies (restocked populations) and undisturbed mainland populations have shown that genetic diversity is severely reduced in the restocked populations (Taylor *et al.*, 1991). This work highlights the need to be cautious when re-establishing locally extinct or depleted populations of endangered species. Thorough characterisation of the genetic variability of remaining colonies will help to ensure that diversity is maximised to give new populations their best hope of survival.

Habitat fragmentation

Many kinds of habitats have become greatly fragmented and, without suitable connecting habitat, the movement of organisms and the mixing of genes are slowed down or stopped. The resulting isolated gene pools may follow different evolutionary paths. If they are too small, they lose genetic variability, especially heterozygosity, more rapidly than if they were connected with each other. Very often small populations isolated from each other are like islands, frequently surrounded by inhospitable 'seas' of urban or rural development (see page 4-13).

Sleepy lizards

The sleepy lizard (*Trachydosaurus rugosus*) is a large lizard that bears live young and is found over much of southern Australia. It experienced a large-scale natural experiment on the genetic effects of fragmentation. As sea levels rose 6000 to 8000 years ago, populations were isolated on offshore islands, preventing gene flow with the mainland. Comparisons of the genetic diversity of island and mainland populations have shown that sufficient time has elapsed to allow significant genetic divergence between them (Sarre *et al.*, 1990). The island populations lack some rare mainland genes — probably because of genetic drift. Changes in the smaller island populations have been greater than those between mainland populations.

Some patterns of genetic diversity

Cryptic species

Animals that look the same may be a complex of cryptic or hidden species distinguished only by genetic means.

Rock wallabies

The rock wallabies (genus *Petrogale*) are found throughout Australia. They live in rocky habitats such as cliffs, gorges and boulder outcrops. Researchers have investigated the problem of distinguishing those that are merely races, between which gene flow can still occur, from those that are true species, where it does not. The most informative genetic technique used so far has been cytogenetics, the study of chromosomes. In mammals, all individuals within the same species have the same number of chromosomes. Also, genes lie along the chromosomes in the same order. Individuals with different chromosome numbers and different gene orders usually cannot reproduce. They may mate but their offspring are sterile. This is one of the mechanisms that separates species.

Rock wallabies inhabiting the east coast were thought to belong to eight races, several of which cannot be distinguished morphologically. Each race has distinctive genetic markers, but where their ranges overlap hybridisation has occurred and hybrid animals can be found. However, genes found in these contact zones have not spread into the populations outside the zone; thus the races are maintaining their genetic identity despite these 'mixed marriages'. In addition, the offspring of the hybrids are sterile. The chromosome and sterility data strongly suggest that these similar-looking animals are, in fact, eight distinct species (Eldridge and Close, 1992).

Two wallabies that look alike but belong to two species: *Petrogale assimilus* and *P. sharmani.*

Velvet worms

The velvet worms or peripatus live in moist habitats such as rotting logs and leaf litter in forests around the southern hemisphere. Peripatus look like caterpillars but they have antennae and 'glue guns' on either side of the jaw with which they capture prey. In Australia they occur in the forests of eastern Australia, Tasmania and south-western Australia.

Velvet worms show little variation in body structure and, until recently, Australia was thought to have only eight species, or eight per cent of the world's species. However, genetic techniques, such as allozyme electrophoresis, have shown that at least 100 species occur in this country (Briscoe and Tait, 1995).

Consequently, the number of recorded velvet worm species in the world has doubled. Almost everywhere a suitable habitat occurs, a new species unique or endemic to that area is found.

Single species

A single species can harbour a broad range of genetic diversity across its range.

Genetic variation within a single species: Sturt's desert pea.

Barramundi

The barramundi perch (*Lates calcarifer*) is a large fish whose range extends from the Persian Gulf in the west to Australia and Papua New Guinea in the east. It inhabits fresh-water ponds and rivers, tidal swamps and estuaries and coastal reefs. It can live to 20 years of age and weigh over 50 kg. In Australia and Papua New Guinea it spends the first six or seven years as a male then metamorphoses into a female, producing 15–45 million eggs per year.

Genetic analysis using allozyme electrophoresis has uncovered great genetic diversity across Australian waters (Shaklee and Salini, 1993). More than 14 genetically distinct stocks have been identified so far: seven in Queensland, six in the Northern Territory and one in Western Australia, and there are likely to be many more as surveys continue into the more remote populations of north-western Australia.

A number of factors have contributed to the high degree of population subdivision. Large gaps between its habitats along the Australian coast have restricted dispersal between isolated breeding populations. Offshore dispersal of larval and juvenile forms is also limited because the females spawn at the mouths of estuaries, not offshore.

The degree of genetic variation across the range of the barramundi and its partitioning into distinct genetic stocks makes it important that local populations are properly managed. If genetic differentiation has occurred, it could well be because the local environment requires that the stock in that region have their own particular set of adaptations. Allowing stocks to mix or moving individuals from one region to another (for aquaculture or restocking, for example) could disrupt a delicate equilibrium brought about by evolution.

Papilionid butterflies

Another important genetic consequence of habitat fragmentation is known as inbreeding depression. As a population declines and numbers dwindle, matings between related individuals become unavoidable. This leads to inbreeding and in many cases a reduction in fitness — inbreeding depression. The decline in health and in reproductive output causes the population to shrink further, often leading to extinction.

Papilionid butterflies in the rainforests of northern Australia illustrate the phenomenon. These northern rainforests persist today as highly fragmented remnants. Normally wide-ranging species of large and beautiful swallowtail butterflies are now often confined within the remnant patches. Experimental breeding between related individuals in captivity showed inbreeding depression in three species (Orr, 1994). The proportion of eggs hatching, the number of caterpillars surviving to adulthood and growth rates were all severely affected in offspring from related parents. These results show that even wide-ranging species can be susceptible to the effects of inbreeding depression and reserves must be large enough to prevent it.

Other processes that lead to steep declines in the size of populations are excessive harvesting, as in some fisheries, and the presence of introduced predators such as foxes and cats and of introduced diseases.

Introduction of exotic genes

Mating between organisms that are related but genetically distinct may result in hybrids and the loss of the identity of the original gene pools. The escape of domestic dogs into the Australian bush and their mating with the dingo has led to a variety of hybrids. Changes in the patterns of genetic diversity of the dingo have probably occurred in some areas of Australia as a result.

Changes in gene-flow vectors

Animals feed on many Australian plants and pollinate them. After brushing against flowers and receiving a dab of pollen, many kinds of animals function as pollen vectors, depositing pollen on neighbouring flowers. Animal pollinators include honeyeaters, pygmy possums, fruit bats and native bees. Likewise, animals that feed on fruit spread seeds in their droppings, and many seeds have special adaptations that prevent them from being destroyed during digestion.

The effects of changes in or losses of pollen and fruit vectors are not well understood. The introduced honeybee appears to compete with native bees, removing nectar and monopolising flowers. However, it is not known whether the effects on native bees are serious. Australian rainforest plants depend heavily on fruit bats for both pollination and seed dispersal, but there is little information on the effects of changes in their numbers and distribution on forest plant species. However, elsewhere in the world, losses of pollen and seed vectors have had drastic effects on plant populations.

Response

Since the 1980s a number of major international agreements (briefly outlined below) have addressed biodiversity. The United Nations has declared an International Day for Biological Diversity (29 December). As a result, biodiversity has become the focus for a number of national policy statements and is now influencing decisions being made at national, State and local levels. This increases the scope for integrating biodiversity with social and economic considerations. It is also being institutionalised in Australia through legislation such as the Queensland *Nature Conservation Act 1992* and the South Australian *Native Vegetation Act 1992*.

In 1987, the General Assembly of the United Nations adopted a report from the World Commission on Environment and Development (1987) (the Brundtland Report or *Our Common Future*). This inspired the National Strategy for Ecologically Sustainable Development, agreed to by all Australian governments in November 1992 (see the box on page 4-40). One of the three core aims of the Strategy is to 'protect biological diversity and maintain essential ecological processes and life support systems'. A key strength of this strategy is its potential to bring together the social, economic and environmental aspects of society's goals and aspirations.

The Brundtland Report fostered the concept of sustainable development and paved the way for the Earth Summit in Rio de Janeiro in June 1992. One of the outcomes of the summit was the presentation for signature of the Convention on Biological Diversity. The key aims of this convention are:

> '*The conservation of biological diversity, the sustainable use of its components and the fair and equitable sharing of the benefits arising out of the utilization of genetic resources, including by appropriate access to genetic resources and by appropriate transfer of relevant technologies, taking into account all rights over those resources and to technologies, and by appropriate funding.*'

The spirit of this convention is reflected in the National Strategy for the Conservation of Australia's Biological Diversity (Commonwealth of Australia, 1996) that has been developed by Commonwealth, State and Territory governments. Its aim is to bridge the gap between current activities and the effective identification, conservation and management of Australia's biological diversity.

The strategy recognises the significant cultural, economic, educational, environmental, scientific and social benefits of biodiversity and the need for more knowledge and understanding of our biodiversity. It contains nine principles to be used for its implementation. These note that biodiversity is best conserved *in situ* and that, although levels of government have clear responsibility for the conservation of biodiversity, the cooperation of conservation groups, resource users, indigenous peoples and the community in general is critical to its successful conservation.

Types of responses

Responses to pressures on Australia's biodiversity are generally based on the assumption that natural resources can be managed sustainably to balance conservation and development. However, much is yet to be learned in order to achieve this ideal (see the box below). Societal responses to the management of biodiversity can be grouped into three broad categories: government, community and industry. Many responses involve feedback and collaboration between different sections of society and there is considerable overlap. Although some of the responses discussed in this section are not specifically targeted at biodiversity, they all affect it.

Government responses

Australia is a party to numerous major international agreements (see the box opposite).

National approaches by governments

In 1989, the Prime Minister committed the Commonwealth Government to preparing a national strategy for the conservation of biodiversity. Subsequently, conservation of biodiversity has become a major focus for policy. It is a core aim of the National Strategy for Ecologically Sustainable Development and a principle of the Intergovernmental Agreement on

the Environment (IGAE, see Chapter 2 for details). A number of policy frameworks for different sectors have included conservation of biodiversity; for example, the National Forest Policy Statement, the draft National Rangeland Management Strategy and the National Ecotourism Strategy.

The National Strategy for the Conservation of Australia's Biological Diversity has been signed by the Commonwealth and all State and Territory governments (Commonwealth of Australia, 1996). It provides an integrated framework of actions to strengthen conservation efforts across Australia with the object of protecting biological diversity and maintaining ecological processes and systems. It covers six main areas:

* Conserving biological diversity across Australia
* Integrating biological diversity conservation and natural resource management
* Managing threatening processes
* Improving our knowledge
* Involving the community
* Australia's international role

The Commonwealth established the Biodiversity Unit in the Department of the Environment, Sport and Territories to coordinate the development and implementation of the strategy. Ultimate responsibility for the strategy at the national level rests with the Australian and New Zealand Environment and Conservation Council. The Biological Diversity Advisory Council has been established to report to governments on issues related to the conservation of biodiversity and the further development and implementation of the strategy. Its aim is to ensure governments receive timely advice from sources such as relevant industries, the scientific community and non-government organisations.

While the strategy takes a broad approach to the conservation of biodiversity, it has also been necessary to provide a specific focus on endangered and vulnerable species, and on ecological communities and the threats acting on them. Commonwealth, State and Territory governments have finalised the Conservation of Australian Species and Ecological Communities Threatened with Extinction: The National Strategy.

Funding

The cost of conserving biodiversity forms part of the overall cost of environmental protection. This includes funds spent on activities as varied as improving water quality, waste management, preventing and mitigating soil erosion, combating weeds, pests and feral animals, rehabilitation of mine and industrial sites, research and education and programs such as Landcare. Australian and overseas studies show a high benefit-to-cost ratio for money spent on the environment, with benefits coming, for example, from enhanced tourism opportunities, improved land values and public health as well as reduced costs of land management and agriculture (Driml, 1994).

Ecologically sustainable development and biodiversity: the need for research

One of the core objects of the National Strategy for Ecologically Sustainable Development (ESD) is to protect biological diversity and maintain essential ecological processes and systems. However, the relationships between ESD and the protection of biodiversity are not well understood and it is widely assumed that, once a human activity appears sustainable, biodiversity will be protected. Some support for this assumption comes from conservation practices such as the reforestation of farmland, the establishment of wildlife corridors and the introduction of forestry and fishery practices sympathetic to wildlife. Although these initiatives are important, we still need research to measure their success in achieving conservation of biodiversity.

Two examples illustrate the need for research. First, some of the species of soil bacteria required to produce leguminous crops are well known, but the contributions of most soil fungi and invertebrate species to the success of these crops are less well understood. In the second case, many native insects are required for crop pollination, but the ecological processes that sustain them and the other species that contribute to their availability are rarely known in any detail. In these contexts, it is likely that once those components of biodiversity necessary to grow a particular crop are working, and apparently sustainable, other components that do not appear to be necessary for growing it will be ignored. This would mean sustainable development without biodiversity conservation.

Sustainability anticipates the needs of Australians many generations into the future. The only way to determine whether new production methods are sustainable is to establish well-designed experiments now, which, when monitored in the future, will identify the improvements and changes required to meet both production and conservation goals. Knowledge of sustainability, by definition, comes only with time. To obtain the baseline data requires an immediate start on the research and monitoring required to ensure that ESD and biodiversity conservation really are achieved together.

International and regional biodiversity agreements to which Australia is a party

- International Convention for the Regulation of Whaling (1946)

- Ramsar Convention on Wetlands of International Importance (1971)

- Washington Convention on International Trade in Endangered Species of Wild Fauna and Flora (CITES) (1973)

- International Convention for the Prevention of Pollution from Ships (1973 and its 1978 protocol) (MARPOL 73/78)

- Convention Concerning the Protection of the World Cultural and Natural Heritage (1974)

- Japan Australia Migratory Bird Agreement (JAMBA) (1974)

- Convention on Conservation of Nature in the South Pacific (Apia) (1976)

- Bonn Convention on Conservation of Migratory Species of Wild Animals (1979)

- Convention on the Conservation of Antarctic Marine Living Resources (1980)

- London Convention on the Prevention of Marine Pollution by Dumping of Wastes and Other Matter (1985)

- China Australia Migratory Bird Agreement (CAMBA) (1986)

- Convention for the Protection of the Natural Resources and Environment of the South Pacific Region (SPREP) (1986 and related protocols)

- Basel Convention on the Control of Transboundary Movement of Hazardous Wastes and their Disposal (1989)

- Convention on Biological Diversity (1992)

- Framework Convention on Climate Change (1992)

However, despite the benefits, only a small proportion of government expenditure is specifically directed to the environment. For example, only 0.52 per cent of the Commonwealth Government's budget in 1994–95 was allocated to the Department of the Environment, Sport and Territories and of the DEST budget, only 29 per cent was allocated to the environment (Commonwealth of Australia, 1994). Even allowing for funds spent on the environment by the States, Territories and other Commonwealth bodies such as the Department of Primary Industries and Energy, the total spent on the environment is a small proportion of the total Commonwealth, State and Territory budgets.

Policy responses

Governments have a range of policy instruments for the conservation of biodiversity. These include legislative, economic and social instruments.

Legislation

No broad Commonwealth, State or Territory legislation for the conservation of biodiversity currently exists. However, governments have, over many years, recognised the need to manage and conserve the environment and natural resources through legislation. Laws relevant to the management of fauna and flora date back more than 100 years, although only since the late 1960s has legislation increased significantly . Few examples, however, specifically identify the conservation of biodiversity. Some key legislation is given in the box on page 4-43.

The effectiveness of legislation varies widely and, in general, depends on the capacity and willingness of the responsible agency to administer the powers provided by the legislation together with the political agenda of the government of the day. The box on page 4-43 clearly shows the central role of the States and Territories in conservation of biodiversity.

One example of Commonwealth legislation to assist in the conservation of particular species and ecological communities is the Endangered Species Protection Act, which came into force in 1993. This Act lists endangered species and commits the government to producing recovery plans for their protection. The total annual budget of the Endangered Species Program in 1994–95 was $5.5 million. This level of funding is inadequate given that the recovery plan for one species alone, the western swamp tortoise (*Pseudemydura umbrina*), will require about $100 000 per year. With 148 recovery plans in operation and more planned (see Table 4.19), the Endangered Species Program is grossly underfunded.

Individual States and Territories have their own protection programs. In Western Australia, for example, the Department of Conservation and Land Management is using poisoned baits to control foxes that prey on native animals as part of a 10-year recovery plan aimed at protecting the chuditch (*Dasyurus geoffroyi*), the numbat (*Myrmecobius fasciatus*) and other native animals.

Research and education

Many organisations are involved in research related to biodiversity. The Commonwealth funds such research through CSIRO and the Environment portfolio. CSIRO has established a multi-divisional program to improve coordination and focus for research on biodiversity. Research on biodiversity has focused on:

- strengthening knowledge of the role of biodiversity in ecosystem function

- planning and management regimes that better integrate biodiversity conservation considerations

Migratory species

Australia's biodiversity includes species of birds, marine mammals and fish that depend on areas outside Australia for part of their life cycle.

Migratory species are subject to pressures in areas other than those under Australian jurisdiction, and so the Commonwealth Government has negotiated several bilateral agreements, for example the Japan and Australia Migratory Bird Agreement and the China and Australia Migratory Bird Agreement. Following meetings between Australia, China and Japan in 1995, the Migratory Waterbird Conservation Strategy and the Shorebird Reserve Network are being developed. A trilateral agreement between Australia, Japan and New Zealand is designed to protect stocks of southern bluefin tuna mainly through imposition of quotas to restrict catches.

Of the 50 species of wading or shorebirds that regularly occur in Australia, 33 (66 per cent) breed outside Australia and migrate in autumn and spring between Australia and places such as central Asia, Siberia and the Arctic zone of North America (Blakers *et al.*, 1984). The red-necked stint (*Calidris ruficollis*), for example, is a small (about 30 g) wading bird that breeds in eastern Siberia and western Alaska, spends the non-breeding season in Australia and migrates to and from its breeding grounds via Indonesia, the South China Sea, Vietnam, China, Japan and probably up the east coast of Asia. About two million waders of various species avoid winter each year by making the 20 000 km round trip to

Australia. Of the 71 species of coastal or seabirds that regularly occur in Australian waters, 25 (35 per cent) breed elsewhere — mainly on islands in the Southern Ocean, in the Antarctic or around New Zealand (Blakers *et al.*, 1984). As well as the migratory waders and sea birds, at least 36 terrestrial species of bird that breed in Australia seasonally migrate between Australia and Papua New Guinea. Large changes in species diversity occur locally on a seasonal basis because of north south migrations undertaken by many birds within Australia.

Migratory species also enhance the species diversity of marine mammals. Southern right whales (*Eubalaena australis*) and humpback whales (*Megaptera novaeanglieae*) migrate from the Southern Ocean to Australian coastal waters to calve (see Chapter 8). Whaling reduced the numbers of whales to dangerously low levels worldwide. Commercial whaling was banned in 1978 in Australian waters and in 1986 in all international waters. Southern right and humpback whales are listed as endangered under the Commonwealth Endangered Species Protection Act and their numbers appear to be increasing. Estimates indicate that nearly 1000 humpbacks migrate to eastern Australia where they have become a popular tourist attraction. Regulations have been introduced to prevent boats and divers harassing them.

Some of Australia's highly diverse fish fauna are migratory species. Marlin (Family Istiophoridae) and tuna (Family Scombridae) are fast-swimming fish that range over large areas in the Pacific, Indian and Southern Oceans and spend part of their life in Australian waters. Southern bluefin tuna for example, spawn in the Indian Ocean south of Indonesia and the young fish migrate southwards along the coast of Western Australia. Some then swim westwards to the waters off South Africa while others swim eastwards across the Great Australian Bight to south of Tasmania and New Zealand. Many species of shark also travel large distances; school shark (*Galeorhinus galeus*) tagged in New Zealand have been caught in Australian waters.

Migratory species are subject to pressures in many of the places they visit. In Australia, migratory birds face a range of pressures including loss of habitat through processes such as urbanisation of the coast, drainage of wetlands and agricultural development. However the main threats come during their breeding migration. Large numbers are shot or trapped in the northern hemisphere each year. Illegal hunting in China is also a serious problem as is habitat loss.

Seabirds also face threats in Australia. Longline fishing in Australian and international waters threatens populations of some species such as the wandering albatross (*Diomedea exulans*) (see page 8-33). Others, such as the brown booby (*Sula leucogaster*), are harvested illegally (Blakers *et al.*, 1984). Most of our pelagic species of fish appear to be in a healthy condition but stocks of southern bluefin tuna have been greatly reduced through fishing by Australian and Japanese fleets (see Fig. 8.13).

occasional sightings

occasional sightings

northern migration

northern migration

southern migration

southern migration

probable mating area

The humpback whale migrates northwards in winter to calve. It feeds in Antarctic waters.
Source: Marsh *et al.*, 1995.

- elucidating trends on the state of Australia's biota resulting from various impacts and the development of geographic information systems to assess terrestrial biodiversity at the scale of landscapes, biogeographic regions and ecosystems

Research by Australian Biological Resources Study concentrates on the systematics of a wide variety of groups of organisms, especially the lesser-known ones. Several universities and most major museums sponsor research and educational programs. These include: the Cooperative Research Centre for Tropical Rainforest Management in north Queensland, the Commonwealth Key Centre for Biodiversity and Bioresources at Macquarie University, and the Invertebrate Biodiversity Conservation Program at the Museum of Victoria.

Biodiversity is increasingly a focus for tertiary and secondary curricula. For example, in Victoria, biodiversity now forms one of the major components of the Year 12 course in environmental studies. Professional development courses for teachers include one on biodiversity and ecologically sustainable development being prepared by the Australian Association for Environmental Education.

A range of information and education materials has also become available recently. These include a primer on biodiversity produced by the Research Unit for Biodiversity and Bioresources at Macquarie University (Beattie, 1995), a CD-ROM on insect biodiversity produced by CSIRO's Division of Entomology, a series of reports on biodiversity, the State of the Marine Environment Report (Zann, 1995; Zann and Kailola, 1995) (see Chapter 8) and other materials produced by the Commonwealth Department of the Environment, Sport and Territories.

Community responses

Over the last two decades, public awareness of environmental issues has grown rapidly. Since the 1970s, many surveys of public attitudes to the environment have been conducted. Of these, only eight have identified specific environmental concerns and the priority that people attach to them (Lothian, 1994). Although a study by ANOP Research Pty Ltd into public attitudes to the environment (commissioned by DEST) found that few people have a good understanding of the term biodiversity, the Lothian study indicates that the public are concerned about many of the issues surrounding it, such as species loss, deforestation, vegetation clearance, logging, harvesting kangaroos and the role of national parks. Overall, biodiversity issues ranked second in the eight surveys behind pollution and waste production/disposal.

In recent years, concern for the long-term ecological sustainability of Australia's natural resources has led to the rapid growth of local and regional environmental community groups. Many groups have changed from being passive recipients and distributors of information to taking a proactive role. They are demanding to be involved

Conservation of biodiversity: relevant legislation
The fragmented approach to the conservation of biodiversity is illustrated by the range of legislation for the management of fauna and flora.

Commonwealth
Environment Protection (Impact of Proposals) Act 1974
Australian Heritage Commission Act 1975
Great Barrier Reef Marine Park Act 1975
National Parks and Wildlife Conservation Act 1975
Antarctic Treaty (Environment Protection) Act 1980
Whale Protection Act 1980
Antarctic Marine Living Resources Conservation Act 1981
Wildlife Protection (Regulation of Exports and Imports) Act 1982
Protection of the Sea (prevention of pollution from ships) Act 1983
World Heritage Properties Conservation Act 1983
Endangered Species Protection Act 1992

Australian Capital Territory
Nature Conservation Act 1980

New South Wales
National Parks and Wildlife Act 1974
Heritage Act 1977
Environmental Planning and Assessment Act 1979
Wilderness Act 1987
Threatened Species Conservation Act 1995

Queensland
Marine Parks Act 1982
Nature Conservation Act 1992

South Australia
National Parks and Wildlife Act 1972
Native Vegetation Act 1992

Tasmania
National Parks and Wildlife Act 1970
State Policies and Projects Act 1993

Victoria
Wildlife Act 1975
National Parks Act 1975
Planning and Environment Act 1987
Flora and Fauna Guarantee Act 1988b
National Parks (Alpine National Park) Act 1989
National Parks (Wilderness) Act 1992

Western Australia
Soil and Land Conservation Act 1945
Wildlife Conservation Act 1950
Conservation and Land Management Act 1984

Northern Territory
Territory Parks and Wildlife Conservation Act 1977

Source: DEST, Biodiversity Unit.

in planning and decision-making processes undertaken by governments for management of biodiversity.

Public awareness of native flora and fauna is symbolised by community support for replacing traditional symbols such as the Easter bunny with native species like the endangered bilby (*Macrotis lagotis*).

Community involvement

The National Strategy for the Conservation of Australia's Biological Diversity relies on community support and community-based actions for its successful implementation. The Commonwealth of Australia House of Representatives Standing Committee on Environment, Recreation and the Arts (1992) stressed the need to develop innovative strategies for increasing community awareness and

Response to species endangerment

A number of legislative and administrative actions have occurred in response to the decline and extinction of a number of species.

The Commonwealth enacted *The Endangered Species Protection Act 1992,* which came into force on 30 April 1993. It introduced a statutory process for the listing of nationally threatened species, communities and key threatening processes. It also led to the establishment of two ministerially appointed committees: the Endangered Species Advisory Committee and the Endangered Species Scientific Subcommittee. These bodies provide advice to the Federal Minister for the Environment on a range of matters prescribed in the Commonwealth Act, including priorities for preparation of plans, scientific advice on listings, and significant statutory responsibilities and powers that affect Commonwealth agencies and areas. The Australian and New Zealand Environment and Conservation Council has developed a strategy titled 'The Conservation of Australian Species and Ecological Communities Threatened with Extinction: The National Strategy'.

Two major Commonwealth programs: the Endangered Species and Feral Pests Programs, address the conservation of threatened species and communities, and the major threats to their conservation. The Endangered Species Program, which was established in 1989, seeks to prevent further extinctions of Australian biota and to restore endangered species and ecological communities to secure status in the wild. Funding for the program in 1994–95 was $5.5 million and cumulative funding since its inception has been $23 million. The Feral Pests Program began in 1992–93 with annual funding of $2.2 million. Foxes, feral cats, feral rabbits and goats are listed as key threatening processes under the Commonwealth Endangered Species Act, as is dieback caused by the fungus *Phytophthora cinnamomi* and longline fishing which poses a threat to albatross.

Action plans have been prepared for birds, marsupials, freshwater fish, reptiles, rodents and vascular plants, and are in preparation for amphibians, bats, cetaceans, seals and dugongs. Conservation overview statements have been prepared for non-vascular plants, and non-marine invertebrates including butterflies. These plans and statements are commissioned to help identify endangered and vulnerable species and nationally agreed priorities for action. The status of recovery action on individual species listed in the schedules to the Commonwealth Endangered Species Act is summarised in Table 4.19. No ecological communities are listed in the Commonwealth Act, however, a discussion paper was published by the Endangered Species Scientific Subcommittee in January 1995 to stimulate comment from the scientific community, conservation organisations, industry groups and interested individuals.

Examples of recovery plans

The western swamp tortoise (*Pseudemydura umbrina*), which is listed as endangered, consists of one wild population of about 48 individuals in Ellen Brook Nature Reserve near Perth and 112 captive individuals that form part of a captive breeding program. The recovery plan aims to increase the population in Ellen Brook Nature Reserve and to establish a second population of up to 40 at a nearby nature reserve that formerly had a population of tortoises. It also aims to maintain a captive breeding population of at least 50 individuals (Burbidge and Kuchling, 1994). The estimated cost of implementation of the recovery plan to 2002 is $1.7 million.

At present, the main threats to the tortoise come from the surrounding agricultural matrix and the availability of water. Conservation of the species depends on the management of two small swamp nature reserves surrounded by extensive agricultural development. Winter water levels need to be maintained. Since 1990, Ellen Brook Nature Reserve has been surrounded by a fox-proof fence and extensive predator control is conducted within the reserve. However, unless the surrounding land is managed to remove any threats to the tortoise population, the plan may still fail in the long term. This case highlights the need for management plans to integrate conservation of biodiversity as an essential part of total landscape management (see page 4-53).

The western swamp tortoise

The Lord Howe Island woodhen (*Tricholimnas sylvestris*) is a flightless rail (Family Rallidae) occurring only on Lord Howe Island, off the coast of New South Wales. Its nearest relative was the wood rail of New Caledonia, which is now extinct. Before human settlement, the woodhen occurred over most of the island. Estimates based on known habitat requirements and the original area of preferred habitat suggest an original population of several hundred. By the mid 19th century the woodhen was confined to the summit regions of Mounts Gower and Lidgbird and between 1974 and 1980 the population had been reduced to about 30 individuals, including at most 10 breeding pairs. The decline was probably initially due to hunting pressure but subsequently to predation and habitat alteration by introduced animals.

An extensive study of the ecology of the remaining birds began in 1971, followed by management to offset the identified threats, and a program of captive breeding and re-introduction. By 1984, more than 80 birds had been bred and released and the wild population had grown to an apparently stable level of 200 individuals, including 50 breeding pairs. Further increases appear to be limited by lack of suitable habitat for more breeding territories. Twice-yearly counting continues for early detection of any further problems, and there is a proposal to establish a captive colony off the island in case of a catastrophic impact on the native population.

involvement. The committee identified several strategies for facilitating community involvement at the grassroots level:

- provision of seeding funds for projects

- government involvement and encouragement in community projects

- extension and dissemination of practical technical advice based on scientific research and knowledge

- bioregional planning and management to coordinate and direct local community actions within a broader perspective

- practical equipment to facilitate on-ground actions

Many mechanisms exist for obtaining input from the community on specific environmental matters. For example, most planning legislation requires opportunity for public comment. Governments have created a range of advisory bodies to better inform people involved in the policy development process. Many non-government groups also provide advice to government through their participation in government committees, such as the Biological Diversity Advisory Council and the Endangered Species Advisory Committee. However, feedback from community groups indicates a need for increased government support, since community input generally relies on the work of volunteers.

Many community groups regularly monitor the environment and undertake field activities to either protect or restore biodiversity (Saunders *et al.*, 1995). Programs such as Worm Watch, Frog Watch and NatureSearch collate information about species collected by members of the public. Community groups monitor organisms as diverse as orchids, birds, frogs, earthworms, dung beetles and butterflies on a nationwide basis. The information collected by these groups has greatly increased our knowledge of the diversity of organisms such as earthworms in Australia. An important example of the role of the community in gathering data is *The Atlas of Australian Birds* (Blakers *et al.*, 1984), which is based on information provided by bird-watchers all over Australia.

The public also donate money towards research related to biodiversity. In New South Wales, for example, the New England community established a 'dieback fund' for research into the causes of dieback killing native trees in the district. Trees for Life provides two million indigenous trees annually free of charge to landholders to assist in the rehabilitation of farmland in South Australia. The restoration of native bush in urban areas is a popular and rapidly growing community activity that seeks to re-establish local species.

Community–government initiatives

Community groups, conservation and government bodies often collaborate on environmental projects.

Table 4.19 Status of recovery action on individual species*

Species	No. of endangered or vulnerable species	Recovery plan prepared or in preparation	Recovery plan being implemented
Mammals	48	13	11
Birds	50	12	8
Amphibians	9	7	2
Reptiles	21	2	1
Freshwater fish	13	6	2
Vascular plants	890	224	124
Total	1031	264	148

*As listed in the schedules to the Commonwealth Endangered Species Act.
No ecological communities are listed in the Act, however, the endangered species scientific subcommittee published a discussion paper in January 1995 to stimulate comment from the scientific community, conservation organisations, industry groups and interested individuals.

Planting trees as part of a community–government initiative, Greening Australia.

A successful example of a community–government initiative is the Land for Wildlife scheme run by the Victorian Department of Conservation and Natural Resources and the Bird Observers Club of Australia. Private landholders are encouraged to conserve flora and fauna and given support by the program. The scheme does not require landholders to enter into agreements with the government, but instead provides a structural framework for the support of voluntary management of wildlife habitat on private land.

Greening Australia and the One Billion Trees campaigns are collaborative projects between government, conservation groups and the community, which have arisen in direct response to loss of native vegetation. The establishment of a number of community networks such as the Community Biodiversity Network, the Marine and Coastal Community Network and the National Threatened Species Network are other examples of collaborative projects. The latter is an active community-based network that increases public support and involvement in the protection of threatened species and their habitats throughout Australia.

The Landcare movement has the potential to be the most important mechanism for integrating conservation of biodiversity into agricultural and

Economic mechanisms for conserving biodiversity

A range of economic instruments is being developed to help achieve the sustainable use of natural resources. However, few 'economic' instruments have been introduced specifically for managing biodiversity.

Environmental pricing

Charges, levies and the setting of prices to fund conservation of biodiversity are rare in Australia — beyond fees for park use, trail access and other uses within reserves. The Great Barrier Reef Marine Park Authority has introduced a one-dollar visitor fee that is expected to raise about one million dollars per year. Most of the funds are being allocated to the Cooperative Research Centre concentrating on the ecological management of the reef.

Some local authorities, such as the Brisbane City Council, have introduced environmental levies on ratepayers. Funds raised in this way are used to buy environmentally sensitive land in order to protect habitats and their associated flora and fauna.

Conservation easements

Conservation easements, like South Australia's heritage agreements, bind owners to a set of conditions, such as prohibition of clearing or cropping an area of land. Other States and Territories also have legislation that facilitates the use of conservation easements, but their limited budgets mean progress has been slow.

Funding arrangements

A revolving fund is one of several ways to maximise the effect of funds for managing biodiversity. This involves buying land and placing a permanent covenant on it to manage part or all of it for conservation. The land is then sold to someone who agrees to abide by the covenant and the money used to buy more land. The Victorian Conservation Trust has established a revolving fund, and other environmental organisations, like the Australian Bush Heritage Fund have expressed a desire to do likewise.

Taxation

Some income-tax deductions are available for control of land degradation, but they are narrowly defined and do not reflect concerns for the conservation of biodiversity. Local governments grant rate relief for land maintained for biodiversity conservation purposes. Rate reimbursements apply in South Australia under the *Native Vegetation Act 1992* and further reductions are available under a heritage agreement. In Victoria, new financial provisions in the *Local*

Government Act 1991 allow local authorities to charge rates under a method of capital improvement valuation. This would reduce the tax burden associated with a decision to lease undeveloped land. However, these provisions have not yet been used.

Transferable development rights

This mechanism is designed to limit development in conservation areas without affecting the underlying value of individual assets. Transferable development rights enable people who own valuable habitat to sell clearing rights to others who own land of lesser biological importance and need a development right in order to proceed with a proposed development. The use of transferable development rights is allowed under the *Resource Management Act 1991* in New Zealand and opportunities exist for its application in Australia.

Performance bonds

Environmental performance bonds are best suited to situations where there is one source of potential environmental damage that can reasonably be estimated. Apart from their use for land rehabilitation in the mining industry, bonds are used as a permit condition for aquaculture in South Australia and New South Wales and by the Great Barrier Marine Park Authority for tourist development.

Financial assistance

Financial assistance forms part of many voluntary management schemes offered by the States and Territories and usually takes the form of payment to assist with the cost of purchasing material associated with the work required. For example, a 50 per cent fencing subsidy exists under the remnant vegetation protection scheme in Western Australia and the Victorian action statement program also pays subsidies.

Land management funding schemes such as the National Landcare Program and the Murray–Darling Management Strategy's Investigations and Education and Integrated Catchment Management programs, while broadly directed towards activities of significance to the conservation of biodiversity, do not contain any explicit biodiversity conservation requirements. The Commonwealth Department of Housing and Regional Development Program (established in 1994 with funding of $15 million over four years) funds three major regional development programs, none of which contains biodiversity conservation criteria.

pastoral production. Landcare was originally established by the National Farmers Federation and the Australian Conservation Foundation in the 1980s (see page 6-43). More than 2000 Landcare groups now exist across Australia, involved in a wide range of activities aimed at improving production, land and water restoration and conservation. Some of these activities contribute directly or indirectly to conservation of biodiversity (Commonwealth House of Representatives Standing Committee on Environment, Recreation and the Arts, 1992).

Industry responses

As the Australian public's attitudes to environmental issues change, so do those of industry. The agricultural sector is increasingly incorporating nature conservation into landscape programs. Some industry responses are driven by consumer demand — that is, the desire to buy more 'environmentally friendly' products. Conservation groups such as WWF, for example, endorse some products. In many cases industries have failed to respond of their own accord to pressures on biodiversity and have only acted in response to government legislation. However, some responses are based on industry's need to preserve natural resources. Commercial fishing organisations in several States employ environmental officers and take active roles in widening the debate on issues such as coastal development and protection of seagrasses. In response to industry lobbying, governments have legislated to protect seagrasses and mangroves over much of northern Australia.

A number of industries have introduced initiatives, such as codes of practice, cleaner production and best practice environmental management, to reduce the impacts of their operations by minimising use of materials and energy and by reducing wastes. These initiatives focus on pollution control and reduction of environmental degradation and do not specifically address biodiversity.

Community actions to integrate the landscape

Technological advances since World War II have made large-scale clearance of native vegetation easier. As a consequence, more land has been cleared in the last 50 years than in the preceding 150 years, bringing associated problems of land degradation and loss of biodiversity. These problems have aroused concern in many sectors of the community, and a number of responses have evolved to address the impacts of clearance.

Since the early 1980s, governments and key community sectors have increasingly recognised the economic, ecological and social benefits of remnant native vegetation and have initiated programs, not only to ensure the protection and management of remnants, but to re-establish large areas of vegetation.

This has resulted in a number of community-based projects supported by programs like the National Landcare Program, whose components include Save the Bush and One Billion Trees. However, while the enthusiasm for 'tree planting' has generated activity on many fronts, this activity has proceeded without any underlying ecological rationale or planning framework. Concern is now being expressed about the planning and preparation for revegetation activities, the appropriateness of plantings and the species selected (Commonwealth House of Representatives Standing Committee on Environment, Recreation and the Arts 1992). Many of the plantings have had a single purpose (such as control of soil erosion or salinisation) and planned solely within the confines of one property. Clearly, it is better to ensure that tree-planting has potential multiple benefits (for example plantings for soil erosion control can be designed to provide habitat) and is placed within a broader planning framework.

Some schemes now advocate the concepts of integrated landscape ecology and bioregional planning as a basis for developing an integrated approach to landscape management. These approaches offer the potential to place previously ad hoc community activities within a regional framework. Using catchments as a basis for designing National Landcare Program projects and planting along corridors — for example Corridors of Green by Greening Australia — are indicative of moves towards an integrated landscape management approach.

The way in which the landscape has been managed is illustrated in a study by Hobbs *et al.* (1993) using a hypothetical section of the Western Australian wheat belt. Individual segments in the landscape, such as road verges, paddocks and patches of remnant vegetation, are usually managed in isolation. The landholder is mainly concerned with individual paddocks and the conservation authority with nature reserves, while the authority responsible for the

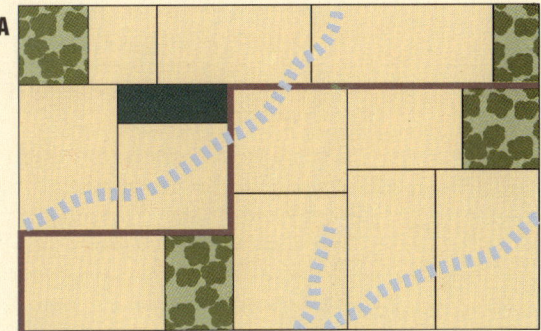

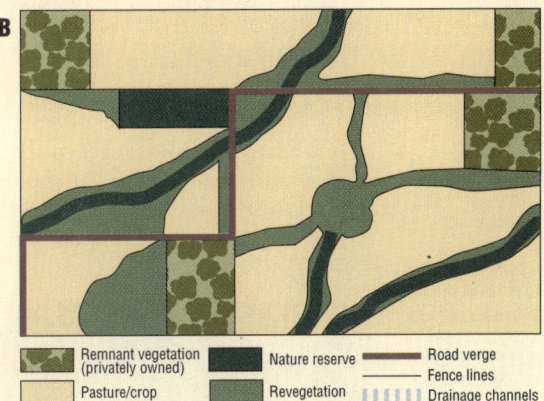

Legend	
Remnant vegetation (privately owned)	Nature reserve — Road verge
Pasture/crop	Revegetation — Fence lines / Drainage channels

Landscape linkages. **A** – A hypothetical section of an agricultural landscape showing remnant vegetation, reserves, roads, paddocks and drainage lines. **B** – Revegetation plan that provides linkages along road verges and drainage lines to connect remnant vegetation.

Source: Hobbs *et al.*, 1993.

roadside verges — usually the local council — is unlikely to share much interest in either of these. While resource managers and landholders manage their own piece of land in isolation, landscape segments are unlikely to operate as an interconnected system. Landscape linkages can be established by planning and planting in line with landscape features. This approach, applied on a regional scale, offers opportunities to enhance the benefits from tree-planting activities at a property/local level.

The benefits of extensive networks of vegetation for the protection of water and soils and for the maintenance of biodiversity now have wide recognition. However, few data are available to either support or refute the value of interconnecting areas of vegetation (corridors) for conservation and maintenance of biodiversity, and their merit continues to be hotly debated in the scientific literature (Noss, 1987; Simberloff *et al.*, 1992). Despite this, it is generally accepted that corridors serve to ameliorate the process and consequences of fragmentation, which constitute a serious threat to biodiversity.

Responses to key issues

On the basis of the preceding sections, the following key issues for management of biodiversity emerge:

- human population patterns
- land clearance
- harvesting of native species
- introduced species
- pollution
- mining
- climate change
- lack of knowledge
- integrated ecosystem-based management of natural resources

The state of Australia's biodiversity is ultimately a consequence of the number of people, our patterns of movement and the resources required to fulfil our desires and to maintain our living standards. Without limiting human population growth and developing management systems that recognise the critical role of biodiversity in human economics the pressures on biodiversity will continue.

Human population patterns

Urbanisation

Australia is the most highly urbanised society in the world (see Chapter 3). The National Strategy for the Conservation of Australia's Biological Diversity has identified a number of goals to minimise the impact of urbanisation on biodiversity, including the following:

- encouraging habitat retention
- improving strategic planning to enhance biodiversity
- reducing fringe development
- encouraging use of indigenous species
- integrating biodiversity conservation into relevant policies and programs such as the Better Cities program

Tourism and recreation

Tourism is a relatively unregulated activity in Australia and consequently government bodies have made little response to the pressures associated with it. However, the Commonwealth Government has responded to growth in the ecotourism industry with the National Ecotourism Strategy designed to provide guidelines to ecotour operators.

All States and Territories have produced policy documents on ecotourism or tourism. Some have also recognised the value of tourism-based activities and contribute some of the revenue gained from using natural resources into the management of those resources. However, the challenge remains to design and implement systems that will link the growth of tourism with conservation of biodiversity (Preece *et al.*, 1995). One example is the Northern Territory Tourism Masterplan, under which the Northern Territory tourism industry and the Northern Territory Conservation Commission are working together to investigate the tourist potential of protected areas.

The growth of ecotourism itself reflects the community's desire to participate in more sustainable types of tourism and to learn more about the natural environment. However, since anyone can call him- or herself an ecotour operator there is no proof that the activities of ecotourists are any more sustainable than other forms of tourism. To overcome this problem a number of accreditation schemes have recently been introduced and the Commonwealth Department of Tourism is investigating the merits of a national scheme (Preece *et al.*, 1995).

Some forms of recreation, such as fishing (see page 8-12), require intensive management to ensure their sustainability. Intertidal rock platforms near large cities such as Sydney and Perth are easily accessible and so shellfish may be over-collected. Recreational abalone fishers near Perth, for example, take between 80 and 100 tonnes each season — about 40 per cent of the available stocks and one and a half times the tightly controlled commercial catch. Controls have been introduced to protect the resource. These include a limit of 20 abalone per person, size limits for common species and regulations on collecting methods (scuba diving, for example, is banned). Fishing is also restricted to two hours a day on eight weekends each year.

Many States and Territories have restrictions on shell collecting — usually a limit on the numbers

French Track, Simpson Desert, SA. Use of four-wheel drive vehicles enables people to appreciate remote parts of Australia but can also lead to disturbance of sensitive ecosystems. ▶

The Bubbler Mound Springs near Lake Eyre South, northern SA. Visitor impact on sensitive sites is often unwitting, highlighting the importance of public education and adequate signage. ▼

that can be collected — to reduce the harvest by amateur collectors.

Recreational fishing is an important leisure activity for more than 4.5 million Australians. However, increasing fishing pressure on inshore stocks from recreational and commercial fishing, coupled with environmental damage, is causing the decline of many fish stocks. A working group set up under the Australian and New Zealand Fisheries and Aquaculture Council has developed a national policy on recreational fishing. The key principles for sustainable fishing include protecting the resource and habitat, and involving government and the community. The policy seeks responsible land use and farming practices, protection of shoreline and floodplain areas and wetlands, and careful use of chemicals and fertilisers that have an impact, direct or indirect, on aquatic habitats or fish stocks (National Recreational Fisheries Working Group, 1994).

Land clearance

The clearance of native vegetation is the single greatest threat to terrestrial biodiversity and a significant threat to aquatic and some inshore marine biodiversity. Thus it is extraordinary that nobody has accurately assessed its extent over the last one to two decades, despite the availability of the appropriate technology (Graetz *et al.*, 1992). The Department of Environment, Sport and Territories is supporting a number of projects to rectify this gap in knowledge. One recently completed project, undertaken by CSIRO, used satellite imagery to assess landcover disturbance at a continental scale (Graetz *et al.*, 1995). The Department has also released a compilation that provides an overview of the pattern of recent vegetation clearance (Biodiversity Unit, DEST, 1995).

Reserves have only limited value as an antidote because most of the land set aside is in small blocks that are infertile and/or steep. Fertile land is less likely to be preserved (see the box on page 4-52). The problem will continue without changes to policies on land clearance and reserve selection.

About 70 per cent of Australia's land mass is under the control of private landholders and resource managers, including indigenous peoples. Vegetation clearance on lands under freehold and leasehold tenure is a major concern. Controls on native vegetation clearance and the provision of incentives for native vegetation retention vary between States and Territories. In South Australia, vegetation clearance is tightly controlled, but other States and Territories have less stringent rules (see page 6-39 and Table 6.8). New South Wales, Queensland and Western Australia are in the process of strengthening measures to control native vegetation clearance. In 1995, the Queensland Government released a package of measures including draft tree-clearing guidelines that demand 30 per cent retention of trees on most leasehold properties throughout the State and prohibit large-scale clearing of mulga (*Acacia aneura*).

Governments have also implemented a range of economic mechanisms to address the issue of land clearance (see the box on page 4-46). For example, all the 1950 taxation provisions designed to encourage it have now been removed from the *Income Tax Assessment Act 1936*, although farmers can still deduct the full cost of clearing in the year of expenditure by using their own equipment and employees.

The removal of incentives that encourage deforestation or land degradation is an effective response to conservation of biodiversity. In the past, it has been argued that drought policy has encouraged land degradation, including the loss of biodiversity values. Recent changes to drought policy at both State and federal levels are making farmers more self reliant in a way that may reduce threats to biodiversity values during time of biological stress.

Environmental lobby groups have focused on issues of native vegetation clearance, especially those resulting from logging operations, conversion to agricultural land and mining activities. The power of large-scale environmental protests over the past two decades to influence government policy and change public perception should not be underestimated. Community groups have also responded to removal of forests by cooperating with government–community initiatives such as Landcare, Greening Australia and Save the Bush programs.

Harvesting native species

Rights to harvest and/or use Australia's flora and fauna vary between States and Territories. In recent years the Commonwealth, several States and the Northern Territory have increased the market orientation of harvesting licences, such as in the kangaroo- and timber-harvesting industries.

The Commonwealth, States and Territories require a permit or licence to take, kill, trade or export protected flora and fauna. These licences, however, are usually issued for a short period with no guarantee of renewal. Beyond these provisions, Australia is a party to the CITES convention which restricts the international trade in endangered species and bans trade in most threatened species. There is continuing debate on the conservation benefit, economics and ethics of using wildlife commercially.

In the fishing industry, measures have been introduced to reduce fishing pressure. These include reductions in the number of trawlers in some prawn fisheries and the number of pots in lobster fisheries, smaller quotas for finfish in the south-east trawl fisheries and controls on mesh size of nets, sizes of animals caught and fishing times. New Commonwealth, State and Territory fisheries legislation emphasises the importance of managing fisheries to achieve ecological sustainability including maintenance of biodiversity (see Chapter 8). Research is being conducted to reduce the quantity of bycatch in prawn trawl fisheries.

Fraser Island

Fraser Island, the largest sand island in the world, supports rainforest, freshwater lakes, coastal heathland and sand dunes. Over the years it has faced a number of threats, including sand-mining, logging, feral animals and (being only four hours drive from a region supporting two million people) tourism and recreation — especially recreational angling.

Numbers of visitors to the island have increased from 50 000 in 1985 to 250 000 in 1994. Following a long public campaign, the Commonwealth Government — against the wishes of the then Queensland Government — refused to issue export licences for minerals produced on the island and so sand-mining was stopped. In response to continuing public pressure, the Queensland Government banned the logging of the island's rainforests. The Commonwealth subsequently obtained listing of Fraser Island as a World Heritage Area and now the island — together with the mainland immediately to its south — has been incorporated into a national park. This park (and areas outside it) are subject to the Great Sandy Region Management Plan, which covers an area of about 840 000 ha.

The plan provides for protection of natural and cultural features of the region, while allowing for provision of services for residents and visitors and ensuring that development is sustainable. It contains a range of strategies, including rehabilitation of degraded areas, removal of feral animals, catchment management, controls on recreational and commercial fishing, rationalisation of vehicle access — including use of four-wheel drive vehicles on beaches — and development and control of tourist areas. If the plan meets its targets, it will provide substantial protection and restoration of the island's biodiversity.

Recreational fisheries have generally had few limits apart from controls on the type of equipment used for fishing and minimum size of fish caught. Most States and Territories now limit the number of fish that can be caught each day. Despite these measures, some species are still declining or their numbers are not recovering and it is likely that controls will be increased in the future.

Introduced species

The devastation caused by introduced plants and animals has elicited a number of government, community and industry responses. The Commonwealth Government established the Cooperative Research Centre for Vertebrate Pest Control, which involves five research groups in developing humane methods to combat vertebrate pests such as rabbits and foxes. Their main approach is to investigate methods of fertility control in vertebrate pests. The Feral Pests Program was set up in 1992–93 to address threats to endangered species posed by feral pests (see the box on page 4-44).

Growing public awareness of the effects of feral animals and introduced plants on native fauna and flora has led to numerous community-based projects. Bush-regeneration groups are common in urban areas, where remnant vegetation is especially prone to invasion by exotic weeds. Individuals and community organisations such as Landcare groups shoot or poison feral cats, rabbits and foxes.

In 1991, the Australian Quarantine and Inspection Service introduced voluntary controls on the discharge of ballast water. These guidelines provide a number of options — the most common being exchange of ballast water offshore to prevent water loaded in foreign ports from being discharged here. In this way, many shallow-water organisms from other ports are dumped at sea in deep water where they are unlikely to survive. Despite modern anti-fouling paints, many organisms are still being transported on the hulls of ships on parts not protected by the paint or on sections where the paint has been scraped off. The Centre for Research on Introduced Marine Pests established by CSIRO under the auspices of the Australian Ballast Water Management Advisory Council, established by the Commonwealth Government in 1994, is researching ways of dealing with marine pests. However, it is unlikely that introduced marine organisms will be eradicated.

Pollution

The National Environment Protection Council is a significant initiative being established jointly through legislation by the Commonwealth, States and Territories to develop harmonised goals, standards and guidelines for the protection of the environment. The areas it covers include: ambient air and water, assessment of contaminated sites, impacts of hazardous wastes and recycling.

Although the amount of government legislation on pollution control has grown steadily in recent years, its effectiveness has been limited by the dispersal of responsibility for enforcing pollution laws across many government departments. This situation has changed recently with the centralisation of responsibility under State environmental protection authorities (EPAs). In New South Wales and Victoria, for example, EPAs can impose fines of up to one million dollars on companies that pollute illegally and in Western Australia the head of one company was gaoled as a result of pollution by the company.

Government legislation generally dictates industry responses to pollution. For example, the Australian Maritime Safety Authority has recently prepared a national plan to combat pollution of the sea by oil. However, industry also reflects changing attitudes towards environmental issues in society generally. Many companies are using a variety of waste-minimisation measures and recycling programs to increase profits and improve their image with an increasingly critical public. Consumer demand for 'greener', 'cleaner' products has generated a flurry of activity and competition in companies to develop products that are seen to be more environmentally friendly. Products are increasingly made from, or packaged in, recycled materials and frequently contain less non-biodegradable chemicals.

The Cooperative Research Centre for Waste Management and Pollution Control involves 18 different research and industry groups and is responsible for research into all aspects of waste management and minimisation. Although it covers all elements of waste, the Centre's research focus is on the treatment of liquid waste.

Mining

When planning new mining developments, companies now must carry out environmental impact assessments that include plans for rehabilitation. In most States, mining leases include a condition requiring the holder to lodge a security deposit, which may be forfeited if the rehabilitation conditions are not met. In 1990, the Australian Mining Industry Council published a handbook that lists standard rehabilitation procedures (AMIC, 1990). These include landscaping, management of topsoil, revegetation, managing waste dumps, acid, alkaline and saline sites, heavy metals and toxic chemicals as well as removal of access roads and constructions. Considerable effort is now applied to rehabilitation of sand-mining areas. However, thousands of abandoned mine sites remain with no rehabilitation apart from natural processes.

In South Australia, mining operators are required under the Mining Act to pay a royalty (presently 10 cents per tonne) on all materials produced. The Extractive Areas Rehabilitation Fund is used to finance approved rehabilitation projects. Some of the larger mining companies employ horticultural advisers or rehabilitation consultants.

Climate change

Commonwealth, State and Territory governments and the Australian Local Government Association on behalf of local governments have accepted the National Greenhouse Response Strategy.

Under the United Nations Framework Convention on Climate Change, the OECD countries are committed to stabilising gas emissions at their 1990 levels by the year 2000. Australia's greenhouse gas emissions in 1990 have been estimated at 572 megatonnes of carbon dioxide equivalent and at present rate of growth would increase by 14 per cent by the year 2000. Vegetation clearing for agricultural purposes in 1990 was estimated to have contributed about 27 per cent of Australia's total net emissions in carbon dioxide equivalent (National Greenhouse Gas Inventory Committee, 1994).

The National Greenhouse Gas Inventory aims to assemble a comprehensive information base of human-induced net emissions of greenhouse gases in Australia. As part of the long-term program to develop an improved capacity to compile this national inventory, about $1.1 million has been directed to scientific research over the three years 1995–97 (see also Chapter 5).

Lack of knowledge

Major research programs are required in three key areas to improve knowledge of Australian biodiversity and its conservation and management.

Inventory

Australian biodiversity consists of hundreds of ecosystems, more than one million species and millions of genes.

Although the country has been divided into biogeographic regions (see Fig. 4.9), each of these contains undescribed ecosystems. Thus, we have a long way to go before we understand ecosystem diversity. Commonwealth, State and Territory agencies are using Geographic Information Systems (GISs) to establish a national inventory of ecosystems. In the future, GISs should enable us to predict, for any given area, how many species live there, species turnover from one habitat to another and numbers of rare, vulnerable and endangered species.

Like the rest of the world, Australia has described only a small proportion of its species (see Table 4.8). While most flowering plants and vertebrates are known, the 'lower' plants (mosses and their relatives) are relatively unexplored. Less than half of the invertebrates are described and the micro-organisms remain little known. We have too few taxonomists to carry out a full inventory of Australian species.

The establishment of the Australian Biological Resources Study (now incorporated in the Australian Nature Conservation Agency) provides a major incentive for the documentation of Australia's species diversity. Funding has been provided, under the aegis of the Commonwealth agency, for the scientific investigation of major groups of animals and plants. Researchers publish their findings in the Flora of Australia and the Zoological Catalogue of Australia and ancillary publications.

The massive task of understanding genetic diversity is being directed towards two goals. The first is to understand the genetics of species in decline or going extinct and to manage their threatened populations to retain the genetic diversity required for recovery. The second is to locate and bring into the laboratory genes that are commercially important, including those from the wild relatives of crop plants (see the box on page 4-5), bacteria and fungi for industrial use and invertebrates for biological control and biological monitoring.

Location and effectiveness of conservation reserves

Conservation areas are parts of the land or sea where special management is applied for nature conservation. Management is intended to separate elements of biodiversity (ecosystems, communities, species, populations, genes) from processes that threaten their persistence. National assessments of conservation areas have been limited to 'strict' reserves (those in IUCN, 1994 Categories I–IV). Regional studies have shown that other types of formal protection measures can be very extensive, but we lack information on these for large parts of the continent. Because of the sheer number and extent of these other formal protection measures and the lack of computerised data bases, it is not possible to examine their distribution and coverage of elements of biodiversity.

It seems obvious that strict reserves should be located so that the elements of biodiversity most needing this type of protection are able to persist in the landscape. However, this does not occur. The location of reserves in Australia has been largely shaped by a process of ad hoc acquisition of land that is determined by its availability for conservation purposes, its cost, and a wide variety of aims, some of which lead to very different priorities for protection. Only recently have more systematic approaches to establishing protected areas had some effect on acquisition priorities. Several decades of ad hoc protection have led to two serious limitations of the national reserve system.

The first is that reserves often do not effectively represent the natural features (such as ecosystems or species) within regions. Some features are represented many times and others not at all. This means that the total area needed to represent all the features in a region, starting with the existing reserve system, is greater than if a whole new reserve system were designed from scratch. An example of this tendency occurs in the Western Division of New South Wales (see figure below). The same results have emerged wherever similar analyses have been done in other parts of Australia. The upward trend in required area as more ad hoc reserves are added to the system indicates a steadily increasing cost of reserving all the natural features in a region and a decline in the likelihood of achieving a fully representative reserve system.

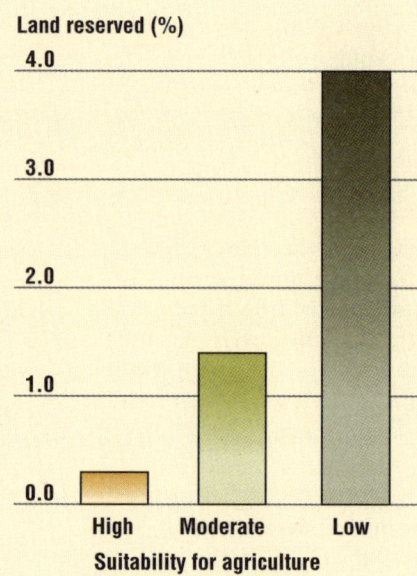

Reservation in relation to three classes of agricultural suitability in the Mt Lofty Ranges of South Australia.
Source: Department of Housing and Urban Development, SA, GIS Unit.

The second limitation is that reserves tend to be a 'residual' land use, with more extensive protection given to land that is least useful for intensive commercial purposes (see the diagram above). In many areas, this means that reserves do not occur where threatening processes are greatest. For example, ecosystems most in danger of outright replacement by crops and pastures often receive zero or minimal protection. The major conservation battles tend to be fought in a small subsection of the environments in need of protection. In eastern and southern Australia, this is typically on areas of Crown land where the main alternatives are forestry or reservation. Crown land itself is usually the residual tenure after freehold land has been released for intensive land use. A residual reserve system is a poor starting point for regional-based planning, which requires a mix of protection measures with the strictest protection applied to those areas least able to persist under any form of extractive uses — particularly intensive ones like agriculture. A residual reserve system provides the opposite starting point.

Ad hoc reservation consumes limited conservation resources and often gives the appearance of conservation progress (for example, increasing area of reserves) without genuinely contributing to the protection of biodiversity. Two critical questions should be asked in a regional context about proposals for new reserves.

• Will the proposed reserve represent features that are poorly protected?

• Will it cover features that most need this form of protection?

If the answer to one of these questions is 'no', conservation resources will not be spent in the most effective way for conserving biodiversity.

Source: Pressey, 1990.

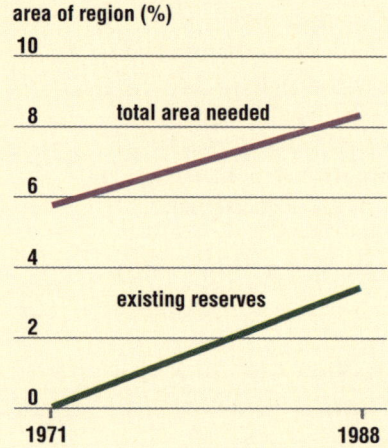

Existing reserves and total area of reserves required to represent all land types in the Western Division of New South Wales. In 1971, before any reserves were dedicated, all land types could have been represented in 5.7 per cent of the region. In 1988, following acquisition of reserves totalling 3.3 per cent of the region, the minimum area of reserve needed to represent all land types, starting with existing reserves, was 8.3 per cent.

Monitoring

Monitoring reveals the success or failure of strategies designed to conserve biodiversity and the sustainability or otherwise of industrial practices. Three types of monitoring correspond to the three levels of biodiversity: ecosystems, species and genes.

Australia is a world leader in ecosystem monitoring by GIS. Satellite imagery can reveal the extent of clearing, deforestation, over-grazing, urbanisation, siltation of rivers or coasts or reef dieback from one period to another. Much work remains to be done on 'ground-truthing'— that is, making sure that the interpretation of satellite data corresponds to what is really happening on the ground or in the sea.

Scientists commonly use the distribution and abundance of flowering plant and selected vertebrate species for environmental monitoring. However, they are also increasingly using invertebrates and micro-organisms. Freshwater and marine environments are routinely monitored by assessing the responses of selected vertebrate and invertebrate species to disturbance such as pollution. Research is required to test the reliability of indicator species when monitoring natural resource management, industrial pollution and conservation practices in all kinds of environments.

Declines in genetic variation need to be monitored in populations of endangered species so that, where possible, they can be managed to stabilise losses and reverse the trend. Australian conservation geneticists have contributed to the management of many endangered plants and animals in the field and to captive breeding programs of species on the verge of extinction in the wild.

Reserve systems

The box opposite highlights the problems and shortcomings of Australia's reserve system. Too often the reservation of a piece of land has been for economic or political rather than biological reasons, with the result that ecosystems that are unproductive — for example, because they are on poor soils — are well reserved while those on more productive land or at highly desirable locations may not be reserved at all. Australian ecologists are developing the criteria and guidelines for establishing reserve systems representative of terrestrial and marine ecosystems. Much information and some software are already available, but much needs to be done to integrate the scientific criteria and guidelines into mainstream political and legislative procedures.

Integrated ecosystem-based management of natural resources

Planning and management should be integrated and regionally based to overcome a number of problems. These include large numbers of public agencies involved in management decisions; administrative boundaries that do not reflect any particular physical, geographic or ecological feature; and the cumulative effects of multiple and continuous developments.

One of the principles of the national biodiversity strategy states that 'biological diversity is best conserved *in situ*'. In view of the limitations of the reserve system, bioregional planning is one way of integrating options for conservation of biodiversity. Bioregional planning has been defined as an ecological and social framework within which government, business and community interests share responsibility for coordinating land use planning and formulating and implementing development options which meet human needs in a sustainable way (World Resources Institute, 1992). However, the boundaries of a region can vary depending on the context for which they are being established. To date, commonly used regionalisations include local government boundaries, catchment boundaries and the Interim Biogeographic Regionalisation for Australia regions.

Traditional planning, while it is effective for controlling site-specific development proposals, has proved ineffective for addressing cumulative, off-site and incremental impacts such as habitat fragmentation and soil salinisation that result from land clearance. Bioregional planning can be used to integrate the often ad hoc approaches in planning decisions, as it facilitates a system-based approach. The move towards greater integration across all levels of government, the community and industry in regional planning is indicative of a move towards a more holistic approach.

It is generally agreed that we will not reconcile production and conservation values simply by expanding the national system of reserves and national parks. Many ecosystems and species will be sustainable only if the environment is managed in accordance with natural rather than economic or political boundaries. For example, it makes little environmental sense if an endangered species is protected under one jurisdiction but not in an adjacent one, or a land-use practice is illegal on one side of a political boundary but not on the other.

Bioregional planning and management is a vehicle designed to overcome problems of fragmented decision-making or the tyranny of small decisions. Many Australians, including landholders, government agencies, industries and conservation groups, are already involved, but much knowledge (not only biological) is required to implement it. While scientific knowledge of the biodiversity of a region is essential, ways of creating the infrastructure to include all interested parties and to resolve their conflicts are, in most cases, still to be worked out. Planners also have yet to evaluate adequately existing knowledge of the effects of one type of activity upon others and upon the environment. Where they do not yet exist, data should be gathered so that facts remain the basis for planning and for dispute resolution.

Table 4.20 Summary

Key issue	State	Adequate Info.	Response	Effectiveness of response
Ecosystem diversity				
Northern rainforests Habitat destruction	Highly fragmented; many areas degraded	✓✓	Listing as protected areas, including World Heritage Register; improved land management	Limited; some unique areas not protected; clearing, fire management, grazing and weeds still problematic
Southern rainforests Habitat destruction	Highly fragmented	✓✓✓	Listing as protected areas	Adequate in some areas only
Tall open forests (wet sclerophyll) Altered fire regimes; land clearance; logging	Extensive losses in area and alteration of species composition	✓✓✓	Improved management; reservation	Reserve system inadequate; management practices still in development
Acacia forests, woodlands and shrublands Clearance; grazing	Habitat loss and degradation; species diversity reduced	✓✓	Improved land management	Locally effective
Eucalypt scrubs and shrublands Clearance; grazing	Extreme fragmentation; possible inability to regenerate	✓✓	Reservation; restoration	Very limited; reserves inadequate
Heathlands Clearance, altered fire regimes, urbanisation, agriculture and sand-mining	Widespread habitat loss; fragmentation	✓✓	Reserves	Limited and only locally effective
Chenopod shrublands Grazing	Widespread habitat degradation; many plant species endangered	✓✓✓	Improved land management; reserves	Locally effective only
Native grasslands Grazing	Many areas highly degraded or altered by introduction of exotic species	✓✓✓	Improved land management/legislation; reservation	Locally effective; reserves inadequate
Alpine and subalpine vegetation Grazing; tourism; predicted to be vulnerable to global warming	Some areas highly degraded	✓✓✓	Reservation; improved land management	Many areas now in national parks; others remain degraded and vulnerable
Salt marshes and mangroves Habitat destruction and degradation	Extensive loss near urban areas	✓✓	Protected areas; development controls; community awareness	Limited; loss and degradation continue in many areas
Species diversity				
Micro-organisms Habitat modification and loss	Unknown but population composition and size likely to be affected	✓	Little direct response	Not known
Marine invertebrates Habitat modification and loss; harvesting of edible species; competition from marine pests	Reduction in population size of exploited species	✓✓	Management plans for exploited species; controls on illegal harvesting	Pressures are continuing; very few successes
Freshwater invertebrates Habitat modification and loss	Insufficient information to assess	✓	Integrated catchment management; waste-water treatment; restoration of wetlands; control of introduced pests	Little known
Land invertebrates Habitat modification and loss	Massive reduction in population size of affected species	✓✓	Little direct response; protected areas	Little known
Marine fish Harvesting of edible species	Many important species overexploited but majority in good condition	✓✓	Management plans for most of the major species	Pressures are continuing on most species
Freshwater fish Habitat modification and loss; competition and predation from introduced species	Generally in poor condition; many species endangered	✓✓	Integrated catchment management; waste-water treatment; restoration of wetlands; control of introduced pests	Very limited effect because of difficulty of controlling introduced species and long time needed to rehabilitate habitat
Amphibians Substantial habitat loss but often pressures not identified	Several species have disappeared or are declining	✓✓	Protected areas; community-initiated protection	Lack of knowledge of causes of declines prevents effective action

Table 4.20 Summary (continued)

Key issue	State	Adequate Info.	Response	Effectiveness of response
Reptiles Habitat loss	Massive reduction in numbers in urban and agricultural areas	✔✔✔	Protected areas; protection of marine and freshwater turtles	Partially effective
Birds Habitat modification and loss; predation from feral animals	Some species disappearing, others threatened; a few increasing their range	✔✔✔	General protection; protected areas	Partially effective
Mammals Habitat modification and loss; competition with and predation by feral animals	Several species lost, others threatened; a few increasing in numbers and range	✔✔✔	General protection, protected areas	Partially effective; pressure from feral cats and foxes continues
Marine plants Habitat modification and loss; pollution; natural events - floods and cyclones	Extensive loss of seagrasses; localised loss of mangroves	✔✔✔	Protection for seagrasses and mangroves but destruction still allowed in some areas by permit	Effective for mangroves; loss of seagrass likely to continue through coastal development and natural events
Freshwater plants Habitat modification and loss	Species threatened	✔✔	Limitation on water licences; protected areas	Little known
Land plants Clearance; habitat modification and loss	Many species endangered or vulnerable	✔✔✔	Protected areas	Effective in some areas
Genetic diversity Habitat fragmentation and loss	Some species show reduced genetic diversity	✔	Protected areas; captive breeding programs; reintroductions; regulation of exploitation	Little known; research in progress.

Conclusion

Australia has a world-class heritage of unique ecosystems, species and genes. Although the country was already occupied by Aboriginal people, recent colonisation by people predominantly from European countries has resulted in the introduction of many practices that, while appropriate in their land of origin, have radically altered and degraded much of the Australian landscape, initiating widespread changes with many flow-on effects still to appear. We are now dealing with a continent-wide ecological experiment, the results of which we are struggling to understand and manage.

Australia has a biodiversity comparable to that of many equatorial countries covered in rainforest but, unlike those countries, is industrialised and has a comparatively small human population. Consequently, we have the opportunity to balance conservation of biodiversity, human population growth and economic development. This will come about only with substantial changes in the way that land and oceans are managed. Clearly, many current practices are not sustainable and

biodiversity-based industries such as agriculture, pastoralism, forestry, fisheries and tourism often erode the resources upon which they depend. Biodiversity is also a source of profound inspiration to Australians and an essential part of our culture. It is a reservoir of resources that as yet remain relatively untapped. All these issues demonstrate the urgency of establishing major, often nation-wide, coordinated programs for the discovery, monitoring, management and sustainable use of biodiversity.

The use of new management strategies, particularly ecologically sustainable development and the precautionary principle, if implemented, would enable this country to provide world leadership for the wise use of natural resources for future generations. Our Commonwealth, State, Territory, and local governments have a central role to play in balancing conservation of biodiversity and economic production. Future generations are unlikely to forgive further losses of biodiversity through bad management or lack of commitment, especially now that its precarious state is recognised.

References

AUSLIG (1990). 'Atlas of Australian Resources (3rd series) Volume 6 - Vegetation, 64.' (Department of Administrative Services: Canberra.)

ABARE (1995). Australian Fisheries Statistics 1995. (Australian Bureau of Agricultural and Resource Economics: Canberra.)

Australian Mining Industry Council (1990). Mine Rehabilitation Handbook. (Australian Mining Industry Council: Dickson, ACT.)

Beattie, A.J. (ed.) (1995). 'Biodiversity: Australia's Living Wealth.' (Reed: Sydney.)

Behn, G., and Campbell, N. (1992). 'Dieback Assessment, Using Multispectral Data, over the Stirling Range National Park.' (CSIRO: Perth.)

Benson, D.H. (1985). Maturation periods for fire-sensitive shrub species in Hawkesbury Sandstone vegetation. *Cunninghamia,* **1**, pp. 339–49.

Benson, D.H., and Howell, J. (1990). Sydney's vegetation 1788–1988 — utilization, degradation and rehabilitation. *Proceedings of the Ecological Society of Australia,* **16**, pp. 115–27.

Benson, J. (1991). The effect of 200 years of European settlement on the vegetation and flora of New South Wales. *Cunninghamia,* **2**, pp. 343–70.

Biodiversity Unit, DEST (1994). Australia's biodiversity: an overview of selected significant components. *DEST Biodiversity Series Paper* No 2.

Biodiversity Unit, DEST (1995). Native vegetation clearance, habitat loss and biodiversity decline. *DEST Biodiversity Series, Paper* No 6.

Biodiversity Unit, DEST (in press). Status of Australia's biodiversity. *DEST Biodiversity series paper* No. 11.

Blaber, S., Battam, H., Brothers, N., and Garnett, S. (in press). Threatened and migratory seabird species in Australia: an overview of status, conservation and management. *Proceedings of the National Seabird Workshop.*

Blakers, M., Davies, S.J.J.F., and Reilly, P.N. (1984). 'The Atlas of Australian Birds.' (Melbourne University Press: Melbourne.)

Briggs, J.D., and Leigh, J.H. (1988). Rare or threatened Australian plants. *Australian National Parks and Wildlife Service Special Publication* No. 14.

Briscoe, D.A., and Tait, N.N. (1995). Allozyme evidence for extensive and ancient radiations in Australian Onychophora. *Zoological Journal of the Linnean Society (London),* **114**, pp. 91–102.

Brown, A.H.D., Grant, J.E., Burdon J.J., Grace J.P., and Pullen R. (1985). Collection and utilisation of wild perennial *Glycine.* World Soybean Research Conference III, Proceedings, Ames IA 12–17 Aug. 1984, ed. R. Shibles. (Westview Press: Boulder, Colorado.)

Burbidge, A.A., and Kuchling, G. (1994). 'Western Swamp Tortoise Recovery Plan. Western Australian Wildlife Management Program No. 11.' (Western Australian Department of Conservation and Land Management: Como, WA.)

Burbidge, A.A., Johnson, K.A., Fuller, P.J., and Southgate, R.I. (1988). Aboriginal knowledge of the mammals of the central deserts of Australia. *Australian Wildlife Research,* **15**, pp. 9–39.

Burbidge, A.A., and McKenzie, N.L. (1989). Patterns in the modern decline of Western Australia's vertebrate fauna: causes and conservation implications. *Biological Conservation,* **50**, pp. 143–98.

Callinan, R.B., Fraser, G.C., and Melville, M.D. (1993). Seasonally recurrent fish mortalities and ulcerative disease outbreaks associated with acid sulphate soils in Australian estuaries. *International Institute for Land Reclamation and Improvement, Publication,* No. 53, pp 403–10.

Castles, I. (1992). 'Australia's Environment: Issues and Facts.' (Australian Bureau of Statistics: Canberra.)

Census of Australian Vertebrate Species (1994). Australian Biological Resource Study. (Australian Nature Conservation Agency: Canberra.)

Cocks, K.D. (1992). 'Use With Care: Managing Australia's Natural Resources in the 21st Century.' (University of New South Wales Press: Kensington, NSW.)

Cogger, H.G., Cameron, E.E., Sadlier, R.A., and Eggler, P. (1993). The action plan for Australian reptiles. *Australian Nature Conservation Agency, Endangered Species Program, Project* No. 124.

Commonwealth of Australia, House of Representatives Standing Committee on Environment, Recreation and the Arts (1992). 'Biodiversity. The Contribution of Community-based Programs.' (AGPS: Canberra.)

Commonwealth of Australia (1994). Budget Statements 1994–95. (AGPS: Canberra.)

Commonwealth of Australia (1992). *Endangered Species Protection Act, 1992.*

Commonwealth of Australia (1996). 'The National Strategy for the Conservation of Australia's Biological Diversity.' (AGPS: Canberra.)

Cracraft, J. (1991). Patterns of diversification within continental biotas: hierarchical congruence among the areas of endemism of Australian vertebrates. *Australian Systematic Botany.* **4**, pp. 211–27.

Dexter, E.M., Chapman, A.D. and Busby, J.R. (1995). 'The Impact of Global Warming on the Distribution of Threatened Vertebrates. (Environmental Resources Information Network: Canberra.)

Driml, S. (1994). Protection for profit: economic and financial values of the Great Barrier Reef World Heritage Area and other protected areas. *Great Barrier Reef Marine Park Authority Research Publication* No. 35.

Eldridge, M.D.B., and Close, R.L. (1992). Taxonomy of rock wallabies, Petrogale (Marsupialia: Macropodidae). I. a revision of the eastern Petrogale with the description of three new species. *Australian Journal of Zoology,* **40**, pp. 605–25

'Endangered Species Protection Act 1992, Schedules 1, 2 & 3.' (Australian Nature Conservation Agency: Canberra.)

Fox, B.J. (1990). Two hundred years of disturbance: how has it aided our understanding of succession in Australia? *Proceedings of the Ecological Society of Australia,* **16**, pp. 521–9.

Fox, M.D., and Adamson, D. (1986). The ecology of invasions. In 'A National Legacy', 2nd edition, eds H. Recher, D. Lunney and I. Dunn, pp. 235–55. (Pergamon Press: Sydney.)

Garnett, S. (1992a). The action plan for Australian birds. *Australian National Parks and Wildlife Service Endangered Species Program, Project* No. 121.

Garnett, S. (ed.) (1992b). Threatened and extinct birds of Australia. *Royal Australian Ornithologists Union Report* No. 82.

Graetz, D., Fisher, R. and Wilson, M. (1992). 'Looking Back: the Changing Face of the Australian Continent, 1972–1992.' (CSIRO Office of Space Science and Applications: Canberra.)

Graetz, R.D., Wilson, M.A., and Campbell, S.K. (1995). Landcover disturbance over the Australian continent: a contemporary assessment. *DEST Biodiversity Series, Paper* No. 7.

Hart, R. (in press). Dieback caused by *Phytophthora cinnamomi*. In: 'The Natural History of Two Peoples Bay Nature Reserve, Western Australia', ed. A.J.M. Hopkins and G.T. Smith (Department of Conservation and Land Management Science Supplement: Perth.)

Hill, B.J. (1992). Minimum legal sizes and their use in management of Australian fisheries. *Bureau of Rural Resources, Proceedings* No. 13, pp. 9–10.

Hill, B.J., and Wassenberg, T.J. (1990). Fate of discards from prawn trawls in Torres Straits. *Australian Journal of Marine and Freshwater Research,* **41**, pp. 53–64.

Hillman, K., Lukatelich, R.J., and McComb, A.J. (1990). The impact of nutrient enrichment on nearshore and estuarine ecosystems in Western Australia. *Proceedings of the Ecological Society of Australia.* **16**, pp. 39–53.

Hobbs, R.J. and Hopkins, A.J.M., (1990). From frontier to fragments: European impact on Australia's vegetation. *Proceedings of the Ecological Society of Australia,* **16**, pp. 93–114.

Hobbs, R.J., and Saunders, D.A. (eds) (1993). 'Reintegrating Fragmented Landscapes: Toward Sustainable Production and Nature Conservation. (Springer-Verlag: New York.)

Hobbs, R.J., Saunders, D.A., and Arnold, G.W. (1993). Integrated landscape ecology. *Biological Conservation,* **64**, pp. 231–8.

Holliday, J.E., and Nell, J.A. (1987). The Pacific oyster in New South Wales. *Agfacts.* F2.1.3: (Agdex 486): pp. 1–4.

Hoskin, E.S., Hindwood, K., and McGill, A.R. (1991). 'The Birds of Sydney, County of Cumberland, New South Wales, 1770–1989.' (Surrey Beatty and Sons: Chipping Norton, NSW.)

Humphries, S.E., Groves, R.H., and Mitchell, D.S. (1991). 'Plant Invasions of Australian Ecosystems. Kowari 2.' (Australian National Parks and Wildlife Service: Canberra.)

IUCN (The World Conservation Union) (1994). 'Guidelines for Protected Area Management Categories.' (IUCN: Gland, Switzerland.)

Kennedy, M. (ed.) (1990). 'Australia's Endangered Species.' (Simon and Schuster: Brookvale.)

Kennedy, M. (1992). 'Australia's Marsupials and Monotremes: an Action Plan for their Conservation. (IUCN: Gland, Switzerland.)

Kirkpatrick, J., McDougall, K., and Hyde, M. (1995). 'Australia's Most Threatened Ecosystems. The Southeastern Lowland Native Grasslands. (Surrey

Beatty and Sons and Worldwide Fund for Nature Australia: Chipping Norton, NSW.)

Lee, A.K. (1995). The action plan for Australian rodents. *Australian Conservation Agency Endangered Species Program, Project* No. 130.

Leigh, J.H., and Briggs, J.D. (eds) (1992). 'Threatened Australian Plants; Overview and Case Studies.' (MacMillan: Melbourne.)

Leigh, J.H., and Briggs, J.D. (eds) (1994). 'Threatened Australian Plants: Overview and case studies.' (Australian National Parks and Wildlife Service: Canberra.)

Lothian, A.J. (1994). Attitudes of Australians towards the environment: 1975 to 1994. *Australian Journal of Environmental Management,* **1** (2), pp. 78–99.

McComb, A.J., and Lake, P.S. (1990). 'Australian Wetlands.' (Collins/Angus & Robertson: Australia.)

Marsh, H., Corkeron, P.J., Limpus, C.J., Shaugnessy, P.D., and Ward, T.M. (1995). The reptiles and mammals in Australian seas: their status and management. In Zann and Kailola (1995) (eds), pp. 151–66.

Morton, S.R. (1990). The impact of European settlement on the vertebrate animals of arid Australia: a conceptual model. *Proceedings of the Ecological Society of Australia* **16**, pp. 201–13.

Morton, S.R., Short, J., and Barker, R.D. (1995). Refugia for biological diversity in arid and semi-arid Australia. *DEST Biodiversity Series Paper* No. 4.

National Greenhouse Gas Inventory Committee (1994). 'National Greenhouse Gas Inventory, 1988 and 1990.' (DEST: Canberra.)

National Recreational Fisheries Working Group (1994). 'Recreational Fishing in Australia. A National Policy.' (Department of Primary Industries and Energy: Canberra.)

Noss, R.F. (1987). Corridors in real landscapes. A reply to Simberloff and Cox. *Conservation Biology* **1**, pp. 159–64.

National Population Council (1992). 'Population Issues and Australia's Future. Environment, Economy and Society. Consultant's Report. (AGPS: Canberra.)

Olsen, P., Fuller, P., and Marples, T.G. (1993). Pesticide-related egg shell thinning in Australian raptors. *Emu,* **93**, pp. 1–11.

Orr, A.G. (1994). Inbreeding depression in Australian butterflies: some implications for conservation. *Memoirs of the Queensland Museum,* **36**, pp. 179–84

Paterson, R.A. (1990). Effects of long-term anti-shark measures on target and non-target species in Queensland, Australia. *Biological Conservation,* **52**, pp. 147–59.

Pender, P.J., Willing, R.S., and Cann, B. (1992). NPF by-catch a valuable resource? *Australian Fisheries,* **51**(2), pp. 30–1.

Preece, N., van Oosterzee, P., and James, D. (1995). Two Way Track - Biodiversity conservation and ecotourism: an investigation of linkages, mutual benefits and future opportunities. *DEST Biodiversity Series Paper,* No. 5.

Pressey, R.L. (1990). Reserve selection in New South Wales: where to from here? *Australian Zoologist,* **26**(2), pp. 70–5.

Pressey, R.L. (1995). Conservation reserves in New South Wales: crown jewels or leftovers? *Search,* **26**, pp. 47–51.

Pressey, R.L., and Adam, P. (1995). A review of wetland inventory and classification in Australia. *Vegetatio,* **118**, pp. 81–101.

Prober, S., and Thiele, K. (1993). Surviving in cemeteries - the grassy white box woodlands. *National Parks Journal,* February, pp. 13–15.

Recher, H.F., and Lim, L. (1990). A review of current ideas of the extinction, conservation and management of Australia's terrestrial fauna. *Proceedings of the Ecological Society of Australia,* **16**, pp. 287–301.

Reid, D.D., and Krogh, M. (1992). Assessment of catches from protective shark meshing off New South Wales beaches between 1950 and 1990. *Australian Journal of Marine and Freshwater Research,* **43**, pp. 283–96.

Reid, J., and Fleming, M. (1992). The conservation status of birds in arid Australia. *Rangeland Journal,* **14**, pp. 65–91.

Resource Assessment Commission (1992). 'Forest and Timber Inquiry, Final Report.' (AGPS: Canberra.)

Resource Assessment Commission (1993). 'Coastal Zone Inquiry, Final Report. (AGPS: Canberra.)

Sainsbury, K.J. (1988). The ecological basis of multispecies fisheries, and management of a demersal fishery in tropical Australia. In 'Fish Population Dynamics' (2nd ed.), ed. J.A. Gulland, pp. 349–82. (John Wiley and Sons: Chichester and New York.)

Sarre, S., Schwaner, T.D., and Georges, A. (1990). Genetic variation among insular populations of the sleepy lizard, *Trachydosaurus rugosus* Gray (Squamata:Scincidae). *Australian Journal of Zoology,* **38**, pp. 603–16.

Saunders, D.A. (1991). The effect of land clearing on the ecology of Carnaby's Cockatoo and the inland red-tailed black cockatoo in the wheatbelt of Western Australia. *Acta XX Congressus Internationalis Ornithologica,* pp. 658–65.

Saunders, D.A., Craig, J.L., and Mattiske, E.M. (eds) (1995). Nature Conservation 4. The Role of Networks (Surrey Beatty and Sons: Chipping Norton NSW.)

Saunders, D.A., and de Rebeira, C.P. (1991). Values of corridors to avian populations in a fragmented landscape. In: 'Nature Conservation 2. The Role of Corridors', eds D.A. Saunders and R.J. Hobbs, pp. 221–40. (Surrey Beatty and Sons: Chipping Norton, NSW.)

Saunders, D.A., Hopkins, A.J.M., and How, R.A. (eds) (1990). Australian ecosystems: 200 Years of utilization, degradation and reconstruction. *Proceedings of the Ecological Society of Australia* **16**, pp. 1–602.

Shaklee, J.B., and Salini, J. (1993). Electrophoretic characterisation of multiple genetic stocks of barramundi perch in Queensland. *Transactions of the American Fisheries Society,* **122**, pp. 685–701.

Short, K. (1994). 'Quick Poison, Slow Poison: Pesticide Risk in the Lucky Country. (Southwood Press: Sydney.)

Simberloff, D., Farr, J.A., Cox, J., and Mehlman, D.W. (1992). Movement corridors: conservation bargains or poor investments. *Conservation Biology,* **6**, pp. 493–504.

Taylor A.C., Marshall Graves, J.A., Murray, N.D., and Sherwin, W.B (1991). Conservation genetics of the koala (*Phascolarctos cinereus*) II. Limited variability in Minisatellite DNA sequences. *Biochemical Genetics,* **29**, pp. 355–63.

Taylor A.C., Sherwin, W.B., and Wayne, R.K. (1994). Genetic variation of microsatellite loci in a bottlenecked species: the northern hairy-nosed wombat *Lasiorhinus krefftii. Molecular Ecology,* **3**, pp. 277–90.

Thackway, R., and Cresswell, I.D. (eds). (1995). 'An Interim Biogeographic Regionalisation for Australia: a Framework for Establishing the National System of Reserves. Version 4.0.' (Australian Nature Conservation Agency: Canberra.)

Tidemann, S.C., McOrist, S., Woinarski, J.C.Z., and Freeland, W.J. (1992). Parasitism of wild Gouldian finches (*Erythrura gouldiae*) by the air-sac mite (*Sternostoma tracheacolum*). *Journal of Wildlife Diseases,* **28**, pp. 80–84.

Tyler, M. (1994). 'Australian Frogs.' Revised edition. (Reed Books: Chatswood.)

Victorian Department of Conservation and Natural Resources (1992), 'Draft Conservation Program for Native Grasslands and Grassy Woodlands of Victoria.' (Scientific Publications: Kew, Victoria.)

Wager, R., and Jackson, P. (1993). The action plan for Australian freshwater fishes. *Australian Nature Conservation Agency, Endangered Species Program, Project* No. 147.

Williams, G. (1990). Invertebrate conservation. In 'Australia's Endangered Species', ed. M. Kennedy. (Simon & Schuster: Brookvale.)

Winter, J.W., Atherton, R.G., Bell, F.C., and Pahl, L.I. (1987). An introduction to Australian rainforests. *Special Australian Heritage Series* No. 7(1), pp. 95–118. (AGPS: Canberra.)

World Commission on Environment and Development (1987). 'Our Common Future.' (Oxford University Press: Oxford.)

World Conservation Monitoring Centre (1992). 'Global Biodiversity — Status of the Earth's Living Resources.' (Chapman and Hall: London.)

World Resources Institute (1992). 'Global Biodiversity Strategy.' (Prepared by WRI, World Conservation Union, United Nations Environment Programme in consulation with FAO and the UNESCO. (World Resources Institute: Washington.)

Zann, L.P. (1995). 'Our Sea, Our Future. Major Findings of the State of the Marine Environment Report for Australia.' Published by the Great Barrier Reef Marine Park Authority for DEST, Ocean Rescue 2000 Program, pp. 1–112.

Zann, L.P., and Kailola, P. (1995). 'The State of the Marine Environment Report for Australia. Technical Annex 1: the Marine Environment.' Published by the Great Barrier Reef Marine Park Authority for DEST, Ocean Rescue 2000 Program, pp. 1–193.

Acknowledgments

The following people reviewed the chapter in draft form and provided constructive comments.

Professor Rhonda Jones (James Cook University)
Dr Sam Lake (Monash University)
Dr Judith Lambert (Community Solutions)
Professor Henry Nix (Australian National University)
Professor Harry Recher (Edith Cowan University)

A number of people from government departments, private industry and voluntary organisations provided information. We especially thank the following who assisted in the preparation of this chapter:

Jim Crennan (Australian Nature Conservation Agency)
Rohan Fernando (Australian Nature Conservation Agency)
Andreas Glanznig (Community Biodiversity Network)
Andrew Taplin (Australian Nature Conservation Agency)

In addition, Commonwealth Government departments and members of the Commonwealth/State ANZECC State of the Environment Reporting Taskforce also helped identify errors of fact or omission. Their assistance is also gratefully acknowledged.

Photo credits
Page 4-1: Richard Weatherly
Page 4-4: (*from top*) CSIRO Division of Fisheries; D. McKillop (GBRMPA); GBRMPA
Page 4-5: (*top*) Kathie Atkinson; (*bottom*) Beth Seviour
Page 4-6: (*top*) Kathie Atkinson; (*bottom*) CSIRO
Page 4-7: Denis Saunders (CSIRO Division of Wildlife & Ecology)
Page 4-8: Kathie Atkinson
Page 4-10: CSIRO Division of Fisheries
Page 4-12: Kathie Atkinson
Page 4-15: Kathie Atkinson
Page 4-18: Ian Morris
Page 4-20: CSIRO Division of Fisheries
Page 4-21: Denis Saunders (CSIRO Division of Wildlife & Ecology)
Page 4-22: Kathie Atkinson
Page 4-27: Land Information Centre, Bathurst; from data captured by ACRES
Page 4-31: (*from left*) Michael Sharp (NSW Parks & Wildlife Service); Wyn Jones (NSW Parks & Wildlife Service)
Page 4-33: Painting by Oldfield Thomas (ANT Photo Library)
Page 4-38: (*left column*) Robert Close; (*right column*) Manfred Jusaitis (Botanic Gardens of Adelaide)
Page 4-44: D. Whitford (ANT Photo Library)
Page 4-45: Greening Australia, Victoria
Page 4-48: Colin Harris (Dept E&NR SA)
Page 4-50: Nick Alexander (Oryx Films)

Chapter 5

The Atmosphere

Thunderstorm cloud near Pt Lookout, NSW.

Prepared by

Bettye Dixon (Chair), Bureau of Meteorology

Willem Bouma, CSIRO Division of Atmospheric Research

Peter Cheng, Department of the Environment, Sport and Territories

Jim Le Cornu, Shell Australia Limited

Phil Morgan, Department of Environment and Heritage, Queensland

Stephen Wilson, University of Wollongong

*Shaun Andrews (State of the Environment Reporting Unit member),
      Department of the Environment, Sport and Territories (Facilitator)*

Assisted by

Roger Beckmann (Consultant science writer)

Contents

Introduction

'On a clear day you can see for ever...' While not strictly true, this phrase nevertheless sums up the sense of well-being that clean, fresh air can give us.

Air is essential for most organisms. Its existence and composition also affect many other parts of the environment. The unique composition of the atmosphere helps to maintain our planet in its present form and make it habitable. 'Atmosphere' is defined as the air environment on all physical scales — from the gaseous envelope surrounding the planet to the air inside a house — and on time-scales ranging from minutes to decades.

This chapter focuses on those aspects of the atmospheric environment where human activities have a detectable effect. The natural processes that operate within the atmosphere need to be understood in order to assess the impact of human activities (see Table 5.1 and Fig. 5.1). The state of Australia's air environment depends on the country's weather and climate as well as on the various human-induced pressures on the natural environment.

The chapter considers the impact of human activities on the atmosphere on three spatial scales — global, regional and local. Although it is convenient to use these scales for this report, in practice, they overlap.

Global-scale impacts encompass the effects of long-lived gases and particles on the global atmosphere.

The enhanced greenhouse effect and stratospheric ozone depletion are examples of phenomena on this scale.

Regional-scale effects involve the dispersion of pollutants within the airshed in which they were emitted and their transport downwind. Examples include motor vehicle emissions in cities and emissions from large industrial plants and power generation.

Local-scale refers to pollutants that are dispersed or inactivated without travelling far from their source or are contained within confined areas — for example, indoor air pollution.

Figure 5.1 The components of the global climate system

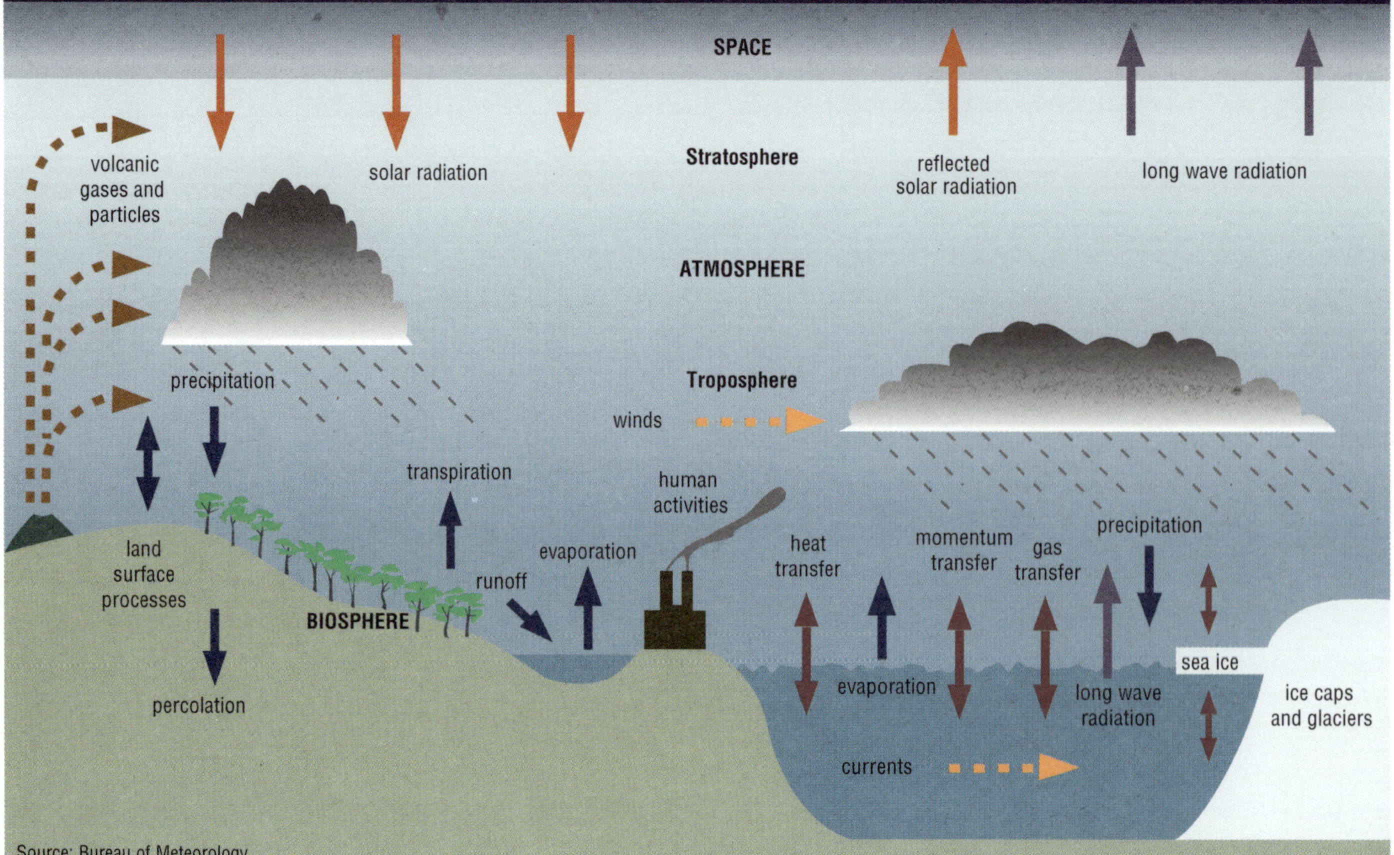

SPACE

Stratosphere

solar radiation

reflected solar radiation

long wave radiation

volcanic gases and particles

ATMOSPHERE

Troposphere

winds

precipitation

transpiration

human activities

land surface processes

runoff

evaporation

heat transfer

momentum transfer

gas transfer

precipitation

BIOSPHERE

evaporation

long wave radiation

sea ice

ice caps and glaciers

percolation

currents

Source: Bureau of Meteorology

Table 5.1 Composition of the lower atmosphere

Gas	Symbol	Percent by volume
Constant Components		
Nitrogen	N_2	78.08
Oxygen	O_2	20.95
Argon	Ar	0.93
Neon	Ne	0.0018
Helium	He	0.0005
Hydrogen	H_2	0.00005
Xenon	Xe	0.000009
Variable Components		
Water vapour	H_2O	0 to 4
Carbon dioxide	CO_2	0.036
Methane	CH_4	0.00017
Ozone	O_3	0.000004*
Carbon monoxide	CO	0.00002*
Sulphur dioxide	SO_2	0.000001*
Nitrogen dioxide	NO_2	0.000001*
Particles (dust etc.)		0.0001*

*Typical value in polluted air
Source: after Crowder, 1995.

Climate of Australia

Our part of the world — the southern hemisphere — enjoys relatively clean air and clear skies. It is mainly ocean, and has a smaller population and consequently a lower level of human emissions than does the northern hemisphere. Australia has the added advantage of being an isolated island, and so is not directly subjected to emissions from neighbouring countries.

We experience a climate quite different from that in Europe and North America. Many distinctive features of Australia make comparisons with other countries difficult (see Chapter 2).

Australia is a fairly flat, sparsely populated island continent located on the western rim of the Pacific in the largely oceanic southern hemisphere. Its geographic location and size mean that it experiences many climate zones. These range from tropical climates in its northern third to temperate ones in Tasmania and the southern parts of the mainland (with a small alpine region occurring in the south-east of the continent and in central Tasmania) to Mediterranean in the south-west and south-central areas. More than 75 per cent of the continent is classified as arid or semi-arid.

Most of the country comes under the influence of the subtropical ridge of high atmospheric pressure. The air above the ridge is in the descending branch of a large 'cell' of air circulation that links the tropics and the middle latitudes (see Fig. 5.2). The air movement of the cell is driven by the temperature contrast between the warm tropical ocean to the north of Australia (the warmest ocean on the planet) and the cold of the Antarctic regions. The other global-scale atmospheric feature exerting a particular influence is the east–west air circulation (the Walker Circulation), with air

ascending over the warm western Pacific and descending over the colder waters off the west coast of South America (see Chapter 2).

Australia's generally low terrain provides little obstruction to the global circulation systems. High pressure cells that travel from west to east dominate the weather and climate over much of the country. In summer they are at a latitude well south of the landmass, and during winter they move northward to become centred over the continent. The high pressure systems ensure stable atmospheric conditions, typically with clear skies, much sunshine, light winds and little precipitation (conditions that favour the build-up of pollution).

In the cooler half of the year (May to October), the high pressure systems pass slowly across the continent, often remaining stationary for several days. Northern Australia is influenced by mild, dry south-east trade winds, while southern areas experience cooler, moist westerly flows. Frontal systems in the westerlies can cause periods of intense rain and abrupt temperature changes, and even snowfalls in the southern higher areas. The coldest temperature ever recorded in Australia was -23°C at Charlotte Pass, New South Wales, on 29 June 1994.

In the warm half of the year (November to April), the highs become centred well to the south. Easterly winds predominate and most of southern Australia experiences fine, warm — often heat-wave — conditions. Marble Bar, in north-west Western Australia, recorded 161 consecutive days above 37.8°C (100°F) between 30 October 1923

Figure 5.2 Large-scale atmospheric circulations affecting our climate

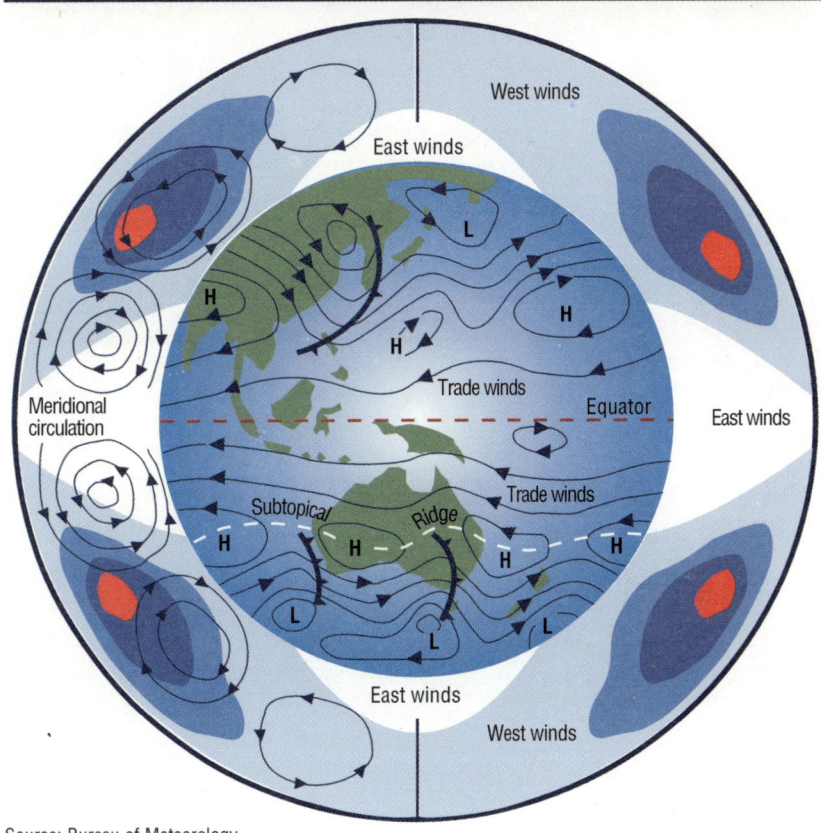

Source: Bureau of Meteorology.

and 7 April 1924. At this time of year northern Australia is influenced by the monsoon lows associated with the southward movement of warm, moist tropical air.

Australia receives less rain than any other continent except Antarctica, and no continent has less run-off from its rivers. Despite its high proportion of arid land, it has a less extreme climate than deserts such as the Sahara. The major features of its climate are: the highly irregular rainfall (see Fig. 5.3), which is closely linked to the El Niño–Southern Oscillation (ENSO) phenomenon (see Chapter 2); the extreme rate of evaporation of available water; and the large temperature ranges.

Between November and April, tropical cyclones develop over the seas to the north of the continent. Both the number of cyclones and their tracks vary greatly from season to season and are related to the ENSO phenomenon. On average, about six cyclones each season affect northern coastal areas, often producing a great deal of rain and strong winds. Some bring widespread heavy rains inland.

Pressure

Human activities exert pressures that may change the state of the earth's natural systems. A 'pressure' does not inevitably cause an environmental or other 'problem'. Whether it does or not depends on the capacity of the system to absorb the pressure — sometimes referred to as the system's assimilative capacity. This may vary in time and place, as well as with the extent of other pressures. Within the air environment, pressures come about mainly through emitted substances. However, it is also possible to create a pressure by altering the assimilative capacity of natural systems (for example, through changing land use) and thereby reducing the capacity of the 'sinks' — the processes or places that remove pollutants from the atmosphere.

Emitted substances may be gases or particulates (fine particles), both of which can remain airborne for considerable periods. Some emitted substances may interact with each other. In this way, or through atmospheric chemical processes and the influence of sunlight, the primary emissions may turn into new, sometimes unwanted, secondary substances. A good example of this is the formation of photochemical smog, which contains ozone, in sunny city air. The ozone is described as a secondary pollutant: it arises almost entirely from the interactions of emitted substances under the influence of sunlight.

However, not all secondary substances are harmful. Chemical reactions in the atmosphere and elsewhere can also change emissions into forms that are less damaging to the environment or to human health. Thus, many emitted substances are degraded and then absorbed by natural systems. Continuous cycles remove many substances — especially naturally occurring ones — from the atmosphere. However, humans may generate a greater volume of emissions than the system can remove.

The period for which a substance remains in its active form in air is called its residence time. Certain emitted compounds, such as chlorofluorocarbons (CFCs), do not occur naturally. They are chemically non-reactive and have a very long residence time. In general, these long-lived emissions are responsible for regional and global pressures, while the short-lived reactive ones lead to local pressures.

How much pressure an emitted substance exerts depends on the type of environment it enters. For example, if emissions are concentrated because of particular air-flow patterns (natural or engineered) in a region or building, concentrations can build up and have an effect that may appear out of proportion to the absolute quantity emitted. Conversely, the atmosphere can also disperse emissions and transport them from one place to another — or even right around the planet. As well, high levels of humidity and frequent rain may 'wash out' particles and certain soluble gases before long-range transport occurs. Sunlight promotes the formation of photochemical smog and so the smog precursors — which are largely emitted from vehicles in urban areas — will create more ozone in sunny weather, and particularly when the winds are light.

How and where a substance is emitted into the atmosphere may also affect its fate. For example, emissions from tall stacks usually travel further than those arising at ground level. Chemical reactions can vary depending on the height within the atmosphere, because factors such as temperature, radiation from the sun, moisture and other gas concentrations vary with height. Air stability also affects the dispersion of emissions.

Pressures on the atmosphere that do not arise from emissions are even harder to characterise accurately. A good example concerns those aspects of the enhanced greenhouse effect that do not involve direct greenhouse gas emissions. In this case, human-induced changes that alter the planet's cover of vegetation may make it harder for the environment to remove additional carbon dioxide emitted by human activity (see Fig. 5.4).

Figure 5.3 Annual rainfall variability over Australia (1890–1990)

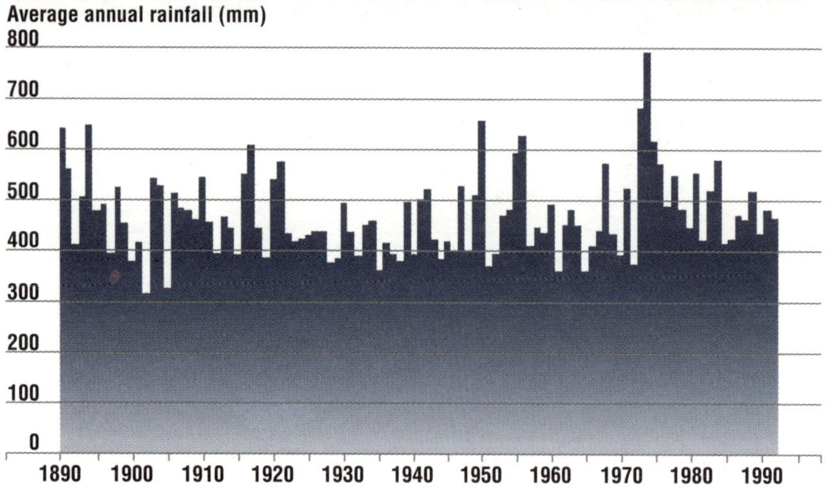

Average annual rainfall (mm)

Source: Bureau of Meteorology.

Emissions

Many substances are emitted into the atmosphere from human activities and from natural sources. Outdoors, the main gases and particulates that are emitted, and that have particular environmental impacts, are:

- carbon dioxide (CO_2)
- carbon monoxide (CO)
- halocarbons, such as halons (used in fire protection), chlorofluorocarbons (of various chemical formulae and known collectively as CFCs) and their replacement products such as hydrochlorofluorocarbons (HCFCs) and hydrofluorocarbons (HFCs)
- lead (Pb)
- methane (CH_4)
- oxides of nitrogen (NOx) including nitrogen dioxide (NO_2), nitric oxide (NO) and the greenhouse gas nitrous oxide (N_2O)
- particles of various compositions and sizes
- sulfur dioxide (SO_2)
- volatile organic compounds (VOCs) other than methane

Emitted gases and aerosols (suspensions of droplets or particles in the air) may contribute towards the greenhouse effect (see the box on page 5-16), the depletion of the stratospheric ozone layer (see the box on page 5-11), the phenomenon of acid deposition, the generation of photochemical smog and, on a local scale, the contamination of air, which then becomes less healthy for humans to breathe. In some cases, the same gases may contribute in different ways to a range of effects. For example, although small concentrations of ozone in photochemical smog in the lower atmosphere are damaging to most living things, small concentrations of ozone in the upper atmosphere all around the globe are necessary for life on earth because that ozone shields the surface from harmful ultraviolet (UV) radiation.

Indoor air is mainly contaminated by emissions generated indoors rather than emissions from outdoors (see Table 5.12).

Sources of emissions

The burning of carbon-containing fossil fuels powers most of Australia's transport and electricity generation and is responsible for a large part of our atmospheric emissions. Many domestic, commercial and industrial processes also emit waste gases. CFCs and halons can 'escape' into the atmosphere by leakage or by the destruction of manufactured items containing them. Agriculture can also be responsible for emissions and for changes to vegetation cover.

Aspects of Australia's society (such as its economy, demography and the lifestyle of its people) underlie pressures on the atmosphere (see Chapter 3). The country has abundant fossil-fuel energy resources, especially coal and natural gas, and is a major energy exporter. The scale of the energy industry is such that it creates considerable pressures on the

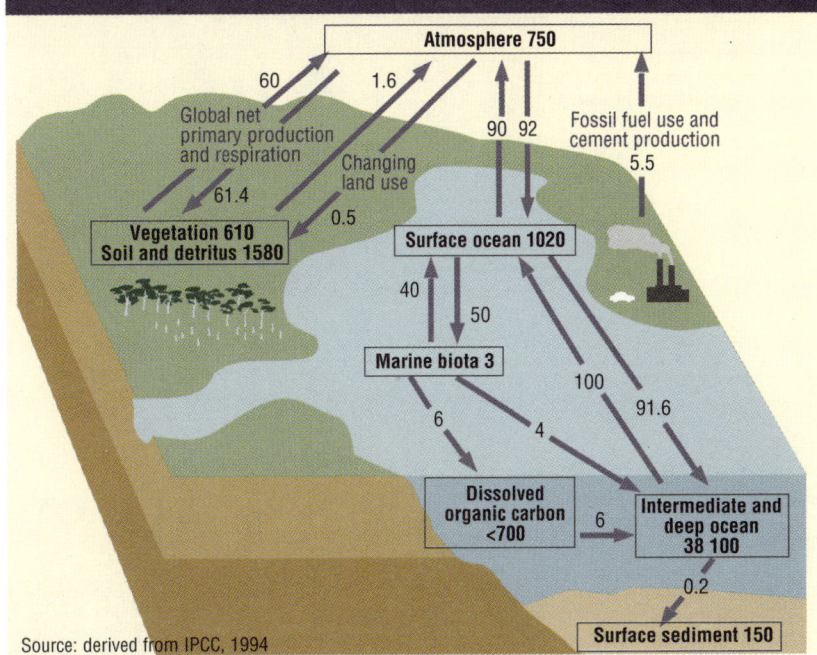

Figure 5.4 The global carbon cycle showing reservoirs (boxed) and annual exchanges of carbon in gigatonnes (Gt)

Source: derived from IPCC, 1994

atmosphere. The main sources of fossil-fuel-derived emissions are thermal power generation, road transport and the use of energy in industrial processes.

Coal is used to generate most electric power. The process produces ash particles, acid gases such as oxides of nitrogen (NOx) and sulfur dioxide (SO_2), as well as carbon dioxide, which contributes to the greenhouse effect. In Australia the amount of SO_2 emitted per unit of energy generated is low by world standards because our coal generally has a relatively low sulfur content. We have a further advantage in that most large power-generating plants are located outside our main cities.

The demand for power is affected by our increasing population, economic growth and, among other factors, the energy-intensive nature of certain major industries such as aluminium production. At present, we have few economically viable alternatives to fossil fuels as our primary energy source. There is no nuclear power generation in Australia, and only a limited capacity for hydro-electricity in certain areas, such as Tasmania and southern New South Wales.

As well as coal, Australia contains rich mineral deposits of iron ore, bauxite, silver, lead, zinc and gold. The country has well-developed minerals, manufacturing and agricultural industries, and a large services sector. Much of its industry is export-orientated — especially the mining and mineral processing sectors, which are important to the national economy. However, the treatment of mineral ores can be a major source of emissions — often of SO_2, but also of other gases, depending on the ore and the type of treatment. Mining and ore-processing are concentrated in small areas, often wherever a suitable ore deposit may be, and these operations are generally remote from major population centres.

▶

Greenhouse gas emissions from a field outside Wagga Wagga, NSW are being measured as part of a collaborative project involving universities and CSIRO. Measurements like this help to quantify Australian greenhouse emissions and are an essential first step in designing strategies for emission reduction.

In urban areas, motor vehicles are the main source of emissions and therefore the main contributors to outdoor air pollution. The design of Australian cities has promoted a high rate of motor vehicle ownership and use. (Our cities have low population densities compared with urban areas elsewhere and are often described as 'sprawling'.) Motor vehicle exhaust contains NOx, carbon monoxide (CO) and volatile organic compounds (VOCs). Together these are responsible for many local air pollution problems. Lead particles from leaded fuel are also present, although the use of such fuel is now declining. Particulate emissions come especially from diesel vehicles, but we have proportionately fewer of these than other countries.

The day-time brown haze that sometimes forms over our major cities is caused by particles and NOx from exhaust fumes. Less obvious forms of urban air pollution also occur. Under the influence of sunlight, NOx and VOCs may react together to form photochemical smog (see the box on page 5-25), containing the invisible gas ozone, a potentially greater threat to human health. In addition, of course, vehicles emit the greenhouse gas CO_2.

Important —but less direct — factors influencing pressures on the atmosphere are the size of the human population, its growth rate and the pattern of consumption of each individual. Australia's population of 18 million lives in relative affluence and is increasing at a rate of about 1.6 per cent per year, a rate higher than in most OECD countries. Expected economic growth and the continued growth of the population mean that the country's demand for energy is also likely to continue to increase.

Global-scale pressures

Enhanced greenhouse effect

The enhanced greenhouse effect is explained in the box on page 5-16. In terms of how human activity affects the global climate system, the pressures that give rise to the effect need to be considered on a global scale. Global pressures are the sum of the smaller pressures coming from individual countries and people.

'Anthropogenic' — a word often used in discussions of environmental topics — means 'brought about by humans'. In the case of the enhanced greenhouse effect, anthropogenic emissions of greenhouse gases are those emissions that enter the atmosphere as a result of human activity — even though some of the gases may occur naturally in the atmosphere.

Human activity has not only caused an increased emission of naturally occurring greenhouse gases such as CO_2, CH_4 and N_2O, but has also released other gases that contribute to the greenhouse effect — such as CFCs and photochemically derived ozone in the lower atmosphere. To assess the impact of anthropogenic emissions from pre-industrial times until the present we need to know the source and characteristics of these emissions.

Because of its particular properties, quantity emitted and long life-time, carbon dioxide (released mainly by burning fossil fuels, changes in land-use and cement production) is the most important anthropogenic greenhouse gas. However, others also make an important contribution. Methane largely derives from the biosphere (for example, livestock, rice cultivation, organic waste and land fills), and from fossil fuels (oil and gas exploration, gas distribution and coal-mining). Ozone in the lower atmosphere is mostly formed from pollutants in urban air. It does not remain for long and is not distributed uniformly — particularly in the southern hemisphere where few

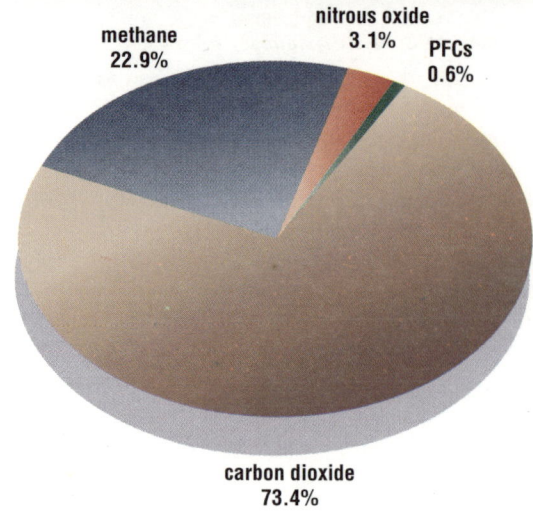

Figure 5.5 Australia's national greenhouse gas emissions inventory by component for 1990

methane 22.9%

nitrous oxide 3.1%

PFCs 0.6%

carbon dioxide 73.4%

Note: This inventory excludes gases controlled by the Montreal Protocol

Source: National Greenhouse Gas Inventory Committee, 1994.

sources occur. Nitrous oxide from agriculture and some industrial processes, and the long-lived halocarbons (such as CFCs and carbon tetrachloride) contribute as well. CFCs have two opposing effects. While they absorb infra-red radiation in the same wave-bands as the naturally occurring greenhouse gases, they also bring about the destruction of stratospheric ozone, a naturally occurring greenhouse gas. In fact, the warming effect of CFCs is reduced by over one-half because of the cooling brought about by their depletion of stratospheric ozone. Because of the impacts of CFCs and some of their replacements, their production is now controlled on a global scale by the Montreal Protocol of the Vienna Convention for the Protection of the Ozone Layer.

As part of its obligations under the United Nations Framework Convention on Climate Change, Australia is required to compile an inventory of its greenhouse gas emissions and sinks (except for gases, such as the CFCs, controlled by the Montreal Protocol) (see Figs 5.5 and 5.6).

Carbon dioxide is by far the most significant anthropogenic greenhouse gas emitted here (see Table 5.2), accounting for about 75 per cent of our total emissions. The main sources of CO_2 are fossil-fuel combustion and fugitive emissions from the energy sector. (Fugitive emissions include leaks, losses of naturally accumulated gases during fuel extraction and emissions from burning a gas or oil 'flare' at drilling rigs.) Slightly more than 30 per cent of net CO_2 emissions are estimated to come from forestry and changes to land use (mainly related to land clearing for agriculture), but this figure is very uncertain. Methane comes mainly from agriculture, waste decomposition (almost entirely landfill) and fugitive emissions from the energy sector, and accounts for about 23 per cent of total Australian greenhouse gas emissions. The remainder are mainly emissions of nitrous oxide from agriculture.

Table 5.2 Greenhouse gas emissions in Australia, 1990

Gas	Emissions (Mt)	Conversion factors (GWP)*	CO_2 equivalent (Mt)	Percent of total emissions
Carbon dioxide (CO_2)	419 807	1	419 807	73.4
Methane (CH_4)	6 243	21	131 115	22.9
Nitrous oxide (N_2O)	60	290	17 444	3.1
Perfluorocarbons (CF_4) and (C_2F_6)	1	(5 100) (10 000)	3 358	0.6
Total			**571 724**	**100.0**

* Global Warming Potentials (GWPs) are conversion factors used to express the relative warming effects of the various greenhouse gases in terms of the carbon dioxide equivalent, IPCC (1990).

Source: National Greenhouse Gas Inventory Committee, 1994.

Figure 5.7 Energy-related carbon dioxide emissions per unit GDP for selected OECD countries, 1970–92

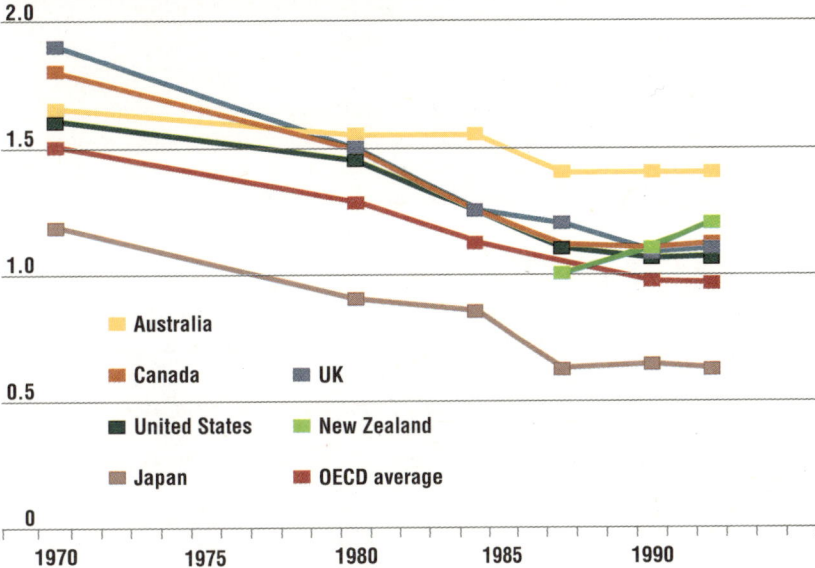

CO_2 per unit GDP (tonnes per 1985 US$1000)

- Australia
- Canada
- United States
- Japan
- UK
- New Zealand
- OECD average

Source: derived from IEA (1994).

Australia's gross greenhouse gas emissions contribute 1–2 per cent of total global emissions.

Based on some criteria, Australia has very high emissions of greenhouse gases relative to other OECD countries. For example, over the period 1987–92, our energy-related CO_2 emissions per unit of GDP declined more slowly than the OECD average (see Fig. 5.7). These emissions have grown over the last 30 years principally because of population growth, industrialisation and continuing electrification (IEA, 1994).
In contrast, in other OECD countries, where economic growth was also high, the effect on emissions was offset by large falls in fossil-fuel consumption per unit of output. Australia has a higher proportion of energy-intensive industry than most of those countries because of its strong natural resource base and competitive prices.
On the other hand, as shown in Fig. 5.8, it has a smaller proportion of CO_2 emissions from the chemicals industry.

Figure 5.6 Australia's national greenhouse gas emissions inventory by sector, 1990

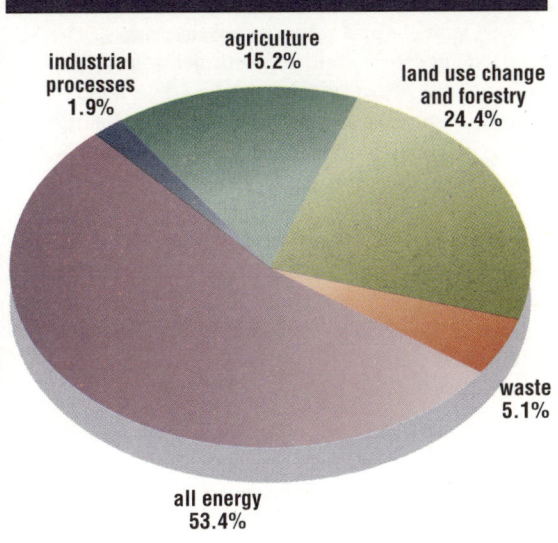

industrial processes 1.9%
agriculture 15.2%
land use change and forestry 24.4%
waste 5.1%
all energy 53.4%

Note: This inventory excludes gases controlled by the Montreal Protocol
Source: National Greenhouse Gas Inventory Committee, 1994.

In 1992 the transport sector accounted for 24 per cent of total CO_2 emissions from energy (just above the OECD average of 22 per cent), industry contributed 18 per cent and electricity generation, using coal as its main fuel, 47 per cent. No other OECD country has such a large fraction of its energy-related emissions accounted for by electricity (IEA, 1994). Additional contributing factors to our high output of greenhouse gases are the level of CO_2 emissions associated with land-clearing and methane emissions from livestock.

Stratospheric ozone loss

The box on page 5-11 summarises what we know about stratospheric ozone depletion. Chlorofluorocarbons (CFCs) are now known to be the main source of anthropogenic chlorine in the stratosphere, and therefore to carry most of the responsibility for ozone destruction (WMO, 1995). Being chemically stable in the lower atmosphere, and not toxic or corrosive, CFCs have a wide variety of industrial and commercial applications. Since their initial synthesis in the 1930s they have been used as refrigerants, aerosol propellants, foam-blowing agents and industrial cleaning solvents. Because they are so chemically stable (part of their initial attractiveness) and do not degrade in the lower atmosphere, they remain airborne for many decades. They are dispersed globally from the point of their release and eventually reach the stratosphere, where exposure to high levels of UV radiation breaks down the molecules, releasing the chlorine that, through a complex series of reactions, causes ozone destruction. A decline in the concentration of global stratospheric ozone has been measured since the late 1970s.

But CFCs are not the only ozone-depleting substances. The fire-fighting chemicals known as halons, which are a class of bromine-containing chemicals, and compounds such as carbon tetrachloride also have this effect. Methyl bromide,

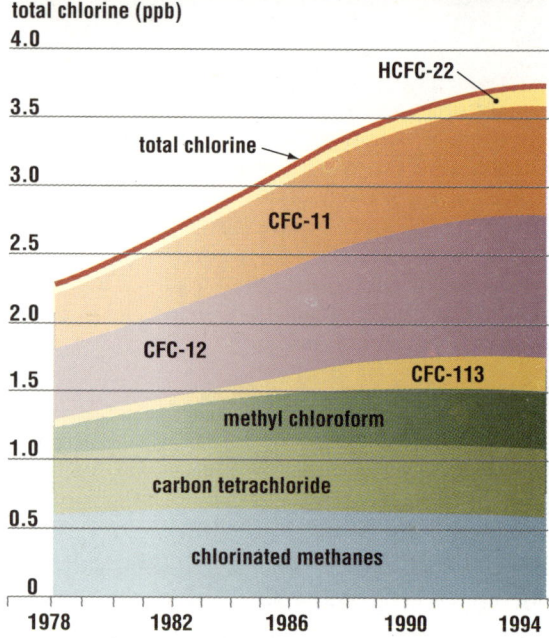

Figure 5.9 Total chlorine in the atmosphere based on Cape Grim data

total chlorine (ppb)

HCFC-22
total chlorine
CFC-11
CFC-12
CFC-113
methyl chloroform
carbon tetrachloride
chlorinated methanes

1978   1982   1986   1990   1994

Source: CSIRO and Cape Grim Baseline Air Pollution Station.

used in Australia and elsewhere as a pest-fumigant, has recently been listed as an ozone-depleting substance, as has methyl chloroform, used as a solvent. (Unlike CFCs, methyl bromide also occurs naturally.)

At the beginning of this decade, concentrations of CFCs were rising at the rate of four to 10 per cent a year, depending on the type. The monitoring station at Cape Grim, Tasmania, has recorded a steady increase in the amount of total atmospheric chlorine (see Fig. 5.9). However, the rate of increase in CFC concentration has now declined, attributable to global action in reducing the emission of ozone-depleting substances.

Regional-scale pressures

On the regional scale, a number of factors affect air quality. These include emissions from human activity and natural sources; the lifetimes of the emitted substances and their interaction with atmospheric chemical processes; and meteorological factors and topographical features. The result may be the transformation, dispersion and/or accumulation of the emissions. The important point to note is that identical amounts of an emission do not always have the same effect, because local conditions play such an important role. Atmospheric scientists usually use the term 'airshed' to describe a body of air marked out by clearly defined geographic features in which a contaminant, once emitted, is contained.

Most of Australia's population is concentrated within a very small proportion of the land area that occurs within about 100 km of the coast. Many human-induced pressures are therefore not distributed evenly across the country but rather are concentrated in major urban areas, making it hard

Figure 5.8 Industry-related CO_2 emissions by sub-sector for selected OECD countries, 1992

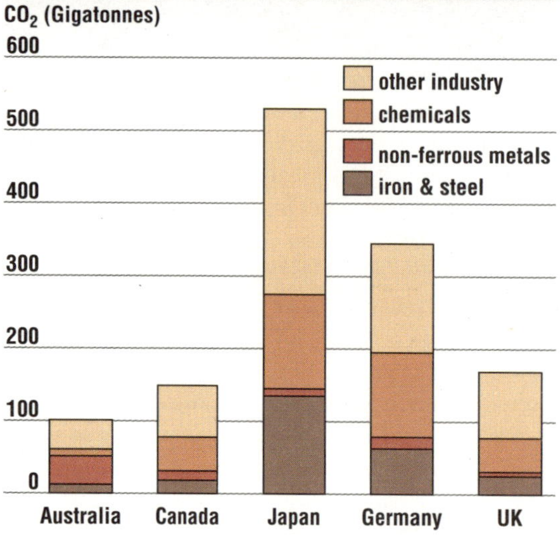

CO_2 (Gigatonnes)

600
500
400
300
200
100
0

- other industry
- chemicals
- non-ferrous metals
- iron & steel

Australia   Canada   Japan   Germany   UK

Source: IEA, 1994.

to define and quantify the pressures on the air environment on a national scale. As well as established centres of high population density, other locations are potential or actual 'hotspots' in terms of air quality, either because of their rapid population growth or because of specific local emission sources.

As mentioned previously, most emissions in Australia derive either directly or indirectly from the use of fossil fuels. A further important source of mainly regional emissions is the metals-processing industry. Processing of ore concentrates containing sulfur produces significant volumes of SO_2 and NOx, and quantities of airborne heavy metals.

Urban airsheds

The Australian Environment Council (AEC) Report of 1988, contains the only national set of published data on urban emissions. It is based on emission inventories up to 1985. Some State government authorities are currently updating capital city inventories.

Based on 1985 data, vehicles emit about 75 per cent of the oxides of nitrogen (NOx) and about 45 per cent of anthropogenic volatile organic compounds (VOCs) in the major Australian urban areas. Vegetation is an important natural source of VOCs, especially in hot weather. Domestic activities, such as house-heating or cooking with solid fuels, oil and gas, can also produce NOx and particles. Both NOx and VOCs are important in the formation of photochemical smog (see page 5-25). Motor vehicles are significant sources of carbon monoxide, particulate matter and lead, and also emit small amounts of a range of other pollutants known as air toxics. These include benzene, 1,3-butadiene, formaldehyde and polycyclic aromatic hydrocarbons (PAH).

More recent data have been published for Melbourne, Perth and Sydney (Carnovale *et al.*, 1991; Weir and Muriale, 1994; EPA NSW, 1995) on the proportion of motor vehicle emissions (see Tables 5.3, 5.4 and 5.5). Motor vehicles remain the major source of NOx and CO and a significant source of VOCs. In Melbourne, vehicles emit about 46 per cent of the particles in summer, but this drops to only 10 per cent in winter when smoke from wood fires dominates (80 per cent).

Most urban airsheds are associated with industrial activities. Industries contributing substantial quantities of emissions include petroleum refineries, chemical and petrochemical plants, solvent-based industries (for example, dry cleaning, degreasing, use of paints and coatings) and mineral-product industries (glass, brick and tile, cement and lime production). In future, the siting of small energy-efficient power stations (local sources of NOx) within urban areas may have detrimental effects on air quality (Cope *et al.*, 1992) but will at the same time contribute to a reduction in carbon dioxide emissions (an improvement from an enhanced greenhouse perspective).

Stratospheric ozone loss

About 15 to 50 km above the earth's surface is the part of the atmosphere called the stratosphere. In the stratosphere — unlike the lower atmosphere (the troposphere) — temperatures increase with height. As a result, it is more stable and does not have the same amount of vertical mixing of air as the troposphere. From the point of view of life on earth, the most important fact about the stratosphere is its concentration of ozone which, although low, is much higher than that in the troposphere.

Ozone is produced when short-wavelength ultraviolet (UV) radiation from the sun acts upon oxygen molecules. Once formed, ozone absorbs UV. But ozone is an unstable molecule, and for every molecule formed in the stratosphere, another breaks down. Thus, the ozone layer is the result of natural processes that both produce and destroy the gas, and its concentration depends on the balance between these processes.

Chlorine, bromine and oxides of nitrogen, in their reactive forms, can catalyse the breakdown of ozone in the stratosphere. They are part of the natural ozone destruction cycle. Human activities — domestic, industrial and agricultural — produce emissions that increase the concentration of these chemicals and accelerate the destruction of ozone. This destruction by anthropogenic chemicals is particularly efficient over Antarctica in spring, as a result of the presence of ozone-depleting substances in the stratosphere combined with the unique temperature, structure and circulation of the Antarctic stratosphere during the long polar night. The seasonal reduction in ozone is referred to as the ozone 'hole', which is a shorthand description signifying that over a large area the stratospheric ozone concentration has fallen below normal (see Figs 5.17 and 5.18). The ozone is replenished at the end of spring, when ozone-rich air from the rest of the stratosphere moves over Antarctica and brings ozone levels almost back to normal.

It is important to note that ozone decline is not just a polar phenomenon — all regions apart from the tropics have shown a decline in stratospheric ozone over the last decade, although not as severe as that over the poles.

Ozone in the stratosphere reduces the amount of damaging UV radiation that reaches the earth's surface. Without the 'ozone layer' life on land would be exposed to such dangerous levels of UV radiation that few life forms would survive. It is because of the ozone layer's essential protective role that any reduction in its concentration gives rise to such serious concern. In general, a fall of one per cent in atmospheric ozone has been calculated to be equivalent to an increase of between one and two per cent in UV radiation at ground level. In humans, exposure to UV can cause sunburn, eye damage, skin cancer and damage to the immune system in susceptible individuals. Fair-skinned people are most at risk. Many other organisms, including plants, are also at risk from increased UV radiation. Too much ultraviolet irradiation reduces plant growth, the sensitivity varying between different species. Whereas humans can avoid exposure, it is more difficult for flora and fauna to do so, although some species can protect themselves by producing UV-absorbing pigments in greater quantities following exposure.

While the existence of stratospheric ozone is so important, increases in ozone in the lower atmosphere (the troposphere) are of concern. Tropospheric ozone has a limited atmospheric lifetime and is not transported to the stratosphere to any significant degree.

Table 5.3 Percentage contributions of major sources to total daily airshed emissions for Melbourne, 1990

	Motor vehicles		Industrial/Commercial point sources		Other sources	
	S	W	S	W	S	W
Nitrogen oxides (NOx)	78	73	17	16	5	11
Volatile organic compounds (VOCs)	50	40	20	16	30	44
Carbon monoxide (CO)	91	70	2	1	7	29
Sulfur dioxide (SO$_2$)	11	11	86	84	3	5
Particles	46	10	29	6	25	84

Note: **S** — summer week day; **W** — winter week day
Source: Carnovale *et al*, 1991.

Table 5.4 Percentage contributions of major sources to total daily airshed emissions for Perth, 1992

	Motor vehicles		Industrial/Commercial point sources		Other sources	
	S	W	S	W	S	W
Nitrogen oxides (NOx)	50.5	51.4	45.9	43.5	3.5	5.1
Volatile organic compounds (VOCs)	47.0	46.1	27.2	12.3	25.8	41.7
Carbon monoxide (CO)	93.4	73.1	2.2	1.7	4.4	25.2
Sulfur dioxide (SO$_2$)	2.7	2.7	96.4	95.5	0.9	1.9
Particles	40.7	9.0	53.1	11.7	6.2	79.2

Note: **S** — summer week day; **W** — winter week day
Source: Weir and Muriale, 1994.

Table 5.5 Percentage contributions of major sources to annual airshed emissions for Sydney and the greater Metropolitan Air Quality Study (MAQS) area, 1992

	Motor vehicles		Industrial/commercial activity		Domestic/commercial activity	
	Sydney	Greater MAQS	Sydney	Greater MAQS	Sydney	Greater MAQS
Nitrogen oxides (NOx)	82	45	13	52	5	3
Volatile organic compounds (VOCs)	49	49	10	10	41	41
Carbon monoxide (CO)	91	69	2	23	7	8
Sulfur dioxide (SO$_2$)	14	2	64	96	22	2
Particles	31	16	36	68	33	16

Note: Greater MAQS area includes Newcastle and the Hunter Valley to the north of Sydney and Wollongong to the south.
Source: EPA NSW, 1995.

Regional airsheds

This chapter will refer to monitored airsheds outside major cities — most of which contain power stations — as 'regional airsheds'. Australian electricity consumption has more than doubled during the past two decades, with most of the power being generated by thermal power stations located near coalfields and away from urban areas. All emit relatively large quantities of NOx. As coal-burning power stations also emit sulfur dioxide, particulate matter and small quantities of toxic organic compounds, monitoring programs have been developed in power-generating regions such as the Latrobe and Hunter Valley airsheds.

The other major emitters to regional airsheds are: mineral-processing operations (for example, copper, lead, aluminium, gold, nickel, iron and steel production); paper and pulp manufacture; extractive industries (mining and quarrying); food processing; and intensive agriculture. Other activities associated with agriculture can also be significant sources. For example, vegetation removal may lead to soil erosion and hence airborne dust. Agricultural burning, bushfires, fuel-reduction burns and aerial spraying can all emit a range of substances, including particles and VOCs. In general, the impact of these activities is poorly known.

Local-scale pressures

Outdoor sources

People's perceptions of air quality are strongly influenced by local emissions, even though these may make only a relatively minor contribution.

Local outdoor pressures can include odours and smoke. The emissions are usually intermittent rather than continuous, with the areas affected often determined by wind direction and atmospheric stability. Important sources of local pressures include traffic, intensive agriculture such as chicken and pig farms, wood stoves, backyard incinerators, spray-painting and even cooking.

Indoor sources

Many pollutants have been investigated in Australian homes and office buildings, but not always in great detail. Few have been studied enough to determine either the existing exposure levels for Australian populations or the most appropriate strategies to reduce exposures. The following factors, alone and interacting with each other, exert pressures on indoor air quality:

- sources of chemical and biological contamination, such as building materials, furnishings or unflued heaters
- combinations of particular levels of moisture and temperature
- building design, ventilation and maintenance
- inflow of outdoor air
- building occupants.

The major pollutants are thought to be tobacco smoke, house dust mites and nitrogen dioxide.

Respirable suspended particles, certain microbes and VOCs may occur at high concentrations but have not been investigated thoroughly. Added to this lack of objective data is the problem of subjectivity — some aspects can be measured but, in the end, results often depend on the perception or symptoms of the individual, and people differ in their sensitivities.

State

Natural variability

A significant feature of the Australian climate is its large year-to-year variability.

Australia's climatic variability comes about partly because the continent lies near the 'centre of action' of the so-called Southern Oscillation. The oscillation, which occurs about once every two to seven years, is a large disturbance in the atmospheric circulation. Every few years the surface waters in the central and eastern Pacific undergo a remarkable warming, known as an El Niño, which leads to substantial changes in the atmospheric circulation throughout the entire Asia–Pacific region. The generic term El Niño–Southern Oscillation (ENSO) is often used to refer to a suite of events that occurs at the time of an El Niño (see Chapter 2).

During ENSO, the area of strongest ascent of air in the Walker Cell moves eastward from Australian longitudes to the central Pacific. It is replaced by a stronger than usual descent of air and accompanying drought conditions over Papua New Guinea and eastern Australia. At the other extreme of the ENSO cycle, the ascent of air over the western Pacific is enhanced and the Australian region is more than usually subject to tropical cyclones and other flood-producing weather systems.

The most widely used indicator of the state of the Southern Oscillation and the strength of the Walker Circulation is the Southern Oscillation Index (SOI). The SOI is derived from the difference in surface air pressure between Tahiti and Darwin. A positive SOI occurs when pressures are higher than normal at Tahiti and lower than normal in Darwin, with above-average rain over eastern Australia. A negative SOI occurs at the other extreme of the ENSO cycle, with drought conditions over much of eastern Australia.

The ENSO phenomenon can be detected in the continuous records of nearly all climatic variables, but has its greatest impact on rainfall and air temperature. The most pronounced variability in the Australian region occurs over the eastern two-thirds of the continent, where the ENSO accounts for 30 to 40 per cent of the variance (a measure of variability) of rainfall. The ENSO is also detectable to some extent in the variability of air quality measurements between years, particularly the measurements of tropospheric ozone.

As well as the ENSO, variations in the sea-surface temperature in the Indian Ocean contribute to climatic variability by their effect on the passage of north-west rain-bearing cloud bands. In some seasons, these cloud bands bring increased winter rainfall to southern and western parts of the continent.

Figure 5.10 Major drought areas during ENSO years since 1972

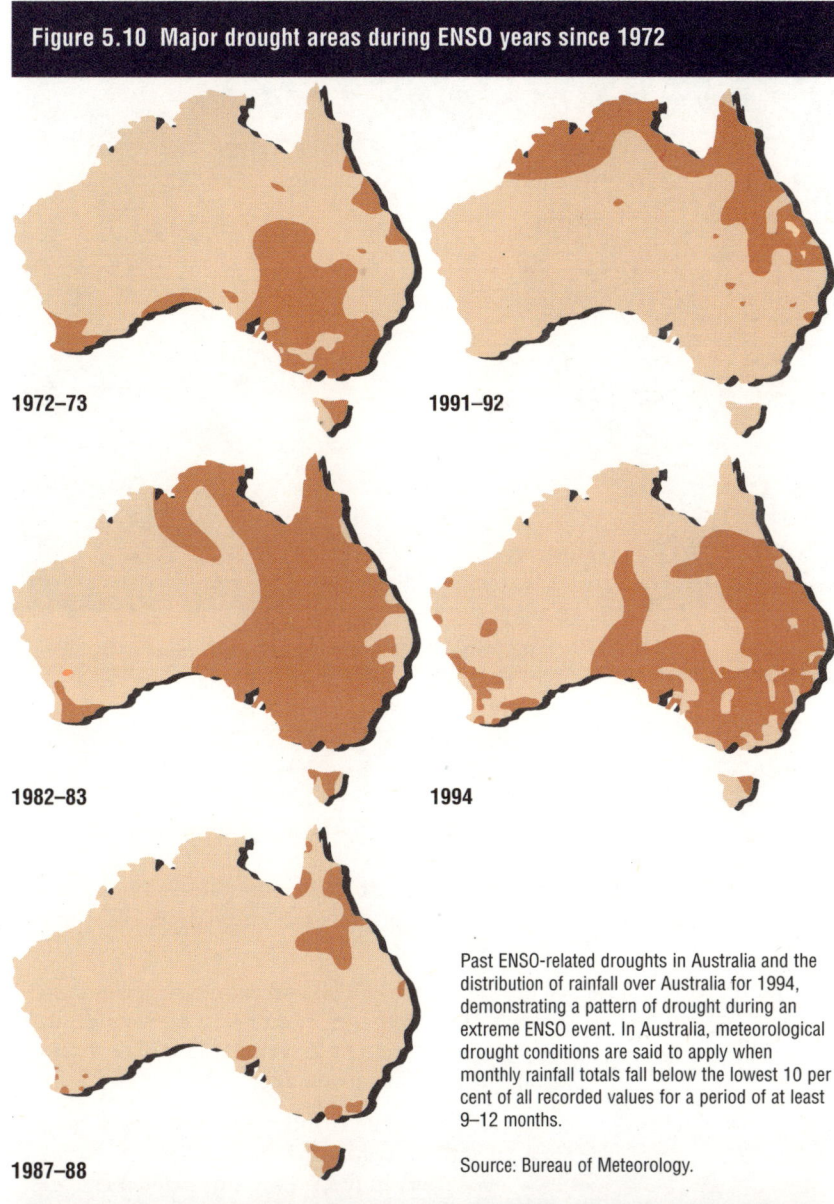

1972–73

1991–92

1982–83

1994

1987–88

Past ENSO-related droughts in Australia and the distribution of rainfall over Australia for 1994, demonstrating a pattern of drought during an extreme ENSO event. In Australia, meteorological drought conditions are said to apply when monthly rainfall totals fall below the lowest 10 per cent of all recorded values for a period of at least 9–12 months.

Source: Bureau of Meteorology.

Figure 5.11 Fluctuations in the SOI since 1972

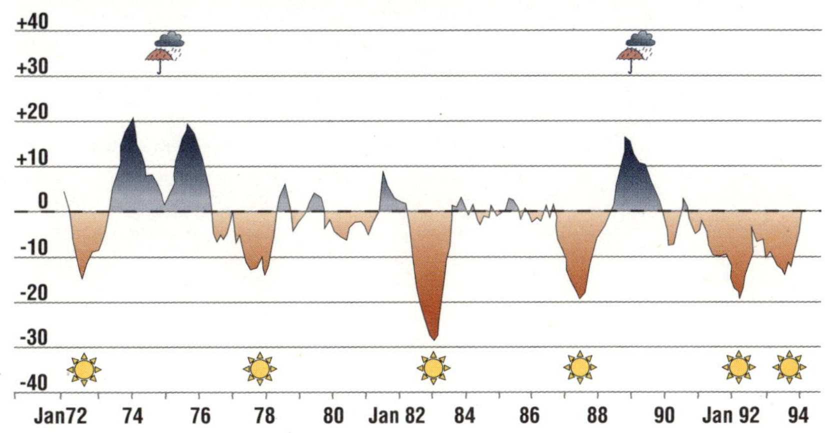

+40
+30
+20
+10
0
-10
-20
-30
-40

Jan72   74   76   78   80   Jan82   84   86   88   90   Jan92   94

Source: Bureau of Meteorology.

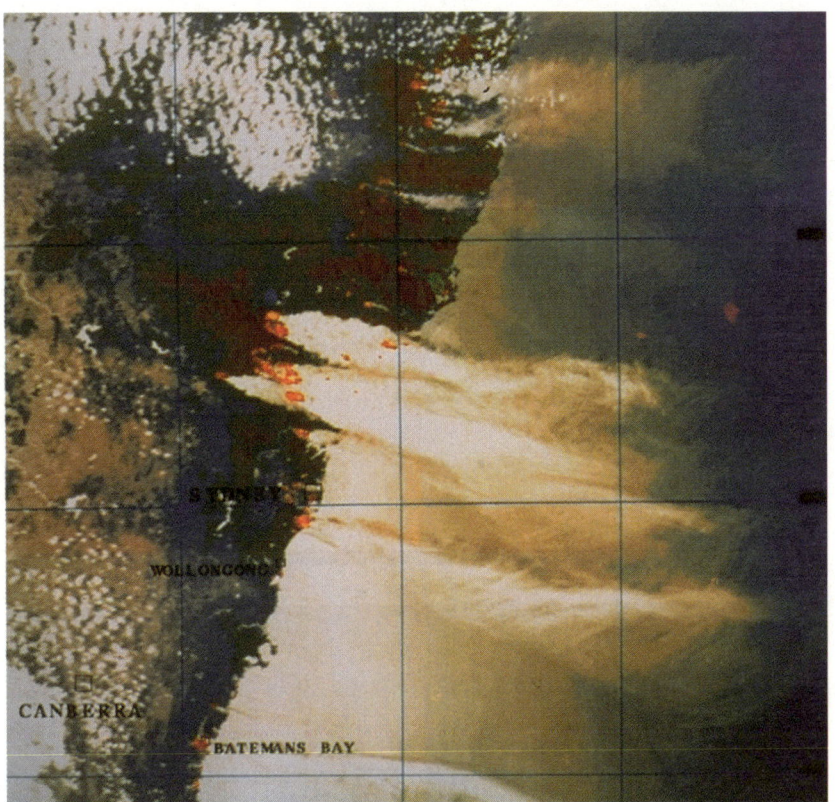

▲

Long range transport of smoke from the 1994 bushfires in NSW during a prolonged El Niño event.

during 1951–1993, with the largest decrease — of about 0.4°C per decade — occurring in the north-east interior. The trend towards night-time warming is consistent with land temperature trends worldwide. Records show an increase in cloudiness over Australia, which is also consistent with a similar trend occurring in North America, Europe and India. After detailed analysis of long-term climate records, Salinger *et al.*, (in press) concluded 'there is a longer-term climate warming trend which is not inconsistent with the enhanced greenhouse effect, and is not directly related to ENSO'.

Atmospheric dispersion and transport

Atmospheric stability is the major meteorological factor controlling the vertical mixing of air pollutants. When conditions are 'unstable', gases and smoke in the atmosphere are quickly mixed (especially in the vertical plane), and so pollutants soon disperse. Such conditions of strong vertical mixing are typical of warm to hot sunny days.

By contrast, vertical mixing is reduced under stable conditions. In some circumstances, the temperature of a layer of air can actually increase with height, rather than decrease as would happen normally. This situation, known as a temperature inversion, generally occurs overnight under conditions of light winds and clear skies when the earth's surface cools rapidly. As very little vertical mixing of polluted air takes place in an inversion layer, pollution accumulates under or within the stable layer of air.

The prevailing wind speed and direction also influence the dispersion of emissions. The greater the wind speed becomes, the greater the volume occupied by the emissions and the greater their dilution. Increased wind speeds also promote increased turbulence and greater vertical mixing. The wind direction determines the shape of a plume and the path it follows. (The term 'plume' covers emissions from a variety of sources, whether backyard incinerators, vehicle exhausts, elevated

The cycle of wet periods followed rapidly by dry ones, increases the risk of fires caused by natural conditions (such as dry thunderstorms) and by human activities. Vast areas of the interior are subjected to bushfires during extended dry periods. The major ENSO-related drought of 1982–83 was marked by severe bushfires in south-eastern Australia; similarly, in January 1994 fires devastated areas of New South Wales. Smoke haze and aerosols from fires can affect large areas downwind.

Highly variable rainfall and extremes in wind can also lead to severe soil erosion and blowing dust. When heavy drought-breaking rains fall on bare soils, further soil erosion may occur.

As well as climate variability linked to the ENSO, longer-term fluctuations can occur. Some studies (Salinger *et al.*, in press) have investigated changes in circulation patterns, particularly in the mid to high latitudes of the southern hemisphere, over the past 100–120 years. Results show that patterns during the periods 1870–1900 and 1950–1990 resembled one another more closely than they did between 1900 and 1950. In particular, the strength of the high pressure systems in the subtropical ridge has increased since the 1950s.

In the period 1951–93, mean temperatures across Australia showed a consistent warming in the range of 0.1 to 0.2°C per decade and above 0.2°C per decade for a broad zone across the country (see Fig. 5.12). By contrast, a cooling trend was observed earlier in the century (1910–51). The observed warming over the past few decades is mainly due to an increase in night-time air temperatures. Thus the diurnal temperature range (the difference between the daytime maximum and the overnight minimum temperatures) decreased over Australia

Figure 5.12 Trends in annual mean temperature (°C) over Australia for the period 1951 to 1992

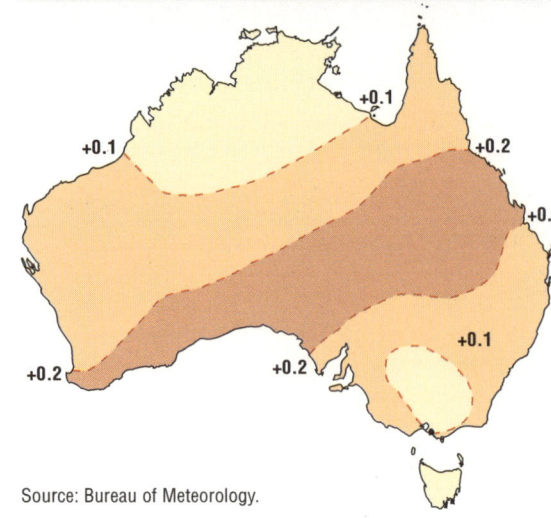

Source: Bureau of Meteorology.

single sources such as industrial chimneys or an entire city.)

In urban areas, assuming no rain falls, the concentration of the urban plume will be determined by the rate of emission of pollutants (from all sources) and the volume of air into which they are emitted. This volume in turn is determined by the average wind speed and the stability of the atmosphere.

Airsheds

In addition to the weather patterns briefly discussed earlier, Australia's long coastline and the large temperature difference between land and sea can greatly influence the patterns of local air circulations and thus air quality.

One of the features experienced to some degree along the entire coastal fringe is a phenomenon known as the sea-breeze/land-breeze regime. During day-time, and under conditions of light or favourable winds, a temperature contrast builds up between the land and the cooler ocean, resulting in a regular onshore sea-breeze. These sea-breeze circulations usually extend about 500–1000 metres vertically.

During the night, stable conditions, light winds and clear skies aid the rapid cooling of the land surfaces. Air in contact with the ground cools faster than the air in the free atmosphere, so cooler (and therefore denser) air near the surface flows down any slope towards lower ground, where it may combine with other cold air flows. Because circulations of this type are typically confined to the lowest 300–500 metres, even relatively low-level terrain can produce such drainage flows of air

(see Fig. 5.13). The ways in which these flows interact is what determines the extent of an airshed. In Australia, the regional airsheds around each major city are discrete and widely separated.

Sometimes, areas along the coastal fringe experience a daily reversal in the direction of air flow (see Fig. 5.14). This usually happens under conditions of a stable atmosphere and light winds. Studies show that, under these circumstances, air carried inland by the afternoon sea-breeze can recirculate in overnight drainage flows. This cycle can continue for some time — up to three days in summer. Pollutants released into the airsheds are trapped within the circulation and concentrations may build up in the lower levels. Some pollutants, mainly those emitted by motor vehicles, can undergo chemical reactions in the presence of sunlight to form photochemical smog (see the box on page 5-25).

By world standards, Australia receives a lot of sunshine. Most of the country experiences, on average, about six hours a day, with some areas receiving more than 10 hours. As a result, the potential for photochemical smog formation is high in all our cities in the summer, and remains so in the winter in northern cities such as Brisbane.

Recirculation of air within urban coastal airsheds is one of the most important events linked to the formation of photochemical smog and ozone episodes monitored over urban areas (Manins *et al.*, 1994). At certain times, major population centres, including the capital cities of Sydney, Brisbane, Perth and Melbourne, are subjected to recirculation and with it, photochemical smog formation (see Fig. 5. 31).

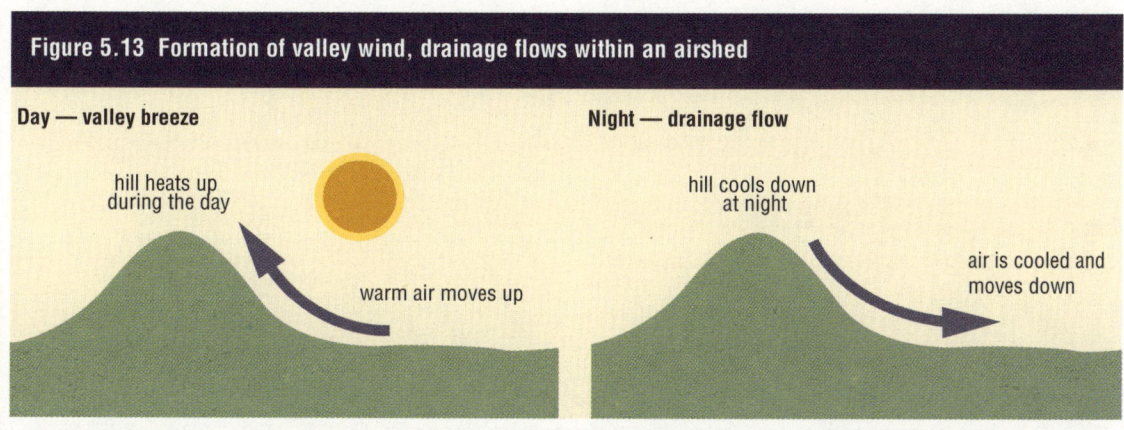

Figure 5.13 Formation of valley wind, drainage flows within an airshed

Day — valley breeze

hill heats up during the day

warm air moves up

Night — drainage flow

hill cools down at night

air is cooled and moves down

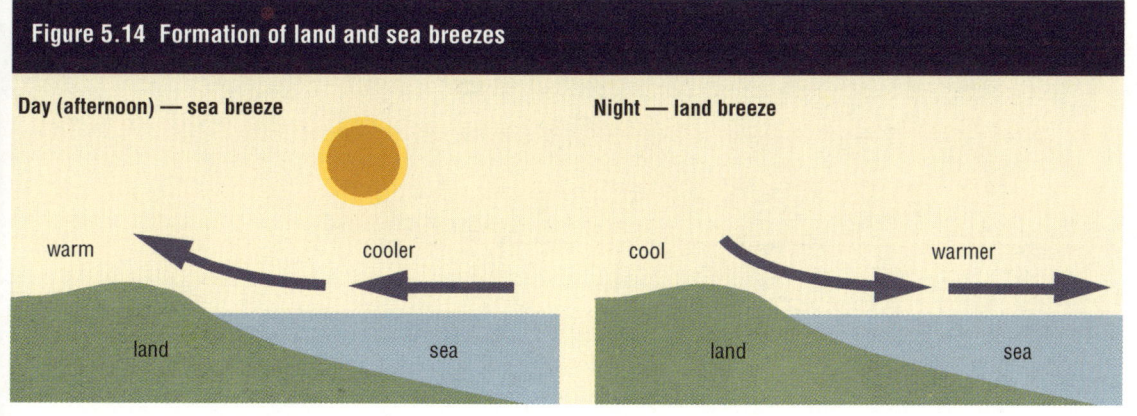

Figure 5.14 Formation of land and sea breezes

Day (afternoon) — sea breeze

warm   cooler

land   sea

Night — land breeze

cool   warmer

land   sea

The enhanced greenhouse effect

Nitrogen and oxygen — the main components of the earth's atmosphere — are almost completely transparent to the sun's rays. The clouds, the oceans, land, snow and ice reflect about one-third of the incoming solar (short-wave) radiation. The earth absorbs the remaining two-thirds of the solar energy, mainly in the tropics, from where large-scale circulations in the oceans and the atmosphere redistribute it. Ultimately, it is re-radiated back to space as infra-red (long-wave) radiation, thus maintaining a balance with the absorbed solar radiation.

Water vapour, carbon dioxide and other trace gases absorb infra-red radiation emitted by the earth's surface and, as a result, have a major impact on this radiation balance. The absorbed radiation is not retained but re-emitted in all directions, thus increasing the temperature of the earth's surface. This warming effect, long recognised as a major element of the climate system, is known as the greenhouse effect. Without clouds, water vapour and these other so-called greenhouse gases (but with no change in the amount of solar radiation reflected back to space) the global surface temperature would average −18°C rather than the present 15–16°C.

We now know that human activity has led, and is still leading, to increased atmospheric concentrations of existing greenhouse gases (carbon dioxide, methane, nitrous oxide and ozone), as well as to the presence of new greenhouse gases such as CFCs. Most of these gases, once released into the atmosphere, persist for tens to hundreds of years, with an associated long-term impact on the background atmospheric levels. Using an understanding of the processes that govern the climate system, and applying this knowledge in computer climate models, scientists consider that the presence of additional greenhouse gases will affect the radiation balance of the atmosphere and lead to a warming at the earth's surface. This is now generally referred to as the enhanced greenhouse effect.

The actual impact on global climate is likely to be complex and involve changes in atmospheric and oceanic circulations, accompanied by possible changes in sea level, diurnal temperatures, rainfall and other climatic variables. While climate model simulations already offer predictions of the global impact of the enhanced greenhouse effect, only broad indications of potential change on a regional scale are currently available.

A further complicating factor has emerged in recent years. This is the realisation that, in addition to raising greenhouse gas levels, human activity is also leading to an increase of aerosols in the lower atmosphere. The most significant are sulfate aerosols that come from sulfur dioxide emissions from power generation and ore processing (see page 5-29). Carbon-based aerosols produced by burning biomass are also important. Aerosols can reflect sunlight as well as change the amount, type and radiative behaviour of clouds, resulting in a lowering of surface temperatures. Due to their short lifetime (days/weeks), their cooling effects are temporary and regional, but for some regions, particularly in the northern hemisphere, the cooling is estimated to be about the same as the warming effects of CO_2.

Since the 1990 Intergovernmental Panel on Climate Change (IPCC) First Assessment Report, considerable progress has been made to distinguish between natural and anthropogenic influences on climate. In its Second Assessment Report (IPCC, 1995) the IPCC concludes that despite uncertainties in key factors, 'the balance of evidence suggests that there is a discernible human influence on global climate'.

IPCC (1995) projects an increase in global mean surface temperature relative to 1990 of about 2°C by 2100 and a corresponding increase in sea level of about 50 cm.

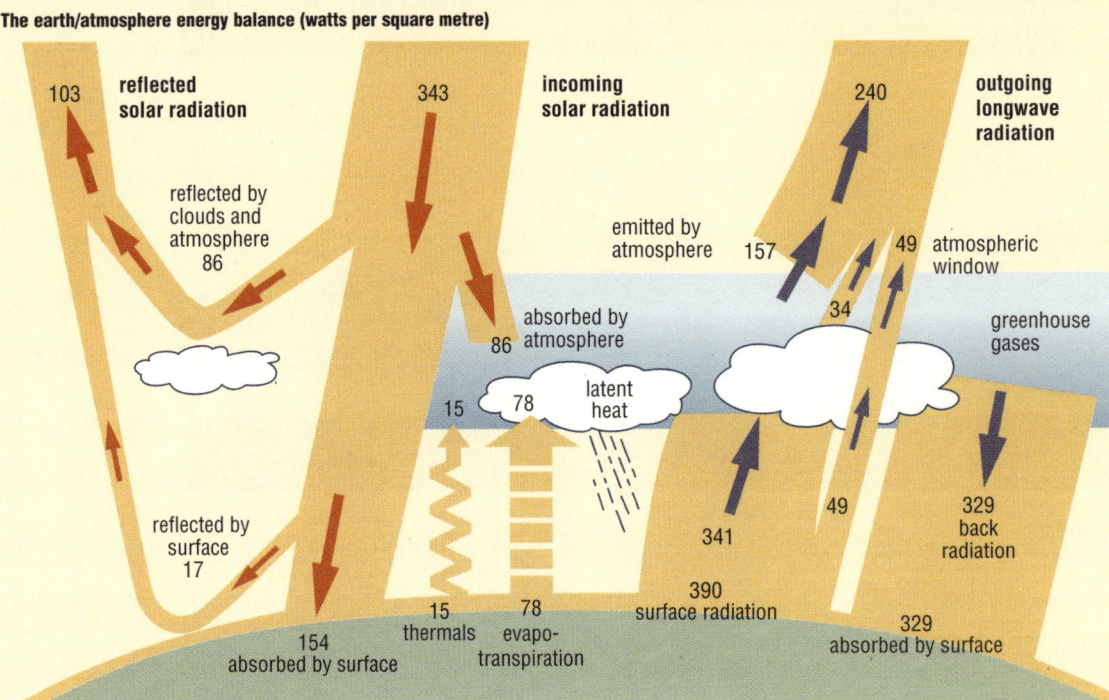

The earth/atmosphere energy balance (watts per square metre)

About half the sun's incoming radiation reaches the earth's surface where some is reflected and absorbed. The atmosphere is not nearly as transparent to the long-wave radiation from the earth. Only a small amount of this radiation is lost directly to space. The remainder is absorbed by the atmposphere and the clouds with enough re-emitted upwards and out to space to maintain the total radiation balance.
Source: IPCC, 1994.

Enhanced greenhouse effect

The major scientific conclusions of the Intergovernmental Panel on Climate Change (IPCC) are that emissions resulting from human activities are substantially increasing the atmospheric concentrations of the greenhouse gases carbon dioxide, methane, tropospheric ozone, nitrous oxide and CFCs. Since the time of the Industrial Revolution (about 200 years ago), the atmospheric concentration of carbon dioxide has increased by more than 30 per cent, that of methane by more than 145 per cent and that of nitrous oxide by about 15 per cent (IPCC, 1995). Climate models indicate that the sensitivity of global surface temperature to a doubling of carbon dioxide is likely to be an increase in the range of 1.5 to 4.5°C. The Panel identified many uncertainties, particularly with regard to the timing, magnitude and regional patterns of climate change. (See the discussion of the enhanced greenhouse opposite). Nevertheless, the IPCC Second Assessment Report (1995) concluded that 'the balance of evidence suggests that there is a discernible human influence on global climate'.

A global network of monitoring stations provides information on the current concentrations of greenhouse gases in the atmosphere. Australia participates in this network through Tasmania's Cape Grim (see Fig. 5.15). Analysis of Antarctic ice cores dating back to the 1300s, collected and analysed by the Australian Antarctic Division and CSIRO, clearly demonstrates the dramatic increase in the atmospheric concentrations of CO_2 since the early 1800s (see Fig. 5.16).

The Cape Grim data, representing the state over much of the southern hemisphere, help us piece together a better global picture, as well as helping us understand the differences between the two hemispheres. The records have made an important contribution towards understanding the science of the atmosphere. It is clear from the measurements that, unless large reductions in emissions occur, the concentrations of most greenhouse gases will continue to rise well into the next century. Indeed, the IPCC (1994, 1995) has reported that, according to a range of models, the atmospheric concentration of CO_2 will more than double unless global emissions are reduced substantially below 1990 levels.

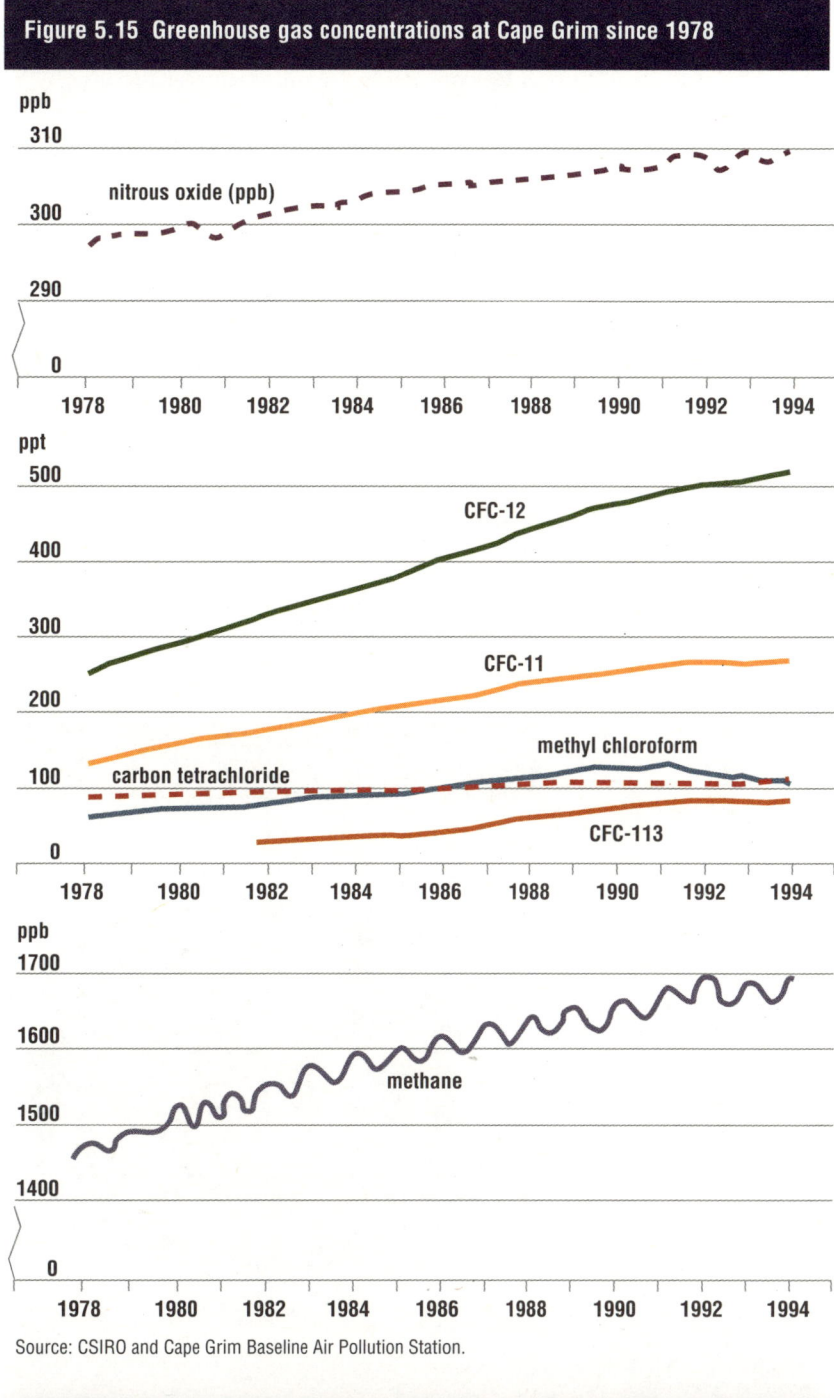

Figure 5.15 Greenhouse gas concentrations at Cape Grim since 1978

Source: CSIRO and Cape Grim Baseline Air Pollution Station.

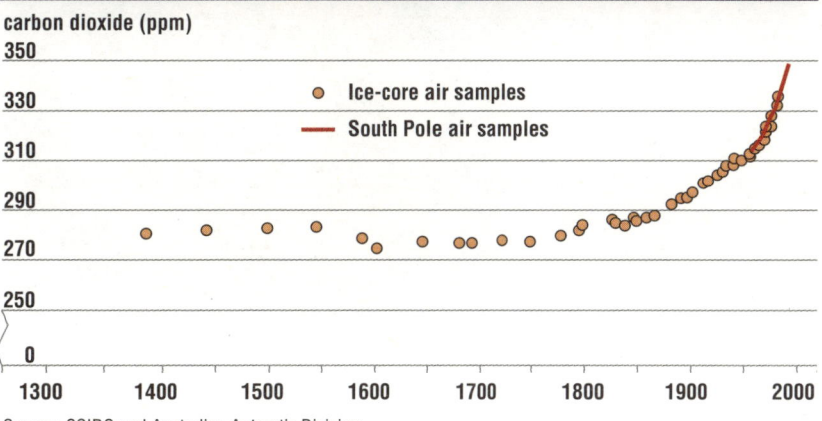

Figure 5.16 Atmospheric carbon dioxide concentrations since 1300 determined from Antarctic ice cores

Source: CSIRO and Australian Antarctic Division.

Stratospheric ozone loss

The ozone layer in the stratosphere protects life on the earth's surface from damaging quantities of UV radiation (see the box on page 5-11). Since the late 1980s, seasonal ozone losses above Antarctica have been severe, with more than 60 per cent of the total ozone being destroyed in spring over a region covering most of the continent as shown by satellite measurements (see Fig. 5.17).

The concentration of ozone in a column of air stretching up from the earth's surface is reported in Dobson Units (DU).

During the middle of an average year in the 1990s, ozone values over Antarctica are about 300–320 Dobson units (DU) (compared with 330 to 350 DU in the early '80s). During spring 1994, total ozone fell to a low of 90 DU in the last week of September, a value similar to the 1993 ozone minimum. Typically, the area of the ozone 'hole' in the 1990s is from 20 to 24 million sq km.

The long term decline in springtime ozone levels over the polar region is also evident from complementary ground-based measurements at Halley Bay in Antarctica (see Fig. 5.18).

Ozone depletion is not just a polar phenomenon. Losses in stratospheric ozone of between two and four per cent per decade have been detected in mid latitudes, including over Australia (WMO, 1995).

The Bureau of Meteorology carries out stratospheric ozone monitoring at five sites in the Australian region. Their results show that ozone depletion now occurs over the most populated parts of Australia all year round (see Fig. 5.19). The occasional passage of ozone-depleted air moving across the south of the country in late spring, following the break-up of the Antarctic ozone hole, may exacerbate the situation.

Stratospheric ozone loss results in an increase in ground-level UV. A few locations worldwide (particularly Antarctica) have recorded an increase in UV-B, the most damaging waveband. The extent of the UV increase correlates well with the measured stratospheric ozone depletion at the relevant latitude (Basher *et al.*, 1994). Accurate

Figure 5.17 Stratospheric ozone levels over the Southern Hemisphere during October for selected years

This series of satellite images shows the progressive loss of stratospheric ozone during the Antarctic spring. The Antarctic ozone hole is defined as the area bounded by the 220DU level.

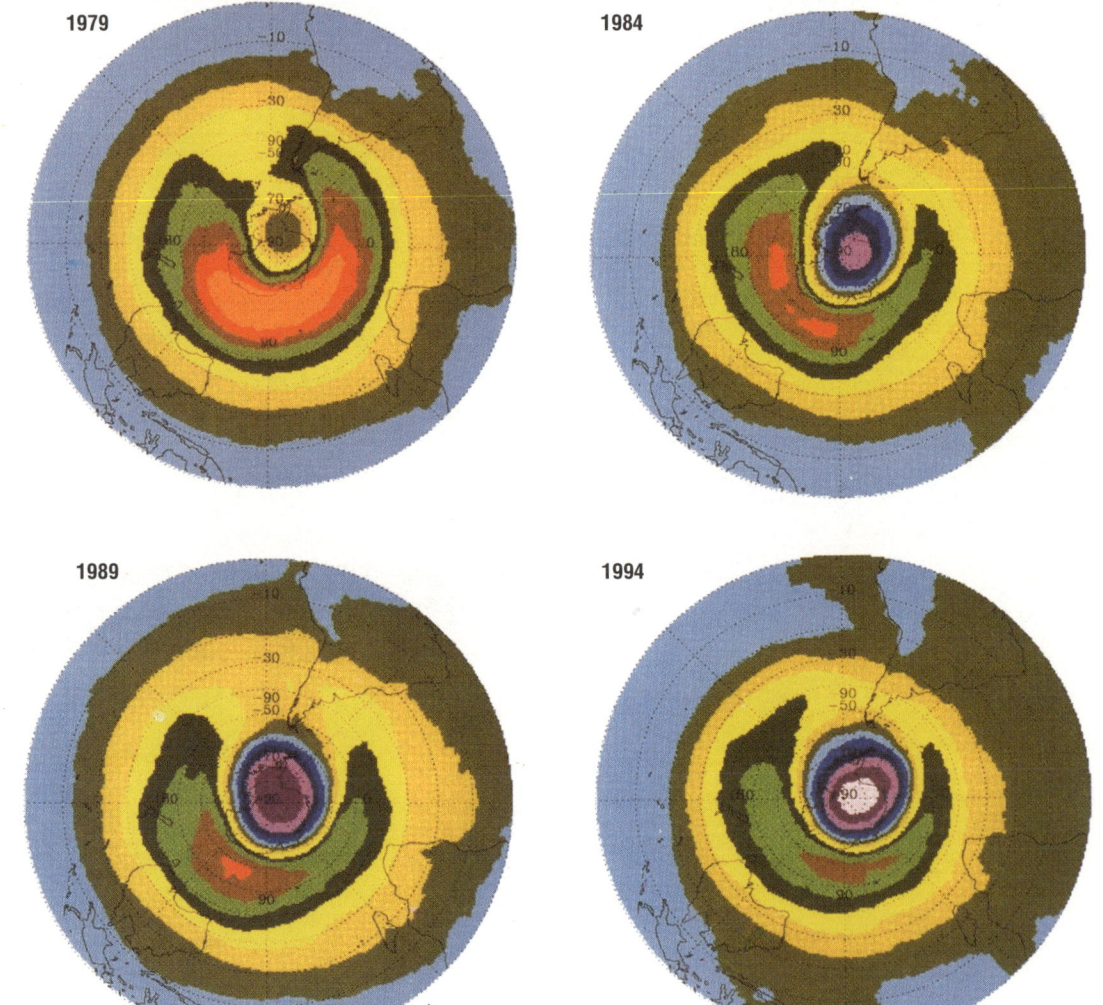

Source: satellite imagery by courtesy of NASA.

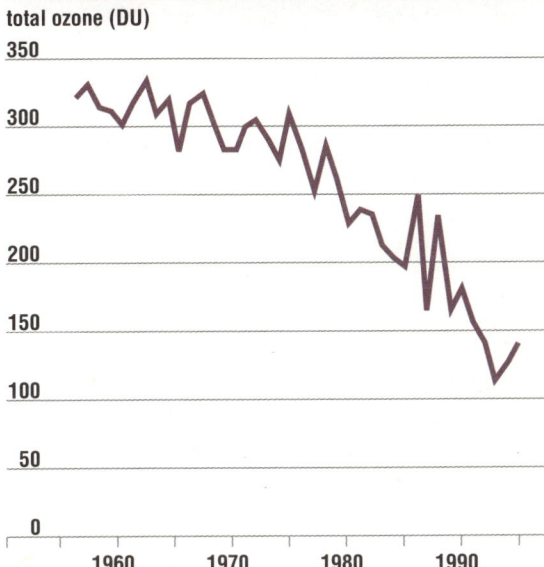

Figure 5.18 Total ozone during October over Halley Bay, Antarctica

Source: British Antarctic Survey.

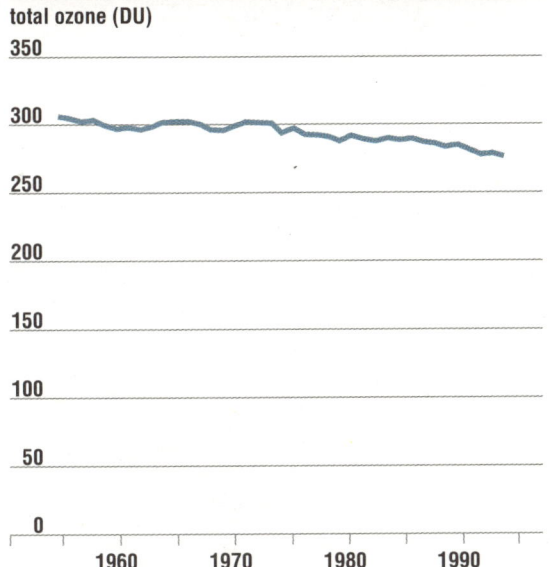

Figure 5.19 Total ozone depletion during January in the air column over Melbourne

Source: Bureau of Meteorology.

monitoring of UV-B radiation is technically difficult. No country has carried out reliable long-term UV monitoring, but Australia and other countries have recently started taking regular measurements. The greatest UV increases in Australia (in percentage terms) are estimated to have occurred in the most southerly (that is, higher) latitudes (Fraser and Bouma, 1990). However, higher-latitude regions, being further from the equator, naturally receive the least amount of total solar radiation (see Fig. 5.20). Although their exposure to UV has increased proportionally the most, southern regions still receive much less UV radiation in total than places nearer the equator.

There is no precise correlation between increased ground-level UV irradiation and human health problems, but there is a clear link between exposure and a range of disorders — for example, the incidence of skin cancer is higher in Queensland than in Tasmania (Marks, 1989). However, recent changes in human behaviour to avoid exposure, brought about by greater public awareness of the issue, make predictions difficult. Information about UV levels is now presented during television weather reports, although this provides broad guidance only. Potential reductions in the yields of some crops, and effects on oceanic plankton, need further research.

Cape Grim Baseline Air Pollution Station, north-west Tasmania, is part of a global network for monitoring background atmosphere.

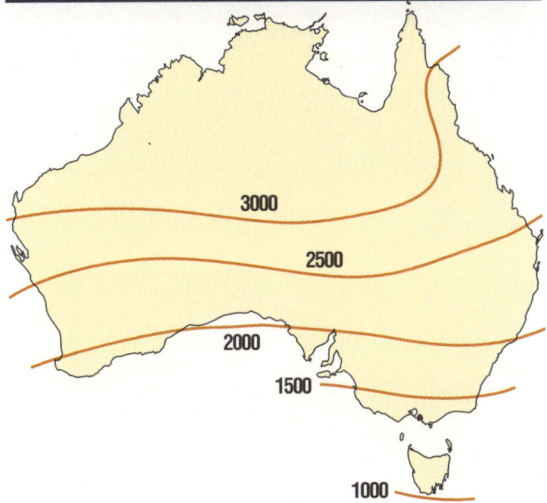

Figure 5.20 Sunburning radiation levels

Note: The calculated average daily dose of the 'sunburning' component of sunlight is measured in erythemal dosage units. Calculations for the above figure include the effects of cloud cover.

Source: Paltridge and Barton, 1978.

Figure 5.21 Australian urban and regional airsheds with ambient air-quality monitoring

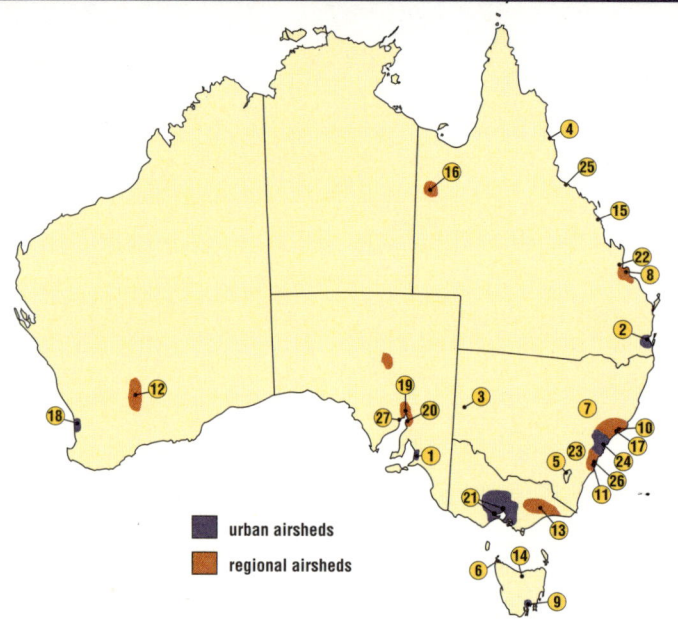

■ urban airsheds
■ regional airsheds

Note: The numbers refer to the table below

National ambient air quality monitoring

While our perception of air 'quality' is often based on how far we can see, the Australian guidelines have largely been based on considerations of what is optimal for human health.

The primary responsibility for environmental management, and thus for the monitoring of air quality, rests with individual States and Territories. Most jurisdictions have implemented some form of routine monitoring program with the goal of assessing the impacts of ambient air quality on human health. Since the late '70s, Sydney, Melbourne and Brisbane (representing a combined population of some eight million people) have had reliable air quality data. Some other jurisdictions have long records for certain selected parameters. Routine monitoring by State and Territory agencies is limited to:

- the major metropolitan areas surrounding most capital cities (except Hobart and Darwin) — areas referred to as 'urban airsheds'

- selected regional areas such as the Latrobe and Hunter Valleys, which contain major thermal power-generating stations linked to coal deposits,

Table 5.6 Number of sites at which air quality indicators are routinely monitored and for which quality assured data are publicly available as at June 1995

Location	O_3	CO	NO_2	SO_2	Visibility	TSP	Pb	Dust	PM10	PM2.5	F	PAH	VOCs	Airtrak
1. Adelaide	2	1	2	1	2	9	9	-	-	-	-	-	-	-
2. Brisbane	8	1	9	3	4	-	5	11	-	-	-	-	-	2
3. Broken Hill	-	-	-	-	-	3	3	27	-	-	-	-	-	-
4. Cairns	-	-	-	-	-	1	-	-	-	-	-	-	-	-
5. Canberra	2	2	2	-	2	5	5	-	-	-	-	-	-	-
6. Cape Grim	1	1	-	1	-	-	1	-	1	1	-	-	-	-
7. Central Tablelands	-	-	-	-	-	2	-	10	-	-	-	-	-	-
8. Gladstone	-	-	3	2	2	-	-	-	2	-	-	-	-	-
9. Hobart	-	-	-	-	-	✓	✓	-	-	-	-	-	-	-
10. Hunter Valley	-	-	-	4	-	20	-	194	-	-	14	-	-	-
11. Illawarra	-	-	-	8	-	9	2	60	-	-	-	16	-	-
12. Kalgoorlie	-	-	-	11	-	-	-	-	-	-	-	-	-	-
13. Latrobe Valley	2	-	2	2	2	-	-	-	-	-	-	-	-	-
14. Launceston	-	-	-	-	-	✓	-	-	-	-	-	✓	-	-
15. Mackay	-	-	-	-	1	-	-	-	-	-	-	-	-	-
16. Mt Isa	-	-	-	1	-	-	-	-	-	-	-	-	-	-
17. Newcastle	2	1	11	10	-	9	7	15	2	-	12	18	-	-
18. Perth (incl. Kwinana)	9	3	11	6	6	3	3	-	4	6	-	-	2	2
19. Port Augusta	-	-	-	✓	-	✓	-	-	-	-	-	-	-	-
20. Port Pirie	-	-	-	✓	-	✓	✓	✓	-	-	-	-	-	-
21. Port Phillip Region	11	5	9	7	10	5	5	-	1	-	-	-	3	-
22. Rockhampton	-	-	-	-	-	2	-	-	-	-	-	-	-	-
23. Southern Tablelands	-	-	-	-	-	-	-	11	-	-	-	-	-	-
24. Sydney	13	8	11	4	8	4	4	10	6	-	-	-	-	1
25. Townsville	-	-	-	-	-	-	-	6	2	-	-	-	-	-
26. Wollongong	2	4	2	2	2	4	5	17	1	-	-	3	-	-
27. Whyalla	-	-	-	-	-	✓	✓	-	-	-	-	-	-	-

Source: Based on Ormerod, in press.

✓ = monitored but number of monitors not specified.

and a few other major industrial centres, referred to as 'regional airsheds'

- some isolated areas around individual large sources of emissions referred to as 'hot spots'

Consequently, we cannot adequately assess ambient air quality over much of Australia because of a lack of data.

Monitored areas

Monitoring covers only about five per cent of the country by area (see Fig. 5.21 and Table 5.6). Although the reporting standards applied to air quality data differ slightly between State and Territory agencies, they are generally subject to well-defined quality control and assurance procedures. The design of the monitoring network and the parameters monitored are a compromise between scientific needs and available resources. With increased knowledge, particularly about the formation of photochemical smog, some monitoring programs, especially in the greater Sydney airshed (Hyde and Johnson, 1990) have been re-evaluated and refocused. The New South Wales government provided substantial resources for the Metropolitan Air Quality Study which has been completed recently. In Victoria, work has started on optimising the design of networks (EPAV, 1994; Ahmet and van Dijk, 1994).

As well as monitoring by government agencies, certain industrial locations undertake extensive self-monitoring. However, the quality of the data in these cases is often unknown and the data themselves are often not in the public domain.

Traditionally, the focus of concern has been on some or all of the general indicators of air quality (SO_2, CO, NOx, ozone, particulates, lead). Recently, however, it has become apparent that some of the more common toxic air pollutants, such as benzene, are present in sufficient concentrations to warrant closer monitoring. So far, investigations have been limited to several discrete surveys rather than long-term programs. Hence the indicators to be routinely monitored should be re-evaluated.

The jurisdictions and organisations that undertake air quality monitoring archive the data themselves. Consequently, no national air quality data set exists.

For the assessment of national ambient (outdoor) air quality, the National Health and Medical Research Council (NH&MRC) and the Australian and New Zealand Environment and Conservation Council (ANZECC) have jointly recommended guidelines that could be used as a basis for comparison across the nation (see Table 5.7). However, some differences occur between agencies in the monitoring, reporting and adoption of these guidelines. As well, some State government bodies have traditionally granted exemptions to specific licensed emitters.

As national ambient air quality guidelines are not uniformly applied, it is difficult to make comparisons between jurisdictions and any such comparisons should only be considered as 'indicative'.

Table 5.7 Ambient Air Quality Guidelines[a] recommended by NH&MRC/ANZECC as at June 1995

Pollutant	Averaging Time	Concentration
Ozone	1 hour	0.12 ppm[b] (under review)
Nitrogen dioxide	1 hour	0.16 ppm[c]
Sulfur dioxide	10 minutes	0.5 ppm
	1 hour	0.25 ppm[b]
	1 year	0.02 ppm
Carbon monoxide	8 hours	9 ppm[b]
Total suspended particulate matter	1 year	90 µg/m^3
Lead	3 months running mean	1.5 µg/m^3
Fluoride (General Land Use)	12 hours	3.7 µg/m^3
	1 day	2.9 µg/m^3
	7 days	1.7 µg/m^3
	30 days	0.84 µg/m^3
	90 days	0.5 µg/m^3
(Special Land Use)	12 hours	1.8 µg/m^3
	1 day	1.5 µg/m^3
	7 days	0.8 µg/m^3
	30 days	0.4 µg/m^3
	90 days	0.25 µg/m^3
Sulfates	1 year	15 µg/m^{3}[d]

Note:
(a) With the exception of fluoride (where the guideline was established to protect vegetation), all these guidelines are based on protecting human health
(b) not to be exceeded more than once per year
(c) not to be exceeded more than once per month
(d) read in conjunction with TSP

Areas not monitored

For the 95 per cent of Australia not covered by routine monitoring, it is believed that the following issues are of major concern:

- sulfur dioxide from industrial point sources, such as coal-fired power stations and major ore-processing plants

- heavy metals (including lead) from ore processing

- particulates from forestry and agricultural activities such as controlled burning, bushfires and the erosion and transport of fine topsoil by strong winds

- pesticides from aerial spraying

- emissions caused by heavy traffic along some roads in rural areas

Although our knowledge about the actual levels of these pollutants and their dispersion is incomplete, it is likely that in the more remote areas of Australia, where there is little human activity, air quality is good (that is, within the guidelines) most of the time. Few studies have been undertaken on the impact that emissions, photochemical smog and ozone may have on vegetation in urban areas.

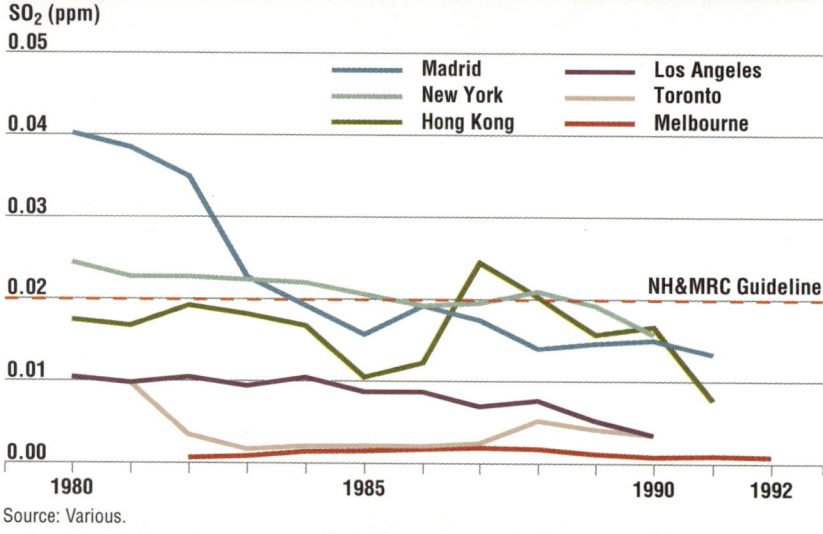

Figure 5.22 Annual average concentrations of sulfur dioxide for selected international cities

SO$_2$ (ppm)

Legend: Madrid, New York, Hong Kong, Los Angeles, Toronto, Melbourne

NH&MRC Guideline

Source: Various.

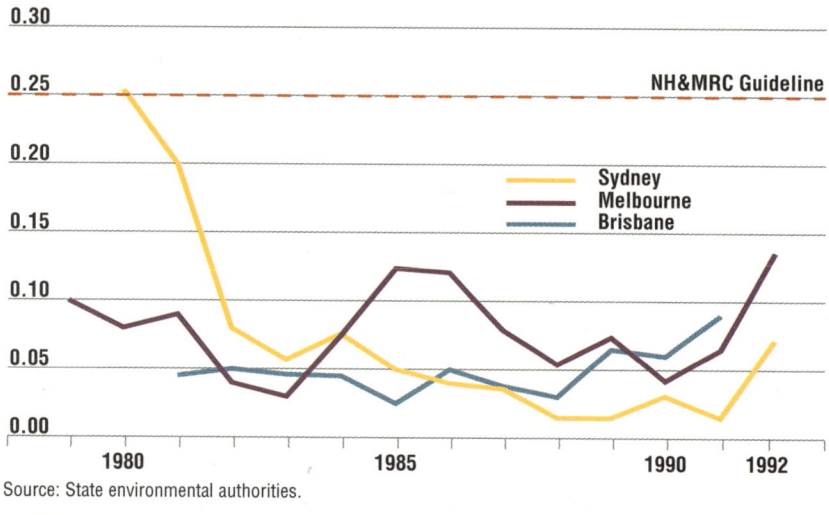

Figure 5.23 Highest one-hour concentrations of sulfur dioxide for selected Australian cities

SO$_2$ (ppm)

NH&MRC Guideline

Legend: Sydney, Melbourne, Brisbane

Source: State environmental authorities.

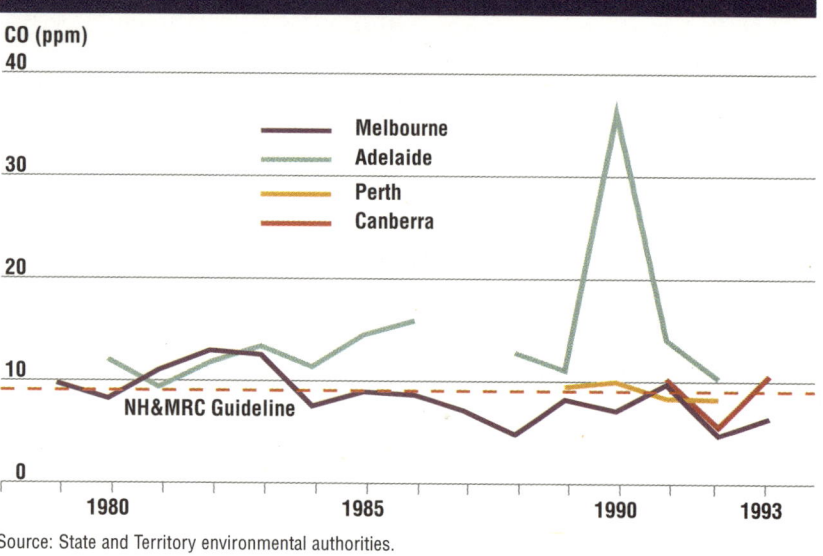

Figure 5.24 Highest eight-hour carbon monoxide concentrations for selected Australian cities

CO (ppm)

Legend: Melbourne, Adelaide, Perth, Canberra

NH&MRC Guideline

Source: State and Territory environmental authorities.

Urban air quality

Motor vehicles are the main source of emissions in urban areas, although industrial sources also contribute. As most power stations are located near coalfields and outside urban areas, their emissions do not generally affect urban air quality.

Sulfur dioxide

By and large, Australian cities do not have a sulfur dioxide (SO$_2$) problem, primarily because fuels used are low in sulfur and also because power stations are not located in urban areas. Some industrial sources of SO$_2$ do exist, such as oil refineries on the outskirts of urban areas, and these can pose local problems.

Annual SO$_2$ averages in Melbourne and Sydney are substantially below the 0.02 ppm guideline. Melbourne's average for the period 1980–84 was 0.002 ppm, which compares well with, for example, Tokyo's average of 0.01 ppm in 1988–89 or New York's of 0.015 ppm in 1986–87 (see Fig. 5.22). Highest annual one-hour SO$_2$ concentrations measured in most Australian capital cities are also well below the NH&MRC guideline (see Fig. 5.23).

Carbon monoxide

A product of incomplete combustion, carbon monoxide (CO), is found in the exhaust emissions of all motor vehicles. As an air quality issue, concern is mainly confined to inner city regions with high traffic density. High winter-time concentrations of CO in some suburban areas can be related to the use of wood fires and combustion stoves. However, recorded values are strongly dependent on the precise siting of the measuring instruments. It is therefore difficult to carry out true comparisons between cities because instruments in some are located very close to high-density traffic flows. In the past ten years, Sydney, Perth, Adelaide and Canberra have all exceeded the NH&MRC eight-hour CO guideline (nine ppm) in locations close to high traffic flows (see Fig. 5.24).

Lead

The major source of airborne lead in Australian urban areas is still from leaded fuel used in motor vehicles. However, the emissions from this source have declined considerably in all cities over the past decade because of the introduction of unleaded fuel for cars fitted with catalytic converters and a reduction in the lead content of leaded fuel (see Fig. 5.35).

Odour

Many of the gases and aerosols responsible for unpleasant odours are volatile organic and sulfur-containing compounds. The extreme sensitivity of the human nose to some of these chemicals means that some people can smell them even at very low concentrations. The fact that certain odours are mixtures, makes it hard to quantify and regulate them.

Industries such as petroleum refining, food processing and tallow rendering can all produce

Figure 5.25 Number of days in Sydney when eight-hour carbon monoxide exceeded NH&MRC guideline

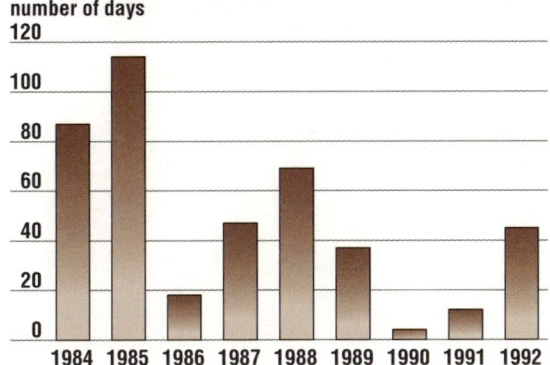

Source: EPA NSW.

Figure 5.26 Annual average concentrations of total suspended particles (TSP) for selected Australian cities

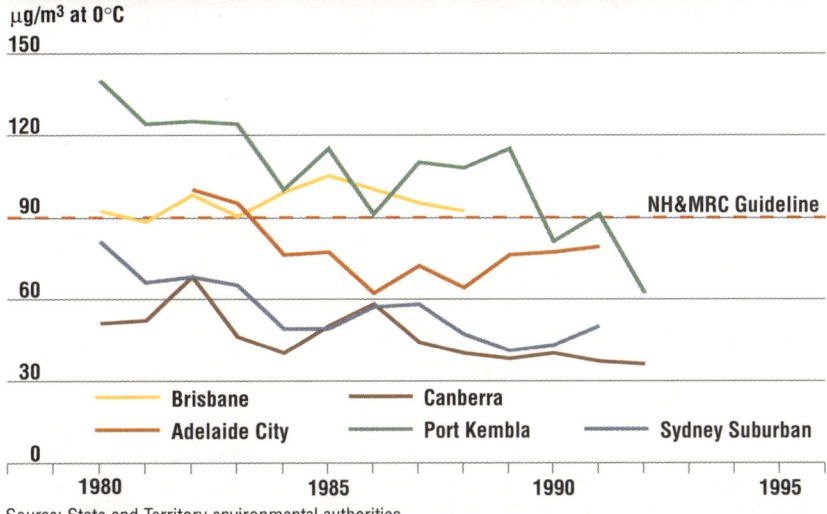

Source: State and Territory environmental authorities.

Figure 5.27 Annual average TSP concentrations in selected international

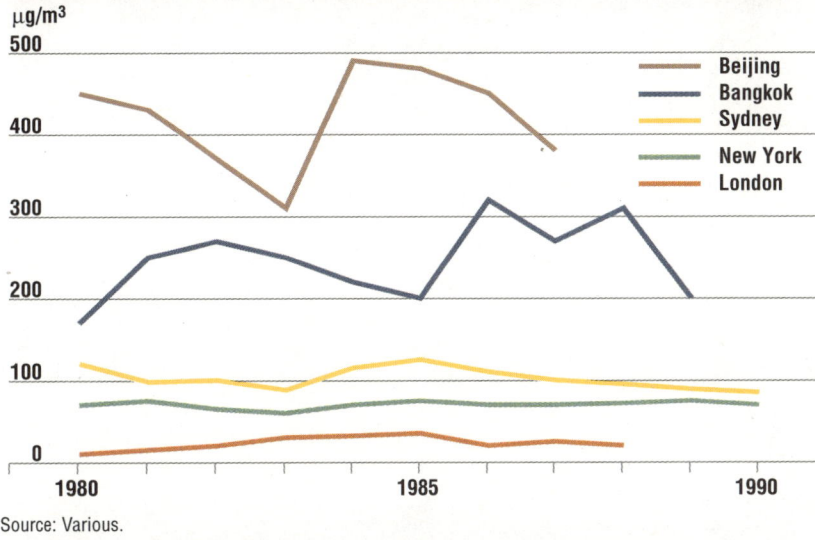

Source: Various.

Figure 5.28 Number of days in Sydney and Melbourne with visibility less than 20 km

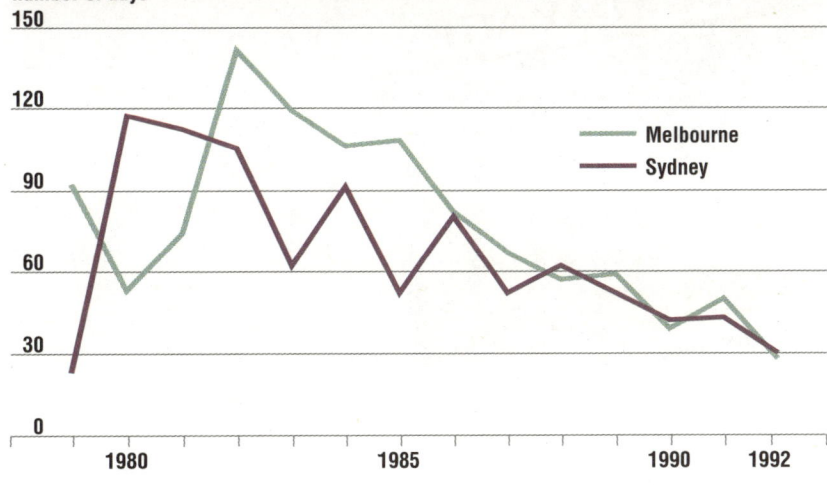

Source: EPA NSW and EPA Victoria.

odours. Other sources are sewage and waste-water treatments, various metal processing and manufacturing industries, and some types of intensive agriculture.

'Bad smells' are an important issue with the public. Available data suggest that from 30 to 60 per cent of all air quality complaints received by State environmental agencies relate to odours.

Air toxics

Air toxics are defined as pollutants present at very low concentrations which are known to cause or are suspected of causing long-term health effects in humans. In the United States, 189 of some 2000 synthetic chemicals emitted to the atmosphere have been identified as air toxics and are regulated by the EPA. Many air toxics are either volatile organic compounds or metallic compounds that could affect health following long-term exposure at very low concentrations.

As yet, Australia has no list of air toxic emissions, but a national pollutant inventory now being developed will include them. The EPA in Victoria has recently carried out studies of air toxic emissions from motor vehicles in Melbourne and from petrochemical industries in the suburb of Altona. Other measurements have shown that Sydney city air contains common air toxics at concentrations similar to those observed in major US cities (Nelson and Duffy, 1994).

Particles

Particles of various sizes are suspended in the air and can reduce its clarity. In Australian cities these particles include: sea salt; sulfate from sea salt and from SO_2 emissions; carbon from combustion processes; silica from soil; and pollen. Lead and other contaminants may also be involved.

Particles are monitored and reported in size-related categories. Total suspended particles (TSP) include all particles from the smallest up to 50μm in diameter: within this range are sub-categories of those less than 10μm in diameter, known as PM10, and those smaller than 2.5μm known as PM2.5.

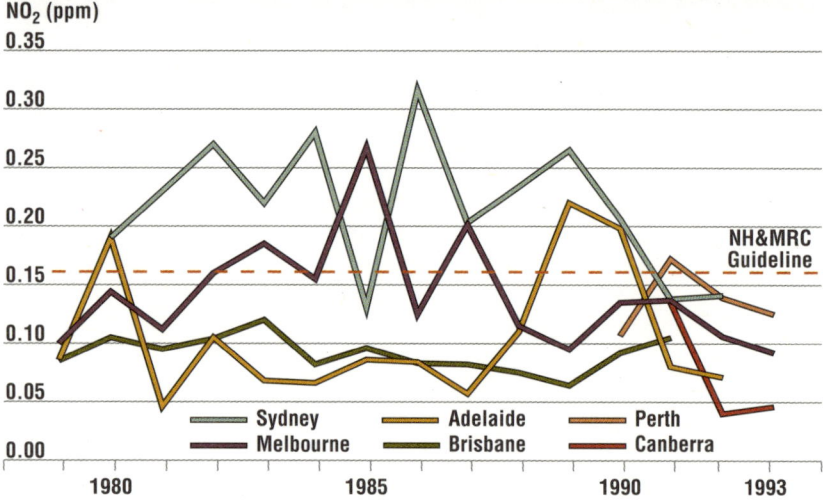

Figure 5.29 Highest one-hour nitrogen dioxide concentrations for selected Australian cities

Source: State and Territory environmental authorities.

their presence is indicated by a reduction in visibility. Existing data suggest that Sydney and Melbourne have experienced the most severe occurrences of urban haze, but improvements have occurred since the early 1980s due partly to tighter controls on industrial and vehicle emissions and to the banning of backyard burning. Monitoring shows that visibility in major cities has improved over the past decade (see Fig. 5.28).

Bushfires and controlled burns

Smoke from bushfires or fuel-reduction burns in the hinterland may drift into major urban airsheds, and under certain weather conditions can reduce visibility for several days. Some of the major air pollution episodes recorded in capital city airsheds have occurred when smoke from controlled burning was trapped during periods of air recirculation.

Photochemical smog and its precursors

When oxides of nitrogen (NOx) and reactive volatile organic compounds (VOCs) are present in the air in sufficient quantities, in the presence of sunlight, chemical reactions give rise to a type of smog. Because these reactions are stimulated by light, it is known as photochemical smog. Both NOx and reactive VOCs are necessary precursors for the production of smog. Ozone is a major product of this photochemical oxidation process (see the box on page 5-25).

The term 'oxides of nitrogen' (NOx) includes nitric oxide (NO) and nitrogen dioxide (NO_2). Motor vehicles are the main sources of NOx in urban airsheds. Nitrogen dioxide — the one of most concern to human health — is also a component of 'brown haze', a visible feature of severe urban air pollution. Current ambient levels in Australian urban airsheds are acceptable, but in the past in some cities have exceeded the NH&MRC one-hour guideline of 0.16 ppm (see Fig. 5.29).

About half the reactive VOCs emitted in Australian urban airsheds come from motor vehicles. Other sources include industrial and commercial plants. As mentioned earlier, vegetation also naturally emits VOCs, especially during hot, dry weather. Native vegetation is probably responsible for most of this. On a typical summer day in the Melbourne airshed, natural emissions of just one common VOC (isoprene) are about the equivalent of 10 per cent of the total anthropogenic VOC emissions (Carnovale *et al.*, 1991).

Ozone is a major constituent of photochemical smog and is the indicator for its presence. As a pollutant, ozone is not emitted directly to the atmosphere, but rather results from natural photochemical processes involving the precursor pollutants NOx and reactive VOCs in the presence of sunlight.

Reported ozone values are affected by the configuration of each monitoring network in our major urban airsheds. The networks are now being reassessed to ensure that they are portraying a true picture of the extent of photochemical smog.

Associated problems include:

- health effects: PM10 and PM2.5 are of most concern to human health because particles up to about 10μm in size can travel deep into the lungs and become lodged there; pollen (which varies in size) can provoke an allergic response in some people

- visibility reduction from particles in the PM2.5 range

- soiling by particles settling on surfaces

For about the last decade, urban airsheds in Australia have recorded falling levels of total suspended particulates. While levels in Sydney are not as high as those of some cities in less-developed countries, they are similar to those of other major cities in developed countries (see Figs 5.26 and 5.27).

The trend in recent years has moved away from monitoring TSP and towards monitoring PM10. The different types of measurement used, and the fact that not all capital cities measure the full range of categories, make it difficult to compare cities.

Under certain weather conditions smoke from bushfires can reduce visibility for several days.

The NH&MRC does not have a guideline for PM10. Available data indicate that average annual levels have been generally below 40 μg per cubic metre (as recommended for Victoria by Streeton, 1990), but that levels at some sites on some days in Sydney, Brisbane, Wollongong and Newcastle would have exceeded the Victorian recommended 24-hour level of 120 μg per cubic metre. However, it is impossible to discern trends, as the data have been collected for too short a time.

Only limited data are available on the concentration of PM2.5. However,

The existing monitoring networks have reported occasional breaches of the NH&MRC ozone guideline over the past decade (see Figs 5.30 and 5.32). The NH&MRC was reviewing this guideline in 1995 and had before it a recommendation for a one-hour average guideline of 0.08 ppm. Standards close to 0.08 ppm have already been adopted in some parts of the world such as California and Japan.

The level of smog experienced in and around major cities depends on the balance between the speed of ozone production and the rate at which the polluted air is diluted by clean surrounding air. The rate of smog production is therefore an important consideration when developing control strategies.

An Australian device, the CSIRO-designed Airtrak, measures the rate of smog formation. The Airtrak samples the air and determines the rate at which the reactive VOCs present can generate smog. By giving a continuous read-out on the parameters controlling air quality, this instrument enables authorities to determine the preferred pollution-control strategies. These data are important because the mechanism controlling ozone episodes (that is, whether it is light-limited or NOx-limited) is often not clear. The diagrams in Figure 5.31 illustrate that, at the same time as the chemical reactions are taking place within the air parcels, the wind is transporting the parcels considerable distances from the source of the original emissions.

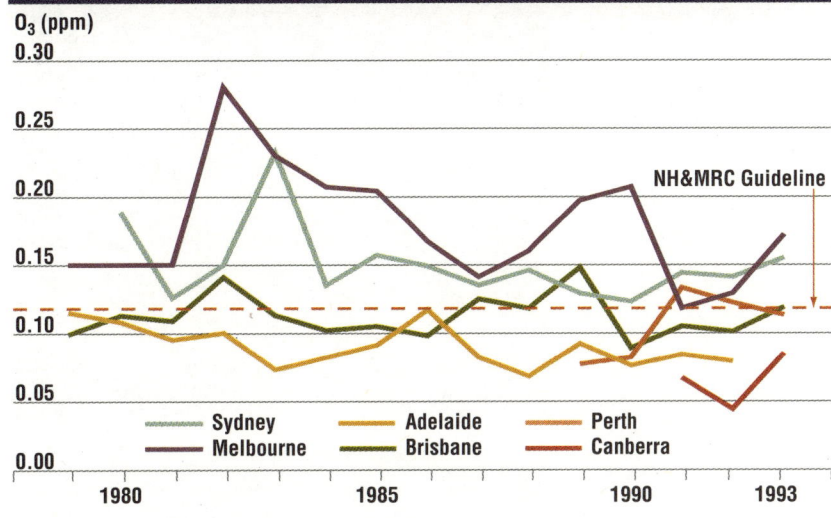

Figure 5.30 Highest one-hour ozone concentrations recorded at individual monitoring sites for selected Australian cities

Source: State and Territory environmental authorities.

Photochemical smog

Photochemical smog is essentially the result of oxidation of nitrogen oxides in the atmosphere. Ozone is the major constituent of photochemical smog.

As air comprises about 20 per cent oxygen, it is not surprising that oxidation is the major chemical reaction occurring in the atmosphere. However, at the temperature range of the atmosphere, oxygen does not react directly with most emissions. In fact, the driving force for most atmospheric chemistry is energy provided by sunlight. (The breaking of chemical bonds by light is known as photolysis.)

The essential ingredients for the formation of photochemical smog are nitrogen oxides, reactive VOCs and sunlight.

Two factors determine the speed at which smog is formed: the concentration of reactive VOCs in the air and the intensity of sunlight available to drive their breakdown.

The process of ozone formation continues as long as both sunlight and nitrogen oxides are available. Nitrogen dioxide also undergoes other oxidising reactions that convert it to nitric acid and organic nitrates, thus effectively removing nitric oxide from the reaction system and ending the production of smog.

Thus, smog formation can be seen as a two-stage process. In the first stage, temperature and radiation are the controlling factors. In the second, the amount of NOx emissions available to take part in the process determines the peak ozone concentration at some distance downwind (Johnson *et al.*, 1990).

These two stages are referred to as 'light limited' and 'NOx limited' respectively.

Production of photochemical smog

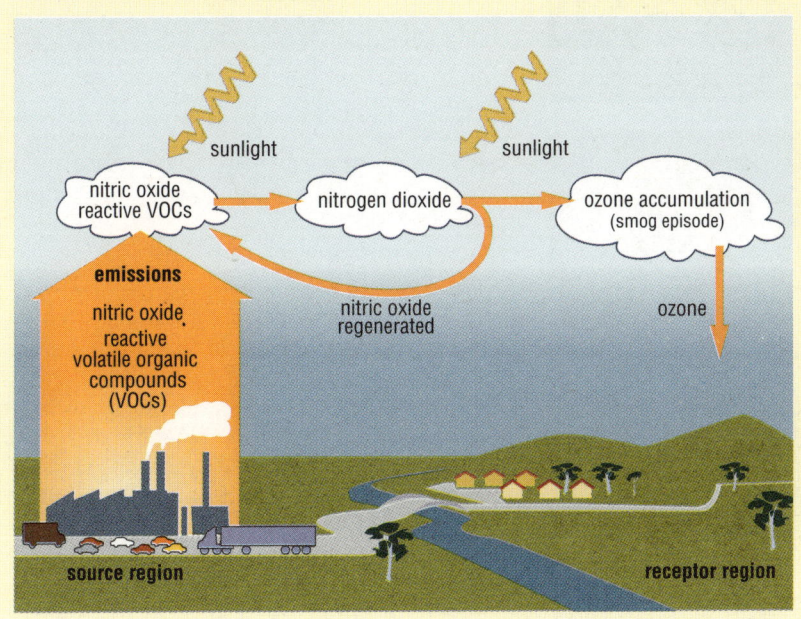

Figure 5.31 Recirculation in Australian airsheds

An early morning satellite photograph. The Melbourne Eddy (see text) is marked by the cloud over the northern part of Port Phillip Bay ▼

Early morning winds over Melbourne on a typical 'smog alert' day. The calculated air trajectory shows how polluted air is trapped and recirculated as the day progresses. The calculations closely match available observations.

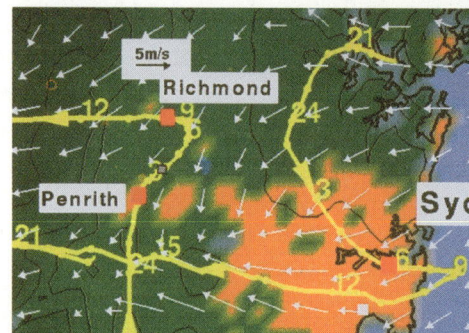

Surface winds at 3 pm: air trajectories over Sydney on high pollution-index days. One trajectory shows how the sea-breeze sends polluted air towards the western suburbs by the late afternoon.

Winds at 2.30 pm, and an air trajectory showing how on days of easterly winds, polluted air is returned from off-shore to the Perth area by the afternoon sea breeze. Overnight the same air is again returned from inland when the land breeze is re-established.

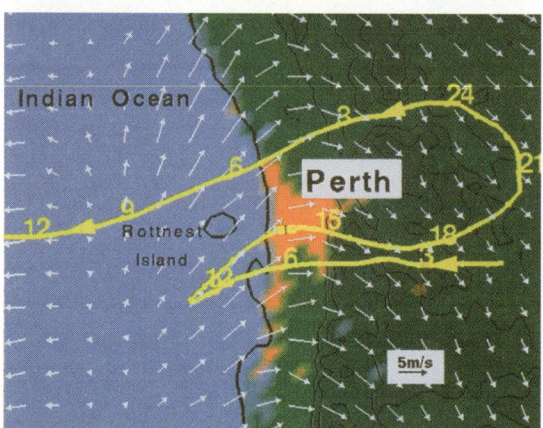

Midday winds on a typical winter's day in Brisbane. The air trajectory highlights how polluted air is trapped and remains over the metropolitan area, allowing emissions to accumulate and photochemical smog to develop.

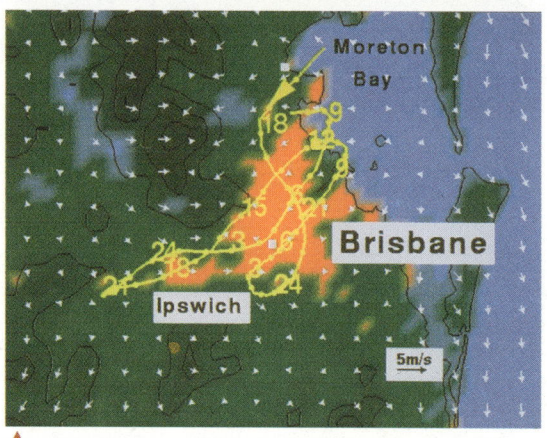

▲
CSIRO computer model simulations of local winds at 100m (white arrows) and associated air trajectories (yellow lines), with time of day marked.

Recirculation in urban airsheds

The recirculation of pollutants over Australia's major capital cities is important for determining pollution levels. Photochemical smog formation under such conditions poses the biggest single threat to clean air in urban Australia.

An important concept when considering the overall air quality of an airshed is its 'assimilative capacity'. This means the degree to which it can 'deal with' emitted pollutants through natural dilution, dispersion and chemical transformation. The build-up of pollutants over the major cities during recirculation episodes often exceeds the assimilative capacity. A proper assessment of air quality requires a knowledge of both the pollution sources and the assimilative capacity of the airshed in which the city is located.

In Australia, only in the last 20 years has attention been focused on the possible role of geographical and meteorological factors in the formation of photochemical smog and its impact on populated areas. Before this, pollutants were usually monitored over central urban areas. However, we now know that where emissions of nitrogen oxides (NOx) from car exhausts are high we expect ozone levels to be relatively low. With the exception of Melbourne, monitoring has rarely been done at locations several hours downwind of the source region, where ozone formation takes place and levels are likely to be higher, although major new studies in Sydney, Brisbane and Perth are addressing this problem.

A new approach to air quality management was needed to mitigate ozone pollution in capital city airsheds. The development of forecasting systems that model the complex air flows within the airsheds, coupled with the development of new photochemical analysis techniques for air sampling (such as the CSIRO Airtrak monitoring stations) have been key factors in increasing our understanding of the major urban airsheds and in the development of new air quality management policies and strategies. The results of recirculation models for major urban airsheds are discussed briefly below (see also Fig. 5.31).

The 'Melbourne Eddy' is a deep, slow, clockwise wind circulation responsible for keeping polluted air over the Melbourne area on some summer days. On days of weak easterly winds and with a low-level inversion, the eddy develops in the lee of the Victorian Alps (to the east of Melbourne), trapping urban emissions. Computer models simulating the Melbourne Eddy have shown that a similar pattern more frequently occurs due to the daily cycle of land- and sea-breezes.

On typical summer days of high pollution in Sydney, night-time air flows from the mountain slopes move cold air to the north towards Richmond in the lee of the Blue Mountains and to the east towards the centre of the city. This latter flow accumulates pollutants as it travels over densely settled and industrial areas in the western suburbs. The air then flows out to sea during the morning. With the onset of the sea-breeze the same air is frequently returned, travelling westward and reaching the Hawkesbury Basin near Penrith or Campbelltown in the afternoon. Under some circumstances air parcels enter the Sydney basin overnight from the Hunter Valley and are caught up in the sea-breeze/land-breeze circulation, which

may carry the air parcel out to sea and down the coast towards Wollongong the next day.

Perth lies on a coastal plain between the Indian Ocean and the 300-metre-high Darling Scarp rising to the east. On summer days of high smog, surface winds are generally easterly. Pollutants from Perth and the southern industrial areas are carried out to sea in the morning, but return as smog in the early afternoon with the onset of a strong sea-breeze.

Mountain ranges rise to the west of Brisbane and extend well to the north and south, while to the east Moreton and North Stradbroke Islands lie less than 20 km from the shore. At midday on a winter's day in Brisbane in light northerly winds, the local winds are also light and very variable in direction. The simulated trajectory of pollutant-laden air parcels near the surface is very confused. The model predicts that the air parcel crosses and recrosses the Brisbane urban area several times, in both the morning and the evening. By the end of the event the whole region will probably have been affected by increased smog levels in the form of ozone, nitrogen oxides or aerosols as the parcels accumulate pollutants and these are changed photochemically to smog.

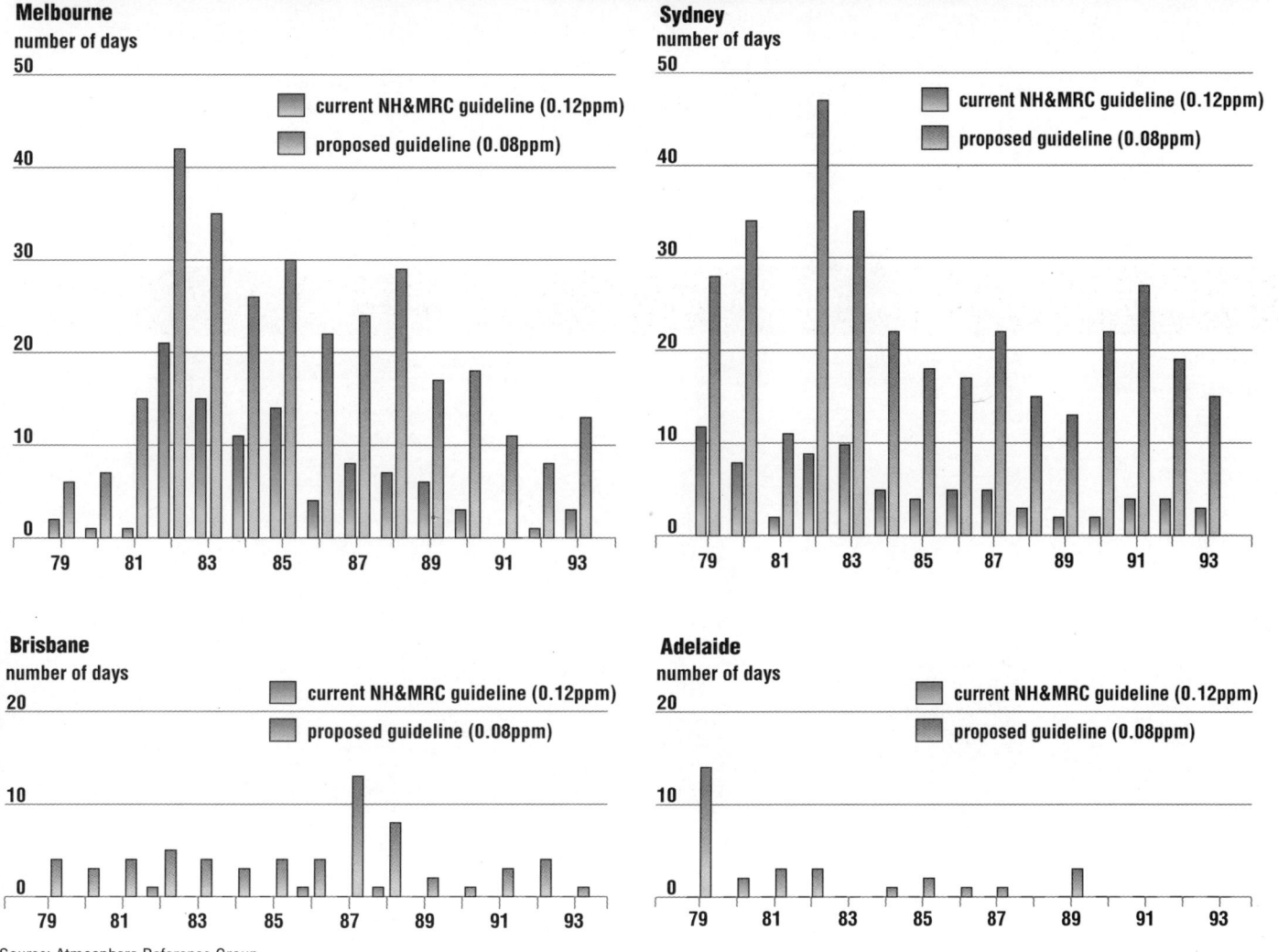

Figure 5.32 Number of days on which peak hourly ozone concentrations exceeded 0.12 ppm and would have exceeded the proposed 0.08 ppm NH&MRC guideline (as at June 1995) in selected capital cities

Source: Atmosphere Reference Group.

Urban air quality trends

In general, the annual mean concentrations of common pollutants within the major Australian urban airsheds are below NH&MRC guidelines, and low by world standards. This is particularly evident for sulfur dioxide. With an increase in the proportion of vehicles using catalytic converters and unleaded fuel, levels of carbon monoxide and lead are likely to decline further. In the short term, a similar trend is expected in smog levels, but this may be counteracted by increasing vehicle usage (brought about both by a larger population and by features of urban design), and the deterioration of existing fitted catalytic converters. Computer models indicate that, based on current trends, the relative contribution by vehicles to the concentration of volatile organic compounds (VOCs) will decline and commercial and industrial sources will instead dominate VOC emissions in established urban airsheds. However, in areas of rapid urban growth (for example, south-east Queensland, Perth, western Sydney) motor vehicle emissions are likely to remain the major concern.

While the annual averages may be low, air pollution levels in Sydney and Melbourne, particularly for ozone, can occasionally approach those in New York or Tokyo. Such air pollution episodes are often associated with recirculation patterns within the urban coastal airsheds. Similar episodes have been recorded in Perth and Brisbane (see Fig. 5.31).

Nevertheless, it is generally agreed that over the past 10 years some aspects of the air quality in major cities such as Melbourne and Sydney have improved following the introduction of stricter controls on motor vehicles, especially the use of catalytic converters on all new cars since 1986 and the simultaneous introduction of unleaded petrol.

Whether this improvement will continue is uncertain. Future air-quality trends may be influenced by:

- the revision of Australian Design Rule (ADR) 37 by the Advisory Committee on Vehicle Emission and Noise (ACVEN)

- a study of the emissions performance of vehicles by type and age, including the deterioration of catalytic converters being carried out by the Federal Office of Road Safety

- long-term projections of vehicle emissions by the Bureau of Transport and Communication Economics.

- responses to the Sydney Metropolitan Air Quality Study with respect to transport and land use planning.

One of the major goals of any air quality monitoring program is the detection of changes or trends in pollution over time. This is essential for assessing the effectiveness of pollution-control strategies and for developing appropriate air-quality management plans, and requires a continuous record of measurements over many years. Data fluctuations in the levels are the result of many factors, including possible trends in the emissions themselves, monitoring programs and any expansions of networks with time. The inherent variability in Australian weather patterns can mask or distort any analysis of trends in air quality.

Assessments of air quality are often based on analyses of extreme or peak events, such as daily maxima or the number of times concentrations exceed hourly guidelines.

In 1995 the NH&MRC was reviewing the ozone guideline of 0.12 ppm with a view to introducing a more stringent guideline. Figure 5.32, based on the re-evaluation of records in four capital cities,

Table 5.8 Summary of Australian urban and regional air quality

Pollutant	Areas of most significance	Measured levels	Trends	Other comments
Ozone	Primarily Melbourne and Sydney	Occasional breaches of guidelines	Signs of improvement may be the result of meteorological variability	A growing problem in Brisbane and Perth as populations increase rapidly
Nitrogen dioxide	Near heavy traffic	Occasional breaches in large cities	No clear trends	
Sulfur dioxide	Near metal ore processing sites	Substantial breaches of guidelines near some sites	Some improvements due to better controls for specific plant	
Carbon monoxide	Areas with heavy traffic, wood fires	Some breaches	Improvements in most cities	Measured levels sensitive to monitor siting
Total suspended particles (TSP or PM10)	Areas with heavy traffic, mining and industrial areas; biomass burning (including wood fires)	Some breaches	General improvements	TSP not as well related to health effects as PM10 or PM2.5
Lead	Near point sources and major roads	Some substantial breaches	Steady improvement in urban areas	Motor vehicles as lead sources declining in importance
Fluoride	Near aluminium smelters and ceramics works	Breaches, often in buffer zones	General gradual improvements	Effects on vegetation are the concern

shows the impact of lowering the guideline to 0.08 ppm. In this case, excess concentrations would have occurred on more days per year in both Melbourne and Sydney and would have been detected consistently in Brisbane and intermittently in Adelaide. The longest consecutive period for which data are available for these cities is 1979–93. (Comparable data for Perth are not available.)

Air quality data show a wide year-to-year variability, reflecting the influence of meteorological factors. In various airsheds researchers have tried to remove the influence of meteorology from the analyses of trends. But this has proved difficult; the results are not always conclusive and appear very dependent on the time interval chosen.

For example, in a detailed study of photochemical smog in the Brisbane region, researchers concluded that, for a few hours per year, the photochemical smog potential was high because of the air recirculation between land and sea (Lunney *et al.*, 1994a; 1994b). The study found few apparent trends, as the inter-annual variability of the weather tended to obscure them.

In Sydney, an apparent decrease in the number of pollution events reported in the late 1980s to early '90s corresponds to an extended period of fast-moving weather systems (Leighton and Spark, 1995), which may explain this trend.

Air quality in regional airsheds

The term 'regional airshed' is used here to describe those areas defined by the local terrain, where monitored emissions come primarily from industrial activity and major thermal power stations, even though some of these areas may contain urban populations. Although there have been few regional airshed studies in Australia, one particularly notable one was the Latrobe Valley Study (Manins, 1988), which ran for a decade from the late 1970s. This study pioneered the development of integrated airshed modelling in Australia.

Regional airsheds are defined both by the characteristics of the source — either ground-level emissions or elevated plumes from industrial stacks — and by the surrounding topography and local wind flows. At times of stable meteorological conditions periodic reductions in dispersion can occur. The key pollutants in Australian regional airsheds are sulfur dioxide, oxides of nitrogen, lead and other heavy metals and fluoride.

Sulfur dioxide

Sulfur dioxide is an important air pollutant in regional airsheds. Both coal-burning power stations and oil refineries emit sulfur compounds. In most cases the SO_2 is emitted from elevated stacks, which may reduce ground-level concentrations near the source but also have the potential to affect areas some distance downwind. High SO_2 levels may also be associated with copper, lead and nickel

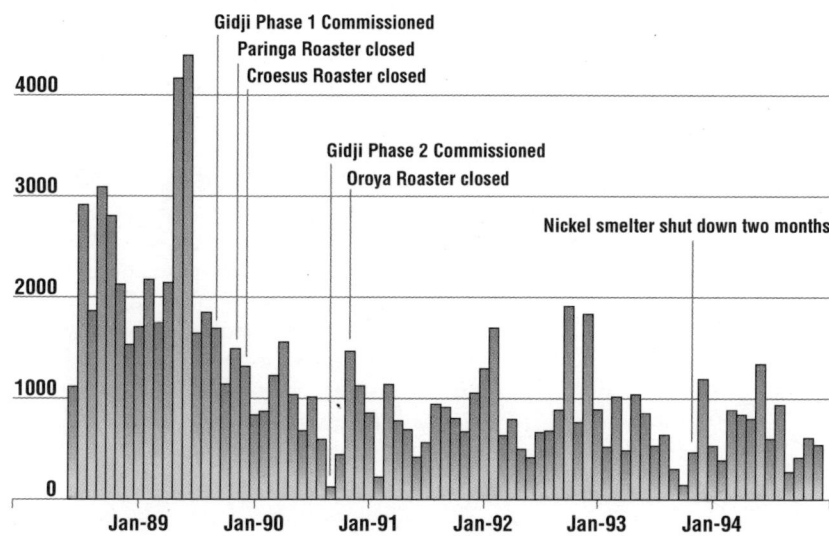

Figure 5.33 Kalgoorlie (hospital) maximum average hourly sulfur dioxide levels for the month

Average SO_2 level ($\mu g/m^3$)

Gidji Phase 1 Commissioned
Paringa Roaster closed
Croesus Roaster closed

Gidji Phase 2 Commissioned
Oroya Roaster closed

Nickel smelter shut down two months

Source: WA Department of Environmental Protection.

smelters and gold-ore roasters, where SO_2 emissions result from the burning of sulfur compounds in the ore feed-stock. In Kalgoorlie, a general reduction in SO_2 levels has occurred following closure of the town roasters and the commissioning of new roasters at Gidji, some 15 km to the north (see Fig. 5.33) although some problems remain. An inventory of major point sources of SO_2 can be found in Steer and Heiskanen (1993).

Nitrogen oxides

The levels of oxides of nitrogen are monitored in some regional airsheds, but other pollutants, such as sulfur oxides and particulates, are usually of greater concern. A comprehensive network of NOx monitors is set up in the Latrobe Valley in Victoria, designed to assess the impact of plumes from the coal-fired power stations, and also around Gladstone, Queensland, to assess the impact of industrial NOx emissions. Data from both networks show that levels are well below the NH&MRC guideline.

Lead

Smelters emit lead and other heavy metal compounds into the air. Industrial areas of major concern are Port Kembla, Broken Hill and Boolaroo in New South Wales and Port Pirie in South Australia.

In the Port Kembla area of New South Wales, airborne lead from a smelter has been a major source of pollution. Monitoring results from three locations indicate that ambient levels near the plant often exceeded the NH&MRC guideline (EPA NSW, 1993). The smelter closed in 1994.

Port Pirie, a city of some 15 000 people, is the site of the world's largest pyrometallurgical lead smelter, in operation since the 1890s. Despite substantial reductions in process emissions since the 1970s, ambient monitoring indicates that lead levels still regularly exceed the NH&MRC guideline in areas close to the works and the wharf stockpiles (South Australia, Department of Environment and Land Management, 1993).

Fluoride

Hydrogen fluoride causes damage to plants at concentrations about 1000 times lower than those that cause detectable human health effects. Certain plants, such as grapevines, are particularly sensitive. Aluminium smelters, power stations and brick and ceramics works are major sources of fluoride, but there is no comprehensive inventory of fluoride emissions for Australia. The NH&MRC and ANZECC have jointly recommended environmental guidelines (based on damage to plants) for fluorides (see Table 5.7).

Limited measurements indicate that fluoride levels are of concern near some industrial sources. Several aluminium smelters have created buffer zones within a distance of about one kilometre, where the concentrations of hydrogen fluoride are likely to be high enough to cause plant damage. In general, brick and ceramics works have a more limited area of impact (Doley and Moller, 1994; Cameron and Rye, 1992).

In the Hunter Valley, New South Wales, some localised seasonal emissions exceeding the ANZECC environmental guideline have been reported near the smelter at Kurri Kurri. The Boyne Island smelter near Gladstone, Queensland, complies with the guideline in nearby residential areas but not in the buffer zone surrounding the plant.

Long-range transport and acid deposition

Apart from global phenomena such as the enhanced greenhouse effect and depletion of the ozone layer, several other processes produce emissions that can survive long enough to be transported many hundreds of kilometres downwind. Unlike many countries in the northern hemisphere, where emissions generated in one country may affect another's air quality, Australia is not currently subject to significant incoming air pollution. (However, this would have occurred during the 1950s and '60s when atmospheric testing of nuclear weapons resulted in worldwide dispersion of radioactivity.) Material injected into the atmosphere from volcanic eruptions may also be carried a great distance, especially if it reaches the stratosphere (for example, the eruptions of Mt Pinatubo and Mt Hudson in 1991).

Some industrial sources (such as ore-smelters and coal-fired power stations) in remote or rural areas of Australia may emit enough acid gases from tall chimneys to justify evaluating the effects of long-range transport. Their major air emissions are SO_2 and NOx. The two isolated locations of Mt Isa and Kalgoorlie, where emissions arise from several ore smelters, account for more than 60 per cent of total national SO_2 emissions.

Assessment of the impact at ground level of long-range pollution transport must include consideration of three major factors: plume dispersion, chemical transformations in the atmosphere and deposition processes — either dry or wet deposition.

It is difficult to assess impacts from large, remote sources as no routine long-distance monitoring takes place downwind of either Mt Isa or Kalgoorlie. Since the early 1980s, CSIRO has conducted a series of experiments, using a specially instrumented aircraft, to investigate the plume characteristics and trajectories from most major sources.

In the arid interior and in the absence of rain, dry deposition of SO_2 is the dominant process. It determines the way in which the emitted pollutants have an effect at ground level. Particles formed by chemical transformations are of lesser significance, although small amounts can be absorbed at the land surface. In the presence of sunlight, NOx is more quickly removed than SO_2.

For remote inland sources, simple assumptions can be made about dispersion meteorology. Using the annual average wind-rose data, researchers can estimate annual average ground-level concentrations and dry deposition rates for distances up to about 300 km downwind from the source. For emissions such as those from Kalgoorlie or Mount Isa and for uniformly distributed winds, it has been calculated that the impacts from the dry deposition of SO_2 are confined to an area about 30 to 40 km downwind from the source. Beyond this, the rate of deposition of sulfur is estimated to decrease to levels similar to those from natural sources (Williams, in press).

The impact of dry deposition on the ecosystems in remote inland areas is essentially unknown.

It is hard to quantify the extent to which rainfall removes both gases and particles from the atmosphere. These processes are more important in the coastal regions than in the drier inland areas.

Any national assessment of the impact of acid deposition must involve quantitative determination of both total deposition (wet acid deposition plus dry acid deposition) and the critical loads for affected areas.

Before 1988, some studies of the composition of Australian rainwater included measurements of rainwater acidity. However, none of them included an assessment of dry deposition and many were conducted for unrelated scientific purposes in regions remote from the major sources. Bridgman (1989) summarised pre-1988 rainwater composition work in Australia while Ayers and Gillett (1988) reviewed what was known about atmospheric acidity and acid deposition in tropical Australia at the time. The Australian Environment

Council (1989) prepared a preliminary national assessment. Several more rainwater composition studies have since been published, but none of these focused explicitly on acid deposition assessment either, or included measurement of dry deposition or evaluations of critical loads. Currently, and contrary to some earlier claims, no evidence suggests that sulfate emanating from the Mt Isa smelters has been detected in rainfall around Jabiru in the Northern Territory, where the composition of the rain has been extensively studied (Noller *et al.*, 1990; Gillett *et al.*, 1990). Most of the sulfate detected there came from natural sources such as vegetation and fires.

One rainwater composition study was carried out specifically to assess acid deposition in the Latrobe Valley, Victoria, between February 1990 and February 1992, but its results are not yet available.

Again, it did not include measurement of dry deposition or determination of regional critical loads. A similar rainwater composition study was carried out at several sites in the Hunter Valley, New South Wales, in two phases between 1984 and 1990, following earlier work in Sydney. Dry deposition and critical loads were not determined in this study either. Two studies that do address the question of dry as well as wet deposition started in New South Wales in 1992, but are not yet complete. The air flows within the airsheds are complicated by day-time sea-breeze effects and night-time valley drainage flows, which invalidate the simple dispersion assumptions described above for remote areas. The prediction of pollution impacts requires the use of complex computer models incorporating terrain and simulated wind fields and the estimation of regional critical loads.

Dry and wet acid deposition

Gases such as SO_2 and NOx are emitted naturally. They are transformed by natural chemical processes in the lower atmosphere into trace levels of sulfuric and nitric acids. These chemicals can be further transformed into small particles known as aerosols. The acids, the gases and their aerosol products are returned to the earth's surface in two main ways:

- through wet deposition in rain or snow (often called 'acid rain')

- through dry deposition, in which atmospheric particles and gases are deposited directly on water, soil, vegetation or other surfaces.

The natural emission, atmospheric transformation, transport and deposition of sulfur and nitrogen are integral and natural parts of the global nutrient cycle.

'Acid deposition' occurs when anthropogenic (human-derived) emissions of acid precursors overwhelm the natural cycles. Anthropogenic activities that emit acid-precursor gases include the combustion of fossil fuels for electricity generation and transport, and industrial processes such as smelting sulfide ores (see figure below). The major gases emitted are SO_2 and NOx. Globally, anthropogenic emissions now dwarf natural emissions three to one.

However, because acids and their precursors typically have atmospheric lifetimes of only a few days, the anthropogenic emissions are not dispersed globally but rather tend to be deposited within some hundreds of kilometres of the source. Acid deposition (mainly in the form of 'acid rain') is high in regions surrounding urban and industrial centres in the mid latitudes of the northern hemisphere, where it has caused environmental problems. In Europe, the long-range transport of acid gases is recognised as a serious transboundary pollution issue. Aerosols also play a key role in the enhanced greenhouse effect.

At present, Australia does not experience long-range transport of acid pollutants from neighbouring countries.

Environmental damage from acid deposition can occur when the extra acid added to surface soil or water exceeds the capacity of these ecosystems to accommodate it. Although arguments persist about the exact causes of damage attributed to acid rain, it is generally accepted that acidification in lakes and streams can cause loss of aquatic life, while on land it can damage leaves and reduce plant growth — most noticeable in the case of trees. In the northern hemisphere the major effects generally arise after some years (typically one or more decades) of excess acid deposition.

It is possible to determine the 'critical load' for a soil or surface water ecosystem below which significant harmful effects on specified elements of the environment do not occur. When total acid deposition (wet plus dry deposition) exceeds the critical load it may start to alter the structure and function of the local ecosystem. As each soil or surface water system has its own critical load, any assessments of acid deposition must include data on the critical load as well as on the rate of deposition. From the limited studies that have been carried out on acid deposition (primarily wet deposition) in Australia, it does not appear to be a serious or widespread problem in this country.

The chemistry of acid deposition

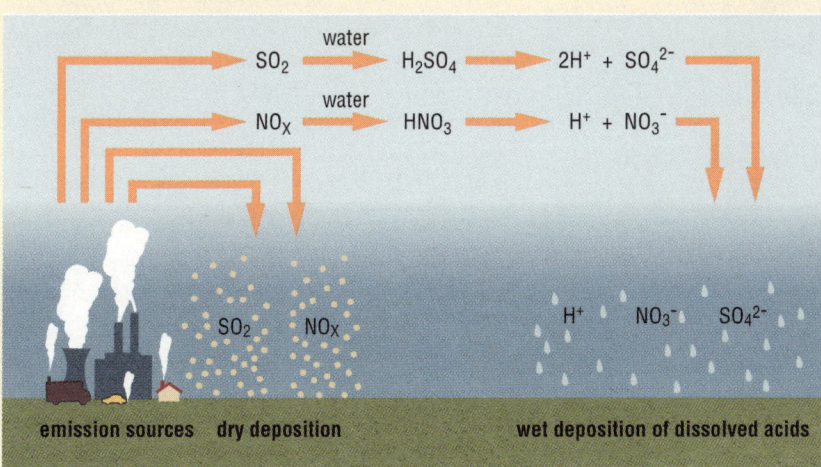

In addition, on certain occasions, the interaction of emissions from power stations with those from nearby cities (such as with the Hunter Valley and Sydney) may occur, bringing the possibility of an extra source of NOx to the urban photochemical smog system, and thereby affecting the distribution and intensity of ozone.

The situation in 1995 has not changed greatly from that presented by the Australian Environment Council (1989), and may be summarised as follows.

- Acid deposition is not a widespread problem here, as sources are generally geographically isolated from each other.

- Unlike many European countries, Australia is not subjected to transboundary transport of emissions from neighbouring countries.

- The major deposition process here is dry deposition of SO_2.

- Nevertheless, Australia has some significant individual sources and regional sources of acid-precursor emissions, and excessive acid deposition in the regions surrounding these sources is a possibility.

- There may be small areas of poorly buffered soils in Australia that could be easily acidified.

Air quality and the environment

Air pollution's potential to damage human health is clearly a major concern, but other effects are also important. These include the possible decline in productivity of agricultural and forestry land (through the impact on trees, pastures, crops and livestock) and the cost of cleaning, repair and maintenance of affected property. In Australia, these effects have only been studied to a limited extent. In Sydney, the annual cost of pollution in terms of cleaning stone buildings was estimated at about \$13 million (Mansfield, 1990). The presence of acidic gases in the atmosphere can accelerate corrosion of steel and aluminium (although natural effects due to the action of sea salt can be more important close to the coastline).

The term 'human well-being' has been coined to refer to areas other than human health that pollution could affect (Doley and McCune, 1993), although human health is the basis of most air quality guidelines. However, for nearly all the common air pollutants, the natural environment is more sensitive to damage than human health (Murray *et al.*, 1992). So human health guidelines alone may not be sufficient to protect the broader concept of human 'well-being'. In other developed countries, guidelines have been developed to protect against impacts such as:

- damage to agriculture and forestry

- disruption to natural ecosystems

- impacts on flora and fauna

- damage to materials, property and buildings

- deterioration of aesthetic aspects of the environment such as landscapes and urban views

Most commonly, people express concern about the potential effects of fluoride, sulfur dioxide and ozone. Although work in Australia has been limited, future guidelines should be based on scientific criteria developed for our conditions. Simply extrapolating from air quality guidelines developed in the major industrial countries of the northern hemisphere is not likely to be valid here (Doley and McCune, 1993).

For instance, some environmental impact assessments for major development projects have assumed, in the absence of suitable data, that eucalypt forests behave like coniferous forests in northern Europe or North America. Similarly, the Australian wheat belt has been modelled on the results of studies in the Canadian prairies (Murray *et al.*, 1992).

We know little about the sensitivity to sulfur dioxide of Australia's 18 000 or so plant species; likewise the effects of any air pollutants on native animals have received little attention. Urban nature parks may need particular study.

The limited work that has been done confirms that many plant and animal species — already naturally restricted in their spread — vary greatly in their response to air pollutants. A range of climatic and

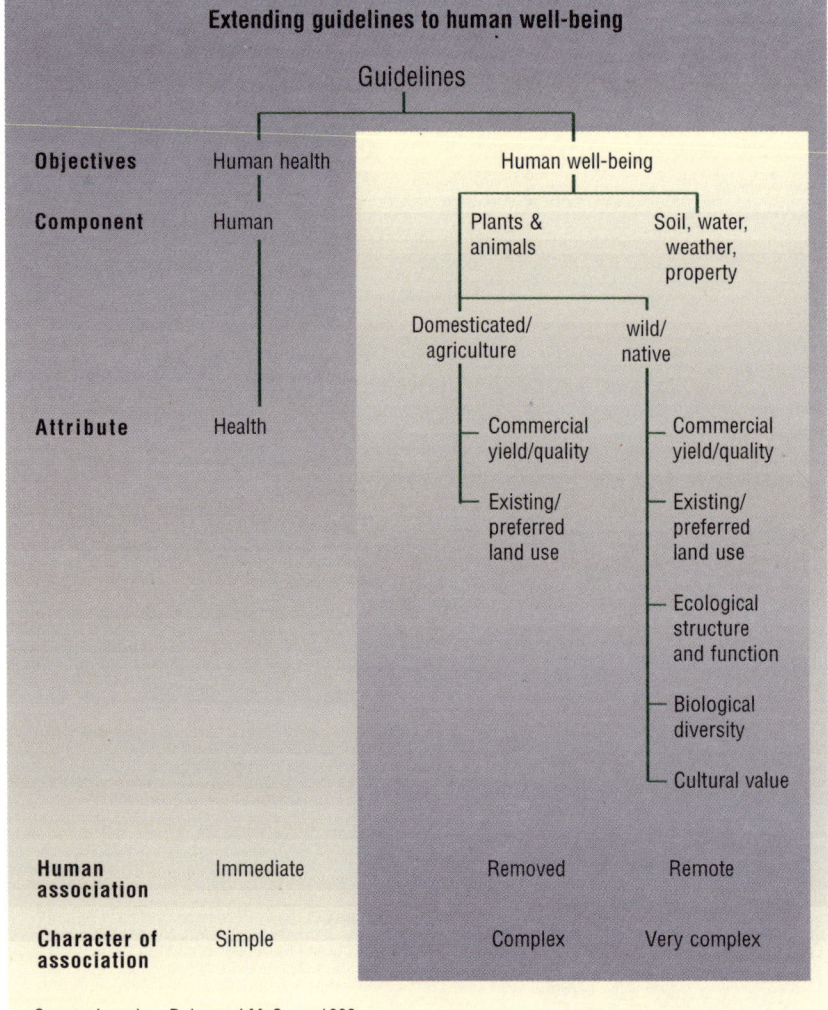

Extending guidelines to human well-being

		Guidelines		
Objectives	Human health		Human well-being	
Component	Human		Plants & animals	Soil, water, weather, property
			Domesticated/ agriculture	wild/ native
Attribute	Health		Commercial yield/quality	Commercial yield/quality
			Existing/ preferred land use	Existing/ preferred land use
				Ecological structure and function
				Biological diversity
				Cultural value
Human association	Immediate		Removed	Remote
Character of association	Simple		Complex	Very complex

Source: based on Doley and McCune, 1993.

management conditions exists for growing any single crop variety, and the responses to airborne pollution are unlikely to be identical for all varieties in all locations.

Sulfur dioxide

Although sulfur dioxide is not monitored widely, it seems from our knowledge of its long-range transport and isolated point sources in remote areas (see page 5-30) that its ground-level concentrations must vary greatly across the country.

Sulfur is an essential element for plant growth. In Australia, the low natural sulfur content of the soils (resulting in crop sulfur deficiency being reported in some regions) means that some plants may actually benefit from airborne sulfur compounds released by human activity (Murray and Wilson, 1989). However, too much exposure can damage plants. As they adapt to other environmental conditions, such as water shortage, their sensitivities to sulfur dioxide may change. However, we know little about the concentrations of sulfur dioxide that would produce adverse responses across the range of Australian native vegetation. According to Murray *et al.* (1992), 'much of the native vegetation in semi-arid areas appears to survive sulfur dioxide concentrations which would devastate the vegetation of more humid areas.'

Until the late 1980s, little attention was paid to the relationship between sulfur dioxide concentrations and effects on important crop varieties and Australian native plants growing under our climatic conditions. In fact, some researchers claim that more work has been conducted on the effects of sulfur dioxide on eucalypts in North America than in Australia (Murray and Wilson, 1989).

Some limited studies (where plants were exposed to SO_2 in open-top chambers in the field) have been carried out on crop species of major economic value — wheat, barley, lucerne, clovers, soybean and peanuts — and on some eucalypt species from native forests near Australia's major coal-fired power stations. Scientists have also studied effects on the major softwood plantation species, *Pinus radiata*. As expected, the results show that individual species have varied and complex responses to SO_2 exposures. Eucalypt species in particular vary greatly in their responses (Murray *et al.*, 1992).

Fluorides

Overseas, the use of fluoride-sensitive plants to monitor the effects of industrial emissions is widely practised. However, in general the Australian environment is too harsh for these methods to be effective (Doley, 1992). An alternative, used for some years in Australia (particularly around several aluminium smelters), involves measuring growth and the appearance of visible injury in the natural vegetation nearby. For gaseous fluoride, the most common visible injury symptoms are yellowing and death around the edges of the leaves. For each species the plant progressively deteriorates with increasing exposure, so we can infer the fluoride dose at various locations around an isolated source from the injury patterns. Despite uncertainties related to both different site conditions and vegetation patterns, it is possible to monitor trends in vegetation condition over time. However, the effects of the pollutants may be masked by:

• seasonal factors, such as the occurrence of drought, variability of rainfall, growth patterns and bushfires

• site factors such as hydrology and mineral nutrition

• salinity.

This masking occurs in the buffer zones that surround industrial point sources, thus complicating the assessment of fluoride impacts.

Photochemical smog

Nobody has assessed the effects of ozone and photochemical smog precursors (such as NOx) on street trees, urban nature reserves and vegetable and fruit production within urban airsheds in Australia. Limited work on the effects of urban ozone on Australian native vegetation (Monk, 1994) has shown great differences in the reactions of various species to ozone exposure.

Indoor air quality

No universal definition describes indoor air quality. The NH&MRC has defined indoor air as 'air within a building occupied for at least one hour by people of varying states of health'. Buildings covered by this definition include homes, schools, restaurants, public buildings, residential institutions and offices — but not workplaces covered by occupational health standards, which are set by the National Occupational Health and Safety Council. (This definition does not include air within vehicles.) The NH&MRC did not define indoor air quality. While many definitions have been proposed, this report uses the following: 'the totality of attributes of indoor air that affect a person's health and well-being'. Under this definition, the chapter uses indicators to determine how well indoor air satisfies thermal and respiratory requirements, prevents unhealthy accumulation of pollutants and enables a sense of well-being.

Many people may be surprised to learn that the quality of indoor air is often poorer than that outdoors. Most Australians spend about 90 per cent of their time indoors (EPAV, 1993) and yet there is little monitoring of indoor air pollutants in homes or recording of individuals' differing exposures to indoor pollutants. Some caravans have been monitored and occasional monitoring in commercial buildings does occur, mainly in response to worker complaints.

Without a systematic program of measurements, it is hard to develop a clear picture of Australia's indoor air quality. Factors affecting it, such as methods of construction and types of building, differ across the nation. Building codes, which specify material used in construction and the ventilation rates of buildings, also differ and have changed significantly over time. National codes have been developed but are not always implemented. Further difficulties can arise if buildings are operated at variance from the original design.

Given the amount of time spent inside, indoor air quality is probably an important environmental factor affecting human health.

In the past, the 'leakiness' of Australian homes, compared with those in the cooler climates of Europe and North America, meant indoor air quality was not a concern. However, changes to building practices that are aimed at improving energy efficiency have substantially reduced ventilation rates, and have probably increased levels of indoor air pollutants. Ventilation rates in our commercial buildings are essentially the same as those in the developed countries of the northern hemisphere.

Major indoor air pollutants can occur inside buildings at much higher concentrations than they are found outdoors. Some are completely different from those found outdoors. Whether a pollution source causes an indoor air quality problem depends on: the nature of the contaminant; the rate of emission from the source; and the ventilation rate of the building.

The following pages provide a perspective on indoor pollutant concentrations found in Australian buildings, including a summary of the major pollutants and possible responses to them. Where possible, measurements are compared with interim indoor air guidelines recommended by the NH&MRC (Brown, in press).

Smoke

Environmental tobacco smoke (ETS) is the term used to describe a complex airborne mixture of gases and particles that tobacco smoking produces. Most measurements of ETS components in Australia have been made in recreational buildings, where smoking is still commonly allowed. Results often showed high levels of ETS pollutants even when ventilation rates satisfied existing guidelines. Further investigation is needed for other building types, to remedy the lack of available information.

Authorities have not determined an exposure standard for ETS because of the complex nature of the mixture, but have used measurements of several components as markers. The most frequently used marker is combustion-derived particulate matter, since it contains a high proportion of respirable size. Such particles are known as respirable suspended particulates (RSP).

In Australia, there appear to be no measurements of typical background levels of RSP in buildings where the occupants do not smoke. Recent measurements of RSP concentrations in Perth buildings with smokers ranged from 150 to 225 μg per cubic metre. and RSP concentrations in the Adelaide casino during peak occupancy ranged from 110 to 430 μg per cubic metre.

In homes, fuel-based heaters and stoves emit RSPs, although their impact on indoor air quality has not yet been well described. Ferrari *et al.* (1988) found an average RSP concentration of 86 μg per cubic metre in eight Sydney homes with wood fires compared with 28 μg per cubic metre in four homes without them.

Legionnaire's disease

Low numbers of the disease-causing (pathogenic) Legionella bacteria are commonly found in soil and water. However, they can multiply rapidly in warm, moist environments such as cooling towers in air-conditioning plants. Inhalation of droplet aerosols containing Legionella bacteria can cause Legionnaire's disease, a pneumonia with a relatively high mortality rate that represents one per cent of pneumonia cases in Australia.

Most Australian outbreaks of Legionnaire's disease have been traced to cooling towers (especially small units) and, to a lesser extent, spa baths. This differs from overseas experience, where large hot-water systems in hotels and the like have been the major sources of the bacteria. Disease outbreaks can be significant and are notifiable in all States and Territories (see Table 5.9).

House dust mites

As their name suggests, house dust mites live mainly in houses — especially in mattresses and carpets — where they feed off human skin flakes and other products. These mites are a major source of the allergy-causing substances (allergens) commonly found in house dust. The main species in Australia is *Dermatophagoides pteronyssinus* (Tovey, 1992), although other species may occur in some buildings. Most of the allergens are found in mite faeces, which easily become airborne when dust deposits are disturbed — for example, during cleaning. It is not easy to measure exposure to house dust mites directly. Instead, indirect methods are used to indicate people's levels of exposure. These involve measuring concentrations of a

Table 5.9 Number of notifications (and incidences per 100 000 population) of Legionnaire's disease for each State and Territory from 1991 to 1994

State/Territory	Number of notifications	Incidence per 100 000 population
New South Wales	251	4.2
ACT	1	0.3
Victoria	149	3.3
Queensland	104	3.4
South Australia	84	5.8
Western Australia	73	4.4
Northern Territory	7	4.1
Tasmania	3	0.6

Source: Brown, in press.

known allergen from the mites or counting the number of mites in accumulated dust (vacuumed from carpets or furniture). Allergen levels above 2 µg per gram of fine dust (equivalent to 100 mites per gram) may increase the risk of sensitisation and of symptoms, while levels above 10 µg per g (500 mites per gram) increase the risk of acute or severe asthma attacks.

As mites require humidities above 40–50 per cent for survival at normal indoor temperatures, the level of mite allergens in very dry or cold regions is generally low (less than 2 µg per g) whereas in regions with one season suitable for growth the mean level is between 2 and 15 µg per g. Coastal areas of Australia, where the climate is suitable for mites for most of the year, have allergen levels of 10 to 40 µg per gram. These data are only broadly indicative, since variations in measuring techniques can lead to two- to five-fold variations in results. Mite allergen levels appear to be highest in coastal regions, become lower inland and are virtually at zero in central Australia. Mites may present a particular health problem in parts of coastal Australia.

Other biological contaminants

Other biological contaminants include microbiological matter such as viruses, bacteria, fungi, protozoa, insect faeces and pollen. The impact of these contaminants on indoor air quality in Australia has received little attention.

Asbestos fibres

Many commercial and industrial buildings in Australia have used sprayed asbestos insulation products that can be a major source of asbestos fibre if they are damaged or deteriorate. The risk of exposure to fibres is particularly high during building maintenance or renovations. Between 1968 and 1979, unbound asbestos 'fluff' was installed as ceiling-space insulation in about 110 houses in the Australian Capital Territory and in some 100 houses in nearby New South Wales towns. In 1988, the Commonwealth Government decided to remove the asbestos fluff from the ACT houses through a $100 million program, which operated from 1989 to 1993.

Asbestos building products have also been widely used in Australia and many of these remain in place. The major items were asbestos cement sheet products for interior and exterior cladding, flooring products (high density underlay sheets, vinyl–asbestos floor tiles and cushion vinyl flooring) and fire, thermal or acoustic insulation products. An estimated 1300 million square metres of asbestos cement building sheets were produced before manufacture ceased in 1983. About half of this product, if still present in buildings, is now more than 30 years old. Emissions of asbestos fibres are greater from those products that are outdoors (particularly roofing), because of surface degradation. Nevertheless, studies have found that asbestos concentrations around such buildings are generally very low and vary little from ambient levels in other urban areas.

Nitrogen dioxide

Although indoor combustion sources produce both nitrogen dioxide (NO_2) and nitric oxide (NO), nitrogen dioxide is of principal health concern, as it is known to cause lung damage at relatively high concentrations. Australian investigations show that its major sources in indoor air are unflued gas heaters and, probably, unflued gas cooking appliances.

Investigations in homes and schools in Australia — particularly in New South Wales where unflued natural gas heaters have been widely used — show high concentrations of NO_2 in many of these buildings. New South Wales government schools are carrying out an extensive program to address the issue, but it is believed that similar concentrations may occur in many buildings in other States that have not yet been investigated.

Volatile organic compounds

VOCs (volatile organic compounds including formaldehyde and pesticides) are organic compounds with boiling points between 50°C and 260°C which are emitted from many materials, equipment and products used in buildings. Generally, the sensitive analytical methods now available can detect from 50 to 150 such compounds in any single building.

Table 5.10 Proportion of private dwellings in each State and Territory in 1981 with asbestos–cement sheet external walls

State/Territory	% of dwellings
New South Wales	19
ACT	1
Victoria	5
Queensland	15
South Australia	9
Western Australia	18
Northern Territory	13
Tasmania	5

Source: Brown, in press.

Figure 5.34 NO_2 concentrations measured in Australian buildings with unflued natural gas heaters during winter

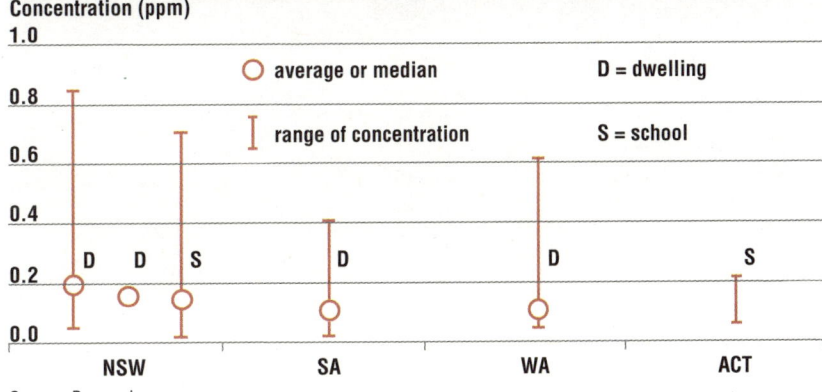

Source: Brown, in press.

The VOCs can irritate the respiratory tract, eyes and nose. They may play a part in 'sick building syndrome' but a clear causal link has not been established.

The major sources of VOCs in indoor air are believed to be wet construction products (paints, adhesives and sealants) in new buildings, and a mixture of wet household products and other materials, such as carpets and soft furnishings, in established buildings.

In Australian buildings — in contrast to those in other countries — VOCs have received very little investigation of any type. Despite this lack of data, the NH&MRC has established an indoor 'level of concern' for VOCs, although it is unknown whether Australian buildings comply with or exceed this level. Further investigation is urgently needed. CSIRO is carrying out a project to characterise VOC concentrations in buildings.

Formaldehyde, an irritating gas with a pungent odour, is one example of a VOC. Its major industrial application is in the production of different resins widely used in indoor materials and in consumer products (particularly pressed-wood and building products such as particleboard and medium-density fibreboard). Small amounts of formaldehyde are also emitted by gas appliances and in tobacco smoke. The NH&MRC guideline for formaldehyde is 0.1 ppm.

Formaldehyde concentrations are low in homes and offices but exceed the guideline in caravans and mobile homes up to five years old, where the high content of pressed-wood products and low ventilation rates appear to increase levels. Concentrations have also exceeded the guideline in homes recently insulated with urea formaldehyde foam insulation, although after several months, when the foam has dried, they generally fall to below the guideline. New or renovated buildings have received little investigation, but Table 5.11 summarises some findings for Australian buildings.

Overall, formaldehyde concentrations in conventional Australian buildings appear somewhat lower than those reported in North America while those in mobile buildings are similar.

In addition, VOCs include pesticides, some of which are widely used in many parts of Australia to protect buildings against termite attack. In homes, people are probably most exposed to pesticides through the use of commercial pesticide products and the infiltration of termiticides from house foundations. Misuse of pesticides is considered to provide the greatest potential for exposure, which can occur by inhalation (including inhalation of previously contaminated house dusts) and by absorption through the skin after contact with treated surfaces. So airborne concentrations are not the only indicators for occupant exposure.

Australia has no guideline for indoor air pesticide concentration. However, concentrations have been investigated recently in small samples of Australian homes treated with termiticides. Studies both here and overseas indicate that homes treated after construction or those with 'leaky' floors have greater indoor air termiticide concentrations. The compounds in termiticides that are of most concern are the organochlorines chlordane and heptachlor. The National Registration Authority recently banned the use of these compounds in the southern States after mid 1995. No investigation appears to have addressed other pesticide sources.

Carbon monoxide

This colourless, odourless gas is produced by incomplete combustion of carbon-containing material. It is a fast-acting poison, as it combines with the oxygen-carrying pigment haemoglobin in the blood, reducing the blood's capacity to transport oxygen. Accidental deaths from CO poisoning have occurred in Australia.

The main indoor sources of carbon monoxide are smoking; unflued gas heating appliances in homes; and vehicles in enclosed car parks in commercial buildings.

In general, indoor CO concentrations are expected to follow outdoor levels except where combustion sources occur in buildings without adequate ventilation. The NH&MRC indoor air guideline for CO is 9 ppm (eight-hour average).

Ozone

A strongly oxidising gas and an irritant, ozone affects the mucous membranes, lung tissues and lung function. Potential indoor sources include electrostatic photocopiers, laser printers, electrostatic precipitators for air cleaning, ionisers and ozone-based sterilisers. Little research into ozone exposure in modern office buildings appears to have been carried out in Australia.

Lead

In residential areas close to some lead-based industries, lead particles may find their way into houses and accumulate in ceiling dust. In general, however, lead in old house paint presents the main health risks to occupants, especially children, due

Table 5.11 Formaldehyde concentrations in Australian buildings

Building Type	No. of buildings	Formaldehyde concentration (ppb) Range	Mean
Conventional home	100	0-97	26
	40	3-73	23
	39	10-33	26
Caravan	20	20-280	90
Caravan/mobile home	24	80-1200	310
Conventional office	3	15-70	21
	4	20-120	66
	8	10-80	40
Mobile office	12	420-830	710

Note: NH&MRC guideline for formaldehyde is 0.1 ppm (=100 ppb)
Source: Brown, in press.

to flaking and chalking of the paint. People are more likely to be exposed during renovations.

Radon

Radon is an inert radioactive gas, given off naturally from most soils and rocks but at widely different rates. The major source of radon in indoor air is therefore the ground under buildings. Radon gives off alpha particles, a very damaging form of radiation. However, alpha-radiation has little penetrating power, so the effects of inhaled radon are generally confined to the lungs. Radon levels in Australian buildings have been widely investigated and found to be well below accepted indoor air guidelines in nearly all locations. This is in marked contrast to some areas in the United States and the United Kingdom, where large numbers of buildings exceed radon guidelines. This difference is believed to result from differences in soil and building designs (such as the more-common use of cellars and basements overseas).

In 1988, the Australian Radiation Laboratory carried out a nationwide survey of radon levels in 3413 homes, and found a measured annual average radon concentration of 12 Becquerel per cubic metre (Bq/m^3)(Langroo *et al*, 1990). The survey estimated that 2000 to 3000 homes nationwide may exceed the NH&MRC recommended level of 200 Becquerel per cubic metre. A subsequent survey in Western Australia reported similar results.

'Sick building syndrome'

Overseas studies have established the existence of a range of building-related illnesses, many with identifiable but diverse causes. The main symptoms are: irritated eyes; irritated, runny or blocked nose; dry or sore throat; dryness, itching or irritation of the skin; and poorly defined feelings such as headache, irritability and poor concentration.

The World Health Organization has termed this cluster of symptoms the 'sick building syndrome'. The symptoms are believed to arise from many causes that, while not clearly understood, are associated mainly with air-conditioned office buildings.

The potential causes include: inadequate ventilation; airborne chemical pollution — with many pollutants probably contributing; micro-organisms and particulates, especially dust from poorly maintained air-conditioning systems and from furnishings; temperatures above 21°C and extremes of relative humidity — both high and low humidities (less than 30 per cent) exacerbate the problem; poor lighting, flicker from fluorescent tubes and absence of windows; and personal and organisational factors.

Assessment of building-related illnesses in Australia has been very limited but research so far suggests a dissatisfaction with office air environments. One study of 228 suburban low-rise office buildings in Melbourne reported that 62 per cent experienced unacceptably stuffy, drowsy conditions, while 82 per cent failed to meet current ventilation guidelines (mainly because of changes in these

Air quality and human health

Researchers in many countries have studied the health effects of the most common air pollutants for some years. These studies were stimulated first by the London smogs of the 1950s, and later by the Los Angeles photochemical smogs. However, as people in different countries rarely experience the same level of exposure to the same set and levels of contaminants, international data are not enough to assess the likely importance of air quality to human health in Australia.

'Proof that air quality has a harmful effect on health depends on showing that deliberately changing the quality of the air leads to an improvement in health' (Peach, in press).

In practice, researchers usually gather data by:

- longitudinal studies, that follow a group of people for many years and compare their history and likely exposure with their patterns of disease
- panel studies, that attempt to monitor the exposure of selected individuals all the time
- epidemiological studies that investigate correlations between exposure and diseases or deaths within a large population
- chamber studies, in which informed, consenting volunteers are exposed to low levels of pollutants for a short time in controlled conditions
- animal studies.

Some Australian studies have found relationships between the levels of pollutants in the air and death rates, attendances at hospital emergency departments or hospital admissions. These relationships are difficult to interpret because the studies failed to exclude the possibility that the people exposed to the pollutants might have had higher death rates, hospital attendances, etc. for other reasons — such as lower socio-economic status or the inclusion of a greater proportion of smokers in the group.

Overseas studies have found a relationship between the levels of several pollutants in the air and death rates or signs of sickness (such as hospital admissions or use of medication for respiratory disease). Some related an increase in signs of poor health with increased levels of sulfur dioxide and total suspended particulates in the air. The levels of many airborne pollutants are generally lower in Australia than in many overseas countries but, despite this, the finding of a possible relationship between the level of total suspended particulates or sulfur dioxide and certain health effects may still have implications for Australia. A detailed critical review of the current state of knowledge of the effects of air quality on human health is provided in Peach (in press).

More knowledge is needed before any clear statements can be made about the impact of air quality on the health of Australians.

guidelines after the buildings were constructed). A similar pattern of incidence has been observed for inadequate ventilation in commercial buildings in Perth.

Within Australian buildings several pollutants have been investigated at varying levels of detail. In general, it is considered that most of these pollutants have not been sufficiently studied to determine either the existing exposure levels for the Australian population or the most appropriate strategies to reduce exposure. In contrast to ambient (outdoor) environments, no regulations have been developed specifically for indoors except for workplaces.

Response

In its definition of the responsibilities of government, the Australian Constitution does not refer explicitly to the environment. This has caused difficulties and has encouraged a fragmented approach to those environmental matters — including the air environment — that cross jurisdictional boundaries.

Under the *Meteorology Act 1955*, the Commonwealth Bureau of Meteorology is responsible for monitoring and forecasting the 'state of the atmosphere'. While the Bureau of Meteorology deals with global baseline monitoring (at Cape Grim) and stratospheric ozone measurements, government decisions in the 1970s gave primary responsibility for air quality to the States.

The Intergovernmental Agreement on the Environment (IGAE), signed in May 1992 by the three levels of government — Commonwealth, State and local — offers the potential for a national approach. Schedule 4 of this Agreement outlines two basic goals regarding national environment protection measures. The first is that people enjoy the benefit of equivalent protection from air, water and soil pollution and from noise, wherever they live. The second is that decisions by business are not distorted and markets are not fragmented by variations between jurisdictions in relation to the adoption or implementation of major environment protection measures. The IGAE has been implemented through legislation which has resulted in the establishment of the National Environment Protection Council (NEPC), a statutory Ministerial Council.

Other basic principles embodied in both the IGAE and the Strategy for Ecologically Sustainable Development (ESD) also form a fundamental part of this national approach. One such, the precautionary principle, has been stated in various ways. In the context of the atmosphere, the principle is as quoted in the IGAE:

'Where there are threats of serious or irreversible environmental damage, lack of full scientific certainty should not be used as a reason for postponing measures to prevent environmental degradation.'

The air environment

Ambient (outdoor) air

The primary responsibility for managing the air environment rests with the individual States and Territories. Clean Air Acts were introduced in most States in the 1960s, with an emphasis on the control of visible emissions from stationary sources. Some States have a record of routine continuous air monitoring dating back to the early 1970s. In 1970, Victoria introduced the first Environment Protection Act followed by a Statewide environment protection policy in 1981.

In recent years, most States have reviewed legislation and regulations to encompass both changing approaches to air quality management and new issues. For example, in Queensland the

Clean Air Act 1963 and Clean Air Regulations (1982) were repealed by the *Environmental Protection Act 1994*.

Although air quality is a State and Territory responsibility, the Commonwealth is involved through joint ministerial councils such as the Australian and New Zealand Environment and Conservation Council (ANZECC). This council of Commonwealth, State and Territory environment ministers provides a national approach to environmental matters. However, its decisions are not binding.

One area where a consistent national approach has been implemented is the setting of national standards for new vehicle emissions. The Advisory Committee on Vehicle Emissions and Noise prepares Australian Design Rules (ADRs) and draft regulations to help control motor vehicle emissions and noise. Before 1989, ADRs were advisory recommendations only; in order to become mandatory requirements they had to be recognised under State and Territory legislation. The introduction of the *Motor Vehicles Standards Act 1989*, administered by the Commonwealth Department of Transport, effectively made ADRs national standards for new vehicles. Australian urban air quality has undoubtedly improved as a result. However, emissions from vehicles already in use remain the responsibility of States and Territories and are less uniformly regulated.

In order to meet the emission requirements of ADR 37, the motor vehicle industry decided that, from 1986, new vehicles would be fitted with catalytic converters, which required the use of unleaded petrol (see Fig. 5.35). A recent further national initiative has successfully encouraged people to use unleaded petrol in older vehicles. This initiative — the 'Lead Round Table' — came about through voluntary agreements between Commonwealth, States and Territories and industry and was coordinated by the Commonwealth. The price differential between leaded and unleaded petrol (in effect since 1 February 1994), the voluntary reduction by industry of the lead content in leaded fuel to 0.3 g per litre (one company offering 'half-leaded' fuel with a concentration of 0.125 g per litre), a further reduction in 1996 and the public-awareness campaign about the lead issue have combined to give significant reductions in airborne lead. Nationally, the use of leaded petrol dropped below that of unleaded for the first time in January 1995.

Under the Constitution, the Commonwealth is responsible for signing and implementing international agreements. In relation to air quality, Australia actively contributes to several international atmospheric monitoring programs, carrying out measurements and interpreting data. The major programs are coordinated by UNEP and WMO (for baseline or clean air monitoring and a rural monitoring network) in conjunction with WHO (for urban air quality — the Global Environment Monitoring System). Australia also provides data on air quality for the OECD State of the Environment reporting process.

Figure 5.35 Levels of airborne lead in selected capital cities showing significant reductions in recent years

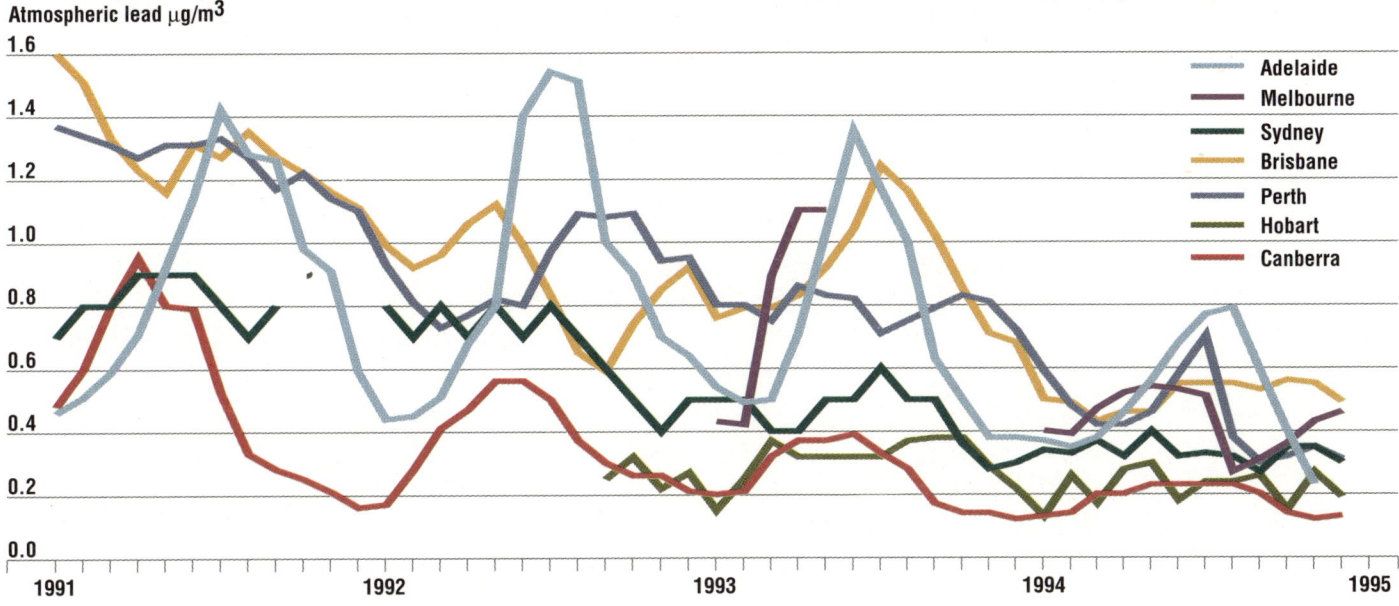

Source: State and Territory environmental authorities.

Within Australia, the NH&MRC used to have sole responsibility for developing guidelines for the protection of human health. Recently, this body has linked with ANZECC to provide joint recommendations covering both human health and the natural environment. However, the States and Territories are not obliged to adopt any of the recommendations. As a result, guidelines vary between each jurisdiction, making comparisons difficult. The establishment of the National Environment Protection Council provides a mechanism to resolve these problems in the future.

Air quality

All three levels of government have in place responses to air quality issues and pressures. Industry and community organisations also carry out some initiatives. The States provide the greatest number of individual responses dealing with air quality. These can be brought about through legislation, regulation, cooperative/voluntary agreements, monitoring, research and education. Some of the approaches to environmental management are listed below.

Regulation

Most States regulate emissions from stationary sources and have inspection and enforcement programs in place. In some cases, responsibility has been devolved to local governments. Environmental agencies can conduct monitoring or it may be required as part of the licence for industrial premises to operate.

Examples of responses by regulation are:

- licensing of emissions from scheduled industrial premises
- setting emission limits
- assessing emissions by testing stacks

- developing technology standards for industry that are either regulated or advised
- enforcing policy standards for major pollutants
- the limited monitoring of vehicles as part of routine vehicle inspections in some States

Cooperation

For many years industry and various State and Commonwealth government agencies have successfully maintained voluntary agreements on air quality management issues. Examples include:

- reducing lead in petrol through a national cooperative program, and the introduction by the petrol industry of half-lead petrol
- voluntary codes of practice developed by industry associations in collaboration with regulators
- voluntary emission reductions of VOCs by some major industries
- the recent move to voluntary cleaner production processes
- the 'Responsible Care program' introduced by the chemicals industry

Prevention

Australian environment protection agencies are changing the emphasis from one of controlling or regulating emissions at the 'end-of-pipe' to one of preventing emissions at their source. While this has potential to provide future savings to industry, it is still largely at the demonstration stage. Examples include:

- voluntary or regulated reductions in emissions through better processes and by minimising waste-producing parts of processes or technologies

- grants and incentive schemes for cleaner production technology

- reductions in licence fees for plants with reduced emissions

- co-generation of power from waste heat (steam)

- power generation from landfill gas (methane) — also a benefit for greenhouse gas reductions

- application of the 'polluter pays' principle

- fuel substitution — through liquefied petroleum gas (LPG) in some taxis, compressed natural gas (CNG), trials of alcohol–diesel mixtures for some buses and petrol–alcohol mixtures for cars in some parts of the country

- use of alternative energy sources such as solar and wind power

Land-use planning

Some land-use planning initiatives can reduce emissions, particularly from road transport, which in turn improves urban air quality. However, as emission reductions and improved air quality are only two of the many aspects considered in land-use planning and decision-making, other initiatives may have the opposite effect. Planning strategies implemented so far include:

- traffic management programs — for example, traffic 'calming' and traffic restriction in city centres

- greater emphasis on, and improved provision of, public transport — for example, the city loop in Melbourne or the eastern suburbs railway in Sydney (the advantages may be partly offset by freeway-building programs, cuts in expenditure on public transport and rises in fares)

- use of buffer zones to protect sensitive areas (may suffer from lack of enforcement in planning schemes)

- the Better Cities Program and urban village projects (scale is too small to prevent continuing urban sprawl)

Monitoring, research, education and information

Education and information programs are common and potentially useful responses. Most of them focus on encouraging waste minimisation or recycling, generally with the aim of changing the behaviour of householders and industry. Programs include:

- reporting air quality data on a daily basis

- air pollution forecasts ('smog alerts')

- airshed studies (including modelling studies)

- 'smoky' vehicle campaigns

- reports of prosecutions in court cases involving air pollution

- research reports on health or ecosystem effects of pollution

- the development of the National Pollutant Inventory

Individual action

Households and individual householders can also contribute to improving air quality by:

- increasing energy efficiency within the home — for example, by using energy-efficient appliances

- minimising vehicle emissions —by using public transport, by car pooling, cycling or walking or by using telecommunications to work from home if feasible

- recycling paper, glass, plastic and cans

- composting rather than burning rubbish

Local governments

Local councils are responsible for several matters that can affect local air quality —backyard incineration control, the management of tips and sewage farms, recycling and community education. They can also have responsibility for broader issues, such as land-use management, traffic management and controlled burning, and may often have sole responsibility for odour control. State and Territory governments may devolve various other responsibilities to local governments.

Commonwealth initiatives

The Commonwealth has only limited powers to respond through legislation and regulation. However, it funds strategic research by CSIRO, universities and Cooperative Research Centres (CRCs). Relevant Commonwealth departments and agencies may be responsible for programs on environmental assessment, on environmental policy and standards, on motor vehicle emissions, on waste management and on environmental protection partnerships. The Commonwealth EPA coordinates and, in some cases, regulates the use of ozone-depleting substances. The Commonwealth can coordinate voluntary initiatives (such as lead reduction in petrol and the National Pollutant Inventory) and, where appropriate, may educate the community and special-interest groups through publications and announcements. It is also responsible for the OECD State of the Environment reporting process.

Indoor air

No single government authority in any jurisdiction has responsibility for indoor air quality. Despite this, a range of responses related to indoor air in Australia is in place. These include State and Territory Government activities, the development of interim guidelines, changes to national ventilation codes, improved building design and community education.

In contrast to workplace and ambient air environments, no regulations have been developed specifically for indoor air (non-workplace) environments, a situation also common overseas. Possible reasons are that:

- the public regards private indoor environments, such as homes, as sacrosanct and therefore not subject to government control

- in practice, it would be impossible to enforce regulations on air quality within homes

- indoor air quality reflects a complex set of factors, including: the effects of building and ventilation system design, construction, operation and maintenance; outdoor climate and pollutant sources; a range and mixture of pollutants and their sources; diverse health effects; and protection of a wide range of people and their sensitivities.

State and Territory government activities

Government departments are responsible for health and environmental regulations in Australia. These departments may also undertake advisory and public education roles.

Environmental protection authorities have probably been the most active in addressing indoor air quality issues. For example, the former New South Wales State Pollution Control Commission carried out research into indoor air quality. However, the Commission was later subsumed by the Environmental Protection Authority, which no longer has this responsibility. The Victorian EPA has responsibility under the State Environment Protection Policy for the ambient (outdoor) environment but has reviewed indoor air quality in residential buildings. Also in Victoria, the Office of the Commissioner for the Environment (OCE) was established in 1986 to identify key environmental indicators and to produce State of the Environment reports. (The OCE has since been disbanded.) The report 'A Review of Air Quality Indicators and Monitoring Procedures in Victoria' addressed both indoor and outdoor pollution.

Standards and guidelines

The boundary between goals for indoor air and occupational exposure standards has became blurred in buildings that act as one person's workplace and another's public place (for example, shopping malls). Indoor-air goals must consider somewhat different factors and risk levels from those in the work environment.

The NH&MRC advises the Commonwealth Government on matters relating to health and also directs research funding. In 1990, ANZEC produced a discussion paper on indoor air quality, which concluded that the issue was not being addressed adequately in Australia and recommended a strategy consisting of three broad approaches: community education and awareness; control of sources of indoor air pollutants; and reduction of the potential for indoor air pollution problems in the future.

Ventilation codes

Australia has no specifications for minimum ventilation (infiltration) rates in residential buildings. In fact, the removal of requirements for fixed wall vents (which occurred in Victoria in 1984 and in New South Wales somewhat earlier), the improvement of construction methods and materials, the use of sheet and concrete slab flooring and the move to improve energy conservation in buildings all appear to have reduced minimum ventilation rates to levels that are in the lower range of those recommended for countries with colder climates than Australia.

The ventilation standard under the Building Code of Australia (1990) is unlikely to be revised for some years. By contrast, ventilation codes overseas are being revised and strengthened with a view to improving indoor air quality. In the Commission of European Communities 'Guidelines for ventilation requirements in buildings' it was acknowledged that not only occupants and their activities but even the buildings themselves could be major sources of pollutants, and that ventilation must be proportional to the total pollutant load. Likewise, the revised United States Ventilation Standard will emphasise the control of indoor air pollution sources. The revision may require 'additional ventilation rates' to be added to the current minimum rates if the building designer does not minimise pollutant sources in the building. The aim of this approach is to encourage the use of low-emission materials rather than increase the level of ventilation.

Reduction of indoor sources

Many indoor air pollutants have clearly identifiable sources. It is now widely accepted overseas that controlling emissions is the most important strategy for achieving improved indoor air quality. In Australia, ANZECC (1990) and NH&MRC (1993) have also recommended this approach. Future Australian Standards for pressed-wood products will probably include formaldehyde-emission limits. The CSIRO Division of Building, Construction and Engineering has recently developed environmental chambers and analytical facilities for research into formaldehyde and other VOC emissions from indoor materials. The gas industry's voluntary initiatives have reduced nitrogen dioxide emissions from new unflued gas heaters.

Public education

Public education is an important tool for improving indoor air quality, especially in homes. Education programs should be based on information from research that identifies potential problem areas, their causes and how to remedy them.

Key issues in air quality assessment

It is impossible to provide a comprehensive, quantitative assessment of national air quality. Some of the difficulties, and possible ways of dealing with them, are detailed below.

Equivalent protection for individuals

Given the lack of data for some 95 per cent of the country, it is impossible to assess whether all people enjoy the benefit of equivalent protection from air pollution — a fundamental goal of the Intergovernmental Agreement on the Environment (IGAE).

Lack of a national approach to the adoption of guidelines recommended by the NH&MRC and

ANZECC has resulted in inconsistencies between States and Territories. Thus, in practice, the current system does not provide, or even attempt to provide, equivalent protection for the 10 million or so Australians whose urban environment is monitored. A consistent national set of ambient air quality standards would be the first step.

The IGAE has the potential to provide a national approach on environment protection matters — particularly air quality. The Agreement includes the formation of a national body — now called the National Environment Protection Council — comprising Ministers representing all signatories to the IGAE. Decisions made by the Council will be binding on all members. However, progress has been slow with the last State, Western Australia, agreeing to join in late 1995. After three years, legislation has been passed in all States, with Queensland the first to enact the necessary legislation in late 1994. When established, the Council will be able to prepare and implement national environment protection measures, including national air quality standards, and so provide the basis for equivalent protection for all individuals. The first meeting of the Council was expected to be held in June 1996.

Monitoring coverage in monitored areas

The current monitoring programs are mainly restricted to the capital cities of Melbourne, Sydney, Brisbane, Perth, Adelaide and Canberra. Recent research, particularly about the formation of photochemical smog under typical Australian conditions, has highlighted deficiencies in monitoring.

Given a reliable emissions inventory, accurate knowledge of pollutant dispersion and a sophisticated representation of air chemistry, studies undertaken by organisations such as CSIRO can be used to forecast the distribution of primary emissions and the formation of secondary

Table 5.12 The main indoor air pollutants and possible response actions

Pollutant*	Typical Concentration Range	Major Sources	Responses
Asbestos fibres	less than 0.002 fibres per millilitre	friable asbestos products	risk management, removal
Synthetic mineral fibres	not characterised	insulation products	unknown
Radon	less than 200 Bq/m^3 per year (the NH&MRC guideline) found 99.9% of time in conventional homes	soil under building	siting of building improved underfloor ventilation
	less than 200 Bq/m^3 per year 91% of time found in earth-constructed homes	earth walls	material selection
Environmental tobacco smoke (ETS)	high in recreational buildings	smoking	prohibition, well-ventilated designated smoking areas, education
Respirable suspended particulates (RSP)	poorly characterised	ETS, cooking fuel combustion	improve ventilation
Legionella spp.	30% of population potentially exposed	water cooling towers	maintenance, siting, regulation
House-dust mites	10-40 µg of mite allergen marker protein per gram of dust in coastal areas.	bedding, carpet, furniture	removal/control of habitat (humidity control)
Microbial	range from hundreds of colony-forming units/m^3 to 18 000 colony forming units/m^3	moist/damp surfaces	control moisture/ mould
Formaldehyde	less than 100 ppb (the NH&MRC guideline) in conventional homes 100 to 1000 ppb in mobile buildings	pressed-wood products	source emission control, ventilation
Volatile organic compounds (VOCs)	poorly characterised	wet synthetic materials	source emission control
Pesticides	median value smaller than 5 mg/m^3. (limited data)	pest control	floor structure, inspection, clean-up, protocols for safe application
Nitrogen dioxide	up to 1 ppm	unflued gas heaters	source emission control, flued systems, improved ventilation
Carbon monoxide	about 10% exceed 9 ppm (the NH&MRC guideline)	incomplete combustion	as for nitrogen dioxide above
Carbon dioxide	poorly characterised	exhaled air	outdoor air ventilation
Lead	poorly characterised	lead paint, the fallout of accumulated roof space dust	clean-up, education
Ozone	poorly characterised	some office equipment	source emission control, ventilation

*Note: no order of priority is implied in the listing of pollutants.
In contrast to ambient (outdoor) environments, there have been no regulations developed specifically for indoors except for workplaces.

pollutants like those found in photochemical smog (see Fig. 5.31). Such studies help in assessing the most appropriate indicators for monitoring programs and for deciding the best sites for monitors.

The Metropolitan Air Quality Study (a study of the greater Sydney airshed) is a recent initiative that includes extending the city's monitoring network. It shows that transport of pollutants can occur between the city and the major industrial areas centred on Newcastle to the north and Wollongong to the south. New, extensive monitoring programs are also under way in the Perth and Brisbane airsheds, and work is scheduled to start in Adelaide soon.

During the next five years, as the population in Australian urban areas increases and city limits extend, motor vehicle usage will almost certainly increase. Vehicles will thus continue to be the major source of urban emissions. Research will continue to focus on understanding the production of photochemical smog and ozone, and policy measures will focus on limiting nitrogen oxides or VOCs. We can also expect an increased emphasis on the need to start routine monitoring of air toxics. This may be underlined by the establishment of the National Pollutant Inventory by the Commonwealth EPA.

In most Australian urban airsheds, authorities may need to reassess the requirement to monitor for sulfur dioxide and carbon monoxide, with a possible change in emphasis to ozone and air toxics.

Current conditions are likely to continue unless a major change in Australian lifestyles occurs. A reduction in motor vehicle use, and thus emissions, is possible. This would be a response partly to urban air-quality problems and partly to the identified need to increase population densities and change transportation modes to minimise our contribution to the enhanced greenhouse effect.

Monitoring for populations outside metropolitan areas

Some remote industrial centres such as Mount Isa, Kalgoorlie and Port Pirie have monitoring programs. In recent years the Queensland government has started monitoring in smaller population centres like Gladstone, Mackay and Townsville. The Melbourne airshed monitoring program has been extended to cover Geelong. Nevertheless, eight million people live outside monitored metropolitan areas.

Some major urban centres, such as the two capital cities Hobart and Darwin and Albury/Wodonga (on the Victoria/New South Wales border), have little data available, which makes it impossible to assess the extent of peoples' exposure to air pollution in those areas. As discussed above, air-quality modelling can be used to identify which population centres need to be monitored and which indicators should receive attention.

National access to air quality data

Although data exist, they are dispersed across different industries and different government agencies responsible for environmental management. This problem has been recognised for some time but remains unresolved. In 1990, a report to the Prime Minister by the Australian Science and Technology Council (ASTEC) stated that 'Australia lacks an integrated national system for measurement of environmental quality, a national database of sufficient calibre to assess and manage environmental quality and appropriate national baseline data to evaluate effectiveness of strategies' (ASTEC, 1990). Initiatives already in place — such as the IGAE Schedule 1, which calls for a national approach to data collection and handling, and the National Pollution Inventory — may provide a more satisfactory situation if fully implemented. A minimum requirement for future SoE reports should be a national inventory of the various data holdings in the States and within industry.

Data quality is also an issue. If industry is expected to monitor emissions — as is currently the case — it is essential that measurements be made using reliable methods, and also that data are independently vetted. In many cases, current practices of collecting and storing data fail to satisfy regulatory goals adequately and may actually prevent industry providing credible evidence of appropriate environmental management.

Geographical coverage of air quality monitoring

Although about 95 per cent of Australia is not regularly monitored for air quality, existing guidelines are probably satisfied over most areas, most of the time. More attention should be given to obtaining data about environmental exposure from major emission sources located in remote areas.

The development of a truly national picture of ambient air quality requires a network specifically designed for this purpose. Such a network would need to include:

- adequate spatial coverage — a sparsely inhabited region probably requires less-dense monitoring than the urban environment
- appropriate choice of well-characterised measurements
- data on the costs and impacts of the pollutants at the concentrations involved
- a pragmatic compromise between ideal scientific standards and the available resources
- coordination with meteorological observing networks.

A design needs to be developed for a representative, up-to-date and cost-effective network.

One possible first step towards a national assessment would be the design of a network to measure appropriate indicators for remote areas — the main ones probably being sulfur dioxide and particulates. This design process would have to

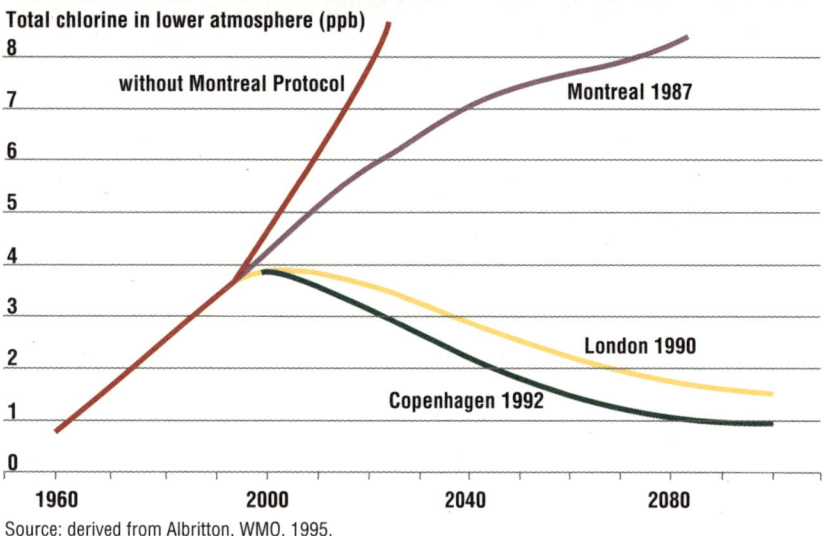

Figure 5.36 Measured chlorine-equivalent concentrations in the atmosphere since 1960 and projections according to the various measures to phase out ozone-depleting substances

Total chlorine in lower atmosphere (ppb)

without Montreal Protocol

Montreal 1987

London 1990

Copenhagen 1992

Source: derived from Albritton, WMO, 1995.

accommodate the sensitivities of different species, different combinations of species and environments and different exposures (so allowing for a range of land uses). However, little research is taking place in Australia and further study of our unique situation would be desirable.

Effects of air quality on flora and fauna

Even though some researchers have studied the effects of air quality on flora and fauna in Australia, the wide variations in climate and soils and the localised ranges of many plants means that their results are rarely applicable nationally. In future, we should aim to acquire more data about the effects of certain contaminants on Australian species. Interdisciplinary studies are important here. Only one research establishment in Australia is conducting controlled open-top chamber experiments to simulate the impact of air pollution under tropical, temperate and arid environmental conditions. Europe and the United States maintain much larger programs, measuring variations across the different zones of those continents.

evolve as we develop our understanding of the interaction between the atmosphere and the rest of the environment (including ecosystems in remote areas) and as we improve our estimates of the cost of impacts of pollutants to society.

Understanding Australian conditions

Overseas researchers have extensively studied air quality and the effect of contaminants on the natural environment. Australia is very different — in weather and climate as well as in soils and ecosystems — from northern hemisphere countries. It is therefore not always appropriate to apply overseas findings here. In our case, it may be appropriate to set more than one standard to

Relationship between air quality and human health

Clearly, more epidemiological studies — properly coordinated and overseen — need to be done. Protocols should be standardised and agreed nationally, along with agreement on confounding factors and how to measure them. More and better longitudinal studies (tracking a group of people and their exposure for many years) are needed, as well as studies on the costs and benefits of changing air quality. Epidemiologists and air-quality experts could be more closely linked and emerging issues — such as levels of air toxics — should receive further investigation.

Indoor air quality and human health

Nobody is adequately addressing this important issue or developing appropriate response strategies. So far, both the measurement of indoor air quality and the assessment of impacts on human health have been fragmented. This undesirable situation will continue until a single body coordinates responsibility for management of indoor air quality.

Stratospheric ozone responses

Depletion of the ozone layer is a global problem that has attracted international attention. By the mid-1980s unusual seasonal reductions in total ozone levels over Antarctica were evident. Scientists believed that these reductions might be caused by the presence of CFCs. In 1985, worldwide concern about ozone depletion led to the Vienna Convention for the Protection of the Ozone Layer, followed in 1987 by the Montreal Protocol on Substances that Deplete the Ozone Layer. By then it had been established that CFCs were not the only industrially produced chemicals that could lead to ozone depletion. Others, such as the fire-fighting chemicals known as halons, were found to exhibit similar activity.

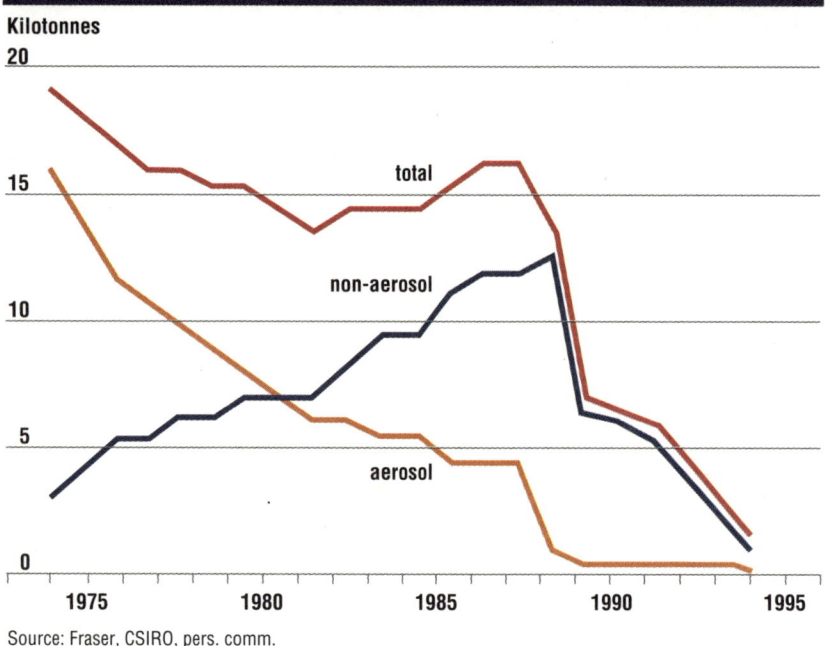

Figure 5.37 Australian consumption of CFCs over the past 20 years

Kilotonnes

total

non-aerosol

aerosol

Source: Fraser, CSIRO, pers. comm.

The 1987 Montreal Protocol introduced a series of measures, including a timetable for action to reduce the release of CFCs into the environment. Australia, the European Community and 32 other countries signed the Protocol. Together the signatory nations were responsible for more than 80 per cent of global CFC usage. The Protocol was amended in 1990 and again in 1992, both times with more stringent controls and the inclusion of more substances recognised for their ability to deplete stratospheric ozone (see Table 5.13). It now has 148 countries as signatories. Figure 5.36 shows the effect of the agreements on slowing the rate of increase of CFCs.

In 1989 — as part of its responsibilities under the Montreal Protocol — Australia adopted a Strategy for Ozone Protection, which was revised and updated in 1994. ANZECC established the Ozone Protection Consultative Committee, chaired by the Commonwealth Government and with representatives from all States and Territories and New Zealand, as well as from industry, science, conservation and community groups.

In Australia, halons were phased out at the end of 1992, a year earlier than the Protocol required. The use of CFCs, methyl chloroform and carbon tetrachloride was due to end by the end of 1995, consistent with the Protocol schedule. The Australian Strategy aims at reducing the consumption of controlled ozone-depleting substances (ODSs) by 95 per cent by the end of 1995; by the end of 1997, all further consumption should have ceased. Figure 5.37 shows the Australian consumption of CFCs over the past 20 years.

The Australian government's approach to restricting ODS consumption is to liaise with the relevant industries and encourage practical responses, backed up with economic incentives where possible. The goal is to change the behaviour of the end-users of the substance, and inform them of alternatives and the means to switch to these wherever possible. States and Territories have addressed usage, emission control and, where necessary, product and equipment bans.

Since 1989 — with the exception of essential uses such as asthma sprays — no aerosol spray cans sold in Australia have contained CFCs. Hydrochlorofluorocarbons (HCFCs) — used as temporary substitutes for CFCs — are to be controlled at a rate ahead of Protocol requirements. Atmospheric monitoring around the world has shown that the previous steady increase in the background atmospheric concentration of several ODSs, particularly CFC-11 (also a greenhouse gas), has levelled off (see Fig. 5.38).

These international measures are aimed at limiting, and ultimately reversing, the most critical impact of stratospheric ozone depletion — that is, the increase in harmful ultraviolet radiation at the earth's surface. The link between UV exposure and a range of human health problems is well established and studies have also demonstrated the UV sensitivity of other biological systems.

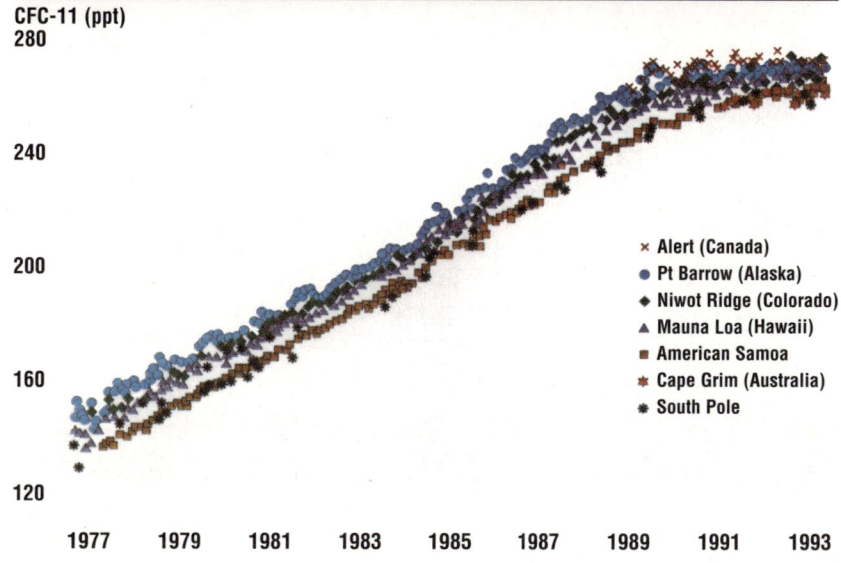

Figure 5.38 Atmospheric concentrations of CFC-11 from 1977 to 1993 from several observation sites

CFC-11 (ppt)

Legend:
× Alert (Canada)
● Pt Barrow (Alaska)
◆ Niwot Ridge (Colorado)
▲ Mauna Loa (Hawaii)
■ American Samoa
✳ Cape Grim (Australia)
✱ South Pole

Source: IPCC, 1994.

Unfortunately, because of the long atmospheric residence time of most of the ODSs, it will take several decades to restore the natural balance between ozone production and destruction. Meanwhile, a large part of the earth will probably be subjected to increased ultraviolet radiation.

Greenhouse responses

The enhanced greenhouse effect is a global problem that requires global responses. Effective global responses will only be achieved by international agreements, such as the United Nations Framework Convention on Climate Change (FCCC), supported by effective national emission reduction and other response programs. Although Australia only produces one to two per cent of total world emissions, we need to play our part in solving the problem.

Interim planning target

Australia has acknowledged the potential impact of climate change resulting from an enhanced greenhouse effect on the nation's natural, social and working environment as well as on the global community. In October 1990, the Commonwealth Government adopted an interim planning target for the emission of greenhouse gases not controlled by the Montreal Protocol. Accordingly, Australia aims to: 'stabilise greenhouse gas emissions (not controlled by the Montreal Protocol on substances that deplete the ozone layer) based on 1988 levels by the year 2000 and to reduce these emissions by 20 per cent by the year 2005....subject to Australia not implementing response measures that would have net adverse economic impacts nationally or on Australia's trade competitiveness, in the absence of similar action by major greenhouse gas producing countries' (Commonwealth of Australia, 1992).

The target is not legally binding but acts as a guideline for action and a standard against which to measure progress (see Fig. 5.39).

National Greenhouse Response Strategy

Late in 1990, the Government recognised that measures to control greenhouse gas emissions should become an integral component of the national Ecologically Sustainable Development process then under way. The interim planning target subsequently formed the basis of the National Greenhouse Response Strategy. In 1992, the Commonwealth, State and Territory governments agreed to the Strategy, which was also endorsed by the Australian Local Government Association. The Strategy recommended a phased approach. First-phase responses will be those of a 'no-regrets' nature — that is: 'a measure that has other net benefits (or at least no net cost) besides limiting greenhouse gas emissions or conserving or enhancing greenhouse gas sinks' (Commonwealth of Australia, 1992).

Many energy-saving measures — such as improved house insulation — are examples of 'no-regrets' actions. The Strategy includes provision for public involvement, a requirement for auditing and reporting and recognition of the need to consider adapting to the impacts (both positive and negative) of climate change. The need for research and analysis to improve knowledge and understanding of the enhanced greenhouse effect is another essential element.

Table 5.13 The 1987 Montreal Protocol* and the 1990 London and 1992 Copenhagen amendments

Ozone Depleting Substances	Montreal (1987)[2]	Control Measures[1] London (1990)[2,3]	Copenhagen (1992)[4]
CFC-11, CFC-12, CFC-113, CFC-114, CFC-115	• freeze at 1986 levels by 1989 • reduce by 20% by 1 July 1993 • reduce by a further 30% by July 1998	• reduce by 50% by 1995 (from 1986 levels) • reduce by 85% by 1997 • total phase-out by 2000	• reduce by 75% by 1994 (from 1986 levels) • total phase-out by 1996
Halon-1211, Halon-1301, Halon-2402	• freeze at 1986 levels by 1992	• reduce by 50% by 1995 (from 1986 levels) • total phase-out by 2000	• total phase-out by 1994 (recycling encouraged)
Other CFCs	not included	• reduce by 20% by 1993 (from 1989 levels) • reduce by 85% by 1997 • total phase-out by 2000	• reduce by 20% by 1993 (from 1989 levels) • reduce by 75% by 1994 • total phase-out by 1996
CCl_4 (carbon tetrachloride)	not included	• reduce by 85% by 1995 (from 1989 levels) • total phase-out by 2000	• reduce by 85% by 1995 (from 1989 levels) • total phase-out by 1996
CH_3CCl_3 (methyl chloroform)	not included	• freeze at 1989 levels • reduce by 30% by 1995 • reduce by 70% by 2000 • total phase-out by 2005	• reduce by 50% by 1994 (from 1989 levels) • total phase-out by 1996
HCFCs	not included	not included, but to be reviewed in 1992	• freeze by 1996[5] • reduce by 35% by 2004 • reduce by 65% by 2010 • reduce by 90% by 2015 • reduce by 99.5% by 2020 • total phase-out by 2030
HBFCs	not included	not included	• total phase-out by 1996
CH_3Br (methyl bromide)	not included	not included	• freeze by 1995[6] (1991 base year) • further study requested • decision on cuts in 1995

Notes:
1. Control measures commence on January 1 of the year indicated.
2. A ten year grace period is allowed for developing nations provided their annual consumption is less than 0.3 kg per capita.
3. Agreement was reached to set up a special fund to provide financial and technical assistance to developing nations to enable them to comply with the Protocol.
4. The special fund was made permanent, but the application of the Copenhagen phase-out dates (plus the ten-year grace period for developing nations) will not be considered until 1995.
5. Based on 1989 HCFC consumption with an extra allowance (ODP weighted) equal to 3.1% of 1989 CFC consumption.
6. Quarantine and pre-shipment treatment is exempt.
* The timetable set by the Montreal Protocol is for bulk consumption of ozone-depleting substances (ODSs) in developed countries (this does not include references to manufactured products containing ODSs. Consumption is defined as the quantities manufactured or imported less those quantities exported in any given year. Percentage reductions relate to the base year for the substance. The Protocol does not forbid the use of recycled controlled substances beyond the phase-out dates.
Source: ANZECC, 1994.

United Nations Framework Convention on Climate Change

More than 150 countries signed the Framework Convention on Climate Change at the 1992 Rio Earth Summit. It came into force in March 1994. At the time of the first session of the Conference of the Parties (March 1995), 127 countries had ratified it. In December 1992, Australia (an Annex I (developed) country under the Convention) became the ninth nation to ratify. The stated aim of the Convention is to achieve:

'...stabilisation of greenhouse gas concentrations in the atmosphere at a level that would prevent dangerous anthropogenic interference with the climate system. Such a level should be achieved within a time-frame sufficient to allow ecosystems to adapt naturally to climate change, to ensure that food production is not threatened and to enable economic development to proceed in a sustainable manner.'

The Convention recognises that developed countries must take the lead in reducing greenhouse gas emissions.

Under its obligations as an Annex I country, Australia provided its national report to the Convention in September 1994 (Commonwealth of Australia, 1994). The greenhouse gas inventory (see Figs 5.5, 5.6 and Table 5.2) forms part of these obligations. The report also outlined the range of policy measures in place to limit emissions, and provided a forward projection of Australia's greenhouse gas emissions for the year 2000 (see Fig. 5.39). However, despite measures in place in 1994 for both emission reductions from sources and increased uptake by sinks, a considerable gap remained between the projections and stabilisation at 1990 levels by the year 2000. Based on current measures, the likelihood of reaching the 2005 interim planning target is quite remote.

Assessment of first phase measures

Several measures identified in the National Greenhouse Response Strategy as satisfying a 'no-regrets' requirement have not been implemented. Mechanisms are in place to address a number of these deficiencies, but only limited progress has been made. However, some companies are using environmental management systems and cleaner production initiatives to reduce energy and commodity consumption with consequent gains. The Government considered, but did not proceed with, the introduction of a small carbon tax (environment levy) on emissions.

Australia's responses are based on, and determined by, our conditions — our high dependence on fossil fuels (particularly for power generation) and the large agricultural sector. The latter contributes significantly to CO_2 emissions from land-use changes and to methane emissions from our large livestock population. Therefore, measures must range across all sectors of the economy; government, industry, agricultural enterprises and individual householders all have a role to play.

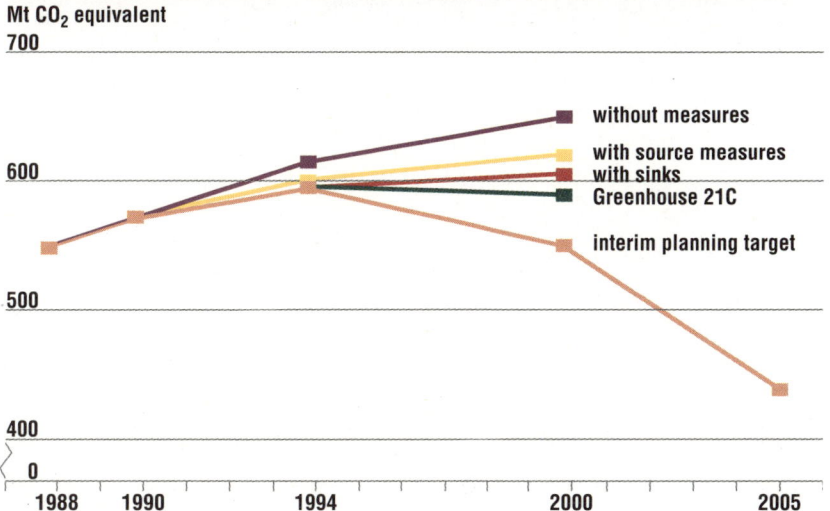

Figure 5.39 Interim planning target and projected greenhouse gas emissions to the year 2000 — these figures indicate that considerable further reductions will be required to meet the interim planning target

Source: adapted from data in Commonwealth of Australia 1992, 1994; DEST, 1995.

A recent joint report of the Australian Academy of Technological Sciences and Engineering, Australian Academy of Science and the Academy of the Social Sciences in Australia compared this country with others in the OECD. It found:

'Australia's relatively high carbon dioxide emissions have become much higher in the last two decades. From 1987 to 1992 energy-related carbon dioxide emissions grew much faster than the OECD average: more than 13 per cent compared with less than five per cent. Australia has had greater increases in population than other OECD countries, but other OECD countries have had greater increases in gross domestic product. Both increasing population and increasing gross domestic product could have been expected to increase greenhouse gas emissions, but carbon dioxide emissions in OECD countries have increased in line with population and gross domestic product while in Australia the rate of emission increase has been significantly higher than the rate of increase in gross domestic product and population.' (Steering Committee of Climate Change Study, 1995).

Our relatively high dependence on fossil fuels (compared with other OECD countries), the lack of nuclear power and the limited opportunities for hydro-electricity generation all restrict the available options to reduce CO_2 emissions. However, we have large reserves of natural gas — the least carbon-intensive of the three main fossil fuels (coal, oil and gas). There is potential for substituting natural gas for coal in some applications. As well, Australia's climate provides abundant supplies of renewable energy sources such as solar and wind power, although their development and use have so far been limited. The conversion of domestic water heating to solar power is one immediate opportunity to achieve worthwhile savings in greenhouse emissions (Commonwealth of

Table 5.14 Scenarios of rainfall change for locations in Australia

Season	Region	Response per degree of global warming	Change in 2030	Change in 2070
Summer half-year (Nov. to April)	Any location	0 – +10%	0 – +20%	0 – +40%
Winter half-year (May to Oct.)	Locations in sub-region A	0 – -5%	0 – -10%	0 – -20%
	Locations in sub-region B	-5% – +5%	-10% – +10%	-20% – +20%
	Locations in sub-region C	0 – +5%	0 – +10%	0 – +20%

Note: Sub-regions are defined in the map below. Values for the years 2030 and 2070 are rounded to the nearest 10%.

low winter rainfall

sub-region A

sub-region B

sub-region C

Source: CSIRO, 1992

Australia, Senate, 1991). The contribution of urban transport to greenhouse gas emissions and energy use can also be potentially reduced.

Although all governments had adopted the National Strategy for Ecologically Sustainable Development, by mid-1995, there has been no indication that they paid close attention to the application of its basic principles in major initiatives such as the operation of the national electricity grid and the extension of reticulated electricity grids to remote areas. Governments appear reluctant to account adequately for environmental costs in the decision-making process when new power stations are constructed.

Greenhouse 21C

In March 1995, the Commonwealth Government announced a package of further measures, entitled 'Greenhouse 21C', which included three main initiatives:

- cooperative agreements between government and industry for reductions in net greenhouse gas emissions

- renewable energy initiatives, including the Commonwealth commitment to establish a Cooperative Research Centre on renewable-energy technology

- continuing emphasis on economic reform in the energy sector, with a greater focus on gas reform and delivery of energy efficiency and renewable-energy programs

Greenhouse 21C also covers: a greenhouse information network to support action by all sections of society; enhanced cooperation with the States and Territories to address issues such as land management and energy-sector reforms; expansion of tree-planting programs to provide greenhouse sinks; and an emphasis on environmental best practice in the Commonwealth Government's own operations.

To date, most effort has been on developing mitigation measures, with particular emphasis on voluntary agreements with industry sectors. Projections in Greenhouse 21C indicate that emission levels in 2000 will be about three per cent above those needed for stabilisation at 1990 levels (see Fig. 5.39). Strong action will be needed in all sectors to achieve these emission reductions. Substantially larger reductions will be needed to achieve the Interim Planning Target for the year 2005. In mid 1995, it was still too soon to judge the likely success of these particular measures.

Climate change research

Australia has made a considerable investment in climate change research and has built up a strong body of scientific expertise both for the country and the region. Australian scientists have also played a major role in the international research effort on the enhanced greenhouse effect, providing a unique southern hemisphere perspective. Research has covered areas such as: the analysis of Antarctic ice-core data; analysis of the global carbon cycle; research in atmospheric chemistry including analysis of long-term background data from the Cape Grim baseline station; climate modelling and validation exercises; impact assessment; and studies of climate variability.

Clearly, a good understanding of the science of climate change should underpin the policy-making process. However, the current uncertainty in the science means that there are no reliable regional predictions of possible climate change over the Australian continent. A precautionary approach is justified in the face of this level of uncertainty. For planning purposes a range of future scenarios has been developed (CSIRO, 1992). Table 5.14 shows an example of a broad scenario designed to help in long-range decision-making, developed by CSIRO for Australian conditions. Such scenarios can help assess vulnerability and develop options for responses (for example, assessing potential costs and benefits, or enhancing the processes that take up greenhouse gases — that is, sinks) and for adaptations (for example, actions to cope with adverse changes and/or take advantage of beneficial changes). In addition, research work designed to help implement adaptation policies will also contribute to our efforts to cope more effectively with the impacts of climate variability, which is an important feature of Australia.

Table 5.15 Summary

Element of the environment/Key issue	State	Adequate Info.	Response	Effectiveness of response
Climate Natural climate variability	Australia's climate displays a high natural variability. In 1994/95 large areas of eastern and northern Australia suffered severe drought linked to a highly abnormal long-lived El Niño episode. Rainfalls in some areas were the lowest on record.	✔✔	An active research program into climate variability is underway.	Seasonal and interannual predictions are available.
Enhanced Greenhouse Effect Global atmospheric concentrations of major greenhouse gases.	Observations at the Cape Grim Baseline Station (Tasmania) show continuing annual increases in global concentrations of all major greenhouse gases.	✔✔✔	Negotiation of the UN Framework Convention on Climate Change (FCCC) including an implied target of developed countries returning emissions to 1990 levels by 2000. Australia has signed and ratified the FCCC. Negotiations continue on the Berlin Mandate.	The first Conference of the Parties of the FCCC, held in Berlin in March 1995, agreed that the global response was inadequate.
Australia's contribution to global emissions.	Australia contributes 1-2% of current global emissions. The NGGI shows that Australia's relatively high contribution to global emissions is mainly due to high fossil fuel use, vegetation clearing and agriculture.	✔✔	1990 : Interim Planning Target - stabilisation at 1988 levels by 2000, 20% reduction by 2005; 1992 : Governments agreed to the NGRS based on a 1995 : Greenhouse 21C including voluntary Industry agreements	Early initiatives will not achieve stabilisation by 2000.It is too soon to judge the success of responses under Greenhouse 21C.
Stratospheric ozone Stratospheric ozone loss	The Antarctic ozone hole has increased in depth and extent since the early 1980s. A decline in stratospheric ozone and a decrease in total column ozone have been monitored over Australia in the past decade.	✔✔✔	Negotiation of the Vienna Convention for the Protection of the Ozone Layer and its Montreal Protocol, London and Copenhagen Amendments. Australia is a Party to the Convention and its protocol and amendments.	The ozone hole will persist but conditions are expected to improve within the next five years, with some expectation that global ozone levels will also begin to increase.
Atmospheric concentrations of CFC-11 & CFC-12	Measurements at Cape Grim (Tasmania) show levels of CFC concentrations are stabilising.	✔✔✔	Australia is meeting all Montreal Protocol obligations ahead of schedule, eg total phase-out of halon imports and use in new equipment, phase-out of CFC production before the end of 1995, most air conditioning units in new vehicles are CFC-free.	Concentrations of several major CFCs are expected to decline globally in the next few years.
Urban air quality Sulfur dioxide (SO_2)	SO_2 levels in major urban airsheds are generally well below levels of concern. Major sources such as power stations are located outside urban areas.	✔✔✔	Low levels result from low sulfur content of Australian fuels.	Effective.
Ozone	Tropospheric ozone concentrations in urban areas result from the chemical interaction of NOx and VOC emissions in the presence of sunlight. Ambient levels are generally low. Episodic high values in Sydney, Melbourne and increases in Perth are of concern.	✔✔	Nationwide introduction of catalytic converters in new vehicles in 1986 reduced emissions of VOCs with a smaller reduction in NOx. Future responses will need to be made in the light of a number of studies currently underway.	Trends are difficult to assess. Increases in new vehicles may be offset by other factors eg Increase in the number of vehicles & the distance travelled. In-service vehicle standards are needed. In future further controls on NOx emissions will be necessary.
Lead	Airborne lead levels have fallen over the past decade.	✔✔✔	Increased use of unleaded petrol following mandatory controls on vehicle emissions (1986) and the installation of catalytic converters in new vehicles. A more recent campaign to encourage increased use of unleaded petrol	Further reductions in lead levels are expected to occur as the number of pre-1986 vehicles decreases.
Toxics	A number of toxic substances have been identified in urban air sheds. However to date only local isolated measurement programs have been undertaken.	✔	Recognition of potential concern. Proposals for a National Pollutant Inventory are under consideration.	Further research is required.

Table 5.15 Summary (continued)

Element of the environment/Key issue	State	Adequate Info.	Response	Effectiveness of response
Particles	Seasonal episodic high values are reported from all major capital cities. Sources include domestic wood fires, backyard incinerators, vehicles (particularly diesel), control burns and bushfires.	✓✓	Elimination or restrictions on the use of backyard incinerators. Improved management of control burns in rural regions upwind or major urban areas; and new emission limits on diesel vehicles.	Improvement in visibility within cities however the lack of controls on domestic wood burning poses a significant problem in some urban areas. Diesel emission standards require further update.
Health Effects	A number of Australian studies have sought to determine whether relationships exist between levels of pollutants in ambient air health/morbidity/mortality but results have been inconclusive.	✓	NH&MRC guidelines for the protection of human health are agreed for a range of individual pollutants but the guidelines are not implemented nationally. A limited number of studies of health effects have been undertaken.	Currently no nationally implemented guidelines. There is a need for more epidemiological studies with standard protocols and agreement on confounding factors. Few longitudinal studies are available.
Regional air quality Sulfur dioxide (SO_2)	A number of major point sources are located in remote inland areas. SO_2 emissions are removed downwind mainly by dry deposition processes. There is no evidence of significant acid (wet) deposition in Australia.	✓	Airshed management of emissions from major sources is used to control ground level concentration mainly for health protection.	Some high levels are occasionally recorded in vulnerable areas. Improved management strategies are being implemented. Effects on native vegetation are generally unknown.
Fluoride	Levels are locally high in the vicinity of aluminium smelters, brickworks, ceramic kilns etc.	✓✓	Buffer zones protect areas adjacent to smelters.	Overall the use of buffer zones is effective however improved emission reduction technology is available.
Indoor air quality Environmental Tobacco Smoke (ETS)	Unknown. Few direct measurements are available.	✓	Designation of non-smoking areas in public buildings. However there are few controls on emissions in private homes.	Non-smoking policies for public buildings are selectively applied.
House Dust Mites (HDM)	Elevated levels are reported in the warm moist coastal areas of Australia.	✓	Nil	Reduced ventilation requirements for energy efficiency may lead to an increase in the concentration of HDMs in private homes.
Legionella	Isolated outbreaks have occurred in mainland Australia with fatal outbreaks linked mainly to poor maintenance of cooling towers.	✓✓✓	Legionella is a notifiable disease. Building codes and NH&MRC guidelines have been developed for the maintenance of cooling towers.	High standards of maintenance must be ensured.
Asbestos	Asbestos has been used extensively as a construction material and for insulation. Asbestos is still present in many old buildings in Australia. The major risk of exposure is linked to disturbances such as renovations and removal.	✓✓	A code of practice for removal has been developed.	The adequacy of the current responses is unknown.
Radon	Levels are generally low in Australian homes.	✓	NH&MRC has defined a guideline.	Effective.
Volatile Organic Compounds (VOCs)	Concentrations can be high in new buildings and caravans. VOCs, particularly formaldehyde, are released by furnishings, carpet and particle board. Concentrations decrease with time.	✓	Apart from ventilation requirements in building codes there are few controls on emissions in private homes.	Ventilation requirements have weakened with the move to increase energy efficiency thus increasing the potential for high concentrations of VOCs indoors.
Health effects and personal exposure.	Limited studies show Australians spend 90% of time indoors where the range of pollutants and personal exposure times are often greater than outdoors.	✓	Small numbers of targeted public education programs.	Limited progress in some aspects but ineffective in regard to many others.

Further reading

Australia, Bureau of Meteorology (1989). 'Climate of Australia.' (AGPS: Canberra.)

Boubel, R.W., Fox, D.L., Turner, D.B., and Stern, A.C. (1994). 'Fundamentals of Air Pollution.' 3rd ed. (Academic Press: San Diego.)

Crowder, R. B. (1995). 'Wonders of the Weather.' (AGPS: Canberra.)

Holper, P. (1994). Atmosphere: weather, climate and pollution. In 'Science and Our Future.' (Cambridge University Press: Cambridge.)

Pearman, G.I. (ed.) (1988). 'Greenhouse: Planning for Climate Change.' (CSIRO: Melbourne.)

Steering Committee of the Climate Change Study (1995). 'Climate Change Science: Current Understanding and Uncertainties.' (Australian Academy of Technological Sciences and Engineering, Australian Academy of Science and Academy of the Social Sciences: Melbourne.)

Swaine, D.J. (ed) (1990). 'Greenhouse and Energy.' (CSIRO: Melbourne.)

References

Ahmet, S., and van Dijk, M. (1994). Monitoring of air quality in the Port Phillip control region, 1979–1991. *Environment Protection Authority of Victoria, Publication* No. 421.

Ahmet, S., and van Dijk, M. (1995). Ambient air quality in the Port Phillip control region, 1979–1993. Compliance with objectives and observed trends. *Environment Protection Authority of Victoria, Publication* No. 468.

Australian Environment Council (1988). Air emission inventories for Australian capital cities. *AEC Report* No. 22.

Australian Environment Council (1989). Acid rain in Australia: a national assessment. *AEC Report* No. 25.

Australia and New Zealand Environment Council (ANZEC) (1990). 'Discussion Paper on Indoor Air Pollution.' (ANZEC: Canberra.)

Australia and New Zealand Environment and Conservation Council (ANZECC) (1994). Revised strategy for ozone protection in Australia 1994. *ANZECC Report* No. 30.

ASTEC (1990). 'Environmental Research in Australia: the Issues. A Report to the Prime Minister by the Australian Science and Technology Council (ASTEC).' (AGPS: Canberra.)

Ayers, G.P., and Gillett, R.W. (1988). Acidification in Australia. In 'SCOPE 36. Acidification in Tropical Countries', ed. H. Rodhe and R. Herrera, pp. 347–402. (Wiley and sons: New York.)

Basher, R.E., Zheng Xiaogu, and Nichol, S. (1994). Ozone-related trends in solar UV-B series. Geophysical Research Letters, **21**, (24), pp. 2713–16.

Bridgman, H.A. (1989). Acid rain studies in Australia and New Zealand. *Archives of Environmental Contamination and Toxicology*, **18**, pp. 137–46.

Brown, S. (in press). Indoor Air Quality. *Report prepared for the Atmosphere Reference Group, DEST.*

Cameron, I., and Rye, P. (1992). Swan Valley fluoride study. *Proceedings, 11th International Conference of the Clean Air Society of Australia and New Zealand, Brisbane, July 5–10, 1992.*

Carnovale, F., Alviano, P., Carvalho, C., Deitch, G., Jiang, S., Macaulay, D., and Summers, M. (1991). Air emissions inventory for the Port Phillip control region: planning for the future. *EPAV Publication SRS* 91/001.

Commonwealth of Australia, Senate (1991). 'Rescue the Future: Reducing the Impact of the Greenhouse Effect.' Senate Standing Committee on Industry, Science and Technology. Parliament of the Commonwealth of Australia, January 1991.

Commonwealth of Australia (1992). 'National Greenhouse Response Strategy.' (AGPS: Canberra.)

Commonwealth of Australia (1994). 'Climate Change: Australia's National Report under the United Nations Framework Convention on Climate Change.' (AGPS: Canberra.)

CSIRO (1992). 'Climate Change Scenarios for the Australian Region.' (CSIRO: Melbourne.)

Cope, M., Carnovale F., and Cook B. (1992). The impact of emissions from gas-fired turbines for cogeneration on ambient air quality in Melbourne. *Proceedings, 11th International Conference, Clean Air Society of Australia and New Zealand, Brisbane, 1, pp. 328–37.*

Crowder, R. B. (1995). 'Wonders of the Weather.' (AGPS: Canberra.)

Department of the Environment, Sport and Territories (1995). 'Greenhouse 21C. A Plan of Action for a Sustainable Future.' (DEST: Canberra.)

Doley, D. (1992). Aluminium production and vegetation injury: an Australian case study 1978–91. *11th International Conference of the Clean Air Society of Australia and New Zealand, Brisbane,.* **1**, pp. 287–96.

Doley, D. and McCune, D.C. (1993) Ambient air quality standards for sulfur dioxide in Australia: 1. Criteria and analysis. *Clean Air,* **27**(3) pp. 122–32.

Doley, D. (1994). Criteria for air quality in Australia: some considerations for sulfur dioxide and multiple pollutants, Clean Air '94: *Proceedings, Clean Air Society of Australia and New Zealand, Perth, WA, 23–28 October, 1994,* **2**, pp. 497–506.

Doley, D., and Moller, I.M. (1994). Vegetation responses to seasonal and operational conditions at an aluminium smelter. Clean Air '94: *Proceedings, Clean Air Society of Australia and New Zealand, Perth, WA, 23–28 October, 1994,* **1**, pp. 113–23.

Environment Protection Authority of Victoria (1993). Indoor air quality in domestic premises in Victoria — a review. *Publication* No. 327.

Environment Protection Authority of Victoria (1994). Review of ambient air monitoring programs. EPAV Publication No. 442.

EPA, NSW (1993). 'New South Wales State of the Environment 1993.' (EPA: Chatswood NSW.)

EPA, NSW (1995). 'New South Wales State of the Environment 1995.' (EPA: Chatswood NSW.)

Ferrari, L., McPhail, S., and Johnson, D. (1988). Indoor air pollution in Australian homes — results of two winter campaigns. *Clean Air,* **22**(2), pp. 68–74.

Fraser, P.J., and Bouma, W.J. (1990). New ozone protocol and Australia. *Search* **21**, 261–4.

Gillet, R.W., Ayers, G.P., and Noller, B.N. (1990). Rainfall acidity at Jabiru, Australia in the wet season of 1983/84. *Science of the Total Environment,* **92**, 129–44.

Hyde, R., and Johnson, G.M. (1990). 'Pilot Study: Evaluation of Air Quality Issues for the Development of Macarthur South and South Creek Valley region of Sydney.' *Report prepared for the NSW Department of Planning, NSW State Pollution Control Comission, Commonwealth Department of Transport and Communications. Macquarie University and CSIRO.*

International Energy Agency (1994). Climate change policy initiatives, 1994 update. Volume 1, OECD countries, *IEA Energy and the Environment Series.*

IPCC (1990). 'Climate Change: the IPCC Scientific Assessment', ed. J.T. Houghton, G.J. Jenkins and J.J. Ephraums. (Cambridge University Press: Cambridge.)

IPCC (1992). 'Climate Change 1992: the Supplementary Report to the IPCC Scientific Assessment', ed. J. Houghton, B.A. Callander and S.K. Varney. (Cambridge University Press: Cambridge.)

IPCC (1994). 'Climate Change 1994: Radiative Forcing of Climate Change and an Evaluation of the IPCC IS92 Emission Scenarios', ed. J.T Houghton, L.G. Meiro Filho, J. Bruce, Hoesung Lee, B.A. Callander, E. Haites and K. Maskell. (Cambridge University Press: Cambridge.)

IPCC (1995). Summary for Policymakers of the Contribution of Working Group 1 to the IPCC Second Assessment Report. (Cambridge University Press: Cambridge.)

Johnson, G.M., Quigley, S.M. and Smith, J.G. (1990). Management of photochemical smog using the Airtrak approach. *Proceedings, 10th International Conference of the Clean Air Society of Australia and New Zealand, Auckland, NZ, 1990,* pp. 209–14.

Langroo, M.K., Wise, K.N., Duggleby, J.C., and Kotler, L.H. (1990). A nation-wide survey of radon and gamma radiation levels in Australian homes. *Australian Radiation Laboratory Report* ARL/TR090.

Leighton, R., and Spark, E. (1995). 'An Investigation into the Synoptic Patterns Associated with Air Pollution in Sydney.' (Bureau of Meteorology: Melbourne.)

Lunney, K.E. Best, P.R. and Anh, V.V. (1994a). Air quality characteristics of the Brisbane airshed determined from historical monitoring information. *Proceedings, 12th International Conference, Clean Air Society of Australia and New Zealand, Perth,* **2**, pp. 283–95.

Lunney, K.E., Best, P.R., and Anh, V.V. (1994b). Synthetic approaches to air quality management problems for urban areas. *Proceedings, 12th International Conference, Clean Air Society of Australia and New Zealand, Perth,* **2**, pp. 107–15.

Manins, P. (ed.) (1988). 'Special issue: Latrobe Valley airshed,' *Clean Air,* **22**, (4) pp. 123–228.

Manins, P.C., Physick, W.L., Hurley, P.J., and Noonan, J.A. (1994). The role of coastal terrain in the dispersion of pollutants from Australia's major cities. *Proceedings, 12th International Conference, Clean Air Society of Australia and New Zealand, Perth,* **2**, pp. 179–88.

Mansfield, T. (1990). The cost of stone building soiling in Sydney. *Clean Air,* **24**(1), pp. 31–33.

Marks, R. (1989). Possible effects of increased UV radiation on the incidence of non-melanocytic skin cancer. In 'Health Effects of Ozone Layer Depletion. A Report of the National Health and Medical Research Council Working Party, Melbourne 1989.' (AGPS: Canberra.)

Monk, R. (1994). Effects of urban ozone on Australian native vegetation. *Proceedings, 12th International Conference, Clean Air Society of Australia and New Zealand, Perth,* **1**, pp. 141–9.

Murray, F., Monk, R., Clark K., and Wilson S. (1992). The relationship between exposure to sulfur dioxide and yield or growth of crops and trees in Australia. *Proceedings, 11th International Conference of the Clean Air Society of Australia and New Zealand, Brisbane,* **1**, pp. 273–86.

Murray, F., and Wilson, S. (1989). The relationship between sulfur dioxide concentration and crop yield of five crops in Australia. *Clean Air,* **23**(2), pp. 51–5.

National Greenhouse Gas Inventory Committee (1994). 'National Greenhouse Gas Inventory 1988 and 1990.' (AGPS: Canberra.)

National Health and Medical Research Council (1993). Volatile organic compounds in indoor air. *Report of the 115th NH&MRC Session.*

Nelson, P.F., and Duffy, B.L. (1994). Air toxics in ambient air and vehicle exhaust. *Proceedings, 12th International Conference, Clean Air Society of Australia and New Zealand, Perth,* **1**, pp. 305–20.

Noller, B.N., Currey, N.A., Ayers, G.P., and Gillet, R.W. (1990). Chemical composition and acidity of rainfall in the Alligator Rivers region, Northern Territory, Australia. *Science of the Total Environment*, **91**, pp. 23–48.

Ormerod, R. (in press). Urban and regional air quality. *Report prepared for the Atmosphere Reference Group, DEST.*

Peach, H.G. (in press). Air quality and human health. *Report prepared for the Atmosphere Reference Group, DEST.*

Salinger, M.J., Allan, R., Bindoff, N., Hannah, J., Lavery, B., Lin, Z., Lindesay, J., Nicholls, N., Plummer, N., and Torok, S. (in press). Observed variability and change in climate and sea-level in oceania. In 'Greenhouse: Coping with Climate Change', ed. W. Bouma, G. Pearman and M. Manning. (CSIRO: Melbourne.)

South Australia, Department of Environment and Land Management (1993). 'The State of the Environment Report for South Australia. (DELM: Adelaide.)

Steer, K., and Heiskanen, L. (1993). 'Options for Australian Air Quality Goals for Oxides of Sulphur. Public Review Document.' (Department of Health, Housing, Local Government and Community Services: Canberra.)

Steering Committee of Climate Change Study (1995). Climate change science: current understandings and uncertainties. (ATS, AAS and ASS: Melbourne.)

Streeton, J.A. (1990). 'Air Pollution Health Effects and Air Quality Objectives in Victoria.' (Streeton: Melbourne.)

Tovey, E.R. (1992). Allergen exposure and control. *Experimental and Applied Acarology*, **16**, pp. 181–202.

Weir, P., and Muriale, O. (1994). Development of an inventory of emissions for the Perth airshed. *Proceedings, 12th International Conference, Clean Air Society of Australia and New Zealand, Perth*, **1**, pp. 457–68.

Williams, D.J. (in press). Long-range transport of pollution and its significance in Australia. In 'Four Air Quality Issues in Australia'. *Report prepared for the Atmosphere Reference Group, DEST.*

World Health Organization (1987). Air quality guidelines for Europe. *WHO European Series* No. 23.

World Meteorological Organization (1995). Scientific assessment of ozone depletion: 1994. *Global Ozone Research and Monitoring Project, Report* No. 37.

Acknowledgments

The following people reviewed the chapter in draft form and provided constructive comments.

Dr John Zillman (Bureau of Meteorology)

Dr Graeme Pearman (CSIRO Division of Atmospheric Research)

Dr John Taylor (Australian National University)

Professor Paul Greenfield (University of Queensland)

Mr Len Ferrari (Consultant)

Dr Clive Hamilton (Australia Institute)

The Atmosphere Reference Group thank the following experts who provided valuable assistance in the preparation of this chapter:

Dr Sabriye Ahmet (Environment Protection Authority, Victoria)

Dr Greg Ayers (CSIRO Division of Atmospheric Research)

Dr Steve Brown (CSIRO Division of Building, Construction and Engineering)

Dr Paul Fraser (CSIRO Division of Atmospheric Research)

Mr Marcel van Dijk (Bureau of Meteorology)

Dr Graham Johnson (CSIRO Division of Coal and Energy Research)

Mr Robin Ormerod (Dames and Moore)

Dr Neville Nicholls (Bureau of Meteorology)

Dr Peter Manins (CSIRO Division of Atmospheric Research)

Professor Hedley Peach (Melbourne University, Department of Public Health and Community Medicine)

Mr David Williams (CSIRO Division of Coal and Energy Technology).

A number of other people from CSIRO, Commonwealth and State government departments, environment protection agencies and private industry provided input. The assistance of Mr Allan Spessa (State of the Environment Reporting Unit) in the final preparation of the Report and Mr James Macnicol in the preparation of a number of the figures is also gratefully acknowledged.

In addition, Commonwealth Government departments and members of the Commonwealth/State ANZECC State of the Environment Reporting Taskforce also helped identify errors of fact or omission.

Photo Credits

Page 5-1: Peter Mackey, by courtesy of the Bureau of Meteorology

Page 5-4: Japanese GMS Image, by courtesy of the Bureau of Meteorology

Page 5-8: David Griffith, University of Wollongong

Page 5-14: Bureau of Meteorology, US NOAA Satellite

Page 5-19: David Whillas, CSIRO Division of Atmospheric Research

Page 5-24: *The Sunday Age* newspaper

Page 5-26: Images by courtesy of CSIRO Division of Atmospheric Research

Chapter 6

Land Resources

Prepared by

Ian Noble (Chair), Australian National University, Research School of Biological Sciences

Michele Barson, Bureau of Resource Sciences, Department of Primary Industries and Energy

Robert Dumsday, La Trobe University, School of Agriculture

Margaret Friedel, CSIRO Division of Wildlife and Ecology

Ron Hacker, NSW Agriculture

Neil McKenzie, CSIRO Division of Soils

George Smith, Queensland Department of Natural Resources

Mike Young, CSIRO Division of Wildlife and Ecology

Mathew Maliel (State of the Environment Reporting Unit member), Department of the Environment, Sport and Territories (Facilitator)

Charlie Zammit (former State of the Environment Reporting Unit member), Department of the Environment, Sport and Territories (former Facilitator)

Contents

Introduction

The word 'land' has different meanings for different people. For many Australians — including indigenous communities, farming families, conservation workers and others who work daily with the land — 'our land is our life'. For Aboriginal people the landscape not only provides a source of sustenance, but also expresses spiritual beliefs and power relationships. The community takes responsibility to actively care for the land and to fulfil ritual obligations. Many non-Aboriginal people also have a strong personal attachment to the care of particular landscapes. For others, especially the 88 per cent of Australians who live in coastal cities, the land is a more distant entity. Nevertheless, acceptance of the land ethic — that is, our individual and collective responsibility for the stewardship of our land — is widespread in the Australian community.

We make many different uses of our land resource and often the same patch of land can serve multiple uses. We are a small population in a large island country with old and relatively infertile soils and low rainfall. The prevailing vegetation is woodland and shrubland. The land is mainly used for extensive grazing by introduced stock, but we also have significant areas of intensively managed land.

This chapter deals with the environmental impacts of the way we use our land resources, the pressures leading to those impacts and the responses taken to the changing state of the land resources. It incorporates the economic forces that encourage or constrain particular actions and also deals with the impacts on agricultural productivity, because losses in productivity are often associated with off-site impacts and pressures to use more land resources. The approach taken is to describe the condition and, where possible, the trends of each of the major land resources (such as agricultural land, which is dealt with mainly in the section on soils, rangelands and forests). The chapter also deals with the sustainability of recent and current usage.

Pressure

General pressures

Australia's land resources are influenced by a number of factors that the OECD describes as 'general pressures'. These are indirect pressures that affect all aspects of our land use and thus the state of our land resource.

Australia has a much lower population density than most other countries: about two people per sq km compared with a world average of 42. By themselves, these figures suggest that Australians need exert little pressure on their land resource. But this simple estimate hides the true situation. The population is not evenly distributed but is concentrated along the eastern, southern and south-western coasts. More than 80 per cent of us live on only three per cent of the land. Australians enjoy a high standard of living and the country supports more than its own population. For example, we feed the equivalent of about an extra 50 million people with cereals (see Fig. 6.1). We also produce one-third of all the world's wool and more than twice our own meat requirements.

Our arable land totals 467 000 sq km or six per cent of our land surface. By comparison, the United States has 20 per cent arable land, the United Kingdom 26 per cent and Spain 30 per cent. And, despite the low fertility of our soils, we also use less fertiliser than comparable countries — only 0.03 tonne of fertiliser per hectare of arable land, or one-quarter of the rate used in the United States and Spain and one-twelfth of that in the United Kingdom. Thus, the effective population pressure on our land resources is far higher than the simple population density would imply.

Australia is a dry continent with infertile soils and high climatic variability (see Chapter 2). These features themselves do not put pressure on our land resource, but human activities that fail to take them into account do. For example, although we should take account of the cycle of droughts caused by the El Niño–Southern Oscillation in our biological and financial planning of land use, individuals and governments often overlook these in strategic planning, leading to responses in the form of crisis management. Whatever its precise nature, climate change will most likely occur as an increase in apparent climatic variability (see Chapter 5). Australia may be in a better position than most countries to deal with this since we are used to coping with variability and have not only an effective research and development infrastructure but the educational and financial resources needed to change agricultural practices.

A sustainable ecosystem must have a set of species to carry out the essential cycle of production (plants), consumption (animals) and decomposition (mostly micro-organisms). Many species perform specialist roles in these cycles — for example, as decomposers of certain substances, as plants adapted to particular habitats and so on. When an ecosystem loses or gains species, the cycles may be disrupted and pressure exerted on the land resource (see Chapter 4).

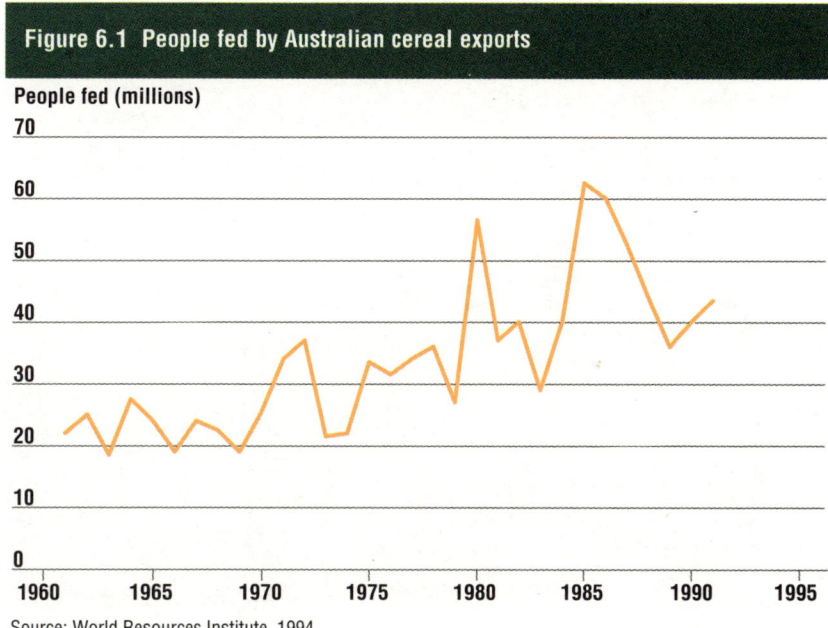

Figure 6.1 People fed by Australian cereal exports

People fed (millions)

Source: World Resources Institute, 1994

The demand for food and fibre imposes general pressures on land resources, which vary depending on production levels and commodity prices. Government actions may exacerbate or relieve the pressures depending on the social and political climate at the time.

Indirect and proximate pressures

Different forms of land-use have a number of indirect and direct (proximate) pressures associated with them.

Agriculture

Most agricultural practices lead to natural vegetation being replaced by plants more suited to the agricultural systems. This occurs either through direct clearing and replacement of native vegetation with crops or pasture species, or through the gradual destruction of native vegetation by overgrazing, by changes in water availability and salinisation, or simply by the failure of the native species to recruit new individuals to replace those that die. Although land clearing was essential if we wished to grow food and fibre, it has opened our land resource to damage by erosion, destroyed soil structure, changed soil chemistry and caused losses of biodiversity and other problems.

Australian soils have very low nutrient contents and much of our agriculture is dependent on inputs of nitrogen, phosphorus and other elements applied as chemical fertilisers or achieved by planting legumes to improve soil nitrogen levels (see page 6-30). For Australia as a whole, the nutrient balance is positive for all the major elements. That is, the natural release of the major elements by weathering, fixation of nitrogen by native and introduced species such as clovers, plus net imports and production of artificial fertilisers amount to a total that exceeds the estimated losses (see Table 6.7). However, on a local level, some areas have a net loss and others a net accumulation. Total fertiliser use per hectare of crops has changed little over the past 15 years. But, the amount of

Intensive agriculture places pressure on the land.

accumulate, or become concentrated in some biological cycles. Organisms can evolve resistance to previously effective chemicals, so higher dosages or new chemicals need to be applied. Chemical usage on farms is increasing rapidly. For example, in Victoria the area treated with insecticides rose from 80 000 ha in 1983 to 226 000 ha in 1989. Many conservation tillage practices require an increased use of farm chemicals for weed and pest control.

Pastoralism

Pastoralism exerts an indirect pressure on the land resource. Its biggest impact is due to increases in the number of large herbivores — domestic, native and feral. These animals can destroy the vegetative cover and break up the soil surface, exposing it to water and wind erosion (see page 6-26). Pastoralists have greatly increased the number of watering points and in sheep grazing areas reduced the population of dingoes, leading to an increase in the density of kangaroos and the maintenance of high densities of herbivores over more of the landscape than before European settlement. They have also modified fire frequencies (usually decreasing them compared with Aboriginal practices), often leading to changes in the vegetation (see pages 6-8 and 6-9).

Forestry

The Resource Assessment Commission Forest and Timber Inquiry of 1992 assessed the pressures of human activities on forest resources. Pressures arise from our use of forests as sources of wood products and for grazing, recreation, conservation and water-catchments. The Inquiry identified areas of concern including: impacts of logging operations on populations of forest flora and fauna, soil compaction and erosion, stream siltation and reduced water quality, and the effects of changed fire regimes in all forests. Its report concluded that human use of forests, including some wood-production activities, affects the habitats and population sizes of some forest-dependent species and increases the risk of invasion by exotic species. However, there was no evidence that these risks posed an immediate threat to the ecological processes on which forest systems depend.
The report highlighted the availability of only a very small amount of scientific information about forest impacts, reflecting the lack of basic research and effective monitoring of many of our land resources.

Mining

Numerous mining sites and petroleum fields occur across Australia (see Fig. 6.3) but the land area they actually occupy is very small (less than 0.01 per cent). Some sites are affected by land clearance and pollution or waste disposal, but the most widespread effects are associated with roads and infrastructure that provide access to remote areas surrounding prospecting leases and mining towns. The major controversies over land use for mining occur where mining priorities coincide with sites of high biodiversity or cultural significance.

phosphate has decreased while that of nitrogen has increased, reflecting changes in cropping practices and the price relativities (see Fig. 6.2).

Cultivation practices often result in land degradation. For example, mechanical fallowing can lead to wind and water erosion, and repeated cultivation for weed control can cause soil compaction, organic matter oxidation and soil structural decline. Realising this, many farmers have changed land management practices — to include rotations with pasture phases, conservation tillage and stubble retention — and reduced stocking density to maintain ground cover.

Most land-use practices modify the hydrology of the surrounding area either directly or indirectly. A common example is heavy grazing, which can lead to vegetation loss and expose the soil to increased water and wind erosion. Another is land clearing, leading to raised water tables and salinisation. Transport corridors often cause inadvertent but severe changes to water flow. Another pressure arising indirectly from agriculture is the construction of water-storages for stock use and irrigation, ranging in size from small farm dams to major reservoirs. They divert stream flow with subsequent effects on downstream environments and redistribute water for irrigation, which can lead to salinity problems (see Chapter 7).

Pollutants, whatever their source, may kill organisms directly or indirectly by changing the environment. Some pollutants — such as pesticides associated with sites of former sheep and cattle dips — have a point source. Others, such as pesticides or herbicides applied over wide areas of land, or gases such as sulfur dioxide released from industrial areas, are described as having a diffuse source (see page 5-33). On a world scale, Australia releases only a small proportion of pollutants and potential pollutants such as pesticides and herbicides. On a national level, a lack of information on the nature and amounts of the materials being released is a matter of concern. Pollutants can gradually become a problem when very persistent chemicals

Human habitation

Although Australian cities and towns occupy only 0.01 per cent of the continent, this area is concentrated in our most important catchments and often on high-value agricultural land. Urban fringe development and coastal development often compete with agricultural and conservation land uses. The impact of human habitations extends far beyond the area cleared for buildings and infrastructure (see Chapter 3).

The tendency for city-dwellers to acquire land for rural retreats or hobby farms has brought about major changes in land use around many cities. These changes can have both positive and negative effects on land and related resources. Hobby farmers often have off-farm incomes that can support improvements such as fencing, weed control, tree-planting, erosion-control measures and improved pastures. Negative effects may include pollution of surface and groundwater by septic effluent, soil erosion through overstocking and destruction of habitats by clearing. Hobby farms in the Adelaide Hills and around Canberra present contrasting examples. While the former have reputedly increased turbidity of urban water supplies through erosion caused by overstocking with horses, the latter have reduced land degradation through increased tree-planting.

Tourists may exert pressure on our land resources (see Chapters 4 and 9). In some places (for example, alpine areas in Kosciusko National Park), concern about camping wastes and trampling pressure of numerous tourists has led to steps to control access by camping restrictions and the use of raised footpaths. In Tasmania there is evidence that fires are more frequent along popular tourist routes and concern that tourists may significantly increase the spread of weeds and diseases.

Corridors for roads, railways, pipelines and powerlines open up access and bring tourists, development, grazing animals, weeds and pests. They can also modify water flows and act as barriers to the movement of native organisms (see Chapter 4). In some areas, corridors are less than a kilometre apart and on a State by State basis separation varies from an average of 1.5 km in Victoria to 71 km in the Northern Territory.

The loss of traditional Aboriginal land-management practices, particularly burning, has led to changes in vegetation. Fire management can be traced through several phases:

- Long-established patterns of burning under Aboriginal management modified the vegetation, but over tens of thousands of years it reached a new equilibrium.

- After European occupation, periods of reduced burning often occurred as Aboriginal practices were lost. These were usually followed by a period of frequent and intense burning during a clearing and settlement phase.

- The new occupants attempted to protect their investment in fencing, buildings, pastures and forests by suppressing fires, but fuel loads built up, leading to years of major bushfires.

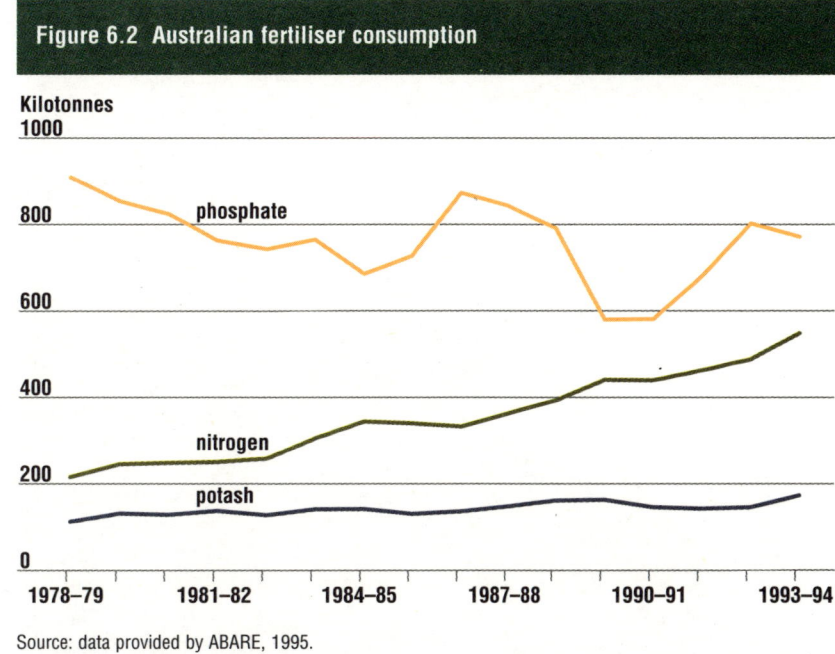

Figure 6.2 Australian fertiliser consumption

Source: data provided by ABARE, 1995.

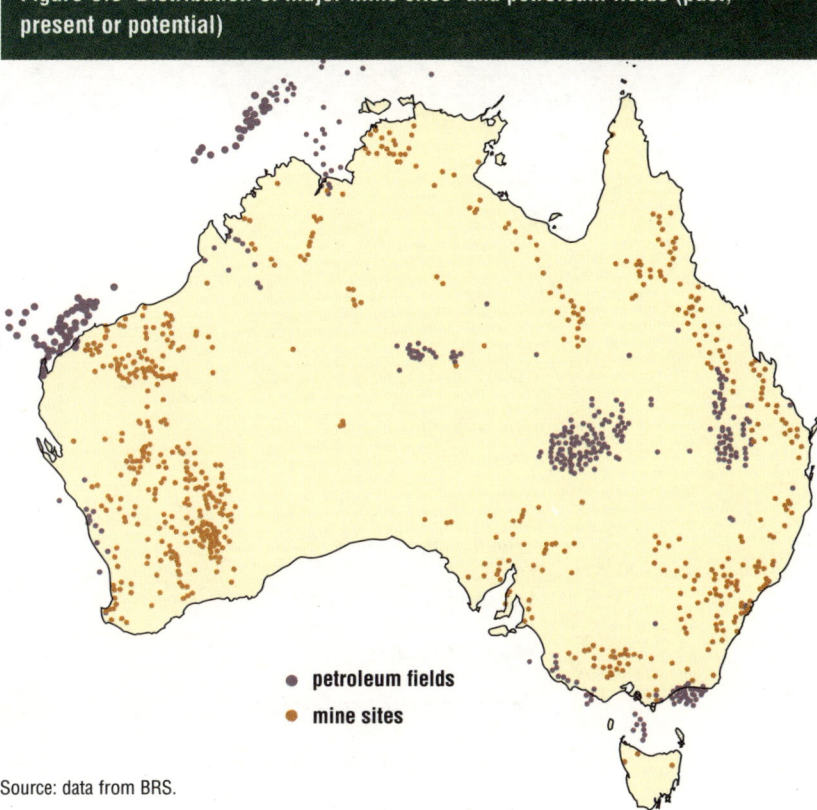

Figure 6.3 Distribution of major mine sites and petroleum fields (past, present or potential)

- petroleum fields
- mine sites

Source: data from BRS.

- In many areas prescribed burning regimes have been introduced — in some places these are designed to reflect Aboriginal practice (although they may no longer necessarily be appropriate in the modified vegetation).

Each of these fire regimes changed the vegetation, animals and even soils. Although the amount and nature of change varies from place to place, it is clear that changes in fire management have imposed significant pressures on the biota and soils.

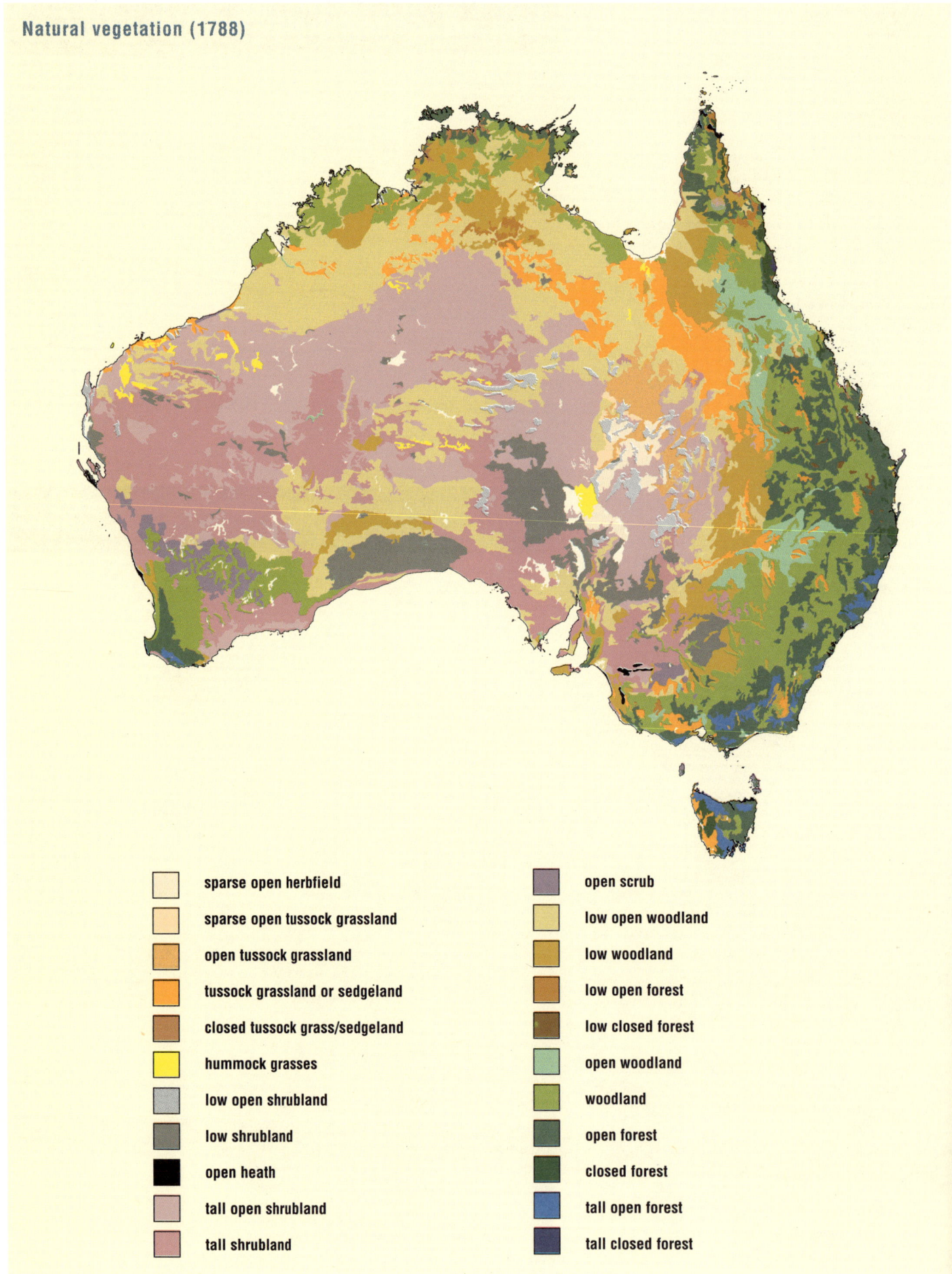

Natural vegetation (1788)

sparse open herbfield		open scrub	
sparse open tussock grassland		low open woodland	
open tussock grassland		low woodland	
tussock grassland or sedgeland		low open forest	
closed tussock grass/sedgeland		low closed forest	
hummock grasses		open woodland	
low open shrubland		woodland	
low shrubland		open forest	
open heath		closed forest	
tall open shrubland		tall open forest	
tall shrubland		tall closed forest	

Source: AUSLIG, 1991.

Present vegetation (1988)

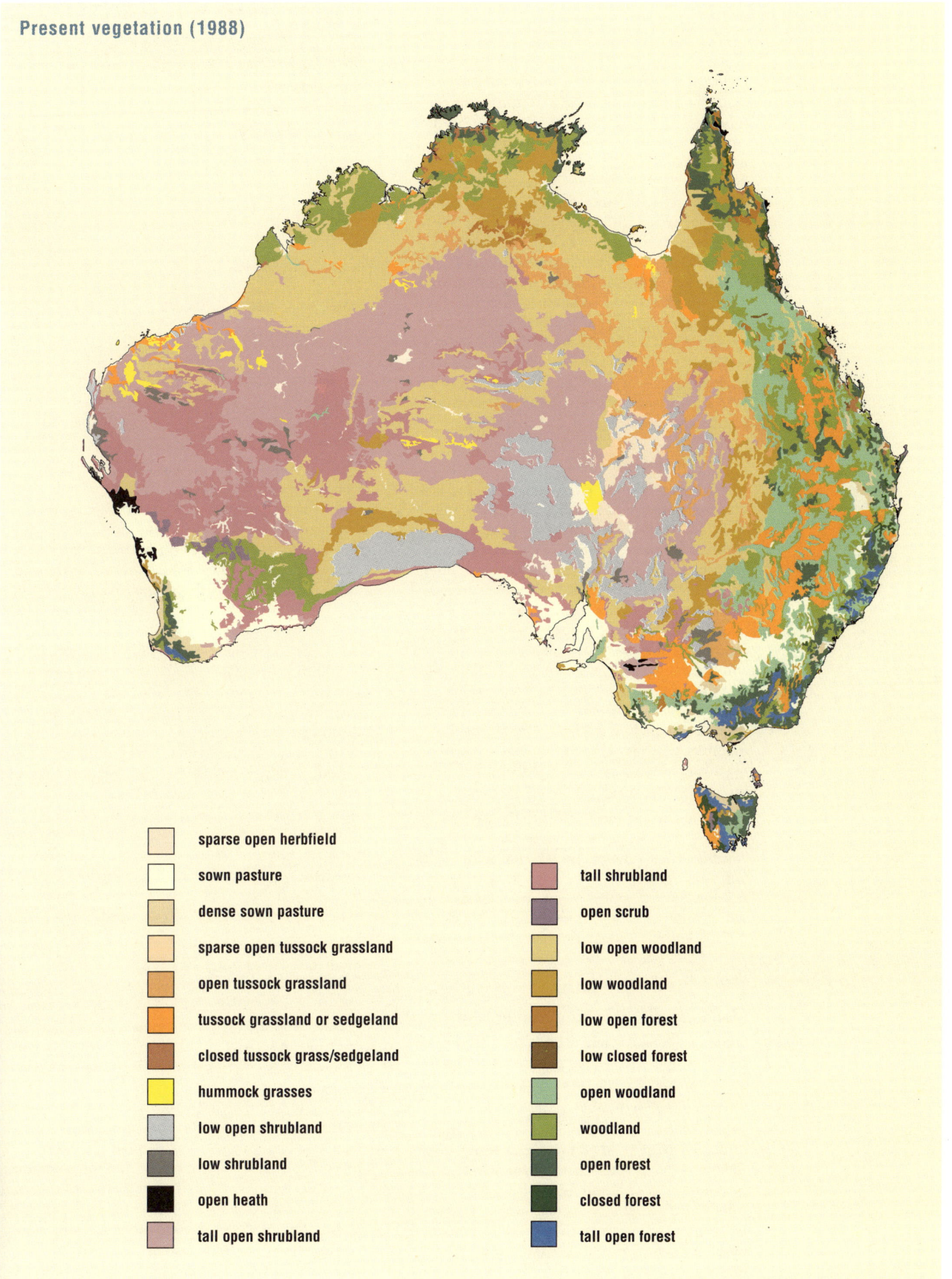

Legend:

- sparse open herbfield
- sown pasture
- dense sown pasture
- sparse open tussock grassland
- open tussock grassland
- tussock grassland or sedgeland
- closed tussock grass/sedgeland
- hummock grasses
- low open shrubland
- low shrubland
- open heath
- tall open shrubland
- tall shrubland
- open scrub
- low open woodland
- low woodland
- low open forest
- low closed forest
- open woodland
- woodland
- open forest
- closed forest
- tall open forest

Source: AUSLIG, 1991.

State

This section emphasises some of the more important aspects of the state of our land resources. Land cover, for example, is important for describing the condition of the land resource, since it both moderates the effects of weather on the land surface and reflects the impact of disturbances. Major variations in land cover warn of some potentially significant changes in the state of the land itself. Other changes may be reflected not by variations in the amount of cover but in more subtle shifts in the balance of certain plant types. These and related changes are discussed under the heading of changes in land condition. The condition of the soil, which underlies the ecological integrity and productivity of land resources, is described in some detail. Many of our responses to the state of land resources are reflected in changes in land use. Thus, this section finishes with a brief summary of current land uses and land tenure, and trends over the past decades.

Changes in land cover

The term 'land cover' refers to the physical state of the land surface and includes vegetation, soil, rock, water and man-made structures. Land cover is the interface between the earth's crust and the atmosphere, influencing the exchanges of energy and matter in the climatic system and the biogeochemical cycles. Land cover changes can affect a wide range of processes — such as the movement of nutrients through plants, soil, water and the atmosphere, emission of greenhouse gases and the movement of soil and water within catchments (see Chapter 7). They can also have major consequences for living organisms; for example, clearing of vegetation can result in an organism with a very limited distribution becoming extinct (see Chapter 4).

Vegetation covers most of Australia. There is little bare soil apart from the three per cent of the continent that is fallow prior to cropping. In the arid zone, vegetation cover may be sparse in some seasons. Surface waters, including salt lakes, occupy some three per cent of the continent (Cocks, 1992). Australia's vegetation cover when Europeans arrived and that prevailing 200 years later are shown on pages 6-8 and 6-9.

Management practices affecting land cover

Some of the management practices affecting land cover include expansion and intensification of cropping, grazing practices, commercial forestry, mining and urbanisation.

In the 1950s and 1960s, long periods of high prices for many rural commodities and general national growth led to an expansion and intensification of farming (see Fig. 6.4). Intensification resulted in substantial clearing of remnant stands of trees on grazing properties to develop sown pastures. In some wheat-growing regions — such as south-west Western Australia, the mallee wheat lands of South Australia and the wheat belt of central New South Wales —

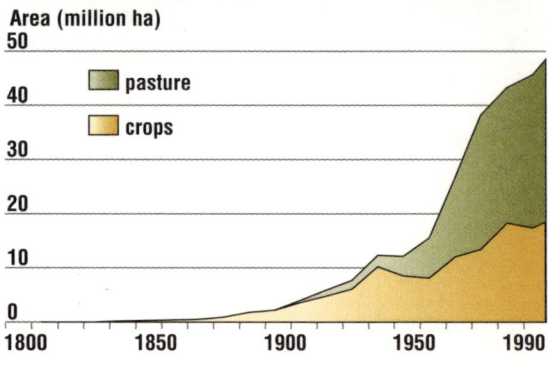

Figure 6.4 Area of crops and sown pastures 1800–1995

Area (million ha)

■ pasture
□ crops

Source: ABS Yearbooks.

expansion and intensification of cropping activities have resulted in major changes to land cover (see Fig. 6.5). Less than 20 per cent of the native vegetation remains in these regions.

Lands used for grazing comprise predominantly woodlands, open woodlands and shrublands. Only a small proportion of grazed land is true grassland, although tussock and hummock grasses form an understorey over very extensive areas. As well as clearing, landholders have thinned trees to promote grass growth, by ringbarking and more recently by using chemicals, over large tracts in the high-rainfall and semi-arid zones. Little thinning has occurred in the arid zone although pastoralists lop scrub, particularly mulga, in the arid and semi-arid zones of eastern Australia to provide drought feed for livestock. Overgrazing of chenopod shrublands by livestock and rabbits has prevented seedling regeneration, which, in conjunction with periodic wildfire, has resulted in the conversion of extensive areas to grasslands and herbfields. In the semi-arid and arid woodlands and shrublands, thinning of perennial grasses due to overgrazing by livestock, and to some extent rabbits, kangaroos and feral goats, together with a reduction in fire frequency, has resulted in an increase in native and exotic woody species or 'woody weeds'.

Overgrazing of grasslands in the semi-arid zone has resulted in some thinning of Mitchell grass communities, and in some instances conversion of grassland to herbfields, although most of these grasslands remain relatively intact. The expansion of exotic and native woody species poses a serious problem. In the tropical woodlands, native grasses are generally of low quality from a pastoral perspective. In the past, this, together with the use of British breeds of cattle (*Bos taurus*), which need relatively high-quality diets, limited the grazing pressure on these communities. The introduction of hardier *Bos indicus* cattle and the use of urea and molasses supplements around watering points have allowed much higher grazing pressures, and consequent degradation of sensitive areas. In southern Australia the temperate native grasslands have largely disappeared, replaced by cereal crops and introduced pasture species.

Commercial forestry involves harvesting and removal of some or all of the land cover in a given area. Each year about 200 000 ha or one per cent of the native forest available for logging is harvested and undergoes various regeneration practices. Additionally, about 30 000 ha of new softwood and over 10 000 ha of new hardwood plantations are established annually — mostly on land previously cleared for agriculture. Historically, harvest operations have ranged from clearfelling, which removes all the timber·in a coupe or patch ranging in size from 10 to 300 ha, to various forms of selective logging, which removes part of the existing growing stock at each harvest to create a stand of trees with a range of ages and sizes (see Fig. 4.3). Clearfelling is the most common method of harvesting in the wet sclerophyll forests of Tasmania, Victoria and Western Australia, the drier forests of south-eastern New South Wales and lower-elevation mixed forests in Victoria (RAC, 1992). No estimates are available of the areas of forest clearfelled in the past or the current annual rates of clearfelling. This practice, which is usually followed by burning of the wood debris to provide an ash bed for seedlings, results in a temporary loss of land cover until natural regeneration occurs or the coupe is replanted or direct-seeded. Clearfelling is often modified by leaving a number of trees for environmental protection or conservation. The site is seldom reharvested until the trees have reached a commercially valuable size — usually after 60 to 120 years.

Selective logging operations harvest part of the growing stock, repeating the harvest at relatively short intervals (decades), resulting in regular regeneration of seedlings. Practices range from single-tree selection to 'heavier' selective logging, where groups of trees are removed. Variants of the systems are used in the forests of Queensland and northern New South Wales, the jarrah forests of Western Australia and some mixed forests of Tasmania and Victoria.

Mining was responsible for destroying local land cover, as smelting was often done on the spot to reduce the bulk of the material to be transported. For example, miners stripped the timber around the Kapunda, Burra and Wallaroo/Kadina copper mines in the 1840s–'60s, around Broken Hill in the 1880s–'90s and around Zeehan, Mt Lyell and Mt Bischoff in western Tasmania in the early 1890s. The activities of gold-miners also laid waste to large areas of woodland — around Kalgoorlie and Coolgardie among others (Williams, 1988).

Although the growth of cities has only affected land cover over a small area, most of the urbanisation has occurred around the coast, sometimes in regions of high biological diversity. Patterns of clearance are usually influenced by slope and elevation, resulting in disproportionately high impacts on the vegetation of gentler slopes and coastal plains. For instance, in coastal south-east Queensland between 1974 and 1989, 33 per cent of the remaining bushland was cleared, predominantly for residential development but also for some plantation forestry and sugarcane production (Catterall and Kingston, 1993).

Land cover change 1788–1993

In 1788 forests covered almost nine per cent of Australia (see Fig. 6.5). Woodlands and open woodlands each covered about 21 per cent, while shrublands — including acacia and mallee eucalypts, heaths and saltbushes — occurred across 40 per cent of the continent. About seven per cent of the continent's surface was grassland (that is, grasses without an overstorey of woody vegetation) and less than one per cent was unvegetated. By the 1980s, about 175 000 sq km of forest had been thinned to woodland or open woodland, and a further 140 000 sq km cleared mainly for grazing, leaving about five per cent of the continent forested. Some 260 000 sq km of woodland were thinned to open woodlands and a further 320 000 sq km cleared for pasture and cropping as well as 50 000 sq km of open woodland. Altogether, clearing led to a decrease in the area of woodland, from 21 to 14 per cent, and a four per cent increase in that of open woodland. While the proportion of shrublands cleared (mostly mallee) has been relatively small (three per cent), about 20 per cent has been thinned. Grasslands now cover almost 16 per cent of the continent, the major changes being an increase of three per cent in the tussock grasslands and a six per cent cover of sown pasture grasses. Eight per cent of the native grassland present in 1788 has become open woodland.

Some people believe that the reduction in burning, which occurred as European settlers displaced Aboriginal people, resulted in extensive eucalypt regrowth and a subsequent expansion of the area of forest and an increase in its density. Anecdotal reports suggest that these changes took place in the

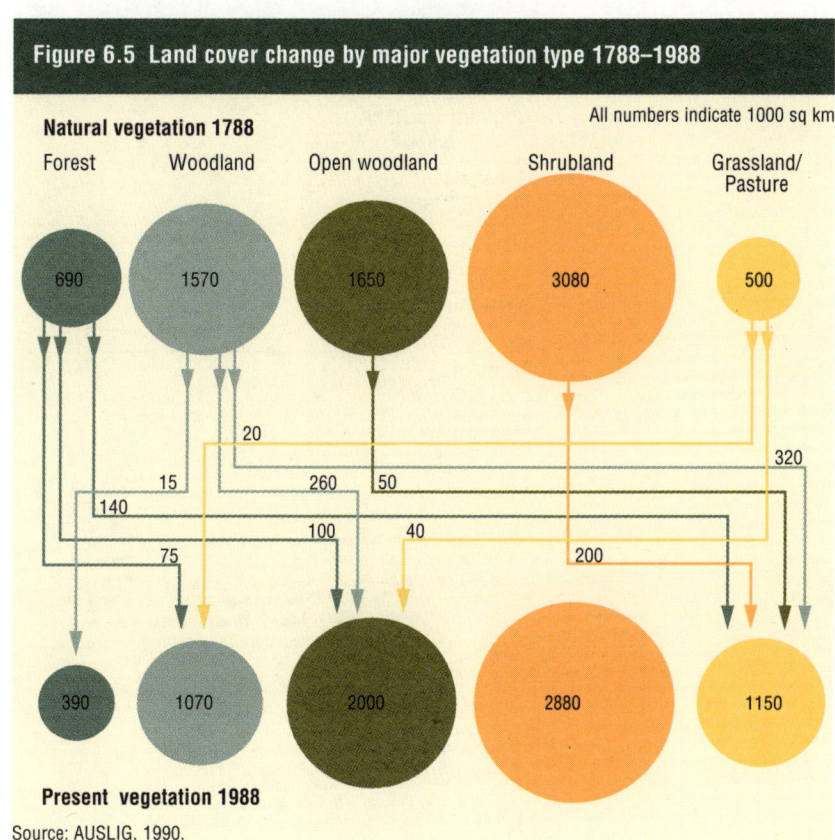

Figure 6.5 Land cover change by major vegetation type 1788–1988

Natural vegetation 1788

All numbers indicate 1000 sq km

Forest	Woodland	Open woodland	Shrubland	Grassland/Pasture
690	1570	1650	3080	500

20
320
15   260   50
140
100   40
75   200

| 390 | 1070 | 2000 | 2880 | 1150 |

Present vegetation 1988

Source: AUSLIG, 1990.

Figure 6.6 Timing of vegetation clearance since 1788

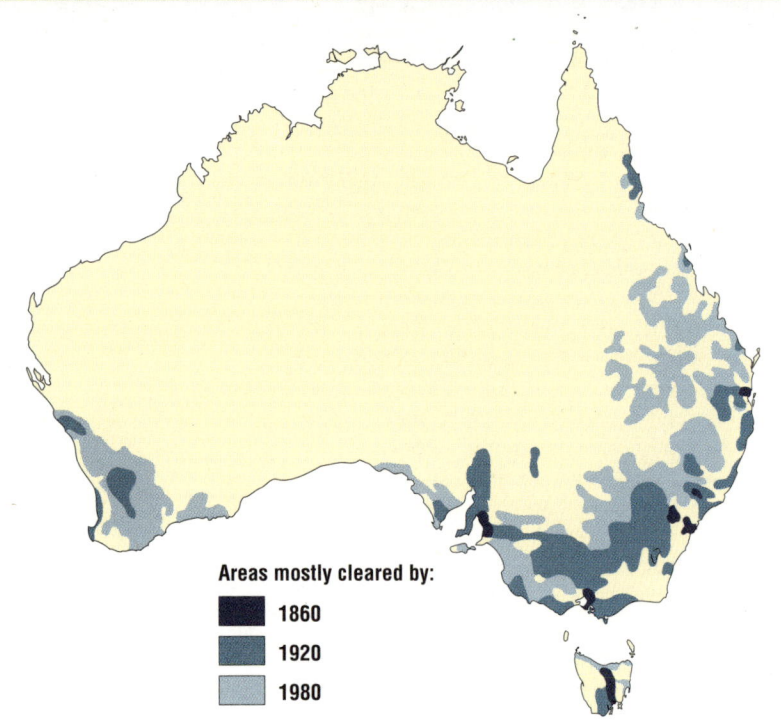

Areas mostly cleared by:
- ■ 1860
- ■ 1920
- ■ 1980

Source: derived from AUSLIG, 1990.

Table 6.1 Estimated rates of clearing on agricultural land 1983–1993

Sources of data used to estimate clearing 1983–93	Estimated area cleared	
	Average 1983–93 (sq km)	1990 (sq km)
ACT 1983–93[1]	0	0
New South Wales 1983–93[1]	1500	1500
Northern Territory 1983–91[2] 1992–93[3]	163	163
Queensland 1983–93[4]	3000	4500
Victoria 1983–88[5]# 1989–93[3]	78	62
South Australia 1983–93[3]	116	45
Tasmania 1983–88[6]#	60	60
Western Australia 1986–89[7] 1990–93[3]	260	311
Total		**6640**

Notes: Estimates based on:
1. Professional opinion
2. Department of Lands, Housing and Local Goverment records
3. Clearing permit data (Estimates of clearing based on the area for which permits have been granted are uncertain as the clearing may not have been done and illegal clearing may have occurred)
4. Survey of clearing contractors
5. Woodgate and Black, 1988
6. Kirkpatrick, 1991
7. Kestel Research and Victorian DCE, 1990
#. Figures based on remote sensing of land cover change

Source: adapted from DEST, 1994.

Victorian Wimmera and right across Gippsland, particularly after heavy rains in the early 1860s and 1880s and in northern New South Wales (the Pilliga forest). The extent of this regrowth is not known. More recently, increases in the density of woodlands in central and south-west Queensland managed for pastoralism have been attributed to increased grazing pressure and reduced burning.

Until recently, many people believed that the last major clearing for agriculture occurred in the 1970s and that relatively little had taken place since (see Fig. 6.6). However, a recent review undertaken for the National Greenhouse Gas Inventory (DEST, 1994) suggests that annual rates of clearing over the last 10 years could be greater than 5000 sq km (see Table 6.1).

ABARE's 1994 survey of broadacre and dairy farms asked farmers how much forest and woodlands they intended to clear in the next five years. At the national level, the response was about 3.28 million ha between 1994–95 and 1998–99 — equivalent to about 6500 sq km a year, and close to the estimated national rate of clearing for 1990 (see Table 6.1). Most of this intended clearing was expected to occur in Queensland (ABARE, 1995). Recent changes in that State (see Table 6.8) will influence the extent of any future clearing.

Researchers (Graetz *et al.*, 1995) have assessed the extent of clearing or thinning of major vegetation types in eastern Australia, south-west Western Australia and the Top End using low-resolution satellite data (see Fig. 6.7). By 1991, about 35 per cent of Australia's vegetation within this region had been cleared and 17 per cent thinned. About 36 per cent is largely unaffected, while the remaining 12 per cent could not be assigned to either category. It is unlikely that these estimates include the smaller-scale clearing of remnant vegetation.

Figure 6.7 Status of vegetation in the regions used for agricultural production in 1991

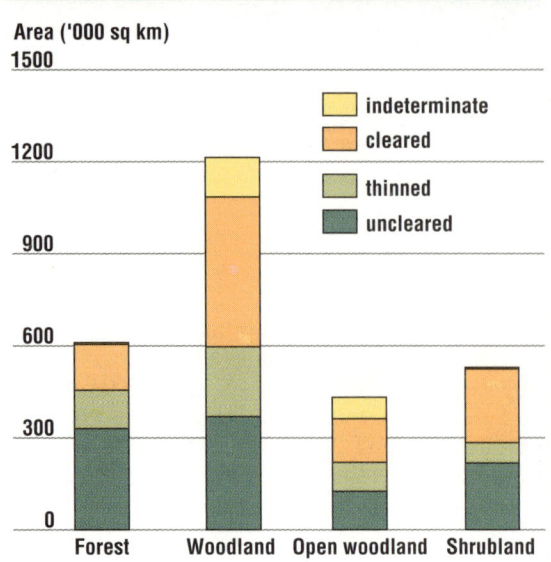

Area ('000 sq km)

Legend:
- □ indeterminate
- □ cleared
- □ thinned
- □ uncleared

Forest Woodland Open woodland Shrubland

Source: Graetz, pers. comm.

Changes in land condition

Land condition may be likened to the 'health' of the land, but how it is assessed depends on what the land is being used for. The condition of agricultural land, for example, is judged by its potential for production, which in turn depends largely on the state of the soil. The state of Australian soils is discussed in detail in a later section. The condition of remnant native vegetation within agricultural areas is usually assessed with reference to its 'pristine' state or, in other words, to its conservation status. People often equate it with the degree of change in structure or species composition. The presence of remnant vegetation may contribute in some degree to the productive potential of the surrounding agricultural land, by providing shelter for stock or by harbouring useful insects or predators of pests, but this contribution is largely unknown for Australian agricultural systems. Rangelands can be evaluated in many ways since they support a diversity of uses. The most common commercial use is pastoralism and so current assessments focus on productive potential. Range condition in this sense depends on the status of both vegetation and soils, and is discussed in greater detail below (see page 6-14).

Agricultural lands

About six per cent of the continent is now used for broadacre cropping (predominantly cereals), intensive grazing of sown pastures, intensive horticulture and field crops such as sugarcane and cotton. Development of the wheat–sheep zone (see Fig. 6.8) resulted in large-scale conversion of native land cover to agricultural land cover. In the high-rainfall region, forests and woodlands have been replaced by pastures sown for dairying, sheep and beef production as well as for cropping. However, substantial areas remain uncleared, particularly on more rugged terrain.

Clearing for agricultural development has been very selective, with the vegetation types occupying the better soils and gentler slopes being used first. As a result, many land cover types are now severely under-represented in remaining vegetation. For example, 85 per cent of Victoria's box–ironbark forests and woodlands have been cleared. Those remaining are located mainly on rocky areas, upper slopes, poorer soils or periodically inundated floodplains (Bennett, 1993).

Today, in the more intensively used grazing and cropping lands, remnants of the native land cover occur as isolated, uncleared patches or narrow strips along roadsides. The condition of the remainder is of considerable concern. Some remnant vegetation is now dying through old age, and is not being replaced because of poor seedling regeneration due to grazing by livestock or rabbits or competition from introduced pasture grasses. Invasion of weeds and fertiliser drift from adjacent agricultural lands will also affect the condition of patches of remnant vegetation.

Implications of clearing for greenhouse gas emissions

Australia, as a party to the United Nations Framework Convention on Climate Change, is required to prepare an inventory of greenhouse gas emissions. The inventory is prepared using an internationally agreed method developed by the Intergovernmental Panel on Climate Change. Using this method, several land use changes that can modify carbon dioxide fluxes between the terrestrial biosphere and the atmosphere were identified — including clearing for agriculture. These fluxes were calculated from estimates of the areas of vegetation cleared over the last 10 years (see Table 6.1) and the likely carbon content of this vegetation (about half the dry weight of this vegetation is carbon) and the soils below. The results suggested that land clearing may be contributing up to 27 per cent of total greenhouse gas emissions. However, the limited amount of information available on rates of clearing, the type of vegetation cleared and the effect of regrowth on carbon fluxes means that such contributions could be anywhere between seven per cent and 45 per cent of total emissions.

Since the mid '60s reports of premature tree decline or 'rural dieback' have been frequent. This affects many species and all ages of trees, and has been reported in most States. The condition is particularly severe in areas used for intensive livestock production and has a long list of possible causes. They include not only direct effects of pastoral management — such as fertiliser application, the introduction of exotic pasture species and damage from livestock and machinery — but more indirect changes in the quality of soil and water resources, as well as increases in the populations of tree-defoliating insects. In most regions it is likely that dieback has multiple causes.

Large-scale clearing for agricultural purposes has occurred mostly on the better soils and gentler slopes.

▼

Figure 6.8 Location of the major agricultural zones

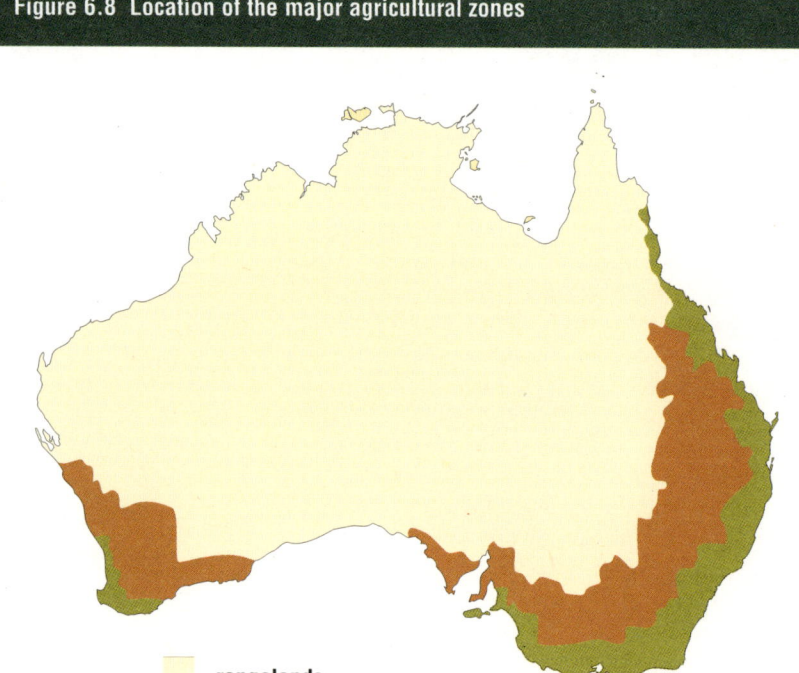

- rangelands
- wheat–sheep zone
- high rainfall zone

Source: ABARE.

Rangelands

Rangelands occupy about 75 per cent of the Australian continent, or approximately 6 million sq km (see Fig. 6.8). About 46 per cent is shrubland and 42 per cent woodland or open woodland. Tussock and hummock grasslands cover only about nine per cent of the rangelands, although grasses occur as an understorey over a further 68 per cent.

Most of the rangelands are leasehold and administered by State agencies, with only small areas privately owned (freehold). Extensive areas remain vacant Crown land. No nationally coordinated data-collection and reporting procedure exists for monitoring the vegetation and soil resources of these lands, although one is being developed. Consequently, their condition, and changes in condition, cannot be reported from a single set of national figures.

Nevertheless, it is widely accepted that rangelands have been degraded by the grazing pressure of domestic livestock, feral animals (such as goats and rabbits), and the increase in population of some native herbivores, particularly the larger kangaroo species, which in some areas has accompanied pastoral development.

Evidence for such degradation may be found in the historical pattern of livestock numbers typical of many semi-arid and arid areas. Livestock numbers commonly rise to a peak some years after settlement, crash during a period of severe drought and fail to recover to pre-drought levels (see Fig. 6.9). Stocking rates during the early years of pastoral development were generally unsustainable and the interaction of unfavourable seasonal conditions with grazing pressure undoubtedly caused much change. Rangelands in many areas

still reflect these impacts. However, overall Australian rangelands are not extensively 'desertified', despite the changes.

The term range condition is used to describe the state of their vegetation and soil. The meaning of this term has been the subject of considerable controversy. Historically, attempts to assess range condition as a basis for management of grazing land have been either ecologically or productivity based. The ecological approach measures condition by the change in species composition from the original vegetation (Dyksterhuis, 1949). An assessment based on productivity defines the productivity of the landscape for its current use (Humphrey, 1949). Both concepts require that assessments are made relative to some ideal state that is specific to each type of land, and are not influenced by short-term seasonal conditions.

Australian usage has generally favoured the productivity approach, either implicitly or explicitly. However, the uncertainty in the relationship between assessed range condition and animal production requires a degree of caution in interpreting survey results. The emphasis on pastoral productivity means that the available data are of limited value in defining other aspects of land condition, such as conservation status.

Recently, a radically different approach to rangeland assessment has been proposed, based on the capacity of the landscape to produce cover in response to rainfall (Pickup, 1989). This method has so far been used successfully only in central Australia. Virtually all other published data are based on assessment of the landscape as a pastoral resource, except where particular forms of land degradation (for example, rilling, sheet or gully erosion) are described.

To date surveys have been conducted over about 51% of the rangelands which includes most of the pastoral lands. In all survey regions most of the land is assessed to be in fair to good condition, with only a relatively small proportion considered

Figure 6.9 Sheep numbers in the rangelands of the West Gascoyne region of WA

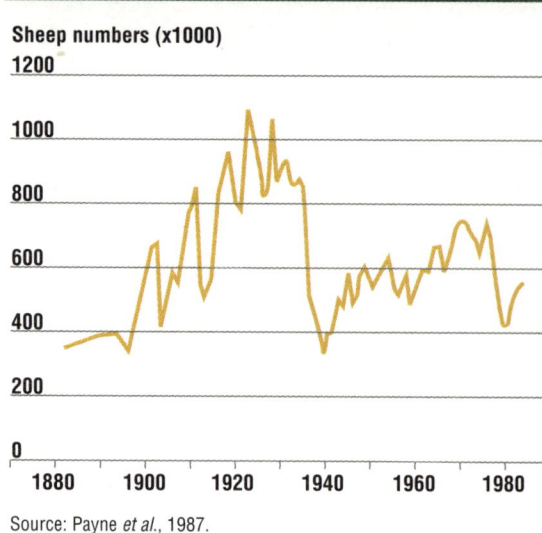

Sheep numbers (x1000)

Source: Payne *et al.*, 1987.

to be in poor condition or unrecoverable by normal property management. Where assessed, areas of severe degradation and erosion, which are probably unrecoverable, amount to no more than 2.3 per cent of the land surface. However, Wilcox and McKinnon (1972) believed that a considerably larger proportion (about 15 per cent) of their survey area required removal of grazing to prevent irreversible degradation. Throughout the Australian arid zone as a whole, about 4.4 per cent of the land used for pastoralism was estimated to exhibit vegetation degradation and severe erosion in 1975, with an additional 8.5 per cent showing vegetation degradation and substantial erosion (DEHCD, 1978).

Regional differences occur in range condition (see Table 6.2). The arid zone of Western Australia, the

Table 6.2 Results of regional rangeland condition surveys

Survey region	Reference	Survey area (sq km)	% severely degraded & eroded	% of survey area Good* or A**	Fair* or B**	Poor* or C**
Western Australia						
Gascoyne*/***	Wilcox and McKinnon, 1972	68 700	1.9#	6	65	29
Carnarvon Basin*	Payne et al., 1987	74 489	0.9	45	32	23
Murchison*/***	Curry et al., 1994	85 890	1.8	21	37	42
NE Goldfields*/***	Pringle et al., 1994	110 570	0.4	39	33	28
Nullarbor*/***	Mitchell et al, 1988	47 400	0	50	10	40
West Kimberley*/***	Payne et al.,1979	89 600	2.3#	19	51	30
West Kimberley**	Tothill and Gillies, 1992	127 270	N/A	73	19	7
Ashburton* (Alienated land)	Payne et al.,1988	61 128	0.9	64	27	9
Roebourne Plains*	Payne and Tille, 1992	10 200	2.3	51	27	22
Pilbara	Tothill and Gillies, 1992	264 740	N/A	79	19	2
North Kimberley	Tothill and Gillies, 1992	88 950	N/A	84	15	<1
East Kimberley	Tothill and Gillies, 1992	73 370	N/A	55	35	10
Northern Territory						
Darwin	Tothill and Gillies, 1992	18 550	N/A	91	6	3
Gulf	Tothill and Gillies, 1992	101 960	N/A	96	4	0
Barkly Tableland	Tothill and Gillies, 1992	150 520	N/A	79	15	6
Victoria River District	Tothill and Gillies, 1992	130 450	N/A	94	6	<1
Central Australia	Condon et al.,1969***	140 595	N/A	48	44	9
	Condon et al.,1969*** excluding spinifex dunes and plains	86 179	N/A	27	59	13
				##%PL<5	%PL 6–15	%PL >15
	Bastin et al., 1993	34 263	N/A	75	21	4
	Bastin et al., 1993 excluding spinifex dunes and plains	19 506	N/A	57	36	7
Queensland						
(compiled from data of Tothill and Gillies provided by the Meat Research Corporation)						
Burdekin (AER 23; 75%)		65 060	N/A	47	34	18
Nth Qld Tableland (AER 25; 100%)		8 060	N/A	67	27	6
Central West & Gulf (AER 26;100%)		353 559	N/A	56	34	10
Cape York (AER 27; 100%)		209 896	N/A	44	41	14
Marraki-Arnhem Land (AER 29; 30%)		41 230	N/A	74	21	5
Maranoa/Warrego (AER 40; 36%)		91 070	N/A	23	52	25
Channel country (AER 43; 60%)		296 042	N/A	36	44	20
South Australia						
Kingoonya Soil Con.Board.*/***	Anon., 1991	65 815	N/A	43	27	30
New South Wales						
Western Division	Soil Con. Service of NSW, 1988	325 000		Nil–Minor	Moderate	Severe
Woody Shrub Infestation				62.4	27.3	10.3
Sheet and Rill Erosion***			N/A	99.9	0.1	0
Gully Erosion***			N/A	91.1	7.4	1.5
Scalding			N/A	81.4	17.4	1.2

Notes:
*   Range condition classes from surveys marked*
**   Degradation classes from Tothill and Gillies, 1992; A — sustainable, B — deteriorating; C — degraded
***   Amalgamation of classes, or change of class names, from original survey; data for Condon et al. refer to watered area only
#   Estimate only, not mapped
% Productivity Loss Index for Bastin et al, 1993
AER   Agro-ecological region after Australian Agricultural Council, 1991. Percentages indicate the proportion within the Queensland rangelands
N/A   Not available

Table 6.3 Condition of major vegetation types in the rangelands

Vegetation type	Survey area	Derived from	Survey area (sq km)	% of survey area Good/A	Fair/B	Poor/C[1]
Chenopod shrublands[2]	Nullarbor	Mitchell *et al.*, 1988	46 856	50	10	40
	Murchison	Curry *et al.*, 1994	15 313	26	30	44
	Gascoyne	Wilcox and McKinnon, 1972	16 123	5	63	32
	NE Goldfields	Pringle *et al.*, 1994	16 881	50	28	22
	Ashburton	Payne *et al.*, 1988	8 231	26	42	32
	Carnarvon Basin	Payne *et al.*, 1987	12 750	35	30	35
	Kingoonya Soil Con. Board, SA	Anon., 1991	38 870	55	25	20
Mitchell grasslands	West Kimberley	Tothill and Gillies, 1992	12 190	60	30	10
	Barkly	Tothill and Gillies, 1992	68 190	78	16	6
	Victoria River District	Tothill and Gillies, 1992	12 680	71	29	<1
	Qld - Rolling downs (north)	Tothill and Gillies, 1992	208 830	70	25	5
	Qld - Stony downs	Tothill and Gillies, 1992	26 320	30	50	20
	Qld - Ashy downs	Tothill and Gillies, 1992	24 110	30	50	20
Acacia woodlands/shrublands	Gascoyne	Wilcox and McKinnon, 1972	34 630	9	68	23
	Ashburton	Payne *et al.*, 1988	28 120	63	31	6
	NE Goldfields	Pringle *et al.*, 1994	39 596	33	37	30
	Murchison	Curry *et al.*, 1994	64 827	22	39	39
	Carnarvon Basin	Payne *et al.*, 1987	41 954	40	39	20
	SW Qld - mulga[3]	Tothill and Gillies, 1992	183 580	20	50	30
	W Qld - gidgee	Tothill and Gillies, 1992	8 040	70	20	10
	SW Qld - gidgee	Tothill and Gillies, 1992	9 400	20	35	45
	W Qld - Georgina gidgee	Tothill and Gillies, 1992	15 980	70	20	10
	Kingoonya Soil Con. Board, SA	Anon., 1991	15 180	20	30	50
Perennial tallgrass pastures	Darwin, NT	Tothill and Gillies, 1992	1 450	40	40	20
	North Kimberley	Tothill and Gillies, 1992	10 700	78	21	1
	East Kimberley	Tothill and Gillies, 1992	2 130	29	44	27
	West Kimberley	Tothill and Gillies, 1992	3 600	85	10	5
	Cape York	Tothill and Gillies, 1992	9 280	90	5	5

Notes:
1. Amalgamation or renaming of condition classes as for Table 6.2 A (sustainable), B (deteriorating) and C (degraded) refer to classifications of Tothill and Gillies, 1992
2. Includes communities in which chenopod shrubs form a mid storey
3. Includes hard and soft mulga, mulga on residuals and mulga-whitewood
   The vegetation types listed are derived from the survey reports and do not correspond to Carnahan's 1990 classification.

Cape York and Burdekin regions of north Queensland, and the Maranoa/Warrego and channel country regions of south-west Queensland are more severely affected than elsewhere. The degradation in south-west Queensland extends into New South Wales and north-east South Australia (Woods, 1983), but a lack of comparable data makes comparison difficult.

Regional assessments generally disguise considerable local variability in land condition. A common finding in rangeland surveys is that those land types or vegetation communities with the highest potential for pastoral production (for example, creek and river frontages) are the most severely degraded. These are frequently the most accessible for livestock and are generally well supplied with watering points. They are also generally limited in area, and surrounded by much larger tracts of less-productive country, so they receive preferential use. Thus, the implications of rangeland degradation for total pastoral productivity may be greater than the regional assessment figures indicate.

Examination of the condition of major vegetation types (see Table 6.3) shows that: extensive areas of Mitchell grassland remain in good condition; perennial tallgrass pastures in the higher-rainfall areas of the west Kimberley, north Kimberley and Cape York are also generally in good condition; and the chenopod and acacia communities in the arid zone with winter or non-seasonal rainfall show the most severe degradation. However, all vegetation types display wide variation in condition between survey areas and few generalisations are possible.

Trends in rangeland condition

Although it is not possible to present a comprehensive account of the trends in rangeland condition on a national basis, some regional trends can be shown. Comparison of data from surveys of the west Kimberley region (Payne *et al.*, 1979;

Tothill and Gillies, 1992) and evidence from long-term photographic monitoring sites (WA Dept of Agriculture, unpublished data) strongly suggest that rangelands in this region have improved over the period 1972–91. The improvement can be attributed to the reduction in stock numbers and improved stock control associated with both the Brucellosis and Tuberculosis Eradication Campaign and changing markets for cattle, together with control of feral donkeys by aerial shooting.

A similar comparison in Queensland (Weston *et al.*, 1981; Tothill and Gillies, 1992) suggests that both positive and negative trends have occurred over the period 1978–80 to 1991. Of 21 rangeland pasture communities, six appear to have remained relatively stable (albeit in some cases in poor condition), eight have apparently improved in condition and eight have deteriorated. Generally the improvements have been relatively minor while the downward trends are often marked. Part of this deterioration may be attributed to the encroachment of exotic weeds, such as *Parkinsonia*, prickly acacia (*Acacia nilotica*) and mesquite (*Prosopis* spp.) in the Mitchell grasslands, and rubber vine (*Cryptostegia grandiflora*) and chinee apple (*Zyziphus mauritiana*) in the Gulf and Cape York Peninsula areas. Increased grazing pressure resulting from low beef prices in the mid 1970s (and consequent retention of stock), widespread adoption of tropically adapted *Bos indicus* cattle and increased provision of feed supplements, combined with poor seasonal conditions during the 1980s, have also contributed to the decline.

Site data collected by the Western Australian Rangeland Monitoring System (WARMS) in the southern rangelands of the State also indicate mixed positive and negative trends. Comparison of data collected over (usually) a five-year interval showed that density of palatable perennial shrubs had increased on 39 per cent and decreased on 50

per cent of 539 sites analysed in 1991–92; the remainder were unchanged (WA Dept of Agriculture, unpublished data). Pastoralists in this region generally agree that range condition is better now than 20 years ago (Wilcox and Cunningham, 1994), but over the shorter period covered by the monitoring data it is evident that changes have not been uniform.

Recent trends in woody vegetation cover in western New South Wales have also been variable. Increases in cover, due mainly to encroachment of 'woody weeds', occurred over substantial areas in the Cobar and Bourke districts in the 1980s. Over some smaller areas cover decreased (Weir *et al.*, 1992). Extensive regeneration of degraded chenopod shrublands and sandhill–claypan complexes in various parts of the New South Wales rangelands has occurred since the 1950s, due to favourable seasons, reduced rabbit numbers and a range of socio-economic factors leading to more conservative management (Condon, 1983).

In South Australia, a similar example of rangeland regeneration has occurred (Nicholson, 1983). However, in another case in that State, substantial declines in chenopod shrub density on three out of four large arid-zone properties were reported over a 22-year period to 1972. No decline was recorded on the fourth property, which was conservatively stocked between surveys (Lay, 1979).

Comparison of central Australian surveys (Condon *et al.*, 1969, Bastin *et al.*, 1993) is difficult owing to differences in methods and the size of the survey areas (see Table 6.2). Also, the earlier survey was conducted under severe drought conditions and, thus, some recovery, as suggested by the data, is to be expected.

It is clear that the present condition of our rangelands and the trends over time are highly variable. Thus it is difficult to make broad generalisations about either.

-Heavy grazing can lead to vegetation loss and expose the soil to increased water and wind erosion.

Forests

When European settlers arrived in Australia, the land carried 69 million ha of forests. Since then, about 40 per cent of the forests have been cleared and about the same area has been affected by logging at some stage (see Fig. 6.5). Only about 25 per cent of the original forest estate remains relatively unaffected by clearing or harvesting (see Fig. 6.10). A separate State of the Forests Report is being prepared by the National Forest Inventory for JANIS, a joint committee of Commonwealth, State and Territory governments established to oversee the implementation of the National Forest Policy Statement. This report is expected to be released in 1996.

Relatively little is known about the condition of larger areas of privately owned forests in the high-rainfall zone (see Fig. 6.8). Undoubtedly, some thinning has taken place on less rugged terrain and grazing has affected forest regeneration (see page 6-11).

Forests in conservation areas are usually selected because they provide good representations of pre-European conditions. However, in some areas there is concern about declining condition. In small reserves invading species from neighbouring pasture, crop or urban areas threaten the forest. In all reserves, changes to the fire regime can alter forest condition. Conservation agencies are faced with problems of identifying pre-European fire regimes, deciding whether they should be maintained and establishing how to do that without threatening other values and property.

Figure 6.10 Forest areas affected by clearing or harvesting since 1788

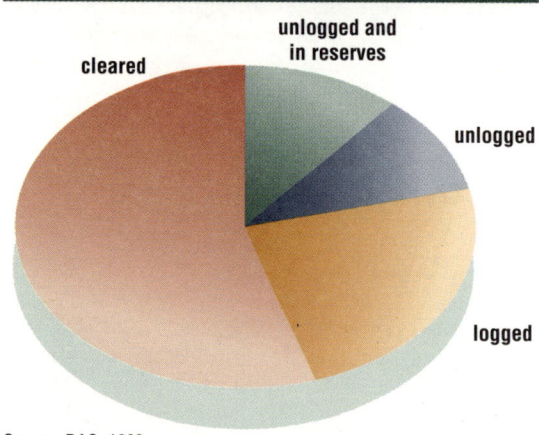

Source: RAC, 1992.

Sustainability of our forests

Forests are renewable resources that can be used to provide timber products, water, recreation opportunities and conservation values. Forest management is a long-term issue. Decisions made at the turn of this century still affect the forest resource available today, and planning for the forests of 2050 and beyond is happening now. Forests are dynamic — the resources and values that they provide change with the age of the forest and with the occurrence of bushfires or cyclones. Human values also change; those we hold now are not the same as those held 50 years ago. Nor are they likely to be the same as those held 50 years hence.

The question: 'Is our use of forests sustainable?' will always have an ambivalent answer. About 75 per cent of the 1788 forests have been cut and the timber and other products used; about half of this land has been converted to other uses such as agriculture. While this has resulted in the irretrievable loss of certain values, such as biodiversity, and some Aboriginal cultural heritage, others have been maintained. Much modern forestry practice is designed to allow near-to-natural forest to regenerate after harvesting. The goal is to retain as many of the multiple values of the forest as possible. In practice, ecosystems tend to be modified so that forests can be harvested on a 50–80 year rotation.

People commonly confuse the concepts of sustainable use and sustainable yield from our forests. Sustainable yield means the output of a specified level of a certain mix of products, under a particular management scenario and with particular assumptions about future climatic conditions and demands for forest products. In its simplest form, sustainable yield can be achieved as an even flow by ensuring that each year's harvest is limited to the net growth during that year. This can be achieved in a fixed forest estate by cutting and regenerating each portion of the estate in a rotation that will allow the return of logging to the same site just at the time it has regrown to its optimum yield (the rotation time). In Australia, forestry is not practised on the basis of even-flow sustainable yield. The existing stock of old-growth forests, competing uses and changing value systems led the Resource Assessment Commission Forest and Timber Inquiry (RAC, 1992) to conclude that sustainable yield was too simplistic a concept to be useful in assessing whether our use of the forests is sustainable.

The following are some important criteria that can be used to assess whether forest use is sustainable:

• no decrease in area

• maintenance of ecological processes

• mechanisms to retain the range of biological diversity

The forest estate includes forests of all types — natural, modified and plantations, publicly and privately owned. To satisfy the first criterion, the total forest estate should not permanently decrease in area. Since the arrival of Europeans 40 per cent of forests have been cleared and only 25 per cent remain relatively unaffected by logging or other significant human activity. Governments throughout Australia have agreed that the permanent forest estate should not decline further.

Second, our forest estate must be sufficient to maintain the ecological processes on which the forests, and many catchments and waterways, depend. Third, it should also be sufficient to protect the full range of biodiversity. In numerous cases, clearing of forests has led to breakdowns in biological processes, erosion, siltation and loss of species and biodiversity. A more contentious issue is whether our use of the remaining forest estate for timber production endangers any of the above values. The RAC concluded that, although inappropriate practices had occurred in the past, there was not enough evidence to conclude that current forest management practices threatened these values. However, it remains possible that some of our patterns of use — and in particular unforeseen problems arising from them or their cumulative effects — are threatening these values. So, any forest use should proceed cautiously, using the best current practices (including conservation practices) and continuous monitoring of the forest to detect emerging problems.

A further issue is maintaining the industries and communities that directly depend on forests. About 10 000 people are employed in log extraction and sawmilling. The number has been declining by more than 500 per year over the past 40 years, mostly due to technological changes. About 30 000 people are employed in the wood products industry, using raw material from native forests, plantation forests and imported timber. It is difficult to attribute the proportion of employment in the recreation and tourism industry that depends directly or indirectly on forests. However, this source of employment is growing.

The RAC considered a number of scenarios of timber production over the next century, including 'current trends' and specific scenarios proposed by industry and conservation groups. These projections dealt mostly with hardwood (usually wood from native eucalypt species) and with sawlogs (logs suitable for processing for sawn timber). The RAC found that, if current trends continued, the availability of hardwood sawlogs would decline slightly, as suitable native forests become scarce, until about 2030 or 2040. Following this, the availability of logs from eucalypt plantations and

Forests: a renewable resource providing timber, water, recreation, conservation and cultural values.

intensively managed regrowth stands would begin to increase and future timber production would become increasingly reliant on this resource. This scenario depends greatly on how industry investment plans for plantations and intensively managed regrowth develop.

Other ways of relieving pressures on native forests and maintaining industry activity include the substitution of softwoods (mostly from *Pinus radiata* plantations) for hardwoods. For many uses (such as housing construction) this could occur, but in other cases it would be difficult to substitute the values of hardwoods. Removal of obstacles to plantation forestry; development of fast-growing hardwoods; and removal of subsidies for the harvesting of native forests are other ways of relieving the pressure on native forests.

Activity at the international level seeks to define indicators for ecologically sustainable forestry (the Helsinki and Montreal processes). Australia is active in these processes, but it will be several years before effective indicators are available.

There is no clear answer to whether our current use of the forests is sustainable. Past practices have not been. All forest management agencies have implemented policies designed to meet sustainability criteria. However, there is no simple path to the destination of sustainable use. Rather it is a compass bearing in a never-ending journey. Staying on that bearing will require a continuing debate by many elements of society with different goals and different value systems for the forests.

Old-growth forests

Old-growth forests have been at the centre of much of the debate about the use of our native forests. The National Forest Policy Statement (1992) defined old-growth forest as forest that is ecologically mature and has been subjected to negligible unnatural disturbance such as logging, road building and clearing (Commonwealth of Australia, 1992). The definition focuses on forest in which the upper stratum or overstorey is in the late mature to overmature growth phases. The Statement acknowledged the significance of these areas to the Australian community because of their very high aesthetic, cultural and nature conservation values and their freedom from disturbance. Their rich understoreys, numerous epiphytes, hollow limbs and fallen logs harbour high biological diversity. They are also a significant timber resource. Old-growth forests are distinguished from regrowth forests that re-establish themselves after logging. They are often rare and scattered throughout the forest estate. They are vulnerable to fire and human disturbance.

The Resource Assessment Commission (RAC) Forest and Timber Inquiry found that across all forest types about 172 000 sq km (40 per cent) remained unlogged but only 24 000 sq km (14 per cent) stood in conservation reserves. The areas protected in conservation reserves range from 64 per cent for mangrove and swamp forest, to five per cent for south-eastern dry forest and woodland.

The RAC concluded that clearing of old-growth forests was potentially a violation of the 'precautionary principle' in that an irreplaceable resource would be destroyed. The National Biodiversity Council and other bodies can see no justification for any further clearing of such forest remnants.

Proper management of old-growth forests is difficult to achieve. It takes many decades or centuries to recreate an old-growth stand. A policy of total preservation is not sufficient to maintain old-growth forests into the future. They are but one part of a cycle of renewal and ageing in native forests. Old-growth stands will inevitably be lost through wildfires and, therefore, we need to ensure that there is a pool of maturing forests ready to replace them.

The Regional Forest Agreement process initiated by the Commonwealth Government will assess the extent of old-growth forests and seek to place at least 60 per cent of existing old-growth forests and 100 per cent wherever practicable (but at least 90 per cent) of rare old-growth forest in a national forest reserve system. The reserve system is also intended to contain 15 per cent of each forest type existing prior to European settlement and 90 per cent or more, wherever practicable, of high quality 'wilderness'.

Impact of pest animals and plants

Pest animals

European settlers brought many animals with them from 'home' — some were brought intentionally, others not. Many animals were domesticated and were intended to provide food, fibre, transport and labour or simply companionship. Some were brought for sport or pleasure, while others arrived as stowaways. Over time, animals escaped or were deliberately released and a number of species became so well established that they now have a significant impact on both agricultural production and conservation values. Some native vertebrates have become pests too. In sheep country, where dingoes are controlled, kangaroos have increased as a result of the greater availability and permanence of watering points and a reduction of natural predators (see Fig. 6.11). Elsewhere, dingoes have flourished because of the availability of food in the form of rabbits and young livestock.

The most significant environmental impact of pest animals and plants is on our biodiversity and this is dealt with in Chapter 4. Nevertheless, pest animals and plants significantly affect our environment by the additional pressures they place on the land resource. Many of these effects occur through losses and control costs to agricultural and pastoral production. Few reliable and comprehensive figures exist to measure the environmental or economic impact of any pest at a national level. Such figures as are available will be affected by year-to-year variation in seasonal conditions. Financial losses in some regions may be counterbalanced by gains in others: for example, a reduction in grain yields in one region may lead to higher prices nationally. Off-site environmental damage is difficult to assess.

Many vertebrate pests are familiar to us, but a number of invertebrates, particularly insects, cause major damage to agricultural production and yet are not known to the general community. The heliothis caterpillar (*Helicoverpa armigera* and *H. punctigera*) attacks cotton and other broadacre crops. Infestations of Queensland fruit fly (*Bactrocera tryoni*) can have a severe impact not just in the short term on local crops but on overseas trade prospects. A variety of insects cause post-harvest damage to stored grain. Australia faces a greater threat from insect pests than countries in higher latitudes, where the cold winters kill pests. Not all insect pests are introduced; indeed, *H. punctigera* and the fruit fly are native to Australia.

Losses due to animal pests

It is difficult to estimate the economic costs of pest species as the result depends so much on what impacts are taken into account and the assumptions that are made about their flow-on effects.

Losses of crop production in one area may be creating opportunities for production in a less affected area. For example, estimates of the costs of rabbits in the agricultural and pastoral industries vary from $90 million to $600 million. A mouse plague in South Australia and Victoria during 1993 cost the affected grain-growers an estimated $54.8 million in reduced yields alone. When on- and off-farm costs were added, the total reached $64.5 million (Caughley *et al.*, 1994).

Kangaroos resulted in estimated losses of $113 million in agricultural production in the commercial shooting areas of Australia in the 1984–85 financial year (1.4 per cent of the gross value of all agricultural production).

No one has quantified the economic damage to the pastoral industry, as a whole, inflicted by large herbivores. It is incurred through competition with livestock for food and water as well as through land degradation, damage to infrastructure, disturbance to mustering and, in some cases, indiscriminate mating among domestic and feral members of the same species, such as cattle.

For invertebrates, costs are also high. The total annual loss due to insects in primary production is about $3.1 billion. The red-legged earth mite (*Halotydeus destructor*), one of Australia's worst pasture pests, causes annual damage of $200 million. The two species of *Helicoverpa* moth mentioned earlier cost the cotton industry over $50 million each year, and failure to control them would lead to the collapse of the industry, worth $800 million per year.

Costs of control

Any control costs will vary widely depending on the enterprise, the pest and its density and the land type. During the 1993 mouse plague in South Australia and Victoria, baits alone amounted to $1.7 million. On a single property in central Australia, ecologists estimated the cost of controlling 2000 feral horses to be about $25 000 and, while the subsequent income from sales was likely to return a profit of about $130 000, follow-up control would have eaten into this return. The reduced carrying capacity resulting from land degradation, and the repair of damage were not costed. In semi-arid South Australia, the cost of controlling rabbits was estimated to be $150 000 for a single property over seven years.

Nationally, sheep blowfly, buffalo fly and cattle tick — all major pests of livestock — are estimated to cost more than $600 million annually to control. The cost of controlling all insect pests in Australia exceeds $1 billion annually, mostly due to expenditure on chemical pesticides. Integrated pest management may reduce dependence on pesticides. No annual estimates of national expenditure on vertebrate pest control are available.

Costs of control per individual pest animal increase as their density decreases, making eradication an unrealistic aim for many species (see Fig. 6.12). As well, any eradication campaign must use techniques that are socially acceptable, do not adversely affect non-target species and are effective against all individuals in the population. It must be possible to detect very low densities of the species and immigration must be zero. In general, practical

Rabbits are a major contributor to environmental damage.

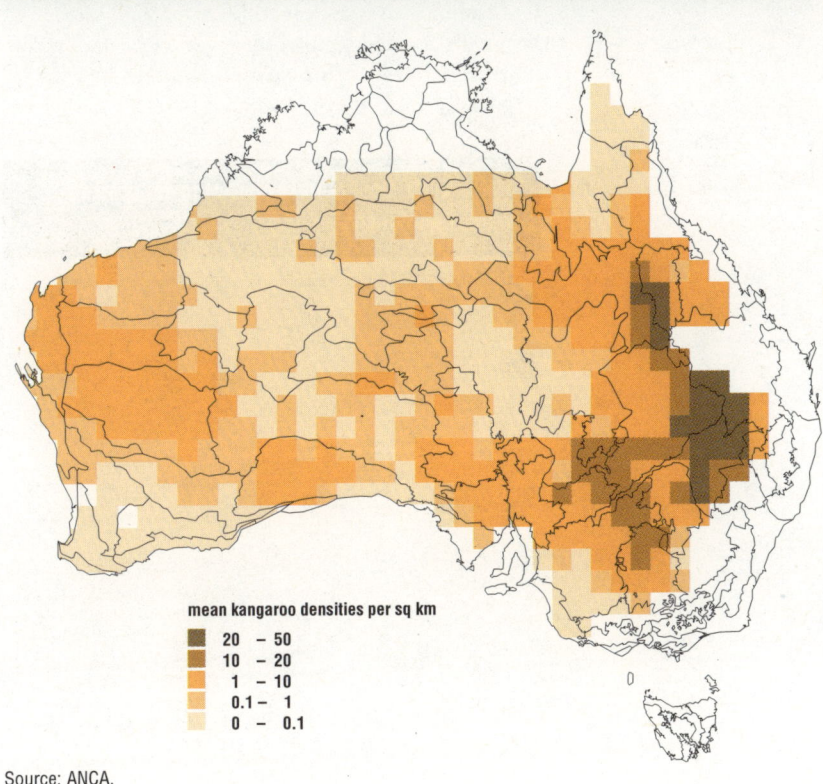

Mice sometimes reach plague proportions.

Figure 6.11 Kangaroo densities

mean kangaroo densities per sq km

- 20 – 50
- 10 – 20
- 1 – 10
- 0.1 – 1
- 0 – 0.1

Source: ANCA.

The cost of controlling all insect pests in Australia exceeds $1 billion annually.

and economic limitations rule out the option of vertebrate pest eradication (Braysher, 1993).

Another economic impact of animal pests involves potential disease outbreak. A 1991 study, for example, estimated that an outbreak of foot-and-mouth could cost $9 billion in lost exports, and losses in the first year alone could be $2 billion. Mammal pests would play a major role in such an event.

The economic benefits of pest animals are modest relative to costs, and are likely to remain so. The estimated annual wholesale value of trade in wild animals and their products is $132–156 million. Commercial harvesting can improve the cost effectiveness of pest management, but can also lead to pests being maintained at commercially rather than environmentally desirable levels. It also has value in improving conservation status and agricultural productivity, but the benefits have not been quantified. The commercial harvesting of native species, whether or not in the context of 'pest control', is a controversial conservation issue and is opposed by many in the community.

Control measures appear to be reducing pest numbers in some regions from earlier peaks. In the Northern Territory, for example, buffalo numbers reportedly fell from more than 340 000 in 1985 to

140 000 in 1989. Now, numbers are limited to 10–15 thousand domestic animals and 20–40 thousand in Arnhem Land. Similarly, over the last decade or so, donkeys in the Kimberley have been reduced from about one per sq km to about one per 10 sq km — that is, to 10 per cent of former levels. In contrast, the number of camels in and around the Northern Territory appears to be increasing. Estimates of 3000 to 6000 in 1969 rose to more than 31 000 by 1984. In the 10 years to 1993, the numbers increased further to around 47 000, due to the animals expanding their range rather than increasing their density in core areas.

Accurate trends for pest numbers are not available and, in any case, such figures would probably be seasonally variable. Trends in damage caused by pest species are likely to be a more valuable indicator, but we lack quantitative information.

Pest plants

Plants are deliberately introduced for their agricultural, forestry or ornamental value and a number of them are of great worth to Australia, providing almost all of our crops and many of our important pastures.

Introductions continue on a significant scale for all our major crop and pasture plants. Since European settlement, more than 1900 new vascular plant species have been added to Australia's complement of 15 000 indigenous species by either intentional introduction or accidental release. About half of these are now regarded as weeds, with more than 220 of them declared as noxious weeds. At least 46 per cent of the noxious species were deliberately introduced.

What is a weed? Common parlance defines it as a plant in the wrong place at the wrong time. The National Weeds Strategy, under development in 1995, defined it as 'a plant which has, or has potential to have, a detrimental effect on financial, social or conservation values'. Nevertheless, many

Figure 6.12 Cost of pest control in relation to population density

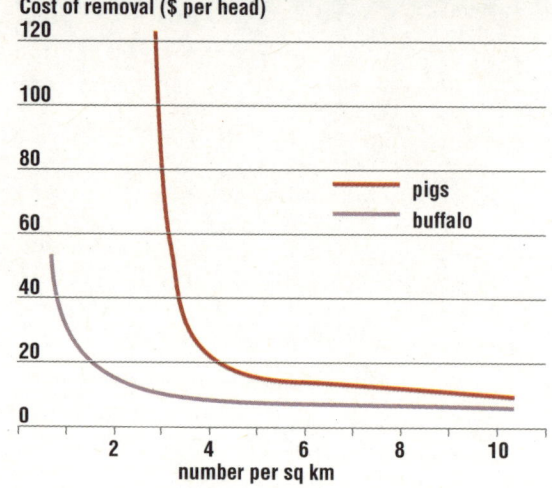

Source: Choquenot, 1994.

Figure 6.13 Distribution of prickly acacia

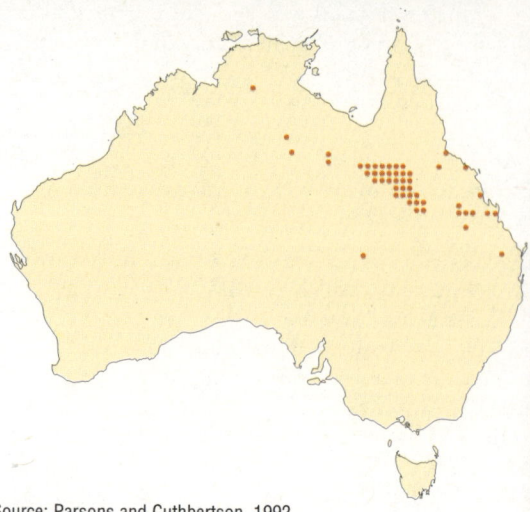

Source: Parsons and Cuthbertson, 1992.

so-called weeds may have some desirable characteristics for particular applications. A definition in financial terms is appropriate for weeds of agriculture but not for those of conservation ('environmental weeds'), which are best defined with regard to their disruption of ecological processes. Weeds in forests (see page 4-18) appear to be of minor economic importance and are not considered further here, where only agricultural weeds are considered.

Weeds can vary from short-lived and herbaceous plants to long-lived trees (for comprehensive descriptions of Australia's noxious weeds, see Parsons and Cuthbertson, 1992). Common short-lived species include wild oats (*Avena fatua*) and skeleton weed (*Chondrilla juncea*) in cereal crops, Bathurst burr (*Xanthium spinosum*) and Noogoora burr (*X. occidentale*) in summer crops and grazing lands, Paterson's curse (*Echium plantagineum*) and blackberry (*Rubus fruticosus*) in temperate pastures and parthenium weed (*Parthenium hysterophorus*) in tropical pastures and crops. Long-lived species such as prickly acacia and various mesquites infest northern grazing lands.

Native species can also become agricultural weeds. For instance, brown beetle grass (*Diplachne fusca*) is a major weed of drill-seeded rice in New South Wales, and a number of native shrubs in the arid and semi-arid grazing lands have reached densities that suppress native pasture production.

A number of pest species remain benign for many years until an environmental trigger like high rainfall, disturbance or a new dispersal agent stimulates a dramatic expansion. Prickly acacia was introduced to grazing lands of western Queensland for shade and fodder in the 1890s, but was only regarded as a weed after its expansion during a series of wet years in the 1950s (see Fig. 6.13). In some cases, expanding populations could simply have been overlooked until the species was entrenched.

Potential weeds are still being introduced. Between 1947 and 1985, 463 exotic pasture grasses and legumes were intentionally introduced into northern Australia (Lonsdale, 1994). Only 21 (five per cent) were subsequently regarded as useful and, of these, 17 were also weeds. Thus, only four species (less than one per cent) were useful without causing weed problems while of the non-useful introductions, 43 became listed weeds, giving a total of 60 (13 per cent) new weeds of either cropping or conservation areas or both.

National quarantine procedures focus on the introduction of potential pests and diseases of agricultural crops and pastures, and do not restrict the importation of plant species that are not on the prohibited list. Thus plants that are not classified as weeds overseas can be introduced here, even though overseas conditions may be a poor guide to potential weediness of a species in Australian environments. Formal procedures exist at State level for screening introductions, but they are more stringent for pasture plants than for horticultural or other crops.

Impacts of weeds

Weeds cost Australia about $3.3 billion annually, excluding the cost of associated health issues and environmental weeds (Combellack, 1989). Annual losses due to crop weeds are usually estimated at 10–15 per cent of crop value (see Table 6.4).

Such estimates must be treated with caution. For pastures, different authorities assess pest impact in different ways and there is little standardisation of methods of data collection or of units of measurement. Rating infestations from heavy to light — a common procedure — does not provide data that can be readily related to pasture or livestock productivity. In the estimates for agricultural crops, costs of cultivation are incurred not just for weed control but to improve soil structure as well. Some 'double counting' of costs probably occurs.

Nevertheless, the cost of weeds to agriculture — in terms of reduced post-production quality, such as vegetable fault in wool, and of herbicide application — is relatively well documented compared with the costs of land degradation and loss of biodiversity.

Primary producers incur costs due to loss of production through reduced crop or pasture yields, poisoning of stock, vegetable fault in fibres, tainted products, carcass damage and animal stress due to physical discomfort. Weeds can also act as hosts for crop diseases, especially if they are closely related. Pathogens cause annual losses estimated at $187 million in agricultural production.

The costs of controlling weeds are not simply those listed in Table 6.4. For example, farmers spend more than $700 million annually in cultivating crops, to control weeds as well as to modify and loosen the soil surface. But added to this is the cost of soil degradation — primarily caused by cultivation. Herbicide use and direct-drilling techniques offer alternatives to cultivation, but these strategies may adversely affect yields in some soils or contaminate soils. The issue of weed control is complex and requires an approach that

Table 6.4 Annual financial losses due to weeds

	Financial loss in $ million			
	Agricultural crops	Horticultural crops	Pastures	Non-crop areas
Direct losses				
Cultivation	710			
Herbicides	263			
Herbicide application	40			
All direct losses	1013	63	71	177
Indirect losses				
Yield losses	713			
Product contamination	143			
All indirect losses	856	240	900	unknown
Total	**1869**	**303**	**971**	**>177**

Source: Combellack, 1989, based on 1986 data.

integrates agronomy and ecology with soil and weed science and engineering design.

Few weeds of crops and pastures have been studied systematically with a view to understanding yield loss, seed production or the decline of seed banks (Cousens and Medd, 1994). As a consequence, weed control using ecologically based management practices has not been widely adopted. Similarly, much of the research on exotic woody weeds has focused on chemical, mechanical and biological control, and there is little information regarding the ecosystem effects of invasions, mechanisms of spread or the effects of grazing and fire on spread. It is conceivable that, once one woody weed species is controlled, another will invade the gap unless follow-up management is ecologically sound.

The benefits, if any, of successful pasture introduction in the northern Australian tropics accrue to the pastoral industry, while the costs of the introduction becoming a pest are largely borne by others (Lonsdale, 1994). Species introductions in arid native pastures are rarely of economic benefit.

Pests and the future

Significant problems to arise in future could include introduction of new animal and plant pests, introduction of diseases, parasites or vectors that interact with pest species, variations in market value of products and complex impacts of climate change. Species already present in Australia — in zoos, plant nurseries and private collections, even some family pets — also have the potential to become pests. The five-lined palm squirrel (*Funambulus pennanti*), an escapee from Australian zoos, and the red-billed quelea (*Quelea quelea*) in private aviaries are pests elsewhere in the world and could become so here.

Changes in the way land is used could alter distributions and densities of pest species. A switch from sheep to cattle, or from grazing to cropping, the introduction of a new crop or a tourism enterprise are a few examples of new sources of disturbance and dispersal.

In anticipation of future invasions, researchers are relating the genetic diversity of pest organisms outside Australia to that in the country and determining suitable management strategies.

The future offers the potential for new technologies, such as genetically engineered infertility, to control pest animals. However, control of one pest, such as the rabbit, could lead to increased predation on native fauna if feral cats and foxes were not controlled at the same time. There is also a risk that existing forms of control will not be maintained, in the expectation of a single 'magic' solution to a pest problem. Any new techniques will have to be integrated with existing methods, and with underlying ecological processes, for the optimal management of pests.

Soils

Australian soils, compared with those of the Northern Hemisphere, commonly have lower organic matter and poorer surface structure. Their clay content often increases abruptly just below the surface (for example, kurosols, chromosols and sodosols) (see Fig. 6.14). Such layers restrict drainage and impede root growth. In these soils, bleached layers with very low nutrient levels are also common. Soils affected by salt, either now or in earlier geological times, cover large parts of the continent (sodosols in Fig. 6.14) and they have various nutrient and physical limitations.

Australia has an unusually large proportion of cracking clay soils (vertosols). Their shrinking and swelling cause problems for engineering structures and farming. Extensive areas in the arid zone are covered by soils formed on aeolian sands (rudosols and tenosols). Finally, the remaining ancient land surfaces have very deep and strongly weathered soils (kandosols), which have very low levels of nutrients.

Soil structure

Structure is a complex soil property that influences physical, chemical and biological processes. It has the following three components (see Fig. 6.15):

- form — the arrangement and packing density of soil particles

- stability — the ability to resist stresses such as cultivation, trampling and rainfall

- resilience — the ability to recover from stress-induced changes

The natural structure in Australian soils is often poorly developed and the impact of European land-use is often not easy to characterise. While changes in soil structure due to land use may be one of Australia's most serious forms of land degradation, the extent, severity and cost have not been well documented.

In many soils, a fall in organic matter after clearing and cultivation has changed their structure. During the first few years of regular tillage, between one-third and half the organic matter may be lost due to increased oxidation. After several decades, organic matter reaches an equilibrium under a

Pest control is a complex issue requiring the integration of agronomy and ecology with soil science and engineering.

given management system. Pastures usually increase organic matter, but the rate of increase is slow. Substantial increases may take at least a decade and a minimum level of two per cent may be needed to maintain soil structure. The effects of management practices on organic matter levels in wheat soils in southern New South Wales and Victoria are shown in Table 6.5.

Significant changes in soil structure can occur in seconds when wet soil is exposed to stresses caused by machinery (for example, tractors, logging vehicles etc.) or stock. One such event can undo years of careful soil management and reduce crop or tree growth. Similarly, exposing bare soil to intensive rainfall can cause slumping and crusting of the soil surface: this is usually a precursor to run-off and erosion.

The degree to which structural decline can be reversed depends on the resilience of the soil and the resources available for land management. For example, on cracking clay soils (vertosols in Fig. 6.14) used for cotton, heavy machines working on wet soil caused soil compaction that reduced yield. This has been remedied by carefully timing operations with respect to soil moisture. Cracking soils are resilient and regenerate structure by shrinking and swelling. In contrast, structural decline in red soils (for example, ferrosols in Fig. 6.14) of eastern Australia is difficult to repair because these soils have no inherent regenerating capability (Bridge and Bell, 1994). Many forest soils commonly have low resilience.

There are few reliable surveys of the extent and magnitude of soil structural changes. One survey of farms on eight major soils in Queensland found adverse changes in soil structure on all farms, with the proportion of individual farms affected ranging from 20 per cent to 80 per cent (McGarry, 1992). Results from other Australian cropping areas could be expected to be similar. The decreases in soil organic matter associated with agriculture shown in Table 6.5 are from southern New South Wales and Victoria and are representative of farming areas elsewhere in Australia.

Soil structure decline can have the following effects:

- reduced porosity and permeability leading to waterlogging, excessive run-off and soil erosion

Table 6.5 Effect of management practices on soil organic matter

Management system	Soil organic carbon level (%)
Undisturbed woodland	8.2
Grazed pasture	4.8
Direct drilled cropping	2.0
Overgrazed pasture	1.6
Tillage-based cropping	1.3

Source: Geeves *et al.*, 1995.

Figure 6.14 Distribution of Australian soils

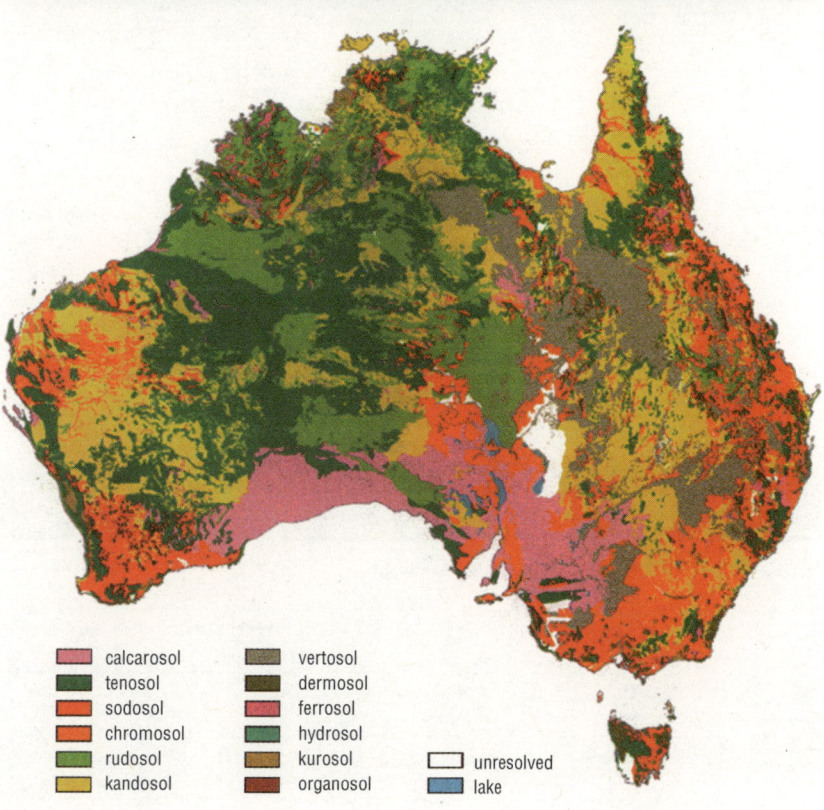

calcarosol		vertosol	
tenosol		dermosol	
sodosol		ferrosol	
chromosol		hydrosol	
rudosol		kurosol	unresolved
kandosol		organosol	lake

Source: CSIRO Division of Soils.

Figure 6.15 Structure characteristics of the upper 0.5 m of soil

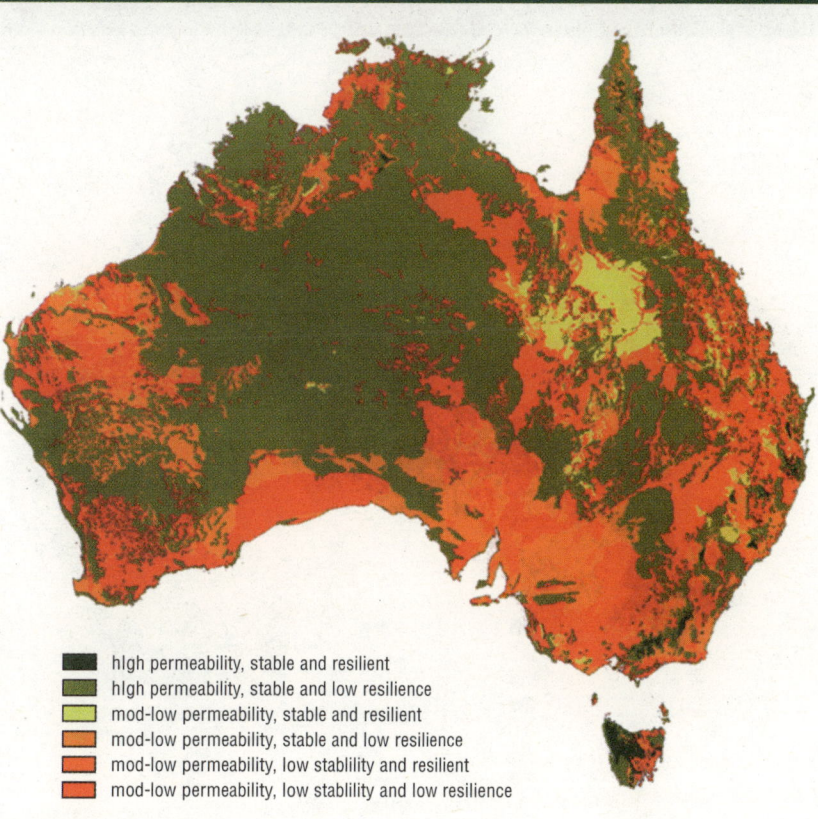

- high permeability, stable and resilient
- high permeability, stable and low resilience
- mod-low permeability, stable and resilient
- mod-low permeability, stable and low resilience
- mod-low permeability, low stability and resilient
- mod-low permeability, low stability and low resilience

Source: CSIRO Division of Soils.

- reduced root vigour leading to plant nutrient deficiencies and restricted water extraction

- reduced yield if management inputs are not increased

- increased management costs (fuel, labour, machinery, irrigation and fertiliser needs)

- fewer management options, and restrictions on the timing of farming operations

The problems of soil structure decline, nutrient decline and erosion are closely related. In general terms, successful management practices maintain a protective cover on the soil surface, increase organic matter levels and minimise damage to soil structure caused by machinery and stock. Various strategies are required for different land uses, and significant improvements have been achieved in some areas.

Table 6.6 Summary of soil loss on sloping lands

Land management system	Region	Soil loss on sloping land (t/ha/yr)*
1. Tropical cropping (eg. sugar, pineapple)	Queensland, NT	100–500
2. Cereal cropping	South Queensland, NSW, Victoria, South Australia	1–50 (higher in northern areas)
3. Forested catchments	South-eastern Australia	0–1 (but 10–50 after bushfires)
4. Pastures (well managed)	Southern Australia	1
5. Bare fallow	Southern Australia	50–100

*Note: Average rates of soil formation are much less than 1 t/ha/yr.

Figure 6.16 Average erosive energy of rainfall events

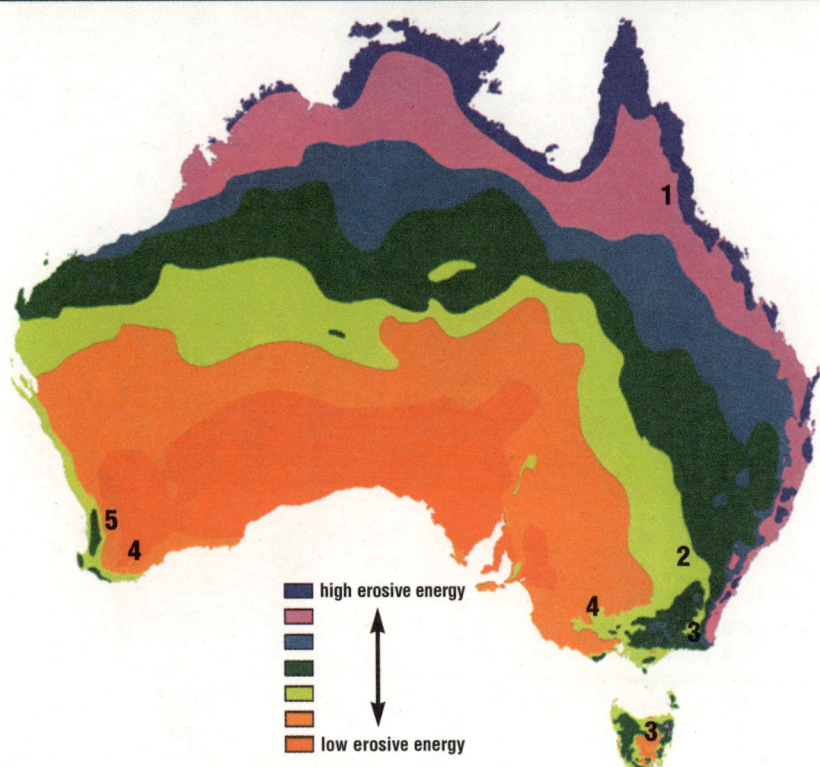

high erosive energy

low erosive energy

Note: Numbers indicate land management systems in the table above.
Source: Edwards, 1991 and CSIRO Division of Water Resources

For example, farming systems that use minimum tillage or permanent beds have demonstrated advantages for improving soil structure. Likewise, the impact of forest operations can be minimised by more careful harvesting layouts, the use of improved machinery and harvesting when soils are relatively dry. Also, adjusting stock rates to avoid overgrazing susceptible soils during wet weather reduces trampling compaction. In addition, soil degradation during engineering constructions can be reduced by controlling traffic and maintaining soil cover.

The degree to which structure limits agricultural production and affects environmental quality is not well documented. Some land management systems are clearly degrading soil structure at present and in other areas structure must be improved to avoid excessive soil loss and improve production.

Waterlogging and salinity

Where changes in land use have replaced deep-rooted vegetation with shallow-rooted species, more water drains past the root system, particularly in southern Australia where winters are wet and cool. Consequently, groundwater levels have risen and waterlogging of the root zone may cause serious yield decline as well as damage to roads and buildings and septic tank absorption systems. In some cases, changed hydrology redistributes salt within the landscape leading to dryland salinity (see page 7-20). Reclamation of saline lands is expensive and not feasible in many areas. Direct costs to landholders result from reduced productivity and erosion. Off-site costs arise with diminished water quality and environmental effects.

Water erosion

The limited data available indicate that rates of soil formation in Australia are very low by world standards. The great age of many Australian landforms and the generally low relief and rainfall differentiate this continent from the very young glaciated landscapes of Europe and North America. For alluvial landscapes, scientists have estimated rates of 0.4 tonne per ha per year (30 mm per 1000 years), but rates for soils forming on bedrock (that is, on hillslopes) would be considerably lower, particularly in the drier parts of the country.

Therefore, given human life spans, soil has to be considered as a finite and non-renewable resource. The few exceptions are in areas where regular deposition occurs (for example, alluvial plains) and where high inputs of organic and mineral materials become affordable (for example, urban areas and some forms of intensive horticulture).

Infrequent but destructive storms are responsible for much of the soil loss even in the gentle-rainfall zones of southern Australia (see Fig. 6.16). The rates of soil loss on sloping lands under different land management systems are summarised in Table 6.6.

Accelerated soil erosion results from clearing, grazing and cultivation. In some regions the rates of soil erosion by water have been much higher than at present; for example, in southern regions, gully erosion developed soon after clearing. The impacts of erosion can be masked for many years, as they are quickly obscured by cultivation or plant growth. The most immediate impact is a loss of nutrients carried away in the soil. This reduces productivity, although it can be offset by applying more fertilisers. A medium- to long-term impact is the reduced water-storage capacity of the soil, which is critical where soil depth, and thus available storage capacity, is already marginal such as on upper slopes in the Central Highlands of Queensland. The reduced water storage causes a loss of productivity and eventually increased run-off.

Most off-site impacts of accelerated erosion are associated with the deposition of sediment (for example, damage to roads, filling of dams etc.) or reduction in water quality (see page 6-28).

The most critical factor in protecting soils from accelerated erosion by water is the maintenance of cover — plant residues, pasture, forest litter etc.— in close contact with the soil surface. Other techniques for erosion control, such as contour banks, reduced tillage and strip cropping, provide supplementary measures. For example, a storm at Cowra in southern New South Wales during 1992 caused soil losses of 342 tonnes per ha for areas under traditional tillage, 362 tonnes per ha under reduced tillage and 65 tonnes per ha under direct drilling (Hairsine *et al.*, 1993).

Despite its demonstrated value, not enough cover is maintained over large areas of the sloping lands of Australia used for cropping and grazing. The reasons include the following:

- Difficulties in destocking and subsequent overgrazing can remove the cover completely, leaving soil unprotected and extremely vulnerable both during drought and when rains (often heavy) eventually fall.

- Systems of minimum tillage that maintain cover can be more difficult to manage. Problems with crop establishment, diseases, pests and machinery may occur and yield may not increase in the short term.

- Landholders often do not recognise that erosion is occurring at rates that are faster than soil formation.

- Many farmers can afford to substitute the costs of other inputs or are able to defer the costs of degradation so they are borne by future generations.

Even in the best-managed systems, the rates of soil loss still exceed rates of formation by an order of magnitude or more across many parts of Australia. It is clear that current farming practices on sloping lands are unsustainable, and the productive lifespans of soils in these areas vary between decades and centuries.

Wind erosion

Natural and human-induced wind erosion

Soil erosion and deposition by wind has had a major role in creating landforms and soils across

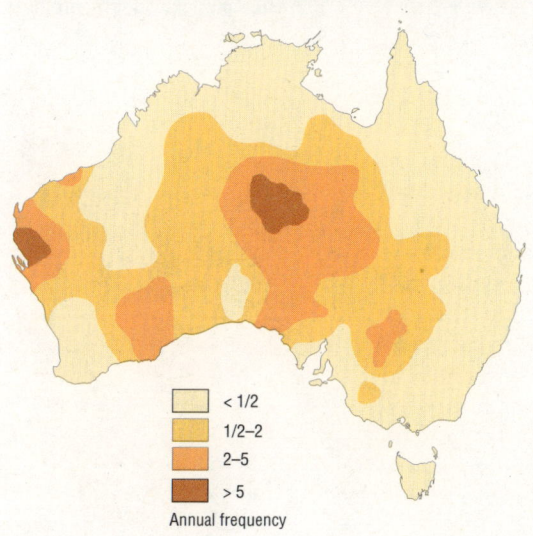

Figure 6.17 Annual frequency of dust storms across Australia

< 1/2
1/2–2
2–5
> 5
Annual frequency

Source: McTainsh and Leys, 1993.

large parts of Australia. The continent experienced a severe arid period between 15 000 and 20 000 years ago, during which dunefields were active and large quantities of dust were deposited around their fringes and in the surrounding oceans. As climates became less arid, vegetation stabilised large dunefields, for example, in the Murray–Mallee, Eyre Peninsula and parts of south-west Western Australia. When these were cleared, the soils became very susceptible to wind erosion. Dust storms provide striking evidence. Most of them occur in the semi-arid and arid lands where annual rainfall is less than 400 mm (see Fig. 6.17). In these areas, large sand drifts can have substantial impacts on agricultural production and local infrastructure.

In many instances, the only evidence that wind erosion is occurring may be atmospheric haze, where the dust consists of fine mineral and organic particles of soil. These particles have nutrient concentrations many times greater than the soil from which they came. Significant quantities of dust are deposited offshore and Australian dust is reported to have been transported to New Zealand on at least seven occasions this century.

The mass of the 1983 Melbourne dust storm was calculated to be about two million tonnes and the cost of replacing nutrients at around $4 million (Raupach *et al.*, 1994). However, the fine haze in previous days, which went largely unnoticed, carried more soil into the Southern Ocean than did the dramatic dust storm.

Landholders can minimise wind erosion by retaining vegetative cover and using windbreaks to reduce wind speeds at the land surface. It is critical to maintain soil structure, particularly in light-textured soils, and this is best achieved by minimal tillage and a low frequency of cropping.

Water erosion and sedimentation

Erosion affects both the soil resource and also the quality of inland waters and aquatic habitats. Erosion occurs as sheets and small channels called rills on bare or poorly vegetated soils, as gullies, and along stream banks and in stream beds. In an average year in Australia, about 14 billion tonnes of soil is moved by sheet and rill erosion (Wasson *et al.*, 1996). This is about 19 per cent of the total soil moved each year globally, even though Australia is only five per cent of the world's land area. Australia, therefore, has a higher than average erosion rate, in places up to 1000 times the rate on undisturbed sites (Edwards, 1991). Sheet and rill erosion is occurring up to about 100 times the soil formation rate on sloping land — a clearly unsustainable situation (Edwards, 1991).

Gully formation, resulting from increased run-off and subsurface flow after clearing, also erodes catchments. Gullies remove soil, reduce access on farms and carry a lot of sediment into water bodies downstream.

The density of gullies over most of the area in the figures below is not changing, because gully networks reach their maximum extent within 50 years of clearing (Stewart, 1968). In areas of recent clearing, such as Queensland, gully formation can be expected. Gully density in areas of stable networks provides a useful indicator of sediment yield from gullied catchments. On the Southern Tablelands of New South Wales and the Australian Capital Territory, annual sediment yield rises on average by 41 tonnes/sq km/year for each increase of 1 km/sq km density of gullies.

The other major type of erosion in catchments occurs along stream banks, but few data exist in Australia to estimate the rate.

Erosion projection for Australia

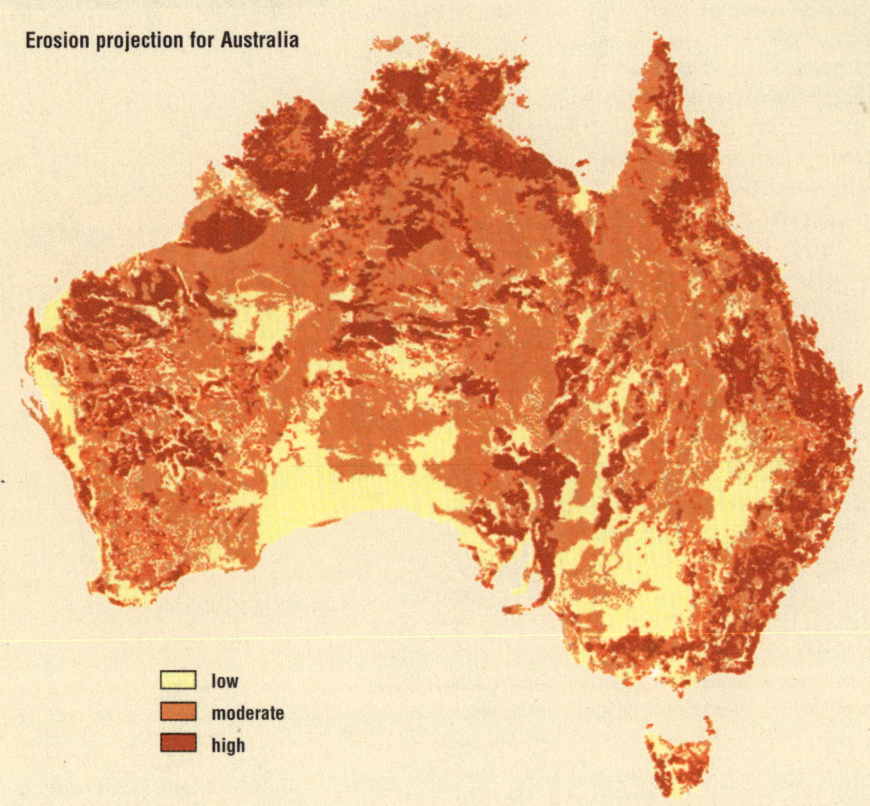

□ low
■ moderate
■ high

An estimate of the spatial pattern of sheet and rill erosion based on a soil erosion equation (Rosewell, in press). It shows high rates (>1000 t/sq km) in areas such as the steep lands of the eastern seaboard, the north, centre and west. This map probably overestimates erosion in the seasonally wet tropics, and in areas of rock outcrop.

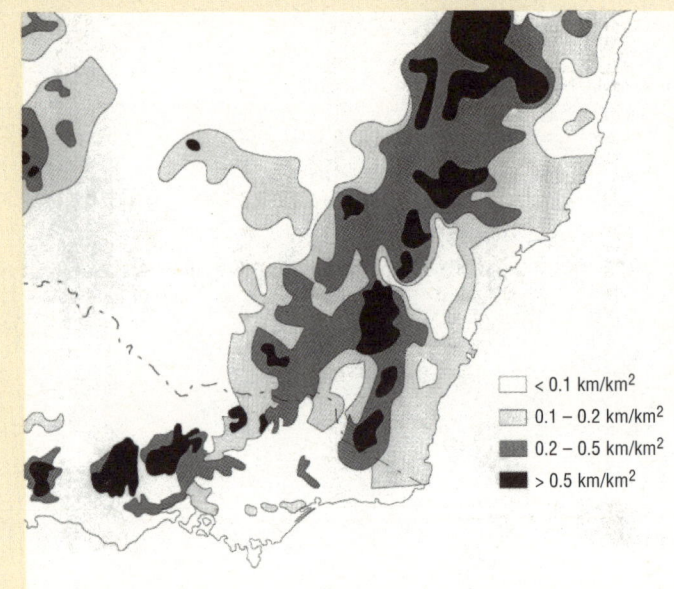

□ < 0.1 km/km²
□ 0.1 – 0.2 km/km²
■ 0.2 – 0.5 km/km²
■ > 0.5 km/km²

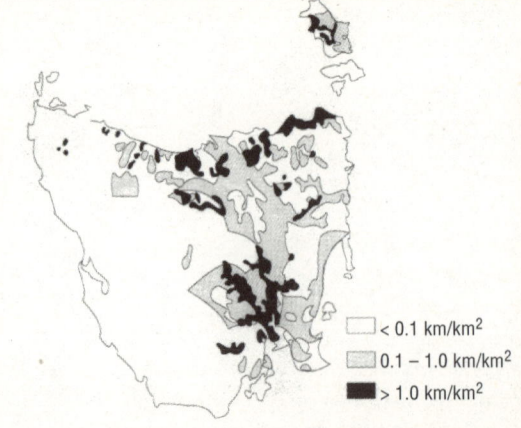

□ < 0.1 km/km²
■ 0.1 – 1.0 km/km²
■ > 1.0 km/km²

Gully density in south-eastern Australia — that is, the total length of gullies per sq km of catchment. The highest density occurs in areas of soils that have high concentrations of sodium salts — a characteristic that promotes collapse of soil structure.

Source: Ford *et al.*, 1993.

Sheet and rill, gully and stream erosion combine to contribute to both turbidity and sedimentation in streams, reservoirs, lakes and estuaries. Sediments have changed the nature of streams and wetland habitats by, for example, filling waterholes that used to be drought refuges for plants and animals, and interfering with the foraging habits of fish. Sediment also carries nitrogen, phosphorus, soil organic matter and possibly pesticides and bacteria. It reduces the storage capacity of reservoirs, and adds substantially to the cost of water treatment.

An indicator of the amount of sediment moving in streams is the sediment yield, which is the quantity lost from a catchment each year.

Channel bank slump in the Ord River, WA.

The highest known rate is for mine waste from the King River in southern Tasmania. Estimates for rural catchments show very low rates in southern temperate areas and much higher rates in northern and arid Australia. In general, these yield estimates match those reported from other continents for similar landscapes which are under similar climatic conditions and disturbed by human activities (Wasson, 1994).

Sediment transport rates in rivers change with time, a long-term measure of which is the rate of accumulation in reservoirs. The rate has been declining this century in eight reservoirs in the uplands of southern and south-eastern Australia. This change reflects improved stock and land management, the use of soil conservation practices, rabbit control and the stabilisation of gullies. Reduced sedimentation in the Laanecoorie Reservoir in Victoria reflects the end of sluice mining for gold in the catchment.

The relative significance of the three types of catchment erosion (sheet and rill, gully, and channel erosion) as contributors to stream sediments is poorly known. In the uplands of south-eastern Australia, Neil and Galloway (1989) showed that gullied catchments yield about eight times more sediment than ungullied catchments — whatever the land use. Radio-cesium, a product of nuclear weapons testing, has been used to show that much less than 30 per cent of the sediment in the Darling River, Murrumbidgee River, upper Condamine, Burdekin and Ord rivers is derived from sheet and rill erosion (Wallbrink and Murray, 1993; Wallbrink *et al.*, 1996). In Victoria, 56 per cent of the sediment in streams has been estimated to be the result of gullying and the rest contributed by the erosion of large channels (Rutherfurd *et al.*, in press). While these results indicate that sheet and rill erosion are not important contributors of sediment to Australian rivers, there may be places where this is not true.

Responses

A lot of effort has been devoted to soil conservation using contour banks, minimum tillage and grazing control. A particularly difficult problem for land managers is that most sheet and rill erosion result from infrequent but destructive storms. Even the best conservation measures can be inadequate in these circumstances, although good ground cover is beneficial.

Gully control — by means of damming, infilling, revegetating and constructing weirs in gully beds — is widespread but limited to small areas due to cost. However, the mounting evidence that gullies are a major source of sediment in Australian streams means that cost-effective control measures are badly needed.

Soil conservation in Australia has mostly been designed to protect the soil resource, and so is primarily targeted at slowing sheet and rill erosion with some attention to gully erosion. While appropriate for its purpose, this form of soil conservation, which emphasises vegetation cover and engineered works to slow water flow, may not be the most appropriate form for water quality protection. As landcare and catchment management become more integrated, truly multipurpose management techniques will be necessary.

Prepared by the Inland Waters Reference Group.

The large dust storm that swept across Melbourne in February 1983 made many urban dwellers aware of wind erosion. However, the fine haze in previous days, which went largely unnoticed, carried more soil into the Southern Ocean than did the dramatic dust storm.

An indication of the severity of wind erosion during recent years in eastern Australia. Note that the data for the 1994–95 drought are not yet available although it is known that the graph will show a sharp increase.

Notable periods of wind erosion included the 1895–1903 drought and the extended drought of the 1930s and early '40s. During these periods, severe wind erosion occurred in the Mallee areas of New South Wales and Victoria and across many of the agricultural lands of South Australia.

Long term, the number of dust storms appears to be falling (see Fig. 6.18), which suggests that control measures have been successful to some extent. For example, rabbits are not in the plague proportions experienced during the 1930s and '40s, and methods of minimum tillage have been

adopted in some areas. In other places, invasion of woody weeds has increased cover and surface roughness.

The highly episodic nature of wind erosion caused by climatic variation indicates that major phases will occur again. For example, in 1994 the number of dust storms increased.

While images of drifting sands and direct damage to properties and infrastructure have dominated media coverage, in recent times it has become apparent that a less visible but serious problem is the loss of nutrients in dust. However, apart from localised case studies, the magnitude, extent and cost of nutrient depletion caused by wind erosion remain largely unknown.

Soil nutrients

Many Australian landscapes are ancient and heavily weathered or have impoverished parent materials. Thus large areas have naturally low nutrient status and only small areas have highly fertile soils (see Fig. 6.19). Acute deficiencies of nitrogen and phosphorus are common and deficiencies of sulfur and potassium and the trace elements molybdenum, copper, zinc, boron and manganese are also not uncommon (Williams and Raupach, 1983). The few toxicity problems that occur are usually related to aluminium or manganese in soils of low pH. Boron toxicity occurs in some of the cropping lands of Western Australia, South Australia and Victoria.

Over the last 200 years, the export of phosphorus, sulfur and nitrogen via food and fibre products has increased. For phosphorus and sulfur, this has at least been matched by inputs. The input/output budget for nitrogen is more difficult to determine because of fluxes associated with erosion, leaching, gaseous losses and fire. Estimated budgets suggest that, at the continental scale, inputs are exceeding outputs (see Table 6.7).

Regional patterns of nutrient increase and decline occur across Australia but reliable evidence is only available from isolated surveys, experiments and monitoring studies with the exception of Western Australia where the main fertiliser company monitors regional trends.

In many permanent pastures or long-rotation pasture systems, soils have accumulated nutrients through biological fixation of nitrogen and fertilisation — most commonly with phosphorus, sulfur, calcium, potassium, molybdenum and zinc. In some areas these gains have to be balanced by costs associated with soil acidification. Nutrient accumulation is probably occurring in areas used for intensive agriculture with high fertiliser use (for example, horticulture and sugarcane production). In some instances, risks of groundwater pollution are associated with excessive fertilisation.

Evidence suggests that nutrient losses are occurring in the nation's most productive soils (Dalal and Mayer, 1986). They have been documented for the cropping lands of south-east Queensland and northern New South Wales, and in the deep red soils formed on basalt (ferrosols in Fig. 6.14) that

Figure 6.18 Total number of dust storm days from 41 eastern Australian weather stations

No. of dust storms

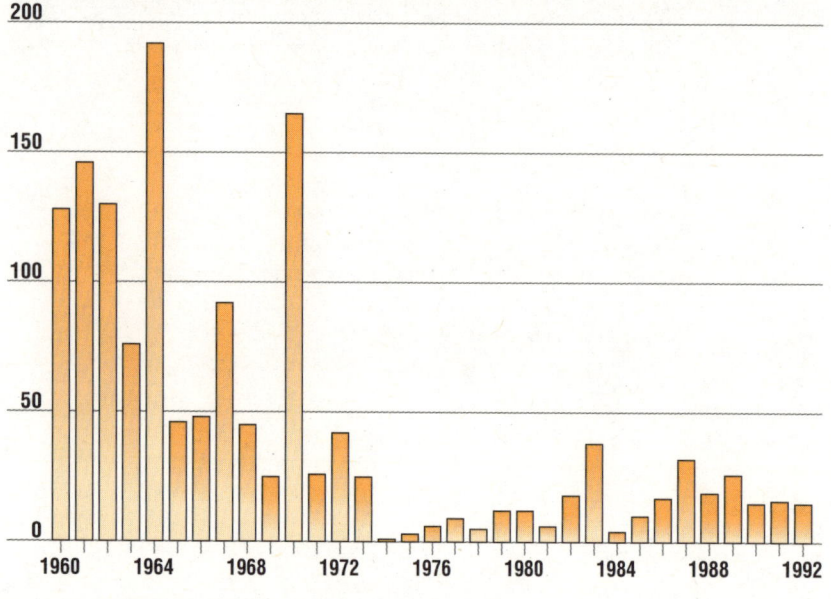

Source: McTainsh, G.H. and Reddan, S.P. 1995 (unpublished data).

are used for intensive cropping in eastern Australia. These declines are caused by inadequate fertilisation, excessive cultivation (resulting in the mining of nutrients in organic matter) and losses associated with soil erosion.

The nutrient status of soils has declined across extensive areas used for wheat production (Hamblin and Kyneur, 1993). Between 1950 and 1990, Australia had one of the lowest average rates of wheat-yield increase in the world (8 kg per ha per year). Although climatic constraints limit the potential gain in grain yield and the likely response from fertiliser and other inputs, this is a disappointing result when the improvements in plant varieties, weed control, tillage methods and fertiliser technologies are considered. However, in one-quarter of the wheat belt, farmers have achieved yield increases of more than 20 kg per ha per year and many have enhanced soil fertility by the use of legume pastures in the rotation.

Without an organised and comprehensive system for surveying, monitoring and predicting the sustainability of nutrient supplies for different systems of land use, it is hard to make conclusions on nutrient status across large parts of Australia. It seems that nutrient levels of soils used for forestry are being maintained except where erosion rates are high or where intensive fires have become more frequent. Some areas used for plantation forestry have experienced nutrient declines and decreases in yield (for example, pine plantations in the south-east of South Australia), but the problem has been rectified by improved land management.

While nutrient losses are probably occurring in those parts of the arid zone prone to wind erosion, their magnitude and significance are unknown. Water erosion causes local redistribution of nutrients in many arid zone ecosystems and substantial degradation can be associated with it.

The off-site effects of nutrient management in agriculture are discussed more fully in Chapters 7 and 8.

Soil acidification

A decrease in soil pH over a period of time — usually decades, but in some cases years — is the most obvious signal of soil acidification. The effect may take place at or near the soil surface, or in the subsoil. As their pH falls below about five, some soils release aluminium and manganese, which are toxic to some plants. Retarded root growth leads to poor water and nutrient use and impaired nodulation of legumes, resulting in poorer yields. By contrast, many calcareous and alkaline sodic soils of southern Australia are likely to benefit from acidification because more nutrients become available to plants and it creates an opportunity for leaching of sodium. Oxidation of acid sulfate soils, predominantly in coastal areas, can have a major local impact, including fish kills in nearby waterways and loss of vegetation.

A number of factors cause soil acidification:

• oxidation of sulfides (acid sulfate soils), usually in tidal areas or in mine tailings

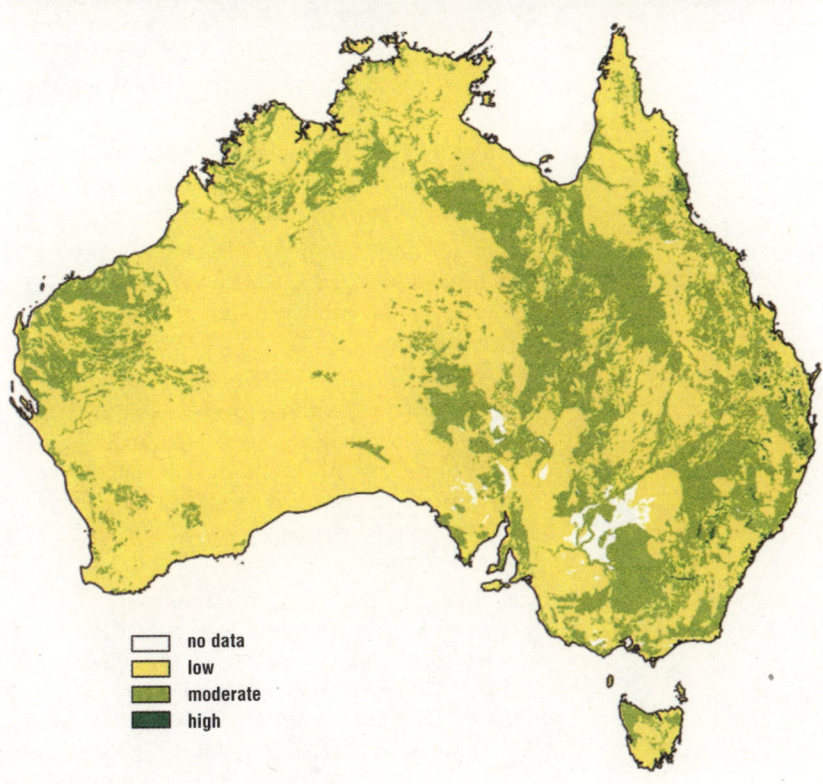

Figure 6.19 The nutrient status of soils

- no data
- low
- moderate
- high

Source: CSIRO Division of Soils.

Table 6.7 A summary of the continental nutrient budgets

	Flux in kilotonnes per year		
	Phosphorus	Sulfur	Nitrogen
Inputs			
Atmospheric deposition	77	769	1150
Food, fish and timber imports	<1	<1	<2
Fertiliser	54	325	380
Phosphate rock	326		
Fixation by sown pastures			1500
Fixation by crops			200
Fixation by forests			200
Total inputs	**458**	**1095**	**3432**
Outputs			
Food and fibre exports	56	52	415
Urban discharges	11	77	32
Soil erosion	6–32	3–17	18–98
Leaching	<3	87	243
Ammonia volatilisation			334
Denitrification			41
Fire	4	120	1200
Biogenic gaseous emissions	-	8–154	-
Total outputs	**80–106**	**347–507**	**2283–2363**

Source: derived from McLaughlin *et al.*, 1992.

- acidifying fertilisers, usually nitrogen fertilisers applied in intensive agriculture, or elemental sulfur

- nitrogen and carbon cycle effects such as nitrogen fixation by legumes and subsequent nitrification and nitrate leaching or increased soil organic acids

- removal of plant products that have more basic cations than inorganic anions and redistribution of nutrients through grazing animals

- acid deposition from the release of oxides of sulfur and nitrogen as gases from industrial sources, sulfide-ore processing or coal-fired power generation

- natural weathering processes driven by rainfall leaching and the action of soil organic acids and carbonic acid; these processes are usually very slow, but many Australian soils on stable landscapes have reached a significant stage of acidification due to extensive leaching and weathering over long periods of time

Soil types vulnerable to acidification are well known (see Fig. 6.20) but the exact distribution of affected soils is not. An estimated 29 million ha of mainly agricultural land are regarded as significantly acidified, at least in the surface layers. Significant areas affected by acidification extending into subsoil layers pose more serious problems for root development and greatly increase remediation costs. Trends in soil acidification are causing concern on some sugar plantations, in intensive horticultural industries and in the northern Australian grazing lands where tropical legumes have been introduced. The extent of acid sulfate soils in Australia is now being assessed.

Soil acidification can be prevented or remedied by applying lime or dolomite as sources of alkaline material. Land managers in areas of acidifying soils have not universally adopted liming practices, despite demonstrated benefits and knowledge of how critical the problem is. Recent estimates in South Australia suggest that only about 10 per cent of the area thought to require liming received applications over the last 10 years. If deep subsoils are acidified the damage will be expensive to repair (Helyar *et al.*, 1988). Other approaches to the problem include the manipulation of management systems (for example, limiting the removal of nutrient-rich products such as hay), and the use of tolerant plant varieties.

Soil contamination

Scientists do not have a good understanding of how contaminants are distributed and move in Australian landscapes. Most of the available information relates to contaminants associated with food (see the box on page 6-49) and water quality (see page 7-34). Researchers have paid minimal attention to the impacts on natural ecosystems.

Impurities in fertilisers and in soil amendments such as lime and gypsum include arsenic, cadmium, fluorine, lead and mercury. Cadmium is the element of most concern because significantly

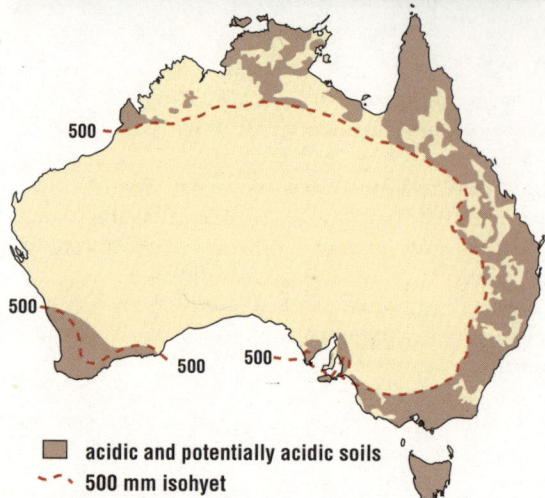

Figure 6.20 Acidic and potentially acidic soils in the agricultural regions

- acidic and potentially acidic soils
- - - 500 mm isohyet

Source: CSIRO Division of Soils.

more of it is transferred from soils to the edible portions of agricultural food crops. Some agricultural produce currently exceeds maximum permitted concentrations. However, levels are in a similar range to, or lower than, those found internationally.

Cadmium concentrations in Australian phosphatic fertilisers have been high by comparison to fertilisers used in other countries. However, lower inputs of fertiliser per unit area than in Europe and less contamination of soils from industrial sources have resulted in similar or lower cadmium loadings to agricultural land. In recent years, people have addressed the problem by lowering cadmium levels in phosphatic fertilisers and developing plant cultivars that restrict its uptake. Acidification and salinisation of soils pose threats because they increase the uptake of cadmium by crops.

In the past, pesticides with harmful residues have been used to control pests and diseases in orchards and on a range of crops, particularly potatoes, tomatoes, cotton, bananas and sugarcane. They have also been used to control external parasites on livestock. Although harmful pesticides (such as organochlorines) are now banned in agriculture, they are persistent and their residues will continue to have an adverse impact on environmental quality.

The available information on the extent of contamination by pesticides and herbicides is patchy. The discovery of residues of organochlorines in beef exports in 1987 led to sampling of farms considered to be at risk because of their agricultural history. In Western Australia, surveys detected the pesticides chlordane, DDT, dieldrin and heptachlor in 18.4 per cent, 39.6 per cent, 39.0 per cent and 18.8 per cent respectively of 11 248 samples analysed (EPA WA, 1989). These samples did not include known hot spots such as soils near wooden power poles with a history of pesticide treatment.

Soils at cattle- and sheep-dip sites are contaminated with organochlorines like DDT and with other pesticides, such as arsenic. Thousands of such sites have been documented in northern New South Wales, where the consequences are particularly serious because of urbanisation and construction of dwellings on or near dip sites.

The extent of soil contamination associated with agrochemical usage in cotton, banana and other intensive agricultural enterprises is under investigation. In northern New South Wales and Queensland, past usage in banana and sugarcane production has resulted in contamination with organochlorines, while copper, arsenic and lead have been problems in orchards and market gardens.

Concentrations of endosulfan and other insecticides or of herbicides like atrazine have exceeded guidelines for the protection of drinking water and aquatic environments in the rivers of central and north-western New South Wales (Preece and Whalley, 1993).

The substantial benefits associated with the use of most pesticides in agriculture outweigh their risks especially since their safety and sophistication have increased during recent decades (Ferris and Haigh, 1993). The irony is that new farming systems such as minimum tillage or conservation farming, which conserve soils by minimising erosion and improving soil structure, rely on agricultural chemicals much more heavily than traditional systems. While the impact of this increased use of chemicals is not fully known, it is clear that land-managers require a substantial level of technical sophistication to use them safely.

Although rural lands surrounding industrial areas and cities are contaminated, serious impacts are usually localised — except where contamination of streams or groundwater is involved. As an indication, dispersed heavy metals from lead smelters at Port Pirie can be detected over thousands of square kilometres, although seriously contaminated areas are restricted to tens of square kilometres (Cartwright *et al.*, 1977). Other urban-generated pollutants such as lead can also be widely dispersed (Tiller, 1992).

Soil biology

Australian soils host a wide range of soil organisms, varying in size from micrometres (for example, bacteria, protozoa and fungi) to several centimetres (insects and earthworms). One square metre of soil may support populations of about 200 000 arthropods and enchytraeids and billions of microbes. One hectare of good-quality soil could contain 1000 kg of earthworms, 1000 kg of arthropods, 150 kg of protozoa, 150 kg of algae, 1700 kg of bacteria and 2700 kg of fungi. Although we know soil organisms break down organic materials in soils, their role in agricultural productivity and the way different management systems affect them remain poorly understood. Thus it is not easy to assess the biological health (increasingly recognised as a key index of sustainability) of Australian soils or predict the

Land resource surveys for good land management

Land managers across large parts of Australia operate with only a rudimentary and sometimes erroneous understanding of the condition, productive capacity and potential hazards associated with the use of land.

Maps of soil and land resources are essential for the development of sustainable systems of land use. Unlike other comparable countries, many parts of Australia have no detailed maps of soil and land resources. Consequently, land use proceeds by trial and error with unnecessary economic and environmental costs, many of which have been documented in this report. The lack of comprehensive survey coverage also severely limits the conclusions that can be drawn about the state of the Australian environment.

The effort directed towards land resource survey by State and Territory agencies has increased during the last decade with funding from the National Landcare Program. However, with the existing level of staff and resources, it would take more than a century to map the arable and forested lands of Australia, which cover only 16 per cent of the continent, at a scale suitable for guiding decisions on practical land management. (For example, 1:25 000-scale mapping or larger is required for farm plans, urban development or forest operations.) Despite this daunting task, land resource agencies have done much to improve the coordination and quality of their mapping programs during the last five years (McKenzie, 1991). This has been achieved through the establishment of the Australian Collaborative Land Evaluation Program and better cooperation between State, Territory and Commonwealth agencies.

A survey scale of 1:100 000 is an absolute minimum for land management except for rangelands, where 1:250 000 is often adequate. These scales provide only a general framework for decisions on land management. It is clear that acquiring the basic information necessary to address contemporary problems of land planning and management will require a long-term commitment by the public and private sector to well focused and strategic land resource surveys.

effects of particular practices. People are starting to recognise the importance of soil organisms and placing more emphasis on balanced ecosystem management within agricultural systems.

Aspects of soil biology are sometimes manipulated to improve production. For example, inoculating legume seeds with *Rhizobium* spp. makes atmospheric nitrogen available through a symbiotic relationship between the bacteria and the legume. Highly productive pasture systems in Australia's temperate regions are based on nitrogen fixation by medics and clovers. While these systems provide the benefit of increased soil organic matter, in the long term they can also increase soil acidity to harmful levels. Some land management practices may introduce or favour pests and diseases. For example, crop sequences and management practices can affect the spread of fungal pathogens (like *Phytophthora* or *Fusarium* spp.) and nematodes, or the population densities of organisms such as wireworms or earthworms. Soil surface conditions, particularly the amount of surface litter and the type of tillage practised, influence the suitability of the soil habitat. Surface litter provides food and shelter for macrofauna whose burrows increase aeration and infiltration, while tillage discourages earthworms.

Land use and land tenure

Land use

Agriculture is the largest land use in Australia (see Fig. 6.21) with the greatest area supporting grazing of cattle and sheep. Most of this grazing occurs in the rangelands in central and northern Australia (see Fig. 6.22), while cropping is concentrated on the coastal fringe. The area of land used for agriculture reached a peak of 500 million ha in 1975 and has since been declining slowly. Most of the change comes from the transfer of grazing lands to other uses including conservation, urban areas, Aboriginal title and unused land such as vacant Crown land. Agricultural establishments comprise about 460 million ha (or 60 per cent of Australia). Of this, crops or sown pastures grow on 47 million ha. Both these uses have increased steadily over the past 40 years from two per cent to six per cent of the land area.

The number of farming enterprises has fallen dramatically over the last 40 years (see Fig. 6.23). Much of this decline is due to smaller enterprises (defined as having an estimated value of agricultural operations of less than $20 000) being excluded from data collection from 1987 onwards. Nevertheless, the number of farming enterprises has fallen by about 1200 per year since 1950. The reduction of numbers has brought a corresponding increase in the average size of individual farms, from 2200 ha in 1951–52 to 3800 ha in 1992–93. The average area worked by each farm worker — including employers, self-employed people and unpaid family help — has increased from 930 ha in 1951–52 to 1270 ha in 1992–93.

Land tenure

Australia is probably unique among Western countries in that a high proportion of our land is still publicly owned (Campbell and Dumsday, 1990). Most agricultural land is operated under long-term Crown leases. Only 13 per cent of the total area of the country has been converted to freehold title (BAE, 1983), and Queensland is the only State where this is still occurring for a significant proportion of land. However, the more

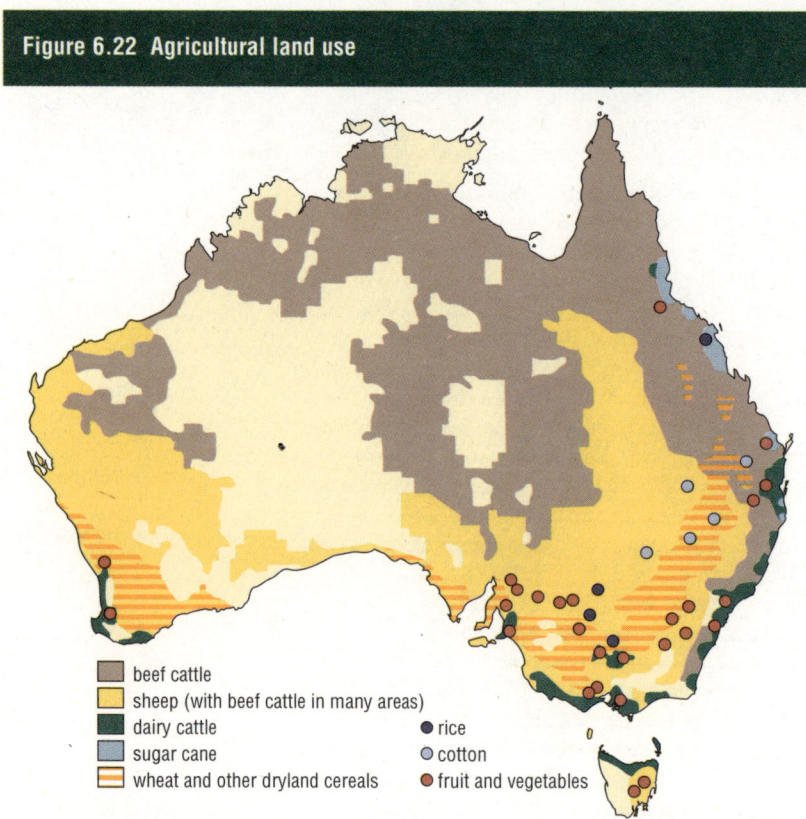

Figure 6.21 Land use in Australia 1990

grazing and fallow

Aboriginal land

other land

urban

protected areas

forest

sown pastures

crops

Source: data supplied by ABARE and ABS.

Figure 6.22 Agricultural land use

- beef cattle
- sheep (with beef cattle in many areas)
- dairy cattle
- sugar cane
- wheat and other dryland cereals
- rice
- cotton
- fruit and vegetables

Source: ABS Yearbooks.

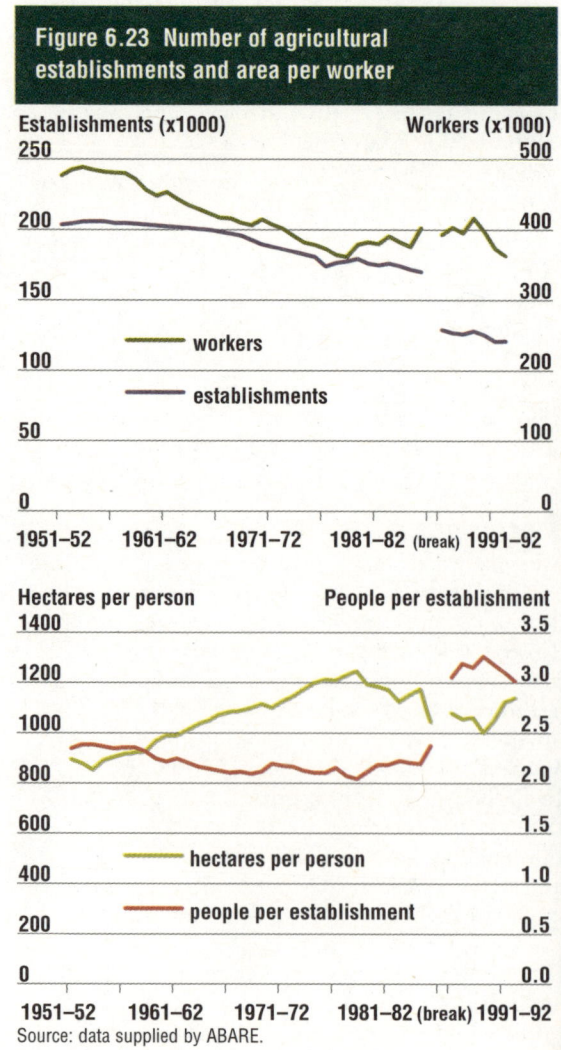

Figure 6.23 Number of agricultural establishments and area per worker

Establishments (x1000)   Workers (x1000)
250   500
200   400
150   300
100   200
50   100
0   0

- workers
- establishments

1951–52   1961–62   1971–72   1981–82 (break) 1991–92

Hectares per person   People per establishment
1400   3.5
1200   3.0
1000   2.5
800   2.0
600   1.5
400   1.0
200   0.5
0   0.0

- hectares per person
- people per establishment

1951–52   1961–62   1971–72   1981–82 (break) 1991–92

Source: data supplied by ABARE.

The evolution of land tenure in Victoria

The historical evolution of land tenure in Australia can be illustrated by the case of Victoria.

"'*Terra nullius*' was the term used to describe land tenure in Australia when James Cook's expedition landed at Botany Bay in 1770. It means land of nothing and, by implication, unoccupied or possessed by nobody, and was used as justification for the British Government taking possession of eastern Australia. By the same stroke that effectively dispossessed the Aboriginal people, Cook established the public land estate of Australia, in its best condition and greatest extent. Over the next 160 years, that estate diminished in extent and much of it deteriorated in condition." (Land Conservation Council, 1988).

Of course, much of the damage to the public land estate was neither deliberate nor recognised at the time. However, even if it had been recognised, some of the damage would have been seen as a cost that could not be avoided in furthering the development of Victoria through industries such as agriculture and gold-mining.

In recent years there has been a steady increase in the area of 'protected' land in Victoria, largely at the expense of commercial agricultural leases on public land. The increase is in line with that for Australia as a whole.

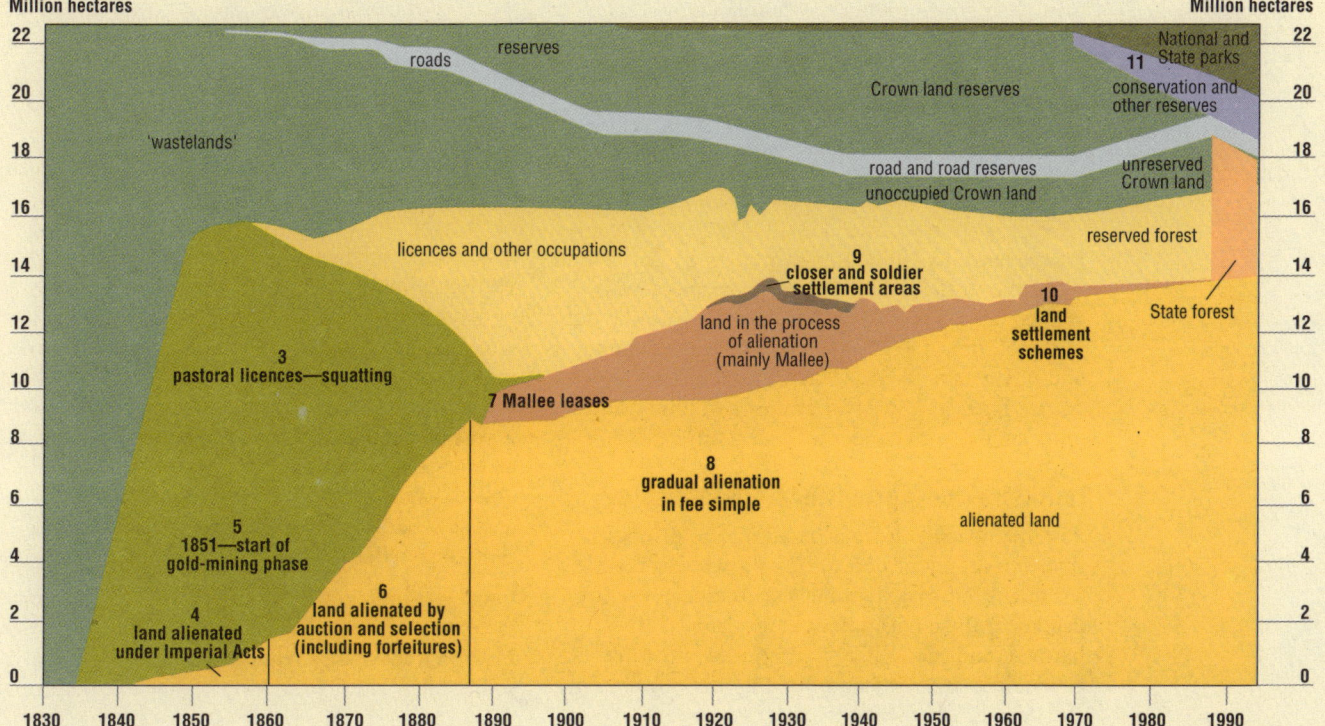

Date		Phase	Pressures on public land estate
1770		Cook's landing at Possession Island	Establishment of the public land estate of New South Wales, including the area later to become Victoria.
1788	1.	First Fleet	First European settlement — clearing of vegetation; hunting of fauna with firearms
Early 1800s	2.	Sealers and whalers	Harvesting of native fauna
1820s to 1890	3.	Pastoral settlement — squatting	Damage to ground flora; competition with grazing fauna; spread of weeds; erosion
1837 to 1860	4.	Alienation under Imperial Acts	Loss of public land; clearing, cultivation, and removal of indigenous flora and fauna
1851 to 1860	5.	Gold-mining and associated settlements	Clearing, timber extraction; mining-sludge deposition; erosion; water pollution; flooding
1860 to 1887	6.	Extensive loss of public land	Extensive loss of public land; widespread clearing, cultivation, and removal of indigenous flora and fauna
1888 to 1970	7.	Mallee settlement and other leasehold	Leasing and later alienation of the mallee country; other new areas opened for selection; effects as above
1888 to 1970	8.	Gradual alienation of public land	Loss of public land and its values as for 4 and 6 above
1919 to 1970	9.	Closer, soldier, and rural settlement schemes	Some areas of public land developed by the State and alienated for farming; effects as above
1960s	10.	Little Desert, Big Desert and other land settlement schemes	Limited areas settled
1970s to 1990s	11.	Withdrawal of agricultural leases from some public land, including Little and Big Deserts	Increase in the area of 'protected' public land

Source: Adapted from Land Conservation Council, 1988.

fertile parts of Australia have a much higher proportion in freehold tenure. For example, about 60 per cent of land in the wheat–sheep zone is privately owned.

Governments reserve about 30 per cent of Australia's land for forests, urban, transport and other purposes, while approximately 53 per cent is made available for commercial use under Crown leases. These are commonly long-term or perpetual and are mostly taken up by agricultural and pastoral interests. Such land is generally secure in tenure and marketable. In management terms, there appears to be little practical difference between freehold and long-term or perpetual leasehold land.

Australian governments have gradually increased the area of 'protected' publicly owned land. Protected areas include land within national or State parks and wildlife or conservation reserves.

The main changes in land tenure in recent times have arisen from recognition of Aboriginal claims. The *Aboriginal Land Rights (Northern Territory) Act 1978* gave freehold title to 240 000 sq km of former reserve land. In 1990, approximately 11 per cent of Australia, in the form of reserves, freehold and granted land, was held by Aborigines, predominantly in the Northern Territory, Western Australia and South Australia (ABS, 1992). The Commonwealth *Native Title Act 1993* will almost certainly lead to more significant changes in land tenure.

Is Australian agriculture sustainable?

Sustainable agriculture is based on the principles that: the supply of necessary inputs is sustainable; the quality of basic natural resources is not degraded; the environment is not irreversibly harmed; and the welfare and options of future generations are not jeopardised by the production and consumption activities of the present generation. There is a further objective, which is to maintain or improve yield. Clearly, sustainability is a complex issue that cannot be easily evaluated for modern agricultural systems.

Sustainability of resources and ecosystems

The use of land for agriculture inevitably changes land qualities and processes from the natural state. Habitats on the site itself are destroyed or altered through land clearing and other sites may be damaged, for example, through sediment deposition in streams. These changes to the basic qualities of the resource, although often imperceptible against the effects of seasonal fluctuations, incur an on-site cost to the producer and an off-site cost to the community. If these costs rise, eventually agricultural practices may need to be adjusted to maintain economic viability of farms and quality of life for communities.

The condition of land resources is fundamental to sustainability since it is the key to the capacity of the land to sustain future uses. Many soil properties, such as organic matter content, soil structure, nutrient availability, water-holding capacity and acidity, influence productive capacity. A major difficulty when trying to assess sustainability in terms of the condition of land resources is the lack of an agreed set of indicators, without which we cannot assess either condition or trend in a consistent and comparable way. National efforts are being made to develop sustainability indicators. However, even when key sustainability indicators are selected and data collected regularly, it will be difficult to separate gradual permanent changes from seasonal variations. We need to be able to predict the accumulated effect of changes and to separate these permanent changes from temporary effects.

Crop yield integrates the effect of many different factors, including changes to basic soil properties. However, it is not a reliable short-term indicator of soil quality. Farmers seeking higher yields will use management inputs (such as new varieties and fertilisers) to improve conditions for plant growth. These inputs change over time and their effect, which is in a sense the sum total of the impact of advances in agricultural technology, may completely mask any tendency for yield to decline due to soil degradation.

The question of sustainability is one of time scales. It is difficult for individuals to see beyond the next few years based on their recent experience. But the threats to sustainability may develop insidiously over many years. For example, in Mesopotamia, an average decline in yield of less than 0.1 per cent per year due to salinisation, over 700 years from 2400 BC, eventually made irrigated cereal production unsustainable (Gelburd, 1985). This was despite adoption of new technologies such as salt-tolerant crops.

One way to assess sustainability is to choose a fundamental measure of soil condition — say, water-holding capacity — and assess the impact of the most powerful degrading process, such as soil erosion. Using computer-based simulation models (see the box on page 6-38), researchers can forecast the likely effects (including the effect on yield) of various management options.

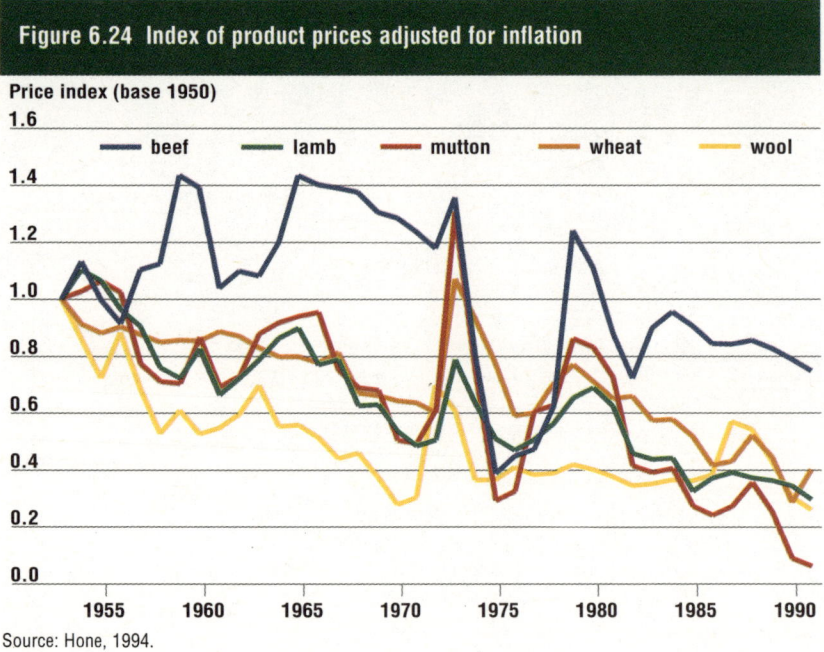

Figure 6.24 Index of product prices adjusted for inflation

Price index (base 1950)

beef — lamb — mutton — wheat — wool

Source: Hone, 1994.

Economic viability of agricultural production

The economic viability of Australia's agricultural production has been under pressure for four decades because of our heavy dependence on competitive export markets. While nominal (undeflated) agricultural commodity prices in these markets have risen, 'real' prices (market prices deflated by the Consumer Price Index) have declined (see Fig. 6.24). Real per-unit prices of inputs have also fallen, but at a slower rate. As a result, the real net income per farm in Australia has fallen steadily over the past 40 years and is now only about 60 per cent of its level in the 1950s. But the number of farms has also fallen, resulting in a fall in the total real value of agricultural production to 40 per cent of its level in the 1950s (see Fig. 6.25).

These changes to economic viability are due to market forces, which have placed enormous pressure on farmers to increase their productivity by applying new technologies that will make production more efficient while not necessarily accounting for any adverse changes in resource condition. As these economic pressures can be expected to continue, we can examine past productivity trends to assess the overall resilience of Australian farming systems, as shown by trends in wheat yield and agricultural productivity.

Trends in agricultural yields

Analysis of mean Australian wheat yields for each decade since 1870 suggests that early agricultural practices were unsustainable (Donald in Hamblin and Kyneur, 1993), but since about 1900 yields per unit area have steadily increased (see Fig. 6.26). Yield data provided by the Australian Bureau of Statistics (ABS) and the Australian Bureau of Agricultural and Resource Economics (ABARE) for coarse grains, oilseeds, rice and cotton also show increasing trends.

However, increasing trends in overall yield may hide negative trends in some areas (Hamblin and Kyneur, 1993). Any such negative trends suggest

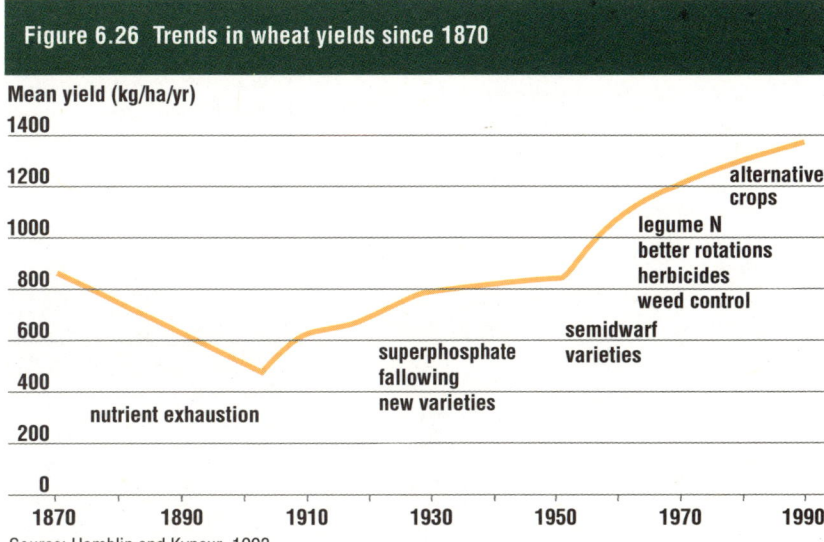

Figure 6.26 Trends in wheat yields since 1870

Mean yield (kg/ha/yr)

Source: Hamblin and Kyneur, 1993.

that enhanced technologies either are not being applied or are not compensating for adverse changes in soil resources. In the sugar industry, for example, ABS and ABARE data and observations by researchers (Henzell, 1992) indicate that a yield plateau has occurred for the last 20 years, despite the adoption of potentially higher-yielding varieties. These negative trends or lack of expected response to yield-enhancing technology inputs may indicate that the system is becoming unsustainable.

Trends in agricultural productivity

Yield data provide one measure of productivity, but they ignore the resources, other than land, required to produce yields. 'Total factor productivity' (TFP) is an index of productivity growth that attempts to incorporate all the resources (or factors of production) used and all products for an industry. Estimates of TFP provide a measure of the overall productive capacity of agriculture.

A viable long-term agricultural system could be defined as one where TFP and output over time are constant or increasing. It would be difficult for agricultural industries to meet this criterion if substantial land degradation was occurring. For example, severe erosion would normally result in reduced crop yields or an increased need for fertilisers, or both, leading to reductions in TFP. On the other hand, the discovery of a trace element deficiency that is severely limiting yields and that can be overcome at little cost would lead to increases in TFP and output.

Over the last 40 years, substantial increases in TFP have occurred in agriculture, both globally and in Australia (Chisholm, 1992). Knowledge is the key resource accounting for the increase (Crosson and Anderson, 1992), which has been accompanied by only a modest expansion in the quantities of land and water devoted to agricultural production (Chisholm, 1994). The increases in global productivity have been the main cause of the decrease in real commodity prices.

Over the 40-year period from 1950 to 1989, farm output in Australia increased two and a half times,

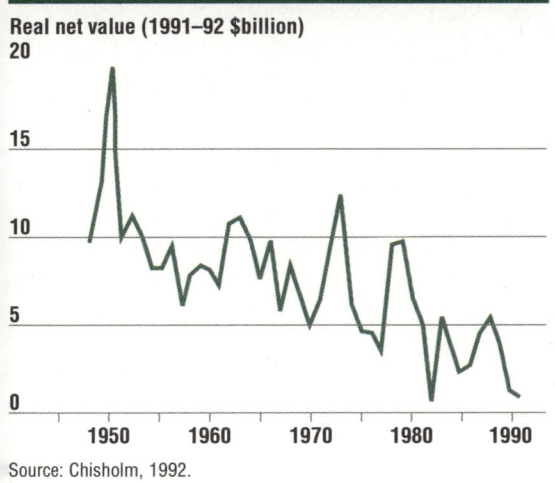

Figure 6.25 Net value of farm output adjusted for inflation

Real net value (1991–92 $billion)

Source: Chisholm, 1992.

Computer models as a guide to resource sustainability

Australia has a strong tradition of scientific research into environmental constraints and innovative agricultural management practices. However, climatic variability means that field experiments have to be run for several years to sample a representative range of seasons. For example, records show that much soil erosion by water occurs in infrequent and episodic catastrophic events. Thus, field investigation of the effect of erosion on sustainability is expensive. In recent years computer-based models have been used to simulate natural systems. This approach integrates data from many experiments, and knowledge from many sources, into mathematical equations. Data such as long-term daily rainfall records are entered into the model, which generates predictions that can be used to compare sustainability of systems under various land management options, without the time and expense of field experiments.

For example, the bar charts generated by the PERFECT (Productivity Erosion Runoff Functions to Evaluate Conservation Techniques) simulation model (Littleboy *et al.*, 1989) show the episodic nature of events and the need to consider more than one aspect of soil degradation. The simulation results show that zero tillage greatly reduces soil erosion. On the other hand, it increases deep drainage, which could increase shallow water tables and salinity, although this is not yet a problem. Opportunity cropping (not shown) is another option in which sowing is timed to optimise use of water stored in the soil: it shows the importance of timing sowing to use available soil water. This option reduces deep drainage but increases erosion slightly. On balance, opportunity cropping is considered to be both more profitable and less degrading of soil resources.

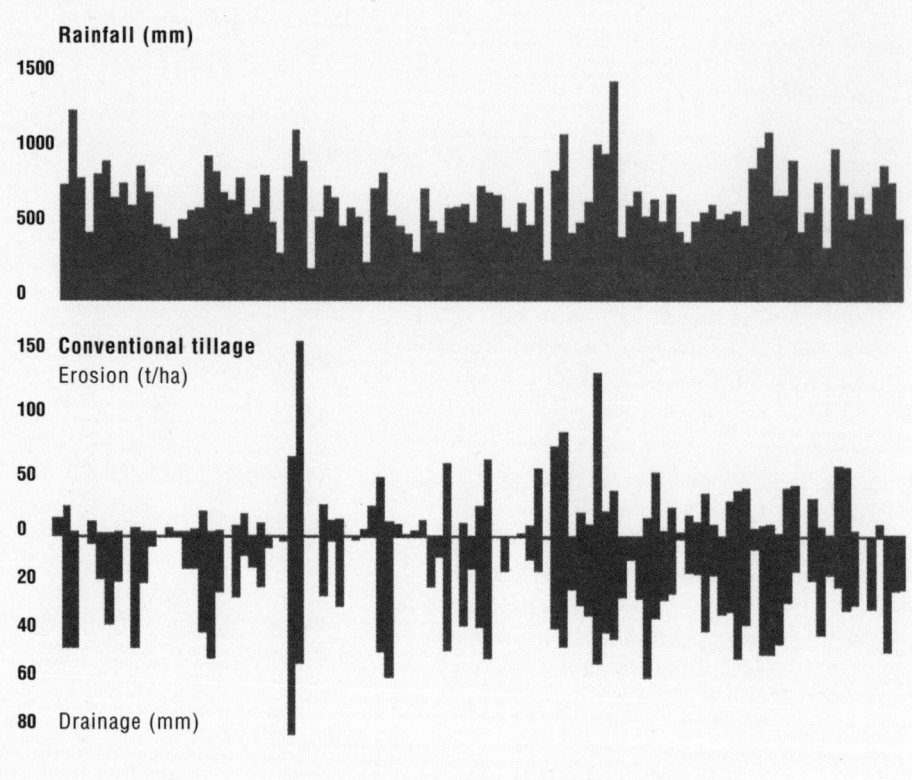

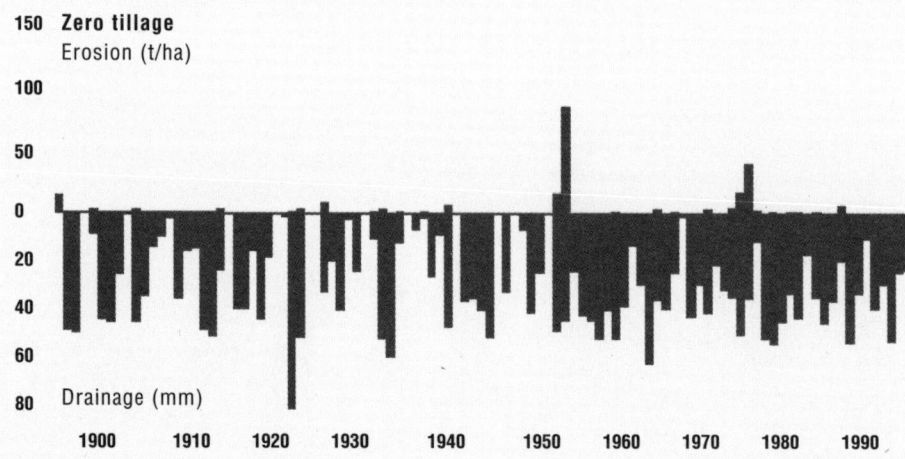

A simulated comparison of two management practices showing effects of erosion and drainage.
Source: Littleboy, M., unpublished.

or about 2.5 per cent per year, while real per-unit costs of production declined. The rate of increase in output per unit of land over the last three decades has been higher than for any other OECD country (Alston *et al.*, 1993), reflecting the skill and effectiveness of Australian scientists and farmers to develop and apply enhanced technologies to sustain agricultural production systems.

The relationship between land degradation, conservation and productivity growth is complex. As productivity increases, land quality may remain unchanged, improve or decline, depending on the nature of the technology. While land is included in the calculation of TFP, it is not measured in a way that fully captures any decline or improvement in its quality. Nor are off-site degradation effects included in the calculation.

Until we know more about the damage to other ecosystems, we should not draw firm conclusions about the sustainability of agriculture.

Research and development will be increasingly important to the international competitiveness and sustainability of Australian agriculture. If the industry cannot retain competitiveness in a sustainable way, we can expect to see large changes in the way rural land is managed. As in the case of mining, we may see some land move in and out of agricultural production as technology and prices change. Other land may be permanently allocated to non-agricultural uses.

Response

The Australian community's responses to the perceived state of land resources reflect public perceptions of national needs and priorities of the times. Earlier concepts of land are part of our culture and heritage. In recent times the community's focus has moved from establishing productive industries in a pristine but harsh environment, and populating a vast land, to concern about sustainability of industries, the land resources they use, keeping land for conservation purposes, and the impact of increasing populations. It is not surprising that this change has occurred as we approach the limits of land development for agriculture and other purposes including conservation. Concern about the state of the nation's land resources has led governments, industries, communities and individuals to respond in a variety of ways.

Responses may be designed either to change the pressures on a particular resource and thus affect its future condition or to directly change its current state. They vary in their effectiveness: some are inappropriate and may become pressures in their own right — for example, the introduction of pasture plants that become weeds and drought-relief measures that encourage inappropriate risk-taking behaviour.

Governments, industries and communities have all responded to problems relating to land resources. The responses include soil conservation measures, codes of practice for mining and forestry operations, educational programs, drought relief and rural adjustment schemes. Some responses have encouraged land clearing while others have restricted it.

In recent years, the overall nature of the 'response' has changed. People are increasingly aware of the need for sustainable resource development and use within the whole community. Whereas in earlier times, government was seen as the custodian of natural resources, a range of other stakeholders now take greater interest in resource use. Individuals and community interest groups wish to be involved in decisions on allocation and use and in monitoring and evaluation. Also, users, including industry, are expected to act responsibly and to develop and apply sound codes of practice. Governments are expected to compile data on resources and resource uses and to ensure that there is accountability in their allocation and use. This results in a 'whole-of-community' involvement, but also means that individual resource-users or groups of users are taking more of the responsibility.

Many of the responses relating to land resources also affect other components of the environment, and are covered by other chapters in this report. This chapter highlights a series of responses, selected because they are important for determining the state of our land resource, because they provide a historical perspective to our attitudes or because they represent current trends in public and private responses.

Land clearing

Throughout our agricultural history, governments have used various means to encourage land clearing. Some land purchase agreements required it, taxation incentives encouraged it and Departments of Agriculture provided advice on how to clear. At the same time, these Departments were encouraging farmers to plant trees for shade, shelter and wood products, particularly in regions that were naturally treeless. By the 1980s concern about land degradation and the decline in native vegetation had become widespread. This resulted in the Commonwealth Government establishing the National Tree Program in 1982 to reverse tree decline by promoting action at individual, community and government levels to conserve, regenerate and plant trees. In 1989, the Commonwealth Government established the 'One Billion Trees' and 'Save the Bush' programs to protect and enhance native land cover. In 1992 these programs were brought under the administrative umbrella of the National Landcare Program.

To date, $12 million has been spent under the 'Save the Bush' program. Of this, $6.4 million has been provided to community groups for projects involving protection and management of native vegetation. A further $31 million ($3.7 million as community grants) has been spent under the 'One Billion Trees' program, with an estimated 550 million trees being planted or regenerated. However, there has been little assessment of how these programs have affected the extent of tree cover or conservation of bushland. At best, they make a relatively modest contribution.

By the mid 1980s to mid '90s, concern about the rate of disappearance of native vegetation, and associated land degradation problems resulted in the establishment of legislative or regulatory controls on land clearing in most States (see Table 6.8). Statistics for South Australia and Western Australia indicate that, for these States, clearing controls have resulted in retention of significant areas of native vegetation that would otherwise have been cleared. For example, in Western Australia over the period 1986–94, notices of intent to clear were received for 271 784 ha; 24 per cent of this area was retained. This includes additional areas retained (other than the area notified) as a condition of clearing. In South Australia, 97 per cent of the applications to clear relatively intact native vegetation with an understorey received under the *Vegetation Management Act 1985* (which provided compensation for those refused consent) were refused. Under the *Vegetation Act 1991* the few applications received for broadacre clearance have been refused. About half the applications to clear scattered trees have been approved, with approvals invariably tied to conditions such as the replanting of at least 10 trees raised from local seed and the removal of domestic stock from areas of native vegetation, to encourage regeneration.

The need to compile information at a continental level for the first National Greenhouse Gas

Inventory in 1994 highlighted the lack of data available on land cover. Victoria, South Australia and Western Australia have mapped their tree cover recently and a major program is under way to map the vegetation of the Murray–Darling Basin. However, in the other States, no land cover data were available, and only Victoria was monitoring changes in land cover. The results of the inventory indicated that recent land clearing for agriculture could be much more extensive than previously recognised, and may make a major contribution to Australia's greenhouse gas emissions (see page 6-13). In responding to the findings of the inventory, the Commonwealth Government has provided funding for a State–Commonwealth program to assess agricultural land cover change across the continent between 1990 and 1995. The information produced from this program will reduce some of the uncertainties in estimating national emission levels and help develop further policy responses to land clearing.

Pest animals and plants

Community, government and industry all address problems caused by pests at a number of levels. The main channel for the wider community to contribute directly is Landcare (see page 6-43). The Australian Trust for Conservation Volunteers provides opportunities for local and international visitors to devote time to special projects. Both groups, for example, have assisted with rabbit control and weed management programs. Clearly, the whole community contributes indirectly through payment of taxes and levies.

The Commonwealth Government is active in many spheres. The Bureau of Resource Sciences runs a vertebrate pest program, which deals with pest animals in agricultural production. The Australian Nature Conservation Agency (ANCA) has a feral pest program, which addresses the problems of pest management for conservation purposes. The government has also established national bodies like the Australian Plague Locust Commission to deal with specific pests.

Commonwealth funding supports many institutions that are researching pest management — for example, the Vertebrate Biocontrol Centre and the CRC for Tropical Pest Management. The CSIRO is funded federally and also, for pest research, by other agencies and industries. Its Division of Entomology, for example, is

Table 6.8 Summary of land clearing regulations

	Most recent significant changes	Summary
South Australia	1985 and 1991	Consent required to clear any native vegetation; heritage agreements with some financial assistance available for areas not approved. Agreements in place for 600 000 ha cost $70 million in compensation for first six years. No broadacre clearing approved since the Act came into force.
Western Australia	1986 and 1995	Permission required for clearing of areas larger than one ha of native vegetation. No compensation provisions. More than 80 per cent of requests approved and conditions tightened in 1995. Some funding available to maintain remnant vegetation for soil conservation. Technically, Conditional Purchase Lease system still applies over 800 properties, requiring them to be half cleared before they can be converted to freehold.
Victoria	1989	Clearing of blocks of native vegetation larger than 0.4 ha is subject to approval. No provision exists for compensation but heritage agreements may provide financial assistance.
Queensland	1995	The Government has released a preliminary policy for tree clearing on leasehold lands which will be proclaimed under the Lands Act, 1994 when local guidelines have been finalised. A satellite monitoring program has also been established. These restrictions are opposed by some land-owner groups. Many piecemeal measures are administered by State and local governments.
New South Wales	1995	A State environmental planning policy was introduced in August 1995 to control clearing on freehold land. Following community consultation, the Government will consider further amendments, options or alternatives to this policy. Previously existing controls include the Western Lands Act, provisions under the Soil Conservation Act and regional and local environmental plans.
Northern Territory	1992	Commercial activity involving the destruction of vegetation on Crown land or leasehold must be licensed. Some provisions exist under the Pastoral Land Act. Few measures are in place for private land.
Tasmania		Few measures exist, other than at local government level and some provision for private reserves.
ACT		There is no freehold land; vegetation protection is usually incorporated in lease provisions.

Source: Derived from DEST.

investigating major insect pests of crops, pastures and livestock, as well as biological control of weed species. ANCA researches the management of environmental weeds, many of which also have an economic impact. The Commonwealth Government provides further support for weed research through organisations like the Land and Water Resources Research and Development Corporation, the International Wool Secretariat and the Meat Research Corporation. Federal and State agencies are finalising the National Weed Strategy, which will provide guidelines for better weed management and will integrate the efforts of governments, industry, landholders and land managers, community groups and the general public.

State and Commonwealth governments have developed legislation relevant to pest management. Some Acts — such as the Commonwealth's *Quarantine Act 1908*, New South Wales *Noxious Insects Act 1934* and Victoria's *Vermin and Noxious Weeds Act 1958* — relate directly to pests. Other legislation important to pest management includes the Commonwealth's *Agricultural and Veterinary Chemicals Code Act 1994* and various State Acts on soil and land conservation. The Commonwealth Government provides taxation relief for management activities like woody weed clearance.

Departments of Agriculture, Primary Industry, Land and Conservation share responsibility for pest management at State government level. Arrangements differ between States. Some have Agriculture or Pasture Protection Boards to promote research and provide management support for pest control. In some cases, control activities are undertaken through cost-sharing with landholders. For example, an agency may provide equipment and skilled personnel while the landholder contributes financial support.

Industries support pest and weed research through organisations like the Meat Research Corporation and the International Wool Secretariat. Within particular industries, land-owners, managers and land users spend much money, time and effort controlling pests and diseases. They work as individuals and within groups like Landcare, and they contribute to research and development through industry levies on their commodities.

Forests

Forests have always been a major source of food and shelter for Australia's human inhabitants. Aborigines made significant use of forested areas before Europeans arrived. With the arrival of the First Fleet, many of the most valuable timber trees were selected for cutting. The colonists sent the first exports from Australia in 1791 and many settlements were positioned to exploit nearby timber resources. In the early 1800s, the government introduced regulations to control timber-getting and to raise revenue.

The gold rushes of the mid 19th century greatly increased the demand for timber and agricultural land, and large areas of forest around major

diggings were clearfelled. South Australia, the State with the smallest forested area, was the first to adopt legislation designed to manage the forest resource. In the 1870s, the State provided incentives for land-owners to plant trees, set up conservation reserves and began establishing radiata pine plantations. Other States soon followed. After Federation, the States retained responsibility for forest resources and by the 1920s all had set up forest management agencies. During the 1930s, these agencies adopted the principles of multiple use. Forests were to be managed to contribute to timber production, watershed protection, recreation, conservation and grazing. World War II was a turning point in forest management. The demand for timber products increased greatly in the post-war years and State governments instructed their forest agencies to service that demand. Large areas of public and private native forest were felled, while agencies increased the establishment of softwood plantations (see Fig. 6.27). New technologies, such as chainsaws and tracked vehicles, became more readily available. The volume of timber extracted from native forests almost doubled in the 20 years following World War II. Since then the volume harvested has increased by only a small amount, some of which is due to new technology allowing a greater log yield per area harvested. The greatest gains have been in harvests from softwood plantations.

Meanwhile, public interest in conservation increased during the 1960s and 1970s and most States established government agencies specifically responsible for conservation. Conflict increased between organisations (both government and non-government) supporting conservation on the one hand and timber production on the other, leading to a series of confrontations over forest-use issues. These disputes often overflowed into the Commonwealth–State arena because, although

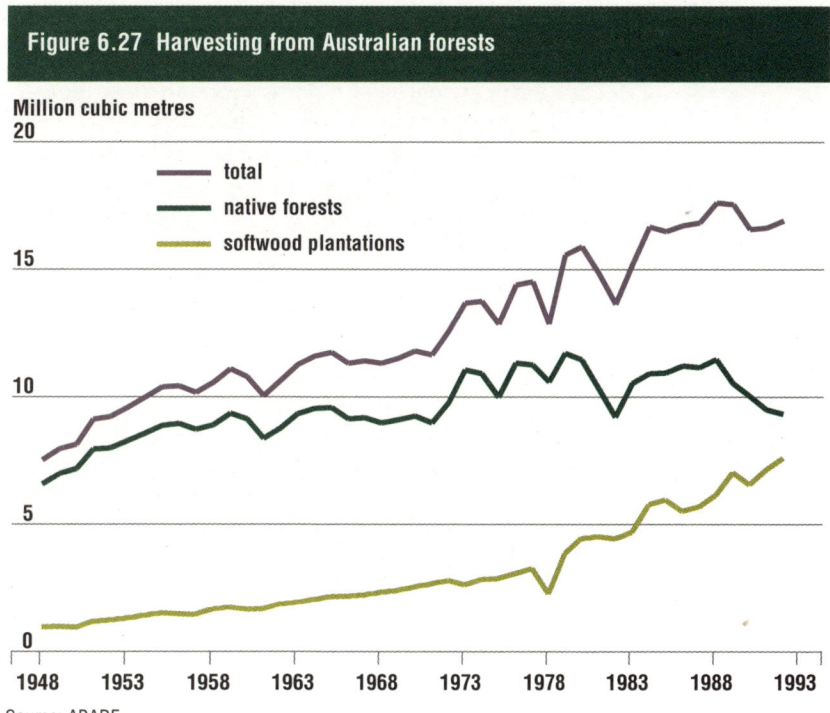

Figure 6.27 Harvesting from Australian forests

Million cubic metres

— total
— native forests
— softwood plantations

Source: ABARE.

The national drought policy

In 1992, Commonwealth and State Ministers agreed to a national drought policy based on principles of sustainable development, risk management, productivity growth and structural development in the farm sector. Its objectives are to:

- encourage primary producers to adopt self-reliant approaches to climatic variability
- maintain agricultural and environmental resources during periods of extreme climate stress
- ensure early recovery of agricultural and rural industries, consistent with long term sustainable levels

The policy includes measures to: encourage farmers to become more self-reliant; ensure the longer term profitability and sustainability of farming; provide welfare assistance for marginally profitable farmers; help re-establish unprofitable farmers; and provide special funding support in extreme drought.

Defining drought, in its various intensities and extent, has long been a contentious issue for those forming drought policy. The Commonwealth Government has now adopted a national system for considering drought declarations by State and Territory governments. The Commonwealth can also make 'exceptional circumstances' declarations for drought. The criteria to be taken into account by both the Commonwealth and the States when considering cases for drought 'exceptional circumstances' declarations are: meteorological conditions, environmental impacts, agronomic and stock conditions, farm income levels, water supplies and scale of the event.

Policy measures available to farmers to deal with drought include:

- Income Equalisation Deposit and Farm Management Bond schemes to increase self-reliance
- a drought-preparedness investment allowance
- Landcare, for maintaining and protecting the resource base
- the Rural Adjustment Scheme to improve farm productivity

Drought is a natural and recurrent feature of the Australian climate.

- the Farm Household Support scheme to improve farmers' welfare
- drought relief payments to assist farmers in extreme drought

A drought research program focuses on three main issues that will allow farmers to anticipate and prepare for drought and respond quickly and appropriately when faced with drying conditions: climate forecasting, drought-risk monitoring and decision support for farm risk management.

Destocking early in the onset of drought is the best way to protect natural resources on grazing lands. However, there are difficulties in providing financial support at an appropriate time to ensure that resource degradation does not occur and that bad management through overstocking is not rewarded.

Large areas of Australia have been suffering extreme drought in recent years. Consequently it will take some time for the country to feel the full impact of the new policy as many producers still have to develop the capability for self-reliance. Even so, the policy, which focuses predominantly on farm business issues, only partially addresses core issues of sustainable land management.

forest management is a State responsibility, the Commonwealth has powers in relation to international agreements such as World Heritage and to export licensing (in particular, woodchip exports), and responsibilities with respect to the National Estate.

Woodchipping, especially woodchipping for export, is a contentious issue in Australia. All States maintain a forest policy that is 'sawlog driven'. That is, logging operations are supposed to be justified only by the sawlogs they produce. But logging waste (branches and unsuitable logs) are also sold for woodchips. The issue provokes such intense debate because it is the value of woodchip sales that makes the logging of many coupes financially viable. Also, whereas almost all timber products are retained in Australia, most woodchips are exported. This allows the Commonwealth to exert an indirect influence over the choice of forests to be logged (usually a State matter) by controlling woodchip export licences.

The disputes over forest use culminated in the Resource Assessment Commission (RAC) Inquiry into the Forest and Timber Industry which reported in 1992. This Inquiry was conducted against a background of opposition by conservation groups and large sections of the community to the logging of native forests, and pressures from other sectors for more secure access to develop and expand forest-based industries. The RAC did not make specific recommendations but identified and evaluated options. Its report addressed such questions as the environmental impacts of wood production on native forests, the sustainability of timber production, the role of plantations, the options for managing old-growth forests and the impacts of pulp mills (see the boxes on pages 6-18 and 6-20). Among the most important outcomes of the Inquiry were its conclusions that there were insufficient grounds to halt logging in native forests in general, but that logging old-growth forests potentially violates the precautionary principle, in that it destroys an irreplaceable resource.

The Ecologically Sustainable Development process occurred in parallel with the RAC Inquiry and included a working group on forest resources. Both processes influenced the contents of the National Forest Policy Statement to which the Commonwealth and States agreed in December 1992 (Tasmania signed in April 1995). The Statement outlines agreed objectives and policies for the future of Australia's public and private forests and sets broad national goals in relation to forest management and the forest industry. In 1991 the National Plantations Advisory Committee prepared a series of recommendations to accelerate the development of plantation forestry. An important conclusion was that while plantation forestry should be subject to appropriate codes of practice, it should not be subject to controls that would not apply to agricultural industries.

In summary, the approach outlined in the National Forest Policy Statement is that governments will set the regulatory framework for the use of native forests in order to achieve social and environmental goals, but within that framework market forces should determine the extent and nature of resource use. Two of the strategies being pursued by governments to meet the conservation goals are the establishment of a 'Comprehensive, Adequate and Representative' (CAR) reserve system, and the development of codes of practice to ensure that both commercial and non-commercial uses of public forests do not adversely affect the forests' ecological base. Each major forest region will have a Regional Forest Agreement between the States and Commonwealth that establishes the obligations and objectives of both governments. The agreement may specify land-use boundaries, forest management guidelines, and consultative arrangements between governments. These agreements will run for 10–20 years subject to regular reviews and are expected to create a more stable forestry environment.

A major goal of the National Forest Policy Statement was that the States and Commonwealth reach agreement over the reserve system, including the protection of old-growth forests. Progress in reaching these agreements has been slow and the deadline specified in the Statement of a completed comprehensive and adequate reserve system by 1995 for public lands was not met.

However, the process is being accelerated; criteria for the reserves are being set and the Commonwealth has made $53 million available in 1995–96 to ensure progress of the regional forestry agreements.

In mid 1995 the Commonwealth recommended as a broad benchmark for a CAR reserve system that at least 15 per cent of the pre-1750 extent of each forest type should be in secure reserves. Old-growth forests are to be given additional protection ranging from 100 per cent for the rarest and biologically most important to 60 per cent in more extensive forest types.

The first stage in the Regional Forest Agreements process was to identify, before the 1996 round of woodchip licence renewal negotiations, interim protection areas (deferred forest areas) to protect areas of high conservation value for a CAR reserve system. The negotiation of Regional Forest Agreements will occur over the following three to four years. States which fail to set aside the necessary areas and to enter into and progress negotiations on the agreements will suffer a 20 per cent per year reduction in export approvals.

The State of the Forests Report, being prepared by the National Forest Inventory, is expected to provide information on the history, ownership, size, composition and biodiversity of the national native forest and plantation estate. Other topics include forest-based industries, water quality, carbon inventory, social values, cultural heritage, research activities and broad management initiatives.

Land conservation legislation

All States established soil conservation agencies before World War II and all have long-standing soil conservation legislation. However, its effectiveness is difficult to assess without established criteria. Some researchers (Bradsen, 1992) conclude that it has generally been ineffective and that legislation revised in the 1980s contains weaknesses.

The role for individual land-users gained new momentum with the formation of land conservation district committees under the 1982 Western Australian legislation. Other States also recognised the importance of increased community involvement, as shown by the Landcare movement (see below).

The focus of concern has moved from 'soil' as a relatively narrow issue to a whole-system approach, where 'land' is used to embrace a range of issues. For example, South Australia passed the *Native Vegetation Management Act* in 1985 and the *Soil Conservation and Land Care Act* in 1989. This trend recognises that use and impacts of all resources are interlinked. The legislation being developed in Queensland (see the box on page 6-50) is another example of this trend.

Landcare — a community response

Concerns over soil erosion by water and wind in the 1930s prompted the development of government soil conservation programs. Farmers were involved in these programs through statutory boards and similar bodies, but the schemes were largely government-driven and focused on single issues such as designing and erecting earth structures to control run-off. Advances in technology — biological control of rabbits, improved pastures, clover ley rotations and machinery for soil conservation works — also reduced erosion. Despite these efforts, erosion continued and more insidious long-term problems — such as soil salinity, acidification and the decline of soil structure and fertility — emerged.

A measure of the progress of responses to past problems can be obtained by comparing some of the recommendations on land resources issues in the 1974 National Estate Report (Commonwealth of Australia, 1974) with the progress of the responses to those recommendations.

Table 6.9 The historical effectiveness of responses

1974 National Estate Report recommendations	Twenty years later
Removal of taxation concessions for clearing.	The Commonwealth Government has removed tax concessions for land clearing but anomalies remain.
Establishment of a system of national parks, especially near the larger cities, together with State parks, nature reserves, urban parks and other classes of reserves.	The number of reserves has increased but progress towards a Comprehensive Regional Assessment program has been limited.
Woodchipping and other operations involving clearfelling of large areas of forest should be discontinued until the environmental effects are better known and properly assessed.	Resource Assessment Commission Inquiry into the Forest and Timber Industry completed in 1992 and National Forest Policy Statement released. Woodchipping from any areas not subject to a Comprehensive Regional Assessment will be phased out by 2000.
There is an urgent need for expanded research into biology of feral species.	An eradication program has substantially reduced the water buffalo numbers in northern Australia and consequently improved the environment there. Government agencies have undertaken surveys of other vertebrate pests and control programs are being developed.
Management of forests to include multiple use and conservation objectives, including rainforest conservation.	Most of the remaining areas of rainforest in Australia are now protected in national parks and listed as World Heritage areas. The Commonwealth and States have signed the National Forest Policy Statement to address issues of sustainable forest management.
A review of Australian Government lands on the coast to see which areas may be included within Coastal Heritage Parks.	The Resource Assessment Commission (1993) has finalised a report into management of the Australian coastline, but few policy decisions have been announced as a result of the report (see Chapter 8).

It became clear that a single-issue approach was inadequate and that government legislation and regulation would not control land degradation unless land-users as individuals accepted responsibility for the effect of their land management practices.

Changing roles of government and community

In the 1980s, the role of State soil conservation programs began to change from one that tried to provide leadership to one that encouraged community-led programs, with agencies involved in a collaborative partnership. For example, in 1982 the Western Australian Soil Conservation Act was revised to encourage greater rural community involvement in land conservation by forming Land Conservation District Committees. Since 1982, 137 committees, encompassing most of the State, have been formed.

During the same period in Victoria, the Garden State Committee and the then Victorian Farmers and Graziers Association sponsored the establishment of farmer-led farm tree groups. In 1986, the Victorian Government, with the support of the Victorian Farmers Federation, established a new land protection program called 'LandCare'. The first Victorian LandCare group was established at Winjallok in the Wimmera in 1986 and by the end of 1987 at least 30 groups existed.

In 1988, The National Farmers' Federation and the Australian Conservation Foundation proposed a national land management program. Its key elements were funding for Landcare groups and for property management planning. The Commonwealth Government designated the 1990s as the Decade of Landcare and supported it with funding for the National Landcare Program. The broad objective of this program is to foster the development and implementation of systems of land use and management that will sustain individual and community benefits now and in the future. It does this by encouraging Landcare and community groups to participate in land management projects. Other programs target increased application of property management planning as a process to build landholders' skills to improve production, reduce degradation and manage risks (such as drought).

The growth of Landcare

While other community groups focus on conservation issues, Landcare groups have a particular emphasis on rural lands and contain people who want to work with others to improve the long-term health of the land. Landcare is voluntary; the agenda for each group is set by the members and each group operates according to its own plan. A facilitator usually helps the group determine priorities. Landcare groups allow people to see that they have the capacity, within their own community, to deal constructively with issues that seem too big for individual families.

The groups have two key areas of influence — immediate impact on people's attitudes, and subsequent impacts on land-management practices as new attitudes lead to change.

The Decade of Landcare plan set a target of 2000 Landcare groups by the end of the decade. However, this target was passed in late 1994. More than 25 000 people are actively involved in Landcare. A 1992–93 ABARE survey estimated that 28 per cent of broadacre farmers and 19 per cent of dairy farmers were members of Landcare groups. Surveys by Charles Sturt University show that the proportion of properties with a Landcare group member varies between States; for example, in Victoria, Western Australia and Tasmania it is about 50 per cent and in Queensland it is about 33 per cent. Studies show that Landcare members have significantly higher levels of adoption of 'best-bet' land management practices than non-members (Curtis and DeLacy, 1994; ABARE, 1994). They also spent an average of $2500 on Landcare activities during 1992–93.

Challenges and prospects

The Landcare movement faces a number of challenges (Campbell, 1994; Sustainable Resource Management Committee Working Group, 1994). It is difficult to find enough people in rural areas with the skills required to lead and facilitate groups, plan and implement farm and catchment strategies, tackle all degradation problems in all areas and develop remedies for difficult technical problems. Groups need to adopt a 'working together to develop sustainable systems' approach rather than a 'fix land degradation problems' approach. They need to develop an effective collaborative partnership to meet the interests of all rural and urban community stakeholders and to develop and adopt practical and profitable sustainable farming technologies and land management systems. The links between social and economic wellbeing of a community and the quality of its land management need greater recognition.

The major challenge for Landcare is to have all land-users adopting best-management practices to ensure sustainable use of resources. This difficult task will involve widespread changes to routine land management practices. Such changes may be resisted, particularly in difficult economic circumstances. There are high expectations for Landcare and evaluations of the Decade of Landcare will assess whether these are being achieved.

The level of success of the Landcare movement will have wide implications for society. It may be used as a model for other 'care' programs to enlist stakeholder commitment to sustainable resource use. If the voluntary approach embodied in Landcare does not succeed, consequences may include continued resource degradation, declining community benefits and a more regulatory approach.

Integrated natural resource management policy

Increasingly, people are recognising that natural resource management issues are interlinked. However, policy and legislation are often based on a single issue. Queensland's initiative to remodel existing legislation to underpin sustainable resource development illustrates the trend towards a more integrated framework.

The initiative recognises that:

- natural resources need to be protected from further degradation

- resource condition influences the economic prosperity and social well-being of the community

- community involvement in the decision-making process is vitally important

- individuals play a critical role in management of natural resources

To help develop the policy, the government has implemented an extensive community consultation process with over 50 public meetings around the State and has also established a policy advisory council.

The initiative addresses three critical areas:

- allocation of resources between productive use and environmental needs

- sustainable management of natural resources using Codes of Practice backed by legislation

- identification of Special Management Areas with specific management needs — degraded environments, critical remnants, run-off control structures and floodplains

The proposed legislation will fold nine existing Acts, covering the management of water, soil and forests, into one. The mandatory provisions of the new Act will be designed to regulate unsustainable management practices that cause rapid resource degradation. Non-mandatory provisions will encourage and support community and whole-of-government involvement in sustainable resource management programs. It will complement integrated catchment management, Landcare, property management planning, Waterwise and Treecare programs.

Table 6.10 Summary

Element/Key pressures	Impact / State	Adequate Info.	Responses	Effectiveness of responses
Clearing of native vegetation/ forest removal				
Agricultural land use	Loss or fragmentation of most of native vegetation, exposing area to many of the soil issues listed below and contributing substantially to greenhouse gas emissions	✔✔	Legislation and controls on tree clearing; monitoring; One Billion Trees program; Landcare	Effective in many areas but not implemented in others; the efficacy of the recovery programs is not tested
Forestry operations	Short term exposure of soil to risks of compaction and erosion and loss or dramatic modification of habitat.	✔✔	Agreement not to decrease the forest estate; Codes of Practice to cover forestry operations and regeneration; Regional Forest Agreements (RFAs)	Mostly effective; RFAs not yet completed
Urban expansion	Permanent loss of habitat; modifications to regional hydrology	✔✔	Local planning regulations	Varied
Forest habitat conservation				
Forestry operations; changed fire regimes	Fragmentation of forest habitat (via roading and areas of operation); alteration of age structure of the forest	✔	National Forest Policy Statement; Codes of Practice to cover forestry operations; Regional Forest Agreements; Comprehensive and Adequate Reserve systems; encouragement of plantation development.	Most important responses still being implemented
Condition of the rangelands				
Pastoralism	Changes in the density and species composition of vegetation; widespread establishment of weeds and feral animals	✔✔	Research and extension; legislation; leasehold conditions (animal stocking rates, soil status); National Strategy for Rangeland Management; inventory and monitoring; Landcare; structural adjustment programs; multiple use policies.	Only limited success through lease administration; localised successes in weed and feral animal control but little progress in many areas; variable trends in vegetation on a regional basis
Tourism	Localised trampling and vehicle damage; infra-structure; pollution	✔✔	Planning regulations; rehabilitation programs; public education	Major re-organisation and rehabilitation in some areas (e.g. Uluru); little effect in others
Soil structural decline				
Agricultural land use and especially tillage systems that lead to loss of soil organic matter	Reduced permeability and increased runoff and erosion; poor root vigour leading to reduced productivity	✔✔	Soil conservation research and extension especially into improved farming systems; Landcare	Increased awareness of the issue, but problem is still widespread
Salinisation				
Clearing of agricultural and grazing land; over irrigating	Rising groundwater levels; in some areas salt is mobilised leading to dryland salinity and stream salinisation	✔✔✔	Landcare and catchment planning, Murray-Darling Basin Commission; salt quotas and Saltwatch; monitoring; expenditure on salinity management	Only minor and localised successes through planting and changed management. Regional-scale responses are inadequate

Table 6.10 Summary (continued)

Element/Key pressures	Impact / State	Adequate Info.	Responses	Effectiveness of responses
Soil erosion				
Agricultural land use and especially those leading to the loss of soil cover	Loss of soil depth; loss of nutrients; off-site effects such as siltation	✔✔	Land management research and extension; expenditure on structural works; Landcare	Uptake of advice still inadequate in most areas
Forestry operations	Localised erosion during harvesting and regeneration	✔✔	Codes of Practice relating to operations and streamside buffers	Significant improvements but continuing debate about effectiveness in some areas
Soil nutrient decline				
Agricultural land use and especially excessive cultivation	Decline in soil organic matter and major nutrients (N and P)	✔	Research and extension on cropping systems, rotations and fertiliser use; promotion of N utilisation, fertiliser application rates, fertiliser imports (t/yr)	Varies from successful farming systems incorporating legume rotations, to reliance on artificial fertilisers, to little effective response
Forestry operations	Loss of major nutrients through harvest losses, erosion or frequent burning	✔✔	Improved forestry practice	Mostly effective but limited information for many areas
Pastoralism	Wind and water erosion leading to widespread redistribution (and loss) of nutrients	✔✔	Improved stock management, fencing and water distribution	Varies from property to property
Soil nutrient accumulation				
Intensive agriculture and horticulture	Accumulation of nutrients with risk of water pollution.		See Inland Waters chapter	See Inland Waters chapter
Soil acidification				
Agricultural land use and especially inappropriate legume rotations, the overuse of fertilisers, removal of plant products and natural processes in some areas.	Increased soil acidity (low soil pH) and the release of toxic levels of aluminium and manganese in some soils; poor root growth. Some calcareous soils benefit from higher acidity	✔	Research and extension; liming and changed fertiliser practices	Very poor uptake of appropriate measures
Soil contamination				
Various agricultural and pastoral land use	Impacts include cadmium (Cd) contamination from phosphate fertilisers; increased herbicide use in minimum tillage systems; pesticide pollution in cattle and sheep dipping sites	✔	Reduction in Cd levels in fertilisers and use of cultivars that restrict its uptake; regulations on pesticide and herbicide use; education	Cd levels are generally low by world standards; much more work needs to be done in identifying and rectifying other contamination problems
Food quality				
Farming practices with respect to hormones, pesticides and fertilisers	Food contamination	✔✔	Legislation; new farming practices, integrated pest management; reduction in pesticide use	Successful in most cases, but risks of accumulation of residues poorly known

Multiple use of rangelands

The dry outback is the best-known part of Australia's rangelands, which cover three-quarters of the country (see Fig. 6.8). After little more than a century of pastoralism, vegetation change and soil erosion are widespread, the mix of native fauna has changed and feral animals and plants are well established. Nevertheless, the rangelands are still largely natural ecosystems, important on a world scale.

By area, about 60 per cent of the rangelands is used for grazing, 15 per cent for Aboriginal homelands and four per cent for conservation reserves; the remaining 21 per cent is technically unoccupied but much of it is under claim by Aboriginal people. Other land uses like mining and tourism may not occupy large areas, but have a disproportionately high economic value (see the Table below). In addition, recreational use is having an increasing impact on more-remote areas.

Our rangelands are also valued for their international, cultural and scientific interest. The outback is a part of Australia's national identity. Its scientific interest lies not only in its unique flora and fauna (which include, for example, the world's greatest diversity of reptiles), but also in the opportunity to develop an understanding of ecological processes in an arid land that has not had thousands of years of agricultural use, as is the case in Asia and Africa.

Although pastoralism has been the most extensive commercial land use since European settlement, its future supremacy is no longer assured. The industry as a whole has largely operated at a loss over the last decade, due to low commodity prices and high costs (Wilcox and Cunningham, 1994). Many landholders seek alternative sources of income, through different land uses as well as off-farm investment.

The wider community wants to maintain the conservation value of these publicly owned rangelands, regardless of how they are used. For large areas of rangelands, conservation and sustainable pastoralism are compatible. However, some key areas for pastoralism can be very important for conservation and need special protection. Decisions regarding land use should be made at the appropriate level; for example, economic production can be optimised at the level of the individual holding, while conservation priorities should be set within a regional context. Further, decisions should be made after considering all the private and public benefits and costs of various uses.

In the foreseeable future, pastoralism is likely to remain the major use for the rangelands. In places where it is not viable in the long term, the emphasis will shift towards alternative or multiple uses and, even where grazing is viable, other uses may prevail. Tourism is rapidly developing as an option across the rangelands, both as a general enterprise and at individual landholder level, and already is worth more than the pastoral industry. Conservation is formally incorporated with pastoralism in areas like the Innamincka Regional Reserve. Mining and pastoralism co-exist on many Western Australian leases, and dryland cropping occurs on the eastern fringe of New South Wales pastoral country. Traditional Aboriginal uses are combined with pastoralism on leases held under Aboriginal freehold title. Other uses being developed include kangaroo- and emu-farming, horticulture and hunting. The 1992 High Court decision in the Mabo case affects the rights of Aboriginal people to land, but the implications for land use are unclear.

Government, industries and the community have responded to changing land-use priorities and will need to make further accommodations in future. Lease tenure and conditions need to be tailored to allow a variety of uses — not necessarily by the titleholder alone — while protecting conservation value. Decisions on land uses that affect interests beyond the landholder need to involve the wider community in order to minimise conflict. Dryland cropping, for instance, which occupies less than 0.4 per cent of New South Wales rangelands, has given rise to a high level of concern about the impacts of land clearing. Structural adjustment will need to continue, and public funding will have to be allocated equitably so as to balance support for commercial enterprises and the public's interest in maintaining conservation values and social equity in a region.

Value of rangeland commodities 1991–92

	$ million	% of Australian total
Cattle sold	527	17
Sheep sold	22	4
Wool sold	274	12
Tourism	3 000	8
Mining	10 000	38

Source: Stafford Smith, 1994.

Food quality

Agriculture depends heavily on inputs of agricultural and veterinary chemicals to enhance productivity and minimise losses caused by pests, weeds, pathogens and post-harvest spoilage. Pesticide contamination, arising from deliberate administration, over-use or chance contact with residues, may compromise food quality. Heavy metals such as cadmium, lead and mercury occur in rural commodities because plants take up trace amounts from the soil, supplemented in some cases by impurities in fertilisers and sewage sludge (see page 6-32). Excessive levels of these also adversely affect food quality.

Australia produces most of its food and exports large amounts. We only import about six per cent by value. The results of Commonwealth, State and industry-sponsored monitoring programs indicate that misuse of agricultural and veterinary chemicals is low. Most analyses record no residues or residues below established limits. The Australian Market Basket Survey (National Food Authority, 1992) determined that dietary intake of agricultural and veterinary chemicals and heavy metals was within the safe limits set down by the World Health Organisation.

A large survey in New South Wales (Edge, 1993) examined 1509 samples of fruit and vegetables for up to 24 pesticides, involving a total of more than 25 000 chemical analyses. Most did not yield any evidence of pesticides and no sample contained more than one chemical residue above the Maximum Residue Limit (MRL). While 25 samples returned a residue above the MRL, 10 of the breaches were almost certainly attributable to traces of the pesticides benzene hexachloride and dieldrin remaining in soil from applications made some years earlier. Thus the violation rate for chemicals intentionally applied to the sampled commodities was 0.7 per cent.

In a 1993 survey, the Victorian Department of Agriculture found that 98 per cent of 448 fruit and vegetable samples were within residue limits for all chemicals tested. Results from the National Residue Survey (NRS, in press), the largest monitoring program for chemical residues and heavy metal contaminants in Australia, are also encouraging.

Among the heavy metals of concern, cadmium regularly exceeds the Maximum Permitted Concentration (MPC) in offal and potatoes. The National Residue Survey reports that eight per cent of offal samples in 1991 and 1992 exceeded the established MPC. More than 10 per cent of samples in the Victorian clean agriculture produce monitoring program of 1992–93 had cadmium levels above the MPC, with violations detected in carrots, potatoes, silverbeet and safflower. None of the samples had lead or mercury exceeding 50 per cent of the MPC.

Contamination of Australian seafood is generally well below permissible levels, although some organochlorines and heavy metals may concentrate in sharks and other predatory fish higher up the food chain (see Chapter 8). The situation should improve with time, since most organochlorines are no longer available for use within Australia.

Australian food quality, as judged by standards for pesticide residues and for heavy metal contamination, compares favourably with that of international produce.

Responses to food quality issues

All agricultural and veterinary chemicals used in Australia must be registered by the National Registration Authority for Agricultural and Veterinary Chemicals. The Authority was established by the Commonwealth in 1993 to consolidate activities previously carried out by State governments. In consultation with the Department of Human Services and Health and other relevant authorities, the National Registration Authority sets MRLs for chemicals in food commodities. After public consultation, the MRLs are adopted into the Food Standards Code of the National Food Authority and individual State governments incorporate them into State food laws. Imported foods must meet domestic standards for both chemical residues and heavy metals.

The National Food Authority also sets standards for MPCs of heavy metals, taking account of both the concentrations that occur in foods and the varying contributions that different foods make to the average diet. Taking cadmium as an example, the MPC for most foods is set at 0.05 mg/kg, while the limits for prawns and lobsters, other shellfish and kidneys are 0.2, 2.0 and 2.5 mg per kg respectively.

Concern about cadmium is such that government authorities and industry have taken steps to limit dietary intake by regulating cadmium inputs from phosphatic fertilisers and sewage sludge, by eliminating the feeding of phosphate supplements containing high levels of cadmium to cattle and by banning the sale of offal from aged sheep and cattle likely to contain elevated levels of cadmium.

The popularity of organic produce could tempt unscrupulous operators to capitalise on premium prices. This prospect underlay the formation of the Organic Produce Advisory Committee and the adoption of the National Standard for Organic and Biodynamic Produce. The Committee reports to the Department of Primary Industries and Energy and the Australian Quarantine and Inspection Service regularly audits organic organisations enforcing the standard.

All States except Tasmania and the Australian Capital Territory and Northern Territory have established monitoring programs for agricultural and veterinary chemicals, and some selectively survey for heavy metals as well. The Australian Market Basket Survey monitors the intake of pesticides and contaminants in food prepared to a table-ready state in the diets of adults, children and infants — both male and female. The survey samples 'market baskets' of over 60 food types for about 50 pesticides, arsenic, cadmium, copper, lead, mercury and aluminium. The National Residue Survey monitors chemical residues and heavy metals in raw food commodities, largely those destined for export. The National Antibacterial Residue Minimisation Program monitors levels of antibacterials in meat and alerts producers to the risks to trade if levels are violated.

Education is the key to developing responsible use of agricultural and veterinary chemicals. It is integral to the Commonwealth's clean-food export program and to a diversity of initiatives by producers, processors, chemical industries, educators and peak industry bodies. Efforts to decrease dependence on pesticides are being made through research and development into integrated pest management systems while genetic engineering is being used to develop plants and animals with resistance to pests and pathogens.

Rural adjustment in the mulga lands

In the Western Division of New South Wales and the mulga lands of south-west Queensland, 2650 pastoral holdings cover 64 million ha of rangeland and in 1991–92 they produced $450 million worth of agricultural products. The issues facing the mulga region illustrate the links between pressure, state and response on a regional and inter-State scale.

South-west Queensland and Western Division of New South Wales

Pressures

Large stations in this region were subdivided under closer-settlement policies, with many of the smaller properties being established after World War I to provide blocks for soldier settlement (Passmore and Brown, 1992). When subdividing these large properties and regulating land management, land administrators did not adequately account for the impact of the variable climate, changing economic conditions and the importance of retaining native pastures.

The main agricultural industries are sheep and wool production from native pastures, with some beef production and cropping in favourable areas. The minimum economically viable property size varies with location, seasonal conditions and wool prices. Because more than 40 per cent of properties are too small to be economically viable, landholders are under intense pressure to extract financial returns through more-intensive land use (Young, 1985). Variable seasons and several droughts between 1957 and 1994 have exacerbated these pressures.

Terms of trade for the wool industry have declined steadily for the past 40 years. In the sheep-grazed rangelands the decline has been faster than for the whole industry and gains in productivity have been slower.

Native species such as mulga, which were used as fodder, helped to maintain excessive stock numbers during drought periods. The discovery of artesian water in the 1880s was important in the development of the region. Producers drilled many bores and established extensive systems of drains. The network of watering points, plus control of dingoes, has increased numbers of kangaroos, goats, pigs and rabbits, and consequently the total grazing pressure.

State

Land condition is inversely related to property size (Young, 1985). Stocking pressures have reduced soil cover, which in turn has increased soil erosion. A 1985 survey showed that in south-west Queensland, bare ground made up 60 per cent of the area and serious soil erosion occurred on 10 per cent. Similar levels of erosion were found in the Western Division of New South Wales (Soil Conservation Service of NSW, 1988). Wind erosion occurs periodically on exposed sandy soils.

Vegetation has also changed due to heavy grazing and altered fire regimes, and woody weeds are increasing. A 1985 survey in south-west Queensland showed that woody weeds had heavily infested 44 per cent of the area. In the Western Division of New South Wales, about 37 per cent is moderately to severely encroached by woody weeds; a further 25 per cent displays low-level infestation but is potentially vulnerable, and only 37 per cent is unaffected (Soil Conservation Service of NSW, 1988).

Kangaroo numbers in the region are estimated at 10–12 million and studies suggest that feral goats number more than one million. Depending on conditions, kangaroos, goats and rabbits can account for 30–70 per cent of total grazing pressure.

Water pressure in the Great Artesian Basin has been declining across the region (see figure below). In south-west Queensland, flow rates from bores have dropped from 600 megalitres per day in the early 1900s to an average of 260 megalitres today. These are projected to fall further, to 80 per cent of current levels over the next 20 years and 60 per cent over 50 years (Barson *et al.*, 1993). Nearly one-fifth of bores in south-west Queensland have ceased to flow and it is estimated that 90 per cent of water is lost to evaporation and seepage from drains. The Great Artesian Basin inter-State

Bore numbers and flow

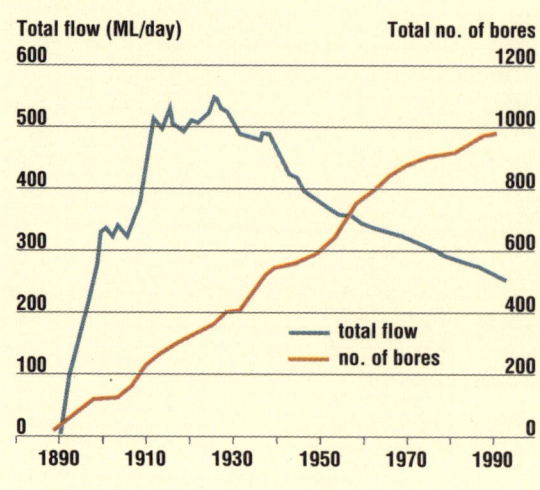

Source: QDPI.

working group of 1992 estimated that capping all bores would result in annual water savings of 250 000 megalitres in Queensland alone.

Many enterprises in the region are unprofitable. For example, projections based on production and cost figures for properties in four areas of the Western Division showed all were non-viable under the wool prices existing in the early 1990s. Although 8000 sheep is generally considered a minimum carrying capacity for long-term viability in this area, up to 40 per cent of properties do not meet this criterion. In south-west Queensland, the minimum property size for long-term viability is 35 000 ha and preferably more than 45 000 ha. Stock numbers vary according to season and pasture conditions, but at least 10 000 sheep are needed for economic viability. About 75 per cent of south-west Queensland properties do not meet these minimum standards.

Mulga is a valuable source of fodder but woody weeds are encroaching on some areas.

Response

The region has suffered economic and resource management problems for many years. The drought and the depression in the wool industry in the early 1990s aggravated the situation. Government agency studies between 1970 and 1990 resulted in some specific initiatives — for example, bore capping and piping projects, research studies of vegetation and fauna, and projects to improve land management. Other responses have varied between the two States.

In Queensland, the Mulga Land Use Advisory Group was formed in 1991 to devise measures to stop land degradation and improve viability of the pastoral industry. It consisted of representatives from industry, government, financial institutions and community interest groups. The group started the 'land degradation voluntary property build-up scheme' in 1992. Eligible applicants could obtain loans at concessional rates to buy land to increase the size of their holding or to finance development in accordance with an agreed property management plan.

In 1993, eight working groups began to devise a strategy for the mulga region. They found that economics, land degradation and social problems were interrelated and should not be considered in isolation. Property viability was the key to ecologically sustainable use, and a regional program was needed to adjust property sizes, improve resource management and increase landholder self-reliance.

In 1992, the New South Wales Department of Conservation and Land Management (CaLM) initiated a community consultation process known as WEST (Working for Equitability and Sustainability Together). It convened a 'search conference' to bring together representatives of all stakeholders in the Western Division to put forward ideas to help achieve a 'desirable and achievable' future. More than 1000 people from pastoral communities and government agencies were briefed on the recommendations developed by the conference and had opportunities to contribute to the recommendations (CaLM, 1993).

The similarity of problems and developments in both States, and recognition of the need for a regional approach, prompted a joint New South Wales–Queensland proposal. The Standing Committee on Agriculture and Resource Management, the Rural Adjustment Scheme Advisory Committee and the Murray–Darling Basin Commission agreed to treat the region as a case study for sustainable rural development. The integrated regional adjustment and recovery program for south-west Queensland and the Western Division of New South Wales will cost government an estimated $91 million over seven years. About $12 million of these funds have already been made available through the National Landcare Program and the Rural Adjustment Scheme.

The program has three components:

- property reconstruction — to assist non-viable producers to re-establish elsewhere, assistance for build-up of viable properties, enhanced business management and planning support, coordinated commercial lending activities, communication and promotion activities and supportive land administration arrangements

- natural resource management — to promote sustainable resource use by reducing total grazing pressure, applying best management and production practices, developing property management plans, applying risk-management practices to minimise pasture degradation, enhancing the network of reserves and voluntary on-property conservation initiatives to secure biodiversity and improving water, stock, pest and pasture management by bore capping and piping

- integrated regional development — to facilitate ongoing adjustment and self-reliance by supporting regional initiatives to develop and manage the strategy, developing a more robust and reliable economic base, coordinating ongoing provision of government services, assisting those who exit industries to find alternative employment, ensuring networks of support and information services are available and balancing the rate of adjustment.

Source: QDPI and NSW CaLM, 1994.

References

ABARE (1994). *Farm Surveys Report 1994* (Australian Bureau of Agricultural and Resource Economics: Canberra.)

ABARE (1995). Australian commodities forecasts and issues, Vol. 2, No. 1. (Australian Bureau of Agricultural and Resource Economics: Canberra.)

ABARE (1995). Trees on farms. Survey of trees on Australian farms: 1993–94. *ABARE Research Report*, 95-7.

ABS (1992). 'Australia's Environment: Issues and Facts.' (Australian Bureau of Statistics: Canberra.)

ABS (1900–1995). 'Year Book Australia.' (Australian Bureau of Statistics: Canberra)

Alston, J.M., Chalfant, J.A., and Pardey, P.G. (1993). Structural adjustment in OECD agriculture: government policies and technical change. *Paper for the Quantitative Analysis and Special Studies Division of the OECD's Directorate for Food, Agriculture and Fisheries.*

Anon. (1991). General summary of lease assessment findings. *Kingoonya Soil Conservation Board District Report.* Rangeland Assessment Unit, Pastoral Management Branch, Department of Environment and Planning.

Anon. (1991). Pastoral regions of Western Australia. *Discussion Paper* No. 3. Select Committee into Land Conservation, Legislative Assembly, Western Australia.

AUSLIG (1990). 'Atlas of Australian Resources. Vegetation.' Vol. 6. (Australian Surveying and Land Information Group: Canberra.)

BAE (1983). 'Rural Industry in Australia.' (Bureau of Agricultural Economics: Canberra.)

Barson, M., Evans, G., Fordham, D., Walcott, J., and White, D. (1993). Opportunities for regional rural adjustment. *Bureau of Resource Sciences Working Paper.*

Bastin, G.N., Pickup, G., Chewings, V.H., and Pearce, G. (1993). Land degradation assessment in Central Australia using a grazing gradient method. *Rangeland Journal*, **15**, pp. 190–216.

Bennett, A. F. (1993). Fauna conservation in box and ironbark forests: a landscape approach. *The Victorian Naturalist*, **110**(1), pp. 15–23.

Braysher, M. (1993) 'Managing Vertebrate Pests: Principles and Strategies.' (Bureau of Resource Sciences: Canberra.)

Bridge, B.J., and Bell, M.J. (1994). Effect of cropping on the physical fertility of krasnozems. *Australian Journal of Soil Research*, **32**, pp. 1253–73.

Bradsen, J.R. (1992). A review of soil conservation legislation in Australia. *Proceedings of the Fifth Australian Soil Conservation Conference, Western Australia. Department of Agriculture, Perth*, pp. 21–35.

CaLM (1993). A progress report on the outcomes of the Western Division consultation and planning process: April 1993. (CaLM: NSW.)

Campbell, A. (1994). 'Landcare: Communities Shaping the Land and the Future.' (Allen and Unwin: St Leonards, NSW.)

Campbell, K.O. and Dumsday, R.G. (1990). In 'Agriculture in the Australian Economy' ed. D.B. Williams (OUP: Oxford, UK.)

Carnahan, J.A. (1990). In 'Atlas of Australian Resources. Vegetation.' Vol. 6. (Australian Surveying and Land Information Group: Canberra.)

Cartwright, B., Merry, R. H., and Tiller, K. G. (1977). Heavy metal contamination of soils around a lead smelter at Port Pirie, South Australia. *Australian Journal of Soil Research*, **15**, pp. 69–81.

Catterall, C. P. and Kingston, M. (1993). Remnant bushland of south-east Queensland in the 1990s: its distribution, loss, ecological consequences, and future prospects. Institute of Applied Environmental Research, Griffith University and Brisbane City Council.

Caughley, J., Monamy, V., and Heiden, K. (1994). Impact of the 1993 mouse plague. *Grains Research and Development Corporation Occasional Paper* No. 7.

Chisholm, A.H. (1992). Australian agriculture: a sustainability story, *Australian Journal of Agricultural Economics*, **36**(1), pp. 1–29.

Chisholm, A.H. (1994). Land use choices in a changing world. *Land Degradation and Rehabilitation*, **5**(2), pp. 153–78.

Choquenot, D. (1991). Pest density, impacts and the cost of control. *Bureau of Rural Resources Working Paper* WPO/20/91, pp. 43–8.

Choquenot, D. (1994). Planned control of feral pigs: optimising costs and efficiency of control in relation to climatic and pasture conditions. *Final Report to Wool Research & Development Corporation, July 1989 to June 1993* DAN 40.

Cocks, D. (1992). 'Use With Care. Managing Australia's Natural Resources In The Twenty First Century.' (University of New South Wales Press: Sydney.)

Combellack, J.H. (1989). The importance of weeds and advantages and disadvantages of herbicide use. *Plant Protection Quarterly*, **4**, pp. 14–32.

Commonwealth of Australia (1974). National Estate. *Report of the Committee of Inquiry Parliamentary Paper* No. 195.

Commonwealth of Australia (1992). 'National Forest Policy Statement — a new focus for Australia's forests.'

Condon, R.W. (1983). What Future for Australia's Arid Lands? *Proceedings of the National Arid Lands Conference, Broken Hill, New South Wales, May 21–25, 1982.*, pp 54–60.

Condon, R.W., Newman, J.C., and Cunningham, G.M. (1969). Soil erosion and pasture degradation in Central Australia. Part IV. *Journal of the Soil Conservation Service of New South Wales*, **25**, pp. 295–321.

Cousens, R.D. and Medd, R.W. (1994). Discussion of the extent of Australian ecological and economic data on weeds. *Plant Protection Quarterly* **9**, pp. 69–72.

Crosson, P. and Anderson, J.R. (1992). Resources and global food prospects. *World Bank Technical Paper* No. 184.

Curry, P.J., Payne, A.L., Leighton, K.A., Hennig, P., and Blood, D.A. (1994). An inventory and condition survey of the Murchison River catchment, Western Australia. *WA Department of Agriculture Technical Bulletin* No. 84.

Curtis, A. and DeLacy, T. (1994). Landcare: does it make a difference? *Johnstone Centre of Parks, Recreation and Heritage Report* No. 12 Charles Sturt University, Albury.

Dalal, R. C., and Mayer, R. J. (1986). Long-term trends in fertility of soils under continuous cultivation and cereal cropping in southern Queensland. I Overall changes in soil properties and trends in winter cereal yields. *Australian Journal of Soil Research*, **24**, 265–79.

Department of Environment, Housing and Community Development (1978). A basis for soil conservation policy in Australia. *Report of the Commonwealth and State Government collaborative soil conservation study 1975–77* No. 1.

Department of the Environment, Sport and Territories (1994). 'Australia's National Greenhouse Gas Inventory 1988 and 1990.' (AGPS: Canberra.)

Dyksterhuis, E.J. (1949). Condition and management of rangeland based on quantitative ecology. *Journal of Range Management* , 2, pp. 104–15.

Edge, V.E. (1993). Pest control and sustainable agriculture — concluding remarks. In 'Pest Control and Sustainable Agriculture', ed. S. Corey, D. Dall and W. Milne. (CSIRO: Melbourne.)

Edwards, K. (1991). Soil formation and erosion rates. In 'Soils: Their Properties and Management.' ed. P.E.V. Charman and B.W. Murphy. (Sydney University Press: Sydney).

EPA WA (1989). Monitoring pesticides — a review. *Environment Protection Authority Bulletin* No. 407.

Ferris, I. G., and Haigh, B. M. (1993). Herbicide persistence and movement in Australian soils: implications for agriculture. In 'Pesticide Interactions in Crop Production: Beneficial and Deleterious Effects.' ed. J. Altman.

Ford, G.W., Martin, J.J., Rengasamy, P., Boucher, S.C., and Ellington, A. (1993). Soil sodicity in Victoria. *Australian Journal of Soil Research*, **31**(6), pp. 869–909.

Geeves, G.W., Cresswell, H.P., Murphy, B.W., Gessler, P.E., Chartres, C.J., Little, I.P., and Bowman, G.M. (1995). The physical, chemical and morphological properties of soils in the wheat-belt of southern NSW and northern Victoria. *CSIRO Division of Soils Report.*

Gelburd, D.E. (1985). Managing salinity: lessons from the past. *Journal of Soil and Water Conservation*, **40**, pp. 329–31.

Graetz, R.D., Wilson, M.A., and Campbell, S.K. (1995). Landcover disturbance over the Australian continent: a contemporary assessment. *Biodiversity Series Paper* No. 7, Biodiversity Unit, DEST.

Greenland, D.J. (1971). Changes in the nitrogen status and physical condition of soils under pastures, with special reference to the maintenance of the fertiliy of Australian soils used for growing wheat. Soils and Fertilizers, **34**, pp. 237–51.

Hairsine, P., Murphy, B., Packer, I., and Rosewell, C. (1993). Profile of erosion from a major storm in the south-east cropping zone. *Australian Journal of Soil and Water Conservation*, **6**, 50–5.

Hamblin, A., and Kyneur, G. (1993). 'Trends in Wheat Yields and Soil Fertility in Australia, (AGPS: Canberra.)

Helyar, K.R., Hochman, Z., and Brennan, J.P. (1988). The problem of acidity in temperate area soils and its management. *Australian Society of Soil Science Inc., National Soils Conference Review Papers.*

Henzell, E.F. (1992). Sustainability of the Australian agricultural system. In 'Environmental Indicators for Sustainable Agriculture', *Report on a National Workshop, 28–29 November 1991.*

Hone, P. (1994). The impact of productivity growth on the management of land in the Australian wool industry, *PhD thesis, La Trobe University, Bundoora.*

Humphrey, R.R. (1949). Field comments on the range condition method of forage survey. *Journal of Range Management* 2, 1–10.

Isbell, R.F. (1996). 'A Classsification System for Australian Soils.' (CSIRO Division of Soils.)

Kestel Research and Victorian Department of Conservation and Environment (1990). Report to the Resource Assessment Commission on forest clearing in Australia. *Resource Assessment Commission Forest and Timber Inquiry Consultancy Series* No. FTC91/02.

Kirkpatrick, J. B. (1991). The magnitude and significance of land clearance in Tasmania in the 1980s. *Tasforests* **3**, pp. 11–14.

Land Conservation Council (1988). Statewide assessment of public land use, Victoria, Australia. (LCC: Melbourne.)

Landsberg, J. and Wylie, R. (1991). A review of rural dieback in Australia. *Growback 91*, pp. 3–11.

Lay, B. (1979). Shrub population dynamics under grazing: a long term study. In 'Studies of the Australian Arid Zone IV. Chenopod shrublands.' Ed. R.D. Graetz and K.M.W. Howes, pp. 107–24. (CSIRO: Melbourne.)

Littleboy, M., Silburn, D.M., Freebairn, D.M., Woodruff, D.R., and Hammer, G.L. (1989). PERFECT, a computer simulation model of Productivity, Erosion, Runoff Functions to Evaluate Conservation Techniques. *Queensland Department of Primary Industries, Bulletin* QB89005.

Lonsdale, W.M. (1994). Inviting trouble: introduced pasture species in northern Australia. *Australian Journal of Ecology*, **19**, pp. 345–54.

Lothian, J. A. (1994). Attitudes of Australians towards the environment: 1975 to 1994. Australian Journal of Environmental Management, **1**(2).

McGarry, D. (1992). Degradation of soil structure. In 'Land Degradation Processes in Australia.' Ed. G. McTainsh and W. C. Boughton. (Longman Cheshire: Melbourne.)

McKenzie, N.J. (1991). A strategy for coordinating soil survey and land evaluation in Australia. *CSIRO Division of Soils Report* No. 114.

McLaughlin, M.J., Fillery, I.R. and Till, A.R. (1992). Operation of the phosphorus, sulfur and nitrogen cycles. In 'Australia's Renewable Resources: Sustainability and Global Change.' Ed. R.M. Gifford and M.M. Barson, *Bureau of Rural Resources Proceedings* No. 14.

McTainsh, G. and Leys, J. (1993). Soil erosion by wind. In 'Land Degradation Processes in Australia.' Ed. G. McTainsh and W.C. Boughton. (Longman Cheshire: Melbourne.)

Mitchell, A.A., McCarthy, R.C., and Hacker, R.B. (1988). A range inventory and condition survey of part of the Western Australian Nullarbor Plain, 1974. *Western Australian Department of Agriculture Technical Bulletin* No. 47.

National Food Authority, 1992. 'The Australian Market Basket Survey.' (AGPS: Canberra.)

National Residue Survey (in press). 'Report on National Residue Survey results 1991–1992.' (Bureau of Resource Sciences: Canberra.)

Neil, D.T., and Galloway, R.W. (1989). Estimation of sediment yields from farm dam catchments. *Australian Journal of Soil and Water Conservation*, **2**(1), pp. 46–51.

Nicholson, D.A. (1993). Pastoralism. In 'What Future for Australia's Arid Lands' *Proceedings of the National Arid Lands Conference, Broken Hill, New South Wales, May 21–25, 1982*. Eds J. Messer and G. Mosley. (Australian Conservation Foundation: Melbourne.) pp. 90–1.

Parsons, W.T., and Cuthbertson, E.G. (1992). 'Noxious Weeds of Australia.' (Inkata Press: Melbourne.)

Passmore, J.G.I., and Brown, C.G. (1992). Property size and rangeland degradation in the Queensland mulga rangelands. *Rangelands Journal,* 14(1) pp. 9–25.

Payne, A.L., Curry, P.J., and Spencer, G.F. (1987). An inventory and condition survey of rangelands in the Carnarvon Basin, Western Australia. *WA Department of Agriculture Technical Bulletin* No. 73.

Payne, A.L., Kubicki, A., Wilcox, D.G., and Short, L.C. (1979). A report on erosion and range condition in the west Kimberley area of Western Australia. *WA Department of Agriculture Technical Bulletin* No. 42.

Payne, A.L., Mitchell, A.A., and Holman, W.F. (1988). An inventory and condition survey of rangelands in the Ashburton River catchment, Western Australia. *WA Department of Agriculture Technical Bulletin* No. 62.

Payne, A.L., and Tille, P.J. (1992). An inventory and condition survey of the Roebourne Plains and surrounds, Western Australia. *WA Department of Agriculture Technical Bulletin* No. 83.

Pickup, G. (1989). New land degradation survey techniques for arid Australia — problems and prospects. *The Australian Rangeland Journal,* 11, 74–82.

Preece, R., and Whalley, P. (1993). Central and North Western Regions Water Quality Program. *Water Resources Technical Services Division 1991/92 Report on Pesticide Monitoring.*

Pringle, H.J.R., Van Vreeswyk, A.M.E., and Gilligan, S.A. (1994). An inventory and condition survey of rangelands in the north eastern Goldfields, Western Australia. *WA Department of Agriculture Technical Bulletin* No. 87.

QDPI and NSW CaLM (1994). An integrated regional adjustment and recovery program for south west Queensland and the Western Division of New South Wales. *QDPI and NSW CaLM internal working document.*

Raupach, M., McTainsh, G., and Leys, J. (1994). Estimates of dust mass in recent major Australian dust storms. *Australian Journal of Soil and Water Conservation,* 7, 20–4.

Resource Assessment Commission (1992). 'Forest and Timber Inquiry Final Report'. Vol. 2B. (AGPS: Canberra.)

Resource Assessment Commission (1993). 'Coastal Zone Inquiry Final Report'. (AGPS: Canberra.)

Rosewell, C. (in press). Potential sources of sediments and nutrients. *Report prepared for Inland Waters Reference Group, DEST.*

Rutherfurd, I.D., Brooks, A., and Davis, J. (in press). Historical stream erosion and sedimentation in Australia. *Report prepared for Inland Waters Reference Group, DEST.*

Soil Conservation Service of NSW (1988). Land Degradation Survey of NSW 1987–88.

Stafford Smith M. (1994). Sustainable production systems and natural resource management in the rangelands. *Outlook 94, Vol 2, Natural Resources.* pp. 148–59.

Stewart, J. (1968). Erosion survey of New South Wales Eastern and Central Divisions, reassessment 1967. *Journal of Soil Conservation Service of New South Wales,* 24(3), pp. 139–54.

Sustainable Resource Management Committee Working Group (1994). *Progress report on evaluation of the National Decade of Landcare Plan.*

Tiller, K.G. (1992). Urban soil contamination in Australia. *Australian Journal of Soil Research,* 30, pp. 937–57.

Tothill, J.C., and Gillies, C. (1992). The pasture lands of northern Australia. *Tropical Grasslands Society of Australia, Occasional Paper* No. 5.

Wallbrink, P.J., and Murray, A.S. (1993). The use of fallout radionuclides as indicators of erosion processes. *Journal of Hydrological Processes,* 7, pp. 297–304.

Wallbrink, P.J., Olley, J.M., and Murray, A.S. (1996). In 'Erosion and Sediment Yield', eds. D.E. Walling and R. Webb. (IAHS.)

Wasson, R.J. (1994). Annual and decadal variation of sediment yield in Australia, and some global comparisons. *Proceedings of the Canberra Symposium on Variability in Stream Erosion and Sediment Transport. IAHS Publication* No. 224.

Wasson, R.J., Olive, L.J., and Rosewell, C. (1996). Rates of erosion and sediment transport in Australia. In 'Erosion and Sediment Yield', eds D.E. Walling and R. Webb. (IAHS.)

Weir, S., Cofinas M., and Tupper, G. (1992). Woody weed monitoring for land assessment. *NSW Agriculture Final Report, Natural Resources Management Strategy Project* N0023,.

Weston, E.J., Harbison, J., Leslie, J.K., Rosenthal, K.M., and Myer, R.J. (1981). Assessment of the agricultural and pastoral potential of Queensland. *QDPI Agricultural Branch Technical Report* No. 27.

Wilcox, D.G., and McKinnon, E. A. (1972). A report on the condition of the Gascoyne catchment. *WA Department of Agriculture and WA Department of Lands and Surveys report.*

Wilcox, D.G., and Cunningham, G.M. (1994). *Land and Water Resources Research and Development Corporation Occasional Paper* No 06/93. pp 87–171.

Williams, C. H., and Raupach, M. (1983). Plant nutrients in Australian soils. In 'Soils: an Australian Viewpoint.' (CSIRO: Melbourne/Academic Press: London).

Williams, M. (1988). The clearing of the woods. In 'The Australian experience. Essays in Australian land settlement and resource management', ed. R. L. Heathcote, pp 115–26. (Longman Cheshire: Melbourne).

Woodgate, P. W., and Black, P. J. (1988). 'Forest cover changes in Victoria 1869–1987.' (Department of Conservation, Forests and Lands: Melbourne.)

Woods, L.E. (1983). 'Land Degradation in Australia.' (AGPS: Canberra.)

World Commission on Environment and Development (1987). Our Common Future. (Oxford University Press: New York.)

World Resources Institute (1994). 'World resources 1994–95: A Guide to the Global Environment.' (Oxford University Press: New York.)

Young, M. D. 1985. The Influence of Farm Size on Vegetation Condition in an Arid Area. *Journal of Environmental Management,* 21, pp. 193–203.

Acknowledgments

The following people reviewed the chapter in draft form and provided constructive comments.

Colin Chartres (CSIRO Division of Soils)

Ann Hamblin (CRC for Soil and Land Management)

Ted Henzell (formerly CSIRO Institute of Plant Production and Processing)

Jamie Kirkpatrick (University of Tasmania)

Judith Lambert (Community Solutions)

Henry Nix (Australian National University)

John Williams (CSIRO Division of Soils)

We especially thank the following individuals who assisted in the preparation of this chapter:

Roger Armstrong (WA Conservation and Land Management)

John Blackstock (Victorian Department of Agriculture)

Tim Boneyhady (Australian National University)

Mike Braysher (Bureau of Resource Sciences)

Julie Burke (Murray-Darling Basin Commission)

Tony Chisholm (La Trobe University)

Joanne Daly (CSIRO Division of Entomology)

Geoff Edwards (La Trobe University)

Graeme Evans (Bureau of Resource Sciences)

Rohan Fernando (Australian Nature Conservation Agency)

Peter Hairsine (CSIRO Division of Soils)

Quentin Hart (Bureau of Resource Sciences)

Peter Helman (Consultant)

Phil Hone (Deakin University)

Rai Kookana (CRC for Soil and Land Management)

Louise Laurence (CSIRO Division of Entomology)

Mike McLaughlin (CSIRO Division of Soils/CRC for Soil and Land Management)

Grant McTainsh (Griffith University)

Richard Merry (CSIRO Division of Soils)

Vicky Passlow (Bureau of Resource Sciences)

Alan Payne (Agriculture WA)

Keith Porritt (Bureau of Resource Sciences)

Sandy Radke (Bureau of Resource Sciences)

Simon Ransome (Victorian Land Conservation Council)

Alison Saunders (Australian National University)

Andrew Taplin (Australian Nature Conservation Agency)

Nick Uren (La Trobe University)

Neil White (Vertebrate Biocontrol Centre)

In addition, Commonwealth Government departments and members of the Commonwealth/State ANZECC State of the Environment Reporting Taskforce also helped identify errors of fact or omission. Their assistance is also gratefully acknowledged.

Photo Credits

Page 6-1: DFAT

page: 6-4: *(clockwise from top)*
Malcolm Paterson (CSIRO);
Ron Hacker (NSW Agriculture); CSIRO;
Kath Bowmer (CSIRO);
Malcolm Paterson (CSIRO)

Page 6-6: W Van Aken (CSIRO Division of Water Resources)

Page 6-13: George Smith (Queensland Department of Natural Resources)

Page 6-17: CSIRO

Page 6-19: Robert Kerton (Waldo Productions)

Page 6-21: Liz Poon (CSIRO Division of Wildlife and Ecology)

Page 6-22: John Green (CSIRO Division of Entomology)

Page 6-24: W Van Aken, (CSIRO Division of Water Resources)

Page 6-29: Sally Robinson

Page 6-30: Bureau of Meteorology

Page 6-42: David Corke (CSIRO)

Page 6-51: Liz Poon (CSIRO Division of Wildlife and Ecology)

Chapter 7

Inland Waters

Prepared by

Bob Wasson (Chair), CSIRO Division of Water Resources

Bob Banens, Murray–Darling Basin Commission

Peter Davies, Freshwater Systems

William Maher, University of Canberra, Water Research Centre

Sally Robinson, Strategic Environmental Solutions

Ray Volker, Department of Civil Engineering, University of Queensland

David Tait (former State of the Environment Reporting Unit member), Department of the Environment, Sport and Territories (former Facilitator)

Stephen Watson-Brown (State of the Environment Reporting Unit member), Department of the Environment, Sport and Territories (Facilitator)

Contents

Figure 7.1 Australia's drainage divisions and surface water resources

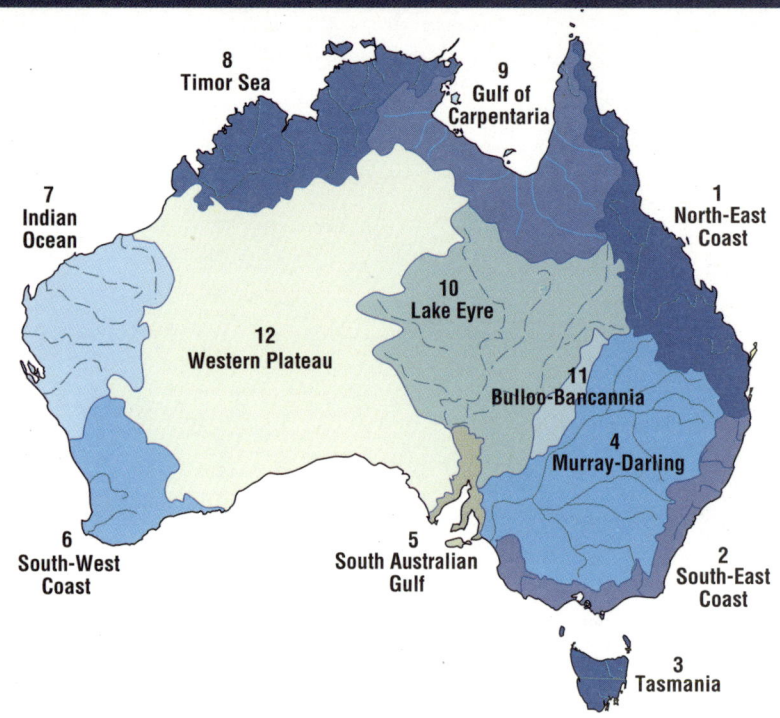

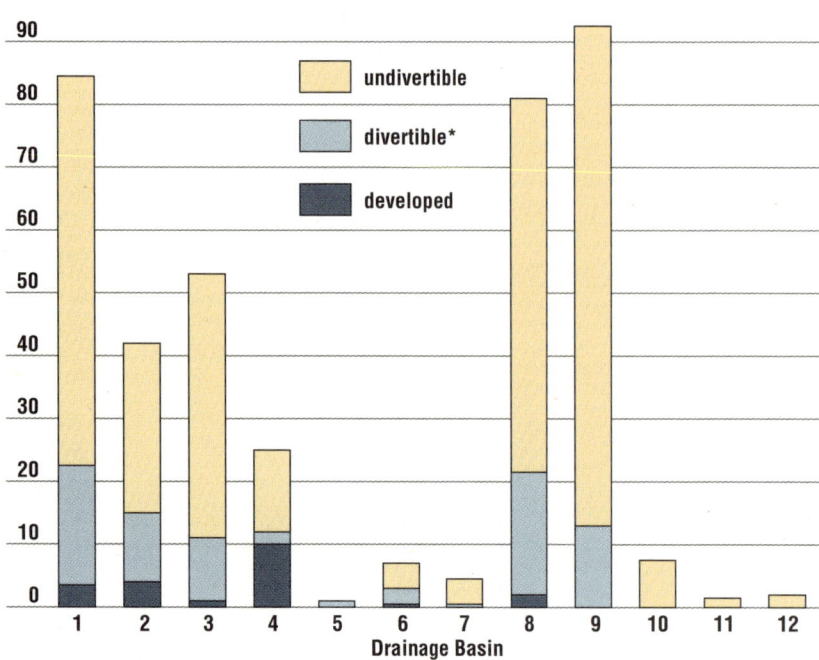

Mean annual run-off in gigalitres

*Note: Divertible run-off represents a proportion that could be used without any regard for economic, social or environmental factors. A responsible level of use may well be substantially less than indicated.

Source: AWRC, 1985.

Introduction

Australia is the driest of all the world's inhabited continents. It has the lowest percentage of rainfall as run-off, the lowest amount of run-off, the least amount of water in rivers and the smallest area of permanent wetlands. The Western Plateau drainage division, which covers 32 per cent of the continent (see Fig. 7.1), produces almost no run-off, and a further 17 per cent does not drain to the ocean. Sixty-five per cent of the continent's mean annual run-off occurs in the northern drainage divisions (1, 8 and 9 on the map), and 10 per cent of the divertible fraction (see note under Fig. 7.1) of this tropical water has been developed for human use. Across all the other drainage divisions, 16 per cent of divertible water has been developed, with a regional high of 81 per cent developed in the Murray–Darling Basin. About 15 per cent of all water used in Australia is groundwater (AWRC and DRE, 1983).

Australia has the most variable rainfall and stream flow in the world (see Chapter 2 and Figs 7.3 and 7.4), and our inland streams have high natural turbidity and salinity (McMahon *et al.*, 1992; Williams, 1982). The Murray–Darling river system is Australia's largest, draining about one-seventh of the continent. It ranks with the world's big rivers in terms of length and catchment area, but has much lower annual discharge. The chemistry of our surface inland waters differs from most waters elsewhere, often being dominated by sodium chloride rather than calcium and magnesium bicarbonates (Williams, 1982). Groundwater is often very old; for example, in the Great Artesian Basin water travels across Queensland, to emerge in central Australia in bores one to two million years after it entered the ground (Torgersen *et al.*, 1991).

The generally arid climate and ancient well-weathered landscape mean that mainland Australia has relatively few permanent and freshwater lakes. Lakes on the mainland are often shallow, dry and salty. Only on the Central Plateau of Tasmania do a number of larger permanent fresh-water lakes occur.

Inland waters include all water inland of estuaries, both in surface features like streams, lakes, wetlands and reservoirs, and in the subsurface as groundwater. The biology of Australian inland waters has many special features (Williams, 1982). Although our invertebrate animal groups resemble those of other continents with a similar environment, many species, and some genera and families, are unique to this country. Our aquatic invertebrates lack several groups that are widespread on other continents. Several families here have adapted to a wider range of environments than is the case elsewhere. The fish of Australian inland waters are represented by few species, many of which have evolved from marine forms and are endemic. Fewer bottom-dwelling creatures live in the lakes, and the fauna of our salt lakes are endemic. Half of the large aquatic plants are also unique, as are some terrestrial forms of the distinctive riverbank vegetation.

The distinctive physical, chemical and biological characteristics of Australia's inland waters have to be managed using criteria that are appropriate to Australian systems and conditions. Rivers and lakes in other parts of the world vary less, are generally less turbid and have fauna with different water-quality requirements, and so do not necessarily provide models for our use. The management of inland waters is increasingly being carried out on a catchment basis, to cover land, water and coasts. All types of waters are included in such an approach, which, in principle, can be applied nationwide.

The condition (or state) of our inland waters encompasses both natural characteristics, such as river flow variability, and the effects of human pressures exerted on the environment.

Much of what follows in this chapter is based on catchments, and their surface, subsurface and groundwater components (see Fig. 7.2). Various features of catchments combine to affect both the amount and quality of water that flows across the surface and as groundwater.

Land use affects soil properties and the amount of water infiltrating to groundwater, the rate of run-off and erosion, and hence the amounts of agricultural chemicals, sediment, and phosphorus and other nutrients reaching water bodies.

Topography and climate combine to influence the risk of salinisation after clearing, and the types of water bodies available for aquatic organisms.

Urban centres and industry produce wastes that can pollute waters, sometimes so badly that almost all life disappears.

The combined effects of topography, climate and land use can produce major changes to coastal waters, and can cause pollution and sedimentation in estuaries, reducing the productivity of fisheries and the diversity of life (Zann, 1995). Some sources of pollution, like sewage and industrial effluent, can be identified easily and dealt with if sufficient resources are available. Others, such as run-off from farmland, are more diffuse and therefore harder to identify and treat. In some cases land management is not the only answer, and water bodies need to be managed by, for example, altering flows or mixing reservoir waters.

Timely reports on the state of inland waters should help target our efforts to restore damaged environments, and to avoid further problems of the kind reflected increasingly in press headlines. However, such reporting is only likely to be successful nationally if databases are available for future reports. Many of the indicators used in this chapter were chosen for pragmatic reasons, and it is to be hoped that future reports will be able to identify the preferred indicators more systematically.

This chapter is based on four key management issues:

- water resources — the quantity of available water, vital for the human environment with significant impacts on the non-human environment

- catchment pollutant sources — increasingly the focus of management to solve water quality problems

- habitat quality — a focus of conservation strategies and some aspects of water quality control

- water quality — another vital component of the human environment, but with greater impacts on the non-human environment.

Figure 7.2 General relationships between catchments, surface waters, groundwaters and the coast

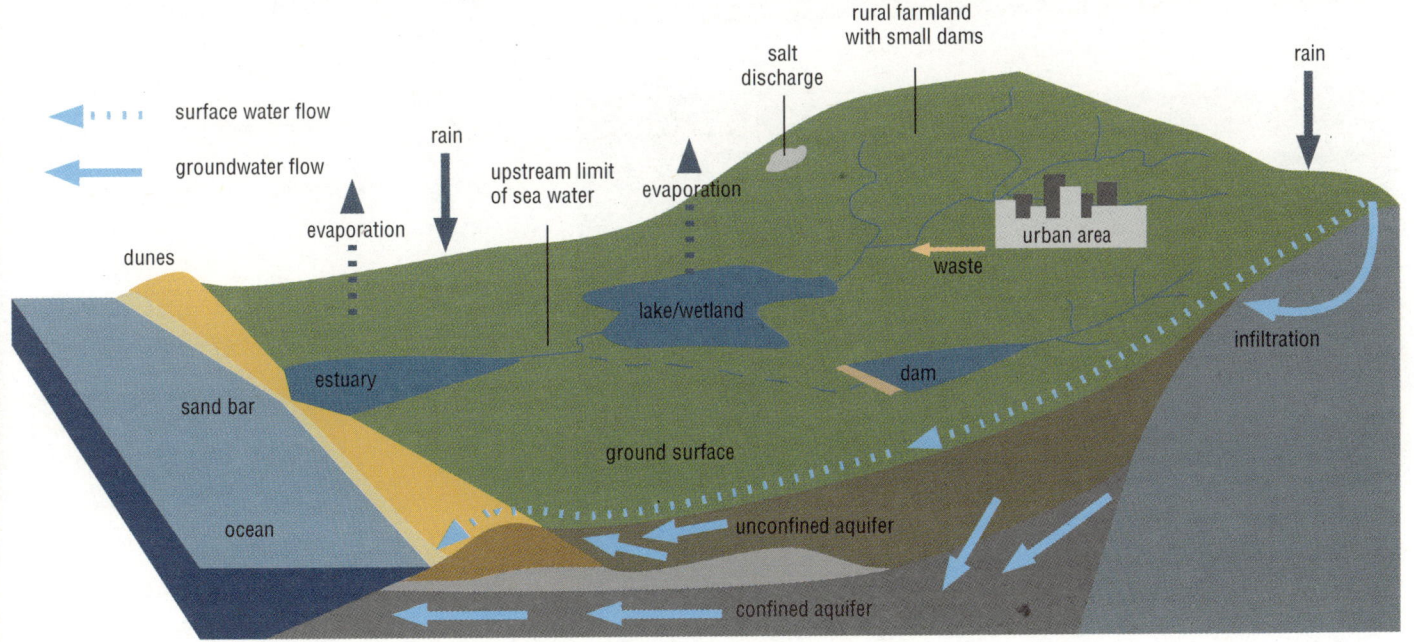

Pressure

Two main types of pressures affect inland waters. The first are intentional changes from direct human modifications of catchments and attempts to reduce the natural variability of systems. The second are unintentional pressures that have been indirectly caused by human actions, as well as inappropriate human responses to pressures and natural constraints of catchments and climate. Pressures usually occur in groups, producing multiple impacts.

Inland waters originate in catchments. Water moving across the surface of catchments moves soil particles, nutrients and chemicals. Therefore, any problem occurring in one part of the catchment can cause another one downstream. In a similar way, water moving through the soil into groundwater takes chemicals with it. This pollution can emerge elsewhere, causing downstream problems. Therefore, it is impossible to consider inland waters in isolation from land and its use, or from near-shore coastal waters and estuaries. The following discussion on pressures and responses recognises these interdependent links.

Direct pressures and their significance

When Europeans first arrived in Australia, they cleared the land and planted exotic species to provide familiar food, and to make the landscape seem more like home. Their descendents and more recent settlers have followed similar lifestyles and so the effects on water bodies and the land have intensified.

To carry out most land uses in the better-watered temperate parts of Australia, settlers replaced the deep-rooted native vegetation with shallow-rooted pastures and crops. This caused a number of environmental problems that affect inland waters today, including: salinity, waterlogging, sedimentation and turbidity, nutrient enrichment, pesticides, and flooding. In northern Australia, where grazing without clearing is common, the amount of sediment and nutrients transported to waters has increased. Clearing and grazing of areas next to lakes and rivers (the riparian zone) has had particularly severe effects on water bodies and on vegetation.

Mining, urbanisation, industrial activity, aquaculture and waste disposal from rural and non-rural industries all pollute waters. The most common pollutants are nutrients, organic matter, metals and pesticides.

Society has tried to accommodate and respond to the variability and seasonality of rainfall by constructing dams to guarantee water supply for themselves, domestic stock and irrigation. But storages affect natural flow, temperature, water chemistry and sediment regimes and replace a flowing water habitat with a still one. The dams themselves stop fish migrating, thus affecting breeding. Flow modification also affects aspects of water quality, like dissolved oxygen levels, and the regeneration and survival of plants.

Other land disturbances, such as urbanisation and agriculture, can greatly increase flow inputs to wetlands, or may result in wetlands being drained. Increasing water use by remote communities — which are now removing more water from inland waters — is also adding to the pressure on the environment.

Lakes, wetlands, rivers and reservoirs are all used for recreational activities, both active (like canoeing and water-skiing) and passive (like bush-walking and picnicking). These activities exert pressures that lead to erosion of river banks, trampling of vegetation and disturbance of habitat and wildlife. Activities in the catchment or the water body may also directly affect water quality in water supply dams.

Indirect pressures

Our use of land and water bodies changes their characteristics and creates indirect or unintentional pressures. These pressures also include inappropriate attempts to correct land- and water-degradation problems.

Removal of native vegetation for agriculture has caused salinisation of land. This process now affects a large area of agriculturally productive land and native bush land in Australia, and will affect more before it stops (see the box on page 7-20). It puts salt into water, with serious impacts on human uses and environmental health. Salinity also occurs in irrigated areas as a result of rising watertables.

Soils are affected by how we use them. In particular, soil structure is destroyed by repeatedly driving large, heavy farm machinery over it, by excessive ploughing and by other land-use practices. This results in compaction and water repellence that reduce both the amount of water penetrating to the root zone of plants and replenishment of groundwater. Repellence can cause or increase sheet erosion, loss of topsoil and gully formation. These processes result in increased amounts of sediment in inland waters.

Soil acidification associated with applying fertilisers to some forms of intensive horticulture and improved pastures can cause acidification of inland waters.

Rice growing in the Murrumbidgee Irrigation Area: expansion in irrigation has been the major factor contributing to the growth in water use in Australia.

Excess nutrients (especially phosphorus and nitrogen) in water bodies arise from eroded soils, fertiliser, septic tanks, discharges from sewage-treatment plants and animal wastes. In most parts of Australia, phosphorus is bound to soil particles and soil erosion will carry it from paddocks to water bodies. In other catchments, phosphorus can move dissolved in water, especially where soils are poor in clay and iron and cannot bind it.

When phosphorus and nitrogen occur in excess, they can result in algal blooms in rivers and creeks, thereby reducing natural biodiversity and increasing risks to human and animal health from toxins.

Monoculture crops require rigorous pest-management programs that may include more chemical applications than mixed crops. In some cases herbicides and pesticides have been shown to cause more damage to the natural organisms in water bodies (for example, frogs) than to the target organism (WA Department of Environmental Protection, 1995).

Like nutrients, herbicides and pesticides can be moved around the environment attached to sediment particles, in solution, and in the atmosphere; ending up in water bodies.

As catchments are cleared, the flood risk increases because water can run off the land more rapidly than when it was covered with vegetation (see the box on page 7-14).

Wetlands are often mined for diatomaceous earth and peat, and rivers for sand and gravel. Other changes include desnagging to remove trees and other obstructions, and river 'improvements' such as straightening.

Our need for water — for ourselves, for stock and for irrigation, particularly during drought (see Fig. 7.3) — drives the demand to manage water supply. Most large water storages do not fill in a single season and so the storage represents several years of stream flow. In river systems it is very easy to get into a situation where too much stored water has been promised (that is, allocated) to users, based on expectations of rainfall and river flows that may not eventuate. Rivers are prevented from flowing at the

appropriate times for their biota to thrive, and reproductive cycles can be seriously affected.

Invasions by introduced animals and plants are causing major problems in Australian waters. Some of the worst include: weed infestation by water hyacinth in lakes, choking of irrigation drains by weeds, spread of woody shrubs across floodplains in tropical Australia and increased turbidity caused by European carp (*Cyprinus carpio*).

Figure 7.4 Monthly flows for selected rivers illustrating seasonal variability across the continent

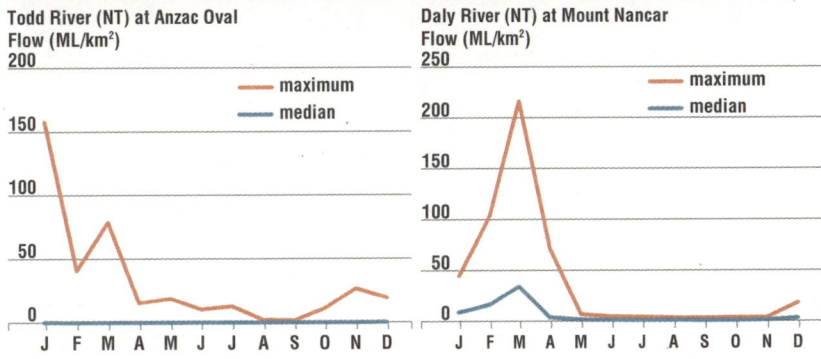

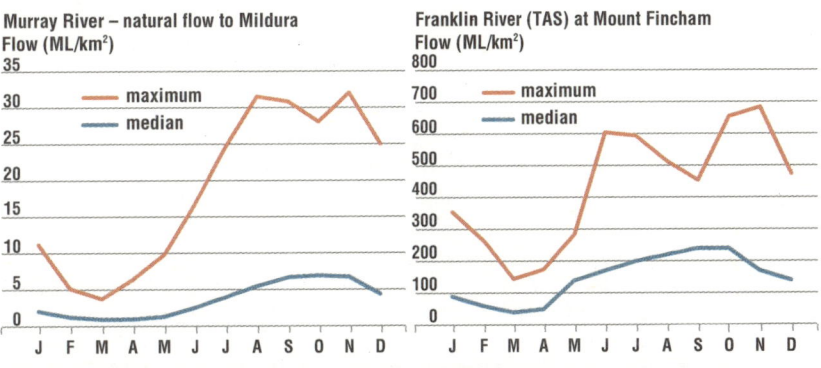

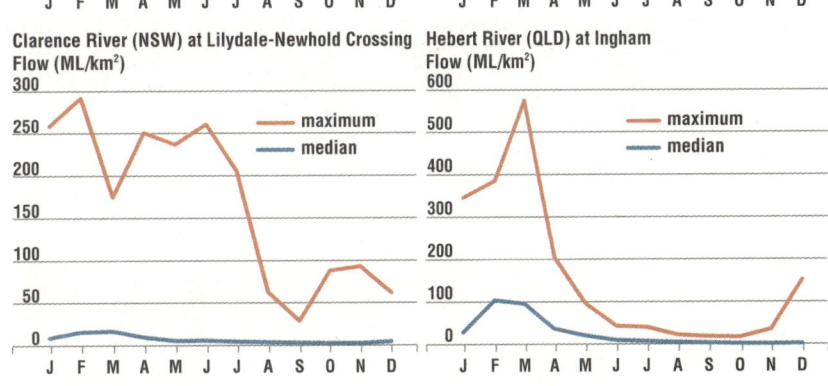

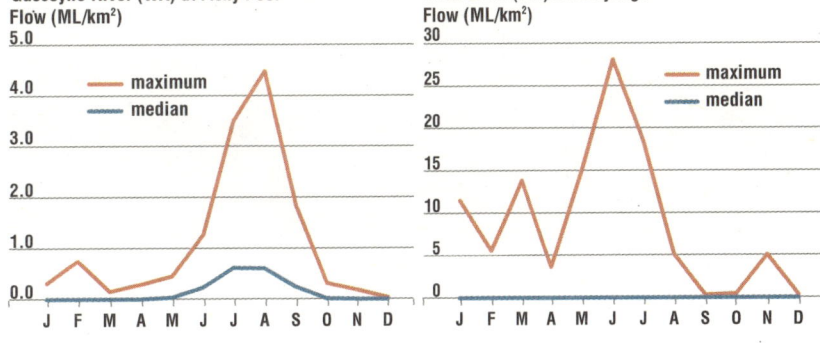

Source: MDBC, NT POWA, QDPI, pers. comm.

Figure 7.3 The incidence of drought 1965–80

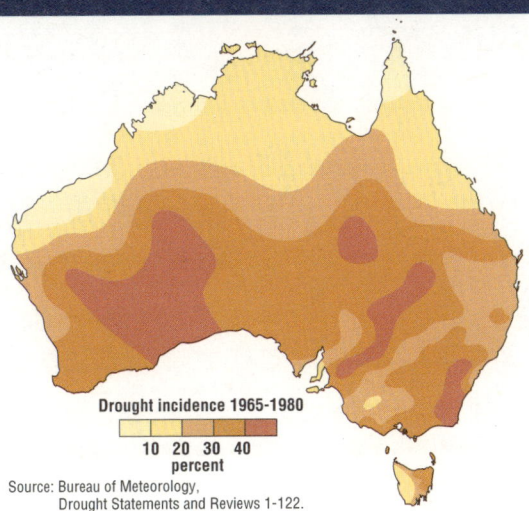

Drought incidence 1965-1980

10 20 30 40
percent

Source: Bureau of Meteorology, Drought Statements and Reviews 1-122.

State

Water resources

The nature of surface water varies considerably through time and across the continent (see Fig. 7.4). Australia's environments range from the wet tropics in the north to cool temperate in the south, with the vast proportion of the inland being arid. Surface water resources are distributed extremely unevenly between Australia's 12 drainage divisions (see Fig. 7.1). Only half these divisions produce a significant level of useable run-off. The Murray–Darling Basin is now facing major environmental problems because of overdevelopment of its water resources.

Groundwater occurs almost everywhere, but is highly variable in quality and useful quantities cannot always be obtained. In large tracts of inland Australia, it is often the only practical source of water supply for the economically important pastoral and mining industries and their associated communities. The Great Artesian Basin ranks among the largest groundwater systems in the world and is of critical importance over a large area of eastern Australia. Where good-quality groundwater is readily available, it is also used extensively in coastal communities. In the Perth region, it constitutes about two-thirds of total water use and about 30 per cent of the water supplied by the Western Australian Water Corporation. Districts such as Bundaberg, Mackay and Ayr–Home Hill along the Queensland coast use large quantities of groundwater for both urban and irrigation purposes. It provides domestic water supplies for over one million Australians in about 600 communities (AWRC, 1992). Of the 15 million megalitres (ML) of groundwater estimated to be available annually for extraction in Australia, about 15 per cent is developed for human use.

Development and state of the resource

After World War II, in an attempt to ensure a reliable water supply for domestic and irrigation purposes, and to drought-proof Australia, a massive nationwide program of dam-building took place (see Fig. 7.5). Australia has the highest per capita water storage of all countries, because we have the world's most variable rainfall. Sydney stores 932 kilolitres (kL) of drinking water for every inhabitant, compared with 250 for New York, 182 for London and 86 for Birmingham. For irrigation water, New South Wales stores 1580 ML per sq km of irrigated land, compared with 760 for the United States, 380 for Egypt and 150 for India (NSW DWR, 1994). Australia's storage capacity in major reservoirs totals some 81 000 gigalitres (GL), or 3.7 times the developed resource. This is equivalent to three Olympic swimming pools for every one of the country's 17.8 million people.

The bulk of our water storage is concentrated in a few very large reservoirs. Australia's 10 largest storages hold about 50 per cent of national capacity. In New South Wales, the 10 largest storages contain 90 per cent of that State's storage volume (NSW SOE, 1993). Most of these large water bodies are concentrated in the south-east of the continent along the Great Dividing Range, and in Tasmania. This reflects availability of run-off, topography suitable for dam construction, and population distribution. The one striking exception is Australia's second-largest storage, Lake Argyle in the north-west of Western Australia. Hundreds of small storages scattered across the rural landscape are also important for water supply and irrigation.

Changes in some farm practices, such as the move to smaller paddocks for better and more efficient management of livestock, have resulted in large increases in the numbers of farm dams (Banens, 1981). For example, in a 10-sq-km catchment in the Yass valley of New South Wales the number of farm dams increased from six in 1959 to 780 in 1985 (Srikanthan and Neil, 1989). Victoria alone is estimated to have some 300 000 small farm dams (VSOE, 1988). With such numbers, although farm dams are individually small, significant reductions

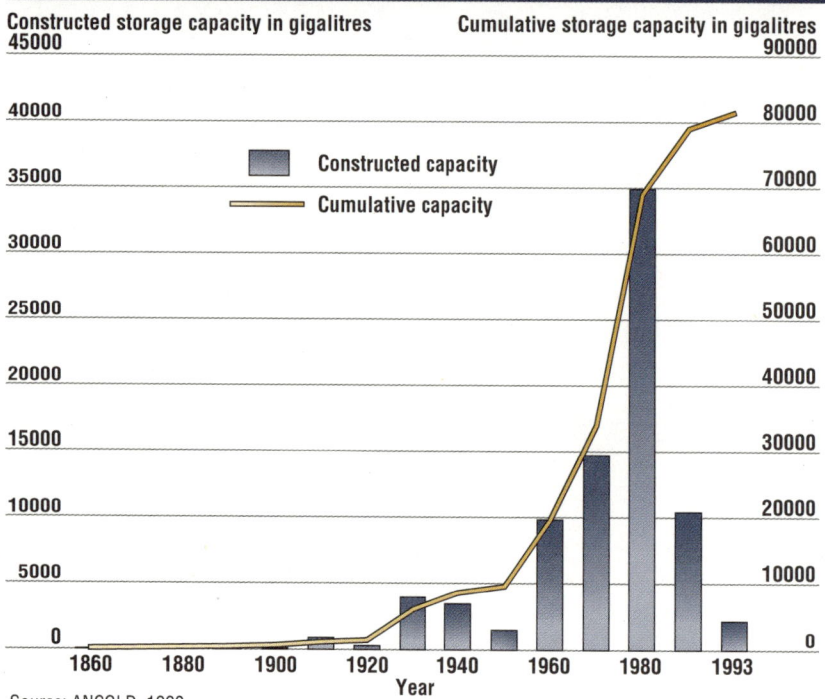

Figure 7.5   Growth in capacity of major storage reservoirs

Constructed storage capacity in gigalitres

Cumulative storage capacity in gigalitres

- Constructed capacity
- Cumulative capacity

Year

Source: ANCOLD, 1990.

Near Colac, western Victoria. Farm dams reduce stream flow but they can become important microhabitats for flora and fauna.

in stream flow can occur, particularly under dry conditions. In the Lal Lal Reservoir catchment in Victoria, for example, farm dams reduce average annual stream flow by seven per cent, increasing to a 50 per cent reduction in drought years (VSOE, 1988). This has implications for large and local-scale water resource planning and management, as well as for the environment. On the other hand, farm dams frequently represent the only near-permanent water in the landscape and therefore can become important microhabitats for fauna and flora.

In most rural areas and for some rural communities, rainwater tanks are used extensively and often represent the sole source of drinkable water. However, in most cities and many rural towns, regulations and disincentives have prevented their wider use. Adelaide is the only major city where rainwater tanks are used extensively, and in this case the spur has been partly the poor quality of the reticulated supply. Augmentation of community water supplies by rainwater tanks can be both cheaper and more effective than building additional large storages (Gippel and Perrens, in prep.).

The development of groundwater reserves is far from uniform. Many groundwater provinces were not being used at the time of the last summary of the resource in 1983–84, while others were being tapped of their entire divertible resource. 'Mining' of old groundwater is undoubtedly occurring, although we have no reliable up-to-date database for all groundwater basins to show the amount of groundwater abstraction compared with its recharge. The Great Artesian Basin has lost an excessive amount of water as a result of the large number of bores left flowing. Groundwater pressures have declined and some bores have

▲ Irrigated pasture — lower Murray

stopped flowing altogether. Bores are now being capped to conserve the resource, but it will take several decades to complete the work.

In 1983, the Department of Resources and Energy listed some of the important groundwater systems under stress (see Table 7.1). Four of these systems are in Queensland which uses a far larger proportion of groundwater for irrigation than the other States. Since then, a surface-water scheme has been built at Bundaberg, thus providing an alternative source, while in the Burdekin delta additional artificial recharge works have reduced the stress on that aquifer.

Aquifers are also affected by increases in recharge from clearing of forests and introduction of irrigated agriculture. These land use changes result in waterlogging and increases in salinity.

Table 7.1 Groundwater systems under stress

System	Aquifer type	Main use	Abstraction (GL/year)	Natural recharge (GL/year)	Induced recharge (GL/year)	Present management	Future strategy
Burdekin Delta, Queensland	surficial	irrigation	263	200	53	extraction limits; recharge to avoid salt water intrusion	further recharge planned and under construction to provide extra 50 GL
Namoi Valley, NSW	surficial and sedimentary basin	irrigation	160	110	0	restricted numbers of bores in part of area	conjunctive use; increased numbers of bores in other areas
Bundaberg, Queensland	surficial	irrigation	100	-	0	replacement of groundwater use by surface water	surface water scheme under construction
Condamine Valley, Queensland	surficial	irrigation	87	13	0	restricted numbers of bores since 1970; metering	surface water scheme planned to partly replace groundwater
Lockyer Valley, Queensland	surficial	irrigation	47	25	1	recharge weirs; control on groundwater use in one area	additional offstream storage (10 million GL) and recharge weirs (5 million GL); restricted numbers of bores

Source: DRE, 1983.

Table 7.2 Area (ha) of irrigated agriculture by major commodity group and State, 1993–94

Commodity	New South Wales	Victoria	Queensland	South Australia	Western Australia	Tasmania	Total Australia
Pastures (incl. dairy)	635 000	555 600	70 000	53 000	14 000	33 000	1 360 000
Cereal crops (non rice)	284 000	23 000	45 000	7 000	1 000	2 000	363 000
Rice	125 000	-	-	-	-	-	125 000
Vegetables	17 000	20 000	27 000	9 000	6 000	17 000	96 000
Fruit	22 000	18 000	25 000	16 000	5 000	3 000	90 000
Grapes	11 000	16 000	–[a]	24 000	2 000	–[a]	53 000
Other crops	176 000	13 000	73 000	4 000	3 000	6 000	275 000
Sugar cane	(<1 000)	-	81 000[b]	-	-	-	81 000
Cotton	189 000[c]	-	48 000	-	-	-	236 000
Total	**1 459 000**	**645 600**	**369 000**	**113 000**	**31 000**	**61 000**	**2 679 000**

Notes: (a) grapes included with fruit; (b) Queensland irrigated sugar cane represents 48% of total cane area; (c) NSW irrigated cotton estimated as 90 percent of total cotton
Source: ABS, Australian Irrigation Council, Cotton Growers Association, pers. comm.

Figure 7.6 Relative water use by key irrigation commodities

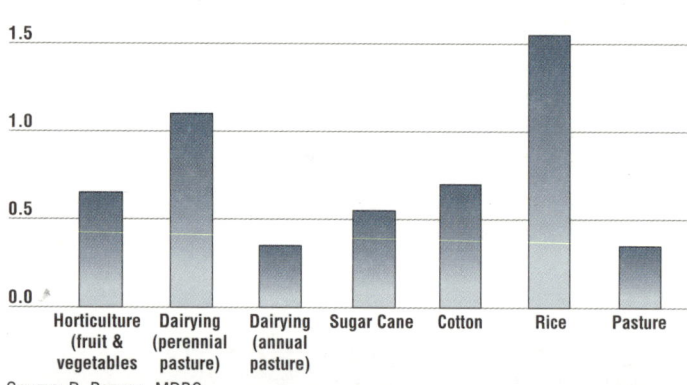

Source: R. Banens, MDBC.

Figure 7.7 Relative profitability of commodities using irrigation

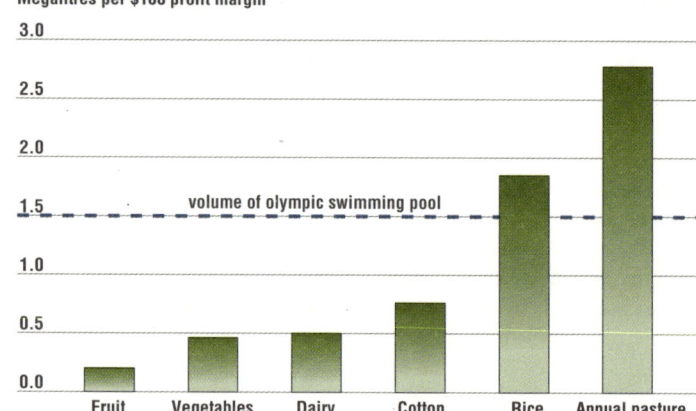

Source: adapted from Hall *et al.*, 1994.

Figure 7.8 Growth in area of irrigated crops in Australia

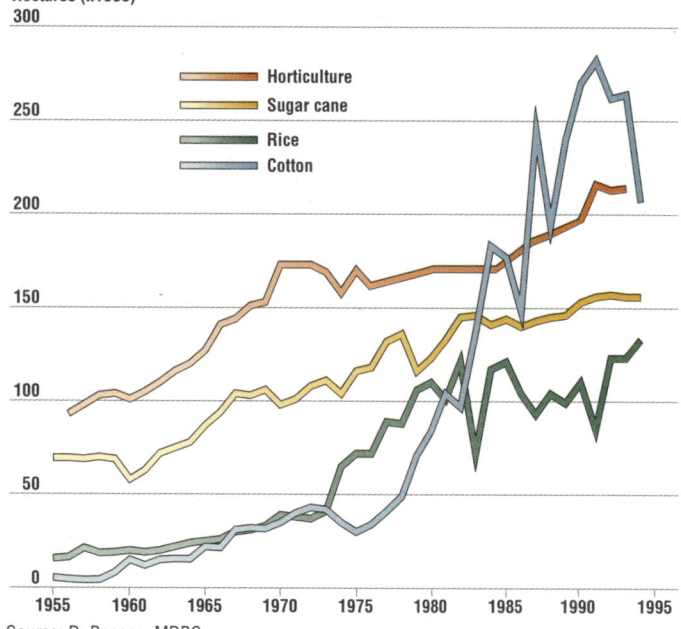

Source: R. Banens, MDBC.

Figure 7.9 Growth in water use in Murray–Darling Basin

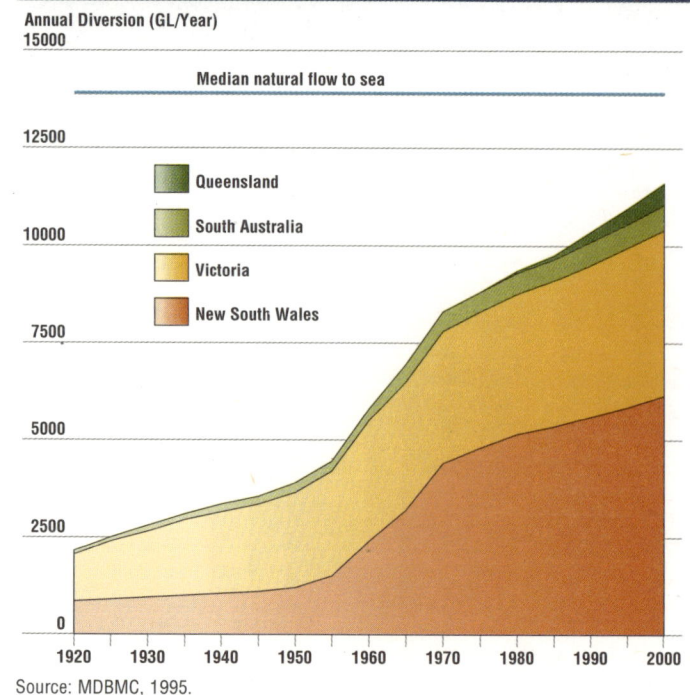

Source: MDBMC, 1995.

The irrigation industry

Of all the developed water in Australia, 70 per cent (15 000 GL) is used in irrigation. This compares with 21 per cent for urban and industrial use, and nine per cent for rural water supply. Expansion in irrigation has been the major factor contributing to the growth in water use in Australia, with consequent pressures on the resource and environment. This growth was particularly evident prior to the early 1970s, with development of the rice and horticulture industries along the Murray and Murrumbidgee rivers, and during the 1980s and 1990s with major expansion of the cotton industry in the Darling River system. About half the irrigated area in Australia is used for pasture (including dairying). This ranges from 14 per cent in Queensland to as high as 85 per cent in Victoria. Two other industries that use a lot of land and water are irrigated cereal crops (including rice) and cotton (see Table 7.2). Water use per hectare and profit margin per ML of water vary significantly between commodities (see Figs 7.6 and 7.7). These factors, coupled with total water use, value of produce and export earnings provide a different perspective on the relative importance of various irrigation commodities (see Table 7.3). Irrigated agriculture makes a large contribution to the Australian economy, with crops such as cotton, rice, wine, sugar and dairy/livestock contributing to a multibillion-dollar export industry.

The area under irrigation is continuing to expand, particularly in New South Wales and Queensland. At the same time, the irrigation industry is undergoing structural reform associated with changes in water marketing and pricing. This has resulted in a shift away from low-value activities such as mixed farming towards high-value crops such as cotton and horticulture (see Fig. 7.8). The development of the cotton industry in the Darling River system has been particularly spectacular, and can be compared with earlier developments in the rice and horticulture industries.

Irrigation in the Murray–Darling Basin is nearly at the limit of the water resource (see Fig. 7.9), indicating that future expansion of irrigated agriculture will be largely through productivity increases and industry restructuring.

Domestic water use

After irrigation, the next highest consumption of water occurs in large urban areas. Although urban use includes industrial and commercial activities, domestic water use is by far the largest component. Domestic consumption has increased significantly over the past 40 years as a consequence of increasing population and rising per capita use. However it varies significantly across the major cities with rainfall, number of rain days, mean temperatures and humidity, availability of water, pricing and education. Annual consumption figures for the

Table 7.3 State of irrigation in Australia

Commodity	Production value ($ million)[a]	Irrigation export value ($ million)[b]	Irrigated area (ha)	Total water use (GL)[c]	Environmental impacts[d]
Horticulture	2600 (Fruit 1490 Vegetables 1110)	496 (wine 247)	214 000	1400 (@ 6.5ML/ha)	slight to moderate impact: localised impact of pesticides, sub-surface saline drainage water
Dairy products	1040	476	280 000	2500 (75% perennial pasture @ 11ML/ha. 25% annual pasture @ 3.5 ML/ha)	moderate impact: regional, localised and downstream impacts of nutrients and organic matter generally as point sources, regional impact on rising water tables and downstream impact of saline drainage
Sugar cane	367	447	183 000	1007 (@ 5.5ML/ha)	moderate impact: localised downstream impact of soil loss and phosphorus, as well as nitrate leaching into estuarine and inshore marine waters, regional impacts of rising water tables and salinisation
Cotton	565	603	264 000	1800 (@ 7ML/ha)	moderate to significant impact: regional and downstream impact of pesticides, large-scale water use in an area of limited water availability and reliability, potential for nitrate leaching to groundwater
Rice	164	228	123 000	1900 (@ 15.5ML/ha)	significant impact: regional impact on rising water tables, salinity, and pesticide and herbicide use, potential for soil acidification in some regions
Irrigated pasture	144 (livestock production)	?	850 000	3000 (@ 3.5ML/ha annual pastures)	significant to severe impact: large scale regional impact on rising water tables and salinisation, and downstream impact of saline drainage

Note: (a) 1992–93 value at farm gate; (b) 1992–93 value free-on-board, source: ABS, pers. comm.; Australian Irrigation Council, pers. comm. (c) Estimated water use rates, source: W. Meyer, pers. comm., for all commodities except sugar cane, source: M. Everson and G. Kingston, pers. comm. for sugar cane; (d) Generally excluding the direct impact of water use.

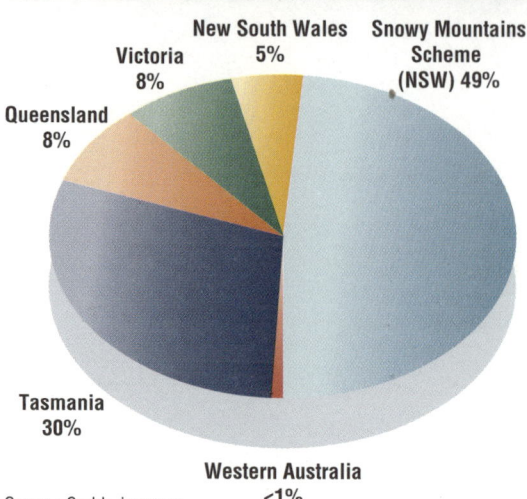

Australians use a lot of water on traditional gardens but native gardens can use less water.

average household of 2.8 people range from 263 kL for Sydney to 700 kL for Darwin (see Table 7.4). With the exception of the Darwin data, these figures are comparable with an average figure for the United States of 397 kL, and a Californian figure of 556 kL (Sydney Water, pers. comm.).

In Australian households, water is mainly used outdoors, with some 30–55 per cent spent mostly watering lawns and gardens. Water use in Darwin is

about 700 kL per year for detached houses but 323 kL for flats without gardens (Northern Territory Power and Water Authority, pers. comm.). Not surprisingly, the level of outdoor water use fluctuates markedly with rainfall and temperature, and this often masks other changes in consumption patterns. Topping up swimming pools and associated backwashing of filters is estimated to average about two per cent of total water use across Sydney, with this outdoor component running as

Table 7.4 Average annual household water consumption in Australian capital cities, 1993–94

	Sydney	Perth	Melbourne	Adelaide	Canberra	Brisbane	Darwin	Hobart
Average household consumption (kL/yr)	263	330	270	265	400	430	700	570
Average annual rainfall (mm)	1227	869	656	451	625	1149	1659	626
Average annual rain days	147	119	147	123	107	123	109	159
Percentage outdoor use	30	42	38	56	55	na	est 45	na

Source: ARMCANZ, 1994.

Figure 7.10 Average residential water use, Sydney, 1993

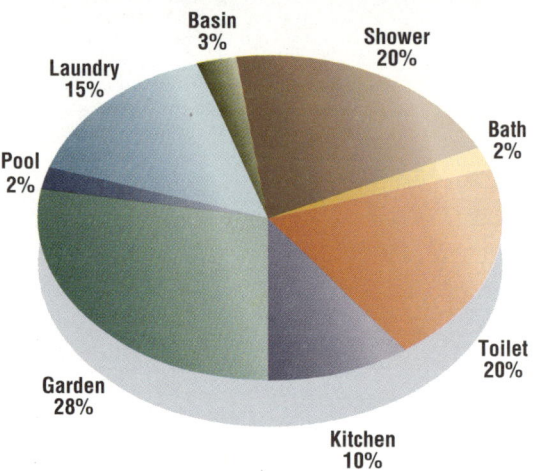

Basin 3%
Shower 20%
Laundry 15%
Bath 2%
Pool 2%
Garden 28%
Toilet 20%
Kitchen 10%

Source: Sydney Water, pers. comm.

Figure 7.11 Australia's hydroelectric power generating capacity

New South Wales 5%
Snowy Mountains Scheme (NSW) 49%
Victoria 8%
Queensland 8%
Tasmania 30%
Western Australia <1%

Source: Crabb, in press.

high as 15 per cent or more of total water use in households with pools (Sydney Water, pers. comm.). Toilets, showers and laundries each account for between 15 and 20 per cent of total domestic water use. The requirement for drinkable water in the kitchen is small — between four and 10 per cent of the total, with the higher figure being more recent and likely to reflect increased adoption of dishwashers (see Fig. 7.10). These figures represent the average for urban households in large cities with secure reticulated water supplies. Country areas with a less secure or no reticulated supply may well have significantly different patterns of domestic water consumption.

Industry and hydroelectricity

Overall, industry does not consume much water. The largest users are thermal power stations, and the petro-chemical, chemical, smelting, pulp and paper, food processing and mining industries. It is difficult to generalise about water use under this category because of the diverse range of quantity and quality requirements as well as considerable opportunity for recycling. The mining industry, often located in the arid parts of Australia, is already embracing recycling, in response to both scarcity and a desire to reduce aquatic pollution. For example, at the Port Kembla smelter, the production of one tonne of steel requires 217 000 L of fresh water and 74 000 L of salt water, yet only about 4750 L are actually consumed (Crabb, 1986). A number of industries have made the innovative move to using 'grey water' (such as sewage effluent), with both economic and environmental benefits.

In some States, the hydroelectric power-generating industry is also a major user of water. For example, Tasmania, which has about 30 per cent of Australia's hydroelectric capacity (see Fig. 7.11), uses about 13 600 GL annually for power generation. This compares with 480 GL used for all other purposes (Department of Primary Industry and Fisheries, Tasmania, pers. comm.). The Snowy Mountains Hydro-electric Scheme discharges some 2700 GL annually in generating its 5100 gigawatt-hours of electricity (Snowy Mountains Engineering Corporation, pers. comm.). For this relatively efficient Scheme, this means the equivalent of three Olympic swimming pools of water are used to generate the annual power consumption of a typical household consuming 23 kilowatt-hours per day.

Altogether, Australia's dedicated hydroelectricity industry uses about 20 000 GL of water for generation annually, a volume comparable to that used for irrigation. Nevertheless it should be noted that this water is not 'consumed', but is considered an 'in-stream use', since it remains available for other uses after passing through the turbines. For example, the Snowy Mountains Scheme was designed for both hydroelectric and irrigation purposes, and many larger irrigation and domestic water supply storages are now being designed or retro-fitted with small hydro-power stations to improve the efficiency of their operations.

Table 7.12 Major dams in Australia

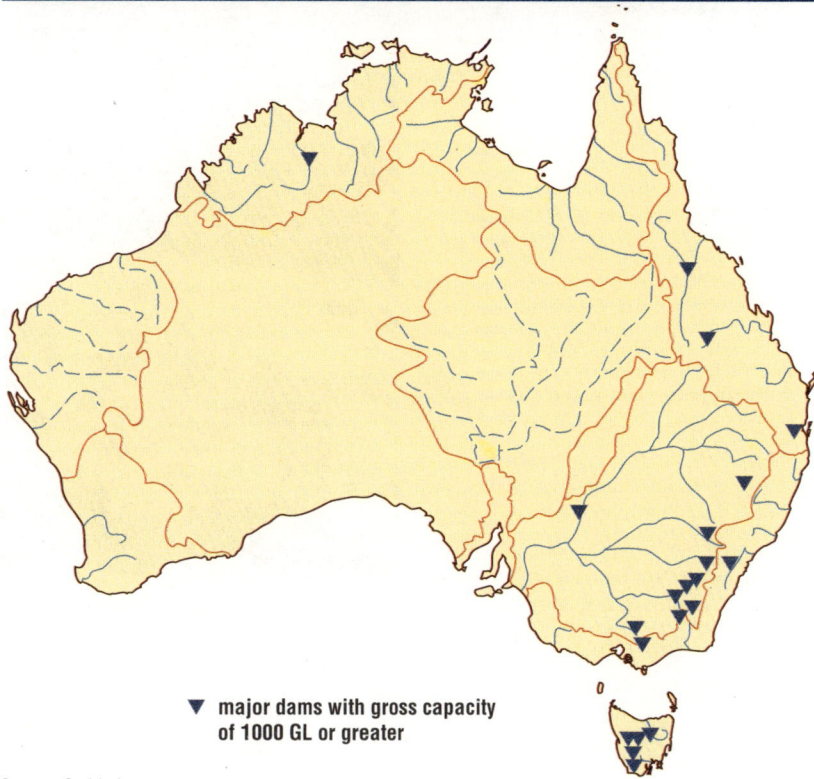

▼ major dams with gross capacity of 1000 GL or greater

Source: Crabb, in press.

Water use and the environment

The amount of water in rivers in irrigation areas has been dramatically reduced during the non-irrigation season because of dam construction (see Fig. 7.12) and poorly controlled abstraction of water from rivers. The use of large amounts of water, often for low-value enterprises such as irrigated pasture and mixed farming (see Table 7.3) has increased the amount of shallow groundwater in irrigation areas.

Darling River; large scale abstraction of water has major implications for inland streams.

Flooding

Australia is a land of extremes — 'of droughts and flooding rains...'. While both occur throughout the country, droughts last longer than floods, which are irregular, relatively infrequent and variable in magnitude. The cost of flooding, in terms of life and property, can be enormous, averaging $400 million a year across Australia. The Brisbane Valley flood of late January 1974, caused by the intense rainfall of cyclone Wanda and a monsoonal air mass, was the most costly flood in Australia, with a damage bill close to $200 million (1974 value) (NDO, 1992). The waters reached 6.6 m at the Brisbane Post Office, affected some 13 000 buildings in Brisbane and Ipswich, and washed away 56 houses. The April–May flood of 1990 was one of the most extensive and dramatic ever recorded, covering more than one million sq km of inland Queensland, New South Wales and parts of South Australia and Victoria. A number of communities, including Nyngan and Charleville, were completely inundated, and stock losses were estimated at one million head. The total cost of the flood exceeded $250 million (1990 value), with the cost at Nyngan alone estimated at $50 million.

By modifying catchments through land clearing and other development, we have increased the frequency and magnitude of floods, particularly in small catchments, as well as our susceptibility to floods. Changes in land use — particularly clearing of vegetation — and increased imperviousness of the soil associated with urbanisation and soil compaction result in increased total volume and peak discharge. In addition, the

Flood waters at St. Lucia, Brisbane, 1974.

time to peak run-off is greatly reduced (see figure below left). For example, in a flash flood in Canberra's Woden Valley caused by severe thunderstorms, the water rose so fast that seven people drowned before any warning could be issued (NSW SES, 1994).

The changes in flooding frequency caused by dams in many regulated lowland rivers have had a significant environmental effect, with the number of small floods on the floodplains substantially reduced. This also affects the biological health of the rivers, because nutrients and organic matter are no longer being exchanged between the rivers and the floodplain; the community structure of billabongs on the floodplain is partly dependent on flood pulses; and fish movement and recruitment depends upon links between billabongs and rivers (Boon *et al.*, 1990; Hillman and Shiel, 1991; Cadwallader and Lawrence, 1990). For their long-term sustainability and health, rivers and their associated wetlands and floodplains will need to be reconnected to allow a return to more natural flooding regimes.

Before the major development of water resources, it was recognised and accepted that 'floodplains are for flooding'. The harnessing of stream flow through numerous weirs and large dams over the last 45 years has resulted in significant changes to the nature of flooding and people have forgotten what floods are like or have never experienced them. This has given rise to a naive perception that river systems can be 'flood-proofed'. Developers then apply pressure to local governments for approvals for developments on floodplains. Around large urban areas this has often led to incremental creep onto the floodplain — sometimes with disastrous consequences when major floods occur. Levees built to protect communities and property may initially reduce flooding, but in the end can actually exacerbate it by presenting ineffective barriers to major floods. Sometimes such banks fail catastrophically causing even greater damage.

As the enhanced 'Greenhouse Effect' starts to exert its influence over the next 50 years or so, an increase in flooding can be expected in some regions (Bates *et al.*, 1994), adding to the effects of land clearing and urbanisation.

Impact of urbanisation: differences in run-off from similar rural and urban catchment areas during the same storm event.

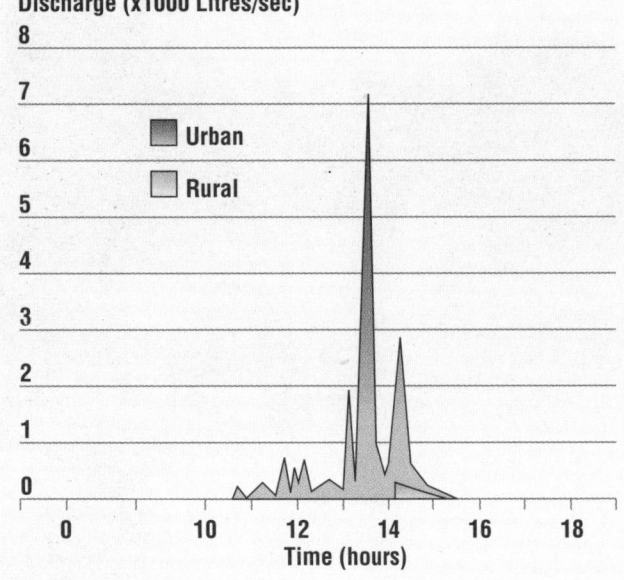

Source: Aitken & Moodie, 1983.

Water for new developments has become scarce, particularly in the Murray–Darling Basin. Some cities are also facing limitations and are now trying to conserve water through strategies such as education, pricing reforms, and recycling, rather than building new dams.

We are starting to realise the environmental consequences of the post-war development of Australia's water resources — among them, disappearing wetlands, and rivers containing too little water at the appropriate times for various fish and bird species to breed. People continue to advance schemes to direct water from coastal or tropical areas to dry areas or those running out of water. The suggestions include piping water from Lake Argyle in the Kimberley to Perth, diverting some northern Queensland and northern New South Wales coastal rivers inland, and towing icebergs from the Antarctic to Adelaide. Restoration is also being considered — the most recent widely publicised example being Lake Pedder in Tasmania. The debate about water supplies in areas of over-allocation and interbasin transfer is now hotting up on environmental grounds. Most governments are not taking the schemes seriously because they are both environmentally and economically unsound and very expensive.

Re-allocating water for environmental purposes, or so-called 'environmental flows', now receives serious consideration. Few resolutions have yet been achieved, but balance between economic and environmental needs is central to the discussion.

The most useful indicator of the impact of water resource use on the non-human environment is the amount of water abstracted each year. Unfortunately, data are not consistent enough to show such an indicator through time.

The message about surface water

- Australia has the highest per capita storage capacity of all countries in the world as a result of dryness and variability of climate.

- Of the developed resources, 70 per cent are consumed for irrigation, followed by domestic, industrial and commercial uses.

- In parts of both the Murray–Darling Basin and eastern seaboard, water is grossly over-allocated and demand continues to increase.

- Over-allocation is placing aquatic environments under severe stress in these regions, but the stress is less severe in the rest of Australia and absent in undeveloped areas such as south-west Tasmania, the wet tropics and parts of the seasonally wet tropics.

The message about groundwater

- Groundwater provides domestic water for more than one million Australians in 600 communities, and 60 per cent of the continent totally relies on it for all uses except drinking water.

- In a number of basins — such as the Great Artesian Basin, Pioneer Valley (Mackay, Qld),

Namoi Valley (Tamworth, NSW) and Burnett Basin (Bundaberg, Qld) — groundwater is being used faster than it is being replenished.

- In other areas, rising watertables from clearing and irrigation are waterlogging and salinising streams and large areas of land. Even with remedial action, this problem will continue to increase for some years.

Catchment pollutant sources

Catchments shed sediment, nutrients (particularly phosphorus and nitrogen), organic matter and agricultural chemicals such as pesticides. Land can also become salinised from extensive clearing or irrigation, and industry and urban areas produce a variety of wastes.

Sediments and river channel change

Rivers provide sources of sediment, places where sediments are deposited, habitats for many creatures, locations for recreation and sources of water for human use. Although their physical state is difficult to report nationally, four categories of channel change can be identified: valley floor incision, dramatic widening, increased meander migration and channel burial. Some channels have not changed (Rutherfurd *et al.*, in press).

Channel incision occurs mostly in uplands in the humid temperate areas, as a result of increased run-off after clearing (see Fig. 7.13). A reconstruction of sediment yield from an incised catchment shows early rapid incision and high sediment yield, then a decline as incision slows (CSIRO, 1992).

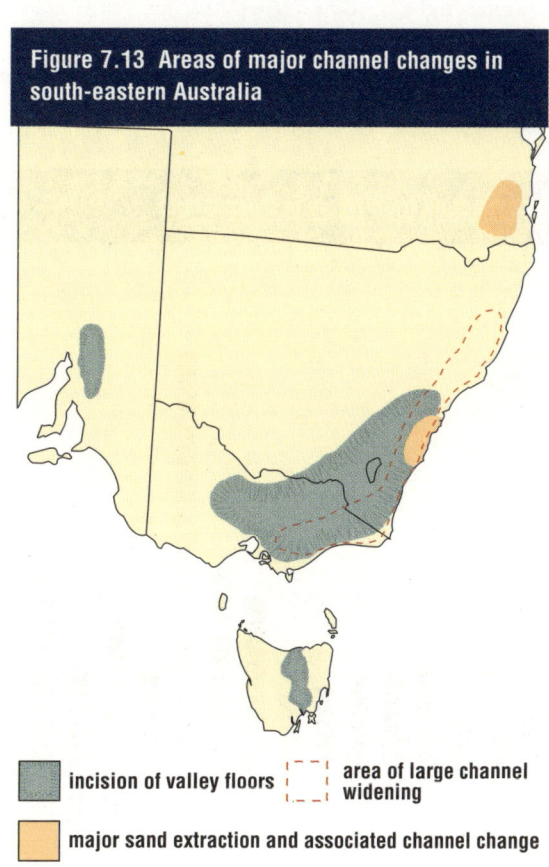

Figure 7.13 Areas of major channel changes in south-eastern Australia

incision of valley floors

area of large channel widening

major sand extraction and associated channel change

Source: Rutherfurd *et al.*, in press.

▲ Incision of small stream resulting from land use.

This pattern began in the 1830s in the south-eastern uplands, and as late as 40 years ago in the south-west of Western Australia. It is probably still being triggered in areas of modern upland clearing and heavy grazing. Channel incision has also occurred as a result of the straightening of rivers in Western Australia, New South Wales and Victoria (so-called 'river training'); and as a result of sand and gravel extraction, particularly in Queensland and New South Wales (see Figs 7.14).

During the early stages of incision, some parts of streams are buried by sediment originating in incisions upstream. This is happening in the south-west of Western Australia. Mining also causes channels to be filled or buried. Examples can be seen in the King and Ringarooma Rivers of Tasmania, in the gold-mining districts of central Victoria and New South Wales, and in the uranium province of the Northern Territory.

Dramatic channel widening has been documented in coastal Victoria and New South Wales. Widening can be a natural process or the result of land use. Almost all valleys in the Southern Tablelands of New South Wales are incised as a result of increased run-off due to grazing. The Avon River in Victoria has increased its width 25-fold, with little identifiable human cause. By contrast, the Latrobe River with large-scale human intervention — 70 meander cut-offs and five desnagging episodes — has widened by only 50 per cent. Sand and gravel extraction can also cause channel widening (Rutherfurd *et al.*, in press).

The rate of bank erosion of large streams is proportional to the amount of water in the channel, which in turn is proportional to the catchment area. Four estimates of meander migration rate for Australian streams conform to the apparent world pattern established by Hooke (1980) (see Fig. 7.15). There is no quantitative evidence that land use has increased these rates, although eroding, cleared and grazed river banks are obvious. However, other Australian rivers do not conform to the world pattern, including the low-gradient, muddy channels of the Murray–Darling Basin, in which the flow of water is very small relative to catchment area. There is no quantitative evidence that migration rates have increased since settlement, although eroded banks are evident. However, some evidence establishes widening of these channels (both of convex and concave banks) at rates between 0.16 and 0.74 m per year, probably as a result of increased flows for irrigation (Rutherfurd *et al.*, in press). Yet, the banks of these muddy rivers contribute less than five per cent of the total sediment in transport (Olley *et al.*, 1995).

Channel change alters some habitats. Generally, the new habitats are less diverse than the original ones, and so are likely to be biologically of lower quality. Using the criteria of rate of channel change, increase of sediment yield and the number of cases, it appears that gullying has wrought the largest

Figure 7.14 Sand and gravel extraction from rivers in Queensland

Extraction volume (million cubic metres)

[Bar chart with y-axis from 0 to 2.0 (marked at 0, 0.5, 1.0, 1.5, 2.0) and x-axis years 1950, 1960, 1970, 1980, 1990, 1994]

Source: QLD Department of Primary Industry Year Books.

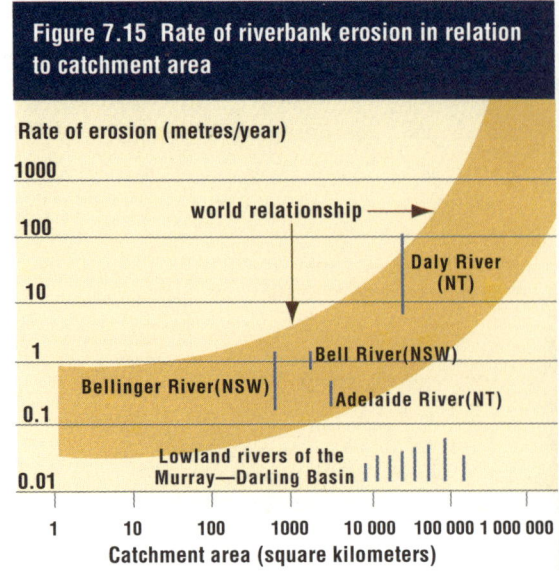

Figure 7.15 Rate of riverbank erosion in relation to catchment area

Rate of erosion (metres/year)

[Graph with y-axis logarithmic from 0.01 to 1000 and x-axis Catchment area (square kilometers) from 1 to 1 000 000]

world relationship →

Daly River (NT)

Bell River (NSW)

Bellinger River (NSW)

Adelaide River (NT)

Lowland rivers of the Murray—Darling Basin

Catchment area (square kilometers)

Source: Hooke, 1980; Rutherfurd *et al.*, in press.

change, followed by channel incision in valley floors, mining, meander migration, urbanisation and river 'training'.

Dramatic channel changes produce large quantities of sediment, the fine-grained part of which moves quickly downstream or onto floodplains. The coarse fraction often moves as a 'slug', burying the riverbed, increasing local flooding and causing additional channel widening as the channel through which it moves adjusts to accommodate flood flows. Mining also produces slugs of sediment, often of enormous size. The biological effects of these coarse-grained slugs are poorly known, but are likely to be significant because they move slowly and so have a prolonged period of impact. Granite catchments yield more coarse grains than others, and so slugs can be natural features, although land use can increase their size (Rutherfurd *et al.*, in press).

Sand and gravel extraction — already noted as a secondary cause of channel change — can directly cause such change. Almost everywhere in eastern Australia these resources are being extracted at a rate faster than they are renewed (Rutherfurd *et al.*, in press). The rate of extraction in both New South Wales and Queensland is steadily rising, to meet demand for construction (see Fig. 7.14).

Phosphorus

Phosphorus, which is an essential nutrient for life, can lead to nuisance growth of blue-green algae and water plants (see the box on page 7-48). Australian catchments have many sources of phosphorus, some related to land use, some derived from animal and human wastes as well as natural sources from the soil (see Table 7.6 and Fig. 7.16).

Estimates of the production (or generation) rate of phosphorus from a land use or land type, as summarised in Table 7.6, tell only part of the story. The area covered by a particular land use and its position in a catchment influence the amount of nutrient that reaches a water body. It is important to know the proportional contributions to nutrient levels in water bodies from different land uses in order to manage eutrophication.

There are few estimates of the relative amounts of phosphorus from different sources entering our streams. Table 7.7 contains a national and some individual catchment estimates, some of which should be treated with caution. The estimate of diffuse sources reaching the Great Barrier Reef is probably too low, given that large floods were not adequately sampled. Floods were well sampled in the case of Lake Burley Griffin, but the sampling period was short. The estimate for the Murray–Darling Basin point-source contribution is likely to be too high, given that measurement-based estimates from the Lower Murrumbidgee and Murray are lower than the Basin-wide estimate of 44 per cent. The Western Australian cases are generally well measured and highlight large variations in the proportion of total phosphorus derived from point sources, with some noteworthy results. For example, the point sources of a city like

Table 7.6 Nutrient generation rates for different land uses in catchments

Land type and use	Total phosphorus generation rate (kg/ha/yr)	Total nitrogen generation rate (kg/ha/yr)
Rural (Australia wide)		
Grazing		
— improved pasture (includes irrigated pasture for dairying)	0.1 – 3.00	0.6 – 10.8
— unimproved pasture	0.002 – 0.4	2.0 – 15.0
Crops — unirrigated	0.2 – 2.0	2.0 – 6.0
Horticulture	15.0	6.0 – 35.0
Forests	0.01 – 0.2	0.9 – 10.0
Urban and industrial (Australia wide)		
Industrial and suburban land	0.1 – 3.6	1.0 – 22.4
Sewage treatment plants	4[a]	35[b]

Note: Broadacre land uses on soils that contain clay and iron contribute phosphorus mainly in particulate form by sheet and rill erosion, and in dissolved form where soils are sandy. An emerging view suggests that erosion of channels produces up to 11 times the maximum amount of phosphorus listed in this table. The urban and industrial category includes point sources (industrial and sewage treatment plant discharges) and diffuse sources from urban stormwater run-off (see Chapter 3).

a. Based on 0.2 kg/person/year; b. Based on 1.0 kg/person/year

Source: Sewage treatment plants; Wasson — various sources; Young *et al.*, in press.

Perth produce no larger proportional contribution than the small towns and intensive industries of the Peel–Harvey catchment. The very high point-source contribution to Princess Royal Harbour (Albany) originated from a fertiliser factory. This has now been rectified.

The Murrumbidgee at Wagga Wagga transports about 620 tonnes of phosphorus each year. Of this, sewage contributes about 44 tonnes — an amount

Phosphorus transport has been calculated by combining measured concentrations of total phosphorus in the surface horizons of major soil types with the sheet and rill erosion estimate of the box on page 6-28 High mean annual run-off occurs along the eastern seaboard, from the eastern uplands, in parts of NT and WA and in a few locations in central Australia.

Figure 7.16 Estimated phosphorus transport resulting from sheet and rill erosion.

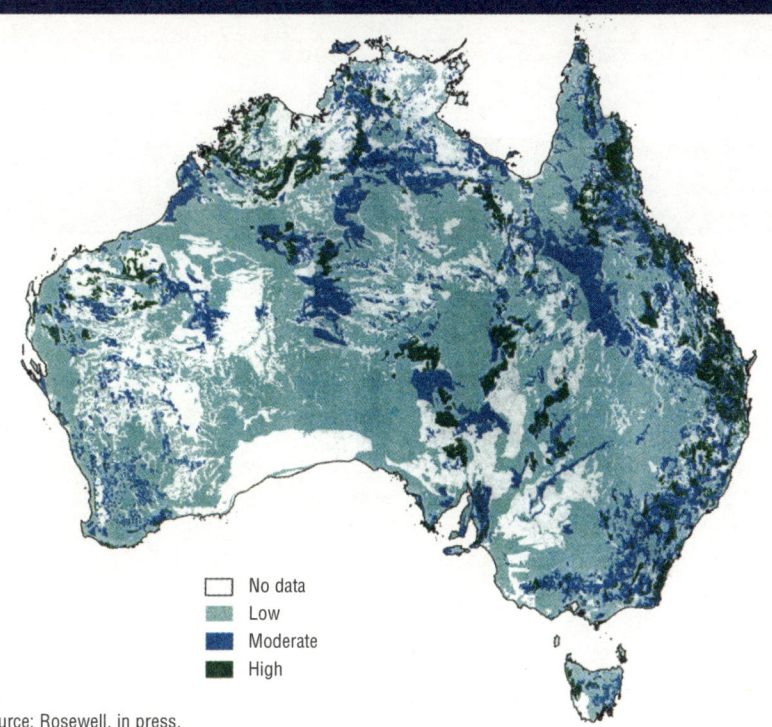

- ☐ No data
- Low
- Moderate
- High

Source: Rosewell, in press.

that has prompted considerable expenditure to dispose of sewage off-river. Of the 580 tonnes from diffuse sources, less than five per cent comes from fertiliser-enriched surface soils. So the two sources of phosphorus assumed by many management agencies to be of greatest significance in this region, namely sewage and fertiliser, actually constitute less than eight per cent of the total load of the nutrient (Olley *et al.*, 1995).

Sewage may be only locally significant as a source of nutrients. In the Hawkesbury–Nepean system

Table 7.7 Estimated proportionate contributions of different sources to phosphorus loads in waterbodies

Location	Point sources (per cent)	Diffuse sources (per cent)	Reference
Hawkesbury–Nepean	15–18 (sewage)	75–82	Cuddy *et al.*, 1993
Murrumbidgee at Wagga Wagga	8 (sewage)	92 (>95 per cent from gullies and streambanks)	Olley *et al.*, 1995
Murray at Albury–Wodonga	32	68	Walker and Hillman, 1982
Murray–Darling Basin	44	56	GHD, 1992
Lake Burley Griffin	29	68	Cullen and Rosich, 1979
Upper Murrumbidgee	23	77	NCPA, 1994
Drainage to Great Barrier Reef	14	86	Furnas *et al.*, 1994
Peel Inlet	35	65	Bott, pers. comm.
Harvey Estuary	15	85	Bott, pers. comm.
Princess Royal Harbour	75	25	Bott, pers. comm.
Oyster Harbour	10	90	Bott, pers. comm.
Swan Estuary	35	65	Bott, pers. comm.
Australian agricultural districts	22	64 (erosion) 6 (leaching) 8 (fire)	McLaughlin *et al.*, 1992

Note: Different methods have been used to derive the percentage contributions from point and diffuse sources. Some are based on measurements, some on models.

Figure 7.17 Change in phosphorus content relative to iron in the Burrinjuck Reservoir sediment core

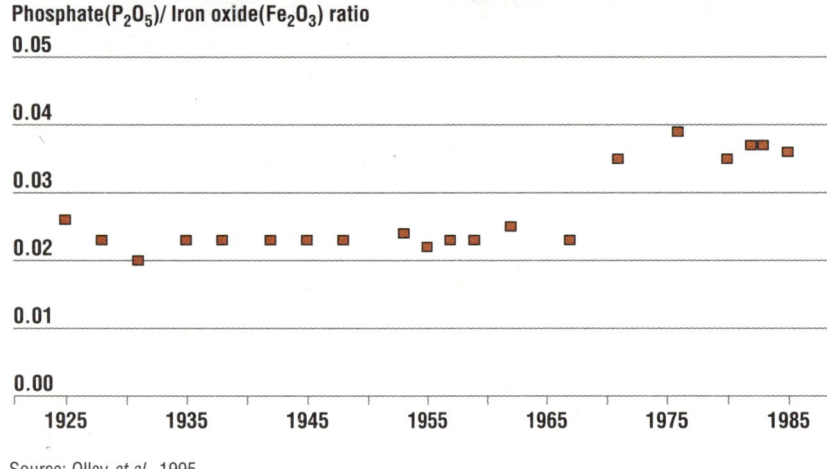

Phosphate(P_2O_5)/ Iron oxide(Fe_2O_3) ratio

Source: Olley *et al.*, 1995.

sewage phosphorus is deposited in the riverbed within a few kilometres of its discharge point (Jones, 1994). In the Barwon River in northern New South Wales, the same process occurs within a few tens of kilometres (G. Hancock, pers. comm.). However, scientists are still trying to determine how much of this phosphorus is subsequently available for plants and algae.

Very few long-term records exist to allow determination of whether the amount of phosphorus reaching Australian inland waters is changing through time. However, a core of sediments from Burrinjuck Reservoir, on the Murrumbidgee River record increasing amounts of phosphorus (see Fig. 7.17). The increase has occurred since the 1970s, even though phosphorus has been removed at the Lower Molonglo Water Quality Control Centre. This suggests that most phosphorus reaching the lake did not come from Canberra.

As the amounts of sediments and phosphorus entering waterways increase, fine sediments in reservoirs and riverbeds can become significant storages for the nutrient. Under various chemical conditions, some stored phosphorus is released and supports blue-green algae (Donnelly *et al.*, 1992, 1994; Harris, 1994).

Nitrogen

In fresh water, nitrogen is not usually the limiting nutrient for algal growth, but it can control the growth of algae in estuaries and near-shore marine environments (see the box on page 7-48).

Nitrogen can also pollute groundwater. In South Australia industrial wastes have increased the nitrate concentrations of groundwater used by people (Lawrence, 1983). In central Australia, natural processes involving bacteria on soil surfaces and in termite mounds produce nitrate (Barnes *et al.*, 1992), which is then leached, and contaminates groundwater used by remote settlements. At high concentration, nitrate is a human health hazard.

Few studies have been done to document the major sources of nitrogen in Australian catchments (McLaughlin *et al.*, 1992), but Table 7.6 summarises the available data. Horticulture and urban land uses have the potential to produce most nitrogen, along with animal-processing industries. An investigation of nitrogen transport to the Great Barrier Reef lagoon shows that rivers draining rural lands contribute about 60 per cent, rainfall about 20 per cent and sewage three to four per cent (Furnas *et al.*, 1994). The rest comes from ocean upwelling.

Other pollutants

A wide range of other pollutants is produced by industry and urban waste disposal. The most important source industries and activities are listed below.

• waste disposal and treatment industry

• electricity generation

- town gas production
- fire-fighting
- transport industry
- agriculture and aquaculture
- chemical and petroleum industries
- mining and mineral-processing
- manufacturing industry
- domestic and commercial developments

The main contaminants of industrial waste streams are metals and inorganic chemicals, organic chemicals such as phthalates, phenols and cresols, aromatics, polycyclic aromatic hydrocarbons (PAH), polychlorinated biphenyls and related compounds (PCB), halogenated aliphatics, pesticides and metabolites, radionuclides and biological compounds (AWRC, 1990).

While these contaminants probably all occur in Australian catchments, few measurements are available even on a local scale and certainly very few at a national level to assess the scale of the problem (O'Loughlin *et al.*, 1992). The accuracy and thoroughness of reporting and recording incidents of point-source pollution varies widely between and within States. It is unlikely that the situation will change for some time. Legal sensitivity makes people reluctant to disclose the nature, extent and location of most actual and potential contamination sites. For more information about waste-generation rates, see Chapter 3.

▲
Pesticides sprayed on crops sometimes end up in rivers as run-off.

Table 7.8 gives one example from each State and the Northern Territory as an indication of contamination incidents.

Of the diffuse sources of pollutants, pesticides are probably the most important. While it should be possible to determine their level of use from sales or import figures, as far as can be determined no one has collated this information nationally, or for most regions. An exception is a pesticide audit conducted by Rayment and Simpson (1993).

Table 7.8 Examples of point source contamination in Australia

Location	Source	Contaminants	Present and future effects
Botany sandbeds, Sydney (NSW)	Manufacturing industry	Petroleum hydrocarbons, organic solvents and heavy metals	Contamination of a major industrial water resource; contamination of wetlands and marine environment
Rum Jungle (NT)	Mine waste	Heavy metals, acid waters, total dissolved solids	Concern for future contamination of surface waters by groundwater discharge; remedial measures in place
Willawong and Kingston hazardous waste sites (Qld)	Industrial waste	Organic solvents, petroleum hydrocarbons	Long-term environmental effects
Mt Gambier region (SA)	Timber treatment	Arsenic, chromium and tin compounds, phenols, cresols and pesticides	Localised contamination of the shallowest aquifer leading to long term resource degradation, environmental effects and some potential public health impacts
Mt Direction and Westbury (Tas.)	Timber industry	Tannins and lignins	Potential future degradation of resource and environmental impact
Western suburbs, Melbourne (Vic.)	Manufacturing industry, landfills	Petroleum hydrocarbons, organic solvents, phenols, heavy metals, nitrate, sulfate, pesticides	Widespread contamination of a vulnerable, brackish water; highly transmissive aquifer with low attenuation, leading to environmental effects Limited potential for public health effects except where highly contaminated groundwater is intersected in excavations
Perth Coastal Plain, Kwinana (WA)	Petroleum, chemical and mineral refining and processing; heavy metals, nutrients, food processing	Petroleum hydrocarbons, organic solvents, pesticides, total dissolved solids	Widespread contamination of surficial sediments and aquifer has led to resource degradation and environmental impact; also poses a substantial threat to public health in the long term

Source: AWRC, 1990.

Dryland salinisation in south-west Western Australia

Before extensive land clearing occurred, native vegetation used most of the water in the soil and unused water leaked into the groundwater. However, the replacement of native plants with shallow-rooted crops and pasture grasses has led to more water leaking into the groundwater, causing watertables to rise. As the watertable nears the ground surface, waterlogging begins and salt stored in the soil is mobilised, creating saline seeps. Water and salt emerge in the floors and sides of valleys, forming salt crusts, killing plants, aiding erosion as plants die, and increasing the salinity of streams.

In south-west Western Australia, about 14.6 million hectares have been cleared for grazing and cropping. This area receives less than 900 mm of rainfall each year and, because it is dry, wind-borne sea-salt has been stored in the soils in large quantities for millennia. Typically, between 20 and 120 kg of salt lie beneath a square metre of ground (Schofield, 1989). This salt is being mobilised as the watertable rises at rates between a few millimetres and two metres each year. Water levels beneath areas of native vegetation in the wheat belt are five to 10 m lower than those in cleared areas nearby. Since clearing, the levels have risen between five and 20 m on hilltops, between five and 10 m in the middle of hill slopes, and about five metres in valley floors (Salama and Bartle, in press).

Estimates of the area of salinised land (see figure below) from the Australian Bureau of Statistics (ABS) show an average increase of 0.07 per cent each year since 1955, which represents an annual increase of nearly 12 000 ha. But these figures underestimate the problem. A recent, more reliable, survey based on aerial photographs and ground surveys (McFarlane, pers. comm.) shows that the area of salinised land in 1994 is nine per cent of the cleared area (equivalent to 393 average wheat farms), not 3.1 per cent as estimated in 1993 by the ABS. The ABS estimates rely on landholders' perceptions — probably reflecting bare land rather than salinised land, which can be either bare or thinly vegetated (WA Select Committee into Land Conservation, 1990).

River of salt, south of Kellerberrin, Western Australia; salinisation destroys agricultural land and native vegetation and increases stream salinity.

The best estimate of the area of salinisation of cleared land in 2010 is 16 per cent or 2.9 million hectares. This is 11 per cent higher than the ABS estimate. This figure will increase before the rising watertables reach a new equilibrium, at which point the area affected is expected to be 24 per cent in the western south-coast region and 29 per cent in the eastern south-coast region, about 4.5 per cent and 13.5 per cent more respectively than in 2010. To make matters worse, most of the ground that will be affected between now and 2010 is the better-quality agricultural land in the valleys.

The salinisation of land also increases the salinity of streams, making them unsuitable for human or domestic stock use, and affecting their biota. The salinity of streams in south-west Western Australia has been declining generally in uncleared catchments since 1940, but rising sharply in cleared catchments (Schofield, 1989). Total soluble salts have risen at an average rate of between one and 76 mg/L/year and this rate is increasing.

In 1985, 43 per cent of the run-off in the south-west drainage division was considered divertible for human purposes. Of this, 52 per cent has since been degraded by salt. In those cases where river salinities have increased, they are continuing to do so and are likely to increase fivefold over the next 30 to 40 years (WA Select Committee into Land Conservation, 1990).

Governments and landholders have been slow to respond to salinisation of land and streams in this area. Land clearing has continued well after the causes of salinisation were identified. In some areas, the government has imposed clearing controls, and these appear to have slowed but not halted stream salinity increases. The only practical way of slowing salinisation is by major revegetation, which involves changing land use significantly over wide areas. Where commercial forestry is possible, the salinity trend may be reversed within a decade. In drier areas, the increase in salinisation will continue if current land uses are maintained. The salt in the wetter areas will be flushed out under current land uses in about 10 years, but, in the dry areas, centuries will be needed if current land uses continue (Peck and Williamson, 1987).

Estimates and projections of the area of salinised land in south-west Western Australia

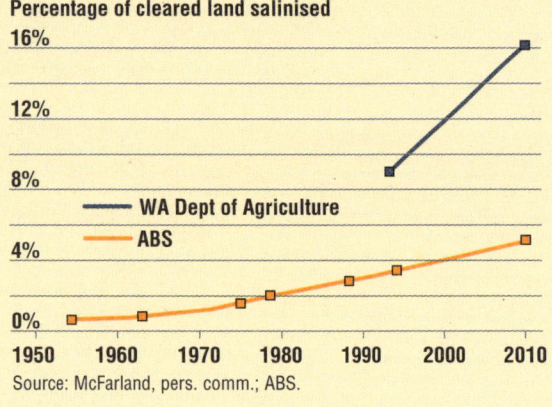
Source: McFarland, pers. comm.; ABS.

Salinisation

In some parts of Australia — most notably the large irrigation areas of the Murray–Darling Basin, which were established at the turn of the century — salinised surface soils, caused by land use, have been present for a long time. The south-west of Western Australia has also suffered from dryland salinity since early this century. In other areas, such as the dryland agricultural areas of central and northern New South Wales, salinity is regarded as a more recent phenomenon. However, it is difficult to assess the scope of the problem, as large areas of the continent lack long-term data. Available data sometimes conflict, indicating that we need standard definitions of relevant parameters. The key indicators of environmental state are rate of rise of groundwater levels, area of land underlain by shallow watertables and rate of change of stream salinity. The best-documented area is in south-west Western Australia (see the box opposite).

Irrigation-induced land degradation

All irrigation practices add water to underlying groundwater. If more water is being added than can move laterally in the aquifer, groundwater levels will rise. Unless the groundwater is highly saline, most irrigation-induced land degradation begins as waterlogging. After time, and depending on evaporation rates and degree of flushing, salt concentrations increase until they affect crop yield. Waterlogging alone also reduces crop yield.

In 1985, the Murray–Darling Basin Commission estimated that 360 000 ha of irrigated land had shallow watertables, and 87 000 ha of land in the Victorian portion of the Basin were visibly salinised (MDBC, 1993a). Victoria's 1991 State of the Environment Report contained an estimate of about 140 000 ha of salinised irrigated land, representing about 30 per cent of the State's irrigation areas. It has been estimated that high watertables lie under some 220 000 ha (41 per cent) of the Shepparton irrigation region. In South Australia, 4600 ha of shallow groundwater exists in irrigation areas (State Dryland Salinity Committee, 1989–90).

The Wakool, Deniliquin and Murrumbidgee irrigation areas of New South Wales were estimated to contain 199 000 ha of land with shallow watertables and 9000 ha of land that is visibly salinised (1985 figures, MDBC, 1993a). By 1991, the area of land overlying high watertables had increased dramatically. For example, the salinised area in Berriquin (part of the larger Deniliquin area) had grown from 22 000 ha in 1985 to 91 300 ha in 1990).

Salinity problems associated with irrigation in other States are much less significant and quite localised at present. However, the prospects of waterlogging and salinisation have caused authorities to reassess the area available for development in the Burdekin River Irrigation Area in north Queensland.

Dryland salinisation

Dryland areas are those that depend solely on rainfall for plant growth. They are susceptible to hydrologic disturbance when deep-rooted native plants are cleared and replaced by introduced shallow-rooted crops that use less water. More water then moves below the root zone, raising groundwater levels to a higher point in the soil profile and remobilising salts in the higher, previously unsaturated zone. Salt usually then appears at the surface after evaporation follows periods of waterlogging.

For Western Australia, the estimated area of salinised land in 1994 was 1.6 million ha or nine per cent of the area of cleared agricultural land in the State (see the box opposite). The Murray–Darling Basin Commission estimated that about 200 000 ha of the Basin suffered from dryland salinity in 1992, although this is now regarded as an underestimate (MDBC, 1993a). In South Australia and Victoria, estimates of areas affected by dryland salinisation stand at about 400 000 and 150 000 ha respectively.

While no one has completely documented the scale of the problem in New South Wales and other States, in 1992 about 20 000 ha were recognised as salinised in New South Wales (MDBC, 1993a). In Queensland, about 10 000 ha were estimated to be affected in 1990, and Tasmania has only recently mapped small-scale scattered pockets of dryland salinisation. In South Australia, 25 000 ha are salinised, and a much larger area has some evidence of salting (State Dryland Salinity Committee, 1990).

Rate of watertable rise

The most comprehensive available information on rates of watertable rise exists for the irrigation areas of the Murray–Darling Basin. Watertables are rising at the greatest rate (about 100 to 500 mm per year) in the developments of the south-eastern parts of the Basin. It is not yet clear whether the rate has slowed as a consequence of improved irrigation efficiency or because of the dry conditions of the last few years. But the rise in salt transport rate in the Murray River at Morgan is slowing (see Fig. 7.29), which may indicate such changes.

Although data on rates of watertable rise in dryland catchments are very limited, long-term observations indicate rises of many tens of metres in some regions since clearing began. Observation indicates that groundwater levels have increased by up to 30 m since the 1880s in parts of south-eastern Australia and by about 20 m in the south-west of Western Australia.

Salt load in streams

In south-west Western Australia, the average rate of rise of stream salinity has ranged between one and 76 mg/L/year over the last few decades (Schofield, 1989). The 1992 State of the Environment Report for Western Australia quotes rates of increase between 11 and 117 mg/L/year for 17 major streams in the south-west of the State for the period 1965–1986. Elsewhere, very few reliable data quantifying the rate of change of stream salt loads exist. In the Murray–Darling Basin estimates of the rate of increase in salinity of the Murray River at

Morgan, made in 1985, suggested a figure of about 1 mg/L/year, compared with an annual fluctuation of about 200 mg/L in river salinity. Preliminary analysis of salt loads in other streams in the Basin shows similar trends.

In South Australia, increasing salinity has made several reservoirs useless for human consumption, and the cost of treatment for Adelaide's water supply is likely to increase as catchments in the Lofty Ranges become salinised. About 21 per cent of the divertible surface water in the State has a salt concentration of about 1500 mg/L, which is above the National Health and Medical Research Council guidelines for human consumption (State Dryland Salinity Committee, 1990).

The message about catchment pollutant sources

- Catchments are the sources of a range of aquatic pollutants, including sediments, nutrients, salt and pesticides.

- The natural character, human disturbance and management of catchments control the nature and amounts of pollutants. Intensive uses produce the most pollutants.

- Sources of sediments and phosphorus include land adjacent to streams, channel beds, banks and gullies. The relative magnitude of these sources varies across the country and through time.

- Fine sediments usually transport other pollutants such as phosphorus, pesticides and pathogens.

- Nutrients also come from point sources like sewage-treatment plants, feedlots and urban and industrial run-off and discharge.

- The relative importance of diffuse and point pollutant sources varies between catchments and with run-off, with point sources being more important in low-flow conditions and in areas of urban development and animal waste disposal.

- The spread of salinised land is increasing salt loads to rivers in many parts of Australia, although the rate of increase of load in the Murray River might have begun to slow.

- Increasing agricultural use of pesticides will result in increasing levels in waters with unknown biological and health consequences.

- Other pollutants such as trace metals and synthetic organic chemicals are important in some localities, although their biological and health consequences are largely unknown.

- The leaching of nitrate to surface and groundwaters from sewage, intensive animal industries, food-processing, fertilisers and natural sources is a potential human health hazard in some areas.

Impacts of forestry

As in all cases where the vegetation cover of a catchment is changed, forestry — both native forests and plantations — inevitably affects soils and water. Although it occurs in lowland regions of Australia, the impacts on water quantity and quality are likely to be most extreme in the uplands where run-off is produced.

Water quantity

Plants draw water from the soil, and transpire some of it to the atmosphere. As well, they catch rain on their leaves, from where it can evaporate. Through these processes (known as evapotranspiration), surfaces covered by vegetation lose more water than those that are bare. When a forest is removed, more water flows into streams, from run-off and soil seepage. Vigorously growing young trees use a lot of water. The amount depends on the number and type of trees, and can be greater than that used by a mature native or plantation forest.

In forested areas the flow of water in streams increases immediately after clearing and logging (Bosch and Hewlett, 1982), and then is reduced as regeneration begins. Fire can produce results similar to clearing and logging.

Some of the differences between initial water-yield increases in different catchments are due to different forestry practices. Among these, the most significant are in the area of logging. Run-off has a tendency to increase as the logged area of a catchment increases.

Native forests in Australia are a mosaic of untouched, logged and regenerating patches, producing a stream-flow response that is difficult to relate to specific forestry operations. Few streams have large entirely forested catchments, and agriculture also influences flow rates in large catchments. The only way of describing the state of water yield as a result of forestry land use is therefore by using small research catchments, of the kind illustrated in the diagrams, where the individual effects can be isolated and measured.

Water quality

Increased run-off and flow through soil after logging can increase the sediment load of streams by increasing slope erosion. It may also increase stream size, and so sediment that moves out of forests can be derived from the streams themselves (Olley *et al.*, 1995). In spite of this, few quantitative data are available to demonstrate relationships clearly. A recent review (Campbell and Doeg, 1989) concludes that timber-harvesting operations have significant effects on stream sediment levels, water quantity, water temperature, nutrients and aquatic biota. The effects appear to be site-specific, with some studies (Grayson *et al.*, 1993; Olive and Rieger, 1987) failing to demonstrate any effect of forestry operations on water quality because of very high natural variability or good forestry practice.

During the establishment phase of plantations, erosion of forest roads, hauling of logs downhill and clear-felling cause increased turbidity and sediment loads (Boughton, 1970; Cornish, 1989). Elevated sediment loads are therefore likely to occur at a number of times during the growth/harvesting cycle of both native and plantation forests. However, few studies in Australia last long enough to detect these changes. Studies in other countries show that the effects are still evident 20–50 years after logging, in the form of elevated stream sediment loads and woody debris (Beschta, 1978). Only one Australian catchment has been studied for more than 30 years, at Coranderrk near Melbourne (Doeg and Koehn, 1990).

Ecological impacts

While changes in flow regime and water quality associated with logging activities have been shown to cause significant impacts on stream ecosystems in other regions of the world, research in this area in Australia is still in its infancy (Campbell and Doeg, 1989). Recent studies of streams in Western Australia, Tasmania and New South Wales have shown that logging and associated road building cause significant changes. For example, spread of tree disease, changes in characteristics of bed sediments, amount of wood debris, water temperature and shading that are associated with declines in the abundance and diversity of macroinvertebrates and fish (Richardson, 1985; Growns and Davis, 1991; 1994; Davies and Nelson, 1993; 1994).

Chemicals are widely used in forests to control competing plants and pest insects. Commonly used chemicals include the atrazine and glyphosate herbicides, and a variety of pyrethroid and other insecticides. These chemicals have been shown to cause disturbance and mortality of stream fish and invertebrates (Barton and Davies, 1993; Davies *et al.*, 1994). In some areas public concern is high over the contamination of untreated rural domestic water with chemicals used in forestry. Concentrations of 1–100 µg/L of atrazine were regularly recorded in some Tasmanian plantation forests (Davies *et al.*, 1994).

Responses

Although people have long recognised the threat that forestry operations pose to water resources, they have rarely considered the value of water or the role of forests in ecological processes, in calculating the value of forests. Responses in some States, particularly Western Australia and Victoria, have been to carry out research into water quantity and quality in relation to forestry operations. Victoria, Western Australia, New South Wales and Tasmania now have codes of forest practice and practical guides for field operations to protect water quality. Education and enforcement of these codes is essential. Where water supply either is at risk (for example, in the mountain ash forests near Melbourne) or could be enhanced by thinning (for example, part of the Perth supply), careful research has played a key role in identifying options. The closed-catchment policy for Melbourne's supply has been maintained because of the risk posed by logging and human recreation (O'Shaughnessy and Jayasuriya, 1987). A similar policy exists for Sydney and Perth.

In general, forest activities and their impacts have not been explained to the public and research has been either non-existent or patchy, with research results often neither analysed nor published. Under such circumstances, it is difficult to be confident that policies and practices are appropriate. Continuing public protest over logging is partly a response to a lack of information, among other concerns.

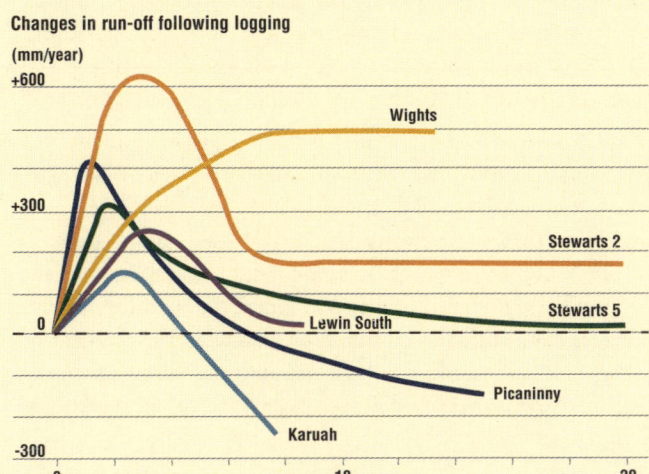

Changes in run-off following logging

Generalised water run-off curves from small research catchments show responses to logging. Stewarts 5 (Vic.), Picaninny (Vic.), and Karuah (NSW) all show flow increases after logging, then a decrease to values lower than the pre-logging flows in the last two cases. That is, streams have less water in them as a result of forest regeneration. Lewin South (WA) shows an increase and then decrease, but groundwater played a more significant role than in the Victorian and NSW cases. Wights (WA) and Stewarts 2 (Vic.) were converted from forest to pasture, but show different responses. Groundwater in the Western Australian case played a major role in maintaining flow at a higher level than would be expected once pasture had established, a factor that is less important in drier areas. Curves are smoothed from the original data.

Sources: Karuah, NSW (Cornish, 1993); Picaninny, Stewarts 2 and 5, Vic, (Nandakumar and Mein, 1993); Lewin South and Wights, WA (Ruprecht and Stoneman, 1993.)

Changes in run-off following wildfire

Long-term water yield following wildfire in mountain ash (*Eucalyptus regnans*) forests of Victoria. After the 1939 fire, an initial rather small increase occurred in stream flow (run-off) in catchments in the Melbourne water supply region, then a period of decreased flow for 20–30 years as regrowth used water that previously replenished streams. Gradually flow increases, and should reach the pre-fire values after about 130 years — assuming that no other disturbances to the forest arise.

Sources: Kuczera (1987), Jayasuriya *et al.*, (1993).

Groundwater: a vital but often neglected resource

People living in about 60 per cent of the Australian continent are totally dependent on groundwater, and those in another 20 per cent use more groundwater than surface water. The volume of groundwater in the upper one kilometre of the earth's crust is 10 times larger than the combined volume of all rivers and lakes on earth. Groundwater allows streams to flow through prolonged dry periods and yet, for many people, it has a certain mystical quality, perhaps because it cannot be seen until it is pumped or flows to the surface. This mystique is reflected in the widely accepted but scientifically baseless practice of water-divining.

Despite its vital role in supporting communities and industries, people often abuse groundwater. They often regard it as free and unlimited in quantity and generally do not appreciate that groundwater can be over-used and become polluted. Not long ago the tendency in waste disposal was to deposit waste underground and to ignore any consideration of leakages from surface facilities, presumably on the premise 'out of sight, out of mind'. The huge groundwater clean-up costs that many countries are now facing provide ample evidence of the folly of these practices.

Groundwater is an integral component of the hydrologic cycle and constitutes the largest terrestrial water store in that cycle. At its source, it is inextricably linked to the surface environment. It sustains many wetlands and supports vegetation. In turn, disturbances to the surface environment affect aquifers. Management of land and water resources must therefore be approached in an integrated way that takes full account of both groundwater and surface water.

At the broadest level, the way in which groundwater occurs in aquifers is not difficult to understand. Aquifers may be unconfined or confined, and may be further classified as 'surficial', 'sedimentary basins' and 'fractured' (see diagram below). They are all susceptible to pollution where they intersect the surface, but surficial ones are at greatest risk in the short term because they are hydraulically connected to the surface over much larger areas proportionally, and through shorter flow paths.

A key factor in groundwater management is the very much longer time period for aquifers to respond compared with surface waters. For example, it takes a matter of weeks for water to travel in surface streams from inland of the Great Dividing Range in Queensland to central Australia after flooding rains. In the Great Artesian Basin it takes a million years for water to travel a similar distance. While surficial aquifers react significantly more rapidly, the slow response time still needs to be recognised in managing them. In particular, cleaning up polluted aquifers is a prolonged and costly process.

It is not hard to find examples of groundwater contamination in other countries. In Bangkok, the large resource underlying the city is subjected to major pollution from human and industrial waste products deposited in the recharge areas above it as well as from salt-water intrusion from the sea because of overpumping for water supply. In Europe, both industrial and agricultural contamination are widespread. In many places, the nitrate level in groundwater has soared above the 50 mg/L permissible drinking water limit and continues to increase. Given the large number of possible contaminants, and the fact that several hundred new organic molecules are fabricated and released into the environment each year, this obviously presents a control problem that can only be tackled at the national or international level.

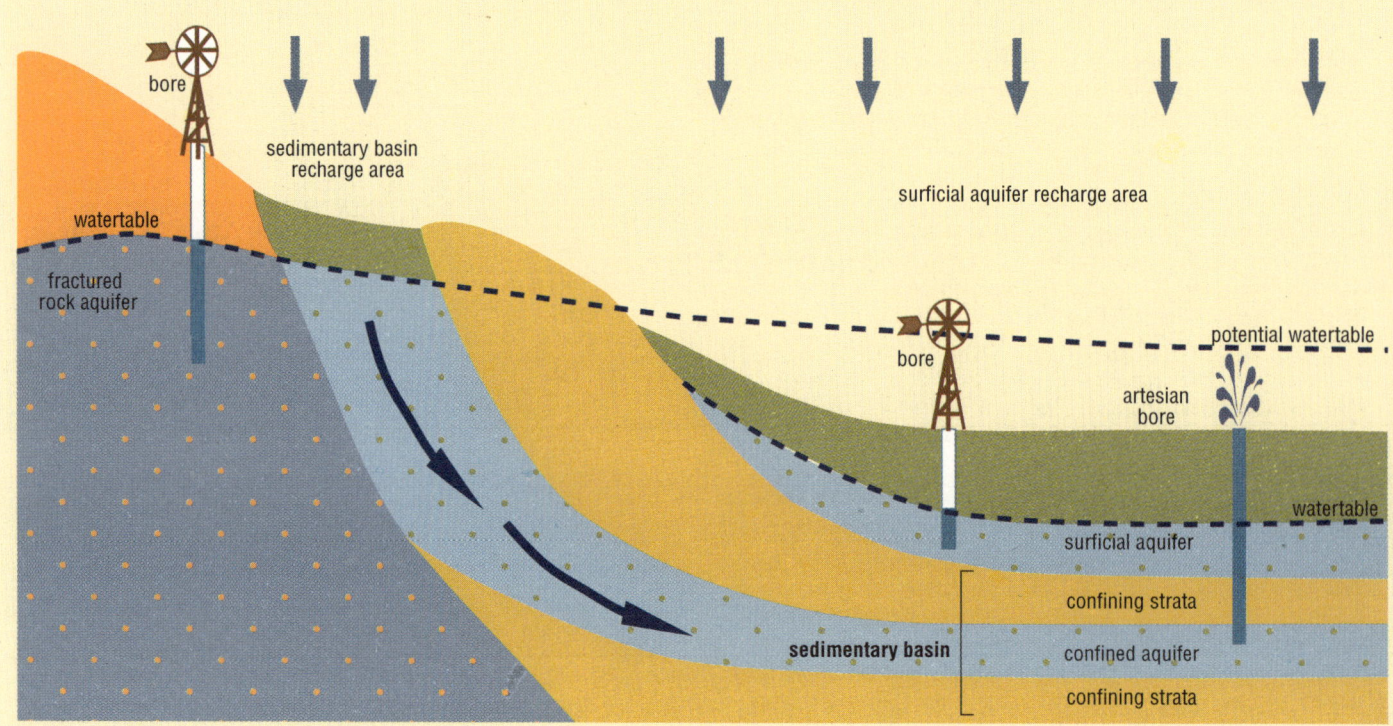

Running bores in Great Artesian Basin

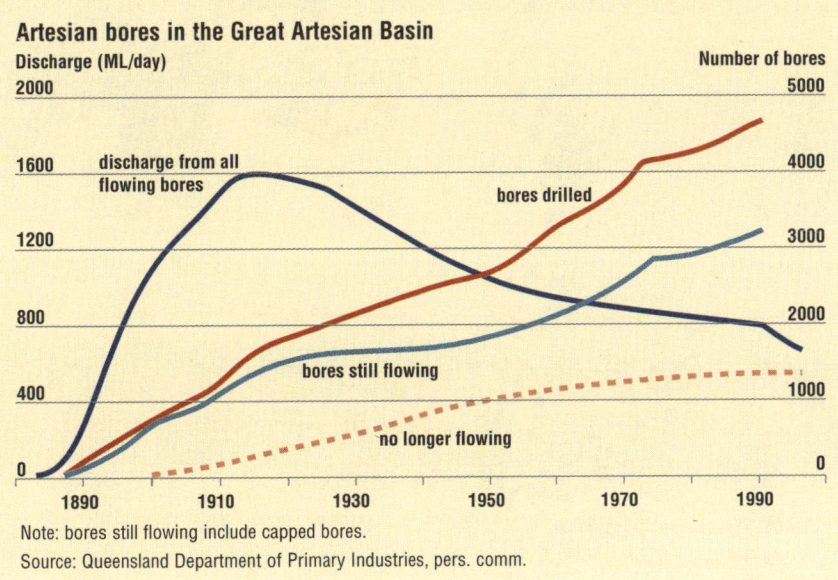

Internationally, the quality of surface water in some locations has improved substantially in the last 30 years — thanks to the construction of waste-water treatment systems and the recognition of the need to protect and rejuvenate catchments, streams and their ecosystems — but the same cannot be said about groundwater. Over the same period groundwater quality has decreased markedly. Its usage has increased in the last 30 to 40 years — doubling in the United States since 1950. In Australia and worldwide, aquifer systems are suffering from dual pressures; the level of exploitation of the resource is increasing at the same time as they are sustaining increased levels of pollution. In Australia there is still time to reverse these trends if action is taken soon.

The Great Artesian Basin — the lifeblood of much of eastern inland Australia — is recharged mainly along the western slopes of the Great Dividing Range running parallel to the New South Wales and Queensland coastlines. It lies under one-fifth of the continent, but at considerable depth (1000–2000 m) in the areas where its water is used. Uncontrolled flow from bores in the Basin has caused major and unnecessary depletion of water resources (see figure below). A program of bore capping is under way, which will gradually stem the water, although at considerable cost.

If pollution were to occur in the Great Artesian Basin, the time scales already mentioned make it obvious that a clean-up would be almost impossible to achieve. It is therefore a matter of concern that in 1990 the Australian Water Resources Council listed the New South Wales recharge area as a possible site of major groundwater contamination. It listed dissolved solids, metals, pesticides, and nitrate as contaminants, and agriculture and mining as sources (AWRC, 1990).

The Burdekin River delta in Queensland is the site of a major surficial aquifer formed as the river changed its course over the centuries and deposited water in alluvial materials near its mouth. The aquifer was first used to irrigate sugarcane in the late 1800s. The area under sugarcane expanded spasmodically in the first half of the 20th century and after several dry years, such as occurred in the 1930s (see Fig. 7.8), the aquifer showed signs of stress. After substantial expansion of the cropped area and corresponding increases in water withdrawals in the early 1960s, it was clearly under threat, with sea-water intrusion on a large scale a potential consequence. This led to a detailed assessment of the water balance and trials of artificial recharge as a means of supplementing natural recharge. The Burdekin is now the site of the largest artificial recharge operation in

Australia, which has sustained agricultural output for more than 25 years.

On the Swan Coastal Plain in Western Australia, groundwater from mainly unconfined aquifers provides about 60 per cent of Perth's water requirements as well as playing a vital environmental role. Rainfall is the main source of recharge while extraction by bores is the largest withdrawal component. About 80 000 private bores are unlicensed and unregulated, and rough estimates of their use for gardens suggest about one megalitre of water per garden per year. Aquifer management attempts to maintain water supply and to safeguard environmental requirements but management to date cannot take account of the unlicensed bore extraction. Authorities have identified and investigated several sources of pollutants, including nutrients from septic tanks and organics from petroleum product storage leaks (Barber *et al.*, 1993).

Artesian bores in the Great Artesian Basin

Discharge (ML/day) — Number of bores

- discharge from all flowing bores
- bores drilled
- bores still flowing
- no longer flowing

x-axis: 1890, 1910, 1930, 1950, 1970, 1990

Note: bores still flowing include capped bores.
Source: Queensland Department of Primary Industries, pers. comm.

Kakadu National Park (above); a healthy wetland and Lake Wyangan, near Griffith, NSW; a degraded wetland

Figure 7.18 The distribution of wetlands in Australia

a. Permanent freshwater swamps

b. Permanent freshwater lakes

c. Intermittent freshwater swamps

d. Episodic fresh lakes

e. Land subject to inundation by fresh water

a. Wet, vegetated with persistent emergent species, water greater than 1m deep

b. Areas of open water generally greater than 1m deep, little permanent vegetation

c. Alternately wet and dry, with emergent vegetation when wet, less than 1m deep

d. Mostly dry lakes

e. Water present for short periods after rain; typical wetland vegetation not present

Source: after Paijmans *et al.*, 1985.

Habitat quality and biota

Wetlands

Wetlands are areas of land that are flooded naturally or are inundated or waterlogged on a permanent, seasonal or intermittent basis. They include marshes, ponds, lakes, billabongs, meadows and swamps. Australia has a wide variety of wetlands, many of which have unique features and are of high ecological value. Numerous species of birds, fish, amphibians and other aquatic life depend on them as habitats for survival (McComb and Lake, 1990). National inventories of wetlands were developed in 1985 (Paijmans *et al.*, 1985; ANCA, 1993), listing the various types and their locations (see Fig. 7.18). The Ramsar Convention lists 567 wetlands of international significance, 40 of them in Australia (Phillips, 1993). Although significant wetland areas are protected by various levels of reservation within national parks and World Heritage Areas, on a national scale wetlands are not well-protected from human impacts.

The extent and condition of Australia's wetlands have deteriorated greatly since European settlement by draining, changes to water regimes and increased sediment and nutrient inputs (Usback and James, 1993).

In Victoria, one-third of natural wetlands have been destroyed, including half of the area of non-permanent fresh-water wetlands. A survey of New South Wales coastal wetlands estimated a 70 per cent reduction of some types (Goodrick, 1970). In a 1992 survey, 38 per cent of New South Wales lakes were found to be degraded by nutrient enrichment and only 18 per cent were considered to be in 'good' ecological condition (Timms, 1992). Some 70 per cent of wetlands on the Swan Coastal Plain, Western Australia, have been lost since European settlement and a survey of those north and south of Perth has shown that most lakes and seasonally wet areas have been substantially or completely altered (Halse, 1989–94). In South Australia, drainage has reduced wetlands in the south-east to 11 per cent of their former area. A 1981 inventory of Tasmanian wetlands found that 63 per cent of the lowland wetlands recorded were slightly to severely disturbed (Kirkpatrick and Harwood, 1981), while the majority of larger Tasmanian water bodies are now artificial and subject to regulation.

Rivers

Australia has a wide diversity of rivers, ranging from upland perennial streams and lowland floodplain rivers to the ephemeral rivers of the arid interior. Most rivers in the lowlands and in agricultural catchments are degraded, with moderate to severe disturbance of riparian and channel habitats as well as increases in salinity, decreases in flow, changes in flow regime and increased nutrient loads. For example, in Victoria, the water quality and aquatic animal and plant life of most rivers and streams are seriously degraded (VSOE, 1988). This is worst in the central, north-west and south-west of the State, where high levels of turbidity, nutrient contamination and salinity occur, and where erosion and riparian vegetation losses are severe. A survey of the state of 27 Victorian river basins revealed that only 44 per cent (12) had more than half of their stream length in an excellent or good environmental category (Mitchell, 1990). A recent survey of the Maroochy River catchment, Queensland, found that only 14 per cent of stream length was rated environmentally as very good and the largest category (26 per cent) was classified as poor (Anderson, 1993). Such results are typical of many Australian rivers on the developed coasts.

By contrast, the northern Kimberley, Central Desert and Nullarbor Plain regions of Western Australia have suffered little disturbance or land clearing and rivers in these areas retain many of their natural values. However, rivers of the south and east of the State are frequently highly modified with multiple problems of high nutrients, sediment inputs and salinity (WASOE, 1992).

Creation and destruction of inland water habitats

Water storages and associated channels in Australia create a substantial amount of artificial wetland habitat, which is often of low ecological value and has an unnatural water regime. It usually also represents lost natural river or wetland habitat. Australia has more than 400 major dams greater than 10 m high (Crabb, in press). In Victoria in 1992, 83 400 ha were flooded for 2430 impoundments. In Western Australia, more than 80 000 hectares have been flooded (WASOE, 1992). In the nation's largest river system, 73 per cent of the length of the River Murray below the confluence with the Darling has been converted into a series of 10 weir pools (Thoms and Walker, 1992).

While irrigation and drainage channels also create aquatic habitat, it is frequently of poor ecological quality. In 1988, the Wimmera–Mallee area of Victoria had 16 000 km of open channels in domestic and stock water systems.

In 1980, the national estimate of farm dams was more than 400 000 (Timms, 1980). The number is increasing, with figures from the late 1980s putting the number of small farm dams in the south-west agricultural area of Western Australia and Victoria alone at more than 400 000 (Lane and McComb, 1988; VSOE, 1988). There are also many unrecorded and unlicensed farm dams.

Changes in wetland and river hydrology

Regulation has drastically altered the flow regime of many Australian rivers. The construction of dams and weirs causes changes in hydrology, and small farm dams and the operation of pumps in the main river channels and their tributaries also often affect flow. Only a few Australian rivers have not been altered hydrologically by human activities. For example, all 22 coastal drainages between Fraser Island (Queensland) and Lakes Entrance (Victoria) are impounded (Harris, 1985). Most unregulated rivers are in sparsely populated areas — that is,

Lake Corangamite

The condition of Lake Corangamite, a saline lake in western Victoria, has significantly declined since its major inflow stream, Woady Yaloak Creek, was diverted in 1980 (Williams, 1992). The diversity of aquatic biota and waterbirds subsequently declined by 1992. The salinity of the lake had increased by 40 per cent, water levels had fallen by about 2 m and many islands, which were previously used as nesting and refuge sites, had become peninsulas or disappeared. The biological community became typical of a high-salinity lake. Some elements of the biota — including species of shrimp, snail, fish and widgeon grass — virtually disappeared, decreasing the lake's value to bird life, and so both the diversity and abundance of birds also declined. Pelicans, ibis, spoonbills and cormorants no longer nested at the lake.

northern and central Australia.

The allocation of river flows for environmental purposes is now a major issue in Australian water management. The hydrological cycle links changes in the catchment, changes in river flow, and in riparian and channel habitats. Most Australian rivers have been degraded by a combination of these factors (Cullen *et al.*, in press).

Changes in Australian river flows may take the form of all or one of the following:

- decreases in the volume of discharge or occasionally, increases (for example, with inter-basin transfers)
- changes in and reversal of seasonal flow patterns (for example, higher summer instead of winter flows)
- reduction or enhancement of the natural variability of flows on scales from hours (in hydro-electric dams) to months and years (for large irrigation or urban supply storages)
- changes in the frequency, typically suppression, of small to medium floods
- changes in the form of floods, especially the rate of rise and fall

These changes affect the channel form, sediment transport, water quality, habitats and biota of Australian streams — from small creeks to large lowland rivers. Consequently, regulation of the rivers has had major impacts on their ecological health: changing biological communities; eliminating species and reducing water quality through erosion; reducing habitat diversity; disrupting biological processes linked to floods; and flooding or de-watering key habitats. For example, the reduced frequency of flooding on the floodplain of the River Murray has severely affected red-gum forests, with reduced growth, greater mortality and minimal regeneration (Cullen *et al.*, in press). With fewer

▲
Downstream of Gogeldrie Weir on the Murrumbidgee River; eroded, degraded riverbank.

flows, forest litter accumulates and floodwaters lose oxygen and can become toxic to fish. Since European settlement, reproduction and recruitment of fish in floodplain habitats of the Murray has been reduced and is now limited to small species such as gudgeons (Cullen *et al.*, in press).

Changes caused by humans in the water regimes of Australian wetlands include changes in water-level fluctuations (size, frequency and seasonality) and changes in water balance (input, output and seasonality), all of which affect their ecological health. Such changes have occurred to the majority of wetlands along the River Murray, the western Victorian lakes and those on the Swan coastal plain (Pressey 1986, Williams 1992, Froend *et al,.* 1993).

Significant environmental degradation commonly occurs when major water inputs are reduced in flow or diverted (see the box on page 7-27). However, prolonging wetland flooding well beyond the natural pattern may also have significant impacts. The wetlands of the River Murray have a total area of about 2200 sq km, including the Coorong and Lower Lakes (Pressey, 1990). Some 35 per cent of the area that used to be flooded intermittently now never dries out, and 11 per cent receives irrigation water and now may be virtually permanently wet.

Changes to the structure of inland water habitats

Wetland and river habitats rely for their ecological health on their physical environment being maintained. In Australia, physical destruction or degradation of inland water habitats is widespread due to the removal of riparian (bank-side) vegetation, increased bank erosion, river engineering — including realignment, straightening and desnagging — and construction of barriers to fish migration. While the impacts of these practices are well demonstrated in the River Murray, their effects are common to most Australian rivers.

Riparian zones influence habitat composition, stability and energy inputs, and act as 'filters' for the exchange of water, nutrients, sediments and pollutants between terrestrial and aquatic systems (Bunn *et al.*, 1993). The vast majority of Australian riparian areas have been degraded since European

settlement. Clearing of native bank-side vegetation for agricultural and urban development has been extensive in lowland sections of most of our rivers.

A recent survey of the Murray and its side-channels found that riparian vegetation was in generally poor condition due to clearing, weed infestation, soil salinisation, grazing and regulation of flow (Margules *et al.*, 1990). At least 30 per cent of the study area was cleared, and introduced weeds constituted 18–63 per cent of plant species. Regeneration of red gum and black box was also affected.

Bank erosion is widespread in Australian rivers, causing increases in sedimentation and changes in channel form, both of which have negative impacts on river health. The Murray suffers conspicuous bank erosion, where slumping of banks follows rapid drops in water level. An estimated 1.8 million tonnes of material fell into the lower Murray over a 153-km section in 1988–89 alone (Walker, 1992) and channel-widening has averaged 16 cm per year since 1977 in the Lake Hume–Lake Mulwala reach (Tilleard *et al.*, 1994).

Storages trap sediment, increasing river erosion immediately downstream. Impoundments trap up to 73 per cent of sediment in the Murray (Thoms and Walker, 1992), and at least 50 km of river could be deepened substantially from its sediment supply being trapped in the Hume Reservoir (Tilleard *et al.*, 1994). The 180-km Hume–Lake Mulwala reach conveys all the regulated flow destined for downstream use. Here, channel changes include lateral migration of bends, channel deepening and widening, and side-channel development. Regulation has also slowed development of key side-channels. The river bed between Hume Reservoir and Albury has deepened by up to 24 per cent since 1977. Natural changes in channel form are also occurring in the Murray — for example, near Mildura.

River engineering works, designed to provide improvements for human use, have often caused river ecosystems to deteriorate. Along the Goulburn and Upper Murray rivers some 870 and 400 stream-management works respectively, have been recorded (Vic. DWR, 1989), resulting in serious environmental degradation. Levee-bank construction separates rivers from their floodplains, as it has done for the Murray downstream of Mannum, and in many other areas of the country.

Desnagging — the removal of logs and wood debris from river channels, usually to facilitate the passage of water or boats — results in loss of biological habitat and increased stream erosion. Extensive desnagging was carried out in lowland reaches of the Murray–Darling until road and rail transport replaced paddle steamers (Pressey and Harris, 1988). Desnagging and willow removal in the Barmah Choke section of the Murray have reduced overbank flows at the Edward River and Gulpa Creek off-takes for a distance of about 20 km, affecting the health, growth and reproduction of trees in the Millewa Forest (Murphy, 1990).

Many species of inland-water fish migrate over long or short distances to complete their life cycles. Blocking of fish migration by dams, weirs and fords has resulted in population declines in many coastal catchments (Harris, 1984). Some structures have fishways, but many of these are not properly designed or maintained. Of 54 fishways in New South Wales rivers, only 10 have been assessed as effective. Twenty are recorded in the Murray–Darling Basin (see Fig. 7.19), but none is on high-level (10 m or more) dams and only two have been assessed as being effective as fish passages (Cullen *et al.*, in press). Golden perch and silver perch were once common as far upstream in the Murray as Lake Hume, but have now disappeared above Yarrawonga Weir. Dams also have the potential to cause genetic isolation of fish populations, with accompanying loss of genetic diversity.

Changes in water quality

While the main aspects of the state of inland water quality are discussed elsewhere in this chapter, several other related issues affect wetland and river health. Metal poisoning of wildlife from sources such as lead shot and mercury in sediments is a problem in several inland waters. Hunters put large amounts of lead shot into wetlands — for example, Bool Lagoon in South Australia receives four to six tonnes of it per year (Lund *et al.*, 1991) and Howard Springs in the Northern Territory receives 330 000 shot per ha (Whitehead and Tschirner, 1991). Lead poisoning from shot at Bool Lagoon killed magpie geese, swans and black ducks. Mercury, a leftover from gold mining, has been recorded from the sediments and biota of the Gippsland Lakes and tributaries, with some fish above safe limits for human consumption. In Lake Eildon, 70 per cent of large trout sampled had mercury levels above those safe for human consumption. Other sites at Walhalla, Wandiligong, Chewton and Maryborough also showed mercury levels in sediments high enough to predict high levels in fish (VSOE, 1988).

Figure 7.19 Location of fishways in the Murray–Darling Basin

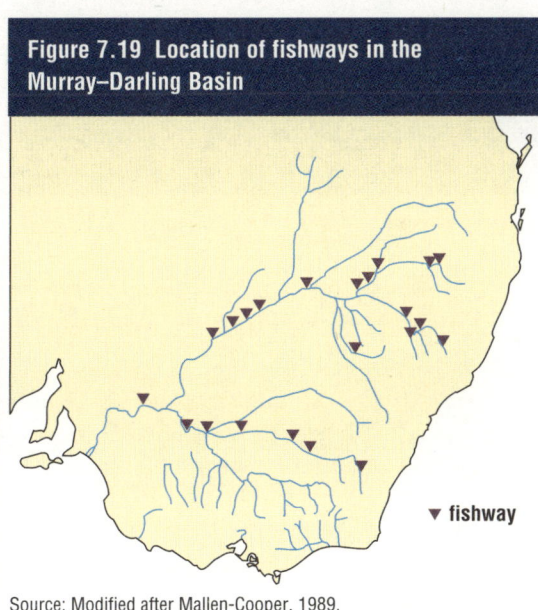

▼ **fishway**

Source: Modified after Mallen-Cooper, 1989.

Figure 7.20 The effects of the Hume Dam on water temperature in the River Murray

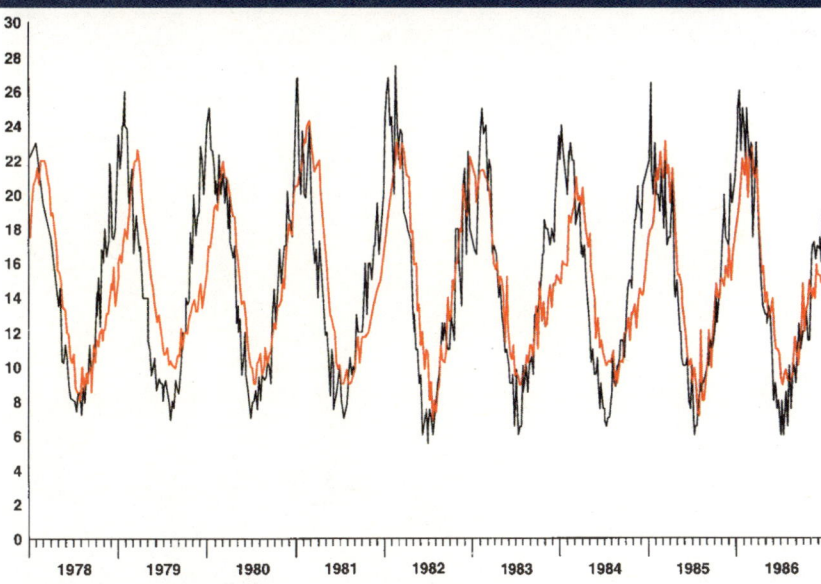

Source: MacKay and Shafron, 1989.

▲

The modified temperature regime in the Murray downstream of Hume Dam (red line) does not vary as much as the natural one (black line).

Other cases of ecological impacts from metals include acid discharges from several Tasmanian mines (most notably Mt Lyell), zinc pollution from the Captains Flat mine in New South Wales, and acid drainage from the Rum Jungle mine in the Northern Territory polluting the Finniss River (Jackson, in press). While its impact is not significant on a national scale, mine discharge has had major local environmental effects on several rivers.

Changes in river temperatures and oxygen levels downstream of storages also affect the health of Australian rivers (see Fig. 7.20). The modified temperature regime in the Murray downstream of Hume Dam does not vary as much as the natural one, and seasonal changes are offset by about one month (Walker, 1985). Such effects are evident for between 40 and 100 km below several dams in the Murray–Darling Basin (Mackay and Shafron, 1989). Summer oxygen levels reflect the discharge of oxygen-poor bottom water. Such changes have caused major disturbances to the biology of several Australian rivers. Examples include the loss of native fish and invertebrate species and their replacement with exotic species such as carp, trout (*Salmo* and *Oncorhynchus* sp.) or redfin perch (*Perca fluviatilis*) (Cullen *et al.*, in press).

Many Australian rivers and wetlands are saltier than they were before European settlement and some are fresher, causing changes in their fauna and flora (Hart *et al.,* 1990, 1991). Salt-affected streams occur throughout half of Victoria and predominantly on the lower-lying plains to the north and west of Melbourne (VSOE, 1988). Salinity levels in the Yass River, New South Wales, are increasing by seven per cent per year (NSW SOE, 1993), while salinity in the River Murray increases downstream (see Fig. 7.28). In Western Australia, many rivers are more saline inland (for example, the Kalgan), and fifty per cent of south-west rivers (those that rise in agricultural areas) have increased salinity (see Fig. 7.21).

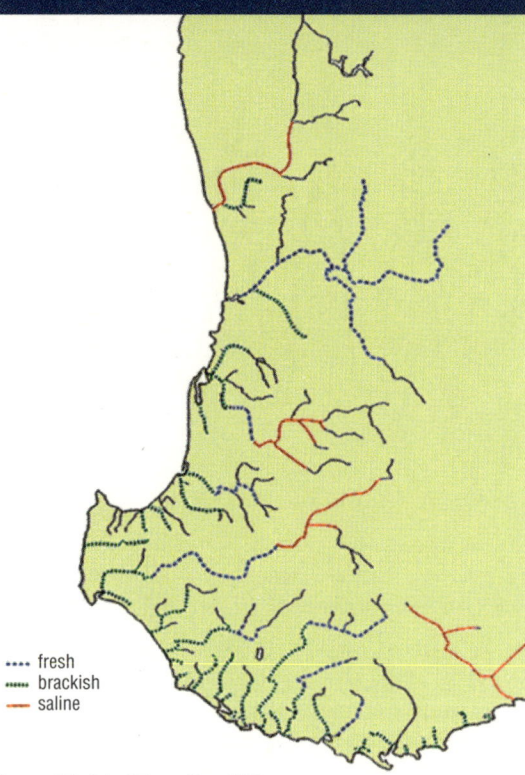

Figure 7.21 Severity of salinity stress in Western Australian rivers

.... fresh
..... brackish
— saline

Source: WA Select Committee, 1990

Inland water biota

The complex and multiple changes to the quality and extent of Australian inland water habitats have caused fundamental shifts in the structure and function of ecosystems and the composition of biological communities. There is certainly scientific evidence for major changes in Australian inland-water-ecosystem structure and function as a result of changes in habitat. But, as yet no published information is available on the extent to which such changes have occurred at a national or even regional level, other than for wetland and riparian vegetation. Indicators of change in the biota of inland waters include changes in the distribution and abundance of native fauna and flora and measures of the impacts of introduced and displaced biota, including aquatic weeds and faunal pests. While many specific examples exist of declines in the biological values of Australian inland waters associated with forestry, pesticide use, sewage effluent, mining, dam construction, increased salinity and river regulation, much effort is still required at a national and State level before the state of wetland and river ecological health can be reported in a systematic and consistent manner.

Although scientists routinely carry out widespread monitoring of water storages for plankton, there is little information on the composition of algae in other inland waters that can be used to assess the impacts of human activity. However, it is likely that the incidence of algal blooms, particularly of nuisance blue-green algae, has increased, suggesting widespread reduction in the diversity and stability of ecosystems in agricultural and urbanised catchments (see the box on page 7-48).

The composition and abundance of planktonic algal communities varies widely over different places and times, making it difficult to measure trends.

Habitat degradation has reduced the range and diversity of many aquatic plant species, especially in eastern Australia. Of our many different aquatic plants, five species are currently considered endangered (ANZECC, 1993).

Macroinvertebrates, which include aquatic insects, are both diverse and abundant in inland waters. They are surveyed to assess human impacts on aquatic ecosystems. Their communities are often affected by changes in water and habitat quality in Australian wetlands and rivers. For example, a study of Gowrie Creek, south-east Queensland, showed that the macroinvertebrate diversity declined by 90 per cent, with only two or three families present below a sewage outfall (Cosser, 1988). Such systems recover downstream, usually with marked changes in the types of species present. Studies of forestry impacts on Tasmanian and Western Australian streams show large changes in the abundance and diversity of macroinvertebrates, mainly associated with sediment inputs from logging and riverbank disturbance (Growns and Davis, 1994, Davies and Nelson, 1994) .

Scientists have also recorded changes in the distribution of many invertebrates. The River Murray crayfish (*Euastacus armatus*) has declined in range and abundance since the 1940s (Walker, 1985) and 13 out of 14 native snails have disappeared from the banks of the River Murray due to artificial changes in water level (Walker, 1994). The Tasmanian giant crayfish, *Astacopsis gouldi*, is also declining in range (see Fig. 7.22).

Native fish species have suffered declines in abundance and diversity in most regions of Australia since European settlement. Surveys in Victoria indicate that only two out of 17 segments of river basins still have high-quality native river fish populations (VSOE, 1988). Similar evidence exists for most of the country. Most species of the lowland river fish, including the murray cod (*Maccullochella peeli*), trout cod (*Maccullochella macquariensis*), Macquarie perch (*Macquaria*

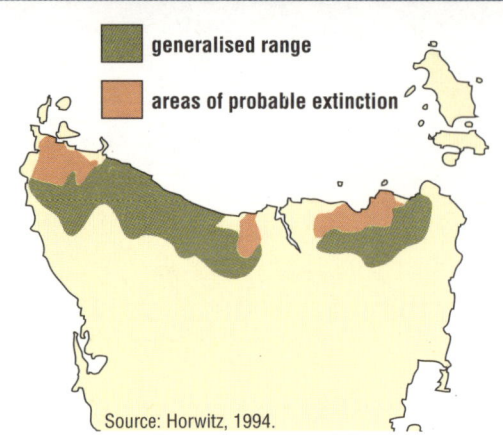

Figure 7.22 Range of the giant Tasmanian crayfish

generalised range

areas of probable extinction

Source: Horwitz, 1994.

australasica) and barramundi (*Lates calcarifer*), have declined in range and abundance (Jackson, in press). Some 33 per cent of inland-water native fish species have undergone major reductions in range (Wager and Jackson, 1993).

In Australia, frogs are declining in number and abundance, a trend also occurring in many other areas of the world (Barinaga, 1990; Tyler, 1991; 1994). Thirty-two species have been recorded as being in decline, with only limited data available for many others. Primary causes are decline in wetland and riverine habitat and water quality. Several species of aquatic tortoise and lizards are currently listed as endangered or vulnerable (Cogger *et al.*, 1993).

The abundance of water birds varies greatly, mainly in response to patterns of rainfall (Norman and Nicholls, 1991). Large variations in population estimates make it impossible to assess trends from current data. Populations are frequently large; Western Australian waterbird numbers were estimated at around 1.5 million in both 1989 and 1991 (Halse, 1989–94).

Three aquatic mammal species are found in Australia: the platypus (*Ornithorynchus anatinus*), water rat (*Hydromys chrysogaster*) and false water rat (*Xeromys myoides*). Of these, only the platypus is restricted to fresh water. Decline in platypus abundance or range indicates severe changes in environmental conditions. Platypuses are still known throughout their original range, but frequently have locally reduced populations — for example, in the lower Murray and Murrumbidgee rivers (Jackson, in press).

Exotic/displaced fauna and flora in inland waters

The introduction of exotic aquatic fauna into Australia, and the movement by people of native species or stocks to areas outside their natural range, have a profound effect on inland water ecosystems (Arthington and McKenzie, in press). Fauna have been introduced mainly for recreational fishing and the aquarium and aquaculture industries. Twenty exotic fish species have established or are likely to establish self-sustaining feral populations. Many of them were introduced in the 1800s and early 1900s and their spread has often been helped by active translocation (McKay, 1984). The aquarium industry imports increasing numbers of exotic aquatic species each year — worth some $2.7 million in 1994 alone. Nine species of these aquarium fish have now established feral populations. Some, like the guppy, *Poecilia* sp., have wide distributions (see Fig. 7.23). Accidental or intentional releases of exotic aquarium species are seen as a principal cause of new introductions into Australian inland waters.

Populations of exotic fish have become established in all States and Territories, ranging from New South Wales with 18 species to the Northern Territory with only one (Arthington and McKenzie, in press). The most widespread are the brown trout (*Salmo trutta*), mosquito fish (*Gambusia holbrooki*) and

Wappa Dam, Queensland, before and after the weevil *Cyrtobagus salvinaea* was introduced to control a *Salvinia* sp. infestation.

several species of cyprinids — the goldfish (*Carassius auratus*), European carp, redfin perch and tench (*Tinca tinca*). All of these species have aggressively expanded their ranges since first introduction, most with human assistance. Trout and salmon have been stocked intensively since the late 1800s, although climate has largely limited the expansion of their range. European carp has rapidly extended its range and continues to do so (Shearer and Mulley, 1988; Arthington and McKenzie, in press). The Boolara carp strain expanded rapidly into the Murray–Darling Basin during the 1970s and 80s and is now well established throughout Victoria (see Fig. 7.24). In the early 1990s, carp established self-sustaining populations in Tasmania. Goldfish are the most widespread of the exotic fish in Australia, being found in every drainage from the Fitzroy River in Queensland to the south-west of Western Australia and in Tasmania (Brumley, 1991).

Figure 7.23 Distribution of guppy and mosquito fish in Australian waters

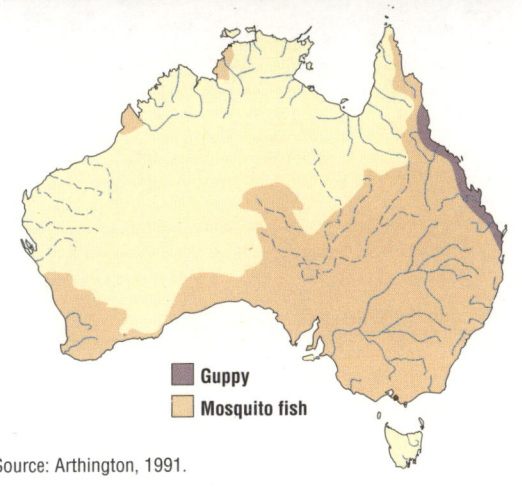

    ■ Guppy
    □ Mosquito fish

Source: Arthington, 1991.

The Lake Pedder *Galaxias,* an endangered native freshwater fish.

Direct evidence from Australia and elsewhere indicates that certain exotic fish species are hardy, opportunistic, readily dispersed, and have an impact on native biota and ecosystems. The conservation status of native fauna and the general state of inland waters partially reflect the establishment of exotic and translocated fish. However, many human pressures also affect aquatic systems. The impacts of exotic fish range from direct competition with and predation of native fauna, through habitat alteration and destruction, to acting as vectors for disease (Arthington and McKenzie, in press). Exotic fish have been implicated in the decline of nine endangered, eight vulnerable and five rare or common native fish species (Wager and Jackson, 1993). Trout are assumed wholly or partly responsible for declines in the abundance and range of nine species, as well as for changes in species composition and abundance of stream invertebrates. The negative impact of European carp on Australian inland waters includes the loss of aquatic plant species and communities in both rivers and wetlands (Roberts, 1993). This has been linked to loss of other fauna, changes in nutrient status, and algal blooms.

Exotic fauna pose a significant disease risk in Australian waters. Eight diseases or pathogens are known to be associated with established exotic fish species, while a further 10 have either been found in Australian aquaria or detected in quarantined aquarium fish, or are known to occur overseas in fish brought into Australia for the aquarium industry (Langdon, 1988).

Native crustaceans have been moved on a large scale, mainly in association with aquaculture or farm dams. At least eight species of crayfish of the genus *Cherax* are actively cultured in Australia, and several of these, most notably the yabby, the redclaw and the marron, have been shifted to large areas outside their natural range (Kailola *et al.*, 1993). The yabby (*C. destructor*) is the most widespread. Risks to inland waters associated with these moves include disease, physical damage to habitats through burrowing, competition with native crayfish and the transmission of symbiotic or parasitic organisms.

Other introduced fauna have also had significant impacts on inland waters. These include several waterbirds, the cane toad, fresh-water snails and other invertebrates (Arthington and McKenzie, in press). The cane toad (*Bufo marinus*) continues to expand its range in eastern and northern Australia and is now known to extend from the eastern end of the Gulf of Carpentaria to New South Wales. The New Zealand snail *Potamopyrgus niger*, first introduced into south-eastern Australia in the 1800s, has replaced native hydrobiid riverine snails in disturbed catchments.

Australian inland waters contain 65 species of aquatic plants regarded as weeds, about 15 of which are significant pests and 13 of which could become so (Arthington and Mitchell, 1986; Humphries *et al.*, 1991) A further five exotic species that are not yet present in Australian inland waters could pose a serious threat if introduced. The established species have been introduced and spread by several routes: intentional plantings for aesthetic, economic or

Figure 7.24 Range expansion of the Boolara strain of European carp up to 1988

- 1970
- 1980
- 1988

Source: Shearer and Mulley, 1988.

Figure 7.25 Known and potential distribution of alligator weed

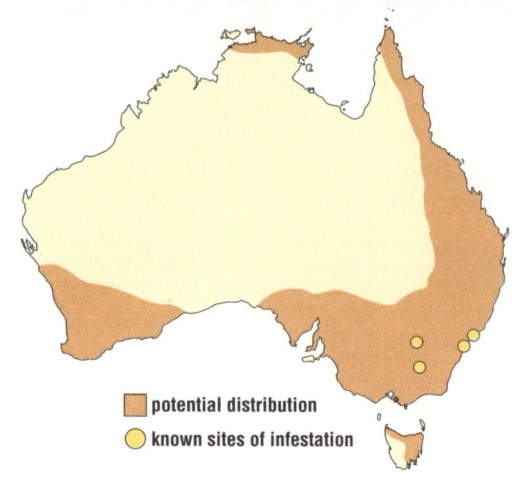

- potential distribution
- ○ known sites of infestation

Source: after Julien, 1995.

management reasons; accidental or intentional releases or plantings associated with agriculture or the aquarium industry; or accidental translocations from other sources.

Exotic aquatic plants are becoming more abundant in all States and Territories. Such increases may be due to recent local introductions or may be a response to deterioration in the quality of wetlands or rivers. Extensive cover of floating or tall shrubby aquatic weeds decreases the amount of light penetrating the water, causing declines in the diversity and abundance of native submerged plants, as well as reducing dissolved oxygen levels and temperatures, which have a negative impact on aquatic fauna (Jacobs, in press). Intense infestations of aquatic weeds also reduce flow rates in wetlands or rivers, decreasing mixing and enhancing sediment build-up, and thermal layering of the water body.

Some of the introduced aquatic plants posing significant environmental risks for Australian inland waters are the giant sensitive plant (*Mimosa pigra*) and the alligator weed (*Alternanthera philoxeroides*). *M. pigra* is causing a major and expanding infestation of Northern Territory wetlands and their riparian zones (Harley, 1992; Lonsdale, 1992). Alligator weed, introduced around 1947 in central coastal New South Wales has been found in locations from southern Queensland to the Australian Capital Territory. The most recent and extensive outbreak occurred at Barren Box Swamp in central New South Wales. A recent study of the potential spread of this species suggests that it is likely to become a more extensive problem over the next decade (see Fig. 7.25), although biological control has been instigated (Humphries *et al.*, 1991; Julien *et al.*, 1992).

There is little information on the impacts and spread of many aquatic weeds in Australia. The proportion of exotic, aquatic plant species in Australian wetlands varies from zero in remote areas to almost complete local dominance in some less-isolated and smaller wetlands (see Fig. 7.26). Where infestations occur, the area covered by exotic species is highly variable (Jacobs, in press), ranging from around two per cent for the lower Macleay and Clarence River wetlands up to 60 per cent for the Botany wetlands (all in New South Wales). The density of some of these weeds ranges up to 15 kg per sq m, and *M. pigra* annually produces about 0.8 kg dry weight per sq m.

Exotic plants also invade the riparian zone and floodplains, the most notable species being the rubber vine (*Cryptostegia grandiflora*) in northern Queensland, hymenachne (*H. amplexicaulis*), *M. pigra* and willows (Bunn *et al.*, 1993) (see page 4-19).

As with animal pests, the impacts of exotic weeds often cannot be separated from other impacts, to which the weed invasion may often be a response. Changes to wetland habitats through altered water

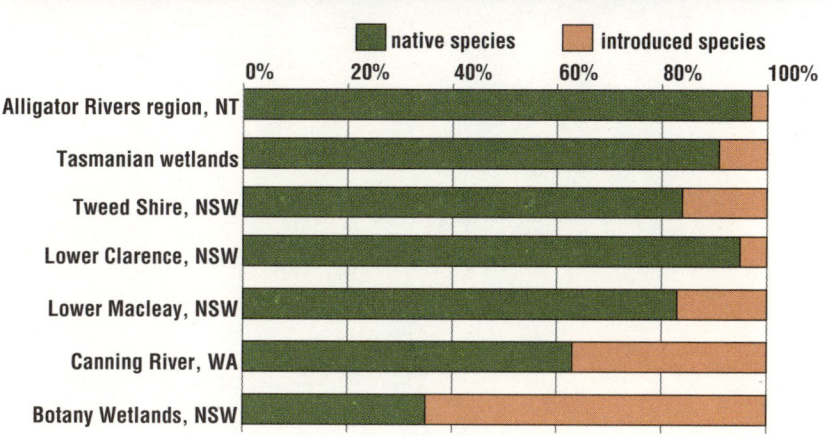

Figure 7.26 The proportion of native and introduced plant species in the various wetlands

Source: Jacobs, in press.

regimes or water quality and through physical disturbance to aquatic habitats and riparian zones can all favour the spread of introduced or translocated aquatic or riparian plants. Their quick growth responses to pulses of nutrients, high dispersal ability and ability to survive stress periods are also important.

The message about habitat quality and biota

- Aquatic habitat quality has deteriorated markedly in areas of agriculture, urban land use and substantial water regulation.

- In many parts of Australia (such as the wet tropics and mountainous areas) where such changes have not occurred, aquatic habitat is still of high quality.

- The area of natural wetlands has been significantly reduced since European settlement.

- Large areas of artificial wetlands (farm dams, sewage treatment ponds, reservoirs, and irrigation channels) have been created, but these are often of low ecological value.

- Regulation, physical barriers, erosion, desnagging, channel modification, introduced species, pollution and algal blooms have all substantially altered and degraded river habitat quality.

- The range and abundance of many species of native aquatic biota have declined significantly, to the point where many are threatened and endangered.

- The introduction, spread and establishment of a large number of exotic biota (such as carp, trout and *Mimosa pigra*) have exerted significant impacts on the biological communities and habitats of inland waters.

- No national or even regional system exists for reporting on the ecological health of Australian inland waters. Systems presently being developed are limited in coverage and developing slowly.

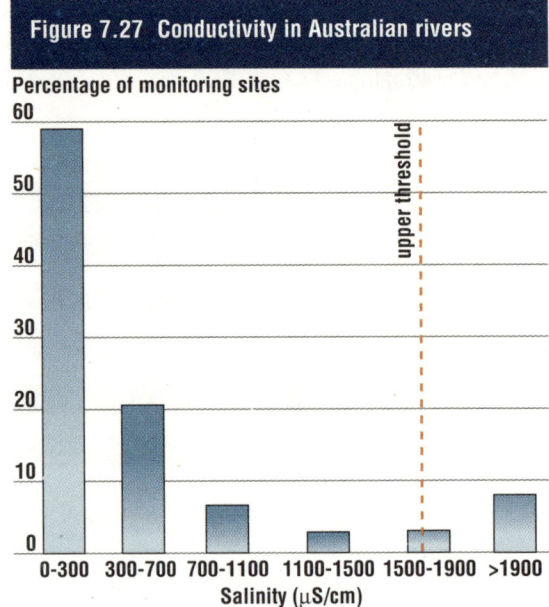

Figure 7.27 Conductivity in Australian rivers

Percentage of monitoring sites

Salinity (µS/cm)

Source: Liston and Maher, in press.

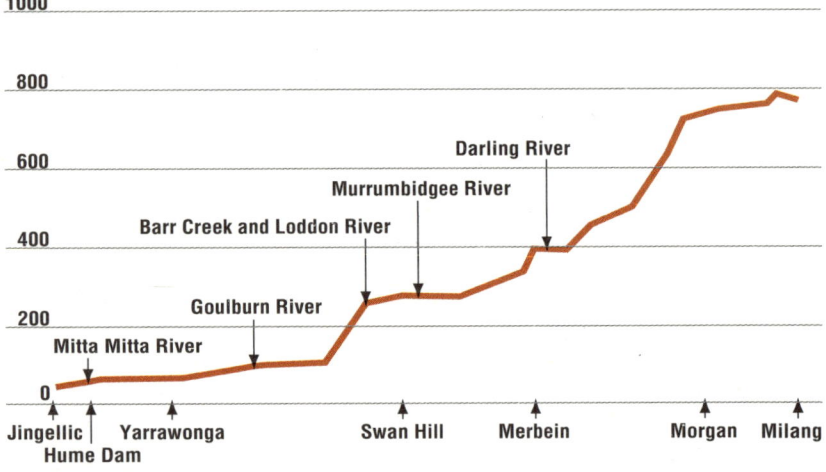

Figure 7.28 Downstream salinity profile of the River Murray

Salinity (µS/cm)

Source: Mackay and Eastburn, 1990.

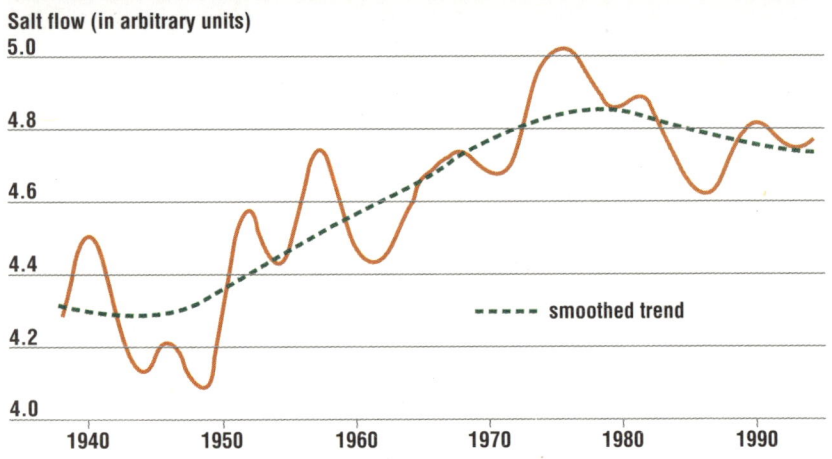

Figure 7.29 Generalised salinity trends in the River Murray at Morgan

Salt flow (in arbitrary units)

- - - - smoothed trend

Salinity increased from 1945 to 1977. Since then it has levelled out, coinciding with the completion of a number of salinity-mitigation schemes. It is not clear how much of the change is due to these schemes.

Source: Morton, 1995.

Water quality

The key environmental water-quality problems are: salinisation, nutrient enrichment, sediments, pesticides, trace metals and organic susbstances.

Indicators of environmental water quality

Indicators of the quality of water needed to maintain aquatic ecosystems have been chosen to reflect potential adverse impacts on water bodies of the problems identified above, and to be scientifically credible and easy to understand (Liston and Maher, in press) (see Table 7.9).

The Australian Water Quality Guidelines for Fresh and Marine Waters (AWQG), produced by the Australian and New Zealand Environment and Conservation Council (ANZECC) and other state of the environment reports provide thresholds for physical and chemical constituents that, if exceeded, are likely to produce an impact on the ecosystem (Liston and Maher, in press; ANZECC, 1992). The Guidelines provide a single guideline value for some indicators, such as conductivity. However, for most of the indicators used in this report the AWQG recommend a range of values, to reflect the marked differences that occur in natural water quality across Australia and the different conditions to which various aquatic ecosystems are adapted. In this section, water bodies are grouped into two categories based on their altitude — namely, upland and lowland ecosystems. For most indicators, upland aquatic ecosystems require water quality of a higher standard than lowland ones.

Australian waters — broad-scale issues

Artificially high salt levels are a major issue for many rivers in Australia. Aquatic ecosystems are adapted to particular salinity regimes and any changes can result in adverse impacts on the biota.

The data presented in Figure 7.27 confirm that salinity (as measured by conductivity) is a cause for concern in some rivers. Most of the sites where conductivity exceeds 1500 microSiemens per cm were on west-flowing New South Wales and Victorian rivers, and on Western Australian rivers. East-flowing coastal rivers in general do not show salinisation. Salinity of water storages has not been identified as a major national issue so no data on it are provided here.

Table 7.9 Indicators of environmental quality

Broad scale issues	Indicators
Salinity	Conductivity
Eutrophication	Total phosphorus,
Eutrophication	Chlorophyll a
Suspended sediment	Turbidity
Local issues	**Indicators**
Toxicity	Pesticides, Trace metals
Eutrophication	Biochemical Oxygen Demand (BOD)

All of these indicators can be related to potential or actual ecosystem impact.

Source: Liston and Maher, in press.

A more detailed examination of salinity in Australian rivers reveals that great differences can occur between rivers, in different parts of the same river (see Figure 7.28) and at different times at the same place (see Fig. 7.29). This variability is partly natural and partly caused by human activities. It indicates that care must be taken when using average levels of salinity or when trying to identify long-term trends.

As more nutrients enter rivers and storages, marked changes in aquatic ecosystems can occur, such as the more frequent occurrence of algal blooms — an indication of ecosystem degradation (see the box on page 7-48).

The key plant nutrients are phosphorus and nitrogen. In nutrient limited waters, small additions of these nutrients can lead to rapid plant growth. Phosphorus is usually the more significant nutrient and is commonly used as an indicator of potential algal problems (see Figs 7.30 and 7.31). Other factors influencing algal growth include light regime, salinity, temperature, the ways the waters mix and levels of nutrients, including trace elements such as molybdenum. A more common way to measure the amount of suspended algae in a waterbody is to measure the concentration of pigment (chlorophyll) in algae. In 75 per cent of lakes and storages, chlorophyll levels exceed those indicative of a healthy ecosystem. The link between the concentration of phosphorus in a lake and the amount of chlorophyll (or algae) is moderately strong. A number of storages have strong average relationships (see the box on page 7-48).

In this section turbidity is used as an indicator of suspended sediment. Increased suspended sediment may smother bottom-dwelling animals and plants, and may reduce light penetration, thereby favouring algal species such as blue-green algae.

Turbidity in Australian waters shows a significant natural variability, much of which can be related to changes in flow. Generally, the higher the flow the more turbid the water and hence the greater the sediment load in transport.

Australian waters — local issues

Although they are thought to affect only a limited number of aquatic ecosystems in Australia, local impacts are likely to be important. Water-quality problems are often linked to particular local activities (Liston and Maher, in press). The most important of these impacts on aquatic ecosystems are due to trace metals derived from mining, urban

▶

Natural differences occur in the concentrations of phosphorus in Australian waters. In general, upland waters should have the lowest natural levels and so should be below the lower threshold. Lowland waters can have higher natural levels of phosphorus and should fall below the upper threshold. For both eco-regions, a large proportion of sites have phosphorus concentrations above their relevant threshold. Of the sites that exceeded the threshold by a factor of two or more, most were lowland ones. Many of these water bodies are in regions used for grazing, intensive agriculture or urban development — activities that increase phosphorus loading to ecosystems. Most storages have phosphorus concentrations below the upper threshold, with some high concentrations, particularly within some upland storages.

◀ Acid mine drainage in the Mt Lyell area of western Tasmania.

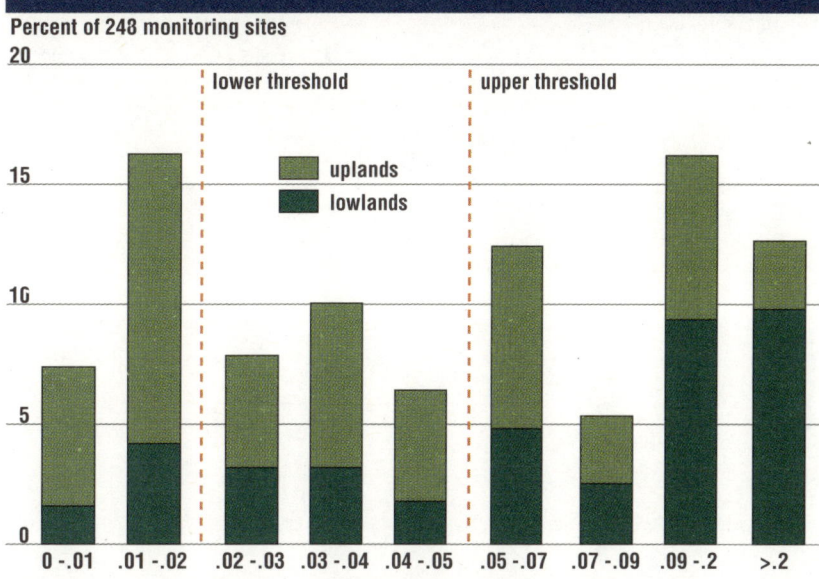

Figure 7.30 Total phosphorus level in selected Australian rivers

Percent of 248 monitoring sites

lower threshold   upper threshold

uplands
lowlands

Total phosphorus (µg/L)

0 -.01   .01 -.02   .02 -.03   .03 -.04   .04 -.05   .05 -.07   .07 -.09   .09 -.2   >.2

Source: Liston and Maher, in press.

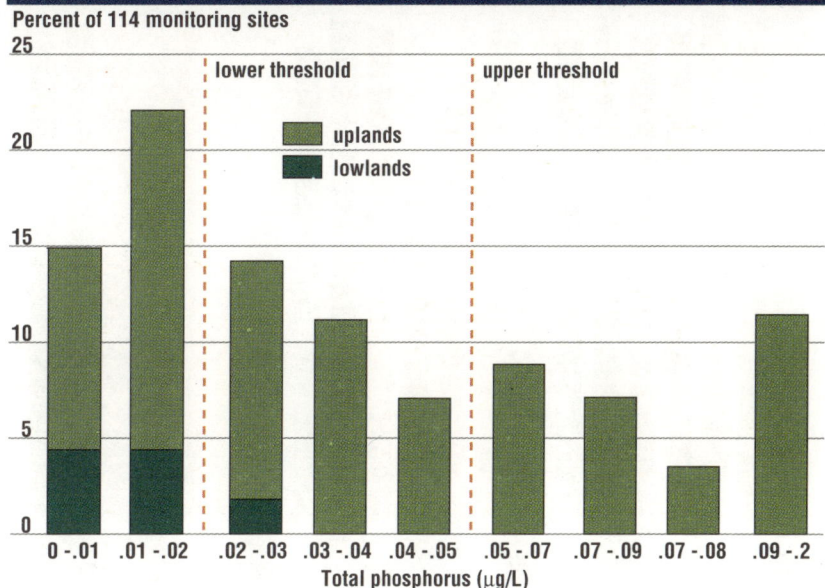

Figure 7.31 Total phosphorus level in selected Australian reservoirs and lakes

Percent of 114 monitoring sites

lower threshold   upper threshold

uplands
lowlands

Total phosphorus (µg/L)

0 -.01   .01 -.02   .02 -.03   .03 -.04   .04 -.05   .05 -.07   .07 -.09   .07 -.08   .09 -.2

Source: Liston and Maher, in press.

Table 7.10 Indicators and guidelines for drinking water quality

Indicator	Guideline level	Reason for concern
Microbiological indicators		
Total coliforms	0 CFU/100 mL	Indicator of faecal pollution
E. coli or thermo-tolerant coliforms	0 CFU/100 mL	
Chemical indicators		
Aluminium	0.2 mg/L	Aluminium flocs. Affects appearance and should preferably be reduced to 0.1 mg/L (potential health problems for dialysis patients)
Chlorination by-products (trihalomethanes)	0.25 mg/L	By-products induce tumours in rats and mice
Nitrate (as NO_3)	100 mg/L (50 mg/L in infants)	Can cause methaemaglobinaemia (blue-baby syndrome) especially in infants
Aesthetic indicators		
Turbidity	5 NTU	High levels affect palatability and appearance
Colour	15 HU	as above
Taste (as salt)	500 mg/L	affects palatability
Supply amenity		
Calcium and magnesium (as carbonate)	200 mg/L	Requires increased use of soaps; causes pipe corrosion or deposits
Iron	0.3 mg/L	Can cause staining of clothing and affects taste
Manganese	0.05 mg/L	as above

Notes:
CFU — colony forming units
NTU — nephelometric turbidity units
HU — hazen units

Source: NHMRC–ARMCANZ, 1995.

Figure 7.32 Bacteria, trihalomethanes (THM), turbidity, iron, manganese and colour in Australian water supplies

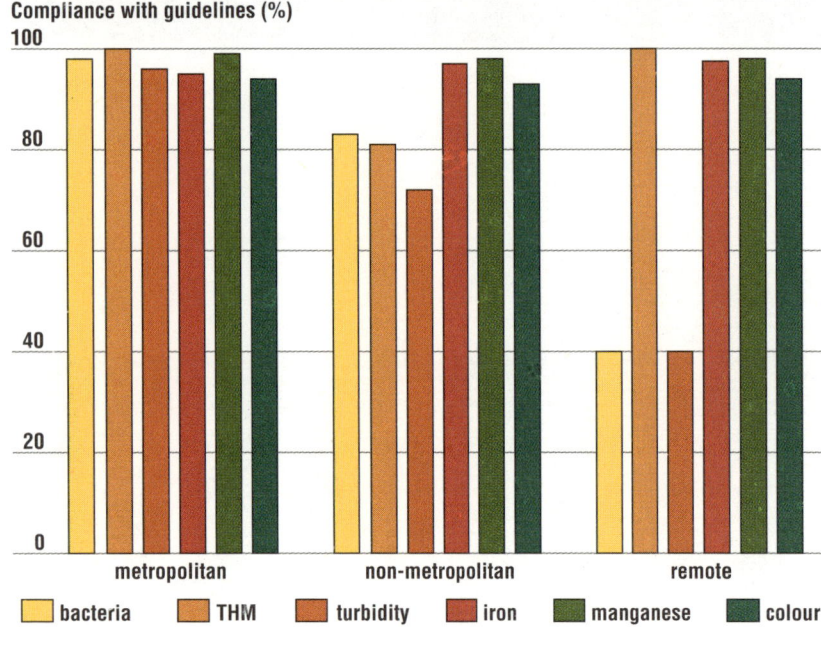

Compliance with guidelines (%)

Legend: bacteria · THM · turbidity · iron · manganese · colour

Source: Maher *et al.*, in press.

run-off or industrial activity, and to organic chemicals derived from industrial or agricultural activity. Examples of the former include trace metal pollution of the South Esk River in Tasmania and the Finniss River in the Northern Territory. Examples of pollution by organic substances include pesticides in the Namoi and Gwydir rivers in New South Wales, sugar-mill effluent in Queensland, milk, dairy wastes and paper pulping effluent. A commonly used measure of organic pollution is biological oxygen demand (BOD).

Stormwater from urban areas contributes a range of potential pollutants (trace metals, hydrocarbons, suspended sediment, bacteria and organics). Because watercourses are often physically modified (dredged, straightened etc.), the precise causes of degraded aquatic systems are difficult to determine.

Tap-water

Water for drinking should be free of disease-causing micro-organisms, harmful chemicals, objectionable taste and odour, and excessive levels of colour and suspended material (WHO, 1993).

Many factors control whether we have a consistently safe supply of drinking water. These include catchment protection and management, coverage of storage tanks, water treatment and maintenance of a reticulation system that prevents the entry, growth and transmission of pathogens. Storing water in reservoirs allows time for particles to settle out and pathogens to die.

Indicators that can be used to measure the effectiveness of these practices focus on microbiological and chemical health and safety, and on the aesthetic acceptability of drinking water.

Drinking water guidelines and indicators

The guidelines for drinking water are based on concentrations of constituents, that is, indicators, below which no harm or offence should occur to individuals. They provide the following benchmarks.

- Health criteria: water used for drinking, beverage preparation and cooking purposes should be free of harmful levels of toxic substances and pathogens.

- Aesthetic criteria: water to be consumed should be free of objectionable taste and odour, and excessive levels of colour and suspended solids.

- Amenity criteria: water used for washing, bathing and other domestic purposes should be free of gross microbial contaminants and should not contain excessive levels of staining, corrosive or scale-forming agents.

While it is possible to ascertain levels of performance against the benchmarks, it is generally recognised that the guidelines are conservative, with high built-in safety margins.

A comprehensive assessment of the 350 substances listed in the guidelines is too costly for many organisations. It is also unnecessary because most of these substances do not occur in Australian drinking water supplies.

The most commonly used indicators and guidelines are shown in Table 7.10. Obviously, this represents only a small fraction of the 350 substances mentioned above.

Assessment of Australian tap-water quality

In the analysis of drinking water quality it is useful to classify settlements into metropolitan, non-metropolitan, and remote settlements. Smaller communities reliant on surface water have inadequate resources to provide and operate water-treatment facilities, and few barriers are in place to ensure the removal of pathogens, turbidity and colour. For Aboriginal and other remote communities it is worth noting that supply of water has been identified as being at least as important as its quality (Federal Race Discrimination Commissioner, 1994).

Data for bacteria, turbidity and trihalomethanes show that, in general, metropolitan areas with protected water supply catchments and full treatment of water (coagulation, filtration and chlorination) perform well against the drinking water guidelines (see Fig. 7.32). The quality of water in rural areas, which rely on rivers and dams and lakes, is lower and more variable, reflecting the catchment land uses and standard of treatment. Settlements classified as remote, such as Aboriginal settlements, and isolated rural communities, show the greatest and/or most frequent non-compliance with guidelines, reflecting poor water treatment and poor source-water quality and availability. However, remote settlements have few excesses for trihalomethanes, a by-product of chlorination because their water supplies are rarely disinfected. For iron, manganese and colour (see Fig. 7.32), all Australian drinking water supplies for which data were made available show levels of compliance greater than 90 per cent.

Closed versus open catchments

In the latter part of the 19th and the early 20th centuries, several water supply authorities adopted a closed-catchment policy to protect water quality, with good effect.

With growth in populations and increased demand for timber resources and recreational opportunities, the policy has come under intense pressure in recent years, and restrictions over activities in some catchments have been relaxed. In some cases this is being compensated for by increased disinfection and monitoring of supplies from these catchments, as an added safeguard.

Adequacy of water treatment

Where a source water varies in quality over time and it exceeds the guidelines, it needs to be stored, managed and treated to ensure a reliable and safe quality of supply. The level and type of treatment are determined in relation to the potential health (see Table 7.12) or aesthetic concerns, and to the level of suspended solids (see Table 7.11).

Water treatment may consist of simple disinfection, or a sequence of flocculation, coagulation, sedimentation, filtration, and disinfection

Table 7.11 Water supply concerns and treatments

Health/aesthetic concern	Cause	Treatment
transmission of disease	bacteria, viruses or protozoa	storage, coagulation/filtration, disinfection
chlorination by-products (carcinogenic compounds)	chlorination and organics	activated carbon filtration, chloramination, ozonation
colour, taste, odour and algal toxins	organics, iron, bacteria and algae	oxygenation, dissolved air filtration, activated carbon filtration
turbidity	suspended solids	coagulation, filtration, storage
salinity	high mineral content	water de-ionisation
hardness	high calcium and magnesium	water softening

Source: Maher *et al.*, in press.

processes. Poor water clarity and high colour affect the general acceptability of drinking water and often result in customer complaints. High turbidity reduces the effectiveness of disinfection, necessitating higher doses of chlorine. Turbidity also increases the likelihood of pathogen survival. For this reason, performance against the guideline for turbidity is being more closely scrutinised.

Because development has intensified in many catchment areas, and controls on the movement of nutrients are inadequate, the incidence of algal blooms has increased within a number of water supply reservoirs in recent years.

Well-designed offtake structures can be used to limit the amount of algal material entering supply systems.

Treatment plants incorporating full flocculation/sedimentation, activated carbon filtration and chloramine-based disinfection can substantially reduce the nuisance by-products of algal breakdown and trihalomethane formation.

These problems highlight the value of having effective planning controls over catchment land uses, and total-catchment management plans as a way of maintaining high water quality.

The message about water quality

- Australia experiences water quality problems caused by salt, pathogens, nutrients, algae and suspended sediments.

- Regional and local water quality problems are associated with nutrient and organic enrichment, pesticides, trace metals and industrial pollutants.

- In particular, high levels of phosphorus, in conjunction with reduced flows in streams, have resulted in extensive and frequent blooms of toxic blue-green algae. These blooms may be increasing in frequency and intensity and can have major consequences.

- Increasing salinity and turbidity in some aquatic environments have contributed to the elimination of sensitive species, resulting in altered ecological communities.

Table 7.12 Main groups of water-borne micro-organisms of concern for human health in Australia

Disease-causing group	Typical pathogens	Health risk
Faecal bacteria	Salmonella excluding *S. typhae*; *Shigella*, *Vibrio cholerae*, enteropathogenic *E. coli*	Mainly gastroenteritis; risks variable; risks from organisms such as cholera now very low
Faecal protozoans	*Cryptosporidium* and *Giardia*	Gastroenteritis; risks uncertain, but possibly quite significant
Viruses	Enteroviruses, Norwalk virus	Risks, if any, extremely low
Parasites (tapeworms)	Zoonoses such as *Echinococcus* (hydatids)	Respiratory and skin infections; risks low
Environmental bacteria	*Legionella, Aeromonas, Pseudomonas, Mycobacterium*	Encephalitis; risks very low but usually fatal
Environmental protozoa	*Naegleria*	Generally low, but always fatal

Source: Maher *et al.*, in press.

Table 7.13 Responses to minimise health risks

Potential type of water contact	Management options	Treatment options
Water supply	Protect catchments, storages and reticulation systems	Clarification and disinfection
Spas and pools	Treatment	Filtration and disinfection
Cooling towers	Hydraulic design and drift prevention	Continuous disinfection
Sewage	Avoid cross connections, discharge of untreated sewage and direct contact with raw sewage	Treatment and disinfection prior to discharge or reuse
Stormwater	Avoid recreational contact	Extended detention where feasible
Recreational use	Avoid contact with sewage-contaminated water	Reliance on natural die-off, sedimentation and predation Avoid recreational contact after heavy rain

Source: Maher *et al.*, in press.

- Domestic water supplies for large cities generally have excellent health and aesthetic qualities.

- Although limited information is available, domestic supplies for rural and remote communities are found to be highly variable, ranging from excellent to poor, with respect to health and aesthetic qualities.

- Where problems in domestic supplies exist they are generally a combination of micro-organisms, chlorination by-products, taste, odour and algal toxins, iron, manganese, turbidity, salts and hardness.

- Increasing development pressures on catchments are resulting in deteriorating water quality and need for treatment.

- Australia's drinking water supplies are generally free of industrial pollutants.

Response

Responses have been of many types, including Federal, State and local government policies, legislation, administrative structures, community education, direct management, changes in community behaviour, setting resource-use targets and so on, as well as initiatives taken directly by communities themselves. They may take the form of broad, strategic solutions (macro-level) or occur at a more local level (for example, a single paddock).

Responses directed towards improving the condition of Australia's inland waters are concentrated in five main areas:

- macro level responses, such as the National Strategy for Ecologically Sustainable Development, State of the Environment reporting, the Murray–Darling Basin Initiative, water market reform, best practice, long-term monitoring and water quality guidelines

- catchment management

- water conservation and management

- biological conservation and management

- water quality control and management

Many other obvious target areas for response are not yet being adequately addressed. These are presented at the end of this section, along with an assessment that indicates the extent of implementation and effectiveness of each response strategy (see Table 7.14).

Macro-level responses

Various State and national legislative instruments and policies can influence the quality of inland waters. Relevant Acts include legislation relating to agriculture, forestry, health, environment, export of various commodities requiring Foreign Investment Review Board approval or Commonwealth government licensing, water supply, irrigation, waste disposal, catchment management, land management, rivers and waterways, industry licensing and approvals, planning and development, local government and industry agreements. All of these tools can be combined to form appropriate responses.

Government and opposition policy statements also provide a basis for the development of responses. However, it is quite common for policy statements to be inconsistent or to conflict across sectors, which can make them less useful unless the boundary issues are resolved satisfactorily. The COAG 1995 Water Reform Agreement provides a national framework to overcome some of these difficulties.

Murray–Darling Basin Initiative

This program promotes and coordinates effective planning and management, for the equitable, efficient and sustainable use of the land, water and other environmental resources of the region. It relies on the Commonwealth and the four State governments concerned — New South Wales, Victoria, South Australia and Queensland —

Pathogens, public health and water pollution

Water polluted by animal and human wastes can seriously affect human health. Pollution barriers, such as closed catchments and water treatment, are intended to protect major Australian water supplies, and increasingly stringent standards are being applied to sewage discharges. However, these approaches can and do fail. Waterborne diseases still far out-rank potential risks from chemicals in water. Even with sophisticated water treatment, in Australia there is still a risk of becoming ill from drinking water, and from water-based recreation (see Table 7.12).

Pressures

Risks arise from many types of water contact, including consumption of drinking water, use of spas and swimming pools, inhalation of cooling-tower drift and swimming in open waters.

Rapid urbanisation and increasing intensity of rural land use are reducing the integrity of many water supply sources, especially where supplies are inadequately treated. Greater volumes of waste water and lack of treatment of stormwater discharged to lakes and streams are issues of increasing concern, as many rural areas use water from these sources without treatment.

State of supplies

Some waterbodies may be improving in quality because of point-source discharge controls for abattoirs, feedlots and sewage. Overall however, general quality is probably decreasing because of increasing land use pressures and lack of management of stormwater.

Recent disease incidents include an apparent viral outbreak affecting some thousands of individuals just outside Melbourne (presumed to be from sewage-contaminated supply). Outbreaks of gastroenteritis in country towns are often attributed to unprotected and untreated supplies but this has not been well documented. The intermittent nature of outbreaks, different types of water contact and the fact that food poisoning and direct person-to-person transmission of disease are also routes for infection makes tracing of waterborne disease outbreaks difficult.

Fully treated water supplies are buffered against any changes in source water quality ensuring consistently high quality at the tap. Recent improvements in the understanding of health effects of microorganisms and disinfection practice have almost eliminated the risk presented by *Naegleria*, but the significance of finding organisms such as *Cryptosporidium* in water supplies, a result of improved detection methods, is not yet clear.

Even though source water quality may be declining, water supplied at the tap is satisfactory microbiologically because of good water management and treatment in cities. There is some evidence of improved drinking water quality in rural New South Wales and Victoria. The number of supplies treated in Australia has increased.

Societal responses

Demands to increase water treatment have arisen because of increased community expectations and also because of contaminated sewage.

Historically, a closed-catchment policy and disinfection of water supplies are the major responses to the risk to public health (see Table 7.13). Chlorination results in the formation of potentially carcinogenic by-products such as THM. However, the risks from disinfection by-products such as trihalomethanes are many times lower than those presented by disease-causing organisms that the treatment destroys.

Catchment protection, especially where supplies are derived from largely unoccupied catchments, should remain a priority. Little opportunity exists for closing catchments already developed. Authorities should resist pressure to open up protected catchments and to urbanise groundwater catchments.

---

working together with the people of the Murray–Darling Basin community to implement and achieve its goals (MDBC, 1990; 1993b; 1995b). The Initiative is one of the largest integrated catchment management programs in the world, encompassing more than one million sq km.

The key to the Initiative is the Natural Resources Management Strategy,(released in 1988) which provides philosophical and organisational structure for governments and communities to coordinate their work.

The Strategy aims to:

- maintain or improve water quality through ongoing research into the problems affecting catchment areas and the run-off entering rivers

- control and, where appropriate, reverse land degradation

- protect, and in some cases rehabilitate, the natural environment

- conserve the cultural heritage

Two funding programs support the Strategy: the first covers investigations and education, and the second integrated catchment management. The former finances the research needed to study the resource management problems facing the Basin and to develop solutions and management tools to treat them. The outcomes are then translated and implemented into practical on-the-ground solutions under the second program, which is based on integrated regional land and water management plans. To ensure commitment from the community, integrated catchment management activities require a significant financial contribution of at least one-third from the local community.

The Murray–Darling Basin Commission is responsible for coordinating the efforts of the governments and communities involved in the management of the Basin. The Commission receives its direction from decisions of the Murray–Darling Basin Ministerial Council, which consists of Natural Resources Ministers representing the various government agencies in the Basin.

The Waterwatch project encourages community participation in long-term monitoring activities.

The Murray–Darling Basin faces grave and worsening degradation of its land and water. The development of an appropriate structure for its management has been a long and painful process. The Initiative brings this process to maturity — as perhaps best demonstrated by the Council's recent landmark decision to place an interim limit on further water diversion from the Basin's rivers (MDBMC, 1995). A final cap on water diversion will be made by 30 June 1997.

Water market reform and irrigation industry restructuring

As well as undergoing recent expansion in some sectors, the irrigation industry is being extensively restructured. This is because of changes in water pricing to seek full cost recovery, the ability to trade in water, the emergence of major environmental problems — such as rising watertables and salinisation — and continuing reductions in farmers' terms of trade. Restructuring will not be easy, will involve difficult decisions by irrigators and agency managers, and may involve considerable pain for individuals and communities. However, restructuring is necessary to ensure a profitable and sustainable future for the industry, based on the realistic cost of water.

In areas of mixed farming, such as the western edge of the riverine plains between Kerang in Victoria and Wakool in New South Wales, broad-scale flood irrigation is now only marginally profitable. As a result of proposed economic reforms and environmental constraints imposed by salinisation, many irrigators may decide to sell their water allocation and retire their properties from irrigation. Regions not suited to sustainable irrigation will decline in prosperity, although alternative dryland agricultural industries may develop. By contrast, irrigators in more suitable

regions may purchase water from the market, invest in upgraded delivery systems and increase more-profitable activities such as cotton-growing, horticulture or dairying. Expansion of irrigation in areas such as the Sunraysia and Riverland regions of the southern Murray–Darling Basin is likely to increase prosperity there. Additional changes are also occurring, including aggregation of farms into larger units to achieve economies of scale, adoption of improved production technology, particularly in relation to inputs of water and nutrients, and training of irrigators in management and marketing.

The large-scale restructuring of the industry should lead to greater water-use efficiency, as well as sustainability and improved profitability.

Best practice

Adoption of best practice methods for all land uses, including urban, industrial and rural, has the potential to bring about substantial improvements in the quality of inland waters. Such methods are being devised through world-wide benchmarking as well as through experimental jointly funded programs like Landcare.

Long-term monitoring

Monitoring provides data that can be used to indicate broad and specific changes to background environmental conditions as a result of both natural and man-made pressures.

Long-term monitoring is critical for providing data that highlight trends in a system — whether it is improving or degrading further. At times, data may need to be collected over long time periods for the significance of change to be fully apparent. Although people who collect and interpret data know this only too well, government agencies and governments themselves often need to be convinced of the role of monitoring and the importance of long time-series data. In a period when many government agencies are undergoing severe financial restrictions and perpetual restructuring, and adopting shorter planning perspectives, agencies are reducing monitoring or even ceasing it in some areas. This is particularly significant for basic hydrographic monitoring — formerly managed by State water authorities. Many have cut back only to areas of immediate interest for possible water resource development. However, the data agencies used to collect are often critical for many other areas of research and monitoring, such as flora and fauna and nutrient loads. Governments at all levels remain to be convinced of the importance of long-term monitoring, which needs to be reflected in guaranteed funding over reasonably long time-frames, not subjected to the annual budget cycle. They also need to understand the differences between monitoring and research and the relative roles of each.

In Australia, long-term monitoring of critical sites provides the only way of telling whether changes are occurring to the environmental state, and of estimating the rate of change and its significance.

Water agencies in all States carry out measurements of stream flows and height on rivers in which they are interested (usually those that are or could be used for water supply). These gauging measurements (gauging) are also carried out in some other areas. Modern digital loggers and automatic samplers, and also remote telemetry, enable more monitoring to be carried out and at less cost. Very little water quality monitoring is conducted often enough and long-term programs are relatively uncommon. Long-term monitoring of the biota of inland waters is rare in Australia. The National River Health Program and Waterwatch are major initiatives in this area.

Community water-monitoring initiatives can be tied in to the critical gauging stations of water agencies and cross-checked to ensure results are comparable.

Water quality guidelines

Guidelines have been produced at a range of levels from local through to international.

The Organisation for Economic Cooperation and Development (OECD) and the World Health Organisation (WHO) have guidelines for various water uses — in particular, human drinking water standards. As well as referring to OECD or WHO levels, Australian States have also used their own guidelines which are generally set by health or water agencies.

At a national level, the National Heath and Medical Research Council (NH&MRC–ARMCANZ) has produced drinking water guidelines (see Table 7.10). ANZECC (1992) has produced environmental water quality guidelines.

These guidelines are useful to various groups (such as the OECD, health managers and industry) providing some consistency between water quality requirements in different States and countries. They simplify planning and management for industries and regulators and save costs, as well as making uniform reporting and internal and external comparisons easier. Until recently, guidelines were variable between States or absent.

Catchment management

Catchment management responses are strategic level responses and include:

- integrated catchment management initiatives
- the National Landcare Program
- research
- educational and monitoring programs such as the Western Australian Ribbons of Blue and the national Waterwatch
- rural adjustment schemes (State and national)
- catchment revegetation programs
- protection/restoration of riparian vegetation
- agricultural extension and advice

Total and integrated catchment management schemes incorporate all of the other responses above to develop general large-scale catchment

improvements. Catchment management ultimately only works if researchers accurately assess the environmental resources of the catchment area and authorities set appropriate management goals. These need to be conservative enough to ensure that whole-catchment improvements — such as reduction of land degradation, salinity and improvements in water quality — can occur.

The total or integrated catchment management approach provides a mechanism to bring together the various parties and interests in a catchment and to focus disparate groups on common goals. Catchment goals are likely to be different from those of individual land-users and there will inevitably be winners and losers if long-term sustainable development is to result. This approach provides a way of identifying the winners and losers, of arranging outcomes that benefit as many interests as possible while ensuring sustainability, and of arranging mechanisms to support change. In some cases it will be necessary to remove some areas from production, at least in the short term. Rural adjustment provides a way of doing so.

The National Landcare Program provides important focus and financial resources to support local community landcare initiatives and field-scale trials (see Chapter 6). These activities often need to be supported by research that must be translated into useful information and guidelines.

Programs such as Waterwatch and Ribbons of Blue are important in helping communities to understand the role and importance of monitoring water quality, as is the National River Health Program for developing protocols and standard biological indicators of water quality and catchment change. Many school-children and adults involved in these programs have discovered that monitoring, although it can be done at low cost, is time-consuming. Similar programs overseas have shown that, despite the greatly reduced cost, the information they collect is often as useful as that collected by water management agencies.

The aim of catchment management is to repair catchments and to prevent undesirable changes in land use. If the aim of Landcare is for the community to become involved, then social indicators such as number of groups and degree of community participation may be valid indicators of success.

In some instances Landcare is an inappropriate response, as its scale is too small to address whole-catchment or off-farm problems such as salinity and excess nutrients. Also, the non-participation of farmers in critical parts of catchments (for example, recharge zones) can undermine the good work of others.

Much relevant research has already been done but is not widely available to the community because it is not in a form that is easily accessible to the community. Considerable scope still exists to turn research results into useful products and guidelines that are easily accessible and are directly relevant to landholders.

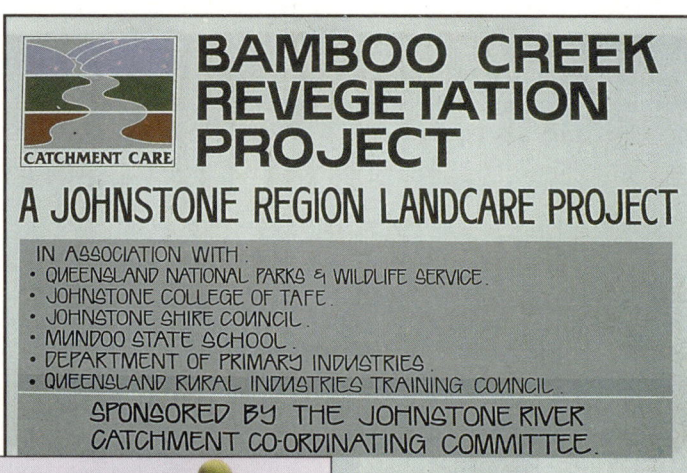

Catchment management is now an important community activity.

The responses to conservation and management of water include water conservation and demand management, water pricing, water rights or transferable water entitlements, education, recycling, monitoring, research, agricultural advice, reservoir management, reservoir redesign, flood control, water-sensitive urban design and regulation.

Water conservation encourages the community to use less water and to become more conscious of levels of use. Several States have imposed water restrictions because of drought. The restrictions commonly include time restrictions on watering gardens or bans on fixed hoses.

The WA Water Corporation has recently proposed a 10 per cent reduction in community water use, with restrictions imposed. So far, consumers have reduced consumption by up to 9.4 per cent and other restrictions are being considered.

As well as restrictions, most agencies are now moving rapidly to a full 'user pays' system, charging for water at close to its true supply cost (taking account of costs such as treatment, pumping, pipelines and reservoirs). The 'free' allowance previously built into the normal water rate is being phased out and all water consumed must be paid for. Pricing is also being altered to progressively remove subsidies, whereby some sectors of the community pay less per kilolitre than others.

The trend towards 'user pays' makes people more aware of the true cost of water and provides real incentives for them to use less and save money. In the long run, the community also saves money by delaying the need for construction of expensive new water resources such as dams and bore fields.

The next logical step is to move to water rights (also known as Transferable Water Entitlements) that can be traded like any other commodity. If water can be sold there is an incentive to improve the efficiency of its use, so there is some left over to sell to someone else. This measure can result in significant improvements in efficiency in large water-users such as industry and irrigators.

In a system of transferable entitlements, water rights can also be bought or reserved to protect the environment. Nowadays, people are starting to think of 'sustainable yield' as the amount of water that can be sold after an adequate amount has first been allocated to the environment. Such allocations need to take account of years of very low flow, when the amount of water available may be severely reduced. Ideally, any water allocation should indicate a range available, not guarantee an absolute entitlement. The actual amount available should range from a minimum (to take account of very-low-flow years), which in some cases might be nil, through to a more generous amount in high-flow years.

Good agricultural advice and practices go together with water rights, demand management and water pricing. Crop selection appropriate to particular types of land, proper selection of best-practice irrigation equipment and sound management of

Total Catchment Management is a New South Wales initiative that is underpinned in legislation. Queensland and Western Australia have State-level integrated catchment management policies. South Australia and Victoria are using the same principles to coordinate various programs and agencies to bring about whole-catchment management.

Whole-catchment management is quite a challenge, even when a river system occurs wholly within one State. Bringing about environmental improvements often involves considerable amounts of money as well as goodwill from landholders. In the case of the Peel–Harvey estuary system, Western Australia, the strategy involved catchment management in the coastal area and the construction of a new channel to the ocean costing more than $50 million (see page 7-46).

The challenge is even greater where a river system crosses State boundaries. The approach taken by the Murray–Darling Basin Natural Resources Management Strategy (NRMS) provides an outstanding example of integrated catchment management. The NRMS provides a mechanism for coordinating management goals between the Commonwealth and State governments and for funding investigations and implementation. However, the Australian constitution gives the States primary responsibility for land management, and hence strong political will and persistence is necessary to cooperate on a common approach and avoid lowest common denominator decisions.

Water conservation and management

Many people will remember the 1000-km algal bloom that occurred in the Darling River in late 1991 and focused world attention on Australia. Such incidents are caused by a number of factors, including: modification and reduction of river flow by reservoirs and direct take-offs; point-source pollution from sources such as town sewage treatment works and industry; non-point-source pollution from farming and horticultural practices; and lower-than-expected rainfall and run-off.

tail-water disposal all have an important influence on water conservation and management, and on downstream environment and other water-users. Tail-waters, if they flow back directly into the main river system, can provide high levels of pollutants, particularly salt, nutrients and pesticides.

Education and consultation are vital components of water management, which help the community contribute to and understand pricing and conservation strategies. Water managers also need to be educated about community preferences and should contribute to community education by exposing the environmental trade-offs involved in various community preferences. A well educated community can ensure that water management agencies carry out thorough investigations, and can decide how to maximise their returns from water that will increasingly cost them more.

Other societal responses include reservoir management — for example, the type and timing of releases of water to the environment — and redesign such as roofing small reservoirs to reduce both evaporation and contamination, with consequent reduced chlorination costs.

Biological conservation and management

Once native species are lost from an area, it can be extremely costly or impossible to reintroduce them. In many cases, the changes in biodiversity are absolute with many species and assemblages being lost permanently. The range of approaches to achieve biological conservation and management includes: promulgation of reserves, policing of protected areas, allocation of water to ensure continuation of environmental flows, education, monitoring, research, restoration of wetland systems, pest control, conservation extension advice and management of threatened commercial, recreational and introduced species.

Monitoring has become more important as an educational tool — with the development of programs such as Ribbons of Blue, Waterwatch and Frogwatch — and as a management tool in the National River Health Program. Managers need to be educated so they understand the usefulness and limitations of data; and changing the management culture towards a more-holistic ecosystem approach is vital.

Pest control is an important response, as removal of feral animals and weeds increases the opportunity for native animals and plants to recolonise areas. Many water bodies have a faunal content that is unique to a particular habitat (Bunn, pers. comm.), suggesting that aquatic biodiversity is even greater than had been thought previously. This is a complicating factor in habitat restoration.

Pesticides in aquatic systems can have severe detrimental effects on native invertebrates that form the predominant food supply for many other animals. Some chemicals, including pesticides, can have long-term effects on breeding of aquatic organisms.

Management of commercial, recreational and introduced species is necessary to protect native flora and fauna. Unfortunately, many native species have already declined or disappeared as a result of inadequate protection and management. In some cases, due to lack of knowledge about breeding habits.

Where dams and weirs provide barriers that fish cannot cross, several States have made active efforts to install and improve the design of fish-passage structures ('fish ladders'), with mixed success.

A number of inland fisheries now use licences, regulations and strict guidelines to protect dwindling stocks of fish (such as barramundi and Murray cod) and crustaceans (such as the Murray crayfish and marron).

Introduced species, unless managed properly, can displace all the native species from an area. Various control and containment programs have been instigated for these pest populations, including attempts to eliminate them, through the release of sterile fish stocks.

Increasingly, water is regarded as a multiple-use resource and catchments and reservoirs are receiving increasing pressure from the community for a range of active and passive uses. Reserve declaration usually nominates a purpose for which the reserve has been created. Reserves require active management unless the community is to be excluded, and in many instances the management resources for generating, implementing and enforcing management plans are not available.

Water quality control and management

Responses aimed at quality control and management include: setting water quality guidelines, standards and objectives for drinking water, recreational water use, stock watering, irrigation, environmental (including aesthetic) uses and various types of waste water; licensing; education; monitoring; research; recycling and re-use, including stormwater management and water-sensitive urban design; water treatment; and best-practice, best available and best-bet technology.

The responses are linked to water supply and treatment costs. Having water of a continually improving standard and responses such as recycling and re-use incur a cost penalty. Although the community wants to consider recycling and re-use as management options, water managers are often reluctant to change established practices. They often cite additional costs and health risks as the reasons, but it may be more an aversion to change that is the problem.

The World Health Organisation has established health standards for human water consumption. In Australia, the National Health and Medical Research Council and the National Water Quality Management Strategy have set guidelines and standards for water for various consumption and environmental purposes (NH&MRC–ARMCANZ, 1995; ANZECC, 1992).

Table 7.14 Summary

Element of the environment/issue	State	Adequate Info.	Response	Effectiveness of response
Water resources Storage and abstraction for human use	Secure supplies but substantial overuse in some regions	✔✔✔✔	Conservation policies; public education in most States; consideration being given to reallocation of water to environment	Good; poor understanding of ecological processes; reallocation could be ineffective
Forest harvesting	An initial increase in water yield is followed by a reduction as regeneration occurs, sometimes dramatically	✔✔✔	Logging restrictions; small-scale and patch logging; fire management	Appropriate, but water issues often not considered in forestry planning
Groundwater	Overuse — use greater than recharge in some areas	✔✔✔	Bore metering and licensing; regulation; capping of bores; water pricing; education; some artificial recharge	Appropriate but limited effect
Farm dams	Proliferation has reduced streamflow particularly during dry conditions	✔✔	No action	Inappropriate — farm dams not considered in water resources planning
Irrigation	Irrigation is the major user of stored water and a large portion is used inefficiently for marginal economic benefit	✔✔✔✔	Water pricing; reform and restructuring of industry; improved irrigation technology	Appropriate but not applied nationally; minimal effect yet
Domestic and urban uses	Supplies adequate but increased demand is leading to more dams	✔✔✔✔	Water pricing; demand management; education; some recycling	Good regional effect; stabilisation of per capita demand where responses have been implemented; minimal national effect
Catchment pollutant sources Agriculture and land clearing	Most waterbodies in areas of agriculture affected by fine and coarse sediment, elevated nutrient loads, and, in some cases, salt; increased volume and rate of run-off; major stream channel changes	✔✔	Strategic revegetation and farm forestry; clearing bans; drainage; broadacre soil conservation and fertiliser management; tree planting to reduce salinity; streambank stabilisation; catchment management and Landcare	Poor — not targeted at water quality; effectiveness of tree planting unknown; streambank stabilisation costly and only partially successful; Landcare working in some areas
Mining	Localised pollution by metals and acid run-off; many sites of disturbance in past areas of coal, alluvial tin and gold mining, sulfide mining and sand and gravel extraction	✔✔✔	Stricter management of all new mines; recycling of water; stabilisation and rehabilitation of some old mine sites	Good in relation to new mines; poor in relation to old mines
Intensive animal industries	Localised but significant pollution by nutrients, organic matter and bacteria	✔✔✔	Guidelines and regulations for effluent discharges, operation, and management; education; implementation of regulations	Inadequate because responses not widely adopted; some local effectiveness
Irrigated agriculture	Localised but significant pollution by sediments, nutrients, pesticides, salt and waterlogging producing serious environmental and social problems	✔✔✔	Guidelines and regulations for effluent discharge and drainage in some areas; improved irrigation techniques; soil conservation; education; industry restructuring; water industry reform	Locally effective but often problem transferred further downstream; guidelines often based on poor biological knowledge; too early to judge effectiveness of restructuring; insufficient land and water management
Urban and industrial development	Localised but significant pollution by sediment, nutrients, oils, organic chemicals and metals	✔✔✔	Guidelines and regulations for effluent management; monitoring; education	Good for trade wastes; poor for general urban runoff; monitoring adequate for some surface runoff; inadequate for groundwater
Forestry	Localised pollution by sediments, nutrients and pesticides	✔	Guidelines and field practice manuals; buffer strips; patch and selective logging; strategic forest and plantation planning	Inadequate because of poor integrated land and water management
Habitat quality and biota (Aspects not already listed above) Drainage	Destruction of wetlands with effects on waterbirds and other biota	✔✔	Almost no response	Inadequate

Table 7.14 Summary (continued)

Element of the environment/issue	State	Adequate Info.	Response	Effectiveness of response
Changed flow regimes	Reduced flows, increased flows, reversal of seasonal flows, reduced medium floods all change habitat quality; reduction and/or extinction of some native species; decline in ecological health	✔✔	Debate, limited trials, and research	In the right direction
Reservoirs and farm dams	1. New habitat and drought refuges 2. Reduced mobility of biota, especially fish 3. Downstream effects, especially erosion, temperature change and changed flow regimes	✔✔ ✔✔✔ ✔✔	1. No active management 2. Fish ladders 3. Little response	1. n.a. 2. inadequate 3. inadequate
Riparian vegetation changes	1. Riparian habitats widely degraded or destroyed 2. Exotic species produce organic inputs to streams different from native species	✔✔	1. Little response; some local fencing and provision of alternative water points; some research and demonstration underway, eg Landcare, Save the Bush 2. Limited control of exotic species	Needs much greater action to protect and repair riparian zones; research and action needs to be in a whole-catchment context
Water quality changes	Changes in water quality affect habitat quality biodiversity and ecological processes, sometimes dramatically; some extinctions in native species; algal bloom enhancement and encouragement of exotic species	✔✔	Catchment management; flow management and point source control	Point source control adequate and effective with increased emphasis on rural and urban catchment management; urban and rural diffuse source control not widely developed
Exotic species	1. Displaced native species 2. Some waterbodies dominated by exotic plants and animals	✔✔	1. Attempts to control import, translocation and spread of potentially damaging exotic species of plants and animals 2. Biological control; management of new outbreaks	Inadequate and ineffective particularly with regard to the control of aquatic plants and fish imports. Education programs required 2. Significant potential effectiveness in some cases
Water Quality (Aspects not already listed above) Chlorination	Produces byproducts that are potentially damaging to human health	✔✔✔✔ (cities) ✔✔ (rural)	Shift to chloramination, filtration prior to chlorination, use of dissolved air flotation, and activated carbon to reduce byproducts	Appropriate and appears to be successful; little knowledge of rural communities
Recreation	Localized but relatively minor increases of bacterial, nutrient and algal concentrations in heavily used waterbodies; some water quality unsafe for some recreation	✔✔	Recreation banned in most drinking water supply reservoirs; provision of toilets; bans and warning notices on water bodies	Appropriate and successful
Sewage disposal	Increased nutrient and pathogen concentrations locally; sewage flow often maintains river flow in dry times	✔✔✔✔	Treatment and land disposal; integrated land and water management and research; community efforts to reduce inputs to sewage plants; discharge licences	Inadequate — level of treatment needs to be reviewed; land disposal in some areas is successful but elsewhere may be unnecessary and unsustainable
Management Short-term thinking	Most governments and organisations focus on the short term	✔✔✔✔	Ecologically sustainable development; national and regional strategies	Appropriate but not widely implemented; too early to judge success
Policy development	Policy development and decision making does not take adequate account of science and does not cope well with scientific uncertainty; reduced skills due to restructuring of major water authorities	✔✔	Appointment of scientific advisory panels; more targeted and integrated science — eg CRCs; economics often used in place of science rather than complementary to science; no response to reduction of skills in water authorities	Inadequate; maybe improving in some areas but not in privatised or corporatised organisations; use of economics often inappropriate
Data and monitoring	Inadequate data sets; fragmentary in space and time; short term; often of limited value; often not interpreted or archived	✔✔✔✔	Very little; National River Health Program; key site monitoring; EPA review	Too early to tell; agency restructuring and privatisation hinders effectiveness
Big picture management, integrated decision making	Lack of integrated decision making	✔✔✔✔	Catchment management and flow management	Many single issue policies developed without consideration of whole policies

Peel–Harvey Estuary System, Western Australia

Prior to European settlement, the catchment of the Peel Inlet and Harvey Estuary (collectively called the Peel–Harvey System) south of Perth was covered in deep-rooted native vegetation. The rivers in the catchment flowed directly across the coastal plain, often via swamps, before entering the estuary. The Murray and Serpentine Rivers entered Peel Inlet and the Harvey River entered the southern end of Harvey Estuary through a birds-foot delta. The single exit to the ocean naturally blocked by formation of a sandbar from time to time (see diagram A opposite). Water entering via the rivers took about 53 days to move through the whole system and out to sea through the exit channel (now called the Mandurah Channel).

The soils in the coastal part of the catchment, beneath the Darling Scarp, were mainly light, sandy soils. The native vegetation was adapted to the naturally low nutrient levels in the soils, and to the relatively dry, Mediterranean climate. European settlement brought about many changes:

- The Serpentine and Harvey Rivers were dammed for water supply and much of the remaining flow of the Harvey River was diverted directly out to sea through the Harvey Main Drain.

- Extensive clearing took place, particularly in the 1960s, to make way for agriculture. The replacement of deep-rooted, native plants with shallow-rooted pastures resulted in a rise in groundwater level, which caused seasonal waterlogging of soils and salinity problems. Extensive artificial drainage had to be installed for agriculture to continue.

- Increased run-off resulting from the extensive clearing more than compensated for the amount being removed by dams upstream. More water reached the estuary in total.

- In order to thrive, the introduced dryland and irrigated crops and pasture required applications of fertiliser — especially phosphorus, potassium and the trace metals: zinc, copper and molybdenum. In some cases they also required irrigation.

- Both the drainage and application of fertilisers resulted in more water, carrying higher levels of nutrients, reaching the estuary faster. This sparked off the estuary nutrient-enrichment problems and subsequent excessive growths of nuisance green algae.

- Urban development and rural living contributed nutrients to the system both from sewage and applications of fertiliser to gardens.

- Intensive horticultural and animal industries became significant contributors of nutrients.

- The algal problems reached a point where management was necessary. Algae were causing severe problems and costing the community money.

- The ocean channel entrance and an area into Peel Inlet were dredged to make a navigable entrance.

Over time it became obvious that management needed to occur in two ways — first, to ensure less nutrient entered the Estuary System; and second, to hasten the removal of nutrients from the System. This provided the foundation for the combined management program, consisting of a catchment management program and the construction of a new channel to the ocean (the Dawesville Channel). Catchment management reduced the inputs of nutrients to the System. The new channel works to reduce the algal nuisance in three ways: by massively increasing the flushing (turnover) rate of water entering the System so more estuarine water is removed; it makes the estuaries more saline for a greater part of the year, thus preventing the germination of one of the algae, which requires lower salinity; and by injecting well-oxygenated marine water into the System. Less phosphorus is released from the sediments as the necessary low oxygen conditions occur seldom if at all. These combined measures have achieved real change (see diagrams A, B and C opposite).

The management approach focused on the nutrient phosphorus, rather than nitrogen, because growth of the blue-green algae was shown to be limited by phosphorus. The improved flushing of the Peel–Harvey System will also remove nitrogen, and people were aware that this would be likely to cause slight increases in nitrogen enrichment of near-shore marine waters. However, a judgement was made that this would have less impact environmentally than the problems that would be fixed by the management strategy (Humphries and Robinson, 1993).

The diagrams opposite illustrate the changes that have occurred to the system in terms of:

- changes in the amount of water entering the system

- changes in phosphorus concentrations over time in the rivers entering the system

- changes in the total amount (load) of phosphorus entering and leaving the system per year

- the length of time that water remains in the system and changes brought about by dredging the Mandurah Channel and by constructing a new channel to the ocean

All numbers used on the diagrams are for an average year for the pre-European settlement, post-European settlement, pre-management, and post management states of the system as assessed by Humphries, Bott and Robinson (in prep.).

Since European settlement, river flows to the estuary have approximately doubled, even with the dams being built on the Serpentine and Harvey Rivers. During the same time, prior to management, phosphorus loads to the system have increased about 20 times, from six to over 140 tonnes per year.

The Peel–Harvey management strategy is one of the largest and the most successful attempts in the world to bring about real improvements to a whole system (Humphries and Robinson, 1995; Humphries and Ryan, 1993).

1. Pre-European settlement: Deep-rooted vegetation in flood plain and headwaters of catchment. No application of fertilisers. No dams on Darling Scarp. No diversions of major rivers (Serpentine, Murray and Harvey Rivers). No artificial drainage in coastal part of catchment. Vegetation adapted to very low phosphorus levels. No problems with nutrient enrichment in estuary.

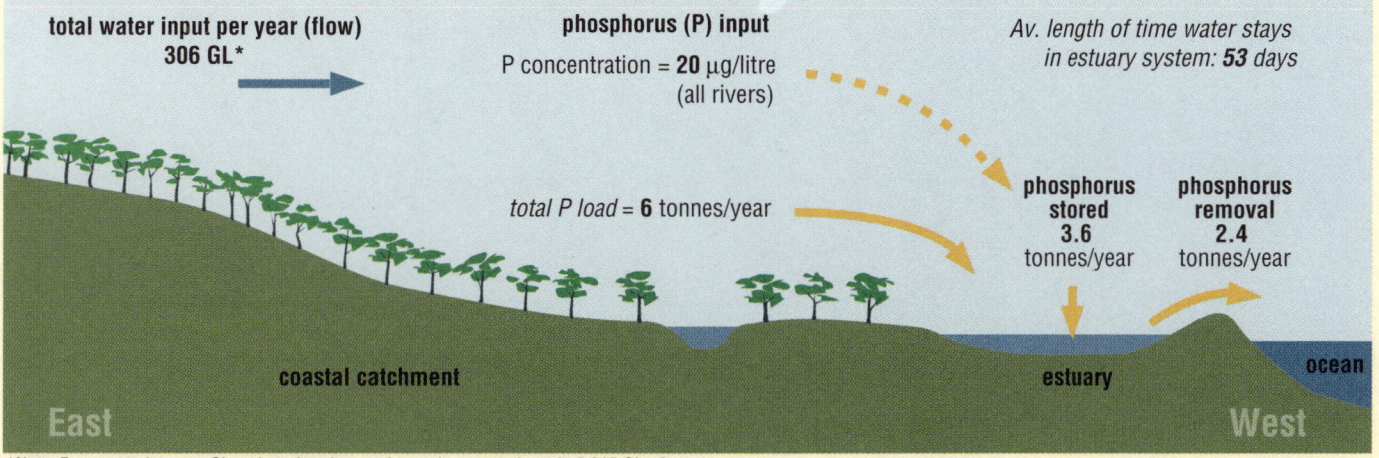

Note: For comparison, an Olympic swimming pool contains approximately 0.015 GL of water.

2. Post European settlement (pre-management): Clearing of deep-rooted vegetation produces water-logging and need for extensive artificial drainage to enable cropping of land. Clearing results in large increases in annual water flows. Increased flow from clearing exceeds water removed by dams on Serpentine and Harvey Rivers. Extensive fertiliser application.

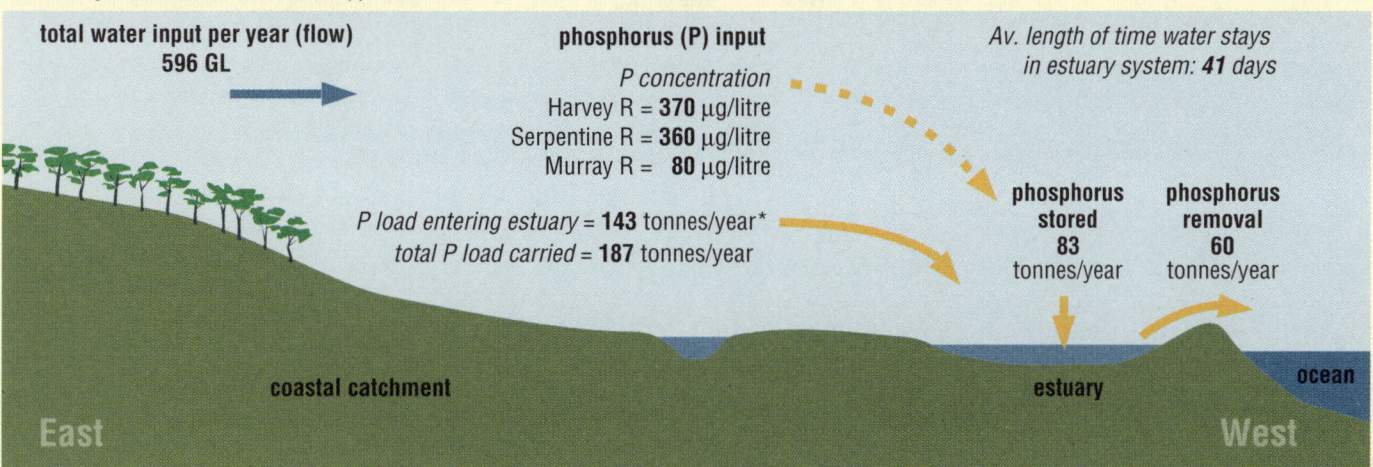

Note: Partial diversion of Harvey River to ocean diverts 44 tonnes of phosphorus a year away from the estuaries and reduces the total amount entering from the catchment to 143 tonnes a year.

3. Post management: Catchment management implemented in urban and rural activities. Less fertiliser applied. Timing of application changed to lessen phosphorus losses. Tree planting on catchment. Drainage into estuary reduced. Activities requiring large inputs of fertiliser not permitted. Design of rural living changed. Additional channel to the ocean constructed to speed up phosphorus removal.

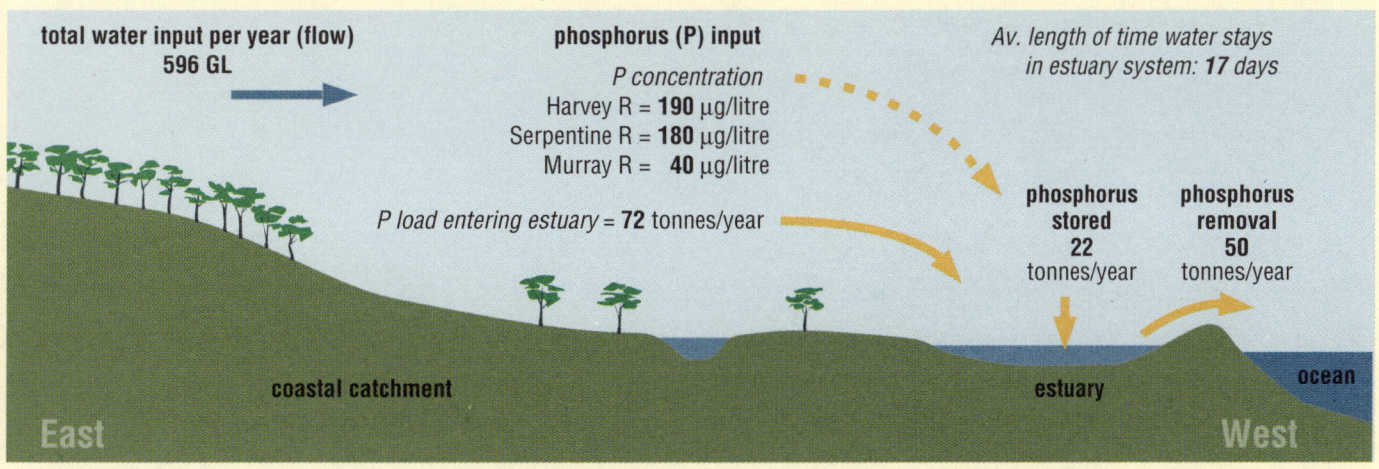

Notes: (1) With the Dawesville Channel, the rate of phosphorus loss to the sea is greatly increased from approximately 40% to approximately 70%.
(2) The phosphorus concentrations and loads are targets required to fix the estuary. These have not yet been achieved but progress has been good. There has been an 80% achievement in the Harvey River and 100% in the Murray River to date. Over the same period, the Serpentine River has barely changed (and may be getting worse) because of inadequate phosphorus controls on intensive animal and horticultural activities in its catchment.

Source: after Humphries, Bott and Robinson, in prep.

Eutrophication of inland and coastal waters

Eutrophication, the process of nutrient enrichment of water bodies by phosphorus and nitrogen, is one of the major water quality problems facing Australia. Human activities accelerate this natural process, resulting in greatly increased amounts of algae, including noxious and sometimes toxic blue-green algae. A number of side effects are associated with the excessive eutrophication of inland, estuarine and coastal waters, including decreased clarity of water, decreased diversity of aquatic plant and animal life and deoxygenation of bottom waters. In terms of water supply, algal blooms are most noticed for their effect on the taste and odour of water, while toxic blooms lead to periodic livestock deaths when animals drink heavily contaminated water.

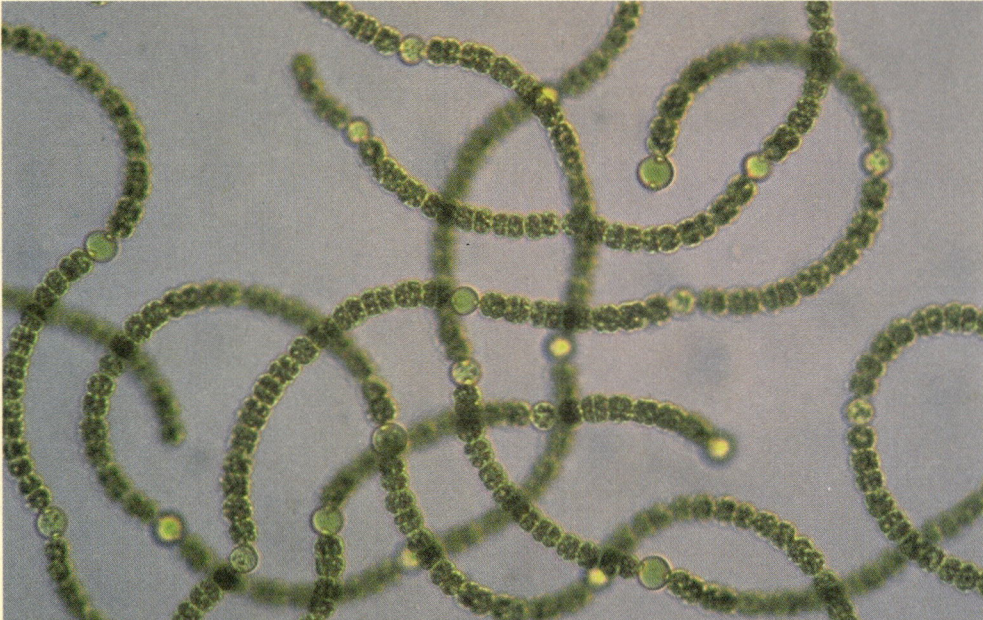

Anabaena circinalis — a common blue-green alga in inland waters

In fresh water, phosphorus typically limits the growth of algae, whereas in marine waters nitrogen is generally the limiting factor. In estuaries, either of the two elements may restrict algal growth. Increased levels of the limiting nutrient are generally associated with larger amounts of algae (see the diagram below). Relative levels of major nutrients are nevertheless important in influencing the composition of algal blooms. For example, high levels of phosphorus relative to nitrogen in fresh waters encourage the growth of blue-green algae, while high levels of silica and phosphorus are required for the growth of the diatom group.

Other factors are also important. Low flow, abundant light, clear water and warmth all encourage algal growth, while the opposite situation — as well as abundant zooplankton and other micro-organisms — discourages the development of algal blooms. In rivers and estuaries, flow is particularly important in regulating algal blooms. The still waters of reservoirs, lakes and slow-flowing regulated rivers are particularly susceptible to algal growth.

Pressure

Virtually all human activity has accelerated the process of eutrophication both directly, through increased nutrient inputs to waterways, and indirectly through modifications to make water available for human use. Nutrient inputs originate from both point and diffuse sources (see Table 7.7). Point sources include sewage effluents, irrigation drains, stormwater run-off and drains from intensive livestock industries. Diffuse sources originate from leaching or soil loss from farmland and stream banks. In some aquatic systems such as reservoirs, stagnant inland rivers or estuaries, nutrients may also be derived from bottom sediments and decaying organic matter.

One of the reasons why eutrophication problems are so widespread, particularly in rural areas where diffuse sources dominate, is the large difference in nutrient concentration between nearby land and water systems. Soils typically have a nutrient status of between 200 and 500 ppm phosphorus, whereas the minimum threshold concentration for algal bloom development in water is between 0.02 and 0.05 ppm — that is, one-ten-thousandth the concentration of phosphorus in the soil. This means that even a small, barely noticeable leakage of phosphorus from well-managed agricultural land may be more than sufficient to stimulate excessive algal growth in a water body.

Studies in Australia and overseas have shown that, in general, the more intensive the land use, the higher the loss of nutrients from the land into the surface waters (see Table 7.6). Typically, phosphorus generation rates increase from 0.1 kg/ha in forests to 1.0 kg/ha in cropped and urban areas, and much higher values under intensive agricultural use.

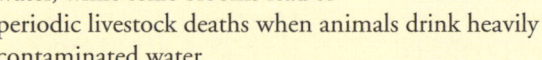

The relationship between maximum chlorophyll-a concentration (a measure of algal levels) and average phosphorus concentration in some south-eastern Australian reservoirs

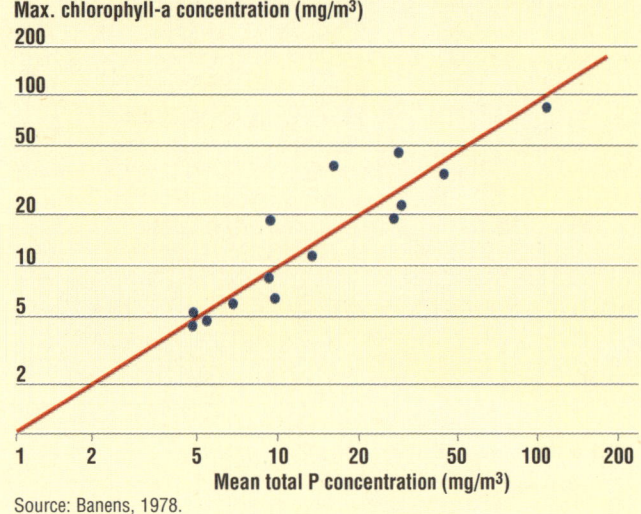

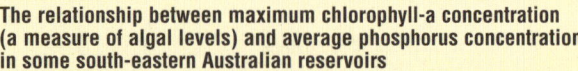

Source: Banens, 1978.

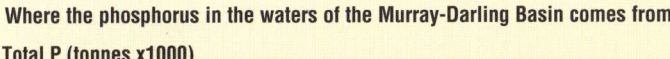

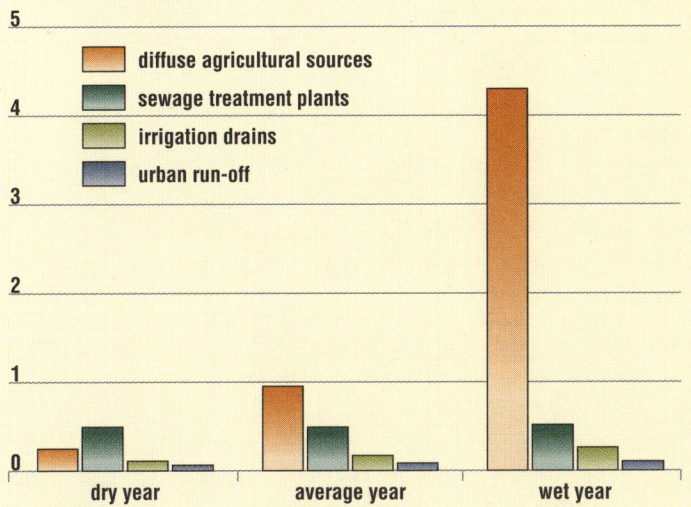

Where the phosphorus in the waters of the Murray-Darling Basin comes from

Total P (tonnes x1000)

- diffuse agricultural sources
- sewage treatment plants
- irrigation drains
- urban run-off

dry year   average year   wet year

Source: GHD, 1992.

The extensive use of superphosphate fertiliser in Australia is often singled out as the major source of phosphorus in inland waters and therefore, by implication, the principal cause of algal blooms. However, research has shown that applied superphosphate rapidly becomes attached to clay particles in the soil, and is not easily mobilised into streams and rivers except through the physical erosion of the soil. Only in sandy soils, such as those of the coastal plain near Perth, is phosphorus leached relatively easily through the soil into streams. Superphosphate is only one of a number of diffuse sources that contribute to the nutrient load of Australia's inland waters.

While diffuse nutrient sources are particularly significant in rural areas, their relative importance in relation to point sources varies according to location and run-off. A nutrient study for the Murray–Darling Basin (GHD, 1992) showed that diffuse agricultural sources greatly dominated the input of phosphorus to the Darling River system during wet years, whereas in dry years sewage treatment plants contributed just over half the total (see the diagram above).

The recent increase in the reported incidence of algal blooms and related water quality problems in Australia may be due to an increased awareness of the eutrophication problem. But it is likely that increased pressure from population growth and human activities has caused a real increase. Ongoing and increasing disposal of sewage effluent, increasing run-off from expanding urban areas and the expansion and intensification of agricultural activities are probably increasing the nutrient loads being delivered to inland and coastal waters. In conjunction with a continuing diversion of flow from our major rivers, the increased nutrient load is believed to be increasing eutrophication. In most cases data do not exist to demonstrate this conclusion.

State

The explorer Charles Sturt noted the vegetable green colour of the largely non-flowing waters of the Murray, which indicated that algal blooms were present in Australia before the impact of European settlers. This suggests that Australia's slow-flowing inland rivers are naturally susceptible to algal blooms. Similarly, both Captain Cook and Charles Darwin noted red-coloured blooms of the blue-green alga *Trichodesmium* on their voyages of discovery, indicating that such blooms also occurred naturally in the marine environment. The first description of a toxic algal bloom (the blue-green alga *Nodularia*) anywhere in the world, and of its impact on livestock, was reported in Lake Alexandrina as far back as 1878 by the South Australian government health inspector (Francis, 1878). While the extent and frequency of algal blooms in Australia have not been documented until relatively recently, numerous rural communities have for many years experienced the taste and odour problems associated with annual summer outbreaks in their water supplies. A national survey of the incidence of major blooms in inland waters in 1992–93 shows the extent of the problem, particularly in the south-east corner of the continent (see the map below). However. it still underestimates the eutrophication problem, as it does not include unmonitored reservoirs, small rural water supply storages, farm dams, natural water bodies, estuaries and inshore marine areas.

Only over the last 25 years have some scientists and water managers become aware of the health and other problems associated with eutrophication. The world's largest toxic blue-green algal bloom along a 1000-km stretch of the Darling River in the summer of 1991, attracted extensive coverage in the media both nationally and overseas. The Darling bloom and the closure of water supplies caused significant disruptions

Location of algal blooms 1992–93

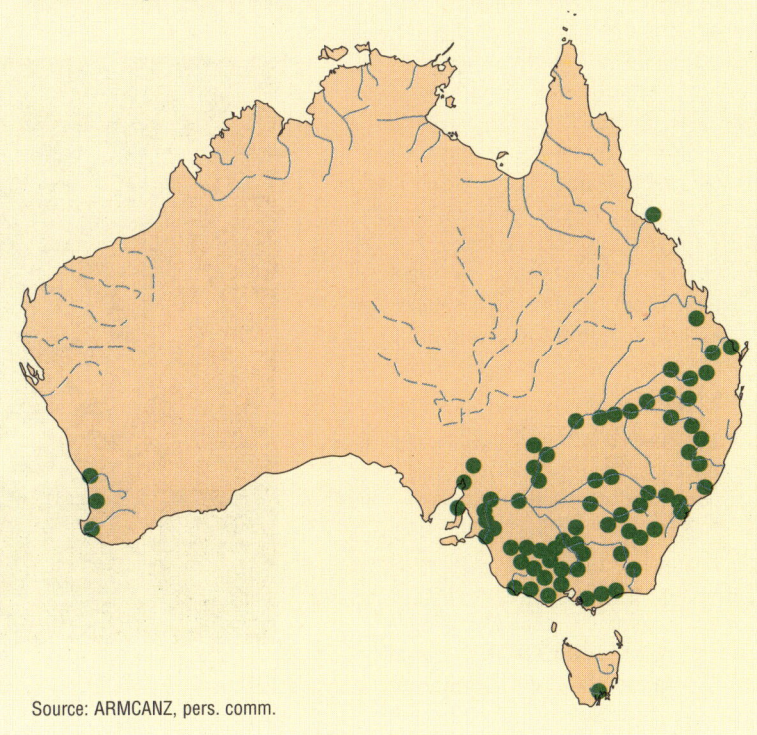

Source: ARMCANZ, pers. comm.

to local communities, as these usually had only a single source of water, and limited and expensive alternative supply options.

Not all major eutrophication problems occur in fresh water. The Peel–Harvey Estuary of Western Australia experienced massive blooms of various species of filamentous algae during the 1970s and '80s. The size and extent of these blooms was such that the commercial fishery was affected, and the hundreds of tonnes of rotting algae accumulated on the shoreline reduced both the recreational value of the estuary and its value for future urban and tourist development. In this particular case the progressive and severe deterioration of water quality in the estuary and the resultant problems were directly attributable to excessive application of phosphorus fertiliser to the estuary's catchment (see the box on page 7-46).

Excessive phosphorus-rich industrial discharges into Albany's Princess Royal and Oyster harbours have resulted in a major dieback of their seagrass beds due to the smothering effect of significantly increased levels of epiphytic algae growing on the seagrasses. Similarly, in a number of areas of the inner Great Barrier Reef, coral degradation and decline has been attributed to increased levels of phytoplankton (floating microscopic algae) associated with elevated nutrient inputs derived from some agricultural activities in the catchments of some coastal rivers.

Societal responses and actions

The world's largest toxic algal bloom occurring in the Darling River with all its associated water supply problems, was the single most important factor in generating or accelerating government and community action. Responses included: algal and nutrient management strategies focusing on both operational and longer-term actions to manage the problem; a Senate enquiry into toxic blue-green algae; major research funding initiatives; and community education programs.

Research into the primary causes of eutrophication, namely nutrient inputs to aquatic systems, resulted in a series of best-bet management activities targeting the reduction of both diffuse and point sources of phosphorus. For example, better farming practices to ameliorate nutrient loss to streams were identified and guidelines for the use of buffer strips and other techniques for sediment and nutrient interception are being developed. Other research explored and developed options for improved sewage treatment, including phosphorus precipitation, off-river disposal and effluent re-use.

Community-based nutrient management strategies are being developed and implemented through catchment management plans. Such plans include the upgrading of sewage-treatment plants and treatment to reduce point sources, and a review of stream-flow management options to reduce the impact and likelihood of algal bloom development. Community education is an

important part of any strategy to deal with the eutrophication problem. This is because long-term management of diffuse nutrient sources requires individual landholders to adopt best-practice methods, while urban dwellers need to change their behaviour in relation to nutrient polluting activities.

One obvious target for nutrient reduction has been the high phosphorus levels in detergents, which form a major phosphorus component of sewage effluent in many cases. As a result of pressure from governments and the scientific community, the detergent industry agreed to a voluntary maximum phosphorus level of five per cent, in conjunction with a labelling program that indicates whether detergents contain phosphorus. This action was followed up by a number of trial community-awareness programs focusing on point source phosphorus reduction from urban areas. Targeting both school and adult groups, these programs educate the community about the benefit of widespread use of low-phosphorus detergents, as well as other individual activities to reduce nutrient input into rivers. Initially trialled in Albury, New South Wales, and since adopted in nearby Wagga Wagga and other places, the phosphorus-reduction campaign has been particularly successful in modifying community behaviour and reducing the phosphorus level in sewage effluent discharged to these rivers.

Research has also been undertaken into the management of algal blooms in inland waters. This has mainly focused on the use of flow and destratification to reduce the likelihood of blooms developing, as well as water-treatment processes to cope with the associated toxins and odours. Contingency plans to cope with the outbreaks of toxic blue-green algal blooms have also been developed. These include emergency water treatment and the identification and development of alternative sources of water. Few management options are available for management of algal blooms in estuarine and marine environments, and prevention remains the best approach.

A 'red tide' at Lake Macquarie, NSW caused by the dinoflagellate, *Noctiluca*, an expression of eutrophication.

References

Aitken, A.P., and Moodie, A.R. (1983). The effect of urban development upon local hydrological regimes. In 'The Effects of Changes in Land Use upon Water Resources', ed. J.W. Holmes, pp. 60–80. (Australian Mineral Foundation: Adelaide.)

ANCA (1993). A Directory of Important Wetlands in Australia. (Australian Nature Conservation Agency : Canberra.)

ANCOLD (1990). 'Register of large dams in Australia, 1990.' (International Commission on Large Dams. National Committee on Large Dams, c/- Hydroelectric Commission: Hobart.)

Anderson, J. (1993). 'State of the Rivers. Maroochy River and Tributary Streams.' (Queensland Department of Primary Industries, Water Resources: Brisbane.)

ANZECC (1992). National Water Quality Management Strategy Guidelines for Fresh and Marine Waters. (Australia and New Zealand Environment and Conservation Council.)

ANZECC (1993). 'List of Threatened Australian Flora.' (Australian Nature Conservation Agency: Canberra.)

Arthington, A.H., and Mitchell, D.S. (1986). Aquatic invading species. In 'Ecology of Biological Invasions: an Australian Perspective', ed. R.H. Groves and J.J. Burdon. (Australian Academy of Science: Canberra.)

Arthington, A.H. (1991). Ecological and genetical impacts of introduced and translocated freshwater fishes in Australia. *Canada Journal of Fisheries and Aquatic Science*, **41**(1), pp. 33–43.

Arthington, A.H., and McKenzie, F. (in press). Review of impacts of displaced/introduced fauna associated with inland waters. *Report prepared for the Inland Waters Reference Group, DEST.*

ARMCANZ (1994). Pricing systems for major water authorities in Australia. *Agriculture and Resource Management Council of Australia and New Zealand Paper FCMC* No. 2.

AWRC (1985). Review of Australia's water resources and water use. Australian Water Resources Council Nov. 1987, Vol. 1, Water use data set.

AWRC (1990). The status of groundwater contamination and regulation in Australia. *Water Resource Management Committee, Occasional Paper* No. 1.

AWRC (1992). 'National Water Quality Management Strategy, Draft Guidelines for Groundwater Protection.' (Australian Water Resources Council : Canberra.)

AWRC and Department of Resources and Energy (1983). 'Consultants Report No. 2 for Water 2000: Australia's Groundwater Resources.' (AGPS: Canberra.)

Banens, R.J. (1981). Changes in land use and hydrology in the Severn Valley. Department of Resource Engineering, University of New England, *Water for the Future Report* No. WWF1.

Barber, C., Barron, R., Broun, J., Bates, L.E., and Locksey, K. (1993). Evaluation of changes in ground-water quality in relation to landuse changes in the Gwelup wellfield WA. *CSIRO Water Resources* No. 12

Barinaga, M. (1990). Where have all the froggies gone? *Zoological Record*, **126** (16).

Barnes, C. J., Jacobson, G., and Smith, G.D. (1992). The origin of high-nitrate groundwaters in the Australian arid zone. *Journal of Hydrology* 137 pp. 181–97.

Barton, J. L., and Davies, P.E. (1993). Buffer strips and streamwater contamination by atrazine and pyrethroids aerially applied to *Eucalyptus nitens* plantations. *Australian Forestry*, 56, pp. 201–10.

Bates, B.C., Charles S.P., Summer, N.R., and Fleming, P.M. (1994). Climate change and its hydrological implications for South Australia. *Transactions of the Royal Society of South Australia*, **118**, pp. 35–43.

Beschta, R.L. (1978). Long-term patterns of sediment production following road construction and logging in the Oregon Coast Range. *Water Resources Research*, **14** pp. 1011–16.

Boon, P., Frankenberg, J., Hillman, T., Oliver, R., and Sheil, R. (1990). The Murray — billabongs. In Mackay and Eastburn (1990).

Bosch, J.M., and Hewlett, J.D. (1982). A review of catchment experiments to determine the effect of vegetation changes on water yield and evapotranspiration. *Journal of Hydrology*, 55 pp. 3–23.

Boughton, W.C. (1970). Effects of land management on quantity and quality of available water. A review. *University of New South Wales, Sydney, Water Research Laboratory Report* No. 120.

Brumley, A.R. (1991). Cyprinids of Australasia. *Zoological Record*, **128**, p. 15.

Bunn, S.E., Pusey, B.J., and Price, P. (eds) (1993). Ecology and management of riparian zones in Australia. *Land and Water Resources Research and Development Corporation Occasional Paper Series* 05/93.

Bureau of Meteorology (1965–80) Drought Statements and Reviews, 1–122. (Bureau of Meteorology: Canberra.)

Cadwallader, P., and Lawrence, B. (1990). Fish. In Mackay and Eastburn (1990).

Campbell, I.C., and Doeg, T.J. (1989). Impact of timber harvesting and production on streams. A review. *Australian Journal of Marine and Freshwater Research* 40 pp. 519–39.

Close, A. (1990). River salinity. In: Mackay and Eastburn (1990), pp. 127–44.

COAG (1995). Council of Australian Governments Water Reform Agreement.

Cogger, H.G., Cameron, E.E., Sadlier, R.A., and Eggler, P. (1993). 'The Action Plan for Australian Reptiles.' (Australian Nature Conservation Agency: Canberra.)

Cornish, P.M. (1989). The effects of radiata pine plantation establishment and management on water quantity and water quality — a review. *Forestry Commission of New South Wales, Technical Paper* No. 49.

Cornish, P.M. (1993). The effects of logging and forest regeneration on water yields in a moist eucalypt forest in New South Wales, Australia. *Journal of Hydrology*, 150 301–22.

Cosser, P.R. (1988). Macroinvertebrate community structure and chemistry of an organically polluted creek in south-east Queensland. *Australian Journal of Marine and Freshwater Research*, 39, pp. 671–83.

Crabb, P. (1986). 'Australia's Water Resources: Their Use and Management.' (Longman Cheshire: Melbourne.)

Crabb, P. (in press). Impacts of anthropogenic activities, water use and consumption on water resources and flooding. *Report prepared for the Inland Waters Reference Group, DEST.*

CSIRO (1992). Towards Healthy Rivers. A report from CSIRO to the Honourable Ros. Kelly, Minister for

Arts, Sport, the Environment and Territories. *Division of Water Resources Consultancy Report* No. 92/44, November 1992.

Community Discussion Document on the Major Issues.' (ASSERT Social Science Unit, CSIRO Division of Water Resources: Perth.)

Cuddy, S., Marston, F., Simmons, B., Davis, R., and Farley T. (1993). Applying CMSS in the Hawkesbury–Nepean Basin. *CSIRO Division of Water Resources Consultancy Report* No. 93/37.

Cullen, P., Decker, R., Gehrke, P., Harris, J., Hillman, T., Humphries, P., and Swales, S. (in press). Impacts of anthropogenic water use and flow modification on natural aquatic ecosystems. *Report prepared for the Inland Waters Reference Group, DEST.*

Cullen, P., and Rosich, R. (1979). Effects of rural and urban sources of phosphorus on Lake Burley Griffin. *Progress in Water Technology*, **11**, pp. 219–30.

Davies, P.E., Cook, L.S.J., and Barton, J.L. (1994). Atrazine herbicide contamination of Tasmanian streams: sources, concentrations and effects on biota. *Australian Journal of Marine and Freshwater Research*, **45**, pp. 209–26.

Davies, P.E., and Nelson, M. (1993). The effect of steep slope logging on fine sediment infiltration into stream beds in ephemeral and perennial streams of the Dazzler Range, Tasmania. *Journal of Hydrology*, **150**, pp. 481–504.

Davies, P.E., and Nelson, M. (1994). Relationships between riparian buffer widths and the effects of logging on stream habitat, invertebrate community composition and fish abundance. *Australian Journal of Marine and Freshwater Research*, **45**, pp. 1289–1305.

Department of Resources and Energy (1983). 'Water 2000.' (AGPS: Canberra.)

Doeg, T.J., and Koehn, J.D. (1990). A review of Australian studies on the effects of forestry practices on aquatic values. *SSP Fisheries Division Technical Report* No. 5.

Donnelly, T.H., Olley, J.M., Murray, A.S., and Wasson, R.J. (1992). Algal blooms in the Darling River. *MDBC Consultancy Report* No. 92/13.

Federal Race Discrimination Commissioner (1994). 'Water: a Report on the Provision of Water and Sanitation in Remote Aboriginal and Torres Strait Islander Communities.' (AGPS: Canberra.)

Francis, G. (1878). A poisonous lake. *Nature*, **18**, pp. 11–12.

Froend, R.H., Farrell, R.C.C., Wilkins, C.F., Wilson, C.C. and McComb, A.J. (1993). 'The Effect of Altered Water Regimes on Wetland Plants. Vol. 4, Wetlands of the Swan Coastal Plain.' (Western Australian EPA and Western Australian Water Authority: Perth.)

Furnas, M.J., Mitchell, A.W., and Skuza, N. (1994). 'Nitrogen and Phosphorus Budgets for the Central Great Barrier Reef Shelf.' (GBRMPA: Townsville.)

Gippel, T.M., and Perrens, S.J. (in prep.). Feasability analysis of a domestic rainwater system as an alternative supply augmentation for Uralla, NSW, unpublished manuscript.

Goodrick, G.N. (1970). A survey of wetlands of coastal New South Wales. *CSIRO Division of Wildlife Research Technical Memorandum* No. 5.

Grayson, R.B., Hayden, S.R., Jayasuriya, M.D.A., and Finlayson, B.L., (1993). Water quality in mountain ash forests — separating the impacts of roads from those of logging operations. *Journal of Hydrology*, **150** pp. 459–80.

Growns, I.O., and Davis, J.A. (1991). Comparison of the macroinvertebrate communities in streams of logged and undisturbed catchments 8 years after harvesting. *Australian Journal of Marine and Freshwater Research*, **42**, pp.689–706.

Growns, I.O., and Davis, J.A. (1994). Effects of forestry activities (clearfelling) on stream macroinvertebrate fauna in south-western Australia. *Australian Journal of Marine and Freshwater Research*, **45**, pp. 963–75.

Gutteridge, Haskins and Davey (1992). 'An Investigation of Nutrient Pollution in the Murray–Darling River System: Report Summary.' (Murray–Darling Basin Commission: Canberra.)

Hall, N., Poulter, D., and Curtotti, R. (1994). ABARE Model of Irrigation Farming in the Southern Murray-Darling Basin. *ABARE Research Report* No. 94.4.

Halse, S.A. (1989–94). Wetlands of the Swan Coastal Plain — past and present. *Proceedings of the Swan Coastal Plain Groundwater Management Conference, Western Australian Water Resources Council, Perth*, pp. 105–112.

Harley, K.L.S. (ed.) (1992). 'A Guide to the Management of *Mimosa pigra*.' (CSIRO: Canberra.)

Harris, G.P. (1994). Nutrient loadings and algal blooms in Australian waters — a discussion paper. LWRRDC Occasional Paper No. 12/94, p. 99.

Harris, J.H. (1984). Impoundments of coastal drainages of south-eastern Australia, and a review of its relevance to fish migrations. *Australian Zoologist*, **21**, pp. 134–250.

Harris, J.H. (1985). Concern over dams' effects on fish populations in south-east. *Australian Fisheries*, April, 37–42.

Hart, B.T., Bailey, P., Edwards, R., Hortle, K., James, K., McMahon, A., Meredith, C., and Swadling, K. (1991). A review of the salt sensitivity of the Australian freshwater biota. *Hydrobiologia* **210**, pp. 105–44.

Hart, B.T., Bailey, P., Edwards, R., James, K., Swadling, K., Meredith, C., McMahon, A., and Hortle, K. (1990). Effects of saline discharges on aquatic ecosystems. *Water Research*, **24**, 1103–17.

Hillman, T.J., Sheil, R.J. (1991). Macro- and invertebrates in Australian billabongs. *Verhandlungen (Internationale Vereinigung feür Theoretische und Angewandte Limnologie)*, **24**(3), pp. 1581–7.

Hooke, J.M. (1980). Magnitude and distribution of rates of river bank erosion. *Earth Surface Processes*, **5** pp. 143–57.

Horwitz, P. (1994). Distribution and conservation status of the Tasmanian giant freshwater lobster *Astacopsis gouldi* (Decapoda: Parastacidae). *Biological Conservation*, **69**, pp. 199–206.

Humphries, R., and Robinson, S. J. (1993). The Peel–Harvey Estuary management strategy: development of the management strategy; *prepared for Macquarie University Coastal Zone Management Conference.*

Humphries, R., and Robinson, S. J., (1995). Assessment of the success of the Peel–Harvey system management strategy — a Western Australian attempt at integrated catchment management. *Water Science Technology*, **32**(5–6), pp. 255–64.

Humphries, R., and Ryan G. (1993). The Dawesville Channel and its effects on the Peel–Harvey Estuary — the predicted changes and their consequences for management. *Paper presented at ANZAAS Congress, Perth, Western Australia, 28–30 September 1993.*

Humphries, R., Bott, G., and Robinson, S. J. (in prep.), An Evaluation of the Effectiveness of the Peel-Harvey Management Strategy to 1994.

Humphries, S.E., Groves, R.H., and Mitchell, D.S. (1991). Plant invasions of Australian ecosystems: a status review and management directions. In: 'Plant Invasions. The Incidence of Environmental Weeds in Australia', Kowari 2, Part One, pp. 1–134. (Australian National Parks and Wildlife Service: Canberra.)

Jackson, J. (in press). State of habitat availability and quality in inland waters. *Report prepared for the Inland Waters Reference Group, DEST.*

Jacobs, S.W.L. (in press). Impacts of displaced/introduced plants associated with inland waters. *Report prepared for the Inland Waters Reference Group, DEST.*

Jayasuriya, M.D.A., Dunn, G., Benyon, R., O'Shaughnessy, P.J. (1993). Some factors affecting water yield from mountain ash *(Eucalyptus regnans)* dominated forests in south-east Australia. *Journal of Hydrology,* **150** pp. 345–67.

Jones, D.R. (1994). Assimilative capacity of creeks in Sydney's N.W. sector. *CSIRO Division of Coal and Energy Technology Investigation Report* No. CET/IR336.

Julien, M.H., Bourne, A.S., and Low, V.H.K. (1992). Growth of the weed *Alternanthera philoxeroides* (Martius) Grisebach (alligator weed) in aquatic and terrestrial habitats in Australia. *Plant Protection Quarterly,* 7, pp. 102–8.

Kailola, P.J., Williams, M.J., Stewart, P.C., Reichelt, R.E., McNee, A. and Grieve, C. (1993). 'Australian Fisheries Resources.' (Bureau of Resource Sciences and Fisheries Research and Development Corporation: Canberra.)

Kirkpatrick, J.B., and Harwood, C.E. (1981). The conservation of Tasmanian wetland macrophytic species and communities. *A Report to the Australian Heritage Commission from the Tasmanian Conservation Trust Inc.*

Kuczera, G. (1987). Prediction of water yield reductions following a bushfire in ash–mixed species eucalypt forest. *Journal of Hydrology* **94**, 215–36.

Lane, J.A.K., and McComb, A.J. (1988). Western Australian wetlands. In McComb and Lake (1988), pp. 127–46.

Langdon, J. (1988). Diseases of introduced Australian fish. In 'Fish Diseases', pp. 225–76. (Post-Graduate Committee in Veterinary Science: Sydney.)

Lawrence, C.R. (1983). Nitrate rich groundwaters of Australia. *AWRC Technical Report* No. 79.

Liston, P., and Maher, W. (in press). Water quality for maintenance of aquatic ecosystems. *Report prepared for the Inland Waters Reference Group, DEST.*

Lonsdale, W.M. (1992). The biology of *Mimosa pigra.* In 'A Guide to the Management of *Mimosa pigra*', ed. K.L.S. Harley, pp. 8–32. (CSIRO: Canberra.)

Lund, M., Davis, J., and Murray, F. (1991). The fate of lead from duck shooting and road run-off in three Western Australian wetlands. *Australian Journal of Marine and Freshwater Research,* 42, 139–49.

McComb, A.J., and Lake, P.S. (eds) (1988). 'The Conservation of Australian Wetlands.' (Surrey Beatty and Sons Pty Ltd: Australia.)

McComb, A.J., and Lake, P.S. (1990). Australian Wetlands. (Angus & Robertson: Australia.)

McFarlane, D. (1995). *WA Department of Agriculture internal report.*

Mackay, N., and Eastburn, D. (eds) (1990). 'The Murray.' (Murray–Darling Basin Commission: Canberra.)

Mackay, N.J., and Shafron, M. (1989). Water quality. *Proceedings of the Workshop on Native Fish Management, 1988, Murray–Darling Basin Commission,* pp 137–48.

McKay, R.J. (1984). Introductions of exotic fishes in Australia. In 'Distribution, Biology, and Management of Exotic Fishes', eds W.R. Courtenay Jr. and J.R. Stauffer Jr., pp. 177–99. (Johns Hopkins University Press: Baltimore, Maryland.)

McLaughlin, M.J., Fillery, I.R., and Till, A.R. (1992). Operation of the phosphorus, sulphur and nitrogen cycles. *Bureau Mineral Resources Proceedings* No. 14, 67–116.

McMahon, T.A., Finlayson, B.L., Haines, A.T., and Srikanthan R. (1992). 'Global Run-off. Continental Comparisons of Annual Flows and Peak Discharges.' (Catena Verlag: Cremlingen-Destedt.)

Maher, W.A., Wade, A., and Lawrence, I. (in press). Drinking water quality. *Report prepared for the Inland Waters Reference Group, DEST.*

Mallen-Cooper, M. (1989). Fish passage in the Murray–Darling Basin. *Proceedings of the Workshop on Native Fish Management, Murray–Darling Basin Commision, Canberra,* 1988, pp. 345–59.

Margules and Partners Pty Ltd, Smith, P. and J., and Dept of Conservation, Forests and Lands Victoria (1990). 'River Murray Riparian Vegetation Study.' (DCFL: Melbourne.)

MDBC (1990). Murray-Darling Basin Natural Resources Management Strategy. Murray-Darling Basin Ministerial Council, Canberra.

MDBC (1993a). Dryland Salinity Management in the Murray-Darling Basin. Report to the Murray-Darling Basin Ministerial Council by the Dryland Salinity Management Working Group, Murray-Darling Basin Commision, Canberra.

MDBC (1993b). The Murray-Darling Basin Commision — Managing Australia's Heartland. Murray–Darling Basin Commision, Canberra.

MDBC (1995a). 'Water Use and Healthy Rivers — Working Towards a Balance. An Audit of Water Use in the Murray–Darling Basin.' (Murray–Darling Basin Commission: Canberra.)

MDBC (1995b). 'Murray–Darling Basin Initiative.' (Murray–Darling Basin Commission: Canberra.)

MDBMC (1995). 'An Audit of Water use in the Murray–Darling Basin.' June 1995 (Murray–Darling Basin Ministerial Council: Canberra.)

Mitchell, P. (1990). 'The Environmental Condition of Victorian Streams.' (Department of Water Resources Victoria: Melbourne.)

Morton, R. (1995). Trends in salinity in the River Murray 1938–1995. *CSIRO Biometrics Unit Consulting Report* ACT 95/3.

Murphy, J. (1990). Watering the Millewa Forest. In Mackay and Eastburn (1990), pp. 245–8.

Nandakumar, N., and Mein, R.G. (1993). Analysis of paired catchment data for some of the hydrologic effects of land-use change. *Hydrology and Water Resources Symposium, Newcastle, June 30–July 2, 1993.*

National Disasters Organization (1992). 'Hazards, Disasters and Survival.' (Natural Disasters Organization: Canberra.)

NCPA (1994). Regional water quality study, upper Murray catchment. Prepared for ACT and sub-region planning committee and upper coordinating committee.

NH&MRC–ARMCANZ (1995). 'Australian Drinking Water Guidelines.' (AGPS: Canberra.)

NH&MRC–AWRC (1987). 'Guidelines for Drinking Water Quality in Australia.' (AGPS: Canberra.)

NSW Department of Water Resources (1994). 'Our Water: a Review of the Current Status of the Water Resources of New South Wales and the Key Issues Relevant to their Future Development.' (DWR: Sydney.)

NSW SOE (1993). New South Wales State of the Environment 1993. (New South Wales Environment Protection Authority: Sydney.)

NSW State Emergency Services (1994). *Annual Report 1993–94.*

Norman, F.I., and Nicholls, N. (1991). The Southern Oscillation and variations in waterfowl abundance in southeastern Australia. *Australian Journal of Ecology,* **16**, pp. 485–90.

Olive, L.J., and Rieger, W.A. (1987). Eden catchment project: sediment transport and catchment disturbance, 1977–83. *Australian Defence Force Academy Department of Geography and Oceanography, Monograph Series No. 1.*

Olley, J.M., Caitchen, G., Donnelly, T., Olive, L., Murray A.S., Short, D., and Wallbrink, P.J. (in press). Sources of and temporal variations in phosphorus in the Murrumbidgee River. *Interim Report to the EPA.*

Olley, J.M., Murray, A.S., and Wallbrink, P.J. (1995). Identifying sediment in a partially logged catchment using natural and anthropogenic radioactivity. *Zeitschrift fur Geomorphologie.*

O'Loughlin, E.M., Young, W.J., and Molloy, J.D. (1992). Urban stormwater: impacts on the environment. *CSIRO Consultancy Report* No. 92/29.

O'Shaughnessy, P.J.O., and Jayasuriya, M.D.A. (1987). Managing the ash type forests for water production in Victoria. *Proceedings, 1987 Conference of The Institute of Foresters of Australia, Perth, Western Australia. September 2–October 2, 1987.*

Paijmans, K., Galloway, R.W., Faith, D.P., Fleming, P.M., Haantjens, H.A., Heyligers, P.C., Kalma, J.D., and Löffler, E. (1985). Aspects of Australian wetlands. *CSIRO Division of Water and Land Resources Technical Paper* No. 44.

Peck, A.J., and Williamson, D.R. (1987). Hydrology and salinity in the Collie River Basin, Western Australia. *Journal of Hydrology* , **94** pp. 1–198.

Phillips, W. (1993). Australia's RAMSAR wetlands. In 'A Directory of Important Wetlands in Australia', pp. 2.1–2.11. (Australian Nature Conservation Agency: Canberra.)

Pressey, R.L. (1990). Wetlands. In Mackay and Eastburn (1990), pp. 167–81.

Pressey, R.L., and Harris, J.H. (1988). Wetlands of New South Wales. In McComb and Lake (1988), pp. 35–57.

Pressey, R.L. (1986). Wetlands of the River Murray. *River Murray Commission Environmental Report* 86/1.

Rayment, G.E., Simpson, B.W. (1993). 'Pesticide Audit.' Queensland Department of Primary Industries Report to the Condamine–Balonne Water Committee.

Richardson, B.A. (1985). The impact of forest road construction on the benthic invertebrate and fish fauna of a coastal stream in southern NSW. *Bulletin of the Australian Society of Limnology* **10**, pp. 65–87.

Roberts, J.R. (1993). Carp and the demise of aquatic plants in rivers and wetlands of south-western New South Wales. *Conference and Symposium of the Ecological Society of Australia, Canberra.*

Rosewell, C. (in press). Potential sources of sediments and nutrients. *Report prepared for the Inland Waters Reference Group, DEST.*

Ruprecht, J.K., and Stoneman, G.L. (1993). Water yield issues in the jarrah forest of south-western Australia. *Journal of Hydrology,* **150** pp. 369–91.

Rutherfurd, I.D., Brooks, A., Davis, J. (in press). Historical stream erosion and sedimentation in Australia. *Report prepared for the Inland Waters Reference Group, DEST.*

Salama and Bartle (in press). Past, Present and Future Groundwater Level Trends in the Wheatbelt of Western Australia. *CSIRO Division of Water Resources Technical Memo* No. 95.10.

Schofield, N.J. (1989). Stream Salinisation and its amelioration in south-west Western Australia. *Proceedings of the Baltimore Symposium, May 1989, IAHS Publication* No. 182, pp. 221-220.

Shearer, K.D., and Mulley, J.C. (1988). The introduction and distribution of the carp, *Cyprinus carpio* Linnaeus, in Australia. *Australian Journal of Marine and Freshwater Research,* **29**, 551–63.

Srikanthan, R., and Neil D. (1989). Simulation of the effect of farm dams on sediment yield from two small rural catchments. *Australian Journal of Soil and Water Conservation,* **2** (1), pp. 40–5.

State Dryland Salinity Committee, (1990) *Annual Report* 1989/90.

Thoms, M.C., and Walker, K.F. (1992). Channel changes related to low-level weirs on the River Murray, South Australia. In 'Lowland Floodplain Rivers: Geomorphological Perspectives', eds PA Carling and G.E. Petts. (Wiley: Chichester.)

Tilleard, J.W., Erskine, W.D., and Rutherfurd, I.D. (1994). Impacts of River Murray flow regulation of downstream channel morphology. *Water Down Under '94,* November.

Timms, B.V. (1980). Farm Dams. In: 'An ecological basis for water management.' ed. W. D. Williams, pp. 345–59 (ANU Press: Canberra.)

Timms, B.V. (1992). The conservation status of athalassic lakes in New South Wales, Australia. *Hydrobiologia,* **243/244**, pp. 435–44.

Torgersen, T., Habermehl, M.A., Phillips, F.M., Elmore, D., Kubik, P., Jones, B.G., Hemmick, T., and Gore, H.E. (1991). Chlorine 36 dating of very old groundwater. 3. Further studies in the Great Artesian Basin, Australia. *Water Resources Research,* **27** pp. 3201–213.

Tyler, M.J. (1991). Where have all the frogs gone? *Australian Natural History,* **23**, pp. 618–25.

Tyler, M.J. (1994). 'Draft of The Action Plan for Australian Frogs.' (Australian Nature Conservation Agency: Canberra.)

Usback, S., and James, R. (1993). 'A Directory of Important Wetlands in Australia.' (Australian Nature Conservation Agency: Canberra.)

Victoria, Department of Water Resources (1989). 'Water Victoria: an Environmental Handbook.' (DWR: Melbourne).

VSOE (1988). 'State of the Environment Report 1988. Victoria's Inland Waters.' (Victorian Office for the Commissioner for the Environment: Melbourne.)

Wager, R., and Jackson, P. (1993). 'The Action Plan for Australian Fishes.' (Australian Nature Conservation Agency: Canberra.)

Walker, K.F. (1985). A review of the ecological effects of river regulation in Australia. *Hydrobiologia*, **125**, pp. 111–29.

Walker, K.F. (1992). The River Murray, Australia: a semiarid lowland river. In 'The Rivers Handbook', Vol. 1, ed. P. Calow and G.E. Petts, pp. 472–92. (Blackwell: Oxford.)

Walker, K.F. (1994). Historical changes related to flow regulation of the River Murray, South Australia. II. The biological environment. *Proceedings of the Fifth International Streams and Regulated Rivers Symposium, Montana, 1991.*

Walker, K.F., and Hillman, T.J. (1982). Phosphorus and nitrogen loads in waters associated with the River Murray near Albury–Wodonga, and their effects on phytoplankton populations. *Australian Journal of Marine and Freshwater Research*, **33**(2), pp. 223–43.

WA Department of Environmental Protection (1995). Acute toxicity of herbicide to selected frog species. *WA Department of Environmental Protection, Technical Series* 79.

WA Select Committee (1990). Select Committee into Land Conservation. (Western Australia Legislative Assembly) South west region of WA. *Discussion Paper* No. 1.

WA SOE (1992). 'State of the Environment Report.' (Government Printer: Perth.)

Whitehead, P.J., and Tschirner, K. (1991). Lead shot ingestion and lead poisoning of magpie geese *Anseranus semipalmata* foraging in a northern Australian hunting reserve. *Biological Conservation*, **58** pp. 99–118.

WHO (1993). 'Guidelines for Drinking Water Quality. Vol. 1 Recommendations.' 2nd ed. (World Health Organization: Geneva.)

Williams, W.D. (1982). Australian conditions and their implications Proceedings of a Symposium Australian Academy of Science, Prediction in Water Quality, pp. 1–10.

Williams, W.D. (1992). The biological status of Lake Corangamite and other lakes in western Victoria. *Report to Department of Conservation and Environment, Colac, Victoria.*

Young, W.J., Marston, F.M., and Davies, J.R. (in press). Nutrient exports and land use in Australian catchments. *Journal of Environmental Management.*

Zann, L., 1995. Our Sea, Our Future: Major Findings of the State of the Marine Environment Report for Australia. Townsville, Qld. Published by the Great Barrier Reef Marine Park Authority for the Department of the Environment, Sport and Territories, Ocean Rescue 2000 Program.

Acknowledgments

The following people reviewed the chapter in draft form and provided constructive comments.

Professor Barry Hart (Monash University)

Dr Richard Pearson (Centre for Tropical Freshwater Research, James Cook University)

Dr John Williams (CSIRO Division of Soils)

We especially thank the following individuals who assisted in the preparation of this chapter.

Professor Angela Arthington (Centre for Catchment and Instream Research, Griffith University)

Dr Peter Crabb (Australian Defence Force Academy, University of NSW)

Professor Peter Cullen (University of Canberra)

Mr Ray Evans (AGSO Canberra)

Ms Lyn Hutchison (CSIRO Division of Water Resources)

Mr Surrey Jacobs (NSW Botanic Gardens)

Mr Trevor Jacobs (Murray–Darling Basin Commision)

Dr Peter Liston (Department of Environment, Land and Planning, ACT)

Mr Ian Lawrence (CRC for Freshwater Ecology)

Ms Jacqui Olley (CSIRO Division of Water Resources)

Mr Col Rosewell (NSW Department of Land and Water Conservation)

Dr Ian Rutherfurd (CRC for Catchment Hydrology)

Dr Allan Wade (ACT Electricity and Water)

Professor Bill Williams (Department of Zoology, Adelaide University)

In addition, Commonwealth and State Government departments and members of the Commonwealth/State ANZECC State of the Environment Reporting Taskforce also helped identify errors of fact or omission.

Photo credits

7-1: Nick Alexander (Oryx Films)
7-06: Kath Bowmer (CSIRO Division of Water Resources)
7-08: Nick Alexander (Oryx Films)
7-09: Murray-Darling Basin Commission
7-12: Bill Van Aken (CSIRO Division of Water Resources)
7-13: Murray-Darling Basin Commission
7-14: University of Queensland
7-16: Bill Van Aken (CSIRO Division of Water Resources)
7-19: Kath Bowmer (CSIRO Division of Water Resources)
7-20: Bill Van Aken (CSIRO Division of Water Resources)
7-25: Ray Volker (University of Queensland)
7-26: Kath Bowmer (CSIRO Division of Water Resources)
7-28: Bill Van Aken (CSIRO Division of Water Resources)
7-31: Peter Room (CSIRO Division of Entomology)
7-32: Wayne Fulton (Inland Fisheries Commission, Tasmania)
7-35: J Johnston (Environment & Land Management, Tasmania)
7-40: Chris Mobbs (ANCA)
7-42: (*from top*) Ray Volker (University of Queensland); Murray–Darling Basin Commission
7-48: Bob Banens (Murray–Darling Basin Commission)
7-50: Gustaaf Hallegraeff (University of Tasmania)

Chapter 8

Estuaries and the Sea

Prepared by

Bernard Bowen (Chair)

Trevor Ward (Deputy Chair), CSIRO Division of Fisheries

Alan Butler, University of Adelaide

Phillip Cosser, Queensland Department of Environment

Nigel Holmes, Kinhill Engineers Pty Ltd (Technical Secretary)

Derek Staples, Bureau of Resource Sciences

Leon Zann, Great Barrier Reef Marine Park Authority

Allan Haines (State of the Environment Reporting Unit member),
    Department of the Environment, Sport and Territories (Facilitator)

Contents

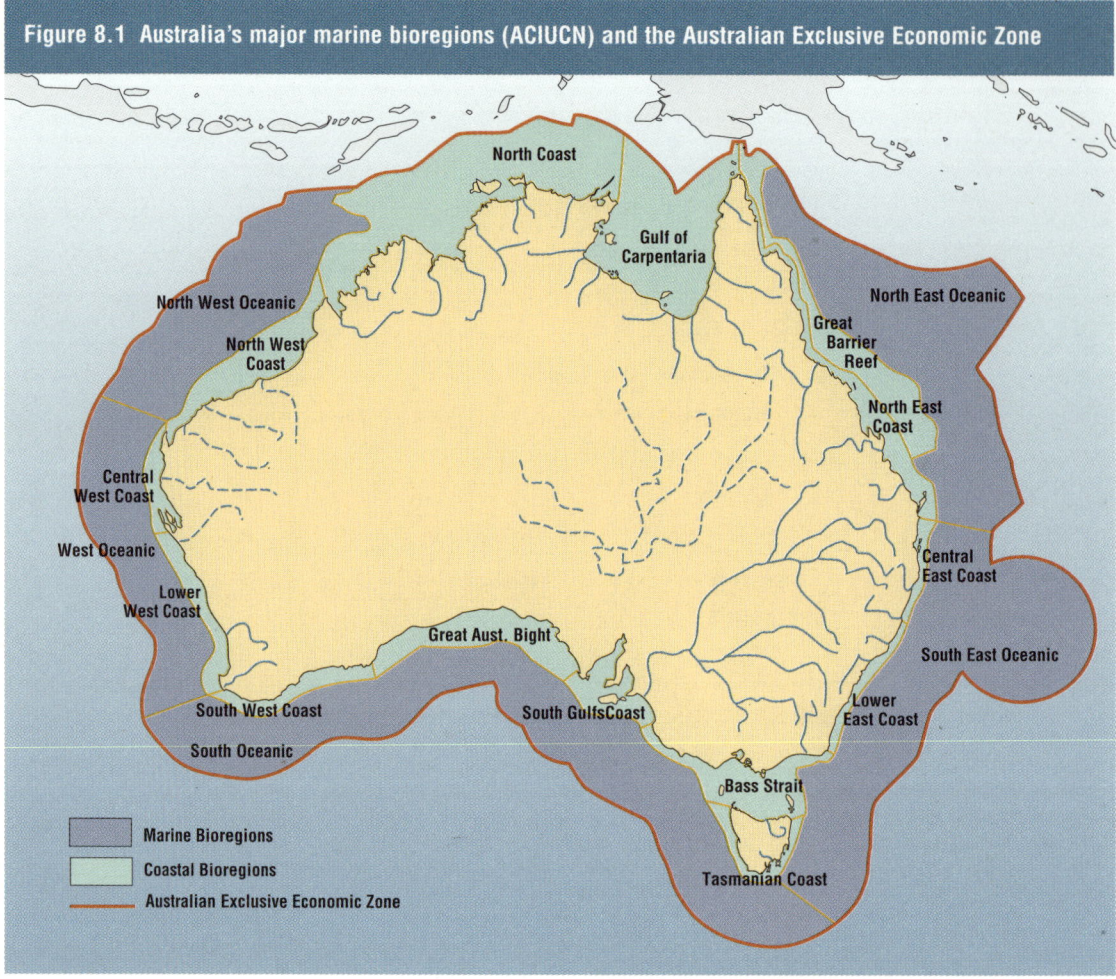

Figure 8.1 Australia's major marine bioregions (ACIUCN) and the Australian Exclusive Economic Zone

North Coast

Gulf of Carpentaria

North West Oceanic

North East Oceanic

North West Coast

Great Barrier Reef

North East Coast

Central West Coast

West Oceanic

Central East Coast

Lower West Coast

Great Aust. Bight

South East Oceanic

South West Coast

South GulfsCoast

Lower East Coast

South Oceanic

Bass Strait

Tasmanian Coast

Marine Bioregions

Coastal Bioregions

Australian Exclusive Economic Zone

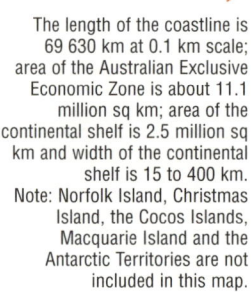

The length of the coastline is 69 630 km at 0.1 km scale; area of the Australian Exclusive Economic Zone is about 11.1 million sq km; area of the continental shelf is 2.5 million sq km and width of the continental shelf is 15 to 400 km. Note: Norfolk Island, Christmas Island, the Cocos Islands, Macquarie Island and the Antarctic Territories are not included in this map.

Introduction

Australia, as an island continent with a long coastline, has many different marine and estuarine environments. Most of these are far removed from the major population centres and are little affected by human activities. The northern, far north-eastern and most of the western coasts of the continent, the Great Australian Bight and Australia's External Territories in the Indian Ocean, South Pacific, Southern Ocean and Antarctica are among the least-polluted places on earth. However, even the most remote regions show traces of persistent global pollutants.

In 1995, the Commonwealth Government published a major report — 'Our Sea, Our Future' (Zann, 1995) — which indicated the growing recognition of the importance of estuaries and the sea. Together with its associated technical papers, the report has provided substantial reference material.

Many different environments

The marine and estuarine environments around Australia's mainland and its many islands cover a considerable area and span a wide range of coastal types, climates, geological and biological regions (see Fig. 8.1). Two major ocean boundary currents — the East Australian Current and the Leeuwin Current — influence the east and west coasts of the continent respectively. (see Fig 2.10). The strength, seasonality and southward extension of both these currents are

highly variable and their flow can influence coastal and ocean conditions along the south of the continent.

Australia also has responsibility for southern subpolar territories, such as Heard Island, and polar (Antarctic) regions. However, these areas are subject to separate management and treaties, and are not included in this report.

The best recognised of Australia's marine environments are those populated near-shore areas, such as rocky shores, beaches and intertidal reefs, that are used for recreation and tourism. The near-shore shallow areas also contain a wide diversity of species and are easily accessible. However, Australia's marine environment extends from these coastal shallows to the boundary of its 200-nautical-mile Exclusive Economic Zone (EEZ). This environment includes large areas of the seabed that are important for fishing, oil and gas production, and possibly mining, and areas of water that, in places, are highly productive biologically. The water provides an important pathway both for pollutants and for early life stages of marine plants and animals. As a result, many habitats, although distant from each other, may be interconnected.

Estuaries are semi-enclosed water bodies at the border of marine and fresh-water ecosystems. They are influenced by the tides and also by fresh water from the land. Australia has 783 major estuaries

(415 tropical, 170 subtropical and 198 temperate) (Zann, 1995). Few exist on the long arid coastlines of southern and western Australia. Estuaries are ecologically important habitats, which usually contain naturally high concentrations of nutrients, high productivity and wide biological diversity. Many species of invertebrates, fish, birds and mammals depend on estuaries for feeding, spawning and/or nursery grounds at some stage in their life cycle.

Australia's coastal population creates significant pressures on estuarine ecosystems from urban, agricultural, industrial, tourist and recreational development.

Water bodies that become increasingly saline as their distance from the open sea increases are called inverse or reverse estuaries. Worldwide, inverse estuaries are uncommon and usually occur in arid areas that do not have sufficient fresh-water inflow, or sufficient sea-water flushing, to compensate for evaporation. Australia has several large and important inverse estuaries: Gulf St Vincent and Spencer Gulf in South Australia; Shark Bay and Exmouth Gulf in Western Australia; and some estuaries in northern Australia. The biological communities in these estuaries are consistently subjected to very salty water and, like those in other estuaries, are susceptible to disturbances such as nutrient inputs, suspended sediments and alterations to fresh-water inflows.

Plants and animals

The variation from tropical to temperate latitudes in Australia has created a vast range of biological communities that live in marine and estuarine environments. These environments can be divided into a number of broad regions (see Fig. 8.1) which include well known habitats such as coral reefs, algal reefs, seagrass beds, mangroves and saltmarshes, but also the less understood midwater, outer-shelf and deepwater habitats.

All major groups of marine organisms are represented in Australian waters, and many of the highly diverse species are endemic here. In southern Australian waters, which have been geographically and climatically isolated for around 40 million years, most of the known species are endemic, or restricted to the area. In the waters of northern Australia, which are connected by currents to the Indian and Pacific Ocean tropics, most species are shared with the Asia–Pacific region. Australia has the world's largest areas and highest species diversity of tropical and temperate seagrasses, largest area of coral reefs, highest mangrove species diversity and third-largest area of mangroves (Zann, 1995).

Australia's marine and estuarine environments are important for fishing, maintaining biodiversity, providing genetic resources and for their aesthetic, social and heritage values.
Clockwise from top left:
Kalbarri Cliffs, Western Australia; Coffin Bay, South Australia; northern Great Barrier Reef, Queensland; and Cottesloe Beach, Western Australia.

These satellite images show the variation in sea surface temperatures around Australia. The subtropical regions experience the greatest changes in surface water temperature from summer to winter. All areas show considerable variation from year to year.

Source: C. Rathbone, CSIRO Division of Fisheries

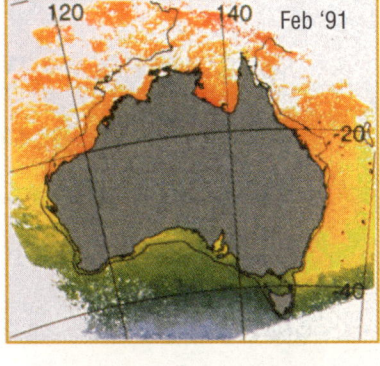
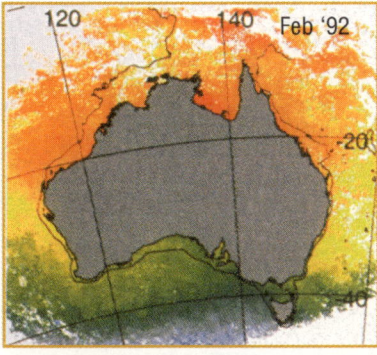
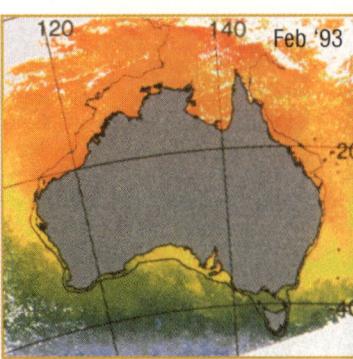
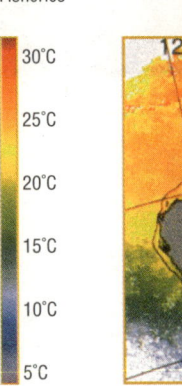
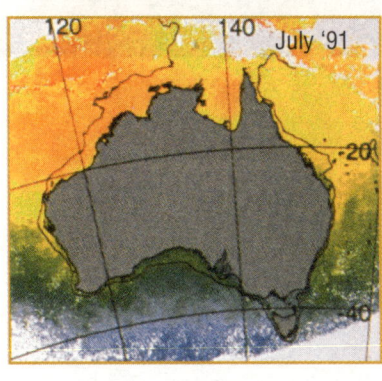
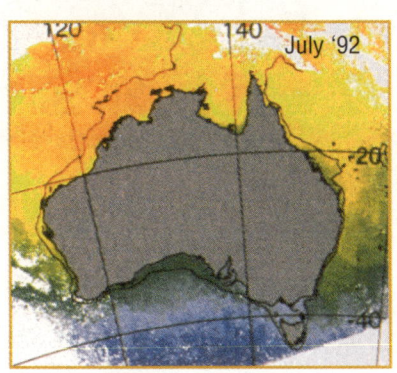
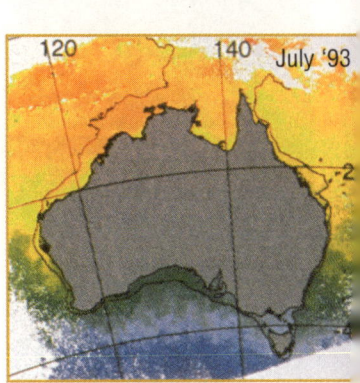

Productivity

Australian near-shore marine waters are generally low in nutrients, which means they have relatively low numbers of phytoplankton and therefore low productivity (see Fig. 8.2). Although upwellings bring nutrients from deep ocean waters to the surface, these are localised and episodic (occurring off parts of New South Wales, South Australia and the North West Shelf). There are no large and continuous upwellings and, historically, run-off from the land has been naturally low in nutrients. Many of the marine species are adapted to low nutrient conditions. Nonetheless, seabed productivity is high in certain habitats (particularly near shore) and in Australia a much greater importance is attributed to the productivity of plants attached to the sea floor (seagrasses, macroalgae, mangroves) and coral zooxanthellae than in some other parts of the world (Zann, 1995).

The great variability in the Australian climate — particularly the variability in run-off from the land — creates pulses of increased nutrient supply to coastal waters from time to time. Important ocean upwellings also occur, although these are generally localised and seasonal. Marine and estuarine communities have evolved under this regime of low, but in some places highly variable, nutrient supply.

Biological systems can absorb occasional high fluxes in nutrients, but constant (chronic) inputs, even at low levels, can cause significant imbalances in some communities. For example, temperate seagrasses and coral reefs, both considered highly productive, are found in clear waters with few phytoplankton. Adding nutrients to these systems over a long period of time (even if the increase in nutrient concentrations is only slight) can cause major changes. The elevation of nutrient levels near sewage or industrial effluent outfalls has favoured the growth of algae attached to seagrass leaves (epiphytes). These shade and weigh down the leaves, reducing photosynthesis, and can eventually kill the seagrass.

In estuaries, nitrogen and phosphorus are the key nutrients supporting growth of plants and phytoplankton. However, in marine systems phosphorus is readily available in high concentrations, and forms of nitrogen are probably more limiting for plant growth. Both elements can cause imbalances when discharged to estuaries and the sea because of their tendency to stimulate the growth of certain algae to the detriment of other species.

This satellite image shows the average distribution of surface water chlorophyll over the period 1978 to 1986. High concentrations of phytoplankton are consistently recorded in southern ocean waters (known as the Subtropical Convergence), in the Gulf of Carpentaria and in other inshore tropical waters. Other areas, particularly northern oceanic waters, are consistently low in phytoplankton.

Figure 8.2 Average densities of phytoplankton, indicated by chlorophyll concentrations measured by satellite

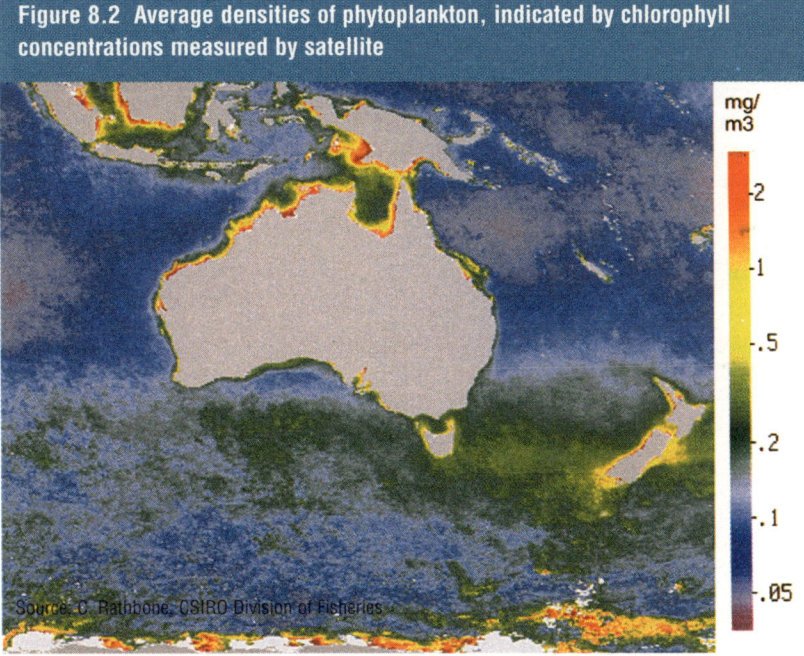

Source: C. Rathbone, CSIRO Division of Fisheries

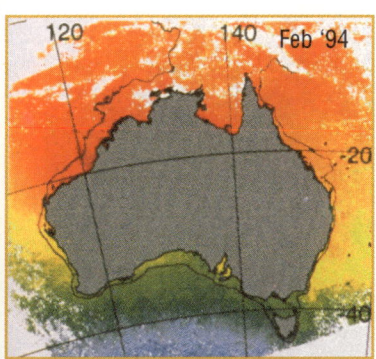

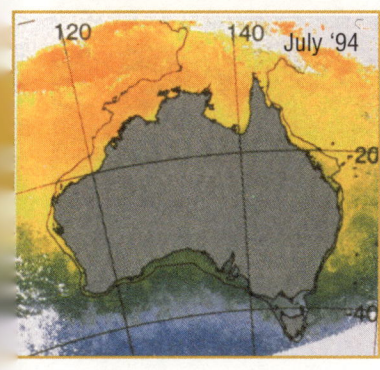

Pressure

Marine and estuarine environments are naturally variable. Large-scale natural processes, like rainfall, wind, wave energy and ocean currents, are continuously causing change. The alterations that we make to the environment need to be considered in the context of this natural variability.

Many aspects of human endeavour in Australia contribute to the pressures on estuaries and the sea. These include both direct pressures and downstream impacts from land-based activities. Some of the heaviest pressures on the environment come from: catchment activities; coastal margin development; contaminants (point and diffuse sources); fishing (commercial, recreational and traditional); mining (coastal and off-shore); and transport (ports and shipping).

It is difficult to identify and measure all of these pressures but, for some, practical indicators may provide an indirect measure of their severity. The size of coastal cities or towns, for example, may provide a broad index of the increase or decrease of pressure on the marine environment. Other useful indicators are: the percentage of catchments cleared; the number of visitors per year; the volume and treatment level of effluent discharged per day; and the number of fishing licences or fishing days.

Coastal margin development

Human settlement in Australia has developed at a series of points around the coast, concentrated on what are now the capital cities of the States and Territories (see the box on page 8-8 and Chapter 3). As the populations in these settlements grow and infrastructure for urban development proceeds,

changes occur to nearby marine and estuarine environments. People also develop adjacent land areas for activities such as farming and recreation (for example, Queensland's Gold Coast). Pressures from development have been particularly concentrated around the east, south-east and south-west of Australia, where most people live. Although coastal development has directly affected a relatively small proportion of the continent's coastline, many changes have taken place in areas of ecological importance with particular aesthetic appeal. About 70 per cent of Australia's coastline remains sparsely inhabited (Zann, 1995).

Land reclamation, modification of habitats and hydrology, tourism, recreational activities and stormwater discharge and other point sources of pollution all exert major pressures.

Coastal engineering structures such as breakwaters and seawalls associated with ports, harbours, airport runways and canal estates and marinas also have an impact. Estuaries and the coastal lakes and lagoons in the south-east have been particularly affected by seawall construction. Shoreline erosion is an increasing problem in many areas due to alteration of coastal morphology. We can expect this problem to accelerate in the future if the sea level continues to rise.

Habitat destruction and modification

Habitat destruction may occur when the environment is altered to create new land or seabed surfaces suitable for development. It may be caused by land reclamation, canal development, dredging for harbours and similar activities. Most reclamations involve the enclosure and infilling of

Habitat destruction at Wingfield, South Australia. A levee road through mangroves and saltmarsh showing the effect of restricted water exchange.

▶ North Haven, South Australia. This housing development, fishing port and base for recreational activities on the shores of Gulf St Vincent exemplifies coastal habitat modification.

former dunes, saltmarsh, mangrove, mudflat and shallow-water areas for port and industrial developments or disposal of waste materials. Such activities cause a direct loss of natural resources, including the plants and animals that would normally live in these habitats.

Human pressures

Since European settlement, humans have changed coastal processes by clearing and developing land, using resources and changing the flow of rivers. The major pressures on our coast come from the increasing pace of human activity.

Australians enjoy the right to live and work on the coast and to use coastal and shore areas for their recreation — for beach sports, boating, bait-collecting, diving, fishing, sunbathing and surfing. In 1993 the coastal zone supported about 80 per cent of the Australian population and 66 per cent of these lived in coastal cities and large towns.

The population of the non-metropolitan coastal zone is growing at an unprecedented rate. Between 1971 and 1991 it grew by 95 per cent (from 2.1 million to 4.1 million), while the total Australian population grew by 32 per cent.

The most severe pressures on the coastal environment are related to residential development in the most accessible 10 per cent of the coast in and around urban areas, particularly those in the south-east. A further 15–20 per cent is subject to increasing development and eventual urbanisation. Between 1983 and 1990 more than three-quarters of all building approvals concerned the one-third of statistical districts that had coastal populations growing at more than double the national average rate between 1971 and 1991. Over 63 per cent of coastal building approvals were for residential developments and more than half of these (34 per cent) concerned non-metropolitan coastal statistical areas.

Many of the pressures have grown gradually, but they usually interact. Urban growth involves land clearing, damming for water supply, flow modification for flood control, pollution (from sewage, urban and industrial waste disposal) and the impacts of meeting a rising demand for fishing, tourism, coastal recreation and a wide range of transport, health, education and community infrastructure and services.

Source: Kenchington, *in press*

There are no firm data on the extent of land reclamation. Although each of the developments only affects a relatively small area, they can have a substantial accumulated effect as in the case of Moreton Bay (see the box opposite).

The effects of land reclamation can extend well beyond its physical boundaries to affect habitats in adjacent intertidal and subtidal areas. Reflected waves and diverted currents may remove sediment, which may be deposited in areas that become more sheltered (Bird, 1994). Many structures alter the shape of the shoreline, its substrates and seabed level. More direct impacts arise from activities such as selective clearance of vegetation. An increasing source of pressure is the disturbance of acid sulfate soils (see Chapters 4 and 6), which results in increasing acidification of estuaries and death of animals and plants. In some cases habitat modification results in species being replaced by others that are better-adapted to the new conditions.

Hydrological modification

Coastal margin development usually entails structural modification of the shoreline to control or modify the environment. Breakwaters and groynes, for example, can alter water movement along the coast. Most of Australia's rivers that feed estuaries are dammed to provide water for urban supply or power generation (see Chapter 7). This has major impacts on the hydrological patterns and occurrence of above-ground and below-ground flows of water. A dramatic example arose in 1981 when the mouth of the Murray, Australia's largest river system, became completely blocked.

Urban and agricultural developments usually reduce the infiltration of rainwater into the ground and increase the run-off of water and suspended sediments to the sea. On a local scale this can alter patterns and timing of surface and underground water availability, thus causing changes in vegetation and availability of habitat resources to wildlife. Changes to the hydrology can also result in changes to physical, chemical and biological processes such as nutrient cycling, soil development, sediment settlement, erosion and movement of salts.

Recreation

Australia's coastal region has important natural tourist and recreational resources that attract many domestic and overseas visitors. The international tourist industry — most of which is based on the coast — contributes more than six per cent of GDP and is a growth industry. In 1994 Australia had more than 3.3 million international visitors. About half of these participated in some sport or outdoor activity: 900 000 surfed or swam, 580 000 snorkelled or scuba dived, 58 000 went fishing and 102 000 sailing (Bureau of Tourism Research, pers. comm.). During 1991, Australians made 49 million trips as domestic tourists, and spent 215 million visitor nights away from home, mostly in coastal areas (Zann, 1995).

On midsummer weekends, individual Sydney beaches attract crowds of up to 50 000, and Melbourne's Port Phillip Bay beaches attract up to 300 000 bathers. Queensland's Gold Coast, one of Australia's most popular holiday destinations, attracts over nine million visitors each year.

All these visitors require a range of public facilities, from accommodation, transport, shopping, toilets and boating facilities to parks, sports grounds, sewerage, water and electricity supply systems and garbage disposal services. These facilities are typically placed on or near the particular attraction, which may be a picturesque and secluded beach, inlet or island, thus changing its natural and cultural environment and reducing the scenic values (Dutton and Luckie, 1994).

Although tourism and recreation are generally considered to be clean industries, they have had significant negative impacts on many parts of Australia's coastal strip. These negative effects may include beach and dune erosion, loss of habitat to facilities, trampling of reef, collecting of intertidal plants and animals, declines in wildlife and fisheries, reduction in water quality, disturbance of animal feeding and breeding behaviour, seabed damage by boat anchors and moorings and the effects of antifouling materials.

Stormwater discharge

Stormwater discharges to estuaries and the sea are a common feature of coastal urban areas. People have only just started to recognise that stormwater carries significant quantities of debris and contaminants, like litter, oil and heavy metals, that may cause environmental harm.

We do not know the level and content of such discharges on a national basis, but in places, stormwater can carry significant quantities of suspended sediments, nutrients and pesticides, chemicals from domestic and industrial discharge and accidents, vehicle-emission wastes from roadways and litter. For example, researchers in Victoria have estimated that drains, creeks and small rivers (excluding the Werribee Complex and the Yarra River) annually carry about 53 tonnes of surfactants (from

Coastal development in Moreton Bay

Moreton Bay is an estuarine bay enclosed by barrier islands. Biotic communities include mangroves, saltmarshes, seagrasses and coral reefs. The bay is recognised as a significant habitat and nursery area for estuarine and oceanic fish, and an important habitat for turtles, dugongs, dolphins and seabirds. It provides about 10 per cent of the value of east coast fisheries and employs about 1000 full-time fishers. The annual recreational catch (1900–2800 tonnes) is also significant.

Central to Brisbane and surrounding areas, Moreton Bay has been subjected to intense pressure from coastal development — particularly through reduced water quality, fishing, coral-dredging, sand extraction and port operations.

Suspended sediments, heavy metals, nutrients and microbial pathogens from the Brisbane River reduce its water quality. These pollutants arise from both diffuse catchment and point sources. In 1987, researchers estimated that 17 per cent of mangroves and 21 per cent of saltmarsh had been lost to coastal development since European settlement, and that there had also been large impacts on seagrass and algal communities (Queensland Department of Environment and Heritage, 1991). The most recent major development has been Brisbane's International Airport.

The government and community have become increasingly aware of issues affecting Moreton Bay with a resultant intense pressure for protection measures. Piecemeal and uncoordinated decision-making on development issues has had a cumulative detrimental effect and is listed as a significant factor in resource degradation and lack of knowledge about the bay. Significant changes have, however, recently occurred in political, administrative, planning and management attitudes at federal, State and local levels. The Moreton Bay Strategic Plan covers areas designated for preservation and habitat conservation, National Estate registration of areas by the Australian Heritage Commission, coordinated management of the Brisbane River and various local initiatives for foreshore management.

The bay's social, economic and ecological importance is being increasingly recognised and the need for conservation and sustainable management has become a significant factor in decision-making about its uses and development.

detergents) into Port Phillip Bay (NSR Environmental Consultants, 1993). In New South Wales, urban stormwater is regarded as a major pollutant of the coastal environment. Between Palm Beach and Cronulla 200 stormwater outlets of about 450-mm diameter discharge water containing high levels of pollutants such as sediments, bacteria, nutrients, trace metals and organic chemicals (Environment Protection Authority of NSW, 1993). These pollutants are derived from pets (dogs and cats) as well as from vehicles, industry emissions, gardens and roads.

Table 8.1 Distribution of commercial tourist accommodation in the coastal zone, 1990–91

	Hotels and motels	Caravan parks	Holiday units, flats & houses	All tourist accom.
Capital cities	960	312	210	1482
Non-metropolitan coastal zone	1824	1248	1050	4122
Total in coastal zone	2784	1560	1260	5604
Total in Australia	4800	2600	1400	8887
Percentage in coastal zone	58	60	90	64

Source: RAC 1992 & 1993

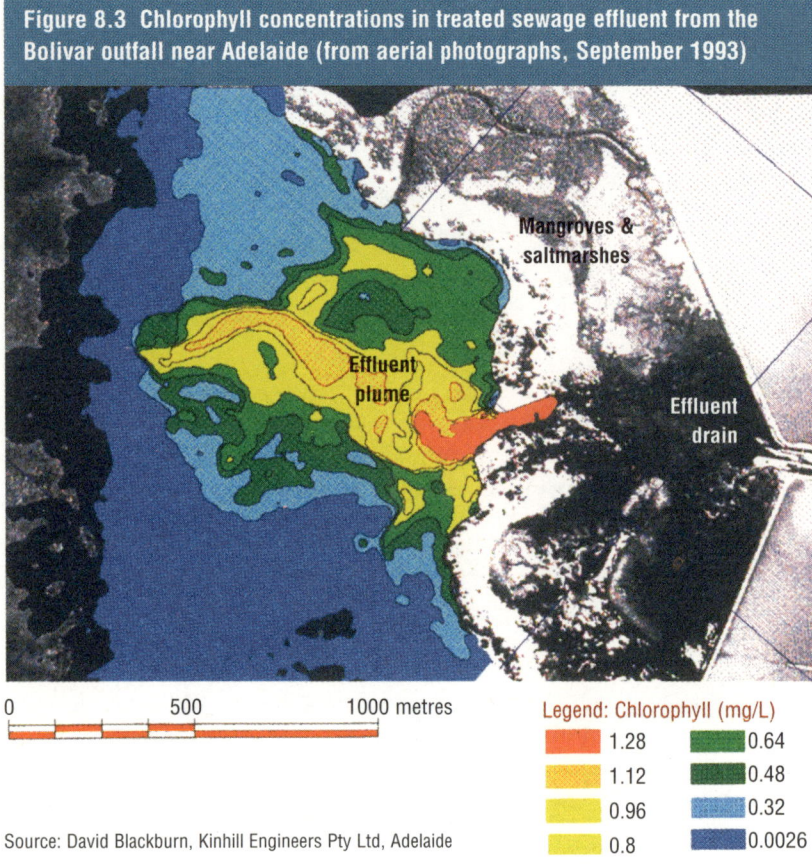

Figure 8.3 Chlorophyll concentrations in treated sewage effluent from the Bolivar outfall near Adelaide (from aerial photographs, September 1993)

0   500   1000 metres

Legend: Chlorophyll (mg/L)

1.28	0.64
1.12	0.48
0.96	0.32
0.8	0.0026

Source: David Blackburn, Kinhill Engineers Pty Ltd, Adelaide

▲

Sewage effluent from the outfall at Bolivar near Adelaide enhances growth of both phytoplankton and benthic algae. The area has lost extensive seagrass beds

Contaminants

About 80 per cent of contaminants in marine and estuarine environments are thought to enter the sea from the land (Zann, 1995), both from direct point sources such as pipes and drains and from diffuse sources such as river and urban catchments, and from the atmosphere. The main sources include: stormwater run-off that carries contaminants like heavy metals, oils and litter; agricultural run-off containing fertilisers, pesticides and suspended sediments; and sewage effluent discharges, which carry organic matter, nutrients, pathogens and industrial wastes.

Some of the contaminants persist in the environment and can become concentrated in fish and other organisms. Pollutants do most damage in coastal and inshore waters close to the sources. However, the more persistent contaminants may also affect offshore waters, although no data are available to confirm this.

Nutrients

Australia's marine and estuarine systems have evolved with relatively low inputs of nitrogen and phosphorus (the main nutrients controlling the natural growth of marine plants) derived from land run-off and ocean upwelling.

One of the most serious large-scale threats to Australia's near-shore marine environment is the input of excessive nutrients (see page 8-13). Tropical coral systems and temperate seagrass beds are highly sensitive to nutrient impacts. Estuaries and coastal lagoons whose upper river catchments have been cleared for intensive agriculture and

whose lower reaches are subject to major urban and industrial developments are also at risk (Zann, 1995). Australia suffers blooms of nuisance microalgae, including toxic species, and macroalgae that can have significant environmental, social and economic impacts. These algal blooms can degrade ecosystems, decrease the recreational value of waterways, affect human health and destroy aquaculture production.

Soil erosion, fertiliser run-off, sewage discharge, intensive animal production and industrial and urban discharges all contribute nutrients to the marine environment. Fertilisers are a significant source of nutrients from agricultural areas, as are sediment leaching and erosion from cleared land (see Chapter 7). Scientists believe that in Queensland the amount of sediments, nitrogen and phosphorus entering the sea each year have increased three to fivefold since European settlement (Moss *et al.*, 1993).

Sewage effluent from urban areas contributes significant amounts of nutrients. Most sewage receives only primary or secondary treatment and the effluent remains high in nutrients (see Table 3.38). Each year Australia's sewerage systems discharge around 10 000 tonnes of phosphorus and 100 000 tonnes of nitrogen (Brodie, 1995), much of which finds its way into the sea. However, as mentioned before, stormwater run-off from cities also contains a large quantity of nutrients from animal and other wastes, which may equal that generated from urban sewage (Zann, 1995).

Suspended solids

Suspended solids can include fine sediments and very small particles of organic debris. Clearing of forests, overgrazing and agriculture have greatly increased soil erosion and, consequently, the amount of sediments and suspended solids entering the sea. Urban development is also responsible for increases in suspended solids associated with run-off, erosion, sewage-sludge discharges and localised activities like construction, dredging and dredged-spoil disposal. As well as contributing nutrients, suspended solids can affect marine and estuarine communities by smothering sedentary plants and animals, clogging gills, reducing the light available to plants for photosynthesis and altering seabed level and sediment grain size (Zann, 1995).

The rivers of Queensland's east coast catchments are estimated to deliver about 14 million tonnes of sediment annually to estuaries and coastal marine waters — three to five times more than before European settlement (Moss *et al.*, 1993). Most of this sediment comes from the large areas of agricultural land in central and northern Queensland. The finest components of these sediments are widely dispersed into estuaries and the sea.

Pesticides, herbicides and industrial chemicals

Organochlorine pesticides are a particularly persistent and toxic group of synthetic chemicals. The characteristics that make them effective pesticides also make them potential environmental threats. Although organochlorines are only present in very low concentrations in the sea, they can accumulate in animal fats up to 50 000 times the concentration in surrounding waters (Richardson, 1995). They may also accumulate through the food chain into predatory fish, seabirds, marine mammals and humans. Low concentrations of these pesticides and polychlorinated biphenyls (PCBs) contaminate whales and dolphins in all the world's oceans.

A range of organochlorine compounds have been widely used in Australia as herbicides (2,4-D and 2,4,5-T), insecticides (DDT, DDE, lindane and chlordane), fungicides (hexachlorobenzene and chlorinated phenyls) and as electrical insulating fluids (PCBs). Organochlorines are also produced in processes such as chlorine bleaching in the making of paper pulp. Now, they can only be used in Australia for specified purposes like protecting houses from termite attack. Although many former agricultural areas still suffer from contamination, little of this migrates to estuaries or to the marine environment.

Other forms of organic pesticides (such as organophosphates) are still widely used in Australia, often as substitutes for the former applications of organochlorines. An example is endosulfan, a widely used pesticide that has replaced several organochlorines for a variety of uses. There is no coherent national monitoring program for residues of these persistent chemicals in estuaries or marine systems and little is known of their present levels (Zann, 1995). Most studies suggest that their levels are low in water, seafood and sediments, except near known point sources.

Metals

Heavy metals such as copper, lead, cadmium, zinc and mercury, as well as tributyl tin (TBT) from antifouling paints (see page 8-17), have become serious contaminants in the world's estuaries and coastal waters in recent years. Heavy metals can be present on particles or as dissolved compounds — the chemical form strongly determines how toxic or available the metal is to organisms (Batley, 1995). Marine and estuarine species may suffer heavy metal poisoning, but the more likely outcome is chronic exposure at lower levels, with the consequent subtle effects including reduced breeding potential or susceptibility to disease. Of particular concern is the potential impact on humans through consumption of seafood with concentrated heavy metals from water or suspended solids (see page 8-37).

In Australia, most concern for the effects of heavy metals is associated with the discharge of leachates, tailings and waste from point sources like smelters and mining operations. Mining is a major industry in Australia and all States and the Northern Territory have smelters, mines or transport corridors with the potential to release metals to estuaries and the sea. Although some release large amounts of waste metals, their effects are limited to the local surrounding area, usually less than a few hundred square kilometres. However, the large

number of these sites is a matter for concern. Australia has no national inventory of the number of point sources for major heavy metal discharges, the amounts discharged or the severity of effects in their local areas. Well-documented examples of affected areas are Spencer Gulf near Port Pirie and the Derwent River near Hobart.

Data on amounts of heavy metals discharged are available only in specific local cases. These figures show a general trend towards reduction in heavy metal discharges from established sources.

Pathogens

Pathogens in the marine and estuarine environments can include a number of natural and introduced soil and faecal bacteria and viruses. Many of these may originate from stormwater (see Table 8.2). Some seafood species, such as oysters and mussels, are filter feeders and can act as concentrators of pathogens. We know little about how pathogens survive in estuaries and the sea or how both endemic and introduced pathogens affect marine and estuarine biota. Viruses in particular are poorly understood, because of difficulties in culturing them, but diseases such as polio and the hepatitis 'A' and 'E' viruses have been associated with swimming (Zann, 1995) and some viruses are reported to be persistent in marine waters.

People come into contact with marine and estuarine pathogens by eating contaminated seafood or enjoying recreational water sports — especially in enclosed water bodies such as bays and estuaries. The microbes can cause illnesses such as gastroenteritis, hepatitis, conjunctivitis and upper-respiratory tract and wound infections (Ashbolt, 1995).

The incidence of pathogens can also have substantial economic implications. In fisheries or aquaculture, for example, contaminated stocks may need to be held in clean or chlorinated water to allow the pathogens to be flushed out before the food is safe for consumption. Contaminated areas may need to be closed to water sports, with consequent losses to recreation and tourism.

Bacteria also cause problems in bathing waters. Beach areas adjacent to stormwater drains can experience increases in the number of indicator bacteria following rain. In Australia, they cause

Table 8.2 Urban stormwater pathogen levels for five Sydney coastal catchments over a two-year period

Percentage of samples with greater than 1000 faecal coliforms per 100 millilitres.

	Dry weather	Wet weather
Whale Beach	98	100
Greendale Creek	98	100
Bondi	98	100
Shelly Beach	44	–
Malabar	11	10

Source: after EPA NSW, 1993.

common wound infections (particularly in people who handle fish) and 'swimmer's ear'. Parasitic protozoans may be transmitted in sea water (Zann, 1995) and ballast water can transport microorganisms.

Some pathogens do kill estuarine and marine organisms. Diseases affecting the oysters in Georges River in Sydney are a well-known recurrent problem.

Litter

Many of Australia's beaches are littered with plastic bottles, plastic bags, tangled fishing lines, nets and other rubbish. This litter is dropped by beach-goers, carried by stormwater run-off, dumped from ships, recreational and commercial fishing boats and also comes as 'drift' from remote sources far across the ocean. Urban beaches are worst affected, but even the most remote coastal and island beaches are not free from litter, often from distant sources (Wace, 1995). However, recent scientific surveys of beaches near Brisbane, Sydney and Melbourne found that most litter came from streets and garbage dumps via streams and drains.

Not only does litter spoil the appearance of the shoreline, it may also endanger marine life. Worldwide, many thousands of marine mammals, turtles and seabirds die each year from swallowing plastic bags and other objects, or get trapped in discarded fishing gear. This gear may also continue to 'catch' fish (ghost fishing). Many seabirds drown each year after becoming entangled in nets of the world's gillnet fisheries. Some seabirds and turtles swallow plastic debris, although the effects of this on their populations are unknown. Overall, researchers have estimated that seven billion tonnes of debris enters the world's oceans each year, 48–99 per cent of which is plastic. While most of this probably occurs in the Northern Hemisphere, the transportation of seeds, spores, eggs, larvae etc, into the Southern Hemisphere on debris may be an issue of concern (ANZECC, 1995).

Litter damages a wide range of marine organisms. Common problems include urban rubbish and lost fishing equipment.

Fishing

Australia makes good use of its diverse fish fauna and Australians enjoy a plentiful supply of fresh seafood. Many also enjoy recreational fishing. Virtually all estuarine, near-shore and off-shore areas support fishing, which may be commercial, recreational or traditional subsistence fishing, or aquaculture. Only commercial fishing exploits the resources of the deeper waters (depths to about 1400 metres), but recreational fishing is also widespread, extending to remote reef areas.

Methods range from the simple hooks and lines of the recreational fisher to the longlines, nets, traps and trawl nets used in commercial operations. Fishing pressure on marine and estuarine environments has a number of impacts. These include: excessive catches of target and non-target species, alteration of food chains, changing species composition, alteration of the genetic composition of fish stocks, introduction of non-indigenous species (aquaculture), habitat modification (particularly seabed disturbance by trawling) and destruction of in-shore reefs by reef-walking and bait-collecting.

The catch and fishing pressure

In the past, people have used indicators such as the number and size of boats in each fishing fleet, number of hours spent fishing and number of lobster-pot lifts to measure fishing pressure. However, the interpretation of these indicators has become increasingly complicated by technological advances such as the use of spotter aircraft, radar, sonar and satellite navigation systems (GPS–global positioning systems) and increased skill of fishing operators.

In the absence of comprehensive data on fishing effort, the increase in total landings of commercial species over time provides a very crude estimate of increased pressure on fish stocks. The total Australian catch, including crustaceans and molluscs, increased from 73 000 tonnes in 1964–65 to 195 000 tonnes in 1994–95 (see Fig. 8.4).

We do not have enough information to assess the impact of recreational fishing on Australian fish stocks. However, one estimate indicates that the total annual recreational catch may be about 50 000 tonnes (Kearney, 1995). A 1984 survey estimated that more than 4.5 million Australians went fishing at least once a year; more than 800 000 went fishing at least 20 times a year.

Recreational activities such as reef-walking, snorkelling, beachcombing and collecting (for food, bait and shells) are also imposing significant pressures, especially near coastal cities, towns and tourist centres. We do not know the overall impact of these activities, but in some local areas it is intense. Collecting overlaps with recreational fishing in terms of its direct pressure and trampling and rock turning can also damage habitats.

Fishers catch many incidental species, which can include a wide diversity of organisms such as

Nutrients: nitrogen and phosphorus in coastal waters

Nutrients — principally nitrogen and phosphorus — are important to the water quality and productivity of marine and estuarine environments for two reasons. First, they have a fundamental role in the functioning of biological systems; and second, their concentrations and availability to aquatic organisms are susceptible to human manipulation.

Nutrient enrichment has adverse impacts on ecosystem diversity, fisheries stocks and the aesthetic and recreational value of coastal waters.

Sources of nutrients

Nutrients continually enter and leave the marine environment and move, or cycle, between the living and non-living components of marine ecosystems. They enter coastal waters through river inflow, upwelled outer-shelf waters, rainfall, regeneration from sediments, localised discharges of sewage and industrial effluent and, in the case of nitrogen, biological fixation of atmospheric gaseous nitrogen. They leave — or become biologically unavailable — largely through deposition and incorporation into the sediments.

While some of these sources are largely beyond the scope of human influence, others, particularly river inflow and waste water discharges, can be changed by human activity.

Nutrient loads

With increasing population growth and urbanisation, the volume of nutrient-rich sewage and industrial effluent entering localised areas of the coast continues to increase. Nationally, sewage effluent contributed some 10 000 tonnes of phosphorus and 100 000 tonnes of nitrogen in 1994. Much of this is discharged to estuarine and coastal waters.

In certain localised areas, waste water discharges are the main source of nutrients entering coastal waters. However, on a national scale, most nutrients (possibly up to 85 per cent) originate from diffuse catchment sources. In the central Great Barrier Reef region, for example, some 39 per cent of all nitrogen and 52 per cent of phosphorus come from river inputs.

While it is difficult to estimate the quantities of nutrients carried by rivers, we do have some indicative values. Mainland catchments annually discharge an estimated average of 77 000 tonnes of nitrogen and 11 000 tonnes of phosphorus to Queensland's coastal waters of which the Burdekin–Haughton catchment alone contributes some 14 000 and 2000 tonnes respectively. Each year between 1977 and 1991 an average of 75 tonnes of phosphorus was carried into the Harvey Estuary in Western Australia, with up to 130 tonnes in wet years. Parent catchments annually deposit an estimated 170 tonnes of nitrogen and 30 tonnes of phosphorus into Tuggerah Lakes, New South Wales.

Many people now recognise the continuing increase in nutrient loads entering estuarine and coastal waters from both river inflow and waste water discharges as one of the most significant pressures on Australia's coastal ecosystems. As nutrient inputs continue to increase, the pressure on coastal waters will escalate further.

Algae

Elevated nutrient concentrations promote excessive algal growth in many estuaries and bays around Australia.

Scientists first noticed increased algal growth in the Peel–Harvey estuary in the 1960s. The problem became progressively worse during the '70s and '80s, with massive blooms of different species of algae occurring at different times. The progressive and severe deterioration in the water quality of the estuary and hence in its amenity, was due to excessive inputs of phosphorus originating from fertiliser applied in the parent catchment (see pages 7-46 and 7-47).

Excessive inputs of both nitrogen and phosphorus from waste water discharges and diffuse catchment sources have contributed to excessive algal growth in the Tuggerah Lakes. The amount of macroalgae (seaweed) in the lake system has markedly increased in recent decades. As a result, the recreational value of the lakes has fallen in recent years.

The discharge of nutrient-rich waste water to shallow waters off the Caloundra coast, Queensland, caused significant changes to the intertidal algal community near the outfall. Some 15 different species of algae that were present at similar, unaffected sites disappeared from the discharge area, while several species grew prolifically, covering the entire intertidal rock platform with a thick algal turf.

In the Great Barrier Reef region, observers have seen signs of reef degradation due to increased nutrient inflows. However, no one knows whether the increased availability of nutrients has increased the frequency or magnitude of algal blooms.

Responses

In some areas monitoring and research are being carried out as the basis for developing remedial action. On the Great Barrier Reef, for example, recognition of the threat of nutrient enrichment has prompted the Great Barrier Reef Marine Park Authority (GBRMPA) to implement a major nutrient monitoring and research program. The program, directed at identifying the major sources of nutrient inputs, their fate once in the marine environment and their effect on reef ecosystems, will provide the basis for the development of appropriate management strategies.

Where very significant adverse effects on coastal systems are already evident, authorities have tried to remedy the problem and rehabilitate the area. In the Peel–Harvey estuary for example, the Western Australian government developed a three-part program following years of research into the cause of the algae problem. The program consists of catchment management to reduce phosphorus inputs by changing agricultural practices, an algae-harvesting component to remove nuisance deposits and a channel-construction program to increase flushing of the estuary. The $57 million Dawesville Channel was opened on 23 April 1994.

In the case of adverse impacts from sewage or industrial effluent outfalls, such as on the Caloundra coast, the response may be to relocate the outfall to a more appropriate position. Responses commonly applied to water quality problems include the relocation of outfalls, improved treatment or the use of different disposal methods. The GBRMPA, for example, now requires nutrients to be removed from sewage effluent before discharge into Marine Park waters.

Source: Cosser, in press.

▶ Fishing: a livelihood and a way of life. Setting rock lobster pots, Western Australia.

sponges, hard and soft corals, molluscs, crustaceans, fish, marine worms and plants. Some of the non-target species are retained for sale and the rest are returned to the sea as discards. Some fishing methods also unintentionally capture reptiles, birds and mammals.

Commercial fishers use a number of methods, which result in the capture of different types and amounts of non-target species. The discards from prawn trawling, for example, include many species that can make up a high proportion of the catch. Very few of these organisms survive when returned to the water. In the northern and southern prawn trawl fisheries the ratio of non-target discards to prawns can be as high as 8:1 by weight. In 1988, for example, 220 prawn trawlers worked the Australian northern prawn fishery landing 7100

tonnes of prawns and more than 30 000 tonnes of bycatch (Ramm *et al.*, 1990).

Turtles and sea snakes are also taken in prawn trawl nets. The flatback turtle is the main one caught in northern Australia, followed by loggerhead, olive ridley and green turtles. A study of prawn trawlers operating in the tiger prawn segment of the northern prawn fishery found that the trawlers caught around 4100 turtles in 1988, about six per cent of which drowned (Marsh *et al.*, 1995).

Hook fishing, traps and pots generally catch fewer non-target species.

Genetic alteration

Fishing can put selective pressure on fish populations to alter their genetic composition. For example, large-scale pressure for a particular size of a target species of fish could result in a shift in the genetic composition of the species. Similarly, trawling or removal of predators may alter habitats, thus favouring particular individuals with a consequent selection for (or against) particular genetically controlled characteristics. This effect is most likely to occur in small populations or populations that are under large-scale pressure (see pages 4-10 and 4-11).

Aquaculture can result in genetic bias when large numbers of juveniles are produced from selected broodstock (for example, oysters or abalone). If these individuals are released to the wild they can influence the genetic composition of natural populations.

Habitat modification

Some kinds of gear used to harvest fish can affect marine and estuarine habitats. Bottom trawling, in particular, has the potential to alter the habitat. It involves towing nets along the seabed, which catch all animals that cannot escape the path of the trawl.

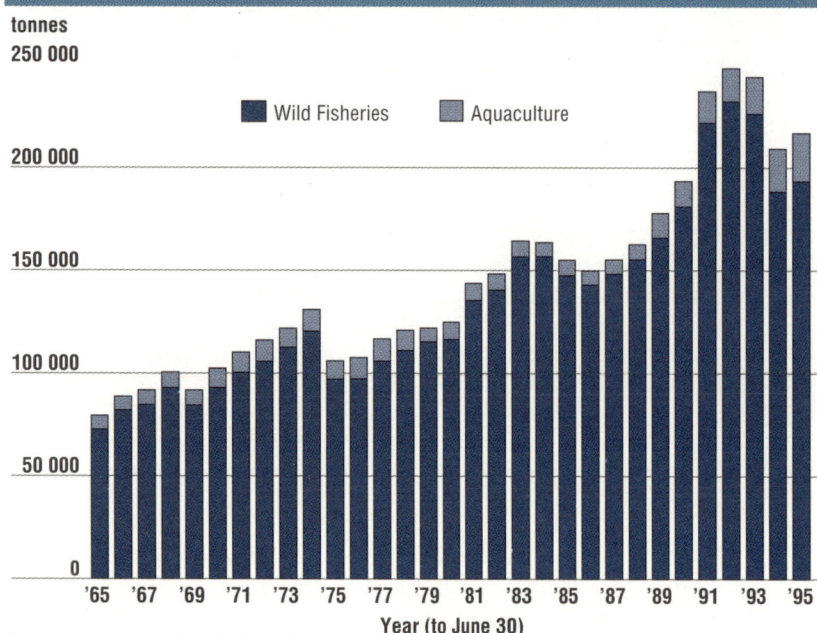

Figure 8.4 Australian fisheries production

tonnes

■ Wild Fisheries   ■ Aquaculture

Source: Kailola *et al.*, 1993; ABARE, 1995.

Year (to June 30)

A number of studies have recently been undertaken and others are in progress to obtain more information on how trawling affects both the habitat and the capture of non-target species. Researchers are also trying to develop trawl gear that is more environmentally friendly.

Mining

Coastal and offshore mining activities that can have an impact on the marine and estuarine environments include: sand and gravel mining for minerals, beach replenishment or industrial uses; oil and gas exploration and production; coral (limestone) mining; diamond dredging; and terrestrial mining.

These activities can cause habitat loss and modification, alter the behaviour of animals, spill hydrocarbon from leaks and flaring, release discharges from platforms and increase concentrations of suspended solids and/or heavy metals (see pages 8-10 and 8-11).

Sand mining and dredging

Sand mining for minerals, beach replenishment and land reclamation occurs in many areas along Australia's coastline, including south-east Queensland, northern New South Wales, Western Australia and South Australia.

It can alter and destroy habitats (see the box on page 8-9), alter seabed level and substrates and generate suspended sediments (see pages 8-10 and 8-17). The effects of sand mining and dredging can potentially alter the species diversity of an area and dramatically change the productivity and composition of seabed-dwellers and associated communities, including commercial fish species. For example, in Cockburn Sound, Western Australia, calcareous sand beneath seagrass beds is mined for cement and lime production, and near Adelaide, near-shore sand dredging is used to replenish sand loss on metropolitan beaches. These activities may have serious effects on the plants and animals living in these disturbed habitats.

Seismic surveys

Seismic surveys are a primary, remote-sensing tool used to explore for oil and gas deposits. Some people have feared that high-energy sound waves could cause mortality or injure nearby marine organisms, or affect their feeding, mating or breeding activities (Zann, 1995; Swan *et al.*, 1994). Seismic surveys are sporadic, of relatively short duration and only likely to have a direct effect on marine organisms that are within 200 metres of the trailing cables (Swan *et al.*, 1994). Such surveys may temporarily affect communications and behaviour of marine mammals in the immediate vicinity of the sound source, but this is unlikely over a wider area (Richardson and Malme, 1994). Survey vessels have a policy of keeping a watch for whales and discontinuing surveys while marine mammals are in the vicinity (Swan *et al.*, 1994).

Given the relatively small scale of seismic activity, the often large areas and the low probability of encountering sensitive populations at critical times and provided seismic surveys are avoided at locations and times of particular sensitivity, the risk for disruption appears to be small for most species.

Oil and gas

Offshore petroleum production has significant economic and strategic importance for Australia. About 86 per cent of our liquid fuels and 75 per cent of our natural gas, worth around $5.4 billion a year, comes from offshore wells in Bass Strait, the Timor Sea and the North West Shelf. Over the past 30 years oil companies have drilled more than 1100 wells offshore and extracted 2800 million barrels of oil. The Australian offshore petroleum industry has a very good environmental record to date and has spilt only about 800 barrels during this period (Zann, 1995).

Crude oil and refined petroleum are complex substances made up of many hundreds of different compounds, including alkanes and aromatic hydrocarbons. The latter, which are toxic, include the polycyclic aromatic hydrocarbons — carcinogens that have been implicated in a wide range of human health problems and diseases in aquatic organisms. Polycyclic aromatic hydrocarbons can occur in small quantities in crude oil and can be produced by incomplete combustion. Most components of oil also accumulate strongly in food chains and bind to organic material in sediments (Connell, 1995).

Dredge spoil dumped at sea has the potential to release large amounts of contaminants ▼

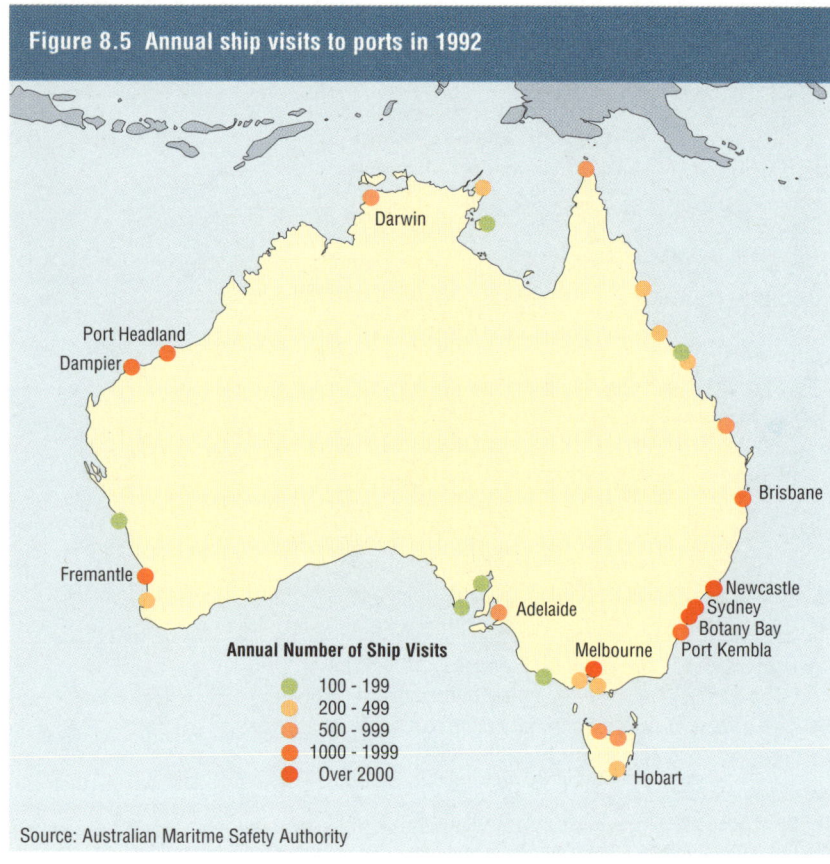

Figure 8.5 Annual ship visits to ports in 1992

Darwin

Port Headland

Dampier

Fremantle

Brisbane

Newcastle
Sydney
Botany Bay
Port Kembla

Adelaide

Melbourne

Annual Number of Ship Visits

- 100 - 199
- 200 - 499
- 500 - 999
- 1000 - 1999
- Over 2000

Hobart

Source: Australian Maritme Safety Authority

Transport and shipping

For Australia, an isolated island continent with a long coastline, shipping represents a major and essential economic use of our seas, estuaries and coastlines. In terms of tonnage carried and distance travelled, Australia ranks as the fifth largest user of shipping in the world. Each year about 12 000 ships arrive from overseas and almost 380 million tonnes of freight are carried in Australian waters. (see Fig. 8.5) (Raaymakers, 1994; ANZECC, 1995).

Shipping operations and the associated facilities of Australia's 68 major ports can impose significant pressures on marine and estuarine environments. These include pollution from oil, spills of hazardous cargoes, release of toxins from antifouling paints, loss of habitat from reclamation and dredging, litter, waste from inadequate port facilities, sewage and introductions of foreign organisms in ships' ballast waters or attached to ships' hulls.

While ports are among the most disturbed marine environments in Australia, technical, engineering and management solutions do exist to prevent or minimise many of the associated environmental impacts (Raaymakers, 1994).

The international nature of the shipping industry and its economic and strategic importance pose particular problems for environmental management. To deal with these problems, Australia has adopted a number of international conventions covering marine pollution such as the International Convention for the Prevention of Pollution from Ships (MARPOL), which covers the carriage and discharge of oil, noxious liquids, packaged harmful substances, sewage and garbage (Zann, 1995).

Ballast water discharged by shipping is the most urgent issue, because of its potential to introduce pest species and pathogens resulting in significant environmental, economic and human health impacts. Hull-fouling can be another vector.

Ballast water and hull-fouling

In 1991, about 155 million tonnes of ballast water were discharged into Australian ports, of which 121 million tonnes came from overseas (AQIS, 1993), mainly from Asia with about half from Japan (Rigby *et al.*, 1993). The other 34 million tonnes was transferred between Australian ports. Ballast water, together with hull fouling, is regarded as a significant carrier of exotic or pest species.

Displaced and introduced species have the potential to replace native species, to cause widespread environmental damage, to endanger human health and to damage fisheries and aquaculture through predation, competition and their impact on seafood quality.

At least 55 species of fish and invertebrates, plus a number of seaweeds, have been introduced into Australia either intentionally, for aquaculture, or accidentally in ships' fouling and ballast waters (Zann, 1995). The principal organisms of concern

The various environmental impacts of drilling for offshore petroleum include: the effect on the seabed of constructing platforms and laying pipes; the disposal of 'produced water', which is present with the oil and contains traces of hydrocarbons; contamination by drill fluids used to lubricate drill bits; and an increase in shipping activity.

Oil drilling is prohibited in marine protected areas like the Great Barrier Reef Marine Park and the Ningaloo Marine Park.

Ballast water is regarded as a significant vector for the introduction or translocation of exotic or pest species. Clockwise from top left: toxic dinoflagellate *Gymnodinium catenatum;* the giant fan worm *Sabella* cf. *spallanzanii;* the mussel *Musculista senhousia;* and the sea slug *Godiva quadricolor.*

▼

are the toxic alga *Gymnodinium catenatum* that causes red tides, the invasive giant fan-worm *Sabella* cf. *spallanzanii*, the Japanese seaweed *Undaria pinnatifida*, which displaces and smothers native kelps and the predatory northern Pacific seastar *Asterias amurensis*.

Blooms of introduced toxic marine algae are a serious marine environmental and fisheries problem in Tasmania and Victoria and may threaten the waters of other States. The toxic dinoflagellate *G. catenatum* can contaminate shellfish and produce toxin harmful to humans; it is of particular concern in areas of shellfish aquaculture (Jones, 1991). Introduced diseases pose a serious threat to our growing aquaculture industry. Organisms that cause human diseases such as cholera have been detected in ballast water, but not in boats entering Australian ports. Outbreaks of the northern Pacific seastar are spreading along eastern Tasmania, threatening marine life, aquaculture farms and scallop and abalone fisheries (Zann, 1995).

Antifouling

Ship-owners employ a range of toxic substances to discourage or prevent fouling growth on the hulls of commercial and recreational vessels. The most notable of these agents is tributyl tin. However, antifouling paints can also contain a range of other metals, such as lead and copper and organic and antibiotic compounds. These paints are designed to leach slowly into the water from the vessel's hull during passage and at docks and moorings. They can also be washed into the water in wastes from slipway operations.

The most significant effect of antifouling has been the impact of the highly toxic tributyl-tin-based antifouling on semi-enclosed water bodies such as bays and estuaries, where some species in marine communities (and aquaculture facilities) have shown significant incidences of deformity. As a result, many countries and most States in Australia, have banned the use of tributyl tin antifouling on vessels less than 25 m in length. Larger vessels are exempt because they do not stay in ports or at moorings for very long (Zann, 1995) and less-efficient antifouling would result in greatly increased operating costs.

Antifouling compounds are of most concern around mooring areas and marinas, which are usually enclosed waters with a high density of boats. Heavy metals, which can accumulate in sediments, are particularly prevalent around slipways and dockyards. The discharges from many of these operations are now controlled and toxins that would otherwise have entered the adjacent waterways must be trapped and removed.

Dredging and spoil disposal

Constructing and maintaining harbours, docks and channels is a significant and regular part of port operations. It often involves extensive dredging of the seabed and disposal of the material dredged up (called 'spoil'). Some spoil is used as landfill on shore, but most is dumped on the sea floor near the dredging area.

Table 8.3 Locations of species introductions attributed to ballast water or sediment discharged from ships

Swan River-Fremantle	**Fish** *Tridentiger trigonocephalus* (Striped goby) *Sparidentex hasta* (Sobaity sea bream) **Crustaceans** *Pyromaia tuberculata* (crab) **Molluscs** *Musculista senhousia* (Asian mussel) *Theora lubrica* (Asian semelid bivalve) *Polycera hedgepethi* *Godiva quadricolor* **Microalga** *Alexandrium minutum* (toxic dinoflagellate)
Port Pirie	**Crustaceans** *Tanais dulongi* (tanaid)
Adelaide	**Crustaceans** *Eurylana arcuata* (slater) *Carcinus maenas* (Euro shore crab) **Polychaete worms** *Pseudopolydora paucibranchiata* *Sabella* cf *spallanzanii* (fan worm) **Microalga** *Alexandrium minutum* (toxic dinoflagellate)
Port Phillip Bay	**Fish** *Tridentiger trigonocephalus* (Striped goby) **Microalga** *Alexandrium catanella* (toxic dinoflagellate)
Bass Strait	**Polychaete worms** *Mercierella engimatica* *Boccardia proboscidea* *Pseudopolydora paucibranchiata*
Hobart-Triabunna region	**Crustaceans** *Carcinus maenas* (European shore crab) **Macroalga (seaweed)** *Undaria pinnatifida* (Japanese giant kelp) **Microalga** *Gymnodinium catenatum* (toxic dinoflagellate) **Echinoderms** *Asterias amurensis* (North Pacific sea-star)
Port Kembla-Sydney-Newcastle region	**Fish** *Acanthogobius flavimanus* (Yellowfin goby) *Tridentiger trigonocephalus* (Striped goby) *Lateolabrax japonicus* (Japanese sea bass) **Crustaceans** *Eurylana arcuata* (slater) *Neomysis japonica* (mysid shrimp)
Brisbane	**Molluscs** *Aeolidiella indica* (sea slug) **Polychaete worms** *Mercierella engimatica* *Pseudopolydora paucibranchiata*

Source: after Manning, *in press*; with updated information from Wells and Zeidler, *pers.comm*; and Allen, 1994.

Dredging usually causes the concentration of suspended solids to rise around the dredging site and especially at the spoil-disposal areas. This can smother seabed organisms, clog fish and invertebrates' gills and reduce the light available to plants. When sediments are disturbed during dredging and spoil disposal, nutrients may be released, toxic trace metals mobilised and dissolved oxygen in the water column depleted.

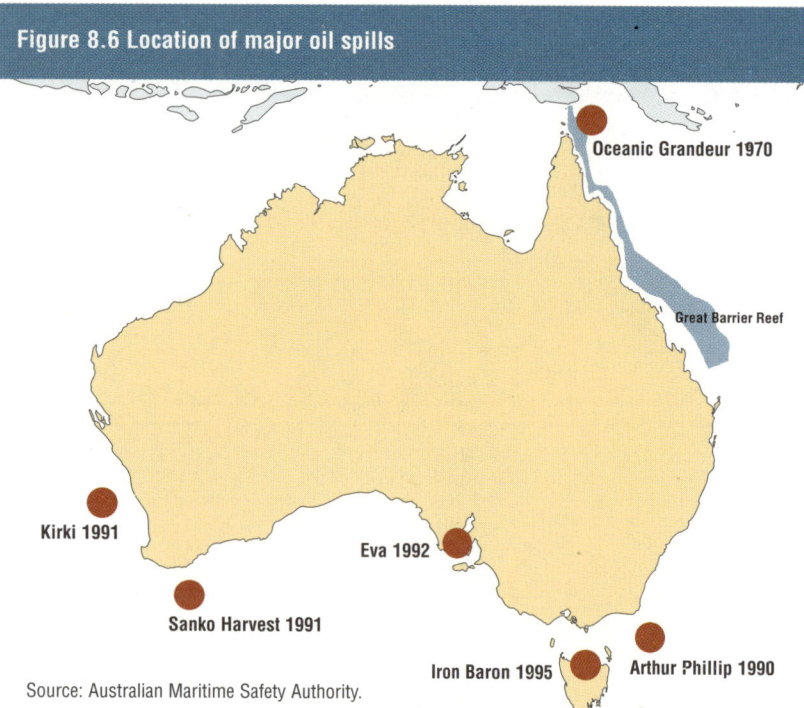

Figure 8.6 Location of major oil spills

Oceanic Grandeur 1970

Great Barrier Reef

Kirki 1991

Eva 1992

Sanko Harvest 1991

Iron Baron 1995   Arthur Phillip 1990

Source: Australian Maritime Safety Authority.

Oil spills

Shipping is a major source of oil spills, both from normal operations and from accidental discharges by oil tankers. Oil spills from ships are of worldwide concern, and their general effects on the marine environment are well known through a long list of environmental disasters such as that involving the *EXXON Valdez* in Alaska. Internationally, vessel operations and tanker accidents contribute 45 per cent of uncontrolled releases of petroleum to the marine environment (ANZECC, 1995). Most spills result from accidents during fuelling of vessels in ports (Zann, 1995).

Historically, few large oil spills from shipping accidents have happened in Australian waters (see Fig. 8.6). But, there is potential for major damage if a spill occurred in a sensitive area. So far, Australia has had only two large spills (over 1000 tonnes): the *Oceanic Grandeur* (1970, 1067 tonnes) in Torres Strait and the *Kirki* (1991, 17 700

tonnes) off Western Australia (Swan *et al.*, 1994). Since 1991, several moderate-sized spills have occurred (South Australia Department of Environment and Land Management, 1993; Swan *et al.*, 1994): the *Sanko Harvest* spill (1991, 570 tonnes) off Esperance, Western Australia, oiled more than 100 km of coast and affected 200 fur seals (13 of which died); the *Arthur Phillip* spill (1990, 100-km slick) off Cape Otway, Victoria, oiled 338 little penguins (273 died); the *Era* spill (1992, 296 tonnes) in Spencer Gulf, South Australia, oiled about 300 birds (most died) and damaged 75–100 hectares of mangroves along 10 km of coastline; and the *Iron Baron* (1995, about 300 tonnes) in the mouth of the Tamar River, Tasmania, oiled more than 2000 penguins (300 died) and 60 cormorants (30 died), and oiled over 15 km of coastline before the vessel was towed off the continental shelf and scuttled in 4000 m of water (Australian Maritime Safety Authority, pers. comm.).

The oil spills from the *Kirki* and the *Sanko Harvest* had minor impacts because of their locations and the prevailing weather conditions. However, the *Sanko Harvest* also spilled its cargo of 30 000 tonnes of di-ammonium phosphate and 300 tonnes of superphosphate fertilisers, which dispersed over a wide area. While the measured effects appeared to be restricted to a localised bloom in phytoplankton, if this spill had been near or in an enclosed bay the additional nutrients could have had a large impact on marine communities (Kinhill Engineers Pty Ltd, 1991).

Risks of a major spill of oil (or other substances) from shipping around Australia are considered to be high. The Bureau of Transport and Communications Economics estimated that the risk of a major spill from shipping was 37 per cent in any five-year period and 84 per cent in any 20-year period (Bureau of Transport and Communication Economics, 1991). Ships travelling inside the Great Barrier Reef (the inner route) are required to have pilotage to reduce the risk of accidents that could damage the reef system.

The Sanko Harvest, wrecked off Esperance, Western Australia in February 1991, spilled 570 tonnes of oil and 30 300 tonnes of fertiliser.

State

We need to know about the state of marine and estuarine environments to define their condition, observe changes over time and assess trends in decline or improvement. Key indicators provide a way to monitor the condition (or state) of the environment. We can use many variables to describe any one element of it, but the difficulties of measurement and interpretation mean that only a few make practical indicators.

In this report, we have chosen three types of indicators to assess the state of marine and estuarine environments. The first type are variables that are directly linked to important elements of the environment of major concern — for example, sea-surface colour as an indicator of phytoplankton biomass. The second type are specific measurements of environmental variables that have wider relevance to the general condition of the environment — for example, the area of seagrasses as an indicator of coastal water quality. The third type are values derived from a weighted aggregation of other estimates — for example, total commercial catch and effort linked with species diversity as a general guide to commercial fish stocks.

An indicator needs to be something that is sensitive to the major pressures on the environment, is related to aspects of the state of the environment in a sensible manner, can clearly identify changes relevant to sustainable resource management and can be easily measured and reported. It is also necessary to monitor indicators of the natural variability to avoid confusion with human-induced changes. An example of this is the Southern Oscillation Index (SOI), which is known to be correlated with the settlement of young western rock lobsters (see Fig. 8.7). Specific indicators of the state of marine and estuarine environments could include the existing area of a resource or habitat (for example, area of coral reefs), the quantities of fish stocks or the abundance/incidence of nuisance species (such as algal blooms or crown-of-thorns starfish). For many important aspects of the marine and estuarine environment, however, appropriate indicators are hard to find and for others few data are available.

Plants and animals

This section outlines, where possible, the state of plants and animals within various important elements of marine and estuarine environments.

Beaches and dunes

Beaches and dunes are the sandy beaches of the open coast and the gulfs and bays (see page 8-21), together with their associated foredune systems including both windward and the leeward slopes (Clarke, 1989).

Vegetation plays an important role in forming and stabilising coastal sand dunes. Dunes provide a

Figure 8.7 The relationship between western rock lobster recruitment, sea level and the annual mean Southern Oscillation Index illustrates natural variability

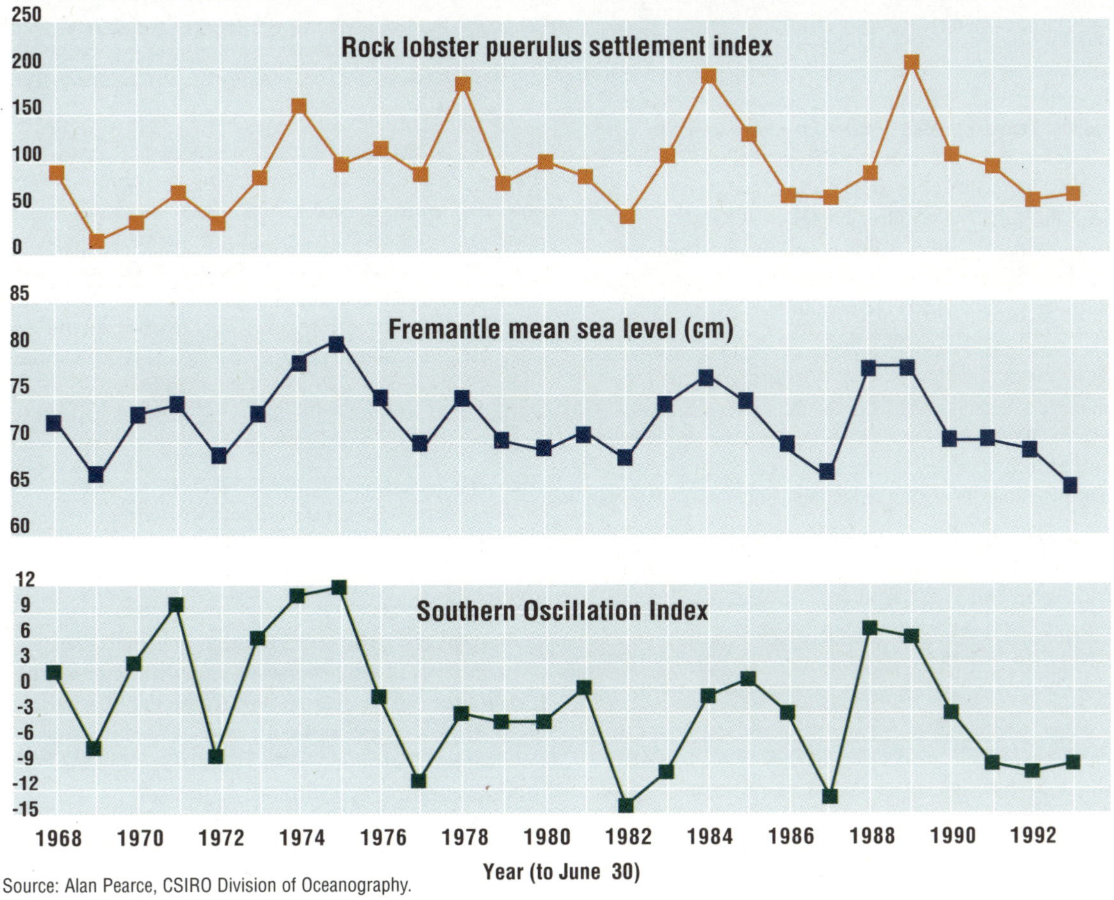

Source: Alan Pearce, CSIRO Division of Oceanography.

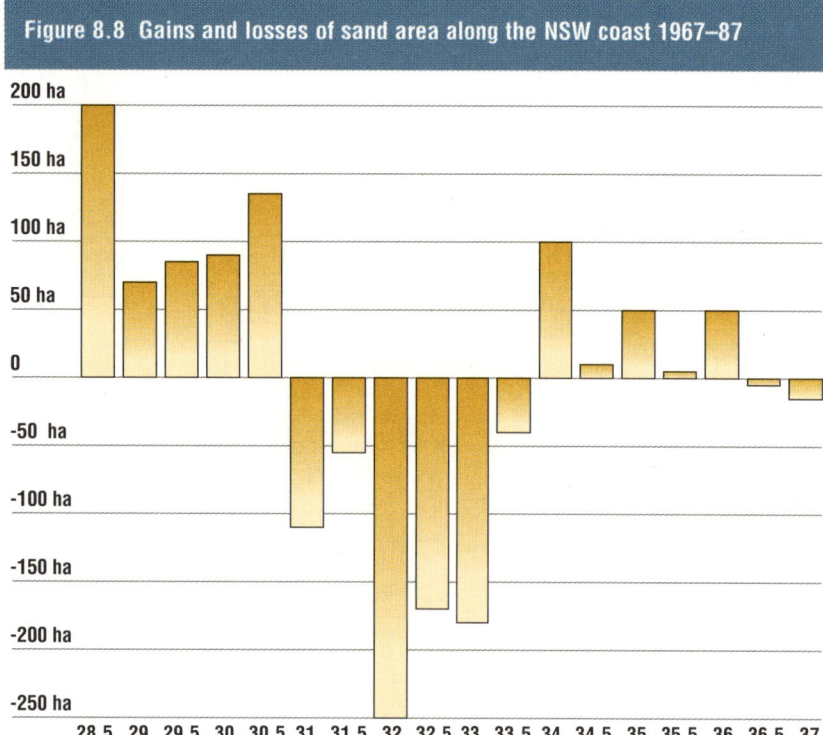

Figure 8.8 Gains and losses of sand area along the NSW coast 1967–87

Latitude (°S)

Source: EPA, NSW, 1993.

Table 8.4 Area of dunes and beaches along Australia's coastline and proportion of coastline by State

	Area (sq km)	Percent of coastline
Queensland	5 109	18.8
New South Wales	1 236	16.3
Victoria	1 653	28.0
Tasmania	984	13.8
South Australia	5 613	53.4
Western Australia	12 057	33.4
Northern Territory	1 242	5.5
Total for Australia	27 897	23.8

Source: Zann, 1995.

reservoir for sand during storm erosion. In the absence of dune vegetation, sand moves inland and is lost from the beach. The vegetation, dunes and beach act as a single dynamic system, essential to maintain the beach and protect inland vegetation and coastal buildings.

Dunes have been particularly subject to urban encroachment because of the strong demand for seafront housing. Most of the dune habitats around coastal cities and towns have been engulfed or alienated and the rest are under threat from erosion by pedestrians, off-road vehicles and introduced weeds. Beaches in turn become degraded by accelerated erosion due to the loss of protective dune vegetation and changes in coastal morphology, which alter patterns of sand movement. Erosion is an increasing problem in many areas and it is expected to accelerate in the coming decades in the event of rises in sea level.

We have estimates of the rate of beach erosion for some areas. In New South Wales, for example, many beaches are eroding at the rate of 0.2 to 1.4 metres per year causing damage to land, property and amenity value. By contrast, sand is being deposited on other beaches (see Fig. 8.8).

Degradation of water quality is an issue for tourism and recreation along some beaches, with turbid water and increased risks from pollutants, especially pathogens, reducing their amenity. Australia's beaches are increasingly littered with plastics, glass, fishing lines, nets and other rubbish.

About 16 000 km (23.8 per cent) of Australia's coastline is dune and beach habitat, occupying some 27 900 sq km. In the more populated areas, beaches and dunes are often the focus of recreation, tourism and residential development and as a result they suffer habitat loss and modification. The best

available indicators for the state of dune vegetation in Australia are the extent of sand dunes (see Table 8.4) and the species composition. In New South Wales, 423 plant species occur on beaches and foredunes up to 500 m inland of the beach strand line. Of these species, 86 have been found on the beach strand zone, 151 on the frontal foredune and 390 on the more protected back-face (lee side) of the foredunes (Clarke, 1989). The number of endemic species increases across the dune zones, from about 20 per cent of species on beach mound areas to the majority of species on the back-face of the frontal dunes.

Estuaries

Estuaries are waterways that are typically marine or brackish, but occasionally are dominated by fresh water (includes rivermouths, deltas and barrier lagoons that may be occasionally or permanently open to the sea).

Australian estuaries occur over a wide range of geological and climatic conditions and consequently display a great variety of form. They are characterised by extremes in conditions and are usually inhabited by species that can withstand variable conditions, particularly salinity. Their sediments and waters can be relatively rich in nutrients from the land so they are potentially very productive. The dominant habitats found in estuaries are saltmarshes, mangroves, seagrass meadows, algal beds, sandflats and mudflats (see Table 8.5).

Estuaries have traditionally been centres of urban and industrial coastal development. Because of their semi-enclosed nature, restricted flushing, shallow water and fine sediments, estuaries readily trap and accumulate pollutants.

Pollution, land reclamation, engineering works, over-fishing, weed infestation and the clearance of catchments all represent major threats to estuaries. Inappropriate developments within catchments may affect water quality by increasing erosion and silt loads and mobilising nutrients and contaminants. In New South Wales, 37 per cent of estuaries have more than half the land area of their catchments cleared. The same applies to 60 per cent of Victorian, 86 per cent of South Australian and three per cent of Western Australian estuaries but in the Northern Territory there are no estuaries with more than half their catchments cleared. There is not enough information to estimate the area of cleared catchment for Queensland and Tasmanian estuaries, but substantial clearing has occurred in central coastal Queensland and in some parts of Tasmania (Zann, 1995).

Generally, Australian estuaries away from the centres of population (in the north and west) have experienced little disturbance.

Gulfs and bays

Gulfs and bays are large (greater than 200 sq km surface area), open (or largely open) and shallow (less than 50 m in depth) coastal waters; typically they are dominated by marine conditions except where rivers flow into them.

The coast of Australia features a number of gulfs and large bays that — because of the protection they afford — are favoured for urban development. Many of them — such as Port Phillip Bay in Victoria, Moreton Bay in Queensland and Cockburn Sound, Western Australia — are important for recreation and transport. Others, like Spencer Gulf in South Australia, Shark Bay in Western Australia, and the Gulf of Carpentaria, Northern Territory/Queensland, are well known for their biological diversity and fisheries. Several of the bays, such as Shark Bay and Jervis Bay in New South Wales, have unique biological and geo-morphological features and are widely recognised for their natural heritage and conservation importance as well as supporting major fisheries. The Gulf of Carpentaria supports Australia's most valuable prawn fishery.

Because gulfs and bays are semi-enclosed or have their water circulation and flushing restricted in some way — by distance from the sea or narrow entrances — cities and towns located on their shores can create many pressures on their natural resources. Fishing, nutrient deposits and land-based run-off like stormwater or farmland drainage may all have detrimental effects on the marine ecosystem.

All the bays and gulfs contain many discrete habitat types, such as reefs, seagrasses and mangroves. Most support a number of potentially competing uses such as tourism and recreation, fishing and mining and, being large, they are typically subject to several levels of government jurisdiction (local and State). They also commonly support many species of high conservation value, such as dugongs and turtles and they may be subjected to various forms of both State and Commonwealth legislation. In this sense, bays and gulfs are diverse and complex systems. However, despite their importance as a class of coastal ecosystem, the only available indicators of their state for most areas are fisheries production data and other indicators of the state of their component habitats, such as mangroves and seagrasses (see pages 8-23 and 8-24).

Continental shelf and slope

The inner continental shelf encompasses the waters and seabed from the shore to the midshelf (about 50 m deep). Beyond this is the outer shelf extending to the 'break' (typically at 150 m to 200 m deep). The continental slope encompasses the area beyond the shelf break into water depths of 2000 m or more.

Most of Australia's marine environment lies in the deeper waters off the coast on the continental shelf and slope, remote from the coastline and human populations. These areas represent the submerged extensions to Australia's continental land mass and the margin with the world's deep ocean basins. Most biological activity is concentrated within the continental shelf and slope zone — in sediments, on the sediment surface and in the overlying water.

Our continental shelf covers about 2.5 million sq km, of which about half lies in less than 50 m of water. At its outer margin lies the continental slope, a further 1.5 million sq km, which starts at depths of 150–200 m and drops down to the abyssal plain at about 4000 m. Very little is known of the deeper-water communities around Australia even though they are subject to trawling and other fishing. The continental shelf surrounding Australia ranges in width from 15 km off the

Table 8.5 Estimated area (sq km) of estuarine habitat types by State, 1988–89

	Open water (incl. subtidal seagrass beds)	Intertidal flats (incl. intertidal seagrass beds)	Mangroves	Seagrass (subtidal and intertidal beds)	Salt marsh	Total
New South Wales	1323	na	107	153	57	**1487**
Victoria	2682	444	41	364	125	**3292**
Queensland	4093	1574	3424	68	5322	**14 413**
Western Australia	17 825	2891	1561	11	2965	**25 241**
South Australia	760	219	111	na	84	**1173**
Tasmania	1825	274	na	na	37	**2136**
Northern Territory	5187	821	2952	23	5005	**13 966**
Total	**33 694**	**6223**	**8195**	**6001**	**13 595**	**61 708**

na = data not available These estimates do not include habitats in bays and gulfs.
Source: Saenger, 1995.

Table 8.6 Distribution and areas of saltmarsh (sq km) by bioregion

	Grass and sedges	Shrublands	Total area
North Coast	4719	3954	8673
North West Coast	1542	1653	3195
West Coast	585	939	1524
South West Coast	21	12	33
Great Australian Bight	21	—	21
South Gulfs Coast	225	951	1176
South Coast	258	1317	1575
Bass Strait	615	45	660
Tasmanian Coast	498	—	498
South East Coast	171	—	171
East Coast	183	—	183
North East Coast	1644	105	1749
Gulf of Carpentaria	3399	465	3864
Total Australia	**13 881**	**9441**	**23 322**

Note: These estimates were derived from aerial photographs taken in the post-war period through to about 1982 and give an approximate estimate of areas in 1982. The estimates for grass and sedges are likely to include a large amount of terrestrial vegetation, but in most places the estimates for shrublands will mainly comprise saltmarsh species.

Source: Galloway *et al.*, 1984.

video. Only three areas around Australia (the Barrier Reef and the south-east and north-west shelves) have been studied in some detail. These studies suggest that hundreds of species are found in shelf and slope areas in surprisingly rich assemblages (Ward and Rainer, 1988).

We know little about the human impacts on Australia's shelf communities. Extensive fishing using otter trawls may remove bottom species and modify the bottom habitat (see page 8-14). The non-target discards returned to the sea may alter food chains by providing more food for carnivores, scavengers and decomposers. Discharged muds and effluent from offshore oil platforms and from dumping of industrial wastes at sea may cause localised damage. Fishing pressure has greatly reduced populations of some shelf and slope fish such as the school and gummy sharks at depths to 200 m and the gemfish at depths 100–700 m. We do not know what effect the removal of these high-level predators has on the ecology of the sea floor. The only easily available indicators of the state of the shelf and slope are fish catches (see pages 8-30 and 8-31).

Saltmarshes

These intertidal salt-water wetland habitats comprise low herbaceous shrubs and grasses, on mainly low-energy shorelines, often behind mangroves. In the tropics they may grade into salt-tolerant terrestrial shrubs and grasslands.

Coastal saltmarshes grow in sheltered intertidal zones. They are dominated by herbs and low shrubs, are highly productive and provide key habitats for many organisms, of both terrestrial and marine origin. They provide roosting sites for many migratory wading birds and breeding sites and habitat for certain endangered or rare species, such as the orange-bellied parrot in Victoria. They are often closely associated with mangroves that lie

south-east coasts to 400 km in the Timor Sea. We share a shelf connection with Papua New Guinea across Torres Strait and with Irian Jaya across the Arafura Sea.

Types of communities found on the shelf are closely associated with the type of sediments, terrestrial inputs from rivers and depth of water. The Australian shelf environment differs from those of other continents, especially in the south, for the dominance of coarse sediment particles and the virtual absence of terrestrial material. These factors contribute to complex microhabitats that may be responsible for high densities of seabed organisms and diversity of species.

The south-eastern to south-western shelves are covered with coarse, calcareous, shelly sands, composed of the remains of bryozoans (lace-corals), molluscs (shells) and foraminifera (calcareous plankton). Terrestrial input of silica and finer silt and clay materials is minimal. The eastern shelf of New South Wales comprises sandy sediments that are terrestrial in origin close to the coast and calcareous below 60 m depth. Muds of terrestrial origin dominate the inner shelf sediments of the Great Barrier Reef and carbonates dominate the mid and outer shelf. The sediments of the Gulf of Carpentaria are very fine and rich in faunal remains. Those of the Arafura Sea and most of the west coast are mostly coarse and calcareous, with very little terrestrial sediment.

Scientific knowledge of Australia's shelf and slope communities is very patchy. Many parts are remote and the water is generally too deep for direct inspection by scuba diving. Research requires time-consuming and expensive shipboard sampling using dredges, grabs and, recently, underwater

Table 8.7 Distribution and areas of mangroves in Australia by bioregion

Region	Area (sq km)
North Coast	4 100
North West Coast	710
West Coast	15
South West Coast	0
Great Australian Bight	0
South Gulfs Coast	40
South Coast	160
Bass Strait	15
Tasmanian Coast	0
South East Coast	65
East Coast	180
North East Coast	2 484
Gulf of Carpentaria	2 307
Offshore islands (estimate)	~1 500
Total Australia	**~11 576**

Note: These estimates were derived from aerial photographs taken in the post-war period through to about 1982 and give an approximate estimate of areas in 1982

Source: Galloway *et al.*, 1984.

to seaward. Along arid and semi-arid coasts, the marshes merge with inland saline habitats and on cliffs and headlands saltmarsh species are found in areas exposed to salt spray. They occur in all States, but are most extensive in the tropical north. Species richness, however, increases with latitude, with southern saltmarshes having the greatest diversity.

Land degradation poses a major threat to saltmarshes in developed areas. In many places they have been filled for ports, marinas, canal estates and urban and industrial sites. Other threats include rubbish dumping, off-road vehicles, invasion by weeds (particularly introduced cord grass, pampas grass, para grass and rushes) and drainage for mosquito and sandfly control. As backwater areas, saltmarshes may accumulate pollutants. They are also highly susceptible to invasion by exotic species; five of the 34 plant species in Jervis Bay saltmarshes are introduced (CSIRO, 1994). In rural areas many saltmarshes are heavily grazed by cattle.

Humans have had a major impact on saltmarshes in the process of land 'reclamation'. This has led to the loss of about 21 per cent of the habitat from Moreton Bay, Queensland (see the box on page 8-9). Although the total loss has not been great in Australia, most of it has been concentrated in the south-east, which had the smallest initial area but the highest biodiversity and incidence of endemic species. Losses are therefore considered to be significant both regionally and nationally. A sea level rise could cause significant contractions in saltmarshes (Adam, 1995; Zann, 1995). Compared to the areas of total loss, much larger areas have been degraded, but data on this are not readily available. The best available indicator for this habitat is the size of the remaining area (see Table 8.6).

Mangroves

Comprising a diverse group of largely tropical trees, shrubs, palms and ferns, mangroves live in the intertidal areas of sheltered marine shores, estuaries and tidal creeks.

Australia has the third largest area of mangroves in the world and the northern communities are among the world's most diverse. They line about 6000 km of Australia's coast. Of the 39 species found here, only one, the newly discovered *Avicennia integra,* appears endemic. Mangroves are most diverse in the wet tropics (some estuaries on Cape York may contain up to 35 species), becoming steadily less so on arid tropic, subtropic and temperate shores. Only one species, *A. marina,* occurs along the southern coastline. The composition and form of the communities varies with temperature, rainfall, river run-off, sediment type, size of tides and coastal structure.

Mangroves have both conservation and economic value. They are generally highly productive and provide important habitats for both bait-fish and table fish. Within various mangrove communities, about 197 fish species are recorded for northern Australia, 65 for Brisbane and 46 for Sydney. Some of Australia's most valuable commercial fisheries are directly or indirectly linked to this habitat.

Seagrasses of Australia

These flowering plants grow in marine and estuarine areas — mainly the sediments of subtidal waters — to 10 m in depth. Some species also grow in intertidal sediments and some in very deep waters, down to 40 m in depth.

Australian waters have the world's highest diversity, with 22 species found in temperate waters and 15 in tropical areas. These grasses stabilise sediment and act as filters to overlying water.

Seagrass meadows provide the nursery areas for many important commercial and recreational fisheries and are critical habitats for turtles and dugongs. They support a diversity of flora and fauna and rotting leaves form the basis of a detrital food web.

Australia has about 51 000 sq km of seagrass meadows, with the largest in Western Australia and Queensland and the most diverse in Western Australia. The major areas occur in the Gulf of Carpentaria, Shark Bay, the southern coast of Western Australia, Spencer Gulf and St Vincent Gulf.

Seagrass meadows are affected by human activities, in particular, nutrient addition from farm run-off, sewage and industrial-waste dumping. The extra nutrients cause blooms of epiphytes and/or phytoplankton, which shade seagrass leaves. Changes to hydrology for harbour or marina development and sediment run-off from terrestrial development have also contributed to a decline in seagrass meadows around the major population centres.

Floods and cyclones have damaged about 1000 sq km of seagrass beds over the past 10 years and their recovery has been slower than expected. About 450 sq km of the beds have been damaged directly by human activities such as dredging and habitat reclamation. However, human-induced changes in water quality — particularly poor land management practices that release sediments and nutrients into estuaries, lagoons and other coastal waters — may also affect the ability of seagrass beds to recover from damage caused by natural events.

Temperate beds are particularly susceptible to degradation because all of the species of the genus *Posidonia* (the dominant temperate seagrass) spread very slowly. They may take centuries to recolonise and regrow in damaged areas.

Recent surveys along the southern coastline have shown major errors in previous estimates of distribution due to seagrass beds often being misinterpreted as algal beds from aerial photographs. More detailed mapping is also needed to accurately estimate losses and the reasons for them.

We do not know the capability of temperate seagrass beds to respond to gradual change in climate. If sea levels gradually rise and if land run-off containing suspended material and nutrients increases as a result of changes in rainfall patterns then available sunlight will decrease. No one has studied the ability of seagrasses to respond to such changes. However, their slow growth rates and their critical role in numerous commercial fisheries, make us concerned about their ability to respond successfully to changing climate.

Most States have various forms of legislation to partially protect seagrasses. We do not know how effective these protective measures are and, generally speaking, do not have enough data to determine whether the loss of seagrasses has been reduced by implementation of government legislation to reduce or control human impacts.

There has not been much research on amelioration and restoration of seagrass beds. Some projects have been successful in restoring tropical seagrass species, but it is unlikely that the key temperate species could be restored without major research. This will be especially important to protect and better manage the remaining vulnerable seagrass beds across much of the Australian coastline.

Source: Kirkman, in press.

For example, the early life cycle of the banana prawn is confined to mangrove-lined estuaries, while many fish targeted by Australian recreational fishers, including bream, grunter, mangrove jack and barramundi, live in tropical and subtropical mangroves.

In the arid and semi-arid tropics, mangroves form the only closed-canopy forest available for birds. A recent study found that 22 bird species in the north of Western Australia were confined to this vegetation for at least part of their range.

Mangroves have been cleared extensively for land reclamation near coastal cities such as Sydney, Newcastle, Brisbane, Cairns and Adelaide. For example, about 17 per cent of stands have been removed from Moreton Bay in Queensland. Many estuaries have been extensively modified by breakwaters, channel dredging, flood mitigation and other engineering works, adversely affecting mangrove communities. Significant areas in south-east Queensland, near Gladstone and near Cairns are under direct threat from development.

Mangrove regrowth can occur by natural recruitment processes. In some places, the community is increasing in area, probably because of increased sediment supply from upland clearing (CSIRO, 1994) or increased sedimentation due to coastal modification (Kinhill *et al.*, 1994).

As with other habitat types, it is difficult to make a national assessment of mangroves at various stages of degradation. However, Table 8.7 shows their estimated area in different bioregions.

Seagrasses

Seagrasses are intertidal and subtidal flowering plants found mainly in the shallow waters of protected estuaries and bays; in the southern temperate regions of Australia they often form dense beds, but in the tropics they may also be found at low densities, widely scattered in near-shore areas.

Australia has the largest number of seagrass species and some of the largest and most diverse seagrass beds in the world (see the box on page 8-23).

Seagrass meadows interact with other habitats. For beaches, their stabilising effect is important in reducing erosion. Seagrasses often occur offshore from mangrove and saltmarsh habitats, where they

Table 8.8 Distribution of seagrass by bioregion

	Central West Coast	Lower West Coast	South West Coast	Great Aust Bight	South Gulfs Coast	Bass Strait	Tas. Coast	Lower East Coast	Central East Coast	North East Coast	Gulf of Carp.	North Coast	North West Coast
Amphibolis antarctica	•	•	•	•	•	•	•						
Amphibolis griffithii		•	•	•	•								
Cymodocea angustata	•												•
Cymodocea rotundata										•	•	•	
Cymodocea serrulata										•	•	•	
Enhalus acoroides										•	•	•	•
Halodule pinifolia										•	•		
Halodule uninervis	•	•								•	•	•	•
Halophila australis		•	•	•	•	•	•	•					
Halophila decipiens		•	•	•				•	•	•	•	•	
Halophila ovalis	•	•							•	•	•	•	•
Halophila ovata								•		•	•	•	•
Halophila spinulosa	•	•								•	•	•	•
Halophila tricostata										•	•		
Heterozostera tasmanica		•	•	•	•	•	•	•					
Posidonia angustifolia		•	•	•	•								
Posidonia australis	•	•	•	•	•	•	•	•					
Posidonia coriacea		•			•								
Posidonia denhartogii			•	•	•								
Posidonia kirkmanii			•	•	•								
Posidonia ostenfeldii			•	•	•								
Posidonia robertsoniae			•	•									
Posidonia sinuosa		•	•	•	•								
Syringodium isoetifolium	•	•								•	•	•	•
Thalassia hemprichii										•	•	•	•
Thalassodendron ciliatum										•	•	•	•
Thalassodendron pachyrizum		•	•										
Zostera capricorni								•	•	•			
Zostera mucronata		•											
Zostera muelleri				•	•	•							

Source: Kirkman, in press.

may provide refuges and nurseries for juvenile fish and prawns or feeding grounds for birds.

Large areas of seagrass have died in recent decades (see Fig. 8.9). A number of cases are still poorly documented and the reasons for the declines unclear. Many losses, both natural and human induced, are attributed to reduced light intensity caused by sedimentation and/or increased growth of epiphytes from nutrient enrichment. Sediment instability, undercutting and erosion can increase the damage. The dominant temperate seagrass (*Posidonia* sp.) has not been known to recolonise from loss or damage and other subtropical and tropical species take a long time (more than 10 years) to recover and then only if the substrate has not been physically disturbed. None of the attempts to replant seagrasses in temperate Australia have been successful.

There are no simple indicators of the health of seagrass beds, but their area can be estimated by techniques such as remote sensing. Table 8.8 shows the currently available information on the distribution of seagrass species around Australia.

Macroalgae

Macroalgae are the larger multi-celled species of algae generally referred to as seaweed. Their growth-form and appearance ranges from tall leathery leaves several metres in length (kelp), to encrusting mats covering coral reefs, and to dense accumulations of filaments which detach from the seabed and become free floating.

Macroalgal communities, composed of the larger red, brown and green algae, predominantly occur in temperate southern Australian waters, where they form a major component of the shallow-water reef communities. The 5500-km coastline from the south-western part of Western Australia to the New South Wales/Victorian border has a wide diversity of macroalgae, with 558 species recorded on the southern coast of Western Australia, 1151 species in South Australia and Victoria, and 398 species around New South Wales and Lord Howe Island. The area is rich in brown algae (Phaeophyta) and particularly rich in red algae (Rhodophyta)

Kelp reef, Gulf St Vincent, South Australia.

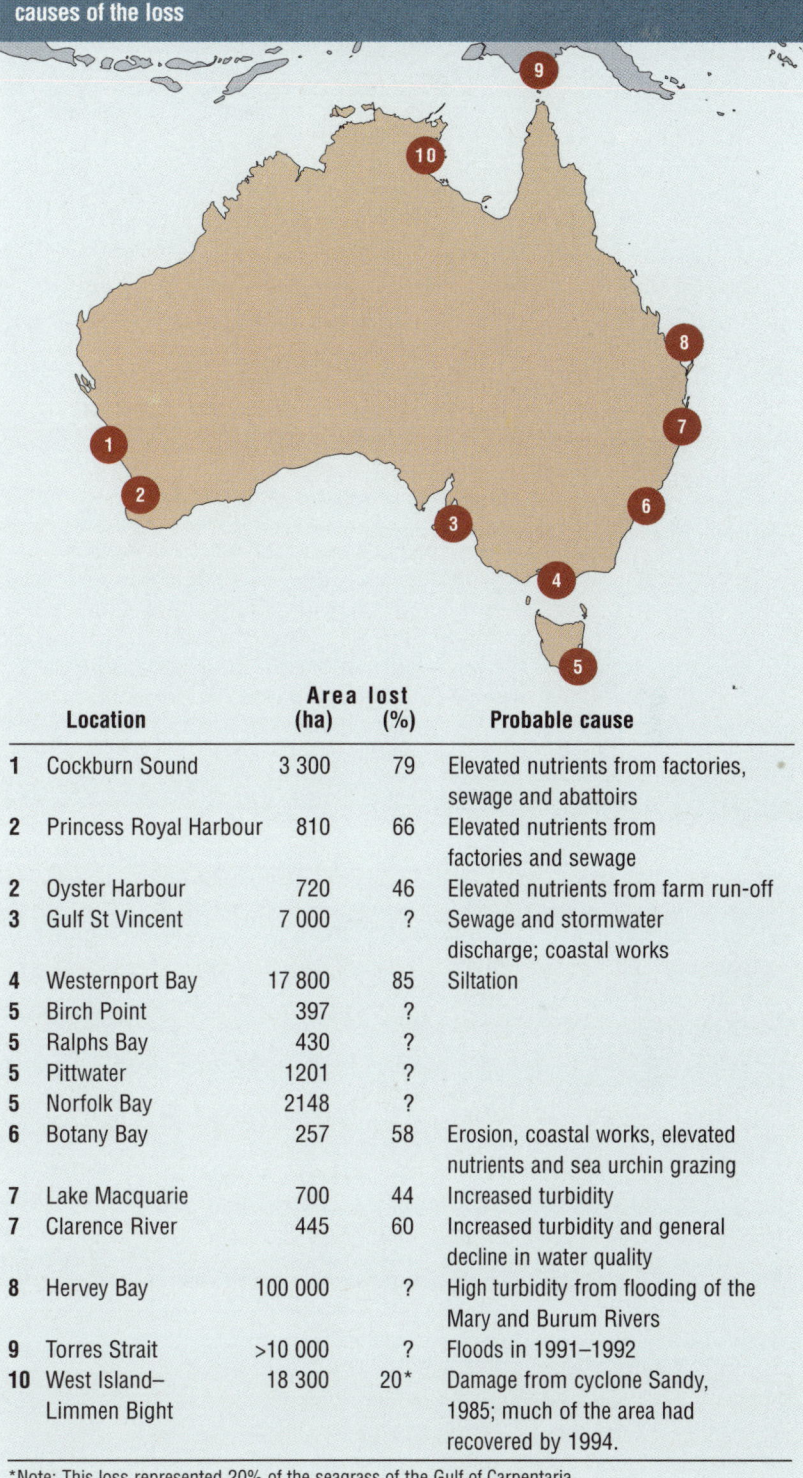

Figure 8.9 Loss of seagrass by area and a brief description of the probable causes of the loss

	Location	Area lost (ha)	(%)	Probable cause
1	Cockburn Sound	3 300	79	Elevated nutrients from factories, sewage and abattoirs
2	Princess Royal Harbour	810	66	Elevated nutrients from factories and sewage
2	Oyster Harbour	720	46	Elevated nutrients from farm run-off
3	Gulf St Vincent	7 000	?	Sewage and stormwater discharge; coastal works
4	Westernport Bay	17 800	85	Siltation
5	Birch Point	397	?	
5	Ralphs Bay	430	?	
5	Pittwater	1201	?	
5	Norfolk Bay	2148	?	
6	Botany Bay	257	58	Erosion, coastal works, elevated nutrients and sea urchin grazing
7	Lake Macquarie	700	44	Increased turbidity
7	Clarence River	445	60	Increased turbidity and general decline in water quality
8	Hervey Bay	100 000	?	High turbidity from flooding of the Mary and Burum Rivers
9	Torres Strait	>10 000	?	Floods in 1991–1992
10	West Island–Limmen Bight	18 300	20*	Damage from cyclone Sandy, 1985; much of the area had recovered by 1994.

*Note: This loss represented 20% of the seagrass of the Gulf of Carpentaria
Source: Kirkman, in press.

(Sanderson, *in press*). An examination of published records shows that, of the nine southern bioregions (see Fig. 8.1), the South West Coast, Tasmanian Coast, Central East Coast and Lord Howe Island regions appear to have fewer species than the others (see Table 8.9). However, this may be simply because some have been much more thoroughly studied than others. Substantial research on the classification of algae is being prepared for publication and the number of known species, especially the red algae, will certainly rise.

Table 8.9 The number of macroalgal species by bioregion for southern Australia and the number of species unique to each region

Bioregion	Green algae		Brown algae		Red algae	
	total	endemic	total	endemic	total	endemic
Lower West Coast	83	(13)	107	(7)	368	(117)
South West Coast	64	(0)	96	(1)	221	(1)
Great Australian Bight	80	(2)	131	(0)	300	(4)
South Gulfs Coast	106	(6)	178	(15)	397	(24)
Bass Strait	87	(3)	174	(8)	381	(18)
Tasmanian Coast	54	(3)	107	(6)	212	(11)
Lower East Coast	67	(7)	107	(9)	272	(49)
Central East Coast	32	(4)	37	(1)	151	(17)
Lord Howe Island	59	(27)	71	(21)	175	(66)

Note: All the numbers need to be treated with caution because some regions have been more thoroughly studied than others.
Source: Sanderson, in press.

We do not understand the effects of human activities on temperate macroalgae and reefs very well. The most serious potential pressures are probably those that affect the habitat-forming species — particularly the large algae. These include discharges and other activities that alter water quality, and direct pressures like fishing and collecting. Victorian and overseas studies claim that sewage outfalls reduce canopy-forming brown algae; and, in California, thermal effluent has reduced kelps. Commercial and recreational fishing may also affect macroalgal communities.

Marine and estuarine algal blooms in Australia

Algal blooms, while a natural phenomenon reported by the first Europeans to explore the Australian coastline, have increased in frequency, intensity and geographic distribution in the past 30 years. Better reporting has increased our knowledge of them. Recurrent and major algal blooms have been reported right around the east coast of Australia from Adelaide to Townsville and around the south-west of Western Australia. Some of those waters affected include: Gulf St Vincent, South Australia; Westernport Bay, Port Phillip Bay and Gippsland Lakes, Victoria; Botany Bay, Tuggerah Lakes, Lake Illawarra, Lake Macquarie, Narrabeen Lagoon, Hawkesbury Estuary and Georges Estuary, New South Wales; Moreton Bay, Queensland; Cockburn Sound, Peel Inlet, Harvey Estuary and Princess Royal Harbour, Western Australia; and Orielton Lagoon and the Derwent and Huon Estuaries, Tasmania.

Blooms of the toxic alga *Alexandrium minutum* were first observed in the Port River estuary near Adelaide in 1986 and seem to occur annually during September–November, making wild mussels from the river highly toxic. Elevated nutrient levels and changing salinity contribute to the formation of these blooms. The alga may have been introduced into Australian waters. In Western Australia's Peel Inlet, observers first noticed enhanced growth of the macroalga *Cladophora* sp. in the late 1960s, which became progressively worse in the 1970s. During the 1980s *Chaetomorpha* and *Ulva* became the dominant genera. In the Harvey estuary, massive blooms of the blue-green alga *Nodularia spumigena* have occurred since 1978. These have affected fish and crab populations and reduced commercial fish catches. Phosphorus inputs originating from fertiliser applied in the parent catchments have enriched the Peel-Harvey system.

In the Tuggerah Lakes, the distribution and density of macroalgae, including *Enteromorpha*, *Chaetomorpha* and *Rhizoclonium* spp., have increased during the past 30 years.

Enhanced growth of various species has occurred in Lake Illawarra since the early 1960s, becoming excessive in recent years. The alga *Scrippsiella trochoidea* has been responsible for major blooms causing red-brown sea-water discolouration and fish kills in several estuaries and coastal lagoons, including West Lakes in South Australia and the Hawkesbury River estuary. In Victoria, mussel beds were closed due to blooms of the diatom *Rhizosolenia* sp.

More open, offshore waters also experience significant algal blooms that, while a natural phenomenon, seem to have intensified in recent years as a result of increased nutrient inputs. In Tasmanian waters, blooms of the toxic alga *Gymnodinium catenatum* occurred in 1986, 1987, 1991 and 1993, causing the temporary closure of commercial shellfish beds. A number of algal blooms were recorded around Sydney and the south coast of New South Wales in 1992–93. In Jervis Bay a substantial bloom of *Gephyrocapsa oceanica* occurred in December 1992, while a bloom of *Noctiluca scintillans* extended along the Sydney coast from Avalon to Garie Beach in January 1993.

The species *Trichodesmium erythraeum* is the most common alga responsible for 'red tides' in tropical Australian waters, producing seasonal (February–April) blooms in the Java, Banda, Arafura and Coral Seas, from where the East Australian Current and Leeuwin Current transport them (covering up to 40 000 sq km) as far south as Sydney and Perth. The algae can accumulate on beaches, causing offensive odours and discolouring water and sand. While some scientists believe that the magnitude and frequency of such blooms have increased in some areas over recent years — notably the Great Barrier Reef lagoons — as a result of increased nutrient inputs, the available data are not decisive at this stage.

Source: Cosser, in press.

Trawling may directly remove plants, while the removal of plant-feeding animals (such as fish, rock lobster or abalone) may alter the balance of grazing, resulting in a shift in the relative abundance of algal species.

Reef habitats are being mapped in South Australia, Western Australia and Port Phillip Bay using satellite images and aerial photography. Recent surveys show that stands of *Macrocystis pyrifera*, believed to be an ecologically important species in Tasmanian waters, are a fraction of the size estimated in the early 1950s.

Massive blooms of macroalgae in the Peel–Harvey Estuary and Princess Royal and Oyster Harbours in Western Australia cause significant problems. All States regularly report minor blooms involving a variety of species, especially *Ulva*, *Enteromorpha*, *Cladophora* and *Hinksia* species. No national data are available, and future indicators should be based on a database of macroalgal blooms that monitors location and extent, frequency, magnitude and type of algae.

A number of macroalgal species appear to have been introduced to Australian waters, principally through ballast waters and on ship's hulls. The actual numbers are unknown, because we are still learning about the taxonomy and ecology of Australian species. A significant introduction is *Undaria pinnatifida* in Tasmania. It blankets the rocky reef bottom for large areas, limiting colonisation by native algae and invertebrates. A relatively recently introduced red alga, *Schottera nicaeensis*, first recorded in Port Phillip Bay in 1972, is now reported from every port and harbour in south-eastern Australia.

The first published record of a macroalgal extinction, that of *Vanvoorstia bennettiana*, has been recorded in Botany Bay. Its disappearance may be due to human activities such as heavy shipping traffic, dredging and urban run-off, all contributing to heavy siltation in the Bay.

Microalgae

These single-celled, microscopic plants predominantly live floating in the water, although they may also live on the sea floor.

Microscopic algae support, either directly or indirectly, the entire production of the open sea. They are usually single-celled and usually inhabit the water column, where they are called phytoplankton, although some (benthic species) live on the seabed and as epiphytes.

Australia's marine phytoplankton comprise representatives of 13 algal classes, including the well-known diatoms (5000 species), dinoflagellates (2000 species), golden-brown flagellates and green flagellates (several hundred species). Our phytoplankton flora are similar to the warm-water and cold-water phytoplankton floras of the Northern Hemisphere. No species are endemic to Australian waters.

We do not know much about the natural dynamics of phytoplankton. However, scientists recognise three distinct assemblages in Australian waters: a

Table 8.10 Recorded algal blooms in the waters of New South Wales from 1890 to February 1995		
Date	**Species**	**Location**
Mar 1890	*Scrippsiella trochoidea*	Sydney Harbour
Jul–Aug 1930–32	*Gymnodinium sanguineum*	Sydney Harbour
Feb 1945	*Alexandrium catenella*	Port Hacking
Dec 1970	*Giffordia mitchelliae*	Port Macquarie
Oct 1972	*Trichodesmium* sp.	Taree and Coffs Harbour
Dec 1972	*Trichodesmium* sp.	Palm Beach and Cronulla
Jan 1980	*Spyridia filamentosa*	Crescent Head
Oct 1980	*Gymnodinium sanguineum*	Lane Cove River
Aug 1982	*Noctiluca scintillans*	Lake Macquarie
Dec 1983	*Trichodesmium* sp.	Newcastle, Narrabeen, Foster, and Bondi Beach
April 1984	*Mesodinium* sp.	Lane Cove River
Nov 1984	*Mesodinium* sp.	Lane Cove River
Dec 1984	*Trichodesmium* sp.	Sydney and Wollongong
Aug–Sep 1985	*Thalassiosira partheneia*	NSW coast
Feb 1986	*Mesodinium* sp.	Lane Cove River
Feb 1986	*Thalassiosira weissflogii*	Alexandra Canal
Jan 1989	*Trichodesmium* sp.	Jervis Bay and Ulladulla
Mar 1991	*Scrippsiella trochoidea*	Hawkesbury River
Jun–Jul 1991	*Gymnodinium galatheanum*	Lake Illawara
Nov 1991	*Heterosigma carterae*	Berowra Creek
Aug 1992	*Noctiluca* sp.	Lake Macquarie
Sep 1992	*Noctiluca* sp.	Berowra Creek
Dec 1992	*Gephyrocapsa oceanica*	Jervis Bay
Jan 1993	*Gonyaulax polygramma*	Darling Harbour
Feb 1993	*Gonyaulax polygramma*	Bate Bay
Jan–Feb 1993	*Noctiluca* sp.	Sydney beaches and Port Kembla
Feb 1993	*Dictyocha octonaria*	Newcastle
Oct 1993	*Alexandrium catenella*	Sydney Harbour
Nov 1993	*Pseudonitzschia multiseries*	Berowra Creek
Apr 1994	*Noctiluca* sp.	Ham and Chicken Bay
Nov 1994	*Alexandrium catenella*	Port Hacking
Feb 1995	*Noctiluca* sp.	Sydney

Source: Hallegraeff, 1993 and Hallegraeff unpublished.

temperate group that inhabit the coastal waters of New South Wales, Victoria and Tasmania; a tropical group confined to the Gulf of Carpentaria and North West Shelf; and a tropical oceanic group in the offshore waters of the Coral Sea and Indian Ocean (Zann, 1995).

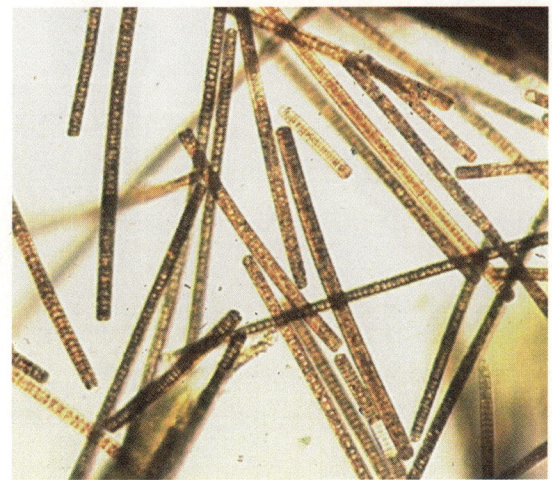

The bloom forming *Trichodesmium sp.*

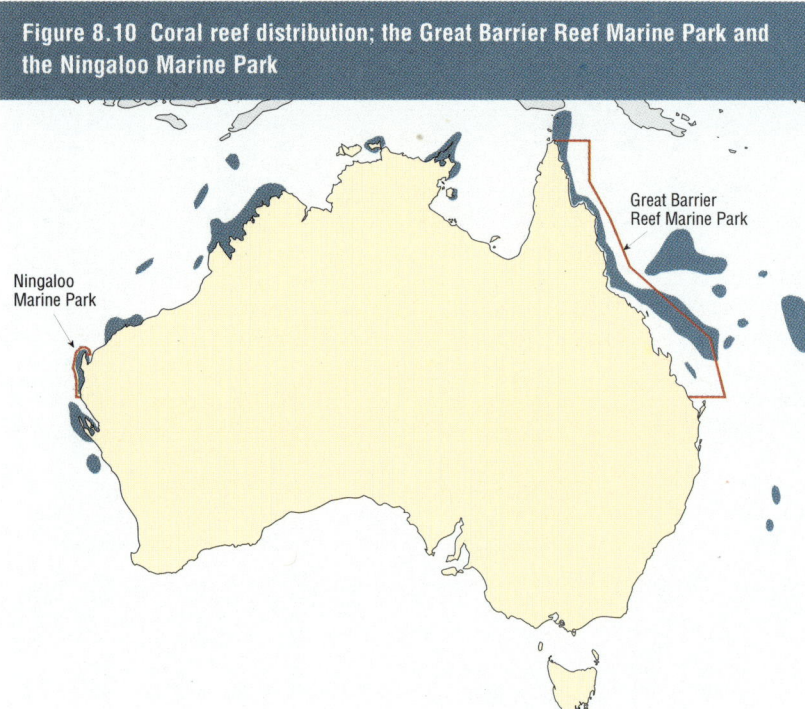

Figure 8.10 Coral reef distribution; the Great Barrier Reef Marine Park and the Ningaloo Marine Park

Ningaloo Marine Park

Great Barrier Reef Marine Park

Source: after Zann, 1995.

Satellite data have enabled mapping of the general patterns of phytoplankton production in the oceans surrounding Australia (see Fig. 8.2) and confirmed that an area of the Southern Ocean known as the Subtropical Convergence has high phytoplankton biomass.

Concern is growing that the deteriorating environment of most major rivers and estuaries near our large coastal cities is affecting the species composition of natural phytoplankton populations, with potentially far-reaching implications for the structure of entire marine foodwebs, which influence fisheries production.

Microscopic algae can 'bloom', becoming very abundant in the right conditions of nutrients, temperature, salinity and light. Some such blooms, notably by diatoms, are a natural part of the cycle of life in certain parts of the sea, but the intensity and frequency of others have increased because of human-induced changes to the ecosystem — especially the addition of nutrients. Some blooms are spectacular because of their red or green colour

Mushroom coral (left) and Soft coral (right), northern Great Barrier Reef

and some (not all) of them include toxic species. They may cause fish kills and threaten human health (for example, by paralytic shellfish poisoning), while others are aesthetically 'nuisance' algae, causing spoiling of beaches, offensive odours and slimy water. In the last thirty years, a large number of Australia's estuaries, bays and coastal lakes have had algal blooms reported for the first time (see the box on page 8-26). The frequency of such blooms is increasing, although the rate is uncertain because of better reporting and public awareness (see Table 8-10).

The best-known case of both phytoplankton and macroalgal blooms is the Peel–Harvey system in Western Australia (see page 7-46). The effects of toxic algae are best known from the closure of shellfish beds in Tasmania and Victoria resulting from blooms of an alga suspected to have been introduced in ballast water.

Coral reefs

Formed from the calcareous skeletons of many species of corals and other organisms, coral reefs support a diversity of fish, invertebrates and plant life.

Australia has the largest area of coral reefs in the world. The Great Barrier Reef (see page 8-44) and Western Australian reefs are well developed and diverse. Most of the world's reefs have been affected by human activity, which threatens the continued survival of some major reef complexes. About 70 per cent of those in the central Indo-Pacific are disturbed to some extent. Australia's coral reefs are extensive and extremely diverse and harbour a number of threatened species. Most reefs are still in a relatively good condition. Only a few other countries have such a low population pressure and/or the capacity to regulate human impacts on their reefs.

However, Australia's coral reefs are now exposed to significant pressures from an array of human activities. Only a few systems remain undisturbed and those close to population centres show the greatest signs of damage. While we have no general assessment of coral reefs, recent photographic comparisons of several sites in the Great Barrier Reef show that major changes have occurred to some of the reef flats over the last 40 to 100 years. In four of the 12 locations studied a significant reduction in the cover of living hard coral has occurred (Wachenfeld, 1995).

Outside the Great Barrier Reef Marine Park and the Ningaloo Marine Park, management and monitoring of Australia's coral reefs are limited.

Fish and fisheries

The terms fish and fisheries are used here to include, as well as the finfish, the sharks and invertebrates, such as crustaceans and molluscs, that are taken by both commercial and recreational fishers.

Australia has an estimated 4000 to 4500 species of finfish, of which around 3600 have been described. About one-quarter of the species are endemic and most of these are found in the south (Zann, 1995).

In 1993–94, production from commercial fishing was approximately 209 000 tonnes, worth about $1.6 billion (ABARE, 1994). Recreational fishing has not been quantified but one estimate indicates that the catch may be of the order of 50 000 tonnes (Kearney, 1995). In 1984 recreational fishers spent about $2 billion on their pastime (Zann, 1995).

Commercial fish markets handle at least 300 species of finfish, sharks and invertebrates. However, about 50 species make up 85 per cent of the catch by weight and only a few of these are widely distributed.

The most important species of commercial fish are either fully exploited and being managed to achieve a sustained yield, or were overexploited and are now being managed in an effort to increase the breeding stock and total catch. Management strategies are based on controls on fishing gear and boats (input controls), limitations on the catch (output controls) or a combination of these.

Population status for a number of the main commercial species is described below and on the following pages to illustrate the state of the population in response to fishing pressure.

Fisheries of Aboriginal and Torres Strait Islander communities

Fish and seafood are significant components of the traditional diet in many communities and are particularly important to Torres Strait Islanders. Torres Strait and northern Aboriginal communities consume a lot of seafood — much more than some other communities in the South-west Pacific. They catch many reef fish species for food and hunt green turtles and dugongs, which form a major part of their diet.

Torres Strait Islanders are increasingly becoming involved in commercial fishing activities, and there is still potential for development in the area. The Islanders sell a growing proportion of their reef fish catch, trolled pelagic fish and tropical rock lobsters.

Table 8.11 Catch, value and status of major commercial fish species (1993–94)

Fishery	Catch (tonnes)	Value ($M)	Catch Trend	Status
Western rock lobster	11 045	287 122	Stable	Fully exploited
Abalone	4 723	176 505	Stable; declining in some areas	Fully exploited; limited by quota
Tiger prawns (two species)	6 062	147 233	Stable overall; some stocks overexploited in 1980s	Fully exploited
Southern rock lobster	5 060	119 505	Stable	Fully exploited
Southern bluefin tuna	6 080	116 348	Stable;recovery uncertain	Overexploited in 1980s
Pearl oyster[a]	-.	96 500	Stable; depressed in Torres Strait	Fully exploited
King prawns	6 056	68 443	Stable	Fully exploited
Northern scallops(two species)	13 445	51 308	Variable	Some over exploited/variable
Oysters (aquaculture)[b]	2 280	48 846	Increasing	Space limitations
Banana prawns (two species)	3 348	36 536	Variable	Fully exploited/variable
Endeavour prawns (two species)	3 056	31 360	Stable	Fully exploited
Crabs	3 551	17 549	Variable	Fully exploited
Southern scallops	9 006	16 604	Variable	Fully exploited/variable
School and gummy shark	5 152	15 536	Stable; some signs of recovery	School shark overexploited in 1980s
Coral trout	1 101	13 212	Stable	Some overexploited/uncertain
Barramundi	861	8 439	Stable	Fully exploited
Mullet and sea mullet	4 464	8 235	Stable	Uncertain
Blue grenadier	3 111	7 079	Stable; quota not reached	Possibly under exploited

Notes: (a) The value given is the value of the cultivated pearls produced.
(b) The value given is that of the cultured product of oysters. The catch is the weight of cultured product.

Source: derived from ABARE and BRS.

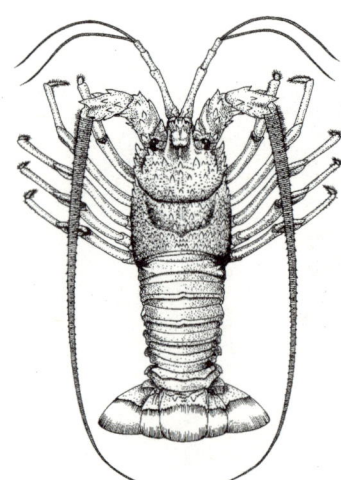

Western rock lobster

The western rock lobster fishery is the most valuable in Australia. The catch has been stable at about 10 000 tonnes over the past 30 years with variations caused by environmental factors. However, heavy fishing pressure reduced the breeding stock to the extent that egg production in 1992–93 had fallen to about 15–20 per cent of its level under light fishing pressure. So far, the reduced level of breeding stock does not appear to have affected recruitment to the fishery. But, as a precaution, fishing pressure is being reduced to allow the breeding stock to increase until egg production reaches about 25 per cent of the original level.

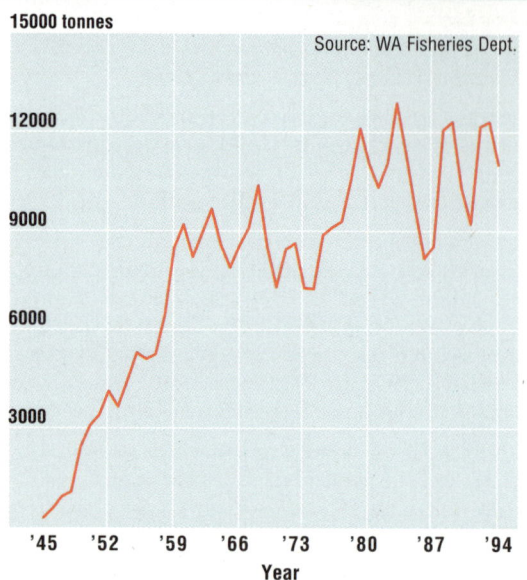

Figure 8.11 Annual catch of western rock lobster

15000 tonnes

Source: WA Fisheries Dept.

12000

9000

6000

3000

'45 '52 '59 '66 '73 '80 '87 '94

Year

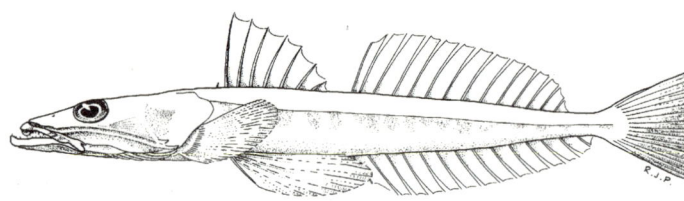

Tiger flathead

Trawl fishing for tiger flathead off south-east Australia has existed since 1915. During the late 1920s the catch rose to nearly 6000 tonnes. It then fell to around 3000 tonnes and subsequent catches have ranged from 1500 to 3000 tonnes. The tiger flathead fishery, which comprises several species, is now an important part of the larger south-east trawl fishery.

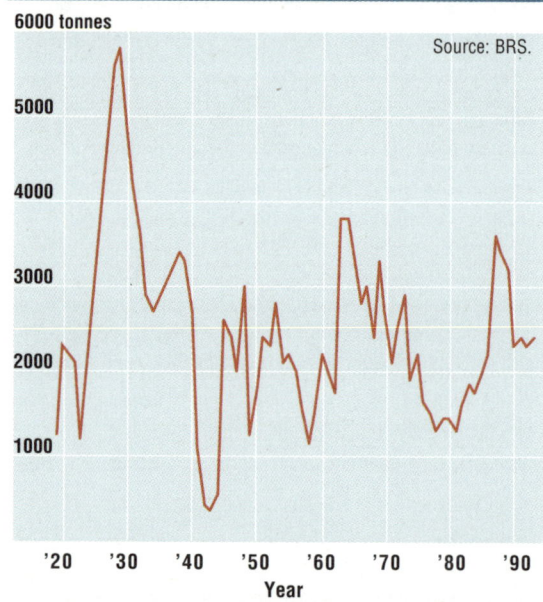

Figure 8.12 Annual catch of tiger flathead

6000 tonnes

Source: BRS.

5000

4000

3000

2000

1000

'20 '30 '40 '50 '60 '70 '80 '90

Year

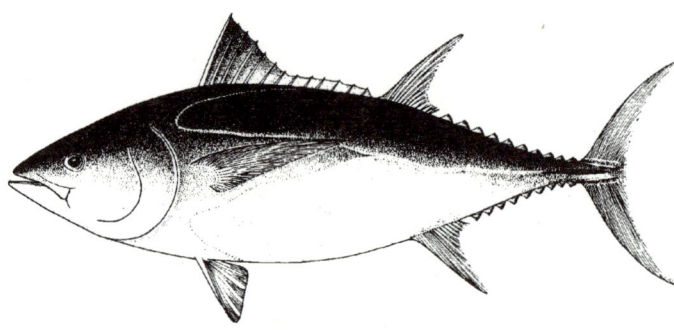

Southern bluefin tuna

Australia's southern bluefin tuna fishery is part of an international fishery. In the past, uncontrolled fishing reduced breeding stock so much that the recruitment of young fish was also reduced. The quantity of breeding fish is estimated to be less than 10 per cent of the original and is considered to be dangerously low (CSIRO, unpublished information). Management agencies have implemented controls on the catch over the past 10 years in an attempt to increase the breeding stock to the early 1980s level, but it is hard to tell if this slowly maturing species is recovering.

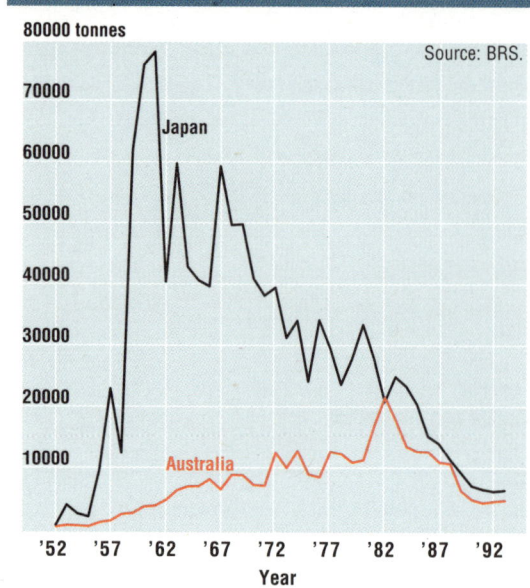

Figure 8.13 Annual catch of southern bluefin tuna

80000 tonnes

Source: BRS.

70000

60000

Japan

50000

40000

30000

20000

10000

Australia

'52 '57 '62 '67 '72 '77 '82 '87 '92

Year

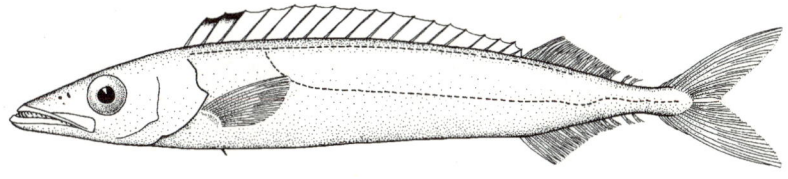

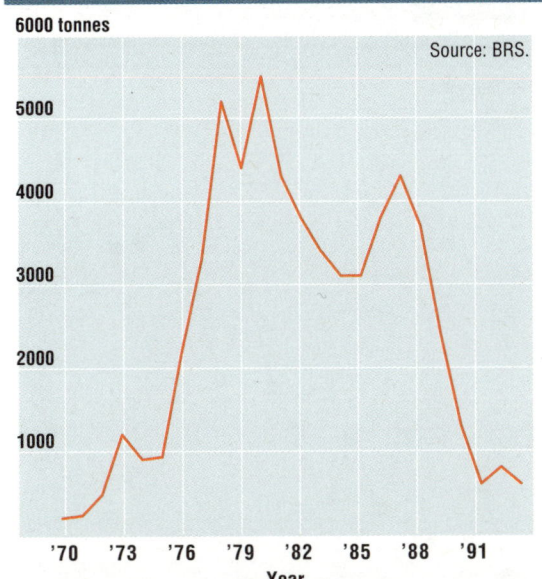

Figure 8.14 Annual catch of gemfish

6000 tonnes

Source: BRS.

Year

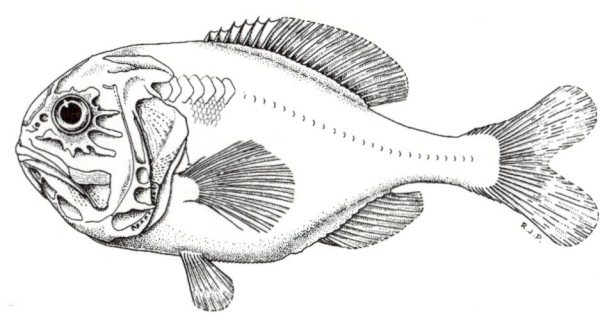

Gemfish

Recruitment of eastern gemfish has been so low for a number of years that the total allowable (targetted) catch was set at zero in 1993 and has remained at zero. However, about 267 tonnes in 1993 and 134 tonnes in 1994 were taken as by-catch during fishing for other species. The biomass of the adult population of eastern gemfish is at a dangerously low level.

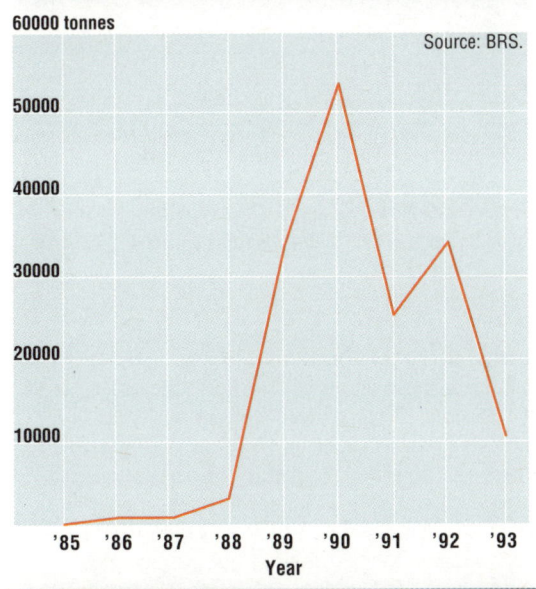

Figure 8.15 Annual catch of orange roughy

60000 tonnes

Source: BRS.

Year

Orange roughy

Fishing of this long-lived species began in the mid 1980s. Catches rapidly increased each year until 1990, when about 44 000 tonnes were taken in the south-east trawl fishery. Research indicated that the breeding stock had been reduced to approximately 30 per cent of its original level. In response, management authorities are now acting to reduce the catch to a target of 3500 tonnes for 1998 and beyond, in an effort to maintain the spawning stock at the current level.

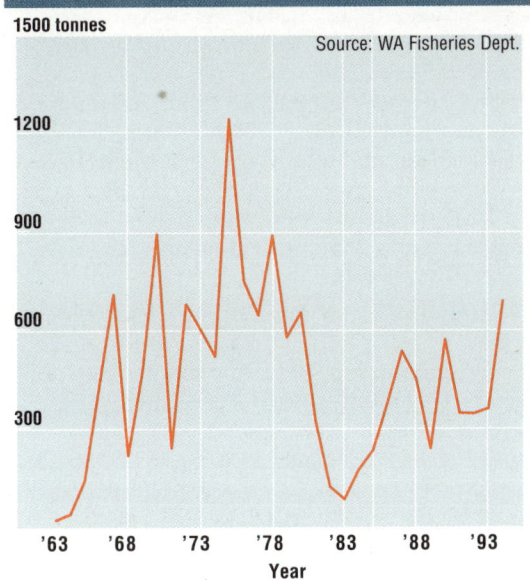

Figure 8.16 Annual catch of tiger prawns in Exmouth Gulf

1500 tonnes

Source: WA Fisheries Dept.

Year

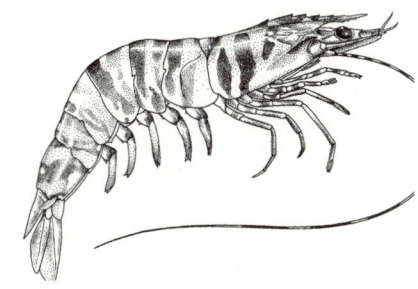

Exmouth Gulf tiger prawn fishery

Heavy fishing during the 1980s resulted in the annual catch of this fishery falling from an average of about 600 tonnes to 100 tonnes. The breeding stock had been reduced to such low levels that, even in those years of a favourable environment, egg production was insufficient to provide high recruitment. Strict controls on the catches planned through cooperative action between the industry and management have allowed the stock to rebuild and catches are again in the order of 600 tonnes per year.

Recreational fishing

Although recreational fishers catch many of the species being fished commercially, they also catch other species. We have few data on the status of the recreational fisheries. However, each of the Australian fisheries agencies is now increasing research on the recreational sector and the species supporting it.

From the limited data that are available it is clear that recreational fishing pressure is increasing. In one case, a study undertaken in New South Wales on boat-based and shore-based recreational fishing on the Clarence and Richmond Rivers found a twofold increase in fishing effort (boat hours) from 1988–89 to 1994–95. Shore-based fishing (angler hours) had also increased, but not to the same extent.

The coral trout fishery in the Great Barrier Reef Marine Park provides another example of increased recreational fishing, as well as increased commercial fishing. A recent study of the private recreational fishing fleet in the region suggests that, since the mid 1980s, the mean fish size has declined by 30 per cent and the catch per unit of effort has fallen by 50 per cent, coupled with a 25 per cent increase in the recreational fishery.

Reptiles

Marine reptiles such as turtles and sea snakes either live in or depend on the sea for their continued survival.

Australia's marine and estuarine reptile fauna includes 30 of the world's approximately 50 sea-snake species, six of the seven known species of turtles and the salt-water crocodile.

Sea-snakes

Of the 30 Australian species of sea-snakes from two families (one inhabiting the reefs and the other the inter-reef areas), 15 are endemic. The first group are well protected because they live in reefs and many of them are found in the Great Barrier Reef

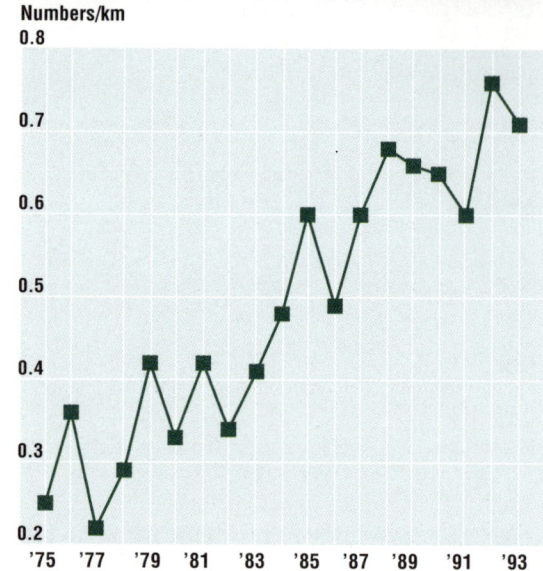

Figure 8.18 Increase in the number of saltwater crocodiles in the Northern Territory since protection began in 1971

Note: These counts are from waterways surveyed by helicopter between 1975 and 1993.
Source: Webb *et al.*, 1994.

Marine Park. The inter-reef species, however, are caught by prawn trawling. In the Gulf of Carpentaria in 1991, the mortality from commercial prawn trawling was estimated to be between 29 801 and 67 051 sea-snakes (Wassenberg *et al.*, 1994), but the impact on the populations is not known.

Turtles

Turtles are large reptiles that live at least 50 years and reach maturity at between 12 and 25 years of age or older. They seldom breed every year, but may wait two to eight or even ten years between breeding attempts. Six species of turtles occur in Australian waters — the leatherback, loggerhead, green, hawksbill, olive ridley and flatback. All six species breed in Australia, although recorded breeding of the leatherback has been infrequent.

The flatback turtle is endemic to Australia. However, the populations of other species move between Australia and neighbouring countries, where they are extensively hunted. In Australia, commercial exploitation of turtles was permitted on the east coast until the late 1960s and on the west coast until the early 1970s. Hunting by Torres Strait Islanders and some Aboriginal communities for traditional purposes continues. Between 1992 and 1994 the annual Australian catch in Torres Strait was about 2500 turtles, mainly green. (CSIRO, unpublished data).

It is hard to assess the status of most sea turtles because of the lack of information, the dispersed feeding and breeding behaviour and the long wait to maturity. However, nesting populations of loggerhead turtles have declined and regional populations are threatened (see Fig. 8.17). Studies suggest that populations of loggerhead turtles will not withstand mortalities of five to 10 per cent above natural levels (Somers, 1994).

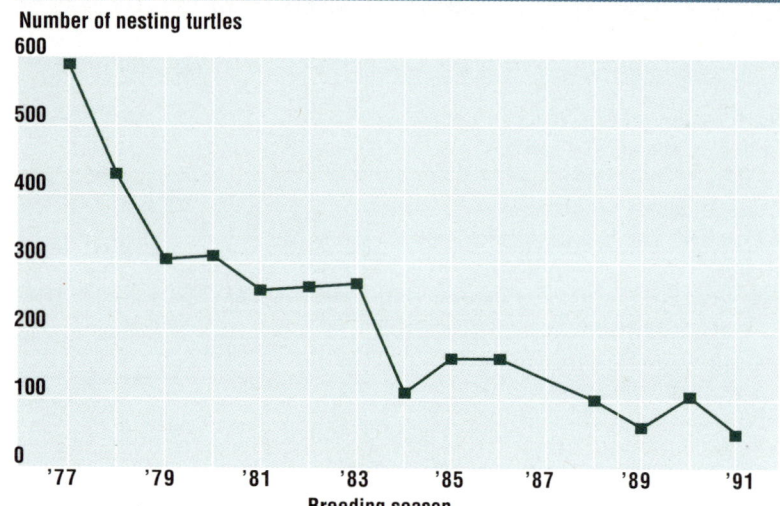

Figure 8.17 Loggerhead turtle population decline at Wreck Island on the Great Barrier Reef

Note: Wreck Island in the Great Barrier Reef is the most significant breeding site for loggerhead turtles in the south-west Pacific.
Source: Marsh *et al.*, 1995.

Olive ridley turtles form massive breeding groups of up to 150 000 in some countries. An estimated 11 500 green turtles were recorded nesting on Raine Island off Queensland on a single night in 1984.

Crocodiles

The saltwater crocodile has a wide distribution outside Australia, from India and the Philippines through South East Asia and Papua New Guinea. In most countries, except Australia and Papua New Guinea, the species has all but disappeared.

It was harvested by shooting in Australia throughout its range from 1945 to 1971 and, as a result, populations fell heavily. When hunting was stopped in 1971, the numbers had been greatly depleted but enough remained to provide the basis for a consistent recovery. Since then, the population status of the species has improved (see Fig. 8.18). Most saltwater crocodiles occur in the Northern Territory, where the original size of the population was estimated to be no more than 100 000 animals, including young of the year. By 1984 the upper estimate was 40 000 animals (including hatchlings, non-hatchlings and captives) and the 1993 upper estimate was 60 000.

Populations have always been smaller in Western Australian and Queensland waters than in the Northern Territory. However, the numbers in those areas are either stable or increasing (Western Australia) or there is evidence to suggest a slow recovery (Queensland), though they may still be declining in populated and agricultural areas.

Seabirds

These are birds such as shore birds and highly migratory birds like albatrosses that depend on the sea for their continued survival.

Australia and its external territories have about 110 species of seabirds belonging to 12 families. Of these, 76 species breed and live in the region and the rest are regular or occasional visitors. The conservation status of most species breeding here appears to be satisfactory (see Table 8.12). In cases where seabirds were previously exploited — for oil, food and bait — most populations have now recovered. The only remaining industry is the harvesting of short-tailed shearwaters for the mutton bird industry in Bass Strait.

Scientists consider 14 species or subspecies of Australia's seabirds (13 per cent) are threatened (Zann, 1995), largely because their colonies on oceanic islands are few in number and are vulnerable to illegal harvesting and natural disasters. Colonies can be subject to human disturbance and predation by introduced and feral animals. The wandering albatross on Macquarie Island, Abbot's booby on Christmas Island and the Australian subspecies of the little tern are classified as 'endangered' under International Union for the Conservation of Nature (IUCN) criteria. Lord Howe's Kermadec petrel and white-bellied storm-petrel and Christmas Island's Christmas frigatebird are considered 'vulnerable'.

The wandering albatross (and other species such as petrels) takes bait used on longlines set for tuna. After taking the bait on the longline hooks, the birds often drown as the hooks drag them just below the surface. This practice has caused considerable mortality and is considered the major factor contributing to the worldwide decline in populations of wandering albatross over the last two decades (Gales, 1993).

Australian and Japanese boats use longlines in the Australian Fishing Zone. The Japanese fleet has introduced measures, such as tori poles, bait-throwing devices and the use of night setting of gear, which significantly reduce the albatross bycatch. Researchers are now collecting data on the effect of the Australian longline fishery on the wandering albatross.

Table 8.12 Estimates of numbers of breeding pairs of seabirds along the Australian coast and Coral Sea

Species	Minimum	Maximum
Little penguin	149 130	249 900
Shy albatross	6 900	8 500
Great-winged petrel	33,050	84 100
Herald petrel	3	3
Black-winged petrel	3	3
Gould's petrel	250	500
Fairy prion	1 055 060	1 682 000
Wedge-tailed shearwater	1 301 150	1 344 400
Flesh-footed shearwater	104 540	310 600
Sooty shearwater	300	1 210
Short-tailed shearwater	12 787 070	16 059 700
Little shearwater	27 060	61 600
White-faced storm-petrel	370 180	396 600
Common diving-petrel	127 220	184 000
Australasian pelican	1 030	1 680
Australasian gannet	5 560	6 140
Masked booby	3 750	4 270
Red-footed booby	1 380	4 990
Brown booby	59 940	73 900
Pied cormorant	13 080	19 120
Little pied cormorant	140	200
Black-faced cormorant	7 740	8 110
Great frigatebird	1 610	1 610
Lesser frigatebird	18 680	19 430
Red-tailed tropic bird	290	380
White-tailed tropic bird	2	2
Silver gull	133 890	163 620
Pacific gull	1 900	1 950
Kelp gull	315	315
Caspian tern	1 160	1 410
Roseate tern	7 220	13 370
White-fronted tern	44	44
Black-naped tern	1 710	2 080
Sooty tern	328 760	383 750
Bridled tern	20 040	57 870
Little tern	560	570
Fairy tern	2 420	2 990
Crested tern	74 350	89 940
Lesser crested tern	4 710	8 170
Common noddy	174 480	214 130
Lesser noddy	79 500	79 500
Black noddy	119 340	130 840

Source: Ross *et al.*, 1995.

Dugong mother and calf. Many of Australia's marine mammals are threatened by human activities.

Mammals

Marine mammals include whales, dolphins, seals and dugongs.

Australia's marine mammal fauna includes one species of dugong, three species of seals and 43 species of whales and dolphins (Zann, 1995). Despite widespread protection, many of Australia's marine mammals are still threatened by dangers such as loss of habitat, pollution, accidental netting, entanglement in litter and disturbance.

Dugongs (sea-cows)

The tropical dugong is the only fully herbivorous marine mammal and the only sea-cow found in Australia. It is listed by IUCN as 'vulnerable to extinction', but is not listed under the Commonwealth *Endangered Species Protection Act*. It is extinct or near extinct in most of its former range, which extended from East Africa to South East Asia and the Western Pacific. Northern Australia has the last significant populations (more than 80 000 animals) in the world (Zann, 1995).

Torres Strait Islanders and some northern Aboriginal communities catch dugongs for food. The Torres Strait take of dugongs by Australians is estimated to be about 1000 per year. The eastern Cape York catch is estimated to be less than 100 per year.

Aerial surveys being carried out along most of the northern Australian coastline are providing

estimates of the dugong population (see Table 8.13). The most important area in Australia and probably the world, is Torres Strait, with an estimated 24 000 animals. Shark Bay, on the west coast, also supports a large population of dugongs and is particularly important as very little human-induced mortality of dugongs occurs in this region.

Dugongs are killed incidentally in commercial gillnets and shark nets set to protect swimmers in northern Australia. A total of 576 dugong deaths were recorded from shark nets in Queensland between 1964 and 1988. These creatures are also potentially threatened by the loss of their habitats to coastal residential, industrial and tourist development, as well as natural causes. For example, many dugongs died in Hervey Bay, Queensland, in 1992 following dieback of seagrasses (Zann, 1995), possibly caused by increased sediments.

Seals

Three species of eared seals breed in mainland Australian waters: the endemic Australian sea lion, the Australian fur seal and the New Zealand fur seal. The New Zealand fur seal also occurs in large numbers in New Zealand and Antarctic waters. The southern elephant seal breeds at Macquarie and Heard Islands. This species did occur (and bred) in Bass Strait, but was hunted to extinction there in the nineteenth century. Non-breeding visitors to Australia's southern shores from the Antarctic and sub-Antarctic islands include the leopard, Weddell, crab-eater and Ross seals.

Australia's seals were badly over-hunted in the last century, until about 1825. Sealing at a few remnant colonies in eastern Bass Strait continued on a regulated basis until about 1923 (Warneke and Shaughnessy, 1985). Seals are now fully protected by legislation and some populations appear to be increasing. Major human threats include entanglement in fishing nets and ocean litter, oil pollution and disturbances by visitors. Fur seals are occasionally illegally killed for shark and lobster bait. They are also killed around fish farms, trawls and nets for 'stealing' fish. Many become entangled in discarded nets and plastic box straps.

The breeding range of the Australian sea lion extends from the Houtman Abrolhos, off the Western Australian coast, across the southern coastline to The Pages, just east of Kangaroo Island. It is the most widely distributed Australian seal and also the least abundant. The total population is estimated to be between 7000 and 10 000 animals, producing less than 2000 pups per breeding cycle (Goldsworthy, 1994).

Australian fur seals — population estimated at between 30 000 and 50 000 animals — have an annual pup production of between 8000 and 9000. The population appears to have stabilised, but remains well below the pre-sealing levels when the annual pup production could have been two to five times higher.

The New Zealand fur seal has an estimated population of 28 000 animals in Australian waters and a pup production of about 7000 per breeding cycle.

Table 8.13 Recent estimates of dugong populations in their 10 principal Australian habitats

Location	Area (sq. km)	Population Estimate	
Shark Bay, WA	14 906	10 529	(±14%)
Exmouth Gulf, Ningaloo WA	28 746	1 974	(±30%)
Northern Territory coast	28 746	13 800	(±19%)
Western Gulf of Carpentaria	27 216	16 846	(±19%)
Wellesley Island	8 848	4 067	(±18%)
Torres Strait	30 560	24 225	(±14%)
Northern Great Barrier Reef	31 288	10 471	(±15%)
Southern Great Barrier Reef	39,396	1 857	(±16%)
Hervey Bay	4 371	600	(±29%)
Moreton Bay	1 400	664	

Source: Marsh *et al.*, 1994.

Whales, porpoises and dolphins

Eight species of baleen whales and 35 species of toothed whales, porpoises and dolphins are found in Australian waters, although none are endemic. They represent almost 60 per cent of the world's cetacean species.

Until recently, hunting exerted the major pressure on them and several species were driven to near extinction. Their populations are slow to increase and are highly vulnerable to overexploitation.

Whalers have caught humpback whales since the early 1800s. However, the greatest effect on the population from hunting began in the late 1940s, from Western Australian land stations and the Antarctic, and during the 1950s off eastern Australia. In 1935, before the major assault began on the humpback whales wintering off Western Australia, the population might have numbered up to 17 000. By 1949 the estimate was down to 10 000 and by 1963 the number of adults had been reduced to only about 500 animals. The east coast population registered a similar decline (Bannister, 1994).

After humpback whaling stopped in 1963, very few were sighted until the early 1970s. Aerial surveys, which began in 1977, have recorded a persistent and encouraging increase in the number of humpback whales sighted off the Western Australian coast. Comparisons with aerial sightings in 1963 indicate that the population now numbers at least 3000 animals and possibly considerably more. A similar 'comeback' has been recorded off the east coast.

By 1845, the population of southern right whales off the Australian coast was sadly depleted. It has grown slowly from small remnants of about 100 animals to number less than 800 today. Encouraging results from aerial surveys undertaken between Cape Leeuwin, Western Australia, and Ceduna, South Australia, show that the number of animals continues to increase, albeit from a very low base.

Whalers caught sperm whales off the south coast of Western Australia from the mid 1950s to 1978. No one has undertaken aerial surveys since then to establish the population status.

Australians have replaced whaling with a new ecotourist industry, 'whale-watching'. Because of

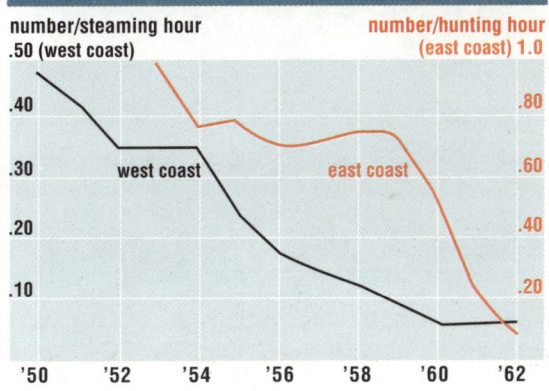

Australian fur seals at Smith Rocks, Joseph Banks Group, South Australia.

concern that boats, aircraft and divers may affect whale behaviour, regulations have been introduced to govern how closely observers may approach whales. An Australian government paper to the International Whaling Commission (Anon., 1994) recognised that the increase in whale-watching activities required better and more coordinated management.

Today whales, porpoises and dolphins within Australia are threatened by gillnets, shark nets, discarded fishing nets, tuna aquaculture nets and the risk of swallowing plastic litter. Protective shark nets off bathing beaches catch considerable numbers. In Queensland, for example, they caught about 520 dolphins between 1967 and 1988. Other pressures, especially on inshore species, may include loss of habitat, reduction in prey numbers, increasing risk of collisions with boats and disturbance of migrating and breeding populations (Zann, 1995). We have little information on a national level on the status of dolphins.

On a worldwide scale, marine mammals face a significant threat of contamination by organochlorine pesticides and polychlorinated biphenyls (PCBs). They are particularly sensitive to reproductive failure due to accumulated PCBs, but we have few data on PCBs in Australian cetaceans (Marsh *et al.*, 1995).

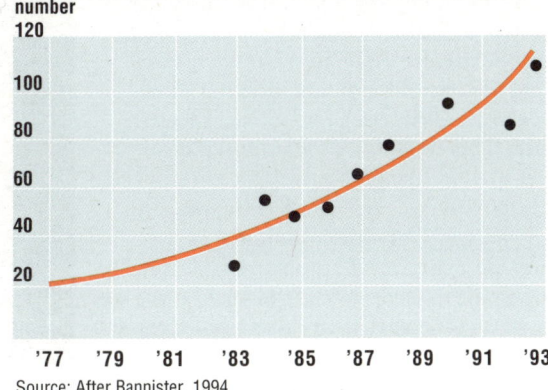

Figure 8.19 Catch rate of humpback whales off the west and east coasts of Australia

number/steaming hour
.50 (west coast)

number/hunting hour
(east coast) 1.0

west coast

east coast

Source: After Bannister, 1994

Figure 8.20 Number of southern right whales off southern WA, 1977–93, from aerial survey results

number

Source: After Bannister, 1994

Species outbreaks

Some species may, for various reasons, undergo population explosions in areas that are part of their normal range.

Problems with pest species are not limited to introductions of species from overseas or movements between localities within Australia. Explosions in the numbers of particular native species may — although we generally lack the evidence — be caused by human-induced changes in their environment. For example, some scientists believe that outbreaks of the coral-eating crown-of-thorns starfish on the Great Barrier Reef might have resulted from overfishing of their natural predators or changes in the nutrient load of the water. Alternatively, the outbreaks may be a natural occurrence, but their frequency may change due to human activity.

Over the past 30 years outbreaks of crown-of-thorns starfish have caused considerable damage to Indo-Pacific reefs, including parts of the Great Barrier Reef and Australia's Tasman Sea reefs. Two such episodes have affected parts of the Great Barrier Reef since 1960 and observers fear that numbers are building up towards a third. The 1979 to 1990 episode affected about 17 per cent of the 2900 coral reefs to some extent, most of them concentrated in the central one-third of the Reef.

Millions of small coral-eating *Drupella* snails have had a significant impact on more than 100 km of the Ningaloo fringing reef off Western Australia. These snails have also caused localised damage in the Cairns section of the Great Barrier Reef (Zann, 1995).

Introduced species

Introduced species are those species which have been spread ouside their natural range.

Complete data on the number of species introduced into Australian waters do not exist.

This is especially true of many invertebrate groups, microorganisms and certain algae and higher plants where the Australian flora and fauna are not well studied. Many introductions will not become established or will not reach numbers that cause any concern, but some have done so.

At least 55 species of fish and invertebrates and a number of seaweeds have been introduced into Australia either intentionally, for aquaculture, or accidentally in ships' fouling and ballast waters. The following organisms are of particular concern.

Gymnodinium catenatum, a toxic alga, causes red tides. Blooms of introduced toxic marine algae are a serious marine environmental and fisheries problem in Tasmania and Victoria and may threaten other States (see the box on page 8-26).

The invasive giant fanworm *Sabella* cf. *spallanzanii* has carpeted areas of Port Phillip Bay and has also been found in the waters of Port Adelaide, in Cockburn Sound and in the Bunbury and Albany port areas of Western Australia. It could have a detrimental effect on local shellfish-farming and on biodiversity of native species.

Of Japanese origin, the seaweed *Undaria pinnatifida* displaces and smothers native kelps. Its spread in Tasmania now supports a small commercial harvesting industry.

The predatory northern Pacific seastar (see the box below) is spreading along eastern Tasmania, threatening marine life, aquaculture farms and scallop and abalone fisheries (Zann, 1995).

Many species of introduced plants have invaded Australian saltmarshes. However, four are of particular concern: *Spartina anglica*, groundsel bush (*Baccharis halmifolia*), pampas grass (*Cortaderia selloana*) and *Juncus acuta*. These species are highly invasive and are capable of substantially altering the ecology of Australian saltmarsh habitats where they become established.

The northern Pacific seastar in Australia

This species, which is a voracious predator on molluscs and other animals, is a pest on shellfish farms in Japan, which is within its natural range. Originally detected in Hobart in 1986, the seastar has spread along about 50 km of the south-east coast of Tasmania and its larvae can potentially be carried across Bass Strait to mainland Australian waters. It is already threatening marine life, aquaculture farms and scallop and abalone fisheries. Although scuba divers have removed many thousands of seastars, it is unlikely that this method can achieve total eradication (Zann, 1995).

The seastar was probably introduced into Australia by ships visiting the Hobart area. Predators occurring naturally in Australian waters have little chance of reducing or controlling numbers. There are usually no practical methods to eradicate such introduced species once they become established. This underlines the importance of effective preventive measures such as stringent controls to minimise the risk of introducing exotic pest species through fouling on ship's hulls and in ballast-water discharges. Notwithstanding the best quarantine measures however, pest species will probably occasionally breach the barriers. It is important to have effective plans in place to predict the spread and manage the impacts of any that may become established in Australian waters.

Seafood quality

Seafood quality is determined by both the environment in which the species lives and its treatment after it has been harvested from the sea. This section deals only with the environmental effects on seafood quality. The major issues concern the accumulation of contaminants in the flesh of seafood species.

In general, Australia's seafoods are low in contaminants. However, exceptions occur in those species that accumulate heavy metals, biotoxins, microbes and pollutant chemicals. Heavy metals, for example mercury, cadmium and lead, can be accumulated either naturally from sea water through the food chain or by species inhabiting locally contaminated areas. Levels of mercury naturally become elevated in top-level predators such as sharks and other long-lived fish.

Microalgae produce biotoxins like ciguatera and paralytic shellfish poisons (PSPs). Coral reef fish can accumulate ciguatera poison (ciguatoxin) by grazing on algae, and filter-feeding shellfish similarly accumulate PSPs. These species can accumulate toxins to such an extent that people who eat them suffer food poisoning.

Shellfish can also accumulate microbes, such as bacteria and viruses, when they are grown in contaminated water. Seafood species can accumulate chemicals in a number of ways — through the food chain, direct from the water or from sediments.

The National Food Authority sets standards to prescribe the maximum permitted limits of contaminants for seafood consumption. The level of risk to people depends on how much and what type of seafood they consume. Recommended public health levels are based on criteria derived for the 'average' Australian community. Some communities may differ from this pattern in both the quantity and parts of the organism they eat. Torres Strait Islanders, for example, are among the highest consumers of seafood in the world (Johannes and Macfarlane, 1991). Fish, turtles and dugongs form a major part of their diet and they also consume more offal than the 'average' Australian community. Appropriate safe levels for seafood intake have not been calculated for such communities.

Ciguatoxin occurrence in seafood is well documented (Lewis, 1994). In French Polynesia outbreaks of ciguatoxic poisoning have been associated with human disturbances to coral reefs (Bagnis, 1994). In Australia, however, field studies have not supported such a causal link (Holmes *et al.*, 1990).

In New South Wales, sewage wastes from outflows have contaminated cultivated oyster crops. Poor water quality has also caused bacterial and viral contamination of shellfish.

The toxic dinoflagellate *Gymnodinium catenatum*, introduced into waters around Tasmania by ballast-water discharges, has contaminated cultivated shellfish with toxins forcing the periodic closure of commercial shellfish farms in the Derwent and Huon River estuaries since 1986 and in Port Phillip Bay since 1991.

In general, Australian seafoods are low in contaminants.

Contaminant data for four species of fish (sea mullet, bream, flathead and luderick) collected by NSW Fisheries in 1991–92 from 10 estuaries along the New South Wales coast showed that fish collected from the Parramatta and Georges Rivers had relatively high organochlorine concentrations compared with fish from other areas. A few areas in that State have been closed for fishing due mainly to the high pollution of waters or sediments and the concern about possible contamination of fish (Environmental Protection Authority of NSW, 1993).

Table 8.14 Maximum permitted concentrations (MPC freshweight) of heavy metals set for seafood in Australian Food Standard A12 (1992)

Arsenic	Fish, crustaceans and molluscs	1.0 mg/kg (inorganic only)
	Seaweed (edible kelp)	1.0 mg/kg (inorganic only)
	All other foods	1.0 mg/kg (total)
Cadmium	Fish (and fish content of products)	0.2 mg/kg
	Crustaceans (and crustacean content of products)	0.2 mg/kg
	Molluscs (and mollusc content of products)	2.0 mg/kg
	Seaweed (edible kelp)	0.2 mg/kg
	All other foods	0.05 mg/kg
Copper	Molluscs (and mollusc content of products)	70.0 mg/kg
	All other foods	10.0 mg/kg
Lead	Fish in tinplate containers	2.5 mg/kg
	Molluscs	2.5 mg/kg
	All other foods	1.5 mg/kg
Mercury	Fish, crustaceans and molluscs (and fish content etc.)	A mean level of 0.5 mg/kg
	All other foods	0.03 mg/kg
Selenium	All other foods	1.0 mg/kg
Tin	Foods not packed in tin	50.0 mg/kg
	All other foods	150.0 mg/kg
Zinc	Oysters	1000.0 mg/kg
	All other foods	150.0 mg/kg

Source: Australian Food Standard 1992.

Table 8.15 Heavy metals in dugong and green turtle from the Torres Strait

		Mean concentrations (mg/kg fresh weight)			
		cadmium	copper	lead	zinc
Dugong	muscle	0.015	0.2	0.03	12.6
	liver	6.43	184.0	0.08	470.0
	kidney	8.17	2.67	0.06	31.0
Green Turtle	muscle	1.14	4.69	0.07	12.5
	liver	10.7	59.3	0.6	38.6
	kidney	26.0	7.4	0.07	23.8

Source: Gladstone, in press.

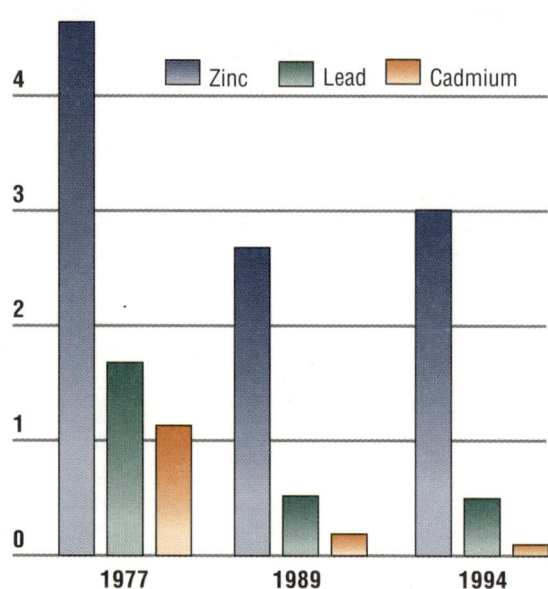

Figure 8.22 Reductions in zinc, lead and cadmium concentration in mussels in Cockburn Sound and Owen Anchorage as a result of pollution controls

5 mg/kg* in mussel flesh

Zinc   Lead   Cadmium

1977   1989   1994

* Note: zinc levels are divided by 10 to fit this scale
Source: WA EPA.

Heavy metals can also be present in seafood in levels above recommended public health limits. Mercury has been subject to detailed investigation (Hancock, 1980) and most States prohibit the sale of shark and billfish over certain sizes because they contain naturally bioaccumulated mercury above recommended limits.

In Torres Strait, high levels of cadmium occur in a range of food species of importance to the Islanders, but levels of other metals are within public health guidelines — depending on which component of the animal is consumed. In general terms, the offal components (such as the liver and kidneys, or their equivalents in invertebrates) of green turtles and dugongs are high in several metals (see Table 8.15). People who eat a lot of these foods may exceed recommended intake rates for some metals.

The high cadmium levels are thought to be due to natural bioaccumulation and not mining operations (Dight and Gladstone, 1993). Preliminary information from other sources suggests that cadmium in prawns may also be of

concern, with concentrations across northern Australia ranging up to 10 times greater than the maximum permitted concentration (Fisheries Pollution Committee, 1991).

In Western Australia, researchers found heavy metal contamination of fish above health safety levels in Princess Royal Harbour, Albany, where mercury was an unsuspected component of an industrial effluent. As a result, authorities stopped the flow of effluent into the harbour and closed part of the fishery. Monitoring of fish and cockle species since 1984 shows that contamination has dropped to a level where it is no longer a health risk. The fishery was reopened in 1992. In Cockburn Sound, a similar trend occurred for the heavy metals cadmium, lead and zinc in mussels, in response to pollution control measures.

Monitoring programs have revealed high levels of tributyl tin (TBT) in a number of marine species around Australia (see opposite). We do not know how significant this is for people who eat seafood, but levels are dropping following the reduction in the use of TBTs as antifouling agents, so it is unlikely to cause concern in future.

We lack information — including baseline data — on levels of substances that may affect seafood quality in Australian waters and sediments. Generally, information is only available from areas with a recognised problem or where organisations have had to provide environmental impact studies for developments such as mining. The National Residue Survey of the Bureau of Resource Sciences surveys the levels of contaminants in seafood from a national perspective; however, the results are not released for public scrutiny.

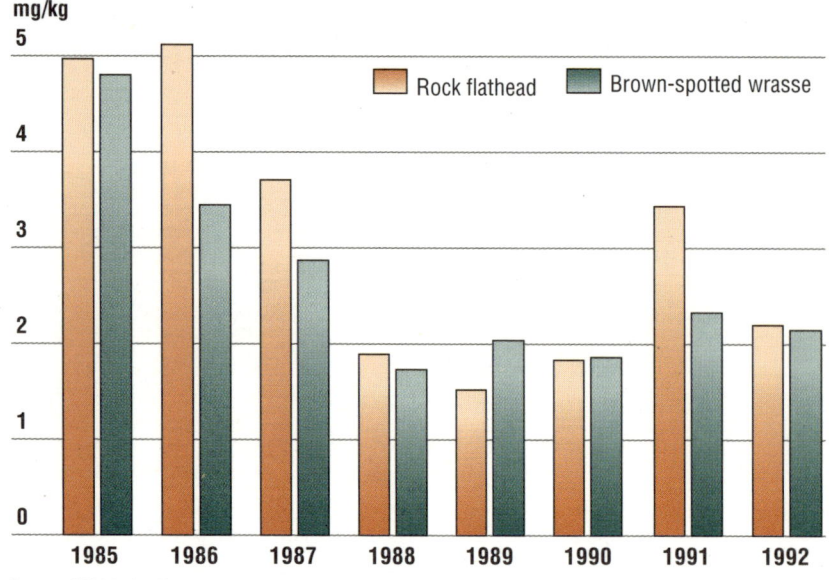

Figure 8.21 Reduction in mercury concentration in rock flathead and brown spotted wrasse in Princess Royal Harbour as a result of stopping an industrial effluent in 1984

mg/kg

Rock flathead   Brown-spotted wrasse

1985   1986   1987   1988   1989   1990   1991   1992

Source: WA Marine Research Laboratories.

Water quality

The quality of water in Australia's marine and estuarine environments is a matter of considerable concern because degraded water quality can have impacts on seafood quality, tourist potential and biological diversity. Degradation of water quality can have many causes, including excess nutrients, sediments derived from land, chemical contaminants, pathogens such as bacteria and litter.

Declining water quality caused by nutrients and sedimentation is the most serious issue affecting Australia's marine and coastal environments. Elevated sedimentation and nutrients are closely linked and largely the result of inappropriate catchment land use practices.

Nutrients

Australia's open, offshore marine waters have low nutrient concentrations. Nearer the coast, concentrations often rise because of the effects of rivers and other land run-off. In some places, near-shore marine systems are periodically exposed to high-nutrient waters derived from upwelling of deep offshore water. In more enclosed, shallow near-shore waters — coastal bays, lagoons and estuaries — nutrient concentrations can rise because of poor flushing combined with river inflows and waste water discharges.

The increased input of nutrients to estuarine and near-shore waters from both diffuse catchment sources and point discharges has significantly altered the nutrient regime of many of Australia's bays and estuaries. An assessment of 22 estuaries in the south-west of Western Australia indicates that seven have low nutrient levels and are in good condition, six have elevated nutrient levels but remain in satisfactory condition and nine have elevated nutrient levels and are in poor condition. Many of the major estuaries in south-east Queensland, including those of the Brisbane, Logan, Albert, North Pine, Caboolture, Maroochy, Mary and Burnett Rivers, have elevated nutrient concentrations, largely due to sewage discharges. In Victoria and New South Wales, most estuaries located near major population centres are — to varying degrees — also nutrient-enriched.

As discussed in the box on page 8-13, elevated nutrient concentrations enhance the growth of marine algae and, in some cases, contribute to loss of seagrass.

Suspended solids

Levels of suspended solids are highly variable, depending on a range of conditions such as particle size, water turbulence, available wave energy, water depth, seabed characteristics and seasonal fluctuations in sources. In Barker Inlet, near Adelaide, for example, the concentration of suspended solids averaged about 20 mg per L, but it varied from 3.5 mg per L during calm conditions to about 300 mg per L during a storm (Kinhill Engineers Pty Ltd, 1993). We do not have long-term data for changes in the concentration in Australian estuaries or the ocean. In Jervis Bay, suspended solids showed great variations over a three-year period, related both to rainfall in the catchment and to nutrient intrusions from the continental shelf (CSIRO, 1994).

Pesticides and industrial chemicals

Organochlorine pesticides like DDT and DDE are thought to be widely present in marine life around Australia, but in very low concentrations away from urban and intensively farmed lands. Organochlorines resist degradation and persist in the environment for decades. They can be passed through the food chain and between generations of organisms.

In Queensland, organochlorines have been detected in very low concentrations on the Great Barrier Reef and in higher concentrations in the Brisbane River. In New South Wales, before Sydney's three deep-water sewage outfalls were built, researchers measured very high levels of organochlorines in marine organisms near Sydney's three major sewage outfalls (Lincoln Smith and Mann, 1989). In South Australia organochlorines were widely present in fish sampled in the early 1970s (Olsen, 1988).

Few surveys of either polychlorinated biphenyls (PCBs) or dioxins have been undertaken in Australia. While PCBs have been detected in offshore waters, increasing towards the coast, their levels are still lower here than in equivalent Atlantic waters: in Port Phillip Bay levels remain high near Melbourne, but are declining in Corio Bay. Dioxins have been found in fish and sediments in Sydney, at Homebush Bay and in Melbourne near sewage outfalls. We do not know the implications of these levels.

Metals

Tributyl tin — widely used in Australia as a poison in antifouling paints since the 1970s — has been a particular water-quality issue. During the 1980s it was found to affect the growth of oysters and other molluscs, and in 1988 the authorities recommended a ban on its use on non-aluminium-hulled vessels smaller than 25 m, which most States have now implemented. Prior to the recommendation, levels of concentration in the waters of Australian dockyards and marinas were frequently 50 times higher than the guideline, but since 1988 have dropped appreciably — to below the guideline. Likewise, in some surface sediments, concentrations are now less than in older sediments below (Zann, 1995).

Table 8.16 Levels of tributyltin in selected waters	
Place	**Level (ng/L)**
Kogarah Bay (NSW)	100 (pre-ban)
Port Phillip Bay (Vic)	3-23 (pre-ban)
Southport (Qld)	45 (pre-ban)
Georges River (NSW)	8-40 (pre-ban)
Georges River (NSW)	1-11 (post-ban)

Note:
The Australian Water Quality Guideline for marine waters is 2 nanograms per litre
Source: Batley, 1995.

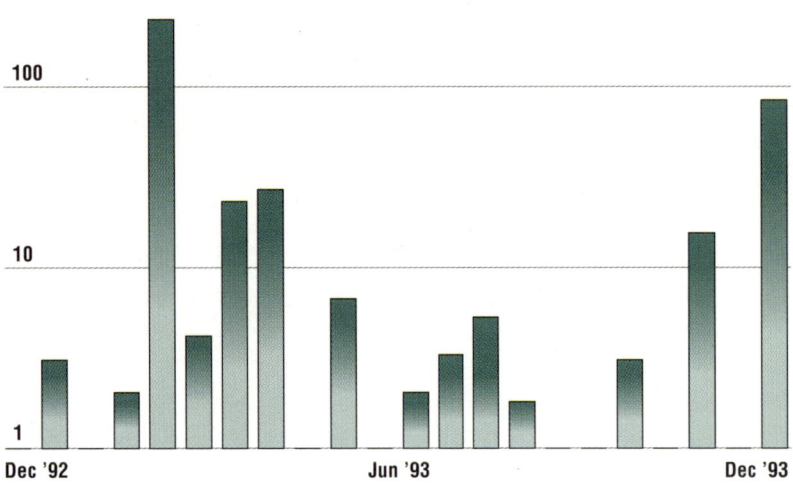

Figure 8.23 Bacteria (*E. coli*) in water from between bathing flags at Surfers Paradise, December 1992 to December 1993 (The Australian water quality guideline is a median value of 150 organisms/100ml)

Source: Qld DEH.

Pathogens

Concern about pathogens in marine and estuarine environments is largely in the context of human health. The areas of most concern are usually those close to urban development, near the sources of pathogens and where the public is most likely to be exposed. Information about the extent of disease-causing microorganisms (pathogens) in most of our marine environment is scarce (Ashbolt, 1995).

In Australia, we usually assess microbial pollution in sea water by measuring the concentration of certain indicator bacteria, primarily faecal coliforms (see Fig. 8.23 and Table 8.2). However, other pathogens, such as viruses and parasitic protozoa, may survive better than the indicator bacteria. Combined with difficulties of sampling and culture methods, this can result in the number of pathogens being underestimated. Nonetheless, local authorities

usually act quickly to close beaches to swimming when they suspect microbial pollution. So, records of numbers and periods of beaches closed to swimming could be used to assess changes in their condition.

Litter

Shipping, beach-goers, recreational boating, commercial and recreational fishing and urban drains all contribute litter to marine and estuarine environments (see Fig. 8.24).

In Australia, a large number of marine organisms become caught in nets, fishing lines and other litter. Researchers estimate that, at any one time, about 500 seals in Tasmanian waters and 45 seals at Victoria's Seal Rocks have 'collars' of plastic litter (see Fig. 8.25). Turtles may die from eating plastic bags, which they mistake for jellyfish (Zann, 1995).

Greenpeace Australia effectively focused public attention on the problem of beach pollution through the 'Adopt-a-Beach' program between 1990 and 1992. Some 123 beaches were 'adopted' by local groups and are now regularly cleaned. The 'Clean up Australia' campaign has continued to focus on the beach litter problem.

Sediments

Sediments form the basis of many marine and estuarine habitats, from sandy beaches, intertidal flats, mangrove muds and seagrass beds to estuarine and deepwater sands and silts. Sediments have several important characteristics that determine what communities inhabit them and how they behave physically. These factors include grain-size composition, organic content, oxygenation and stability. Changes in any of these characteristics usually result in alteration of the associated seabed communities.

Australia's coastal sedimentary environments have changed significantly since European settlement due to the effects of urban, industrial and agricultural development. The greatest impact has been increased sediment input to coastal areas through land run-off caused by erosion from land clearing and agriculture. Associated with these

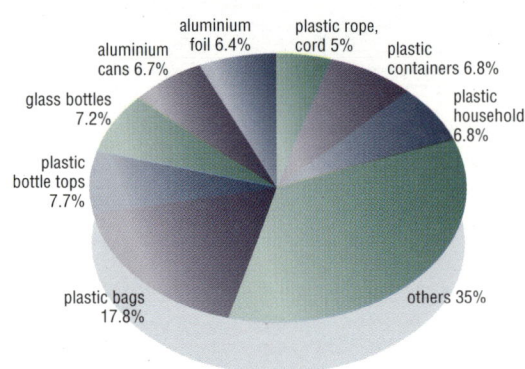

Figure 8.24 Composition of beach litter

Source: Zann, 1995.

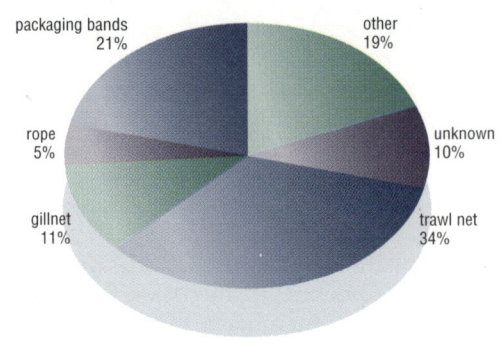

Figure 8.25 'Neck collars' on seals

Source: Zann, 1995.

deposits, the sedimentary environments usually experience changes in grain-size composition, stability and level of sediments, through siltation or erosion. Although beach stability and profiles are increasingly recognised as important aspects of coastal foreshores, no coordinated national programs exist to document changes.

Run-off may include contaminants, like nutrients or heavy metals, from land-based sources. Many such contaminants entering the marine and estuarine environments become chemically bound to sediments and can be released later in response to disturbance or changes in water chemistry (Brodie, 1995; Batley, 1995). There are few programs to monitor subtidal sediments. Disease-causing pathogens and their spores may also survive longer in the micro-environment between sediment grains than in the water, thus increasing the risk of infection (Queensland Department of Environment and Heritage, 1991).

Contaminants in sediments

A large number of contaminants stick to fine sediments. Those bound to surface sediments can be taken up by sediment-dwelling organisms, such as detritus-feeding worms and transferred further along the food chain by predators. Contaminants can also accumulate and be released later when the sediments are disturbed or the local water chemistry changes.

Contaminants originate from many diffuse sources, but most of the problems with sediments remain localised and near the source, such as a stormwater, industrial or sewage discharge. In Australia, potential problems affect up to a dozen major cities located on the coastal fringe. These include the State capitals, a number of other highly industrialised cities — for example, Wollongong and Newcastle — and places near mining or mineral-processing industries, such as Macquarie Harbour, Tasmania.

Heavy metals are of particular concern. On entering the sea they attach to suspended particles and ultimately accumulate in bottom sediments, where they are not lost but may become diluted or covered by newly settling sediments that are less contaminated. Those contaminants that do degrade often exhibit very different behaviour in sediments from that in the water. For example, tributyl tin has a half-life of only a few days in water, but its half-life in sediment has been estimated at about three and a half years. Although some sources of heavy metals have been reduced, concern remains about the high levels in some sediments and the long-term effects of moderate but sustained levels accumulating in the food chain through fish, molluscs, algae and seagrass (Batley, 1995).

During dredging of ports and estuaries the problem is exacerbated. Large volumes of sediment are disturbed and often dumped at sea with the potential to release large amounts of contaminants, including sulfides and organics that naturally occur in anaerobic (deoxygenated) sediments. Although there are no existing standards for heavy metal levels in dredged sediments in Australia, an

Table 8.17 Heavy metals in sediments of selected estuaries in south-eastern Australia

| Place | Levels in micrograms per gram (µg/g) dryweight | | | |
	Copper	Lead	Cadmium	Zinc
Corio bay offshore	2-50	2-210	0.1-13	4-400
Corio Bay mid	4-35	14-100	0.2-9	14-166
Port Phillip offshore	8	22	2	40
Port Phillip near shore	1.5	8	0.8	21
Port Phillip Werribee	<5-75	<20-140	<5	9-300
Lake Munmorah*	70	40	-	150
Tuggerah Lake*	20	40	-	110
Lake Macquarie north*	170	1200	160	2400
Lake Macquarie south*	20	68	4	150
Blackwattle Bay, Sydney#	180	520	3	1150
Quilbray Bay, Botany Bay#	3	10	0.5	25
Port Kembla Harbour#	113	113	2	380
Port Adelaide River#[a]	8 - 38	14 - 38	< 0.5	8 - 118
Sydney (100m water depth)	14	15	-	60

Notes: * 5 cm depth; #10 cm depth; [a] Kinhill unpublished data
Source: after Zann, 1995.

ANZECC Task Force is currently developing guidelines for heavy metals in dredged spoil.

Other contaminants that associate with sediments include hydrocarbons and organochlorine compounds. Hydrocarbons have a strong affinity with sediments; but, while most only exist for a matter of days before being broken down, some persist for a long time. Problems with hydrocarbons usually occur around large or persistent discharges. Organochlorine compounds are of particular concern in marine and estuarine environments because of their toxicity, persistence, ability to accumulate in organisms and concentration via the food chain.

Between 1981 and 1992, different authorities extensively studied the concentrations of organochlorine residues in Australian marine waters, sediments and biota. Most investigations examined concentrations in seafood or biota in localised areas around potential sources — mainly near major eastern seaboard cities. Only a few data exist for sediments and background levels in relatively pristine environments. A survey of organochlorine contamination in waters and sediments in Western Australia (1991 and 1992) indicated that river flushing following rainfall contributed relatively high loadings of some organochlorines and that 63 per cent of waters sampled exceeded Environment Protection Authority criteria for the maintenance and preservation of marine aquatic ecosystems (Richardson, 1995).

In Victoria, measurement of DDT and its breakdown products in sediments from Port Phillip Bay and the Yarra River showed relatively low concentrations in comparison with international data, although concentrations in Australian fur seals were at least as high as those reported in Northern Hemisphere species (Richardson, 1995).

▲
Our beaches are a major cultural icon. A quarter of the population lives within three km of the sea.

Response

Managers of the marine environment need to take account of the links both within and between aquatic ecosystems, and between aquatic and terrestrial ecosystems. Many marine organisms are highly mobile at certain stages of their life cycles, with water currents carrying larvae over long distances. This dispersal stage may be a critical process in maintaining many species. Hence, the concepts of resource management and conservation developed for terrestrial systems are not likely to be directly transferable to marine systems (Kenchington, 1993; Jones and Kaly, 1995).

Estuaries, connecting the land to the sea, add another dimension. Actions and events on land can have far-reaching consequences for the marine environment. Nutrients and pollutants move from the land or the atmosphere to the sea, and the life cycles of some species, such as turtles, involve movement between the sea and estuaries or between the land and sea.

Strategies used to manage the marine environment in Australia

- Controlling disposal of wastes and emissions entering catchments, estuaries and the sea
- Prohibiting or regulating destructive and unsustainable activities
- Protecting important marine and estuarine areas
- Zoning for particular activities, to separate and control incompatible uses
- Requiring environmental impact studies to assess and minimise effects of developments
- Monitoring to assess the effectiveness of management policies
- Protecting particularly vulnerable species or groups of species
- Providing positive and negative economic inducements to encourage sustainable resource use
- Regulating fisheries to conserve the resource
- Undertaking education to promote greater awareness of environmental issues and responsibility
- Establishing industry codes of practice
- Maintaining community action programs
- Undertaking research to provide the essential data on which better environmental management can be achieved

Coastal management

Broadly speaking, the States and Territories have jurisdiction over marine areas to three nautical miles from the coast, and the Commonwealth has jurisdiction beyond those waters to 200 nautical miles. The Offshore Constitutional Settlement (OCS), which established jurisdictions between the Commonwealth and States over marine areas, allows for the application of Commonwealth or State legislation to be varied where there is agreement. Most marine fisheries are now managed under an OCS arrangement. Oil and gas, other seabed minerals, the Great Barrier Reef Marine Park, other marine parks, historic shipwrecks, shipping and marine pollution are managed under 'agreed arrangements'.

Marine environmental conservation in Australia involves a large number of different management strategies and international, regional, Commonwealth, State, Territory and local government agreements, arrangements and agencies (Zann, 1995).

Long-term strategic planning includes the National Strategy for Ecologically Sustainable Development (see Chapters 1, 2 and 10). The strategy seeks to bring together government, industry, unions, environment and community groups and experts in various fields to address 'sectoral' issues, which include fisheries, tourism and transport.

No country can manage its marine environment and resources in isolation from other countries in its region, or from activities on the high seas. International shipping operates under international law, with the rights of innocent passage strongly defended by maritime nations. Numerous international treaties, conventions and agreements are relevant to the coastal zone. Many of these deal with general global issues such as climate change, biological diversity and world heritage.

The United Nations Convention on the Law of the Sea allows nations to claim territorial seas (which extend 12 nautical miles from the coastal baseline), a 200-nautical-mile exclusive economic zone (EEZ) and a legal continental shelf. Australia proclaimed its EEZ on 1 August 1994.

Australia's three tiers of government (local, State and Commonwealth) are all involved in managing the coastal and marine environments. Government inquiries have identified the fragmented and often overlapping responsibilities in the coastal zone as impediments to effective planning and management (Kenchington, 1994; Zann, 1995). The draft Commonwealth Coastal Policy (Commonwealth Department of the Environment, 1992) lists 73 Commonwealth programs, 14 strategies, 49 pieces of Commonwealth legislation and 25 treaties pertaining to the management of the coastal zone. The situation within State/Territory administrations is also complex.

At the local level, the complex administrative procedures involved in managing the many aspects of marine and estuarine environments — from catchment management to policing of recreational fishing — make coordination cumbersome and ineffective. A review of government interests in Jervis Bay, New South Wales, showed that at least 22 Acts of Commonwealth parliament and 29 Acts of State parliament applied directly to the management of the Bay's resources (Environmental Management Services, 1991). Although few places are as complex as Jervis Bay, which combines Commonwealth, State and indigenous ownership of lands and waters, nowhere are the three tiers of government completely coordinated to effectively manage marine and estuarine habitats.

Catchments

Catchment management programs to reduce land degradation are in place throughout Australia. Generally these projects do not have specific marine and estuarine goals, but they have the potential to provide considerable benefits. Communities are becoming increasingly aware of the links between land use and the marine and estuarine environments. For example, this chapter has stressed nutrients as a major issue in estuaries and coastal waters. With the exception of point outfalls of sewage, the source of those nutrients is run-off from catchments (See Chapter 7).

Coastal margins

The Resource Assessment Commission (RAC) Inquiry of 1993 into the coastal zone concluded that coordination and integration between institutions responsible for managing the coast was inadequate. Environmental impact assessments and monitoring programs for coastal margin developments usually concentrate on specific areas of concern and do not help comprehensive regional planning, which crosses the boundary between land and sea.

Although almost 60 government reports and inquiries have been undertaken on Australia's coastal zone since 1960, the RAC Inquiry identified the need for an integrated national approach involving local, State and Commonwealth governments to address the problems of Australia's coastal zone. Major findings of the Inquiry were: no single sphere of government can manage the zone alone; issues of national significance and great public concern are involved; the socio-economic development of the coastal zone is of profound importance to the nation; and Australia has international obligations in the zone that necessitate coordination between the spheres of government.

The Inquiry's report contained many recommendations on the management of Australia's coastal zone, the major one being that the Council of Australian Governments should adopt a national Coastal Action Program to manage the resources of the zone. It recommended that the three spheres of government, in consultation with community and industry groups with responsibility for and interests in coastal zone management, implement the plan.

In May 1995, the Commonwealth government announced a coastal policy, 'Living on the Coast', in response to the RAC recommendations.

Aims of the National Coastal Action Program*

- Reduce degradation caused by urban sprawl and activities in urban and remote locations in the coastal zone
- Provide better facilities for recreation in the coastal zone
- Provide better management and preservation of natural processes in coastal areas
- Achieve more effective and rational use of land in the zone for building, development, tourism and other uses
- Improve recognition by the community of the value of the resources of the zone
- Improve recognition of indigenous peoples' interests in management of the zone
- Improve water quality in streams, estuaries and coastal seas
- Improve management of fisheries through more effective management of sea-based resources of the zone

Proposed by Resource Assessment Commission

Source: Kenchington, 1994.

The policy included funding for the Coastal Action Program, focusing on increasing support for community and local government participation in coastal management, increasing the capacity and knowledge of coastal managers and developing integrated solutions to problems such as urban sprawl and coastal pollution. Other initiatives in coastal zone management include the Queensland government's Coastal Protection and Management Bill, which aims to integrate all management and planning activities in the coastal zone, and a Western Australian report reviewing coastal management, which was released in May 1995.

The Great Barrier Reef is already managed in an integrated way by the Great Barrier Reef Marine Park Authority (see page 8-44). The cost of this increases as pressure from activities in the area, including tourism and recreation, grows. Annual expenditure by the environmental impact management section of the Authority increased from $829 960 in 1988–89 to $2 524 318 in 1993–94.

Public education and environmental awareness of the condition of and pressures on coastal margins are important parts of any response process. Throughout the country, community groups have initiated a number of programs to protect and clean beaches and the marine and coastal environment. One such national program is the Marine and Coastal Community Network, which forms the community component of the Ocean Rescue 2000 program (see page 8-46). The role of this program is to provide a community information network.

Plants and animals

Conservation

Biological conservation and management are largely directed at particular species, areas or sectors — for example, endangered species, marine park areas and commercial fisheries.

Much of the effort in conserving terrestrial biodiversity is based on protecting rare and

Great Barrier Reef Marine Park

The Great Barrier Reef — the largest system of coral reefs in the world — is about 2500 km in length and comprises 2900 separate reefs and 940 islands. Its high species diversity includes more than 400 species of coral, 4000 species of molluscs, 1500 species of fish, six species of turtles, 35 species of seabirds and 23 species of sea mammals.

Internationally recognised as one of the world's best tourist attractions, the region is Australia's premier marine tourism destination. The combined value of tourism and fishing on the Great Barrier Reef is estimated to be around one billion dollars per year.

Pressure

The main pressures on the Reef include: declining water quality in inshore areas — due mainly to elevated sediments and nutrients from changes of land use in coastal catchments: fishing (particularly trawling of the sea floor and overfishing of reef species); coral mortality caused by outbreaks of the crown-of-thorns starfish (the causes of which still remain unknown); storm events; the threat of oil and chemical spills and ballast water introduced from shipping; and the effects of tourism.

About two million people visit the Reef and its adjacent coast annually and the number of visitors is increasing by 10 per cent each year (30 per cent in the Cairns area).

State

The Great Barrier Reef is one of the least-disturbed coral reef systems in the world and much of it is still in a relatively good condition.

Evidence is growing of coral mortality on the tops of some inshore reefs but the evidence is patchy and not consistent on all reef tops for which we have historical photos. Possible causes include cyclones and increased sediments and nutrients.

Recent evidence also suggests that the Capricorn Bunker sector of the Great Barrier Reef is showing signs of degradation: hard coral and fish appear to be reduced in this sector compared with other parts of the Reef (Oliver *et al.*, 1995). The causes of this are unknown, but do not appear to be related to the crown-of-thorns starfish or to cyclones.

Over the past 30 years the crown-of-thorns starfish has damaged nearly 20 per cent of reefs, largely in the central one-third of the Reef. Damage in affected areas ranges from slight to very severe. The causes of the outbreaks are still unknown.

Response

The Great Barrier Reef was inscribed on the World Heritage List in 1981 and is protected under the Great Barrier Reef Marine Park Act. The Marine Park, a multiple-use, protected area of 344 000 sq km, is the world's largest marine protected area. It is managed by the Great Barrier Reef Marine Park Authority, with the Queensland Department of the Environment responsible for day-to-day management. Oil drilling and mining are prohibited in the Marine Park.

The establishment of the Great Barrier Reef Marine Park Authority coincided with a significant period of expansion in tourism, particularly around Cairns. As a result, development and access for visitors has been well planned and regulated within the context of a marine protected area (see Fig. 8.26). Use of the Great Barrier Reef is regulated primarily under a zone-based planning scheme intended to achieve sustainable use of coastal and marine resources. However, better coordination and strategic planning are necessary to maintain environmental protection, particularly in relation to pressures by coastal development, and increasing resident populations and tourist numbers. State and local government bodies and industry need to be involved to achieve this broad management goal.

While the agency approach has worked well on a case-by-case basis and has proved effective even in extremely complex projects — such as the now-defunct floating hotel — it has sometimes proved difficult at a local level. For example, the development of strategic management plans for the Whitsunday and Cairns subregions has highlighted the problems of coordinated management and demonstrated the need for cooperative planning of regional tourism. These cases also reveal the complexity of issues that must be addressed if tourism is to be a sustainable industry in sensitive areas such as the Great Barrier Reef.

The Authority recently coordinated the development of a 25-year strategic plan, involving over 70 user groups, for the Great Barrier Reef World Heritage Area. Variables such as coral cover, crown-of-thorns starfish, dugongs, turtles and nutrients are monitored in parts of the Great Barrier Reef, and the Authority is now developing indicators and a monitoring program for the Area. In 1996 it will publish a report on the state of the Great Barrier Reef World Heritage Area.

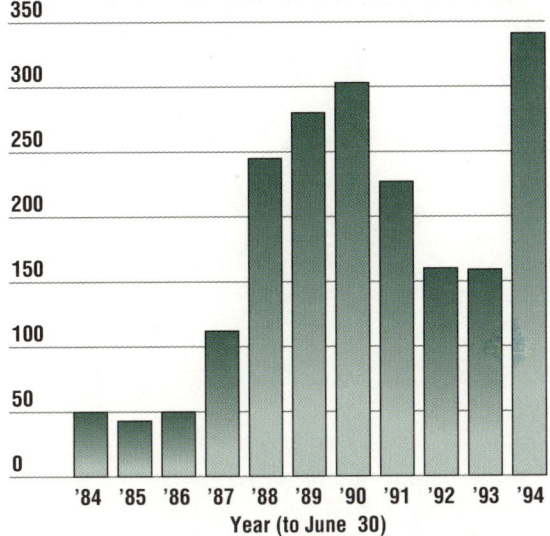

Figure 8.26 Number of tourism permits issued for the Great Barrier Reef Marine Park

Year (to June 30)

Source: Great Barrier Reef Marine Park Authority annual reports.

endangered species. However, the endangered species concept is applicable only to a few kinds of marine animals with certain characteristics, such as those with unusually restricted breeding sites, or species that are highly susceptible to environmental stress (Jones and Kaly, 1995).

Many marine species remain undescribed, and relatively little is known about most of the described ones. Among other things, this adds to the problem of determining whether a species is introduced or a natural inhabitant of any area. An enormous taxonomic and monitoring effort would be required to describe all the marine species in Australia and determine their status.

Given this lack of knowledge, precautionary management strategies are important to conserve marine biodiversity. These strategies include: establishing protected areas for endemic species with small geographic ranges or restricted breeding sites; protecting long-lived, large and wide-ranging species; enhancing populations of excessively exploited species; and establishing protected areas containing representative samples of common habitats.

Some conservation measures apply to specific species while others, like marine protected areas, cover a range of species. An example of a species-specific measure is the listing of the leatherback, green, hawksbill and olive ridley turtles as 'vulnerable' and the loggerhead as 'endangered', under the *Endangered Species Protection Act 1992*, and the flatback as 'potentially vulnerable' by the IUCN. Because some species are highly migratory, a regional and international approach to the management of turtles is important (Zann, 1995).

The establishment of marine protected areas (MPAs) is an important mechanism for conserving and managing a range of marine species and their habitats. An MPA can be any area of intertidal or subtidal terrain — together with its overlying water and associated flora, fauna, historical and cultural features — that has been reserved by law, or other

effective means, to protect part or all of the enclosed environment. These areas may serve many functions, such as conservation of endemic fauna and flora, protection of commercial fisheries resources, protecton of human heritage and provision of opportunities for tourism, recreation, education and research.

Australia is a world leader in using this mechanism for marine conservation and management and has 24 per cent of the total number of MPAs in the world. In 1992, Australia, including its External Territories, had 303 of them covering 463 200 sq km. However, 74 per cent of this area is within a single multiple-use MPA — the Great Barrier Reef Marine Park — which is zoned for multiple uses ranging from preservation to general use. Although the number of Australia's MPAs has increased by almost 60 per cent in the past decade, some are very small and large sections of our marine environment have few or no protected areas. Nearly all (92.7 per cent) of the MPAs in Australia are for multiple-use. Only a small number are dedicated to sanctuaries and preservation (Bleakley *et al.*, 1994).

Many of the MPAs are on the east coast, especially along the Queensland coast. The area protected in the tropics is more than 10 times greater than that in the south. However, more MPAs lie south of the tropics than in the north (175 as opposed to 98). Most of those in the southern and eastern half of the continent are small, even though this is where human activity is greatest and the demand for conservation action is highest. The largest ones tend to be away from the areas of highest human activity (Zann, 1995).

Some MPAs are in the estuaries. Of these, 62 per cent are subject to some form of administrative classification that restricts their uses and 28 per cent have some form of conservation designation, such as marine park, national park, game reserve, flora/fauna reserve or fisheries sanctuary

Table 8.18 Marine Protected Areas (MPAs) by bioregion, excluding external territories and oceanic regions

Region	No of MPAs	Area of MPAs (sq km)	Area of bioregion protected (%)
North Coast	9	3 056	0.55
North West Coast	0	0	0
Central West Coast	4	11 047	22.96
Lower West Coast	4	163	0.26
South West Coast	6	< 2	< 0.01
Great Aust. Bight	2	< 3	< 0.01
South Gulfs Coast	24	361	0.39
Bass Strait	38	572	0.38
Tasmanian Coast	17	555	1.82
Lower East Coast	18	49	0.23
Central East Coast	32	7 500	26.46
North East Coast	95	110 185	97.22
Great Barrier Reef	6	183 280	100.00
Gulf of Carpentaria	6	4 569	1.11
Overall	**261**	**321 347**	**15.00**

Source: Bleakley *et al.*, in press.

▲ Overfishing of southern bluefin tuna during the 1970s and '80s has severely depleted stocks.

(Zann, 1995). The protection varies in name and intent between States and rarely encompasses an entire estuary. Controls are carried out by a range of different administrative bodies within each State. Generally, the protection reflects local interests and does not constitute a national system of estuarine reserves. The level of surveillance and enforcement is likewise highly variable (Zann, 1995) (see Table 8.18).

The Ocean Rescue 2000 program, which the Commonwealth initiated in 1991, aims to establish a national network of MPAs around Australia. The program is an important national response to promote the conservation and sustainable use of Australia's marine and coastal environment. It builds on existing marine conservation and management programs and is part of the National Strategy for Ecologically Sustainable Development.

Other aims of the program are to: develop and implement a marine conservation plan to guide the use and management of Australia's marine resources; ensure adequate baseline and monitoring information on the marine environment, activities and management and ensure its accessibility to decision-makers and managers; and foster an educated, informed and involved community.

The program comprises six elements: the national representative system of Marine Protected Areas; the Australian marine conservation plan; the State of the Marine Environment Report (SOMER); the national marine education program; the national marine information system; and the marine and coastal community network. The Commonwealth is working with the States and Territories in relation to the national representative network of MPAs.

As well as Ocean Rescue 2000, other responses of a broad nature include the formation in 1995 of the Australian Marine Industries and Sciences Council and the publication by the Australian Committee for IUCN (World Conservation Union) of 'Towards a strategy for the conservation of Australia's marine environment' in 1994. A national fisheries policy is being prepared and an oceans management policy is being discussed within government.

Fisheries management

Australia has responsibility under the United Nations Convention on the Law of the Sea for fisheries management within its Exclusive Economic Zone (EEZ). This responsibility is shared between the Commonwealth and the States, depending on the individual fishery, through a cooperative arrangement developed under an offshore constitutional settlement.

In recent years all of the States and the Commonwealth either have introduced or are introducing new Fisheries Acts to manage resources effectively in the face of increasing fishing pressure. While the precise wording of the Acts may vary, they are all directed towards sound resource management, within the principles of ecologically sustainable development and economic efficiency. An example of the aims can be found in the Commonwealth *Fisheries Management Act 1991*.

Liaison between fisheries management and the fishing industry has been strengthened throughout Australia by the development of management advisory committees, which comprise managers, scientists, industry members and in some cases representatives from the recreational sector.

Functions of the advisory committees include helping to develop management plans for each fishery. Once a plan is proposed, interested parties can comment on it. Community groups, including conservation and indigenous groups, are using this avenue for submissions, which government then takes into account. In the Torres Strait Islands, the community is represented on the Torres Strait Fisheries Management Committee.

The advisory committees try to gain a sound understanding of the status of the stocks being fished, and the effect of fishing on the resource, on non-target species and on the ecosystem generally.

One of their priorities is to produce assessment reports for each fishery. Where data are available, this has included a comparison of the status of the breeding stock now compared with the level when fishing was very light. Where breeding stocks have been reduced to levels likely to affect recruitment, management action is taken to redress the problem. This has led to a more detailed consideration of the management strategies required to ensure that the allowable catch is at a level appropriate to the stock status.

Some of the fisheries, such as the South East Fishery, now have total allowable catches, while for others action has been taken to control the effective fishing effort by a combination of a reduction in boats, fishing gear and the number of days fishing (see page 8-47).

As well as management responses to the state of the various fisheries, community groups often express their concern through petitions to the federal Minister for the Environment. For example:

• In 1993 an unsuccessful bid was made to have southern bluefin tuna and orange roughy listed in the Appendices to the Convention on International Trade on Endangered species of Wild Fauna and Flora (CITES).

Managing the fishing resource

Australia is responsible for managing the living resources in its Exclusive Economic Zone, as required by the United Nations Convention on the Law of the Sea (UNCLOS).

All our major fisheries are managed under either Commonwealth or State legislation. Highly migratory species are managed under multilateral arrangements with other countries. An underlying objective of fisheries management is to conserve the limited resources to ensure their long-term sustainability.

The following management goal is taken from the Commonwealth *Fisheries Management Act 1991*:

'...ensuring that the exploitation of fisheries resources and the carrying out of any related activities are conducted in a manner consistent with the principles of ecologically sustainable development, in particular the need to have regard to the impact of fishing activities on non-target species and the marine environment.'

Four examples illustrate responses to a number of fishing pressures. In each case — except for the tropical rock lobster — the pressure on fish stocks has been caused by a rapid increase in fishing during the 1970s and '80s coupled, in some cases, with unfavourable environmental conditions.

Each year a scientific assessment is made of stocks of the four fisheries: southern bluefin tuna, gemfish, tropical rock lobster and barramundi. In the case of southern bluefin tuna, this involves a multilateral effort with Japan and New Zealand. The status of the other species is based on long-term data including information on the size (and age) of fish. In Torres Strait, divers annually survey the tropical lobster stock.

Southern bluefin tuna were overfished throughout the 1960s, '70s and '80s. While numbers remain dangerously depleted, the latest assessment indicates an end to the long-term decline, but the degree to which the stock will recover is uncertain.

Gemfish were overfished in the 1980s and the recruitment of young fish into the fishery declined alarmingly.

Tropical rock lobsters in the Torres Strait are assessed as underfished. However, local areas may be depleted and the catch rate in some heavily-fished areas has declined.

Barramundi monitoring in 1978 revealed that Northern Territory stocks had been reduced to such a low level that recruitment failure was likely. Recruitment of barramundi is highly variable and appears to be strongly influenced by the amount of rain in the early wet season. These environmental effects are now being incorporated into assessments.

Responses

Controls on fishing activity, through fisheries management, constitute the main response.

Southern bluefin tuna — a range of management strategies were introduced during the 1970s, including voluntary seasonal closures in areas off New South Wales and a freeze on further entry of purse seiners and additional pole-fishing vessels around Australia. However, these measures did not contain the fishing effort. In 1979, following the creation of the 200-mile Australian Fishing Zone, arrangements were made for the Japanese to continue fishing in these waters. In 1983, the Australian catch was limited to 21 000 tonnes with a limit imposed on fish size. However, it was not until

1986–87 that Japan limited its catch to 23 150 tonnes. In subsequent years, the catch limit was reduced and has remained stable at 11 750 tonnes (combined Australia, Japan and New Zealand) since 1989–90. The cooperative approach between the countries has been formalised by the establishment of the Convention for the Conservation of Southern Bluefin Tuna.

Eastern gemfish — a management plan was introduced in 1988 to address the problems of declining catch rates and a shift in the population to smaller fish. It set a total allowable catch (TAC) of 3000 tonnes for the winter fishery. This figure was lowered to 1750 tonnes in 1990, 500 tonnes in 1991, and 200 tonnes in 1992 and set at zero in 1993 and 1994 (trip limits allowed fishers to take gemfish as a bycatch). To return stocks to a sustainable level requires protection of both the spawning stock and new recruits. This is obviously difficult where gemfish are also caught as an incidental catch to other species.

Tropical rock lobster — within the Torres Strait Protected Zone, a joint authority involving Queensland and the Commonwealth manages the fishery, with the goal of protecting the traditional way of life and livelihood of the Islanders by encouraging their participation and restricting that of non-Islanders. The catch is shared each year between Australia and Papua New Guinea.

Barramundi — the resource is managed under State and Territory legislation with the aim of providing a balance between commercial and recreational fishers. In the Northern Territory, it is managed under the *Northern Territory Fisheries Act 1988* and the Barramundi fishery management plan 1991. Aborigines who have traditionally used the resource of an area can continue to do so. Commercial fishing has been drastically reduced with the help of a scheme where fishers are charged a levy to compensate those leaving the fishery. For recreational fishers, the time allowed for fishing is controlled and bag limits have been set.

In Queensland, a management plan covering two sub-fisheries imposes closed seasons, size limits, limited licensing, gear restrictions, closures to fishing in some areas, environment habitat protection and compulsory commercial-catch data collection. Education programs also increase public awareness of the issue.

Adequacy of the response

For the three overfished species, responses have successfully reduced fishing effort. For barramundi, allocation of the resource is now possible between commercial and recreational fishers. For southern bluefin tuna and gemfish, time will tell whether reductions in fishing were sufficient and soon enough to allow stocks to rebuild. Because of the large amounts of capital invested in these fisheries, considerable time elapsed between the degraded state of the stock being recognised by scientists and adequate responses being implemented. This applies particularly to the southern bluefin tuna fishery, where a number of countries are involved. Through its membership of the management advisory committees and the Fisheries Management Authority, industry is becoming increasingly involved in decision making for fisheries management.

Source: D. Staples, Bureau of Resource Sciences, Canberra.

- The incidental catch (bycatch) of seabirds during oceanic longline fishing operations has been listed under the Commonwealth *Endangered Species Protection Act 1992* as a 'key threatening process'. The seabirds include the wandering albatross.

- The Minister is considering a request under the same Act to have prawn and scallop trawling listed as a 'key threatening process' to turtles and two species of fish.

- Community groups have also been responsible for having the wandering albatross listed under the Act as an endangered species.

- Requests to have gemfish listed under the Act as an endangered species and southern bluefin tuna listed as a vulnerable species were unsuccessful. A further request to have eastern gemfish listed as a vulnerable species is still being considered.

Long mesh-nets (drift nets) have been banned in waters controlled by Australia because of their harmful effect on non-target species such as birds and marine mammals.

A fisheries response of a different kind is the action necessary to protect the fisheries resources within the Australian Fishing Zone from illegal fishing by operators from other nations. This response has particular significance for the waters off the northern part of Australia.

Exotic species

Governments and industry have taken a number of steps to prevent the introduction of exotic species. These include increased research, voluntary controls on shipping and promotion of international action through the International Maritime Organisation.

Two of the strategies are: to encourage ships to exchange ballast water in the ocean or flush it en route; and to set up quarantine inspection of ballast water prior to discharge. Ocean exchange is the most practical method but it is limited to smaller vessels because of safety (stability) considerations. The quarantine inspection of ballast water is limited by delays in identifying organisms (Zann, 1995). Some research has been conducted into possible treatment of ballast waters during loading and discharge (AQIS, 1993), but effective methods are costly and not easily implemented.

The Australian Quarantine and Inspection Service (AQIS) has extensively studied the issue of introducing exotic species in ballast water and possible management strategies. The International Maritime Organisation has adopted AQIS's voluntary guidelines, and Australian guidelines are now being formulated. Port and quarantine authorities have the power to prohibit discharge of ballast water when they know or suspect that a ship's ballast tanks contain potentially harmful organisms. Scientists at CSIRO's recently established Centre for Research on Introduced Marine Pests (CRIMP) are investigating various options to reduce and manage marine pests. The recently formed Australian Ballast Water

Management Council will coordinate these activities and implement principles to ensure adequate quarantine and to reduce the risk of the accidental displacement of species. Some States already have legislation to control translocation of aquaculture stocks.

Seafood quality

The usual management response to environmental contamination of seafood has been to establish agencies at Commonwealth and State level to directly monitor quality in food, including seafood, and to set safety standards for contaminants.

In 1991, the National Food Authority was established as an independent expert body to make recommendations on the development, variation and review of food standards. Commonwealth, State and Territory governments have agreed that the States and Territories would adopt, without variation, food standards recommended by that Authority and adopted by the National Food Standards Council as prescribed in the Australian Food Standards Code.

Water quality and sediments

Managers of marine and estuarine water quality need to consider all the significant diffuse and point sources of contaminants that affect it. Point sources, such as sewage outfalls, are relatively easy to identify, quantify and manage, whereas diffuse sources such as an urban catchment, are much more difficult to characterise and control. Controlling diffuse inputs of pollutants often means managing sources in areas very distant from the affected marine and estuarine environments but linked to them through the water flow.

Nutrients

Management of nutrient loading to coastal environments and of any subsequent problems of eutrophication, is based on reducing the flow of land-based effluents or better-dispersing existing discharges. Responses aimed at reducing the impacts of nutrients on marine waters include: controlling soil erosion; changing the use or nature of fertilisers; re-using instead of discharging nutrient-rich effluent; diverting discharges into less-sensitive or better-flushed environments; building engineering works to improve flushing; and removing nutrients from effluent. In some places people have tried to rehabilitate existing eutrophic systems by replanting seagrasses or removing nutrient-rich sediments.

The introduction of slow-release fertilisers and improved application regimes is expected to have a major impact on the amount of phosphorus leached from catchments like the Peel–Harvey. Controls over water movement, tree-planting and creation of small wetlands on drainage lines are also being used to slow the movement of drainage water to some estuaries.

In recent years authorities have commonly responded to sewage problems by relocating sewage outfalls into waters with better dilution and dispersion characteristics. While this may be effective in the short term, the full implications for offshore environments are not fully understood and tertiary treatment of sewage to reduce nutrients is preferable in the long term.

Pollution at sea/navigation safety

The Australian Maritime Safety Authority (AMSA) has the national responsibility for marine oil spills. As it is not possible to control a major spill effectively, the primary aim of management is to prevent spills through safe operations and navigation of ships. For example, since the MARPOL Convention took effect, discharges from ships have been reduced by about 60 per cent worldwide. Also, navigation technology has greatly improved, which has helped to reduce the risk of accidents, and oil tankers built after 1993 have double hulls or equivalent construction to reduce the likelihood of oil spillage in the event of accidents such as groundings or collisions (Zann, 1995).

Australia has developed oil-spill-response plans as part of the National Plan to Combat Pollution of the Sea by Oil, which is managed by AMSA and funded by a levy on the shipping industry. The National Plan is a collaborative arrangement between AMSA, the States and Northern Territory governments, the shipping, oil and exploration industries and also includes the Australian Marine Oil Spill Centre, at Corio Quay, Victoria. The Centre was established by the oil industry to assist in responding to major oil spills around the Australian coast and in adjacent areas where Australian-based companies operate (Swan *et al.*, 1994).

Pollution-response equipment is stockpiled at strategic ports and oil terminals, with a response capability for an oil spill of up to 10 000 tonnes (Swan *et al.*, 1994). While fully laden tankers typically carry 60 000 tonnes of oil, most collisions result in only one or two of their internal tanks being ruptured so they spill much less than their fully laden capacity. Usually the oil is lost progressively rather than instantaneously. For example, the *Kirki* spill happened over two weeks.

If a spill larger than 10 000 tonnes occurs, Australia may need to seek international assistance through arrangements under the international Oil Pollution Response and Cooperation Convention. Australia has concluded a memorandum of understanding with New Zealand under this Convention, which will provide assistance in cases of pollution incidents in either country. Similar agreements are currently being negotiated with Papua New Guinea and Indonesia.

Australia has been a leader in regulating international navigation to protect the marine environment. In 1990 the Great Barrier Reef was the first area in the world designated as a 'Particularly Sensitive Area' by the International Maritime Organisation. All ships more than 70 m in length, or carrying oil, chemical or liquefied gas cargoes, must now carry Australian-licensed pilots when using the designated routes within the Great

Barrier Reef Marine Park (Zann, 1995). Australia and Papua New Guinea are cooperating to develop and provide specific preventive and response measures to protect the Torres Strait area from oil spills from tankers and other sources. Improved navigational aids and charting and management of ship passages through the Torres Strait and the Great Barrier Reef are being developed.

Overview

Australian marine and estuarine environments and habitats are generally in good condition to the extent that this can be measured. However, only a few can be regarded as pristine, because of wide-ranging pressures such as nutrient loading, pollution with persistent chemicals and fishing, which have some effect on nearly all parts of our marine and estuarine systems. Near many of our cities and in other localised inshore areas, the condition of some habitats is poor. Of particular concern are the coral reefs of the north-east coast and the temperate seagrass beds of southern Australia. Other signs of deterioration include the increasing incidence of algal blooms, recognition of an increased number of introduced species, continuing loss of coastal habitats such as mangroves and saltmarshes and intense pollution of some estuaries and other near-shore areas.

Poor water quality, caused mainly by high nutrient and sediment loads, is a feature of many Australian estuaries. Together with urban and other coastal development and intensive fishing activities, these exert the major pressures on Australia's marine and estuarine systems. However, for many systems, particularly those offshore or distant from population centres, we do not have enough information to make even a first estimate of their state of health. We believe remote areas are in good health because they face few pressures, but the data to confirm this are scarce.

Governments, industry and the community have recognised the lack of coordination and effective integration of the many levels of government and private-sector interests in management of coastal and marine activities, to the detriment of these ecosystems. In the government sector, many reviews and inquiries have investigated the weaknesses in the coordination of coastal and marine management practices. In the private sector, increased environmental awareness of companies and industry-wide codes of practice are improving approaches to environmental management, both to reduce the inputs of pollutants to waterways and to minimise the unintended consequences of resource utilisation.

At present we lack any integrated framework for management of our marine and coastal systems based on a set of ecosystem-based goals and environmental performance indicators, although steps towards such a system are being taken under the national ESD strategy. Likewise, there is no agreed and common understanding of the ecosystems, their status and the issues affecting them. The development of indicators is essential to this process.

Table 8.19 Summary

Element of the environment/ Pressure	State	Adequate Info.	Response	Effectiveness of response
saltmarshes habitat destruction and degradation	extensive loss near urban centres	✔✔	protected areas, development controls, community awareness	limited – loss and degradation continues in many areas
microalgae algal blooms associated with nutrient enrichment of coastal waters	increasing frequency, magnitude and extent of algal blooms in estuaries, lagoons and bays	✔✔	improved waste disposal treatment and technology: various catchment management initiatives; research on causes	ineffective at present – limited application of technologies
macroalgae blooms due to nutrient enrichment	blooms reported as common; extent and frequency appears to be increasing	✔✔	improved waste disposal treatment and technology: engineering works; changes in land use practice	ineffective at present – limited application of technologies
seagrasses loss of temperate seagrass due to nutrients, sediments, habitat destruction	area diminishing in the south	✔✔✔	protected areas; catchment management; research; protection from trawling and coastal development; improved sewage treatments	inadequate response measure; inadequate at present; inadequate; effective; partially effective;
beaches and dunes habitat destruction from coastal development and mining	generally good, except around human settlements, some areas modified by mining	✔✔✔	zoning; protected areas; impact assessment; rehabilitation	improving but ineffective in addressing cumulative impact
estuaries degradation by pollution, habitat destruction, intensive recreation, introduced pests	most are degraded, particularly south-east	✔✔	pollution controls; catchment management; impact assessment; protected areas; harvesting controls	effective; ineffective at present; limited effectiveness; limited effectiveness; limited effectiveness
gulfs and bays loss of seagrass, algal blooms caused by nutrients, fishing pressures, introduced pests	many degraded especially those near urban centres	✔✔	catchment management; impact assessment; protected areas; pollution controls; fishing controls; management plans	not adequate to cope with existing pressures
continental shelf and slope fishing, oil and gas exploration and production	fisheries fully developed with some over-fishing; habitat conditions unknown	✔	fishing controls, impact assessment; industry codes of practice	partially effective limited effectiveness limited effectiveness
mangroves habitat destruction and degradation	extensively cleared near coastal centres	✔✔✔✔	protection of species; protected areas; development controls; community awareness	mostly effective and improving – rate of decline reduced
coral reefs fishing pressures, recreation and tourism, nutrient enrichment, sedimentation	signs of degradation	✔✔✔	integrated planning and management; protected areas; impact assessment; controls on activities	partially effective – pressures exceeding current response measures
fish and fisheries fishing pressure, both commercial and recreational, impact on habitat and non-target species	most stocks fully exploited; several over-exploited; few under-exploited	✔✔✔✔	fisheries management and planning within the ESD context; research;	partially effective and improving
reptiles previous hunting of crocodiles; threats to turtle population	crocodile populations rebuilding; status of turtle population unknown, susceptible to over-harvesting and disturbance	✔✔✔	national and international protection; croc farming 5 species listed as endangered or vulnerable	effective suspected ineffective because of international nature of issue
seabirds 14 species/subspecies are threatened	populations of most species appear satisfactory	✔✔✔	3 species/subspecies listed as endangered under IUCN; changes in fishing methods;	more information needed
mammals past exploitation of whales and seals; potential habitat destruction and accidental capture; worldwide overexploitation of dugongs	whale and seal populations seem to be recovering although some from a very low base Australian populations of dugongs appear stable	✔✔✔	prevention of hunting; preparation of action plans; increasing community interest; research;	whale and seal population recovery effective; international arrangements are needed for dugong populations

Table 8.19 Summary (continued)

Element of the environment /Pressure	State	Adequate Info.	Response	Effectiveness of response
nutrients high nutrient loads from point and diffuse sources	widespread occurrence of elevated nutrients and associated adverse biological consequences	✓✓✓	improved waste disposal treatment and technology: various catchment management initiatives; research on causes	not effective at present – limited application of technologies
species outbreaks lack of knowledge and understanding of causes, resulting in lack of clear management objectives	crown of thorns major problem in the GBR; Drupella snails a problem at Ningaloo;	✓✓✓	research; localised control of crown of thorns;	inconclusive at present; locally effective
introduced species damage caused by species introduced or translocated by shipping and aquaculture practices	increasing numbers of species documented as introduced; several species recognised as serious pests	✓✓	research; surveys; monitoring of cultured species – closures where necessary; ecological management; codes of practice for aquaculture; control of imports; quarantine and management of shipping practices; pursuit of international agreements	management of established pests ineffective; quarantine and management of shipping practices partially effective; protection of human consumers effective
seafood quality maintenance of high quality seafood; some groups at high risk because of dietary habits	generally good except for heavy metals in large long-lived animals and in seafood from areas where local pollution is high	✓✓	establishment of safe levels in food; closures where levels of contaminants are high; monitoring for export and domestic consumption; community awareness; marketing restrictions	mostly effective, under review; mostly effective; inadequate; inadequate; effective
metals detrimental effects of tributyl tin (TBT); local heavy metal contamination	levels of TBT declining rates of input declining but locally high residual levels	✓✓✓	restrictions on TBT; emission controls; impact assessment	effective but still affecting molluscs locally; control of inputs effective
suspended solids elevated concentrations in many estuaries and near shore waters; impact on reefs; impact on off-shore reefs	trends difficult to determine but thought to be increasing	✓✓	catchment management; dredging controls	ineffective
pesticides accumulation through the food chain with possible impacts on mammals and birds	levels of persistent organochlorines are low, except from some point sources	✓✓	control of use of organochlorine pesticides improved sewage and stormwater management;	usage effectively reduced, changes in residual levels unknown
pathogens threat to public health and recreational amenity	recurrent problem near urban centres	✓✓	closure of beaches when necessary	partially effective
litter impact on amenity and aesthetic appeal; entanglement of and ingestion by marine animals	all parts of the Australian marine and estuarine environment are littered, especially near population centres; impact of litter on marine animal populations unknown	✓✓✓	public awareness programs; waste management for urban areas and shipping (MARPOL)	limited but improving
sediments local pollution and effects on biodiversity and marine resources	local hot spots of contamination but otherwise sediments are thought to be in good condition	✓✓	control of point sources; development of quality criteria; management of marine dumping; catchment management	effective; effective; mostly effective; ineffective at present
marine management fragmentation of management	mess	✓✓	integration of management	ineffective because not widely applied

References

ABARE (1994). *Australian Fisheries Statistics 1994.* (Australian Bureau of Agriculture and Resource Economics: Canberra.)

Adam, P. (1995). Saltmarsh. In Zann and Kailola (1995), pp. 97–105.

Allen, T. (1994). *Habitat Australia* November 1994 pp 6-8.

Anon (1994). Australia — dolphin and whale watching in Australian waters. Paper to International Whaling Commission.

ANZECC (1995). 'Maritime Accidents and Pollution: Impacts on the Marine Environment from Shipping Operations.' (ANZECC: Canberra.)

AQIS (1993). 'Australian Coastal Ballast Water Guidelines, Discussion Paper.' (AQIS: Canberra.)

Ashbolt, N.J. (1995). Human health risk from microorganisms in the Australian marine environment. In Zann and Sutton (1995), pp. 31–40.

Australian Committee for IUCN (1994). Towards a strategy for the conservation of Australia's marine environment. *ACIUCN Occasional Paper* No. 5.

Bagnis, R. (1994). Natural versus anthropogenic disturbances to coral reefs: comparison in epidemiological patterns of ciguatera. *Memoirs of the Queensland Museum,* **34**(3),pp 455–60.

Bannister, J. (1994). 'Western Australian Humpback and Right Whales: an Increasing Success Story.' (W.A. Museum: Perth.)

Batley, G.E. (1995). Heavy metals and tributyl tin in Australian coastal and estuarine waters. In Zann and Sutton (1995), pp. 63–72.

Bird, E. (in press). Coastal modifications and their impacts. *Technical Paper for the State of the Marine Environment Report for Australia.*

Bleakley, C., Ivanovici, A., and Ottesen, P. (in press). State of Australia's marine protected areas. *Technical Paper for the State of the Marine Environment Report for Australia.*

Brodie, J. (1995). The problems of nutrients and eutrophication in the Australian marine environment. In Zann and Sutton (1995), pp. 1–29.

Bureau of Transport and Communication Economics (1991). Major oil spills — risk and response. *Bureau of Transport and Communication Economics, Report* No. 70.

Clarke, P. J. (1989). Coastal dune vegetation of New South Wales. *University of Sydney Coastal Studies Unit Technical Report* 89/1.

Commonwealth Department of the Environment (1992). A draft policy of Commonwealth responsibilities in the coastal zone. (Department of the Arts, Sport, the Environment and Territories: Canberra.)

Connell, D.W. (1995). Occurrences and effects of petroleum hydrocarbons on Australia's marine environment. In Zann and Sutton (1995), pp. 47–52.

Cosser, P. (*in press*). 'Nutrient Technical Paper.' Report for DEST, Canberra.

CSIRO (1994). 'Jervis Bay Baseline Studies, Final Report.' (CSIRO Division of Fisheries: Marmion, W.A.)

Dight, I.J., and Gladstone, W. (1993). Torres Strait baseline study: pilot study final report June 1993. *GBRMPA Research Publication* No. 29.

Dutton, I., and Luckie, K. (in press). Impacts of recreation and tourism on marine and coastal systems. *Technical paper for the State of the Marine Environment Report for Australia.*

Environmental Management Services (1991). Planning the plan: identifying information requirements for the Jervis Bay Environmental Management Plan. *Environmental Information Discussion Paper* 3/91.

Environment Protection Authority of NSW (1993). *New South Wales State of the Environment* 1993.

Fisheries Pollution Committee (1991). 'Cadmium in Fisheries Products.' (Bureau of Rural Resources: Canberra.)

Gales, R. (1993). 'Co-operative Mechanisms for the Conservation of Albatross.' (ANCA: Canberra.)

Galloway, R.W., Story, R., Cooper, R., and Yapp, G.A. (1984). Coastal lands of Australia. *CSIRO Division of Water and Land Resources, Natural Resources Series* No. 1.

Gladstone, W. (in press) Trace Metals in Sediment, Indicator Organisms, and Traditional Seafoods of the Torres Strait — Final Report of The Torres Strait Baseline Study. (GBRMPA, Townsville.)

Goldsworthy, S.D. (1994). 'Status Report on Eared Seals (Otariidae) in Australia and its Territories. Report prepared for the Endangered Species Program.' (ANCA: Canberra.)

Hallegraeff, G.M. (1993). Assessment of the impact of algal blooms on the water quality of Sydney coastal waters. *Report for the Environmental Protection Authority of New South Wales.*

Hallegraeff, G.M. (1995). Marine phytoplankton communities in the Australian region: Current status and future threats. In Zann and Kailola (1995), pp. 85–96.

Hancock, D.A. (Chair) (1980). 'Report on Mercury in Fish and Fish Products to Coordinating Committee on Metals in Fish and Fish Products.' (AGPS: Canberra.)

Holmes, M.J., Lewis, R.J., and Gillespie N.C. (1990). Factors affecting the toxicity of the dinoflagellate *Gambierdiscus toxicus* and the development of ciguatera outbtreaks. *FIRDTF Project* No. 86/10.

Johannes, R.E., and Macfarlane, J.W. (1991). 'Traditional fishing in the Torres Strait islands.' (CSIRO Division of Fisheries: Hobart.)

Jones, G.P., and Kaly U.L. (1995). Conservation of rare, threatened and endemic marine species in Australia. In Zann and Kailola (1995), pp. 183–91.

Jones, M.M. (1994). 'Fishing Debris in the Australian Marine Environment.' (Bureau of Resource Sciences: Canberra.)

Kailola, P.J., Williams M.J., Stewart P.C., Reichelt R.E., McNee A., and Grieve C. (1993). 'Australian Fisheries Resources.' (Bureau of Resource Sciences: Canberra.)

Kearney, R.E. (1995). 'Recreational Fishing: What's the Catch?' *Proceedings, Australian Society for Fish Biology workshop, 30–31 August 1994.*

Kenchington, R.A. (1990). In 'Managing Marine Environments,' p 248. (Taylor and Francis: New York.) .

Kenchington, R.A. (1994). Findings and recommendations of the Resource Assessment Commission Coastal Zone Inquiry. *Technical Paper for the State of the Marine Environment Report for Australia.*

Kenchington, R.A. (in press). Human pressures in the coastal zone. *Report for DEST, Canberra.*

Kinhill Engineers Pty Ltd (1991). 'A Preliminary Assessment of Marine Impacts Derived from the Wreck of the *Sanko Harvest* at Esperance, WA, February 1991.' *Report for WA EPA.*

Kinhill Engineers Pty Ltd (1993). 'Barker Inlet Monitoring and Revegetation Study, Progress Report No. 1.' *Report for Pipelines Authority of SA.*

Kinhill, Metcalf and Eddy (1994). Mapping of historical changes in the region of the Bolivar wastewater treatment plant outfall. *Report for the Engineering and Water Supply Department, South Australia.*

Kirkman, H. (in press). 'Seagrasses of Australia.' *Report for DEST, Canberra.*

Lewis, R.J. (1994). International workshop on ciguatera management. *Memoirs of the Queensland Museum* 34(3), preface.

Lincoln Smith, M.P., and Mann, R.A. (1989). 'Bioaccumulation in Nearshore Marine Organisms II.' (NSW State Pollution Control Commission: Sydney.)

Manning M., Kerr S., and Staples D. (in press). Ballast Water — Technical Overview Report — An overview of the completed Ballast Water Research Projects 1989–1994. *AQIS Ballast Water Series.*

Marsh, H., Breen, B., Lawler, I., Lindoy, N., and Morrissette, N. (1994). 'Dugong Action Plan: Background Document. Report to ANCA Endangered Species Program.' (ANCA: Canberra.)

Marsh, H., Corkeron, P.J., Limpus, C.J., Shaughnessy, P.D., and Ward, T.M. (1995). The reptiles and mammals in Australian seas: their status and management. In Zann and Kailola (1995), pp. 151–66.

Moss, A.J., Rayment, G.E., Reilly, N., and Best, E.K. (1993). A preliminary assessment of sediment and nutrient exports from Queensland coastal catchments. *Queensland Department of Environment and Heritage, Environment Technical Report* No. 5.

NSR Environmental Consultants Pty Ltd (1993). Surface inputs into Port Phillip Bay (excluding the Yarra River and WTC). *CSIRO Port Phillip Bay Environmental Study, Technical Report* No. 12.

Oliver, J., De'Ath, G., Done, T., Williams, D., Furnass, M., and Moran, P. (1995). Long-term monitoring of the Great Barrier Reef. *Australian Institute of Marine Science Status Report* No. 1.

Olsen, A.M. (1988). Pesticide levels in some marine and freshwater fish of South Australia. Department of Fisheries South Australia, *Fisheries Research Paper* No. 19.

Poiner, I.R., and Peterken, C. (1995). Seagrasses. In Zann and Kailola (1995), pp. 107–17.

Queensland Department of Environment and Heritage (1991). 'Moreton Bay Strategic Plan.' (QDEH: Brisbane.)

Raaymakers, S. (1994). Shipping and ports. *Technical Paper for the State of the Marine Environment Report for Australia.*

Ramm, D.C., Pender, P.J., Willing, R.S., and Buckworth, R.C. (1990). Large-scale spatial patterns of fish caught by prawn trawlers in northern Australian waters. *Australian Journal of Marine and Freshwater Research*, 41, pp. 79–95.

Richardson, B.J. (1995). The problems of chlorinated compounds in Australia's marine environment. In Zann and Sutton (1995), pp. 53–61.

Richardson, W.J., and Malme, C.I. (1994). Man-made noise and behavioural responses. In 'The Bowhead Whale', ed. J. Burns, J.J. Montague, and C.J. Cowles. *The Society for Marine Mammalogy, special publication 2*, pp. 631–700.

Rigby, G.R., Taylor, A.H., Hallegraeff, G.M., and Mills, P. (1993). Progress in research and management of ship's ballast water to minimise the transfer of toxic dinoflagellates. *Proceedings, Sixth Annual Conference on Toxic Marine Phytoplankton*, pp. 18–22.

Ross *et al.*, (1995). The status of Australian seabirds. In Zann and Kailola (1995).

Saenger, P. (1995). The status of Australian estuaries and enclosed waters. In Zann and Kailola (1995), pp. 53–9.

Sanderson, J.C. (in press). 'Subtidal Macroalgal Assemblages in Temperate Australian Coastal Waters.' *Report for DEST, Canberra.*

Somers, I.F. (1994). Modelling loggerhead populations. In *Proceedings, Australian Marine Turtle Conservation Workshop, Brisbane.*

South Australia, Department of Environment and Land Management (1993). State of the Environment Report for South Australia.

Swan, J.M., Neff, J.M., and Young, P.C. (eds) (1994). 'Environmental Implications of Offshore Oil and Gas Development in Australia—the Findings of an Independent Scientific Review.' (Australian Petroleum Exploration Association: Sydney.)

Wace, N. (1995). Ocean litter stranded on Australian coasts. In Zann and Sutton (1995), pp. 73–87.

Wachenfeld, D. (1995). A century of change on coral reef-flats: or not? *Reef Research*, 5(3).

Ward, T.J. and Rainer, S.F. (1988). Decapod crustaceans of the North West Shelf, a tropical continental shelf of north-western Australia. *Australian Journal of Marine and Freshwater Research*, 39, pp. 751–65.

Warneke, R.M., and Shaughnessy, P.D. (1985). *Arctocephalus pusillus*, the South African and Australian fur seal: taxonomy, evolution, biogeography, and life history. In 'Studies of Sea Mammals in South Latitudes', ed. J.K. Ling and M.M. Bryden, pp. 53–78, (South Australian Museum: Adelaide.)

Wassenberg, T.J., Salini, J.P., Heatwole, H. and Kerr, J.D. (1994). Incidental capture of sea-snakes (Hydrophiidae) by prawn trawls in the Gulf of Carpentaria. *Australian Journal of Marine and Freshwater Research*, 45, pp. 429–43.

Webb, G.J.W., Manolis, S.C., and Ottley, B. (1994). Crocodile management and research in the Northern Territory: 1992–94. *Paper presented at 12th working meeting of the IUCN-SSC Crocodile Specialist Group, Pattaya, Thailand, May 1994.*

Zann, L. (1995). Our sea, our future: major findings of the State of the Marine Environment Report for Australia. (Great Barrier Reef Marine Park Authority, for the Department of the Environment, Sport and Territories, Ocean Rescue 2000 Program: Townsville.)

Zann, L.P., and Kailola, P. (eds) (1995). 'State of the Marine Environment Report for Australia: Technical Annex 1 — The Marine Environment.' (Great Barrier Reef Marine Park Authority, for the Department of the Environment, Sport and Territories, Ocean Rescue 2000 Program: Townsville.)

Zann, L.P., and Sutton, D. (eds) (1995). 'State of the Marine Environment Report for Australia: Technical Annex 2 — Pollution.' (Great Barrier Reef Marine Park Authority, for the Department of the Environment, Sport and Territories, Ocean Rescue 2000 Program: Townsville.)

Acknowledgments

The following people reviewed the chapter in draft form and provided constructive comments:

Dr Russ Reichelt (Australian Institute of Marine Science)

Ms Diane Tarte (Ocean Rescue 2000 Community Network)

Professor Alistair Gilmour (Macquarie University)

Dr Gustaf Hallegraeff (University of Tasmania)

Dr Nick Harvey (University of Adelaide)

Dr John Volkman (CSIRO Division of Oceanography)

We also acknowledge the comments, suggestions and data on various aspects of the chapter from many others in the Australian academic and government communities: the chapter has been greatly improved by their contributions.

In addition, Commonwealth Government departments and members of the Commonwealth/State ANZECC State of the Environment Reporting Taskforce also helped identify errors of fact or omission.

Photo Credits

Page 8-1: Nick Alexander (Oryx Films)

Page 8-5: *(clockwise from top left)*
Jiri Lochman (Lochman Transparencies);
Ports Corporation, South Australia; Nigel Holmes;
Len Stewart, (Lochman Transparencies)

Page 8-7: Nigel Holmes

Page 8-8: Ports Corporation, South Australia

Page 8-12: *(from left)*
Clay Bryce (Lochman Transparencies);
Tasmanian Parks and Wildlife Service

Page 8-14: WA Marine Research Laboratories

Page 8-15: Nigel Holmes

Page 8-16: *(clockwise from top left)*
Gustaf Hallegraeff (University of Tasmania);
Nigel Holmes; Clay Bryce (WA Museum);
Clay Bryce (WA Museum)

Page 8-18: Nigel Holmes

Page 8-23: Nigel Holmes

Page 8-25: Nigel Holmes

Page 8-27: Gustaf Hallegraeff (University of Tasmania)

Page 8-28: Nigel Holmes.

Page 8-29: Geoff Dews, CSIRO.

Page 8-30: *(illustrations from top)*
FAO; Bureau of Resource Sciences; FAO

Page 8-31: *(illustrations from top)*
John Paxton; Bureau of Resource Sciences; FAO

Page 8-34: Geoff Taylor (Lochman Transparencies)

Page 8-35: Nigel Holmes

Page 8-36: CSIRO Division of Fisheries

Page 8-37: Nick Alexander (Oryx Films)

Page 8-42: Nick Alexander (Oryx Films)

Page 8-44: Nigel Holmes

Page 8-46: CSIRO Division of Fisheries

Chapter 9

Natural and Cultural Heritage

'Barmah Forest' by Lin Onus shows a natural river red gum forest on the River Murray, significant to local Aboriginal communities. The painting suggests irreparable changes to the ecosystem since European settlement.

Prepared by

Rosemary Purdie (Chair), Australian Heritage Commission

Jane Gilmour, Earthwatch

Kieran Hotchin, Aboriginal and Torres Strait Islander Commission

Isabel McBryde, Department of History, The Faculties, Australian National University, and Australian Institute of Aboriginal and Torres Strait Islander Studies

Alex Marsden, Australian Heritage Commission

Ian Robertson (State of the Environment Reporting Unit member), Department of the Environment, Sport and Territories (Facilitator)

Contents

Introduction

Australia's natural and cultural heritage is integral to the environment. Our natural heritage is the physical landscape — the biological and physical elements such as plants, animals, mountains, rivers, deserts and oceans. This landscape is also imbued with human associations, stories, myths, personal histories and emotions.

People have lived in Australia for at least 50 000 years. Over the last 200 years, the first Australians have been followed, initially by British settlers and convicts, and later by immigrants from many countries. All helped shape our physical environment and left tangible evidence in the form of archaeological remains, material objects, structures or remnants of infrastructure. They also left an intangible legacy — the stories of places and people, the meanings attached to places and objects and cultural practices and traditions. This cultural heritage, which provides the fabric, context and web of history, is as much a part of the Australian environment as our natural heritage.

Heritage provides the cultural and physical links with the past, with the history of human habitation and settlement in Australia and with the evolution of biota and the physical landscape. It is integral to our 'sense of place', an element central to the cultural identity of any nation and a source of spiritual well-being. Natural landscapes, with their biological and physical diversity, and cultural landscapes, with their diversity of cultural records and layers of meaning, objects and stories, collectively give us our uniquely Australian sense of place.

A society that values its heritage will want to retain it for future generations and act to do so. The state of our heritage is just as important as the state of Australia's atmosphere, its water, oceans, land, plants and animals.

This chapter covers Australia's natural and cultural heritage as described in the box on page 9-5. Natural heritage comprises significant places and objects that are part of the biophysical environment. As previous chapters focus on the biophysical aspects of the natural environment, this chapter only covers those aspects specifically related to heritage.

Australia's cultural heritage includes places and objects significant to Aborigines, Torres Strait Islanders and non-indigenous Australians. It contrasts strongly with that of countries whose heritage is displayed most visibly by ancient monuments. Yet places significant to the Aborigines may have even greater antiquity. For example, at Lake Mungo in New South Wales, archaeological sites date to at least 40 000 years ago. Australian Aboriginal culture is one of the oldest continuing cultural traditions in the world and remains a vital and creative force in modern Australia. Its ancient record makes places such as Lake Mungo a focus today for Aboriginal groups celebrating this continuity of cultural survival. Only relatively recently has the heritage of Aborigines and Torres Strait Islanders been widely accepted as part of Australia's overall cultural heritage. However, not all Aborigines believe their heritage belongs to all Australians.

The cultural heritage of the last 200 years or so is a complex record of immigration, settlement and dramatic modification of the biophysical environment. It can be seen in skyscrapers and suburban villas, in farms, forests and factories, in ruins in the landscape and relics in museums. People who arrived here as migrants after World War II may perceive heritage as just relating to their own history, in the places left behind or the traditions brought with them. However, the places where they have continued their cultural life and practices in Australia are indeed of heritage value. That these places are relatively young — as is all of our cultural heritage since European occupation — does not make the heritage any less valid. 'It is just more elusive and less easily understood' (Armstrong, 1994).

For both indigenous and non-indigenous communities the value of cultural heritage is not just a matter of age. Local communities value places for their current roles and ongoing uses, as well as their ability to symbolise the past and provide tangible links with it.

The state of Australia's natural and cultural heritage and associated pressures were first described in detail in the 1974 Report of the Committee of Inquiry into the National Estate (the Hope report),

Rottnest Island, Western Australia — a place with significant natural and cultural heritage values.

with the situation reviewed in 1981 (Yencken, 1985). The state of museum collections was described in 1975 in the Report of the Committee of Inquiry on Museums and National Collections (the Pigott report), and in the 1987 report of the Committee to Review Australian Studies in Tertiary Education (CRASTE) (Daniels, 1987).

The following sections outline current human pressures affecting Australia's heritage, summarise its present state and describe responses to its condition particularly over the last decade. Where relevant, the above reports are used as benchmarks to measure changes in pressures and trends in state. Severe constraints were imposed by the very short time available to collate relevant data or initiate studies. Mostly indirect measures were used to assess the condition of the heritage environment. Often it has been possible only to raise relevant issues without being able either to quantify them or assess their importance with confidence.

The inclusion of Australia's natural and cultural heritage adds a new dimension to state of the environment reporting. This chapter concentrates on heritage places and heritage objects. People could well argue that heritage objects should include library and archival material, as they are one means by which heritage meanings are transmitted and also have heritage value in their own right. However, this chapter concentrates on objects with a direct physical relationship to place (see page 9-7) because of their interconnectedness.

Wilderness

The term 'wilderness' is often understood to mean wild and remote areas — that is, land to which people are alien. However, Aboriginal communities have lived in every part of Australia for many thousands of years; they have managed the land, and their stories and songs testify to their relationship with it. For indigenous people today, areas that may be 'wild' in the eyes of many people are cultural landscapes and rich in meaning and law. Because of this long association of indigenous peoples with the land, 'wilderness' is now defined as remote areas that remain substantially undisturbed by the activities of colonial and modern technological society and that are large enough to ensure the long-term protection and integrity of their natural systems and biological diversity.

Land remote from settlements, that has been little disturbed by non-indigenous land-use practices is important for conservation. The protection of these areas may also be important to maintain indigenous values and life-styles, to reaffirm cultural heritage and traditional social relationships and to foster traditional ecological knowledge and land management practices.

Sources: 1994 Environment Policy of the Aboriginal and Torres Strait Islander Commission; Robertson *et al.*, 1992.

Defining natural and cultural heritage

Places

Heritage places are those natural and cultural sites, structures, areas or regions that have 'aesthetic, historic, scientific or social significance or other special value for future generations as well as for the present community' (*Australian Heritage Commission Act 1975,* Section 4). Many places have both natural and cultural heritage values.

Objects

Heritage objects are those which provide material evidence of Australia's natural and cultural environments or its historical and cultural life and biophysical evolution. They may be *in situ* at significant sites or held in collecting institutions — archives, libraries, museums, galleries, zoos, herbaria or botanic gardens — or historic buildings.

Living collections of flora and fauna are also included because of their relevance to biodiversity.

Values and meanings

Places and objects have heritage significance because of the meanings that people attach to them. They reflect the values of their times. These intangible aspects underpin natural and cultural heritage and so are discussed before a description of the state of places and objects, and associated pressures and responses.

It is likely that future generations will value nearly all the places identified today as having heritage significance, but it is also certain they will value other types of places that our society does not. The reasons for this are threefold. Firstly, places (and objects) that seem to be quite commonplace today will have different significance with the passage of time as historical assessment of them changes or as they become rarer through attrition. Secondly, new places and objects will be created and will in time have their own significance. Thirdly, attitudes are constantly changing — in the last couple of decades many new dimensions of our heritage have been recognised and valued. These changes result in a broader awareness of the strong attachments of

The community responded forcefully when Melbourne's Fitzroy swimming pool was threatened with closure — an example of people's strong attachment to places of social value.

Table 9.1 Australia's overseas-born population

	Date of Census		
	1971	1981	1991
Total Australian population ('000)	12 755.6	14 576.3	16 850.5
Overseas-born			
Total population ('000)	2 579.3	3 182.5	4 125.8
Per cent of total Australian population	20.2	21.8	24.5
Total number of birth-place countries[1]	c. 87	c. 102	c. 238
Per cent born in English-speaking countries	48	44	42
Per cent from the top 20 overseas birth places[2]	89.7	80.4	82.1
Per cent from birth places other than the top 20 birth places	10.3	19.6	17.9
Overseas-born as a per cent of the total Australian population, by region of birth place			
Oceania and Antarctica (excluding Australia)	0.8	1.5	2.1
Europe and the former USSR	17.3	15.5	13.6
Middle East and North Africa	0.6	0.8	1.0
Asia	0.8	1.7	4.1
Northern America	0.3	0.3	0.4
South and Central America & Caribbean	0.1	0.3	0.4
Africa excluding North Africa	0.3	0.4	0.6
Not stated	-	1.2	2.2

Notes:
1. The increase in number of birth places from 1971 to 1991 is a mixture of re-classification of birth place categories as well as a real increase in the actual number of countries.
2. Top 20 birth places are the 20 countries with the highest number of overseas born Australians

Source: Unpublished census data from the Australian Bureau of Statistics, January 1995.

Australia's cultural diversity

The cultural diversity of the Australian population is increasing, with the proportion of overseas-born rising steadily over the last two decades. In 1991, people born overseas comprised 25 per cent of the total population, and the percentage from English-speaking countries had declined (Table 9.1). People born in Europe and the former USSR comprised the majority of overseas-born in 1991. Their proportion of the whole population, however, had declined over the last two decades, while that of people from Asia had increased. The proportion of people from countries other than the top 20 overseas birth-places in 1991 had almost doubled since 1971.

different groups of people to particular places and objects. For example, Australians of European origin long ignored the relationship of indigenous peoples to their land in so-called wilderness areas (see the box on page 9-5), and only recently have included indigenous interests in discussions about such areas (Robertson *et al.*, 1992).

Values and meanings related to natural and cultural heritage have rarely been explored in community attitude surveys (see Purdie, in press). However, surveys on attitudes to the environment indicate poor awareness and appreciation of cultural

Heritage registers

In Australia, governments have passed major Acts designed to protect our natural and cultural heritage. Many of these require lists (registers) of places that fall within their definitions of heritage. Some require a judgement about whether places meet a certain level of significance before they are included. Heritage registers include information about the location of places, their characteristics and their significance. The Register of the National Estate is the only one covering significant natural, historic and indigenous places in all of Australia's States, Territories and External Territories.

heritage, both indigenous and non-indigenous. For example, in one study in 1990, preservation of Kakadu National Park (a World Heritage area) and of historic buildings generally were both ranked very low compared with matters such as forest management and conservation of flora and fauna (Imber *et al.*, 1991). In one survey specifically targeted at heritage rather than the environment (Elliott & Shanahan Research, 1993), heritage was most commonly perceived to relate to historic places, although 87 per cent of respondents considered it was important to protect Australia's natural and cultural heritage. Attitudes varied with age, gender, region, education level and ethnic background. People of non-English-speaking background felt their heritage was generally excluded.

Australia's cultural diversity has increased over the last two decades (see the box), placing pressure on Commonwealth, State and Territory agencies to ensure that their heritage registers represent the heritage of all ethnic groups. Such places must also be managed in a way that allows groups to maintain the cultural traditions and continuity that give the places their heritage value. However, this may be difficult where differences in attitudes towards heritage between people from different cultural groups result in conflicting heritage values (Armstrong, 1994).

The homogenising effect of mass media and other global influences on contemporary culture also affects attitudes to places and objects. For some people, these factors strengthen the value they place on their local heritage and they wish to see it retained. For others, local heritage is seen of little importance and moves to replace it may be welcomed.

Differences in community and personal values and in the meanings that people attach to natural and cultural heritage increase the complexity of pressures on heritage places and objects, and complicate decisions about conserving and managing them.

Why are objects part of Australia's natural and cultural environments?

'However we conceptualise culture, in contemporary society or in the past, it is entangled with the objects which give it tangible ... expression ... It is impossible to consider one in isolation from the other.'

(Anderson, in press).

Cultural heritage sites often include objects. These may be: the 'contents' of places, such as fittings and furnishings of historical buildings or machinery of industrial sites; or archaeological material, like shells, fragments of bone and stone artefacts in middens; or broken glass, ceramics and the remains of metal implements in sites of European settlement.

The contents of a place often reveal far more about the owners and their society than the place alone (for example, **Calthorpes' House**). 'Without their contents, places are empty shells, stripped of the primary evidence for their function and use' (Anderson, 1994), their heritage significance thus diminished. The interpretation of archaeological sites is largely based on the objects recovered from them (for example, **Leichardt Billabong**), their position relative to each other and the depositional context.

Individual plants, animals, fossils or rocks are integral parts of the natural environment not generally viewed as 'objects' until 'collected' and removed for scientific or other purposes. The natural heritage value of many places depends on the 'objects' located there, whether animate (such as the **Wollemi Pine**) or inanimate (such as **Australian mammal fossils**). Removal of numbers of 'objects' can threaten the quality of a place's significance.

Natural history and cultural institutions across Australia house vast collections of objects removed from their places of origin (see page 9-28). The millions of biological specimens represent an irreplaceable record of Australia's past and present biota. The specimens of now-extinct plant and animal species and of 'living fossils' (for example, the **Wollemi Pine**) provide an essential basis for understanding historical changes in Australia's natural environment. Biological specimens collected as part of environmental assessment or for management purposes are critical aids to understanding the state of our environment today and for monitoring changes over time. Captive breeding programs in zoos and botanical gardens may ensure the survival of threatened or endangered animal and plant species. The living collections are thus vital to maintain Australia's biodiversity.

Artefacts (such as **toas**) housed in museums are similarly essential for understanding Australia's cultural environment. Many aspects of history and cultural experience cannot be interpreted fully through either the physical fabric of sites or written records.

Objects still *in situ* retain both their physical and cultural context and are an important physical and heritage component of a place. Those included in public and private collections remain significant, although physically and culturally displaced, and if adequately documented help people understand the significance of their places of origin. Objects now stored in collections are therefore an important part of Australia's natural and cultural environments.

Despite their close links, places and objects are often treated separately in legislation, administration and management.

Calthorpes' House, Canberra

This house in Canberra is a fairly typical middle-class home of the 1920s. It was decorated and furnished at the time of building, the fittings chosen and obtained by mail order. Over the years, the owners introduced few technological innovations, and in most cases they kept original implements in store rooms, even if obsolete. The house remains substantially intact, together with the layers of domestic technology. Oral histories from the Calthorpes' long-serving maid and family members provide a record of much of the routine of the household.

Calthorpes' House has been listed in the Register of the National Estate largely because of the heritage significance derived from its intact interior and contents, which allow the place to be interpreted as both home and work site. It is possible to experience it almost as the Calthorpes knew it — a unique opportunity for Australians to glimpse their past — which would have been impossible had the house been stripped of its contents.

Source: Anderson, in press.

Excavation in the limestone hills at Riversleigh, Queensland, (above) revealed treasures such as this 20 million-year-old bandicoot skull and jaw (left).

Australian mammal fossils

Fossils from an area of rugged limestone hills at Riversleigh in north-western Queensland and from caves at Naracoorte in the south-east of South Australia, were first discovered early this century and specimens placed in museums. More recent studies have revealed the immense scientific importance of the sites, which contain fossils providing significant insights into key stages of the evolution of Australia's fauna in prehistoric times. Although many fossils have been collected for research, the sites themselves received the 'ultimate' heritage recognition in December 1994, when they were inscribed on the World Heritage List.

Research on these fossil collections provides information not only about Australia's environment in the past, but also for the development of conservation strategies for animal species and communities living today.

Source: Boden, in press.

Leichhardt Billabong, Northern Territory

At Leichhardt Billabong, on the black clay floodplains of the South Alligator river in Kakadu National Park, the traditional Aboriginal owners and archaeologists have been systematically recording and collecting all the artefacts from an occupation site covering four hectares. This exercise is unusual, as such total collections are rare in archaeological practice. However, special circumstances led to the traditional owners requesting that archaeologists assist them in collecting and recording the artefacts, which they considered to be under severe threat from the increasing numbers of vehicles and visitors using the area, a popular recreation spot. An archaeologist has been employed to make a detailed analysis of the collection so that the maximum amount of information can be retrieved from it.

In this carefully planned exercise, thousands of artefacts have been collected from the surface of the site. Analysis has shown they include a range of implements (such as axe blades, spear points, scrapers, and grind stones), ochre and worked glass. The finding of glass suggests that the site's use continued into the post-European contact period. Cores, flakes and hammer stones bear witness to tool-making there. The site is interpreted as an annual dry-season hunting and fishing place used by Aborigines for at least the last 7000 years. The artefacts are already providing insights into ceremonial life, trading patterns, manufacture of tools from stone and wood, and the collecting and processing of food.

Once the analysis is completed, all the artefacts will be stored in a local Keeping Place to be designed by the traditional owners.

This decorated wooden sculpture (toa) refers to features of an island in Lake Gregory (left), an important place in the Aboriginal Swan history tradition handed down through traditional owners over the generations. The map (left) was prepared at Killalpaninna in 1905.

Source: Behr *et al.*,1994, and M.Grant, pers. comm., January 1995.

The Wollemi Pine, New South Wales

In December 1994, scientists announced the discovery of 40 trees of a previously unknown type of native conifer in a remote area of Wollemi National Park in New South Wales. Initial studies suggest their closest relations are fossil *Araucarites* known only to live in the Jurassic and Cretaceous periods about 65 to 200 million years ago. Botanists from the New South Wales Herbarium are researching specimens of the trees, and horticulturists of the Mt Annan Botanic Gardens have started propagation trials.

The Director of the Royal Botanic Gardens in Sydney considered the find comparable to the 1941 discovery of living plants of *Metasequoia glyptostroboides* in 'Metasequoia Valley' near Shui-Se-Pa in western China. Undoubtedly, the remote small valley in Wollemi where the trees occur will become equally botanically significant. For most people, however, their first opportunity to see the living tree will be in the Botanic Gardens. Plants propagated at the Gardens will also be critical in ensuring continued survival of the species by guarding against possible loss of the population in its natural habitat.

Source: Boden, in press.

Toas

Toas are small, beautifully decorated wooden sculptures between 15 cm and 45 cm in height, shaped to suggest they should be placed upright in the ground. They appear to have been made by one generation of Aboriginal people in the Diyari (Dieri) country east of Lake Eyre and were collected from them by Pastor Reuther of the Killalpaninna Mission between 1900 and 1904. The South Australian Museum acquired them in 1907.

Reuther described toas as direction markers or 'sign posts'. The symbolism of the decoration indicates both the topographic features of a place and its mythological or spiritual associations with the activities of the Muramuras, the creator ancestors. Reuther interpreted the toas as having the capacity to direct observers with knowledge of the country and its Dreaming stories to specific locations. Although toas had religious meaning, they were not sacred but public objects. Recent research suggests more complicated interpretations reflecting interactions between the local Aboriginal people and the missionaries. However, the toas still illustrate well the vital links between people, place, landscape and beliefs in the Lake Eyre region, as do the numerous Aboriginal place names recorded for this part of Lake Gregory.

Source: Jones and Sutton, 1986.

Pressure

The physical condition of natural and cultural heritage places and objects *in situ* is affected by a wide range of 'natural' factors — ageing of cultural structures and materials, soil erosion, storms, cyclones, floods, droughts, sea level changes and earthquakes — many of which are characteristic of the Australian environment. Although human activities may exacerbate the effect of these natural phenomena on the condition of our heritage, they have not been considered in this chapter.

Pressures on heritage places

Over thousands of years indigenous people are thought to have modified the biophysical landscape of Australia (see Chapter 2). The last two centuries of occupation and settlement have caused additional widespread and severe change (see Fig. 9.1). Those regions that remain least disturbed are likely to contain many significant natural and cultural heritage places, although few have been listed on heritage registers (see the box on page 9-6).

Society places a wide range of pressures on heritage places across Australia (see Table 9.2). These may affect the identification, evaluation and conservation of places, or just their physical condition. Some have positive and negative effects, but many have only a negative impact. The pressures are frequently interrelated: they often act in combination, or one may be a consequence of another.

Earlier chapters of this report describe major pressures affecting the natural environment, such as forestry, mining and pastoralism. Many of these have a direct or indirect impact on sites of cultural significance as well as affecting the heritage of natural places.

Progress has been made since 1981 on some pressures identified then on cultural places (Yencken, 1985). Some pressures, like ageing

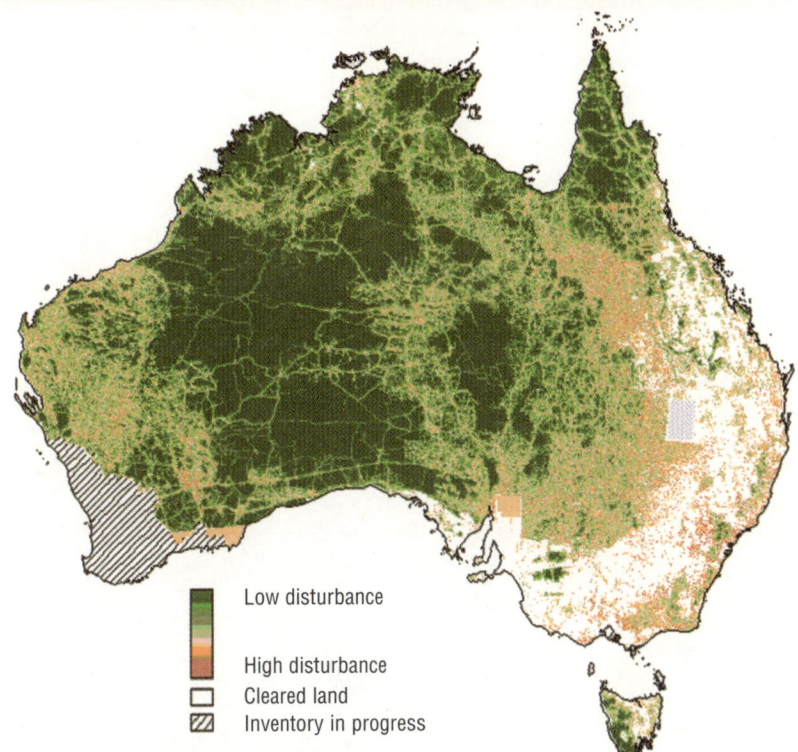

Figure 9.1 Broad levels of disturbance of Australia's natural environment since European settlement

Low disturbance

High disturbance
☐ Cleared land
▨ Inventory in progress

Source; National Wilderness Inventory, 1995.

infrastructure and demolition, are specific to built heritage structures. Others, such as alienation of people from their traditional lands and custodial roles, and loss of languages, apply specifically to indigenous heritage places (see page 9-42).

Many pressures on natural and cultural heritage occur at the three levels of government. In general, information was more readily available for Commonwealth and State/Territory governments than for local governments, although the last are frequently responsible for management decisions affecting heritage places (see Table 9.18).

◄

Australia's major cities contain many significant buildings that are affected by urban redevelopment. The Bow Truss woolstore, Geelong, Victoria, was demolished in 1990 despite professional advice about its heritage value and possible World Heritage significance due to its unique form of construction.

Heritage in Australia's population 'hot spots'

Population growth in Australia has been concentrated in metropolitan areas, especially in coastal areas in south-eastern Australia. Outer metropolitan areas have expanded and the number of dwellings in core and inner areas have increased (see Chapter 3). These areas of highest population growth correspond with regions that contain the highest proportion of listed national estate places because of their pattern of occupation over time (see Figs 9.4 and 9.5) — particularly historic places (see Fig. 9.7).

Pressure

Demographic changes in inner metropolitan areas create strong pressure for redevelopment and consolidation. These impose direct pressures on historic places through demolition or re-use of buildings and precincts, and loss of sympathetic surroundings. In many cities, demolitions anticipating future developments that have not eventuated, have turned heritage buildings into derelict sites. Urban expansion and development of associated infrastructure in outer metropolitan areas create pressures on natural heritage places, such as bushland, which may be destroyed or altered. Sites significant to indigenous communities, especially in coastal areas, are often destroyed or lose their natural and cultural context. Rezoning and altered patterns of land use impose significant pressures on surrounding rural landscapes of heritage significance.

Provincial cities and boom towns frequently lose their historical fabric and sense of place. Heritage planning may be integrated into local and regional planning requirements. However, cultural mapping or main street programs, which have been seen as major ways of achieving this, are not necessarily effective — the former is in its developmental stages, while the latter often focus on economic benefits (Marshall and Pearson, in press).

Population decline in rural areas imposes different pressures on the local heritage. The numbers of places listed in heritage registers are often much lower than in metropolitan regions (see Fig. 9.5) for a number of reasons. However, many places in rural areas have heritage significance because of their historical associations or vernacular architecture, or because of their social value to local communities. Few of these places are likely to have been documented or considered for heritage registers, and hence are not eligible for conservation funding. Empty buildings, which often occur as populations decline, promote physical decay and often invite vandalism. It is often not feasible to maintain the buildings' heritage values by re-using them, because of reduced rural economies.

Some Commonwealth government initiatives focusing on employment or development affect both metropolitan and rural areas. Heritage matters often have a low priority in these programs, and sometimes are addressed only in response to the concerns of the community or heritage agencies.

State

Almost 4900 heritage places have been identified in metropolitan areas in Australia (Australian Heritage Commission, in press). Of these, 91 per cent are places of predominantly architectural or historical significance. Places of social value, those demonstrating modern architectural techniques and styles, cultural landscapes and places significant to ethnic communities in these areas are all under-represented in heritage registers. Equally, in rural regions, such registers under-represent places of social significance to both indigenous and non-indigenous people.

National and State and Territory data do not provide a comprehensive picture of the physical condition of heritage places.

Response

The establishment of local environment plans and city-specific heritage studies have resulted from increased rates of inner city redevelopments. However, heritage conservation has not kept pace with the increased level of identification and evaluation.

Heritage legislation enacted in Queensland, Western Australia, the Australian Capital Territory and the Northern Territory has increased the level of protection of heritage buildings. Developments in metropolitan areas that affect such places are required to meet more stringent conditions for approval. In 1994, a review of government demolition policies by the chairs of Commonwealth and State heritage agencies showed that generally places on their registers are adequately protected. However, many places are demolished or radically altered before their heritage value is assessed or before they can be placed on the registers.

Some industry bodies have responded to heritage issues in metropolitan areas. The Royal Australian Institute of Architects surveyed 20th century buildings in inner city areas as a direct response to redevelopment pressures. This resulted in recognition of the heritage value of many buildings and their inclusion in heritage registers. In 1994 the Building Owners and Managers Association produced a draft document addressing a broad policy view of reforming Australia's system of planning and development control. The 'Heritage and Conservation' chapter looked at streamlining key areas such as legislative controls, duplication, registration processes, appeals and the economic effects of listing. Many of these matters are being addressed through the national coordination program of heritage officials (see page 9-38).

Urban development activities — such as building demolition to make way for new, 'better' developments, inappropriate use of places or rezoning areas for new types of use — often provoke a strong community reaction. The National Trust has continued to be a major force channelling community support for the retention of heritage places under threat. Such support is often strengthened in the face of developments and leads to the formation of local action and lobby groups. In areas of demographic change, new layers of social meaning are added to old and new places, creating the heritage of tomorrow.

Prognosis

Changes in the patterns of Australia's population growth and distribution have a direct impact on Australia's cultural heritage, particularly in cities and associated metropolitan areas experiencing rapid population growth. While heritage registers remain unrepresentative of many types of places, the effectiveness of government heritage legislation will be limited. Community groups will continue to lobby for heritage protection to counter inadequate integration of heritage in government policies and programs developed in response to changing demographic patterns.

Local government appears to accord low priority to heritage matters — particularly indigenous heritage (Brown, 1994), although some councils have taken positive action.

The following sections outline the major pressures affecting the identification, evaluation and conservation of Australia's heritage places.

Population patterns

Regional demographic variations create pressures on natural and cultural heritage (see the box opposite). Areas with the greatest population — the major capital cities and associated metropolitan areas — also have the highest number of historical places listed in heritage registers, and contain other natural and indigenous heritage places. Development, consolidation and expansion in urban areas resulting from population changes create direct pressures on heritage places through either demolition or re-use, although a range of other pressures interact at the same time (see Fig. 9.2).

Some major government programs, such as the Better Cities program (see Chapter 3), have accelerated redevelopment in city and metropolitan areas. These programs are providing future models for the urban environment, including good examples of heritage conservation and re-use. However, they have the potential for major adverse impacts unless heritage matters are addressed early in the planning stages and fully integrated into each program.

▲
Unoccupied buildings are prone to physical decay and vandalism. This cottage in Yass, New South Wales, contributes to the historic character of the town, but it has been allowed to deteriorate.

Table 9.2 Generic pressures on the identification (I), evaluation (E) and conservation (C) of heritage places

Knowledge base
• incomplete basic knowledge about and inventory of heritage places	I, E, C
• research needs exceed available funding	I, E, C

Government management and administration
• policies and programs affecting heritage places, including	I, E, C
— privatisation of government agencies	C
— disposal, leasing of government assets	C
• inadequate development of integrated strategies within governments	I, E, C
• poor coordination between levels of government	I, E, C
• low level of heritage identification and conservation in urban and regional planning	I, E, C

Heritage expertise
• demand for expertise exceeds supply and/or the number employed	I, E, C
• traditional conservation skills disappearing	C
• low level of heritage expertise in many local governments	I, E, C

Community issues
• attitudes and perceptions of heritage generally	I, E, C
• misconceptions/misunderstanding about heritage values	I, E, C
• inadequate community involvement in planning decisions and their implementation	I, E, C

Development from changing population patterns
• urban consolidation, development and expansion	C
• re-zoning and land use planning	C
• land clearance	C

Commercial use of specific heritage places
• use of natural resources such as forests, minerals	C
• industry attitudes to heritage conservation	C
• tourism	C

Degradation arising from general human resource use
• degradation arising from air and water pollution	C
• accelerated natural degradation (eg erosion)	C

Conservation and management
• national economic state	C
• high costs of conservation works	C
• management inappropriate for heritage values	C
• management for conflicting heritage values	C

Monitoring
• inadequate systems to monitor change	C

Source: adapted from Marshall & Pearson, 1995.

Cultural values in natural landscapes

Forest ecosystems, which historically have been managed primarily for their natural values, contain a large number of significant cultural places (see page 9-13). Other natural landscapes are also likely to contain areas with significant cultural values that have been neither documented nor listed in heritage registers, and are not managed to retain their cultural values. Within conservation reserves, experience has shown that active management is often required to conserve cultural places. Specific management to conserve natural values has sometimes destroyed cultural features — for example, the removal of historic buildings to restore the 'naturalness' of remote areas (Griffiths, 1991).

Figure 9.2 Major pressures on historic places resulting from changed population distribution in metropolitan areas

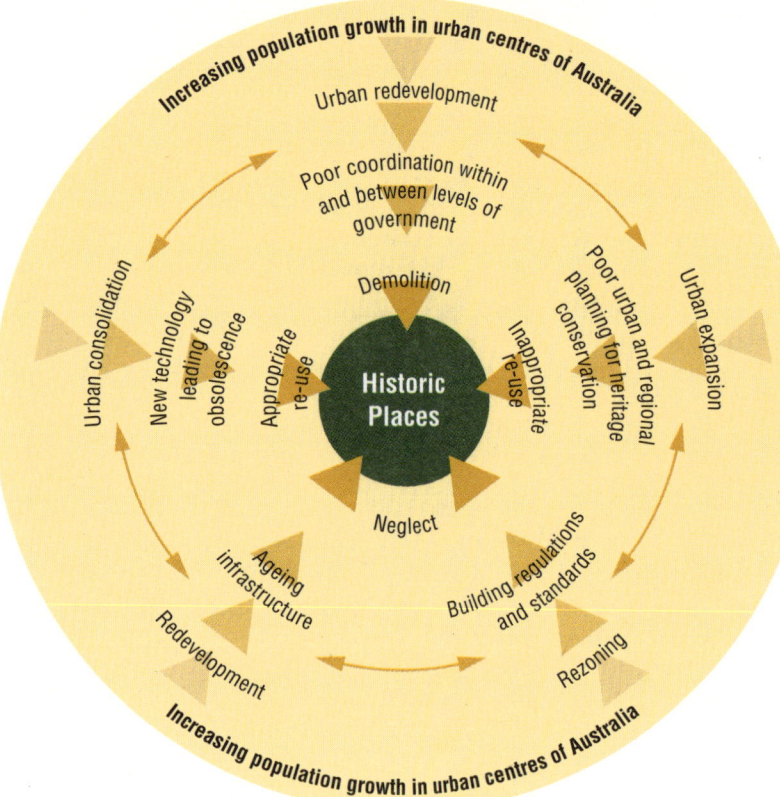

Increasing population growth in urban centres of Australia

Urban redevelopment

Poor coordination within and between levels of government

Urban consolidation

New technology leading to obsolescence

Appropriate re-use

Demolition

Historic Places

Inappropriate re-use

Poor urban and regional planning for heritage conservation

Urban expansion

Neglect

Ageing infrastructure

Redevelopment

Building regulations and standards

Rezoning

Increasing population growth in urban centres of Australia

Level of demand for heritage assistance

The National Estate Grants Program is the Commonwealth's annual funding program specifically to assist in identification, conservation and presentation of the National Estate. In 1994–95, a significant proportion of applications could not be funded in the areas of greatest demand for assistance.

For natural places:

• identification projects comprised more than two-thirds of all applications and funds applied for

• one-third of the projects could be funded, meeting 21 per cent of the funds requested

For historic places

• conservation projects comprised 57 per cent of all applications, representing 62 per cent of all funds applied for;

• 30 per cent of the projects could be funded, but this met only 13 per cent of the funds requested

• community organisations applied for three-quarters of these funds; less than 10 per cent of the funds requested could be met

For indigenous places

• identification projects comprised about three-quarters of all applications and funds applied for

• almost half of the projects could be funded, but this met only 32 per cent of the funds requested.

Source: Australian Heritage Commission, in press.

Tourism

Many of the least disturbed areas of Australia are the focus for specialist nature-based ecotourism and cultural tourism. Museums, historic sites and heritage buildings also attract large numbers of visitors. Tourism has the potential for a wide range of positive and negative pressures, and can affect both the physical fabric of places as well as intangible aspects of heritage (see page 9-32). This is particularly relevant for sites of cultural significance, where physical damage and culturally inappropriate activities by visitors represent a direct assault on the cultural values of people with strong links to the places.

The main negative effects of tourism include: physical damage to sites and their associated objects; a lack of funds required to provide facilities and appropriate management; inadequate data on visitation rates and visitor attitudes essential for sound planning and management decisions; insufficient involvement of community groups in developing and implementing cultural tourism strategies and activities; and inadequate assessment of the social and cultural impact of new tourism projects.

Community involvement in heritage identification

Community groups value many places for their role in social practice and tradition (Blair, 1994). People can contribute to decision-making through the environmental planning process of local governments. However, it is often not until places are visibly threatened that people speak out about their significance, taking planners and decision-makers by surprise.

Heritage studies involving community represent-atives often reveal large numbers of places of social significance that are not identified by heritage professionals. For example, in the Central Highlands forests of Victoria almost half of the significant historic places were identified only by local and regional residents, not as part of expert studies (see the box opposite). More heritage practitioners are recognising the importance of social value in heritage identification, but many studies to date have lacked adequate community input. This is often because funding authorities or professionals have not appreciated the need to determine places of social value or have not accepted the way it is done (Blair, 1994). Language differences exacerbate low levels of community involvement, and put some types of heritage at particular risk, for example, places significant to migrant groups (Armstrong, 1994) and indigenous places (see page 9-13).

While poor consideration of social value and inadequate community involvement in heritage studies continue, heritage registers will remain unrepresentative and important parts of Australia's cultural heritage will remain vulnerable to destruction or degradation through ignorance of their existence.

Cultural values of natural landscapes — forests of the Central Highlands, eastern Victoria

The Central Highlands region north-east of Melbourne covers an area of about 0.7 million hectares. Its forests have long been recognised as a valuable natural resource for timber and as a water catchment for metropolitan Melbourne. Their nature conservation values were given added recognition in 1977 when the Victorian Land Conservation Council (LCC) recommended additional national parks and other conservation reserves in the region. However, cultural values were not considered comprehensively in the LCC study.

A later detailed heritage assessment (Australian Heritage Commission and Department of Conservation and Natural Resources Victoria, 1994) recorded diverse natural values in the region. These included landscapes essentially undisturbed by European activity, areas of old-growth forest, rainforest and remnant vegetation communities, rare, endangered or endemic species of plants and animals, a rich biota, areas acting as biological refuges and important geological and geomorphic features.

The 1994 heritage assessment revealed a diverse assemblage of cultural sites for the first time. Although often not grand or beautiful, these sites are of immense value as reminders and evidence of a rich and complex human past in the region. Places with significant Aboriginal values included 140 sites associated with traditional beliefs and 100 prehistoric archaeological sites. More than 200 significant historic places were identified, including mining landscapes and associated transport routes of goldfields, and an extensive network of sawmills and tramways characteristic of early bush sawmilling in Victoria's mountains. An additional 194 places, including some of Victoria's earliest national parks, were important to local and regional communities for their social and aesthetic values. These places were identified at community workshops, which drew on the extensive local knowledge of the region and its history, as well as allowing local groups to be involved in the assessment of heritage values.

The Central Highlands forests have not been managed consistently in the past to conserve their cultural values. Pre-logging surveys of public forests required under the Victorian Code of Forest Practices often focus on identifying discrete cultural remains, which are then avoided during road-building or logging. This has protected only some aspects of the very limited number of documented sites.

In response to the more detailed knowledge and understanding of the cultural values, those sites identified in the 1994 assessment are being integrated into decision-making and planning processes for State forests in the region, including Forest Management Plans. The information is available for consideration in new plans being prepared for park management and the Melbourne Water catchment, and for planning by local councils.

Throughout the long history of management of the region, the emphasis was almost entirely on natural values — a situation typical of forests across Australia. In the 1970s and 1980s, little systematic information was available about Aboriginal or historic sites in forests. Until recently it was commonly thought that indigenous people did not use such areas extensively in the past. Increasing emphasis on undisturbed natural values and wilderness has seen forests often viewed as devoid of both people and history. At a national workshop in Canberra in 1992, participants concluded that, despite the growing research into Australia's native forests, the human history and cultural significance of these areas continued to be poorly documented and understood.

Identification, evaluation and conservation

Places need to be recognised and their heritage significance evaluated and documented before the most appropriate way of conserving them can be determined. This often does not happen to an extent that matches either the need for heritage information in planning and decision-making at the three levels of government or conservation needs commensurate with the places' heritage values and physical state. For cultural places, the shortfall appears to result from both the high costs involved in identifying and conserving heritage places and the limited number of professionals trained to do such work.

Data from the National Estate Grants Program indicated that many heritage places remain unassessed or in poor condition. In 1994–95 the Program received 866 applications totalling $28 million for the $4.7 million available. Only 30 per cent of the applications, representing 17 per cent of funds requested could be met, with major shortfalls for natural, historic and indigenous places (see the box opposite).

Oral history is an important tool for identifying and assessing social value. For indigenous heritage 'knowledge of oral history, folklore and traditions is often the only way of ascertaining, assessing and assigning significance; it is the only way of dealing with spiritual significance and with the heritage of communities with no written traditions' (McCarthy *et al.*, in press). Over the last decade, the support for oral history research in Australia has increased, reflecting a growing interest in Australian history, although most of the projects have had a biographical emphasis. The work is severely hampered at the local and regional level by inadequate access to training, the required technology for conservation, storage and access to collections and personnel (McCarthy *et al.*, in press).

Loss of languages within cultures

Cultural heritage values and meanings are most fully expressed within the relevant traditional languages. The continued loss of languages of indigenous Australians is being exacerbated by the death of remaining speakers or by other languages replacing them in daily use.

Aboriginal English is now spoken throughout Australia, while two major creoles are spoken by a significant number of indigenous people in northern Australia. Aboriginal English and creoles are necessary for communication in contemporary communities, but their use entails a significant loss of detail of certain types of traditional knowledge (Henderson and Nash, 1995). For this reason and the potential loss of cultural traditions, some Aboriginal people are concerned about the continued spread of creoles in the northern part of Australia, at the expense of traditional languages (see page 9-23).

Indigenous communities in Australia have inadequate resources, including training, equipment, funding and specialist assistance, to maintain their traditional languages and cultures (Henderson and Nash, in press). Education and electronic media are the most powerful influences promoting language shift. However, while education has included some bilingual learning programs (see page 9-40), many indigenous communities have had few opportunities to use electronic media to help maintain their languages. The way in which various government departments and other organisations deal with indigenous languages suggests that they are given relatively low priority (Henderson and Nash, in press). The costs of language maintenance activities are typically under-estimated by government and other agencies, which often have little experience in this area.

It is possible that loss of language will diminish the heritage value of places significant to communities of non-English-speaking background in Australia in a similar manner.

Cultural insensitivity

Many government agencies and industry groups either manage or use land that is likely to contain places significant to indigenous peoples. These organisations appear to be increasingly willing to understand and accept the importance of indigenous cultural values associated with this land (Australian Association of Environmental Education, in press; Council for Aboriginal Reconciliation, 1993) although they often have a poor understanding of these values. With some exceptions, it appears there is little formal provision for staff training about indigenous culture (Australian Association of Environmental Education, in press).

The Distributed National Collection

In 1991, the Heritage Collections Working Group, set up by the Cultural Ministers' Council and the Council of Australian Museum Associations, reviewed heritage collections in Australia. The Group defined these collections as 'those objects or specimens which together constitute the material evidence of Australia's environment and of its historical and cultural life'. Objects of cultural significance included 'not only those judged in some way "unique", but those which provide evidence of a style, trend or movement, or of a political, social, cultural or economic process of significance to Australia' (Anderson, 1991). The Distributed National Collection refers to the aggregate of those objects located in major Commonwealth, State and Territory collecting institutions as well as those held in community, regional and specialist museums, libraries, schools and private collections.

Pressures on heritage objects

Physical decay affects material objects located in their original context as well as those removed to private and public collections, their maintenance requiring some form of conservation. Other pressures may differ, depending on the context. For example, objects located in collecting institutions may be the subject of strong pressure for their return to more culturally appropriate settings, such as a community keeping place in the area of origin. Similar objects *in situ* would not be subject to this pressure. In many cases it is culturally appropriate that indigenous objects left *in situ* are subject to natural decay.

Major pressures described in the 1975 Pigott report and the 1987 CRASTE report still apply. These and other important pressures are described below.

National policies and coordination

Inadequate national policies and inadequate institutional coordination of activities relating to the Distributed National Collection (see the box) can adversely affect its representativeness and physical condition. These problems arise because people do not appreciate the significance and cultural roles of objects and scientific specimens compared with the built environment and natural heritage (Anderson, in press).

The 1975 Pigott report recommended that a national body should be responsible for conservation standards and policies on collecting across Australia. In 1991 the Heritage Collections Working Group identified a wide range of needs for a national approach to the Distributed National Collection. However, national policies and strategies are still not in place, although a national body — the Heritage Collections Committee — was set up in 1993–94 (see page 9-41).

Collecting policies of major museums in Australia have in the past tended to omit aspects of historical or contemporary Australian material heritage related to women, migrants and working-class people. Collections relating to indigenous Australians have tended to focus on scientific concerns rather than on Aboriginal history more generally. Although policy has shifted somewhat in recent years (see page 9-44), continuing bias in collecting policies and inadequate national strategies will inhibit the Distributed National Collection from becoming representative of Australia's cultural heritage.

Museums, zoos, botanic gardens and herbaria have coordinated their efforts, exchanged ideas and shared resources to some extent through 'heads of institution' organisations (Boden, in press). However, individual agencies in the States and Territories appear to operate largely independently within their own charters.

Conservation facilities

Objects are often fragile, requiring special conditions for survival when removed from their original environment into collections. Maintaining the physical condition as well as heritage significance of objects requires appropriate facilities for housing, storing and displaying collections, and the employment of appropriately trained scientific and technical staff to document, maintain and conserve them.

The last two decades have seen a significant decrease in the level of technical and other support for biological collections in Australia's major government-funded museums and herbaria (see Table 9.3). The size of both herbarium and zoological collections approximately doubled, while the number of technical and other support staff decreased by 34 per cent in museums and increased marginally in herbaria. Scientists employed on taxonomic studies dropped by 15 per cent in major herbaria, a reduction related to funding constraints (Richardson and McKenzie, 1992).

In 1994, the Council of Heads of Australian Herbaria stressed the urgent need to do something about pressures on major biological collections caused by funding shortfalls. The Council also called for increased funding of the Australian Biological Resources Study. These measures were seen as essential to implement the National Biodiversity Strategy.

Most State, regional and local museums were unable to provide suitable care for their material culture collections in 1991 (Anderson, 1991). Surveys by the Heritage Collections Working Group indicated that sub-standard storage conditions in some institutions were still causing highly significant material to deteriorate and were posing a threat to the continued existence of the collections. In many cases the situation for institutions with nationally significant collections still warranted concern, despite some problems being identified as early as 1968 (Anderson, 1994) and recommendations for improved facilities in the 1975 Pigott report and the 1987 CRASTE report to redress 'the profound threat of physical decay and loss of vital collections'.

In 1991, only major museums in Australia employed conservators (Anderson, 1991). Local and regional museums experienced difficulty gaining access to conservation services (Anderson, 1993), although they badly needed them. The Pigott and CRASTE reports also raised this situation. State government assistance to regional and local museums is inadequate for long-term curation and preservation of objects, much less research on their heritage significance. The major museums have limited resources from which to offer conservation training programs.

Until recently, material culture studies were largely neglected in formal anthropology courses in Australian universities, so major museums have only a small pool of trained professional staff to document and curate ethnographic collections. These are probably at greater conservation risk than the archaeological ones because of the nature of the materials involved. Relatively few conservators exist in these specialist areas, and opportunities for formal training are still limited (see Table 9.10).

Table 9.3 Diminishing staff support for expanding biological collections

	1975	1991	Change (%)
Herbarium collections (information from six herbaria)			
Number of specimens	2.4 million	4.1 million	+71
Technical & other support staff	55	58.1	+6
Scientific staff	55.5	47.2	-15
Specimens/staff member	23 000	39 000	+70
Zoological collections (information from seven museums)			
Number of specimens	14.6 million	29.4 million	+101
Technical and other support staff	154	101.1	-34
Scientific staff	74.5	90	+21
Specimens/staff member	64 000	133 000	+108

Source: Richardson and McKenzie, 1992.

Major agencies responsible for the protection of indigenous heritage have paid more attention to places of significance than to indigenous objects (Ward, in press). Apart from the established collecting institutions, agencies do not appear to regard indigenous objects as having a high priority for assistance or have no statutory responsibility for them. Thus a heavy burden rests on State museums — particularly in relation to conservation. Yet these museums are already subject to considerable pressure from heavy backlogs of documentation and accessioning, poor storage facilities and heavy demands for specialist services and advice in areas such as curation and conservation. They also have to meet research commitments to interpret and present materials, and to return human remains and return or lend other items of cultural heritage. They have limited resources to deal adequately with such requests. The consequent perceived lack of curation effort has contributed to scepticism among indigenous groups about the value of keeping their material in collections.

◄ Natural history collections, such as the Australian National Insect Collection, are an important part of our natural heritage. Reduced levels of technical and curatorial support threaten the state of many collections.

The ultimate accolade — Australia's World Heritage Areas

Inscription on UNESCO's World Heritage List signifies that a property has been judged to have outstanding universal value. In becoming a State Party to the 1972 World Heritage Convention, the Commonwealth accepted the obligation to the world community to identify, protect, conserve and present World Heritage properties in Australia (see Chapter 2). To this end, the government passed the *World Heritage Properties Conservation Act 1983*.

Only the Commonwealth Government, as a State Party to the World Heritage Convention, can nominate places to the World Heritage List. Decisions about whether places are inscribed are made by the World Heritage Committee with advice from the World Conservation Union (IUCN) for natural nominations, and the International Council for Monuments and Sites (ICOMOS) for cultural nominations. All nominations are evaluated against criteria specified in the UNESCO Operational Guidelines for the Implementation of the World Heritage Convention. Places that fail to meet these criteria, although deemed not to have outstanding universal value, may still have outstanding value to a particular nation. In Australia, many such places (for example, Old Parliament House) are listed in the Register of the National Estate and may be listed on State and Territory heritage registers.

Major pressures

World Heritage areas in Australia have become a major focus for tourism. From 1990–91 to 1993–94 the total number of visitors to the Great Barrier Reef and Willandra Lakes Region grew by about 50 per cent, while visitor levels to Uluru rose by 17 per cent (Hyde, in press). Three other World Heritage areas experienced a 4–8 per cent increase in visitor numbers over the same period. This trend increases the management pressure to satisfy tourist expectations while maintaining the quality of the natural and cultural heritage resources.

Until recently, the applicability of the cultural World Heritage criteria to places in Australia appeared to be limited, especially for places whose significance is related to the living culture of Aborigines and Torres Strait Islanders. Aboriginal cultural practices associated with areas nominated to (and later inscribed on) the World Heritage List for their natural values were not always considered at the same time. Areas of possible historic significance have also not been considered, partly because of perceptions that Australia's historic places are not of sufficient merit. Consequently, current listings favour areas inscribed for their natural values.

Since the early 1980s, some World Heritage listings in Australia have been controversial. Despite processes to assist intergovernmental negotiations, some State governments still oppose further listings, while particular industry and community groups also continue to lobby strongly against them (Reid, 1995).

State

By December 1994, 11 properties in Australia had been inscribed on the World Heritage List for their natural values (see Table 9.4). All met one or more criteria in the Operational Guidelines. The Tasmanian Wilderness Area and Willandra Lakes Region were also inscribed for cultural values associated with their Pleistocene archaeology, while Kakadu was inscribed for its Aboriginal cultural values related to archaeology, rock art and traditional beliefs. Uluru-Kata Tjuta National Park was accepted as a cultural landscape seven years after its inscription for natural values. So far no places are listed solely for their cultural World Heritage significance.

Australia has gained international recognition for its management of Uluru-Kata Tjuta National Park and the Great Barrier Reef. However, management arrangements have sometimes been a source of contention between Commonwealth and State governments. Agreed management arrangements are in place for all but four properties, where they are under negotiation. The Commonwealth and the traditional Aboriginal owners jointly manage Uluru-Kata Tjuta and Kakadu National Parks on Aboriginal lands. The Great Barrier Reef and the Wet Tropics World Heritage Area in Queensland have joint State–Commonwealth management arrangements, and State government agencies manage the remaining areas. Eight of the inscribed properties have management or equivalent plans in place. The Willandra Lakes Region, inscribed in 1981, still had no management plan by early 1995. However, by mid 1995 plans for it and the remaining areas were being actively prepared.

Nourlangie Rock in Kakadu National Park provides impressive examples of the rich and complex rock art of the region. These paintings belong to traditions established over many millenia.

Responses

To facilitate the nomination of cultural areas in Australia, Australian experts have worked with ICOMOS and the World Heritage Committee to change the Operational Guidelines. They were influential in modifications to the cultural World Heritage criteria approved at meetings of the World Heritage Committee in 1992 and 1994, including the addition of cultural landscapes and the applicability of relevant criteria to living cultures.

Australia has also worked closely with the IUCN, the non-government organisation responsible for the evaluation of nominations of natural sites to the World Heritage Committee, about natural areas in Australia.

The InterGovernmental Agreement on the Environment (see Chapter 2) included a schedule to improve World Heritage nomination, community liaison and management arrangements. Since then, the States have approved all nominations submitted. However, limited progress has been made on some potential World Heritage nominations, such as the Nullarbor and the Lake Eyre regions.

In 1993, the Prime Minister announced that the Commonwealth would work with the State governments to develop a 'World Heritage Indicative List of Cultural Sites', as required by the World Heritage Committee. This list currently only includes existing properties and previous nominations that have been deferred (Sub-Antarctic Islands) or foreshadowed (Sydney Opera House and environs). In 1991, a conceptual framework was prepared for assessing places worth including on such a list (Domicelj *et al.*, 1992). The Commonwealth and State governments are continuing to negotiate about its development. National conservation bodies that have strong associations with IUCN have prepared their own indicative lists of natural areas for future consideration by the Commonwealth and States.

Community attitude surveys carried out for the Tasmanian Wilderness and Wet Tropics World Heritage areas have shown strong and positive support for the listings and subsequent management (Purdie, in press). However, some sectors of the community oppose World Heritage listing which they view as giving control to overseas bodies, losing rights and jobs and forgoing development opportunities (Reid, 1995). Management arrangements for most existing places now include avenues for community input, a requirement under the Convention. The Commonwealth is establishing a framework for community involvement in future nominations.

Prognosis

Australia's World Heritage properties are a focus for tourism and receive wide community support. However, misunderstandings about the implications of listing and the uncertainty created by the lack of an indicative list of potential cultural and natural World Heritage nominations exacerbate opposition to listing by some industry and community sectors. Disagreements between the Commonwealth and State governments currently prevent some natural areas being nominated. The revised World Heritage cultural criteria pave the way for the nomination of more cultural sites, especially as components of cultural landscapes and particularly Aboriginal places representing living traditions.

Table 9.4 Australia's World Heritage properties

World Heritage Property	Date inscribed	Values recognised	Management or equivalent plans	Commonwealth-State management arrangements
The Great Barrier Reef, Qld	1981	Natural	In place	Agreed
Willandra Lakes Region, NSW	1981	Natural, Cultural	Being prepared	Agreed
Kakadu National Park, NT — stage 1 — stage 2 — stage 3	1981 1987 1992	Natural, Cultural	In place	Agreed
Tasmanian Wilderness World Heritage Area — stage 1 — stages 2, 3	1982 1989	Natural, Cultural	In place	Agreed
Lord Howe Island, NSW	1982	Natural	In place	Under negotiation
Central Eastern Rainforest Reserves, NSW & Qld — original — extensions	1986 1994	Natural	Most plans in place	Agreed
Uluru-Kata Tjuta National Park, NT — inscribed for natural values — inscribed for cultural values	1987 1994	Natural, Cultural	In place	Agreed
Wet Tropics World Heritage Area, Qld	1988	Natural	Being prepared	Agreed
Shark Bay, WA	1991	Natural	Being prepared	Being re-negotiated
Fraser Island World Heritage Area, Qld	1992	Natural	In place	Under negotiation
Australian Fossil Mammal Sites (Riversleigh, Qld & Naracoorte, SA)	1994	Natural	In place	Under negotiation

Source: Department of the Environment, Sport and Territories, 1995.

▲
Archaeological excavation ahead of development in inner city areas may provide significant information on the past, as did this 1994 dig at The Rocks, Sydney. The dig revealed unrecorded details of inner Sydney's domestic and working life in the early nineteenth century.

Documenting collections

Nationally, in 1991 institutions had a considerable backlog of herbarium specimens and museum faunal collections awaiting processing. Many museums considered that 60 per cent or more of their total collection were well documented, although there were substantial differences between taxonomic groups (Anderson, 1991). While there did not appear to be a shortage of taxonomists in 1991, few of them specialised in some large, poorly known groups of biota such as arthropods, molluscs, and non-vascular plants (Richardson and McKenzie, 1992). This may affect documentation in the future. Many institutions are moving towards computerising their records, but in 1991 few were providing adequate support staff (Richardson and McKenzie, 1992).

Within museums across Australia, documentation vital for assessing the significance of objects related to history and anthropology was judged inadequate (Anderson, 1991). Documentation of the cultural collections was also hampered by a lack of common classification systems and a lack of standard nomenclatures, which requires research and program coordination by institutions. This pressure does not affect natural history collections.

Removing objects from their original context

Objects removed from their context lose heritage value (see page 9-7). The heritage significance of the places from which they are removed is also diminished. Continued demolition and/or sale and re-use of heritage buildings results in pressure for the removal of objects. This is greatest in urban areas subject to redevelopment (see page 9-10). Even historic buildings selected for presentation as house museums may be denuded of their original contents, which diminishes the values for both (Anderson, in press). Removing archaeological objects from sites can have a similar effect unless it forms part of a professional investigation arising from mitigation or research concerns.

Collections of indigenous cultural material

Many Aboriginal and Torres Strait Islander communities wish to assume control of their own cultural material (see page 9-42). While the return of indigenous cultural property to them will ensure this material is kept where it has most meaning, adverse effects may arise from the fragmentation and dispersal of collections. Growth in cultural tourism also results in more Aboriginal communities developing cultural centres, and seeking loan material from major museums for displays in these centres (Hyde, in press). However, as noted earlier for local museums, many indigenous communities also lack facilities, expertise and access to management advice for the on-going preservation of their collections in either keeping places or cultural centres.

The Register of the National Estate — criteria for inclusion

- importance in the course, or pattern, of Australia's natural or cultural history

- possession of uncommon, rare or endangered aspects of Australia's natural or cultural history

- potential to yield information that will contribute to an understanding of Australia's natural or cultural history

- importance in demonstrating the principal characteristics of a class of Australia's natural or cultural places, or a class of Australia's natural or cultural environments

- importance in exhibiting particular aesthetic characteristics valued by a community or cultural group

- importance in demonstrating a high degree of creative or technical achievement at a particular period

- strong or special association with a particular community or cultural group for social, cultural or spiritual reasons

- special association with the life or works of a person, or group of people, of importance in Australia's natural and cultural history

Source: Australian Heritage Commission Act.

The heritage listed garden of Rippon Lea, Victoria is a fine example of a nineteenth century homestead garden. It has aesthetic qualities valued by the community and demonstrates a high level of creative and technical achievement in its design.

State

Any report on the state of Australia's natural and cultural heritage first needs to establish how much of our heritage is recognised and documented. This information is critical to assess the physical state of our heritage and ultimately to make informed choices about which parts will be consciously retained for future generations.

Heritage registers and collections of objects will always be 'open-ended', as they will continue to change in response to evolving community perceptions of what is significant. This chapter applies the concepts of 'representativeness' and 'comprehensiveness' as measures of the current state of knowledge about places (as reflected in heritage registers) and collections of objects. Comprehensiveness is defined as the extent to which the registers or collections include all significant places or objects of a particular type. Representativeness is defined as the extent to which each significant type of place or object is represented in heritage registers or collections.

Large areas of Australia are cultural landscapes with various layers of cultural significance (see page 9-20). Individual places are an integral part of both natural and cultural landscapes, giving meaning to and deriving meaning from them. Their significance is often diminished when treated in isolation from the surrounding landscape. To date, however, heritage registers have focused on individual sites because these are often the focal points of significance and are more easily documented and listed than the whole landscape. This assessment of the state of Australia's cultural heritage is thus biased towards individual sites, particularly for historic places.

State of heritage places

Knowledge

A primary measure of the state of knowledge is the number of places listed in various heritage registers.

The World Heritage List is a global register of places of outstanding universal value (see page 9-16). In Australia, 11 properties had been inscribed on this List at December 1994 (see Table 9.4) compared with none in 1980 (Yencken, 1985). All have been included because of their outstanding natural heritage values and four also for their outstanding cultural values. No formal assessment action has been taken for most other places noted in 1981 (Yencken, 1985) as possible World Heritage nominations.

The Australian Capital Territory, Northern Territory and all States except Tasmania now maintain registers for historic places, while both Territories and all States have registers of

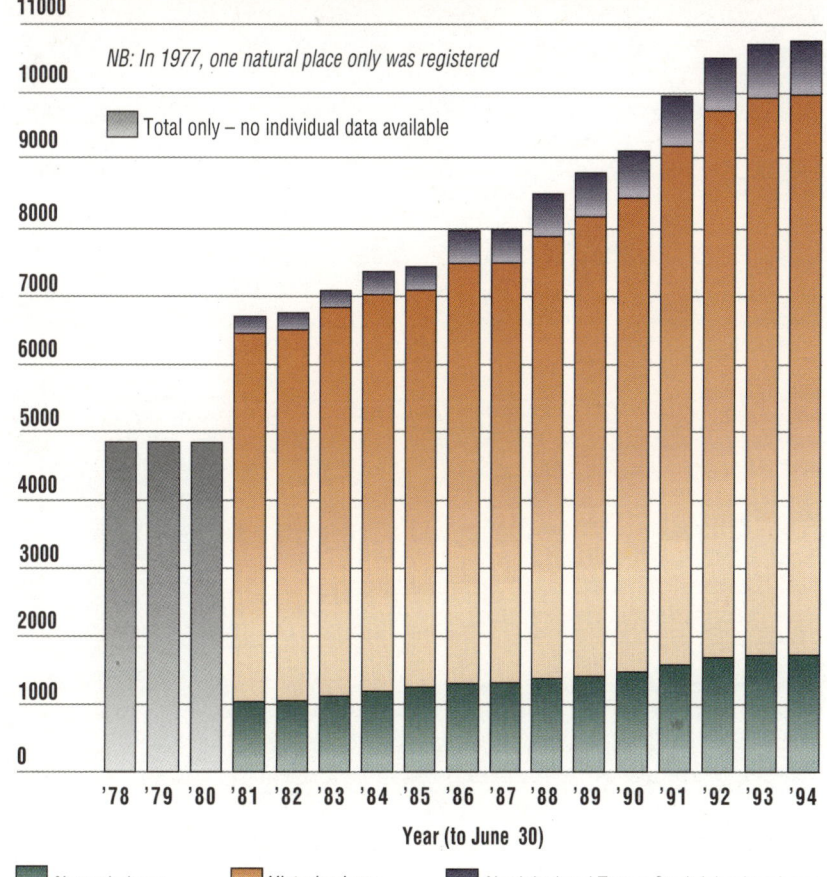

The Earp Gillam bond store, Newcastle, a masterpiece of the German immigrant architect Frederick Menkens, was completed in 1888. It was unused for many years and deteriorated. After the 1989 earthquake (left), part of its front wall had to be demolished . The store was authentically restored (above) and converted for sympathetic re-use as offices at a cost of $2.5 million.

Figure 9.3 Cumulative number of places listed in the Register of the National Estate to June 1994

NB: In 1977, one natural place only was registered

Total only – no individual data available

Year (to June 30): '78 '79 '80 '81 '82 '83 '84 '85 '86 '87 '88 '89 '90 '91 '92 '93 '94

Natural places   Historic places   Aboriginal and Torres Strait Islander places

Source: Australian Heritage Commission, in press.

Cultural landscapes

The cultural landscape is the 'tapestry in which all other artefacts are embedded and which gives them their sense of place'.

(Lowenthal, cited by Lennon in press).

'Cultural landscape' is the term applied to those parts of the environment that have been significantly modified by human activity to distinguish them from natural landscapes, where evidence of human intervention is less apparent. They include cultural and natural elements of the ordinary, familiar, everyday landscape. A cultural landscape is therefore an expression of human attitudes, values and interactions with the environment.

The relationship between people and place creates distinct visual and spatial patterns in the landscape additional to those created by biophysical systems. Landscape is seen not only as a natural system but as a cultural artefact, consisting of the tangible remains left on the land by cultures past and present (Blair and Truscott, 1989). Through these tangible remains the landscape carries — more or less visibly — a record of history where memory, symbolism and actual physical vestiges of the past are held. These meanings are at the heart of community attachment to places and to the development of cultural heritage values (Taylor, cited in Lennon, in press).

Much of Australia may be regarded as cultural landscape because of the traditions and practices of indigenous peoples over thousands of years. Immigrants since the first European settlements have added further layers of historical evidence and social significance to the natural landscape.

Some landscapes are less easy to read than others. In many instances, the layers of historical evidence may not be immediately recognisable. For example, it is difficult for non-Aboriginal people to perceive the layers within the landscape significant to Aborigines. In other instances, physical evidence may no longer be present. The historical evidence of the hand-hewn stone of the first roads that wound their way over the Great Dividing Range has been lost as four-lane highways level their way through the landscape.

The Central Victorian Goldfields region is a landscape which can be more easily read (Lennon, in press). Gold rushes from the 1850s had a great impact on this area. However, the landscape tells a more complex story than just that of the gold rushes. Many people left their marks — Aborigines, European explorers, squatters, travellers, road-makers, surveyors, alluvial-gold diggers, company miners, farmers, foresters and town-dwellers. They created impacts and patterns which can still be identified today in a mosaic of public forested areas, cleared land, abandoned mine workings, archaeological sites, buildings, roads and other signs of human interaction with the land. It is through an understanding of the history of occupation of the area that the cultural landscape can be interpreted and the heritage value of the landscape understood.

Conservation of the cultural landscape raises many management issues. The physical land system is constantly in flux. The natural cycle of decay and renewal changes the cultural landscape. New land uses, township expansion, physical decay of surviving elements and tourist developments can all alter the physical evidence of previous activities as well as adding new layers of meaning.

Many people left their marks on the Central Victorian goldfields region. Abandoned mines, such as the Wattle Gully mine at Chewton, tell just part of the story.

Conservation management and the management of change are more complex because of the number of components and their different inherent characteristics. In areas such as the Central Victorian Goldfields, it may be these inherent characteristics that attract more residents, developers and tourists, thus placing more pressure on their survival.

Conservation of the cultural landscape requires a comprehensive understanding of the landscape, not one which is directed towards preserving one aspect of it. The coordination of public and private effort to conserve components of the cultural landscape becomes more difficult. For example, a farmer would be loath to spend resources stabilising mining ruins on his property, while a pensioner in a historic dwelling may not be able to afford the restoration of historical components.

The adoption of heritage terminology that acknowledges the concept of cultural landscapes and their inclusion on the World Heritage List and on the Register of the National Estate are important preliminary steps towards establishing benchmarks to determine how their features are being managed. Presently there are no indicators to assess their state. Suggested measures (Lennon, in press) include:

• number and range of landscapes listed in heritage registers

• number of planning permits issued for new forms of land use

• number of permits issued for re-use of existing cultural elements

• number of building repair orders

• condition of key natural and cultural places

• number of visitors, and the nature and number of visitor facilities

indigenous places. The Commonwealth's Register of the National Estate, which began in 1976, provides the most comprehensive heritage picture. It covers both natural and cultural heritage of all States, Territories and External Territories, and includes places solely on the basis of criteria specified in the *Australian Heritage Commission Act* (see page 9-18). The values of places listed in this Register range from those of recognised international and national significance to those valued by local communities. The places vary enormously in area — a few natural places cover millions of hectares, but most are hundreds or thousands of hectares in size; many listed Aboriginal places cover hundreds of hectares, while most historic buildings occupy much less than one hectare.

In June 1994, 10 772 places were listed in the Register of the National Estate. Of these, 77 per cent were historic and 16 per cent were natural ones. Only seven per cent were places significant to Aborigines and Torres Strait Islanders, but many of those contain hundreds of individual sites. The number of places listed in the Register has grown steadily (see Fig. 9.3) since the 'first generation' at the end of 1980 (Yencken, 1985), and has increased by 61 per cent since 1981. Over this period, natural listings have increased by 67 per cent, historic places by 52 per cent, and the small number of indigenous places by 222 per cent. While listings to the end of 1980 reflected work carried out in previous decades, many since then have been based on new studies.

Distribution of places listed in the Register

Listed places occur across Australia (see Fig. 9.4). However, they are concentrated in metropolitan and associated regions, particularly in south-eastern Australia, and distributed sparsely through remote and rural areas (see Fig. 9.5). Considerable regional variation in distribution occurs (see Figs 9.6–9.8), reflecting the predominance of assessment and documentation effort in regions of highest population density, and the high proportion of historic places associated with major cities. Comprehensive knowledge of distribution patterns will only be possible when all Australia has been systematically surveyed. This process is being addressed through a range of studies, especially under the National Estate Grants Program.

Changes in the types of places listed in the Register

The number of natural places with significant forests, woodlands, wetlands, geological features and habitats of endangered species has substantially increased since 1981 (see Table 9.5). Most natural places listed in the Register are still terrestrial, with only five per cent comprising or including marine areas. Not all ecosystem types are represented, as many of Australia's biogeographic regions (see Chapter 4) contain no or few listed natural places. Listed places cover five per cent or less of the area of about half these regions (Australian Heritage Commission, in press).

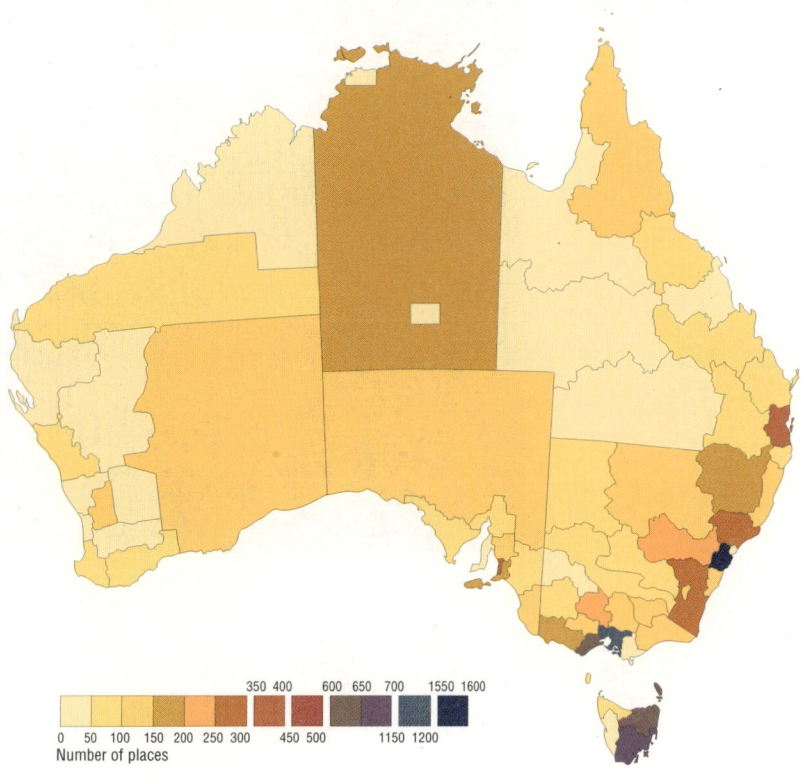

Figure 9.4 Number of all places listed in the Register of the National Estate at December 1994 within Australian Government Regions

Number of places
0 50 100 150 200 250 300 350 400 450 500 600 650 700 1150 1200 1550 1600

Source: Australian Heritage Commission, in press.

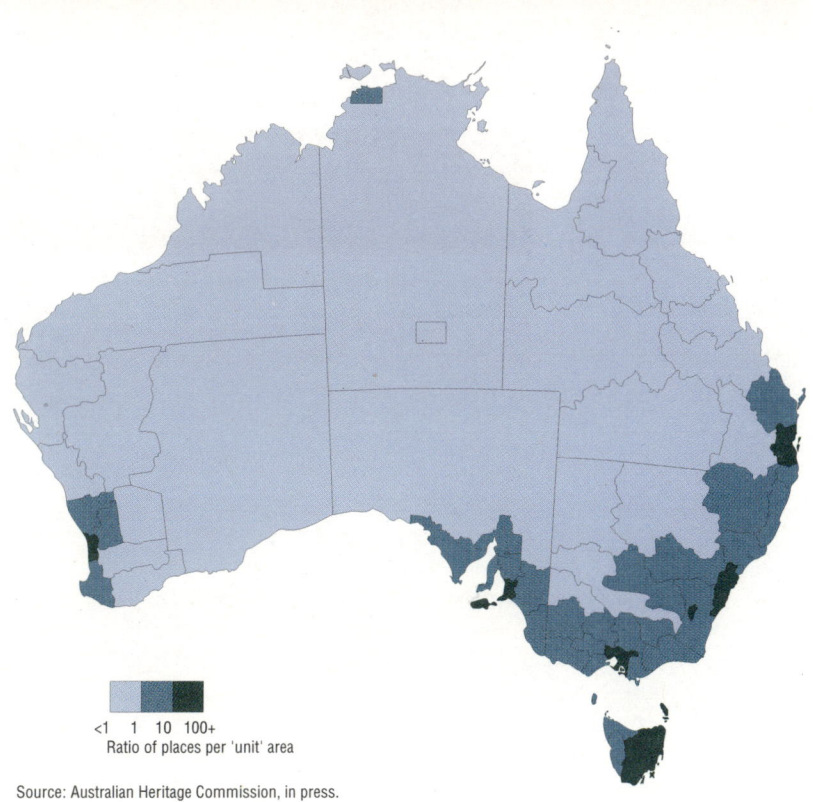

Figure 9.5 Density of all places listed in the Register of the National Estate at December 1994 within Australian Government Regions

Ratio of places per 'unit' area
<1   1   10   100+

Source: Australian Heritage Commission, in press.

Many biogeographic regions in Australia such as the Kimberley, Western Australia, have few places listed in the Register of the National Estate for their natural heritage values.

Since 1981, listings of historic places related to industrial archaeology, gardens, administrative buildings and commercial buildings have increased substantially (see Table 9.6). However, there are still many themes and types of places that are represented in the Register either poorly or not at all (Domicelj, 1992). These include rural landscapes, transport sites and routes, farm complexes and places for community gathering, leisure and recreation. Places representing the history of multiculturalism and minority groups, including 19th and 20th century sites, post-war immigration and women's sites are also poorly represented. Several factors combine to cause these imbalances: some themes or types are not

Table 9.5 Major types of natural places listed in the Register of the National Estate, June 1981 and December 1994

Type of ecosystem/feature present	Number of listed places 1981	1994
Forest	622	784
Woodland	591	690
Shrubland[1]	857	729
Grassland/herbland[1]	522	451
Wetland	164	281
Geological	93	254
Endangered species habitat	184	288
Marine	79	86
Wilderness[2]	72	101
Total number of listed places[3]	1034	1728

Notes:
1. The lower number of places in 1994 is due to different classification of places.
2. See the box on page 9-5
3. Many places contain more than one major ecosystem type/feature.

Source: Yencken, 1985 and Australian Heritage Commission, in press.

Figure 9.6 Number of natural places listed in the Register of the National Estate at December 1994 within Australian Government Regions

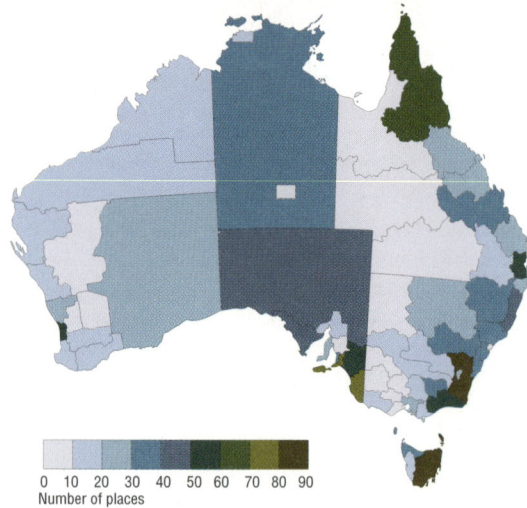

0 10 20 30 40 50 60 70 80 90
Number of places

Source: Australian Heritage Commission, in press.

recognised as heritage and hence are under-studied; survey work is lacking; and appropriate methods of assessment need to be developed.

Among indigenous places listed in the Register in 1994, the largest groups are art sites, complexes that contain many individual sites, places of spiritual or mythological significance and occupation sites (see Table 9.7). Some sites such as wells and modified trees will always be low in number. However, others, like spiritual/religious and historic contact sites or those of contemporary significance, are under-represented. The current representation reflects the focus of non-indigenous professionals on rock art and archaeological sites and a low level of nominations to the Register from indigenous communities.

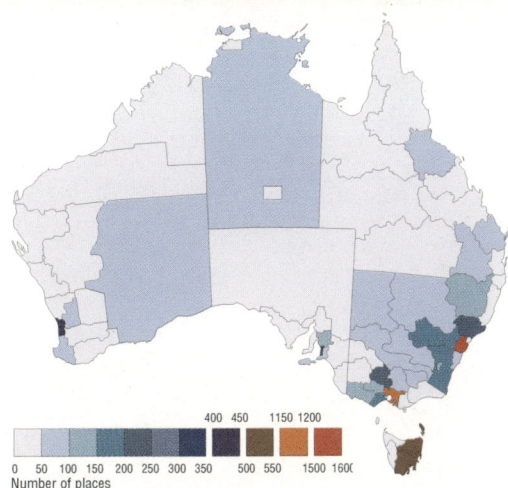

400 450   1150 1200

0 50 100 150 200 250 300 350   500 550   1500 160(
Number of places

Source: Australian Heritage Commission, in press.

The heritage significance of many indigenous places is intimately linked with traditional languages and the knowledge they transmit. Speakers of these languages strongly believe that they are the best vehicle for traditional knowledge (Henderson and Nash, in press). The state of these languages is thus a critical factor relevant to state of knowledge about indigenous places.

Of the 250 indigenous languages thought to be spoken at the time of European settlement, only 90 are still spoken today (see Fig. 9.9), mainly by people in central-western and northern Australia. Just 34 languages have 200 or more speakers today and many have but a handful of elderly speakers (see Fig. 9.10). Only about 20 traditional languages are being passed on to children who use them as their main language. All of the traditional ones being used as the primary means of communication could be lost within a generation. The associated decline in traditional knowledge is not quantifiable but must be profound.

Table 9.6 Major types of historic places listed in the Register of the National Estate, June 1981 and December 1994

Type of place	Number of listed places 1981	1994
Residential houses	**1867**	**2432**
Administrative buildings	**1180**	**1978**
• government functions, including		
– government buildings (parliaments, customs, town halls etc)		129
– court houses, police stations, prisons		416
– libraries, hospitals, civic structures etc		240
– military barracks, bases, fortifications		122
– scientific research facilities		7
• transport and communications, including		
– rail, road, and air transport places		145
– harbour facilities, ports, piers, docks etc		27
– light stations (water transport)		93
– post offices, telegraph stations etc		173
– bridges		243
– shipwrecks		56
• schools and places of education		327
Commercial buildings	**967**	**1540**
• shops, offices etc		690
• hotels, motels, inns		406
• places of recreation (theatres, halls, race courses etc)		260
• banks and financial institutions		184
Industrial archaeology	**162**	**832**
• primary industry (agricultural, pastoral, processing, forestry etc)		664
• industrial sites and buildings		103
• mines and mineral processing works		65
Religious buildings (churches and other places of religion)	**634**	**696**
Conservation areas, historic towns, precincts & groups	**277**	**387**
• towns, precincts, conservation areas		378
• historic landscapes		9
Gardens (parks and gardens)	**61**	**180**
Monuments and other building types[1]	**193**	**132**
• monuments and memorials		69
• cemeteries and graves		73
Historic sites (historic and miscellaneous places)	**76**	**92**
Total	**5 417**	**8 279**

Note: 1. The lower number of places in 1994 is due to different classification of places

Source: Yencken, 1985 and Australian Heritage Commission, in press.

◄ Rural landscapes and farm complexes, such as Gulf Station, owned by the National Trust of Victoria, are not well represented in the Register of the National Estate.

The historic Alice Springs Telegraph Station played an important role in early telecommunications in Australia. Many Aboriginal people also have a special association with the station due to its subsequent use as an Aboriginal Reserve for many years.

Knowledge of our natural and cultural heritage places has grown steadily since 1981, with a considerable body of it gained through targeted heritage studies carried out by professional experts as well as interested members of the community. However, the Register of the National Estate is still far from comprehensive. Nor is it fully representative of Australia's heritage.

Although the types of places representing our current understanding of heritage are generally known, the total number of eligible places is not. New types of places that should be included are likely to emerge as community values change and concepts of heritage expand.

Table 9.7 Major types of Aboriginal and Torres Strait Islander places listed in the Register of the National Estate, December 1994[1]

Type of place	Number of listed places
Art sites	196
Site complexes	106
Spiritual/mythological sites	83
Occupation sites	82
Shell middens	58
Stone arrangements	53
Historic contact sites	41
Modified trees (scarred and carved)	41
Quarries	34
Burials/cemeteries/graves	29
Grinding grooves	19
Ceremonial sites	18
Fish/eel traps	18
Wells	11
Hunting hides/traps	3
Organic resource areas	1
Total	**793**

Note: 1. Comparable data for 1981 were not available.
Source: Australian Heritage Commission, in press.

Figure 9.8 Number of Aboriginal and Torres Strait Islander places listed in the Register of the National Estate at December 1994 within Australian Governement Regions

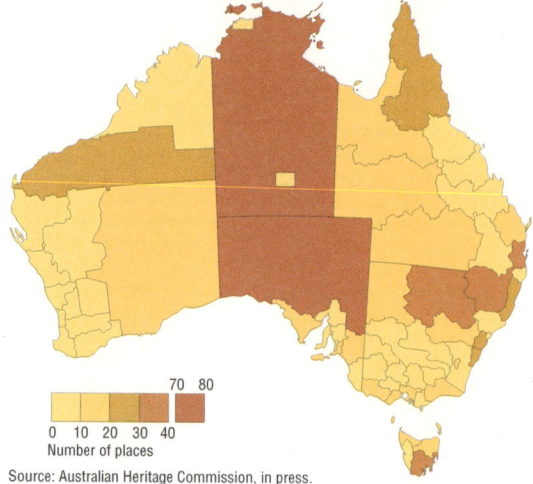

70 80

0 10 20 30 40
Number of places

Source: Australian Heritage Commission, in press.

Legislative protection

All Australia's World Heritage properties are protected under international obligations and Commonwealth legislation (see page 9-16) and are managed to protect their identified heritage values. A wide range of administrative and managerial structures are in place to achieve this, although some arrangements have yet to be agreed (see Table 9.4).

Places listed in the Register of the National Estate receive only a limited measure of protection through the *Australian Heritage Commission Act 1975*. Section 30 of this Act requires Commonwealth bodies to avoid damaging national estate places, unless there is no 'feasible and prudent alternative', and to seek conservation advice from the Australian Heritage Commission. Other bodies make non-statutory requests for advice.

Figure 9.9 Vitality of the original 250 Australian indigenous languages in Australia at the time of the first European contact

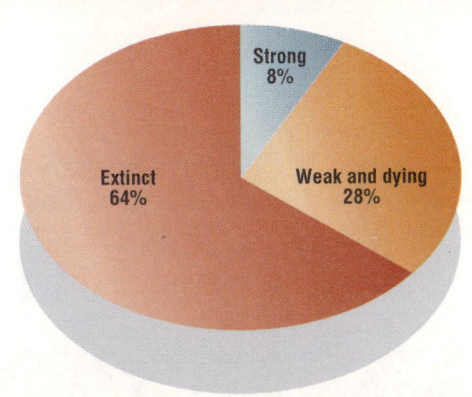

Source: Schmidt, as modified by Henderson and Nash, in press.

Figure 9.10 Geographic distribution and numbers of speakers of traditional languages still used as a primary means of communication

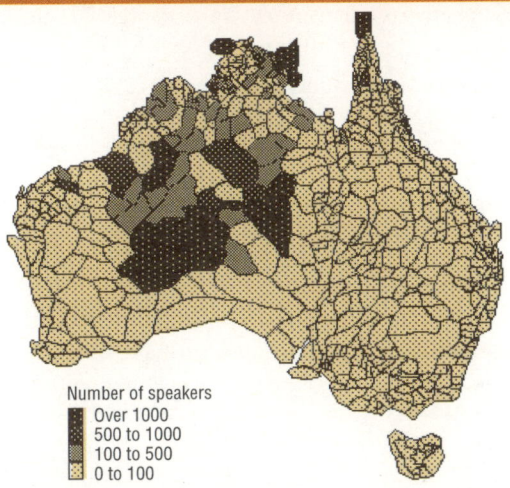

Number of speakers
- Over 1000
- 500 to 1000
- 100 to 500
- 0 to 100

Note: The lines represent the boundaries of language areas identified by Tindale in 1974.

Source: Australian Institute of Aboriginal and Torres Strait Islander Studies, in Henderson and Nash, in press.

In the six years for which records are available, the Commission has received few referrals for conservation advice compared with the total number of places listed in the Register (an average of 5.4 per cent) (see Table 9.8). The average number of Section 30 referrals over the same period compared with the number of listed places was only 3.4 percent. There could be many reasons for this. Relatively few places in the Register are owned by the Commonwealth and, because of the nature of the Register, many other listed places are never subject to Commonwealth activities. Some Commonwealth agencies have a poor understanding of their obligations under the *Australian Heritage Commission Act 1975* (Marshall, 1993) and thus may fail to comply with them.

Referring bodies are not obliged to adopt the Commission's advice. However, it appears to have had a positive influence on many Commonwealth decisions, although it has been followed less often

for decisions affecting natural heritage places than for cultural places (Marshall, 1993).

All States and Territories have legislative and other provisions to protect natural heritage both inside and outside protected areas. Of the natural places listed in the Register of the National Estate in 1994, about 65 per cent were managed as part of some protected area (Australian Heritage Commission, in press), compared with 87 per cent in 1981 (Yencken, 1985). This difference reflects both the focus of State government agencies nominating parks and reserves for the Register in the 1970s, and the growing body of knowledge since then about the conservation values of land outside protected areas.

Apart from Tasmania, all States and mainland Territories now have heritage legislation applying to historic places (see page 9-37), compared with only three States in 1981. By June 1994, approximately 5500 places (mainly historic) were listed in State and Territory heritage registers (Marshall and Pearson, in press) and hence receive the protection afforded by the Acts. Most States have delegated the responsibility for conservation to local councils through heritage, development or planning legislation (see Table 9.18). As the Acts in Queensland, Australian Capital Territory, Northern Territory, South Australia and Western Australia have been passed or significantly revised since 1990, it is too early to assess their effectiveness, while the older acts in New South Wales and Victoria are being reformed.

About 30 major Acts specifically apply to places and objects significant to indigenous Australians (Ward, 1995). Four Commonwealth and 11 State and Territory Acts (see Table 9.17) provide the main protection, although they vary in their scope and provisions. The legislation has grown from four State and two Commonwealth major Acts in 1981. Indigenous archaeological places receive blanket legislative protection under Acts in all States and Territories — that is, places are

Table 9.8 Referrals for conservation advice on places listed in the Register of the National Estate

	Financial Year					
	88/89	89/90	90/91	91/92	92/93	93/94
Number of places[1] (000s)	9.1	9.6	10.3	10.8	10.9	11.0
Number of referrals						
Section 30 (statutory)[2]	-	c. 340	267	354	395	351
Other (non-statutory)[3]	-	c. 110	180	218	374	262
Total	c. 500	c. 450	447	572	769	613
Referral rate[4] (%)						
Section 30 (statutory)[2]	-	4	3	3	4	3
Other (non-statutory)[3]	-	1	2	2	3	2
Total (%)	6	5	4	5	7	6

Notes:
1. Places listed in the Register of the National Estate or on its Interim List.
2. Advice sought by Commonwealth bodies about listed or interim-listed places.
3. Advice sought by Commonwealth bodies about places neither listed nor interim-listed, and advice sought by non-Commonwealth bodies.
4. Number of referrals/number of listed places.

Source: Australian Heritage Commission, in press.

The Richmond main colliery near Newcastle, New South Wales, the largest and most historically significant mine in the South Maitland coalfields, is now part of a historic park. The cooling tower on the left has been restored to resemble its original condition.

protected even where they are unregistered, unrecorded or unknown. Most States now offer legislative protection for sites that are important to contemporary indigenous communities, although some still emphasise archaeological sites. They also provide blanket protection of associated objects, either by specific mention or indirectly by the protection afforded to places. By December 1994, some 85 000 indigenous sites had been recorded across Australia by the relevant government agencies (Ward, in press) — more than twice the number recorded in 1985.

Physical condition

Since 1992, the Commonwealth has provided annual monitoring reports on Australia's World Heritage properties to the World Heritage Committee. However, there is little precise or readily available information on the physical condition of most of Australia's heritage places in general, let alone in relation to their identified heritage values. Information about the state of the natural environment in previous chapters cannot be directly applied for this purpose.

A project is being funded under the 1994–95 National Estate Grants Program to develop and test a pilot method of auditing the condition of listed heritage places. This is intended to provide the framework for a rolling audit program, which will provide this information in the future.

Conservation practice

Measures such as the extent to which heritage places are managed for conservation goals, the adequacy of assistance programs for them and the level of trained expertise available all provide indirect indicators of their condition.

In 1979, Australia International Council for Monuments and Sites (ICOMOS) adopted a charter for the conservation of places of cultural significance (the Burra Charter). Since the early 1980s this has been widely accepted by governments and conservation practitioners as the standard for conservation philosophy and practice in Australia. Cultural places managed in accord with an approved conservation management plan and with adequate resources could be expected to have a good conservation status.

No natural heritage charter comparable to the Burra Charter yet exists. However, if natural places are managed with adequate resources under approved management plans meeting accepted professional practices for nature conservation, they should also have a good conservation status. Such plans will not, however guarantee the condition of cultural places within natural areas if their management is neglected or a low priority.

Heritage assistance programs

A range of funding programs for heritage conservation and associated works were available in 1994; many for natural places have been described in earlier chapters.

Table 9.9 Heritage assistance provided in 1993–94 under annual funding programs

	Funding level ($ million)
Queensland	0.248
New South Wales	1.923
ACT	0.308
Victoria	0.375
Tasmania	not applicable
Northern Territory	0.130
South Australia	0.300
Western Australia	0.199
Commonwealth	4.440

Source: State data from Marshall and Pearson, in press; Commonwealth data from Australian Heritage Commission, unpublished, 1995.

The Commonwealth allocated about $50 million in 1994–95 to manage Australia's 11 World Heritage areas (DEST, 1995). The 1993–94 National Estate Grants Program provided approximately $4.4 million to the States and Territories for the identification, conservation and presentation of national estate places (see Table 9.9). Under current policy, these funds are allocated about equally between natural, historic and indigenous places. Over the same period, $3.5 million was available for mainly historic places under annual assistance programs of the States and Territories (see Table 9.9). In December 1994 more than 30 Commonwealth and State agencies across Australia had the potential to implement specific programs to identify and protect indigenous heritage, but details of funding levels were not readily available (Ward, in press).

Training

Over 80 undergraduate accredited university and TAFE courses on management of natural and cultural heritage were available in 1994 (see Table 9.10). It appears that the total number of courses in management of the built environment had increased slightly since 1981.

Community participation

Many heritage places in Australia are conserved because of community pressure. Community support for and participation in heritage activities are thus important measures of state.

The level of community involvement in environmental and heritage organisations appears to be low. National public attitude surveys in 1991, 1993 and 1994 showed that 0.4 per cent of the population were involved in heritage organisations, 4–6 per cent belonged to an environmental group and 28 per cent donated time or money to protect the environment (Purdie, in press). Comparative data showing trends were not available.

National Trusts, nature conservation bodies and other community organisations, and individuals have actively nominated places to the Register of the National Estate since its inception (see Table 9.11). From January 1991 to December 1994 these groups comprised over 65 per cent of the nominators who were submitting nominations for the first time. Private or government heritage professionals comprised 21 per cent and local governments 15 per cent. Nominations from community groups and individuals comprised 46 per cent of all nominations received over the four years; most were for historic places (Australian Heritage Commission, in press). National Trusts contributed more nominations than any other nominator group over this period: three-quarters were for historic places, one-fifth for natural places and the remainder for places with indigenous significance.

The number of objections to the inclusion of places in heritage registers provides an inverse measure of community support. For interim listings in the Register of the National Estate (see

Table 9.10 Number of higher education undergraduate courses of study on offer in 1994 for management of heritage places and objects, as outlined in institution handbooks and directories

Type of course	TAFE 1981	TAFE 1994	University 1981	University 1994	Total 1981	Total 1994
1. Cultural environment management						
(a) Aboriginal and Torres Strait Islander heritage	n/a	7	n/a	8	n/a	15
(b) Cultural heritage		13		13		26
— directly related to conservation	5		4		9	
— indirectly related to conservation	3		10		13	
(c) Materials conservation[1]	n/a	2	n/a	1	n/a	3
2. Natural environment management						
	n/a	27	n/a	15	n/a	42
Total		**49**		**37**		**86**

Notes:
1. Other courses listed under 1(a) and 1(b) may also include aspects of cultural material conservation.
n/a = not available.

Source: Australian Association of Environmental Education, in press; Yencken, 1985.

Table 9.11 Major types of nominators submitting nominations for the Register of the National Estate 1991–94

Nominator group	Number of nominators from 1976 to January 1991	Number of nominators from 1976 to December 1994	Number of nominators submitting nominations 1991–1994: who had also submitted prior to 1991	Number of nominators submitting nominations 1991–1994: submitting for the first time from 1991 to 1994	Number of nominations submitted 1991–1994
Commonwealth Government	18	19	-	1	1
Commonwealth & State Government	-	1	-	1	63
State Government	56	65	6	9	135
Local Government	149	177	8	28	160
Australian Heritage Commission consultants	58	76	4	18	217
Professional bodies	36	47	-	11	97
National Trusts	14	15	6	1	273
Other community organisations	335	412	2	77	137
Private nominators (individuals)	242	288	6	46	166
Totals	**908**	**1100**	**32**	**192**	**1249**

Source: Australian Heritage Commission 1995.

Table 9.12 Level of objections for interim listings in the Register of the National Estate

Years of gazettal	Number of gazettals	No. of places interim-listed	Places for which objections received	
			Number	Per cent
1977–1978	5	6200	539	8.7
1979–1980	5	680	76	11.2
1981–1982	3	472	97	20.6
1983–1984	4	244	37	15.2
1985–1986	4	770	171	22.2
1987–1988	2	414	64	15.5
1989–1990	5	776	113	14.6
1991–1992	5	631	72	11.4
1993–1994	2	155	22	14.2
Total	**35**	**10 342**	**1 191**	**11.5**

Source: Australian Heritage Commission, in press.

Table 9.12), the two-year objection rate has dropped from its peak of 22% in 1985–1986 to only 14 per cent in 1993–1994. Although objections were often related to the concept of heritage listing, many objectors misunderstood its implications.

In 1994–95, 320 community groups applied for approximately $10 million to carry out heritage identification, conservation and education projects under the National Estate Grants Program. This represented 36 per cent of all funds applied for (Australian Heritage Commission, in press). Community groups submitted 26 per cent of all applications relating to natural places, 39 per cent for historic places and 51 per cent for indigenous places.

State of heritage objects

To discuss the state of Australia's heritage objects is a new initiative in state of environment reporting. Measures relevant to the state of objects are similar to those for places, although it is harder to apply some concepts.

Article 10 of the Burra Charter espouses the principle that objects of cultural significance should not be removed from their original locations unless it is the only way of ensuring their protection and preservation. This stresses the importance of connection with place (see the box on page 9-7). Documentation of objects removed to collections is probably as important as conservation of their physical state: poor physical condition and lack of adequate documentation greatly diminish their heritage value.

This section includes information relevant to living collections of plants and animals in botanical gardens and zoological gardens, and to collections in major museums. Collections located in overseas institutions are not included, although they contain many significant items. Australian university collections, which were the subject of a major study still in progress at the time of preparing this chapter (see page 9-44), are not included either.

Knowledge

The Distributed National Collection (see the box on page 9-14) includes many significant large collections (Anderson, 1991). However, the total number of specimens in a collection is not necessarily a measure of the state of knowledge in the same way as the size of heritage registers. The size of a collection may be more relevant as an indirect measure of physical condition — the greater the number of items, the greater the resources required for their curation and preservation. For living collections, absolute numbers may be relevant for captive breeding programs of endangered species, especially if associated with re-introductions to natural habitat.

Representativeness of collections is relevant to the state of knowledge as it is for places. For example, every species of Australia's biota should be represented in the Distributed National Collection at least once, to provide reference material for taxonomic identification and hence assessment of Australia's biodiversity. Cultural collections should be representative of both Australia's indigenous and non-indigenous communities and history.

For all objects, the level of documentation associated with them is also a significant aspect of the state of knowledge. An object with a richly documented social or natural history may be priceless. Without it, it may be nearly worthless.

Biological collections

Excluding university collections, Australian museums and herbaria held almost 40 million preserved specimens of 'natural' objects in 1991 (see Table 9.13). The number of living specimens in major botanic gardens and zoos was in the order of tens of thousands, but the proportion of native species in all these collections remains unknown.

As well as the eight major gardens (see Table 9.13), at least 100 other local and regional botanic gardens and arboreta existed in 1994 (Fagg and Wilson, 1994). In 57 per cent of these, at least half of the living collections were native species, while 30 per cent grew only natives. In most of the major gardens, less than 40 per cent of the living collections were native plants. In 1993, 61 per cent of the 234 native plant species endangered nationally were cultivated in botanical gardens, compared with only 32 per cent of 203 species in 1984 (Boden, in press).

Some 13 of Australia's major zoos specialise in native fauna (see Table 9.13). In recent years they have tended to hold fewer species with larger numbers of individuals (including small potential breeding populations) rather than single specimens (Boden, in press). By December 1994, Australian zoos participating in the Australasian Species Management Program had captive breeding populations of 12 of the 75 native vertebrate species listed as endangered under the *Endangered Species Protection Act 1992*. Captive-breeding populations of at least a further eight endangered species occurred at zoos or other institutions as part of local programs (unpublished data, Australian Nature Conservation Agency, 1995).

The size of collections in major State and Commonwealth herbaria in 1991 had almost doubled since 1974 (see Table 9.13). The distribution of specimens did not reflect the number of species thought to occur in nature, with some groups, such as fungi, being grossly under-represented. The number of non-fossil fauna specimens in major State and Commonwealth museums or equivalent institutions also approximately doubled between 1974 and 1991. The representation of specimens across taxonomic groups reflected ease of capture rather than the number of species in nature (Richardson and McKenzie, 1992).

Cultural collections

In 1991, the major museums collectively held about 5.5 million items in the fields of science and technology, anthropology, archaeology and history (see Table 9.14). No comparative data for the previous two decades are available. Aboriginal and Torres Strait Islander ethnographic collections, which constituted slightly less than half of the anthropological collections, were not fully representative, as they predominantly related to male cultural artefacts. This was probably because of the circumstances of their collection by male researchers (Anderson, in press) and the durability of many of the items involved.

Of the 1.39 million historical artefacts held in museums, only about 200 000 may be relevant to historic places in Australia — a small number to reflect the material culture produced by 200 years of colonial and post-colonial peace-time activities (see Table 9.14). Many of these collections often have a technological bias, and the experience of women, migrants and working-class people is seriously under-represented (Anderson, 1991). The role of collections in local museums throughout Australia is crucial, given the paucity of history collections in the major museums (Anderson, 1993).

Legislative protection

Objects are covered by State and Commonwealth legislation, although the extent of protection varies with the type of object, and most Acts are designed to protect objects *in situ*.

All States and the two Territories now have legislation (compared with only three States in 1981) to protect indigenous objects associated with archaeological sites or places of religious or other significance (see page 9-25). The blanket protection provided to places affords a high level of legislative protection for those *in situ*, although the actual effectiveness of the protection is not known. In Queensland, New South Wales and the Northern Territory, the heritage Acts relating to historic places (see page 9-25) also cover associated objects, while the Australian Capital Territory has a separate Act to protect all cultural objects. In 1981 only New South Wales had such legislation.

The *Australian Heritage Commission Act 1975* does not specifically cover objects, although the heritage

Table 9.13 Summary of living and material biological collections in major State- and Commonwealth-funded institutions

Type of object	Date	No. of Taxa	No. of specimens
Living collections — plants[a] (1994, information from 8 botanic gardens)			
Total	1994	*c.* 33 400[1]	Unknown[2]
Living collections — animals[b] (1992, information from 13 zoos)			
Total	1992	Unknown[2,3]	20 400
— vertebrate			83%
— invertebrates			17%
Material collections — plants[c] (1991, information from 13 herbaria)			
Total	1991	Unknown[2,3]	5 020 000
— vascular[4]			87%
— non-vascular[5]			13%
Total	1974	Unknown[2]	2 562 000
Total	1965	Unknown[2]	1 954 000
Material collections — animals[c] (1991, information from 9 museums/institutions)			
Total	1991	Unknown[2,3]	33 700 000
• palaeontological			13%
• modern (non-fossil) fauna			
— vertebrates			5%
— invertebrates			82%
Total modern (non-fossil) fauna	1974	Unknown[2]	14 600 000
— vertebrates			6%
— invertebrates			94%
Total plant and animal material	1991/1992	Unknown	**38 740 400**

Notes:
1. Includes native and exotic species but not cultivars; relative proportions variable.
2. Figures not determined; presence and/or accessibility of data not known.
3. Includes native and exotic species; relative proportions unknown.
4. Vascular plants include angiosperms, gymnosperms and pteridophytes.
5. Non-vascular plants include bryophytes, lichens, algae and fungi.

Source: **a**–Boden, in press; **b**–Olney and Ellis, 1992; **c**–Richardson and McKenzie, 1992.

◄ The Leadbeater's possum, *Gymnobelideus leadbeateri,* is an endangered species in Australia. Breeding programs at zoos are an important way of ensuring the survival of the species.

Shipwrecks — part of Australia's underwater cultural heritage

Throughout Australia's history, the sea and inland waters have played a vital role in communications and transport. Aboriginal settlers, who came at least 50 000 years ago, used the coasts and inland rivers intensively. After 1788, European settlers relied on shipping networks (long-distance, local and riverine). Australia's underwater heritage includes ships lost on these voyages. They and their contents are archaeological time capsules, providing incomparable information about the past. Many wrecks have also become significant habitats for marine biota.

More than 5000 shipwrecks in Australian waters have been recorded, although only about 15 per cent of these had been located by 1994, compared with about 10 per cent in 1975 (Pigott, 1975). Thousands of relics removed from wreck sites are now in museums and private collections. They are also part of Australia's cultural heritage and demonstrate aspects of the nexus between object and place.

The Commonwealth *Historic Shipwrecks Act 1976* provides protection to sites (wrecks and their contents) in Commonwealth waters declared under the Act and to material removed from them. An amendment to the Act in 1993 provided blanket protection to all wrecks more than 75 years old. As a result, the number of sites protected under the Act rose from about 150 in 1992 to more than 5000 in early 1995. In 1993, an amnesty for people holding relics from wrecks resulted in more than 3000 registrations of material over a 12-month period.

Shipwrecks located in State waters also receive some form of protection under State legislation, except in Tasmania. However, these Acts (with two exceptions) do not align with the amended Commonwealth Act in providing blanket protection for wrecks over 75 years old. Legislative mechanisms range from 50-year blanket heritage protection (New South Wales) to a situation where individual sites need to be declared (South Australia). Queensland legislation contains no precedent for the declaration of the sites.

The Commonwealth Shipwrecks Program sponsors major activities, which are implemented through State and Territory agencies under cooperative funding arrangements. An Australian Shipwrecks Database has been established, guidelines developed for shipwreck management and public access and a national research plan instigated to develop research priorities and facilitate exchange of information.

Although they are protected by legislation, wreck sites continue to be destroyed. Development activities such as dredging, cable-laying, development of marinas, reclamation, seismic testing and resource extraction can all affect them. The diving community has helped locate and record many wrecks, and have been major lobbyists for their protection. However, the dramatic increase in the numbers of recreational divers has also placed stress on wreck sites through anchor damage and unlawful interference and removal of relics.

Divers, such as those at the Cheynnes (above) and the Vergulde Draeck (right) have helped to locate and record many wrecks. However, the growth in recreational diving places pressure on wreck sites through damage and removal of relics.

Relics removed from wreck sites, whether illegally or through approved archaeological surveys and excavations, quickly deteriorate unless subject to correct conservation measures.

The public is increasingly demanding the underwater interpretation of wreck sites, museum exhibitions of relics from them, publications and community involvement in restoration projects. Such activities are all important means of promoting awareness about wrecks and the need for their conservation.

Australia has high international standing in shipwreck investigation, management and protection, and in 1991 was selected to host and chair the ICOMOS Scientific Committee on Underwater Cultural Heritage. Limitations in shipwreck heritage conservation arise from differences between Commonwealth and State protective legislation, the general lack of enforcement of infringements under the Acts, the level of resources required for monitoring wrecks and collating data on both wrecks and relics and the inadequate coordination of research. Despite the excellent training available, there are still too few marine archaeologists and conservators to meet these demands.

The pressures affecting shipwrecks reflect both their multi-jurisdictional administration and the diverse values they hold for the community, complicated by the underwater environment in which they are located. No comprehensive national picture of the state of wrecks in Australian waters and associated relics was possible from data available at the end of 1994.

Source: Kenderdine, in press.

Table 9.14 Summary of cultural collections in major State- and Commonwealth-funded museums in 1991

Type of objects	Numbers of items (approx.)	Proportion of items well documented	Notes
Major government funded institutions (information from 18 museums)			
Science and technology	*c.* 67 000	Variable (0–70%)	
Anthropology/ archaeology[1] (total)	4 055 000	Mostly poor (0–70%) Many archaeological site collections await processing	Includes *c.* three million unaccessioned archaeological specimens which are bulk site collections
— anthropological[2]	*c.* 524 000		
— ethnographic[3]	*c.* 251 000		
History	*c.* 1 391 000	Variable, often poor (0-100%)	Only *c.* 200 000 are not part of the Australian War Memorial or philatelic and numismatic collections
Total for major museums	*c.* 5 513 000		
Local history museums (information from *c.* 1800 museums)			
Predominantly history, but also includes technology	*c.* 1 800 000	Unknown	

Notes:
1. Archaeological collections relate predominantly to items from Australia
2. Anthropological collections include non-Australian items
3. Ethnographic collections relate just to Australia

Source: Anderson, 1991; tables I and II.

significance of places may be related to their contents. Objects thus indirectly receive a measure of protection through the listing of places in the Register of the National Estate. To date, heritage collections have generally received little attention in the Register. Of the 91 museums listed by December 1994, none had their collections considered a part of their heritage significance (Australian Heritage Commission, in press). The Register recognised the heritage value of living collections of plants and/or animals for about half of the listed botanic gardens and zoos, but did not include any of the museums with collections of major national significance (Anderson, 1991).

Two separate Commonwealth Acts — the *Historic Shipwrecks Act 1976* and the *Protection of Moveable Cultural Heritage Act 1986* — now cover objects of cultural significance. The latter is designed to control export of significant cultural objects no longer *in situ*. Under the Act, Aboriginal secret/sacred objects and scientific Type Specimens are prohibited from export. Separate legislation covers the export of living flora and fauna (see Chapter 4).

The *Historic Shipwrecks Act* covers both shipwrecks and their contents (see the box opposite). Five States (South Australia, New South Wales, Western Australia, Victoria and Queensland) and the Northern Territory have complementary legislation; four of these Acts were passed in 1981 or later (Kenderdine, in press). They are all designed to prevent the removal of objects from wreck sites. The legislative provisions and their implementation appear to provide shipwreck material with a level of protection superior to that for any other type of non-indigenous cultural object *in situ*.

Physical condition and conservation practice

In 1991, about nine per cent of all herbarium specimens still needed to be incorporated into the collections (Richardson and McKenzie, 1992). Computerised databases contained about 21 per cent of vascular plants and seven per cent of non-vascular plants, and covered 19 per cent of all collections. Museums had not yet processed about 33 per cent of all non-fossil fauna specimens. Computer databases included eight per cent of the total modern (non-fossil) fauna collections, which comprised 88 per cent of modern vertebrate collections, but less than 5 per cent of invertebrates. No data were available on the physical condition of the natural history collections, but the reduced level of curatorial and support services for both herbaria and museums in 1991 (see page 9-15) suggests that their collections may be at risk of deteriorating.

The level of documentation on cultural collections in museums varied in 1991 but was less than 10 per cent for a number of institutions (Anderson 1991, in press). A high proportion of specimens of all fields awaited accession. Only a small proportion of artefact collections in some major museums had even been assessed by conservators, and less than 10 per cent had ever been treated. It is highly probable that the physical condition of many artefacts is continuing to deteriorate, despite the conservation crisis identified in the 1975 Pigott and 1987 CRASTE reports. The condition of collections in local museums is likely to be poor as these collections are usually displayed for too long, often in unsuitable physical environments, and the museums generally lack access to curatorial and conservator expertise (Anderson, in press).

Tourism and the environment

Australia's natural and cultural heritage underpins tourism — one of our fastest-growing and most economically significant industries. In 1994–95 international tourism to Australia generated export earnings of $12 billion, accounting for 12 per cent of our total export earnings. More than 3.3 million visitors came here in 1994, three times the number of a decade ago. If growth projections are realised, the tourism industry is expected to generate between $15 and $21 billion (at 1992 prices) annually in export earnings by the year 2000.

Heritage buildings, such as the store at Tilba Tilba, New South Wales, are often used to promote tourism. Their historic context in the landscape may be diminished or lost through the use of unsympathetic signs.

Pressure

Growing interest in and promotion of cultural and natural heritage tourism has increased the number of tourists, particularly to major attractions such as World Heritage areas (see page 9-16).

Tourism exerts both positive and negative pressures (see Table 9.15). It may create an increased awareness and appreciation of Australia's natural and cultural heritage and hence a greater desire for its protection. However, tourism can have negative impacts on both the physical and the non-physical heritage. Physical pressures range from the impact on the landscape of roads, signs and other facilities to wilful acts of vandalism. For community groups, such pressures can also be an assault on the cultural significance of sites. Intrusion on privacy, pressures to conform to stereotypes and payment for 'performances' endanger the cultural integrity of indigenous and non-indigenous communities and reduce the cultural significance of sites. Demands for 'authenticity' that do not recognise the dynamic and evolving nature of indigenous culture and its present relevance turn that culture into a commodity.

Tourism entices more people to become involved because it generates income for a range of ancillary services, such as accommodation and travel. The economic incentive exerts a pressure for natural resource managers to 'open up' their protected areas for tourism, for increased promotion of cultural heritage sites and for provision of tourist services.

Expenditure for tourism contributes to the management costs of sites, creating a further economic pressure in the form of the user-pays principle. However, the percentage of management budget derived from user fees varies significantly. For example, in 1991–92, 64 per cent of the management budget for Uluṟu was derived from user fees, nine per cent for Kakadu National Park and less than five per cent for the Great Barrier Reef, Wet Tropics and Tasmanian Wilderness World Heritage areas (Driml and Common, 1995). In 1995, rates of cost recovery were estimated to be 57 per cent at Uluṟu–Kata Tjuṯa, 30 per cent at Kakadu, 31 per cent for the Great Barrier Reef and 10 per cent for the Tasmanian Wilderness Area (DEST unpublished data, 1995). Figures available from the National Trust of South Australia indicate that, in 1992, entry fees generated less than 20 per cent of the management and maintenance budgets for their properties.

Table 9.15 Pressures created by tourism on natural and cultural heritage

Negative	Positive
Pressures on place	
• disruption of ecosystems	• revegetation programs
• pollution	• protection of wildlife
• waste disposal problems	• development and implementation of management plans based on zoning
• graffiti and vandalism	
• collection of souvenirs	
Pressures on values and meanings	
• overcrowding/over-pricing of host community facilities	• sharing and increased understanding of other cultures
• invasion of privacy	• renewed cultural activity
• privatisation of public space	• stimulus to art and craft activities
• loss of access to traditional land	• promotion of the conservation ethic
• debasement, commodification and exploitation of culture	• re-invigoration of communities with a knowledge of traditional skills and values
• rapid changes in traditional lifestyle	

Source: Hyde, in press.

State

Apart from economic data on tourism expenditure and gross visitor rates, there are virtually no data to measure either positive or negative impacts of increasing tourism on either natural or cultural heritage sites in Australia (Hyde, in press). Impacts at major tourist destinations are usually managed through plans of management that describe the biological and cultural values of areas and detail strategies for ensuring their protection. Zoning systems are often used to regulate levels of use. While various codes of practice have been developed for tour operators and tourists, Australia does not have a national system of accreditation or regulation. Although the concept of carrying capacity is being debated within the industry, it has rarely been applied to establish any limits to visitor numbers, even in terms of measuring environmental degradation. No one has carried out comprehensive surveys to measure the extent to which tourism may generate either positive or negative attitudes among the host population, in terms of access to and affordability of resources and facilities.

Response

The growing demand for tourism based on Australia's natural and cultural heritage has resulted in a rapid increase in the number of ecotourist operators and accommodation establishments. It appears that few of these businesses contribute to the maintenance of the natural environments they utilise (Hyde, in press). The industry, in partnership with the Australian Conservation Foundation and the Commonwealth Department of Tourism, is addressing the need for a national accreditation system for ecotourism operators.

The Commonwealth and all State and Territory governments developed ecotourism strategies or draft discussion papers in 1994, and most are preparing cultural tourism strategies. The Commonwealth committed $10 million over four years to implement its 1994 National Ecotourism Strategy. Since being established in 1991, its Department of Tourism has initiated programs for Forest Ecotourism, Sites of National Tourism Significance and Regional Tourism Development, and has developed a National Rural Tourism Strategy.

In 1994 the Aboriginal and Torres Strait Islander Commission released draft tourism industry and cultural industry strategies to address the specific needs of indigenous communities. These highlight the urgency of giving indigenous people an effective role in decisions about training, delivery of products and services, determination of tourist markets and marketing material and the control of cultural material and its presentation.

Indigenous communities have shown an increasing desire to control their own tourism enterprises. Tourists inspect Aboriginal rock art with the traditional owners, at Udnirr Ingita, Northern Territory (above).

Prognosis

While cultural tourism can have many positive benefits, these will only occur where communities can control its nature. Indigenous communities, in particular, have shown an increasing desire to control their own tourism enterprises as a way of maintaining cultural integrity and ensuring that they share in the economic, social and cultural benefits of the industry. Given the projected exponential growth of tourism in Australia, more sophisticated quantitative and qualitative data on its effect on the natural and cultural environment are needed, in terms both of their capacity to provide sustainable recreational and economic value and to protect their intrinsic values.

Response

Australia's heritage has received increased recognition over the last decade. The 1988 Bicentennial in particular provided a focus for domestic and international attention on the natural and cultural environments, highlighting awareness of cultural identity and inadvertently raising the profile of Aboriginal culture and history.
The importance of protecting our heritage assets — places, objects and associated meanings — has also received more widespread acknowledgement. This has been due to factors such as government and industry recognising the economic importance of heritage through tourism; better understanding by society of the central role of Aboriginal and Torres Strait Islander heritage as part of global moves towards social justice for indigenous peoples; and communities responding to changes in their physical surroundings that affect their sense of place.

Over the last 10 years, schools have included more subjects with an emphasis on heritage. In 1989–90 (the most recent national data available), their environmental education curricula offered a wide range of relevant topics which covered Aboriginal and Torres Strait Islander studies, Australian

Over the last decade, many schools have included subjects with an emphasis on heritage, particularly at primary level.

studies, beliefs, the environment, Australian history and multi-cultural matters (Australian Association for Environmental Education, in press). However, these studies were not generally part of the basic curriculum in schools and, while they were well represented at primary school level, few Year 12 courses encompassed issues relevant to Australia's heritage. The number and variety of specialised tertiary courses associated with cultural heritage has also increased significantly, with 65 courses on offer in 1994 (see Table 9.16). The 1987 CRASTE report highlighted a need for cultural relevance in all areas of study and professional training.

Many sectors of the Australian community have responded to the state of Australia's heritage. However, places have generally received more attention than objects, reflecting their links with the quality of our environment. Objects are rarely perceived as being part of 'the environment', probably because of their less obvious link with the quality of our surroundings.

Governments have initiated a range of responses to protect places, but only indigenous objects appear to have received the same level of attention. Community groups have continued to promote Australia's natural and cultural heritage and its conservation through lectures, walks, publications and other activities. Some corporations and industry groups have also made significant contributions to improve the state of our heritage.

The following pages concentrate on those responses for which qualitative information relevant to natural and cultural heritage across Australia was available. Inappropriate responses also affect the state of heritage and act as pressures on top of those described already.

Table 9.16 Numbers of award courses on offer in 1994 related specifically to Australian cultural heritage, as outlined in institution handbooks and directories

Type of course	TAFE	University undergraduate	University postgraduate	Total
Aboriginal studies (general)[1]	1	15	16[2]	32
Australian studies	-	29	-	29
Applied history	-	2	2	4
Total	1	46	18	65

Notes:
1. Includes language/linguistic studies
2. Includes two courses on Aboriginal Affairs Administration

Source: Australian Association of Environmental Education, in press.

Responses — heritage places

Since the 1981 review of the state of our heritage places (Yencken, 1985), Australia has made substantial progress in the standards of documentation, and methods of evaluating and implementing systematic heritage surveys. More professionals are involved in heritage identification, evaluation and conservation. For example, in 1994, there were 203 heritage professionals in Australia ICOMOS (Marshall and Pearson, in press). A major advance in consistent and creditable evaluation occurred in 1986, when the Australian Heritage Commission adopted specific assessment criteria. These criteria (see the box on page 9-18) were included in the 1990 amendments to the Australian Heritage Commission Act and provided the basis for subsequent State and Territory heritage legislation.

Non-government responses

Community groups have frequently responded to perceived threats to Australia's natural heritage and places of cultural significance with organised protests and demonstrations. Issues such as World Heritage listing (Reid, 1995), mining in national parks and timber harvesting for woodchips in national estate areas (Toyne, 1994) have all prompted actions that epitomise disputes about resource use in heritage areas. Parties from opposite perspectives promote very different views in such campaigns and aim for very different outcomes.

People in the community express their concern for the protection of heritage places by joining interest groups, such as national heritage or conservation organisations, residents' action groups and local history societies. These groups also provide a focus for community involvement in other activities, including oral history projects, legislative and planning processes, ethical investment schemes and educational activities to help maintain and enrich community heritage (Blair, 1994).

The National Trusts have continued to play a significant role in promoting heritage and its protection, and government funding to them has increased in recognition of their role. Heritage Week, which was initiated by the Trusts in 1980, is now a major event in all States and Territories. Donations to the Trusts are also increasing again (see Fig. 9.11), although the level is still far short of that reached during the Bicentennial.

Other non-government, non-profit organisations continue to receive community support through membership and donations.

Protests by environmentalists (left) and the logging community (above) about harvesting old growth forests for woodchip export typify conflicting community views over resource use.

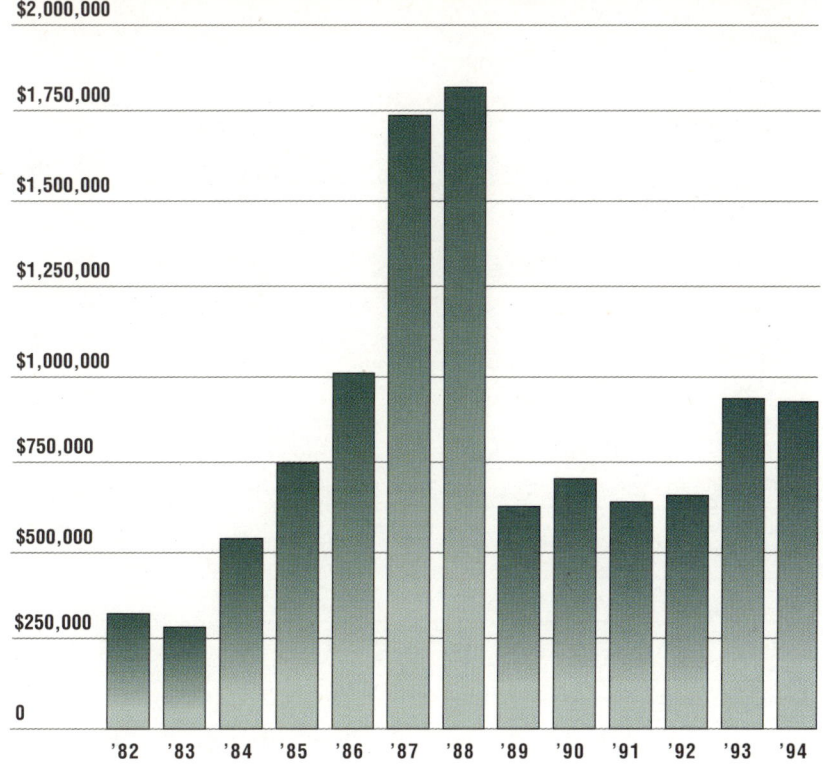

Figure 9.11 Community and industry donations to the National Trust from 1982 to 1994 (a total of $11,008,879 was donated over this period)

Source: National Trust Annual Reports.

For example, the Victorian Conservation Trust has used community donations to acquire 100 properties since 1972. The State government has added more than 40 of these to conservation reserves or other protected lands. Over the last seven years the Trust has used a revolving fund to buy properties then resell them with a protective conservation covenant. The Australian Bush Heritage Fund was set up in 1990 to purchase private land of outstanding conservation value and by mid 1995 it had purchased four blocks.

A number of corporations have policies to manage, conserve and promote their own heritage assets. Others provide sponsorship funding to a range of heritage projects such as the restoration of heritage buildings, the upgrade of facilities at botanic gardens and zoos and research programs. The 1988 Bicentennial provided an important focus for industry support of heritage.

Professional bodies such as the Australian National Committee for the International Council for

Table 9.17 Summary of main State and Territory heritage acts for conserving indigenous places

	Qld	NSW	ACT	Vic	Tas	NT	SA	WA
Date Act(s) passed	1988	1974 (amended 1984, 1992)	1991; 1991	1972 (amended 1973, 1980 1984) Commonwealth Act 1987	1975	1989;1991	1988	1972 (amended 1980)
Consultation with indigenous communities?	To some extent	To some extent	Yes	To some extent (1972 Act) Yes (1987 Act)	To some extent	Yes (1989 Act) To some extent (1991 Act)	Yes	Yes
Advisory Committee requires indigenous representatives?	No	Yes	Yes	Yes (1972 Act only)	No	Yes (1989 Act only)	Yes	Yes
Interests of indigenous people protected?	No	No	Potential	Yes (1987 Act only)	No	Yes (1989 Act only)	Yes	Yes
Provisions of Acts for places								
Main type of place covered? (Significant places include archaeological sites, religious sites, etc.)	Significant places	Significant places	Significant places	Archaeological sites (1972 Act) Significant places (1987 Act)	Archaeological sites	Sacred sites (1989 Act) Archaeological sites (1991 Act)	Significant places	Significant places
Blanket protection provided for archaeological places[1]?	Yes	Yes	Yes	Yes (1972 Act only)	Yes	Yes (both Acts)	Yes	Yes
Sites register required?	Yes	No	Yes	Yes	No	Yes	Yes	Yes
Able to declare protected areas?	Yes	Yes	Potential	Yes (1972 Act)	Yes	Yes (1989 Act)	Yes	Yes
Destruction of places an offence?	Yes	Yes	Yes	Yes	Yes	Yes	Yes	Yes
Penalties (High: $10 000 or more; Low: less than $10 000)	High	High	High	Low (1972 Act) High (1987 Act)	Low	High	High	Low
Provisions of Acts for objects								
Objects covered?	Yes	Yes	Yes	Yes	Yes	Yes (1991 Act)	Yes	Yes
Objects register required?	Yes	No	Yes	Yes	No	No	Yes	Yes
Destruction of objects an offence?	Yes	Yes	Yes	Yes	Yes	Yes (1991 Act)	Potential	Potential
Provisions of Acts for human remains								
Human remains covered?	Yes	Yes	Yes	Yes	Yes	Yes (1991 Act)	Yes	Yes

Note:
1. Other types of significant places must first be identified to receive legislative protection.
Source: Ward, in press.

Monuments and Sites (Australia ICOMOS) have continued to play an important role in raising the standards of heritage identification and protection. In order to increase the accessibility of the Burra Charter adopted by ICOMOS in 1979, an illustrated version was published in 1992 and distributed to all local governments. Australia ICOMOS has continued to influence concepts of cultural significance, assessment methodologies and conservation principles and practices across Australia as well as internationally.

Government heritage legislation and policies

Significant progress has been made in the legislative protection of heritage, with many Acts now addressing major deficiencies identified in 1981 (Yencken, 1985).

Governments have passed seven indigenous heritage Acts since 1987 (see Table 9.17). In general, these Acts have increased the focus on protection of places and objects significant to contemporary indigenous communities, although

Table 9.18 Summary of State and Territory heritage acts for conserving historic places

	Qld	NSW	ACT	Vic	Tas	NT	SA	WA
Date Heritage Act passed	1992	1977	1991	1981 (under review)	(Provisional Bill 1994)	1991	1993	1990 (under review)
Date of any previous Act	1990	-	-	1974	-	-	1978	-
Appointment of staff included under Act?	No	No	No	Yes	(No)	No	No	Yes
Identification & evaluation provisions of Acts								
Assessment criteria included?	Yes	No[1]	Yes	No[1]	(Yes)	Yes	Yes	Yes
Assessment criteria compatible with National Estate criteria	Yes	Yes[1]	Yes	Scope more limited	(Yes)	Yes	Yes	Yes
Heritage Register required?	Yes	Yes[2]	Yes	Yes	(Yes)	Yes	Yes	Yes
Composition of Heritage Council	Interest groups and experts	Largely experts	Largely experts	Largely experts	(Experts and interest groups)	Largely experts	Experts	Largely experts
Who makes final listing decisions?	Heritage Council	Minister	Legislative Assembly	Governor in Council	(Council)	Minister	Heritage Council, but Minister may direct removals	Minister (private places); Heritage Council (crown places)
Conservation provisions of Acts								
Conservation management plan required?	No	No	No	No	(No)	Yes	No	No
Provision for conservation order or equivalent?	Yes	Yes	No	Yes	(Yes)	Yes	Yes	Yes
Provision for heritage agreements?	Yes	No	Yes (only for Aboriginal places)	No	(Yes)	Yes	Yes	Yes
Permit/approval required for developments affecting any place on the register?	Yes	Yes	Yes (for a 'controlled activity')	Yes	(Yes)	Yes	Yes (under the *Development Act 1993*)	No[3]
Financial penalties	up to $1 million	up to $20 000	up to $20 000	up to $150 000	(Yes)	up to $200 000	up to $60 000	up to $10 000 plus daily penalties
Powers of local government under heritage or planning Acts								
Powers delegated for identification and/or conservation?	Yes, for development application approvals	Yes	Not applicable	Yes	Yes for, development approvals only	No	Yes (identification and conservation)	Yes
Maintain some type of register for historic places?	Yes	Yes	Not applicable	Yes	Provision, but no obligation	No	Yes	Yes

Notes:
1. Non-statutory criteria have been implemented in NSW and Victoria
2. NSW list is restricted to places with Permanent Conservation Orders
3. Advice of the Heritage Council must be sought .

Source: data from James, 1993.

The historic Moore's building in Fremantle, Western Australia, was restored by the Fremantle City Council and serves as a model for the adaptive re-use of heritage structures.

in Tasmania the current Act is still based on archaeological value and hence does not recognise historic or contemporary significance. The more recent Acts provide stronger avenues for community consultation and increased involvement of indigenous people in developing and implementing protection measures. Only the South Australian one gives indigenous people control in its implementation. Five States now have major penalties for damaging indigenous places. However, some States such as South Australia (Ward, in press) have not provided adequate resources to implement the legislation effectively.

Three States and both Territories have passed heritage legislation to identify and protect historic places since 1990 (see Table 9.18). Tasmania introduced a draft Bill in 1994. The Acts in the Australian Capital Territory and Northern Territory also cover natural and indigenous places. Each of these Acts requires the compilation of State heritage registers (although ministers rather than appointed councils make the final listing decisions in several States). They all used the 1990 national estate criteria as their basis, and thus compatible concepts of significance now apply across Australia for historic places; all include penalties for damaging historic places; and all except the Australian Capital Territory provide for conservation orders (see Table 9.18). All except Western Australia require approvals for activities affecting listed places.

The passing of the new Acts for historic places has generated some confusion in the community about listings and their implications. However, their common basis has greatly increased the potential

for coordination between governments. In May 1992, heritage ministers across Australia agreed to a program of national coordination of documentation, assessment, listing and provision of conservation advice for historic places. This has been advanced through meetings of officials from the Commonwealth and all States and Territories and of the heads of heritage authorities. Pilot programs are in place for joint heritage assessments and listings and cooperative information-gathering. The various agencies have agreed on standard information requirements and begun work on linking Commonwealth and State heritage databases.

Since 1989, the Commonwealth government has increased its focus on environmental issues and State/Commonwealth relationships. The 1992 InterGovernmental Agreement on the Environment (see Chapter 2) contained specific schedules to improve intergovernmental arrangements about World Heritage and the National Estate.

The Commonwealth government statements: 'One Nation' (1992), Statement on the Environment (1989, 1992) and National Forest Policy Statement (1992) all contained specific initiatives directed at Australia's natural heritage. Some have also helped to identify and protect cultural places. Relevant initiatives have included: the development of a national protected area system, comprehensive regional assessments in forests, identification of significant marine areas through Ocean Rescue 2000, identification and management of wild rivers, completion of national coverage for the National Wilderness Inventory and conservation of sites of national tourism significance. It is too early to assess the effectiveness of these programs nationally in terms of heritage conservation.

The 'One Nation' (1992), 'Distinctly Australian' (1993) and 'Creative Nation' (1994) statements contained specific initiatives to protect Australia's cultural heritage. The 'Creative Nation' statement listed preserving Australia's heritage as one of five principal roles of the Commonwealth Government in cultural development. The policy announced a Committee of Review to examine the management of Commonwealth-owned heritage properties and to provide for better conservation. The work of this Committee will be crucial to developing a model of best practice for the conservation of government-owned heritage across Australia.

Most State and Territory governments have been promoting heritage conservation at the local government level. They have supported the placement of heritage advisers and increased the extent to which local governments must include heritage considerations in their planning (Marshall and Pearson, in press). The involvement of the latter in heritage identification and conservation appears to have grown in some States. Local governments in some areas have also promoted good conservation and adaptive re-use of their heritage resources and used them to promote local and regional tourism.

Funding for conservation

The Commonwealth government introduced the National Estate Grants Program in 1973–74 to provide assistance to the States. The program has remained a significant funding source to help identify and conserve natural and cultural heritage places across Australia (see Figure 9.12). However, its effectiveness is limited as funding has not kept pace with the 61 per cent increase in the number of places eligible for assistance since 1981. All the States and Territories with heritage legislation for historic places have introduced programs to assist their conservation, although the annual level of funding is often low (see Table 9.9).

The Commonwealth government initiated a Heritage Properties Restoration Program from 'One Nation' and the Tax Incentive for Heritage Conservation scheme from 'Distinctly Australian' in response to the high costs of maintaining heritage buildings. The move followed lobbying over a number of years by many groups (Yencken, 1985). The government passed legislation to allow implementation of the tax incentive in 1994. In its first round of operation, 53 of the 91 applications received were approved within the $1.9 million cap on tax rebates. The one-off nature of programs such as the Heritage Properties Restoration Program reduces their effectiveness in conserving Australia's built heritage. It is too early to assess whether the cap on tax rebates will limit the effectiveness of the Tax Incentive program. The current rebate level will remain in place until 1996–97, when it will be reviewed.

The Australian Institute of Aboriginal and Torres Strait Islander Studies has provided special-purpose grants for rock art conservation annually since 1987. The government also provided funding in 'Creative Nation' to establish a Cultural Heritage Protection Program that recognises the special importance of Aboriginal rock art and increases the

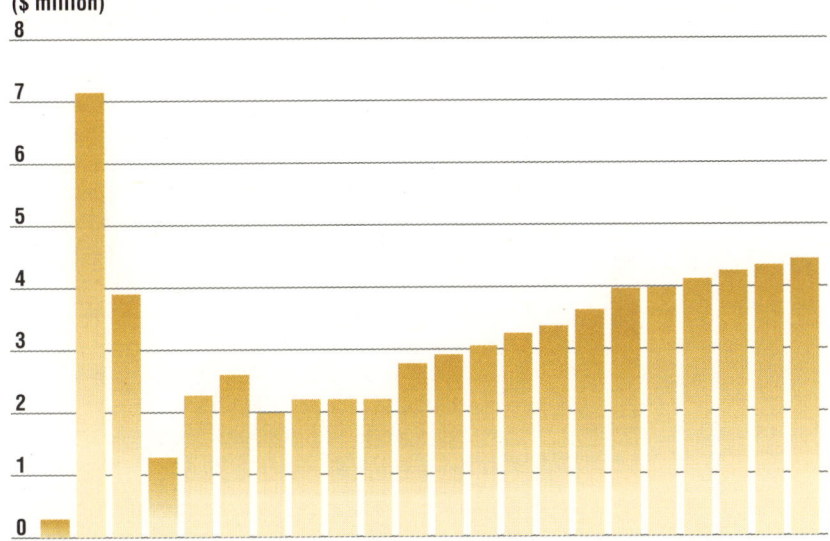

Figure 9.12 Annual funding to the States and Territories under the National Estate Grants Program

Note: Increase in the level of funding from 1980 largely reflects adjustments for inflation.
Source: Unpublished data — Australian Heritage Commission, June 1995.

involvement of indigenous people in management decisions. Increased funding was also provided for rock art research, conservation and management. A pilot workshop for rock art management was held in 1994 to test resource materials developed under the program.

Many heritage buildings need considerable resources for their restoration. The original Cordillo Downs woolshed, South Australia, (left) was restored (below) with financial assistance from the Heritage Properties Restoration Program.

Indigenous heritage

Philosophies and attitudes to indigenous heritage values and the involvement of indigenous people in cultural heritage management, have changed markedly since 1984. Land rights Acts passed by Commonwealth and State governments have facilitated significant change in heritage policy and administration through bodies such as local and regional land councils. Legislation passed in the last eight years (see page 9-36) has increased avenues for indigenous communities to participate in heritage management. Bodies such as the Council for Aboriginal Reconciliation and the Aboriginal and Torres Strait Islander Commission have produced major reports, with recommendations directly relevant to indigenous cultural heritage identification and management. Some local governments identify Aboriginal sites as part of their routine work and promote Aboriginal heritage (Galla, 1993).

The number of technical and professional indigenous staff employed permanently by State and Territory agencies concerned with indigenous heritage has increased, except in one department where the number dropped (Ward, in press). In 1985, each of the 11 major government departments employed an average of 3.8 indigenous people in the relevant sections, representing 25 per cent of those sections' permanent staff. Comparable figures in 1994 were 5.3 indigenous people, representing 30 per cent of the permanent staff. Given the special needs of indigenous heritage, this proportion is still relatively low. Further improvement will probably depend, at least in part, on increased training of indigenous people (Galla, 1993).

More indigenous people have been appointed to policy and decision-making bodies in the area of indigenous heritage since 1985, and many new advisory bodies have been set up (Ward, in press) reflecting changing public attitudes and political will. However, the level of recognition of Aboriginal concerns about custodianship of their heritage and the formal mechanisms provided for achieving this varies between States. Sometimes Aboriginal people may also be reluctant to be involved in European-style advisory bodies.

Over the last decade, government departments and tertiary institutions appear to have provided more training for indigenous people. The number of heritage officers employed and sponsored in their training — often at a tertiary level — by indigenous communities has also increased. Programs such as the Contract Employment Program for Aboriginals in Natural and Cultural Resource Management, which annually receives around $3.8 million and provides opportunities for training through employment for about 1300 indigenous people.

More and more indigenous communities have been applying for funds under the National Estate Grants Program to document sites significant to them. Indigenous communities comprised 51 per cent of all applications for projects applying to Aboriginal and Torres Strait Islander places in 1994–95 (Australian Heritage Commission, in press).

Attitudes to indigenous languages and cultures among non-indigenous people have changed over the last decade. More people recognise the uniqueness of Aboriginal culture and accept it as part of Australia's cultural heritage. A variety of community responses reflect these changes (Henderson and Nash, in press). They include the growth in Aboriginal cultural tourism and the inclusion of more information about the indigenous history and significance of places in interpretive material in parks and in tourist literature. People have become more interested in learning about Aboriginal language and culture, as seen by the growing number of general and educational publications about these subjects. The community is increasingly recognising the wish of indigenous people to be known by their local title — Anangu, Yolngu, Koorie, Nyoongar, Murri and Nunga — rather than 'Aboriginal'. The appropriateness of using their own names for places of special significance to indigenous people is also gaining support, although some name changes have been controversial and have been overturned by authorities.

Retaining indigenous cultural knowledge

The 1994 'Creative Nation' statement recognised the importance of preserving and protecting intangible cultural heritage, particularly that of indigenous Australians. It expressed full support for 'indigenous people in their efforts to retain their material, intellectual and spiritual heritage' and included initiatives to increase their access to material such as oral histories, genealogical data and languages.

The loss of traditional indigenous languages (see page 9-23) has prompted a range of responses. From 1987–88 to 1994–95 Commonwealth funds for indigenous language-related research increased six-fold. The number of Regional Aboriginal Language Centres has increased, and a Federation of Aboriginal and Torres Strait Islander Languages was established in 1992. The Australian Bicentennial Authority provided one-off funding for the development and publication of documents such as a national dictionary of Aboriginal languages — 'Macquarie Aboriginal Words: a Dictionary of Words from Australian Aboriginal and Torres Strait Islander Languages' — and the AIATSIS 'Encyclopaedia of Aboriginal Australia'.

In the 1993–95 triennium, the Commonwealth government allocated specific funding for indigenous languages in the education system. It has been developing a national approach to such courses at the senior secondary level since 1992, and courses are expected to be offered in six places in 1995. The number of bilingual and other indigenous language education programs in schools has increased, as has the number of general market publications with a significant proportion of indigenous language text.

Professional groups have played an important role in identifying Australia's heritage. The Geological Society of Australia identified many geological sites such as Piersons Point foreshore geological monument, Blackmans Bay, Tasmania (below). The Institution of Engineers carried out a national survey of bridges, including unusual ones such as the Adelaide River railway bridge, Northern Territory (left).

After many decades of neglect, the crisis for indigenous languages is now more widely recognised. It is, however, too early to assess how effective the response of recent years will be for maintaining these languages in the long term.

Imbalances in heritage registers

A 1991–92 review of historic places in the Register of the National Estate (Domicelj, 1992) identified a wide range of themes not represented, involving many different types of places. Many of these were subsequently given priority for identification in the National Estate Grants Program. This has assisted in some areas (for example, war memorials, gardens and some types of industrial heritage), but much work remains to redress imbalances. In 1993 national heritage officials began a study to develop a framework of principal Australian historic themes with the aim of providing added focus for future systematic survey work. The framework has been tested in part of the Murray–Darling Basin.

The emphasis on systematic heritage studies has increased in the National Estate Grants Program. In 1994–95, 60 per cent of all applications for heritage identification were for this type of work — 46 per cent of natural projects were for regional studies and 42 per cent of historic projects were for thematic studies (Australian Heritage Commission, in press).

To help redress the under-representation of migrant places in heritage registers, a guide book has been developed (Armstrong, 1994) to assist communities from migrant backgrounds nominate places significant to them. Considerable work has also been carried out to develop appropriate methods for assessing social value (Blair, 1994), with recent regional heritage studies including strong community involvement (see the box on page 9-13).

Some industry and related professional groups have also taken an active part in heritage identification over the last decade. For example, the Institution of Engineers has carried out studies of significant bridges, the Victorian Chamber of Mines and Small Miners and Prospectors Association have been involved in studies of historic mining sites in Victoria and the Geological Society of Australia has continued a nation-wide program of identifying sites of geological significance.

Responses — heritage objects

Despite the fact that objects have a lower profile than places as part of natural and cultural heritage, there have been significant positive responses to assist their protection, largely by governments. Professional bodies such as the former Museums Association of Australia and the Australian Institute for the Conservation of Cultural Material have played important roles in raising standards and providing services. Progress has been made in some areas since the 1975 Pigott and 1987 CRASTE reports, although many pressures remain.

Commonwealth government initiatives

Following recommendations in the Pigott report, the Commonwealth passed the *Historic Shipwrecks Act* in 1976 and the *Protection of Moveable Cultural Heritage Act* in 1986. The latter guards against export of important items of our portable heritage to ensure they remain part of Australia's natural or cultural environments. These include Aboriginal secret/sacred objects, scientific Type specimens, fossils and geological specimens.

The Cultural Ministers' Council and the Council of Australian Museum Associations established a Heritage Collections Working Group in 1990, as part of the National Cultural Framework, in part to define the nature and extent of Australia's heritage collections. The Group adopted the notion of the Distributed National Collection (see the box on page 9-14), and made recommendations towards remedying its existing biases. It was replaced by the Heritage Collections Committee in

Indigenous Australians: protecting their own cultural heritage

The archaeological record demonstrates the great antiquity and continuity of indigenous societies in Australia. Just as they adapted to often dramatic changes in the natural environment over thousands of years, more recently they have had to adjust to massive changes in the social and physical environments brought about since European colonisation. While indigenous people today may not live as their ancestors did, the continuation of many cultural traditions highlight a culture and heritage extending back over vast periods of time.

Indigenous communities feel growing concern to preserve this heritage and the knowledge and values that give it meaning. They are also concerned that they have largely been denied their responsibility of protecting and managing indigenous heritage places and objects. Addressing this need requires attention not only to the responses of governmental agencies, but also to the impediments facing indigenous communities themselves as they attempt to implement their concerns and gain a voice in formal heritage management.

Pressure

Many indigenous people remain isolated from their ancestral lands, within which country, places and objects of significance to them are located. The dominant cultures in wider society and the homogenised culture represented by the international media continue to impinge on indigenous languages and cultural understandings.

The decline of languages and the diminishing numbers of people actively using many of the surviving languages (see Figures 9.9 and 9.10) jeopardise the transmission of knowledge and cultural practices related to indigenous places and objects. The same occurs if indigenous people do not have ready access to the results of archaeological and anthropological research and other relevant studies (Henderson and Nash, in press).

Museums in Australia and overseas contain vast collections of indigenous artefacts and human remains, but it is difficult, if not impossible, for most indigenous people to reach them. While many collections were made with the authority and guidance of local indigenous people, many others were acquired and retained without authority from traditional custodians. Numbers of items were collected, stored and/or displayed in ways inconsistent with indigenous beliefs.

State and Commonwealth agencies responsible for indigenous affairs, museums, heritage agencies and other departments administer about 30 separate Acts specifically relating to indigenous places and objects (Ward, in press). These Acts are often not implemented effectively because of inadequate resources or the use of procedures that are not necessarily compatible with indigenous cultural values and practices. This is exacerbated by a failure to involve indigenous people effectively in administration and management. For the most part, governments still need to develop legislation to protect the traditional knowledge, intellectual property and the integrity of indigenous cultural information.

Many areas in Australia protected for their natural heritage values contain places significant to indigenous people but the management and conservation requirements of the indigenous cultural heritage are generally neglected or given a low priority. Indigenous people have a majority role in management decisions in a few protected areas, but no involvement in many other places.

Because they have little control over matters concerning them, including their own heritage, indigenous people sometimes choose not to have places, objects, traditions and cultural property controlled, or listed in heritage registers, by Commonwealth and State authorities.

Alf Neal, an elder of the Kuku Djungan tribe working with a botanist to extract a pollen core from a lake at Ngarrabullgan in North Queensland in order to investigate changes in the environment and in patterns of Aboriginal land use over time.

State

The cultural context in which places and objects significant to indigenous people are managed may sometimes be a more significant measure of state than their physical condition. Many of the objects removed from their original physical location are stored, conserved or presented in inappropriate cultural contexts, while the management regimes of many protected natural areas do not adequately address the need for culturally appropriate management of significant places.

Response

Over the last decade indigenous people have increased their demands to be more equitably involved in formulating heritage policy and identifying, protecting, managing and repatriating their own heritage. Governments have responded to these demands in a number of significant ways.

The passage of the Commonwealth *Native Title Act 1993* recognising indigenous rights and interests in land was a landmark. It signalled the end of the *terra nullius* concept denying the land tenure of indigenous people in Australia.

Establishment of the Aboriginal and Torres Strait Islander Commission in 1990 gave responsibility for both policy and administration in indigenous affairs, including heritage and cultural concerns, to a structure over which indigenous people exercise control.

The Commonwealth passed the *Aboriginal and Torres Strait Islander Heritage Act* in 1984 to address inadequacies in existing State and Territory legislation for the protection of indigenous sites and artefacts, particularly those of contemporary significance. Sixteen declarations have been made since the Act was passed, nine of them in 1993–94 — three for the protection of significant objects, and six for the protection of significant places.

A number of States have reviewed legislation affording protection of indigenous places over the last decade and passed new Acts that increase the involvement of indigenous people. Over this period, the number of policy-making and high-level advisory bodies related to indigenous heritage has also increased, with an increased proportion of indigenous people appointed to them (Ward, in press). The granting of land rights has allowed indigenous people to re-assert control over places and objects located in many areas.

Prognosis

Increasingly, people are realising and accepting the relationship of indigenous peoples with the physical landscape, and the importance of their intangible cultural heritage. Protection of indigenous culture and its heritage requires that places and objects be protected in a culturally appropriate manner and that traditional languages and cultural traditions be maintained with continuity in their use. The last decade has seen significant improvements in some of these areas. However, indigenous communities do not have sufficient control to protect this critical aspect of Australia's heritage in the long term.

1993–94 as an initiative under the 'Distinctly Australian' (1993) and 'Creative Nation' (1994) cultural statements. The committee was given seed funding to develop and implement collection, organisation, preservation and communication strategies for the Distributed National Collection. Although funding this Committee was seen as a major Commonwealth government role in cultural heritage, ongoing funds were not specifically allocated in the 1993 and 1994 statements.

These recent cultural statements included other initiatives relating to heritage objects, although total funding for material cultural heritage was comparatively low due to the predominant focus of the statements on the performing arts and associated cultural activities. 'Distinctly Australian' included a Bequests Program to encourage private owners to donate significant cultural items to museums. Like the Tax Rebates scheme for heritage buildings, tax concessions under the Bequests Program will be capped (at $2m per annum), which may inhibit its effectiveness. The government passed legislation to implement the scheme at the end of 1994 and was developing the necessary regulations in early 1995. Other initiatives announced in 'Creative Nation' included the development of information networks on collections, national conservation standards and assistance from national institutions to individuals and communities to preserve and present material of cultural significance in their original locations or regions. However, none of these received specific funding allocations in the statement.

Over the last decade the Commonwealth government has supported the return of significant cultural objects to indigenous control. The Aboriginal and Torres Strait Islander Commission implemented a national Heritage Program in 1993–94 for indigenous communities to establish keeping places. It has also provided funds for the return to Australia of human remains from overseas institutions — although State agencies have often experienced subsequent difficulties in returning them to relevant custodians. A major program for the return of indigenous cultural property from institutions in Australia and overseas was announced in 'Distinctly Australian', with the return of human remains a priority. The initiative included a consultative framework and pilot programs to facilitate a primary role for indigenous people in the protection, safe keeping and return of significant cultural property.

The return of cultural items is very complex. For example, in many cases it is

Many botanical gardens in Australia, such as the Kings Park garden in Perth, Western Australia, have programs to cultivate endangered species of native plants.

difficult to identify custodians, while objects considered 'ordinary' at the time of collection may have taken on greater cultural significance to contemporary indigenous Australians. Some projects result in more objects being deposited in museum custody for safe keeping. These, and related issues such as the impact that return programs may have on scientific research, are the subject of continuing debate.

Coordinating living collections

Collections of living plants and animals have not received the same level of support in government policies as material collections. Major zoos in Australia and New Zealand established the Australasian Species Management Program, which was incorporated in 1992, to promote regional cooperation in the management of species in captivity for conservation purposes. The program publishes an annual summary of species management and recommendations for planning collections. The Australian Network for Plant Conservation, established in 1991, includes a range of corporate bodies, organisations, individual members and international associates in its membership. Its aims include helping species recover in their natural habitats and cultivating endangered species as an 'insurance' against extinction in the wild. In 1993, the Network published a list of endangered Australian plants in cultivation.

In 1989, the heads of botanic gardens formed a council to help coordinate activities between their institutions. In 1994 it compiled the first list of all plants in cultivation in major Australian botanic gardens. This provides an aid for cooperation in developing collections and helps reduce pressures on wild populations by minimising the duplication of collecting activity.

The conservation activities of major Australian botanic gardens and zoos are based on recently adopted international conservation strategies, themselves based on the principles of the World Conservation Strategy. The strategies aim at achieving sustainable use of natural resources and

biodiversity, and include a focus on captive breeding populations. About 44 per cent of all botanic gardens/arboreta recorded in Australia in 1994 were engaged on some type of program relating to threatened native Australian plants (Boden, in press).

Reviews of the collections

Since 1991, Australian museums have undertaken major surveys of their collections for the Heritage Collections Working Group (Anderson, 1991, 1992, 1993), partly to determine their state (including size, characteristics and conservation conditions). This work provides the most up-to-date picture of historical, archaeological and anthropological collections in Australia. The survey of herbarium collections and museum faunal collections undertaken in 1991 by the Australian Biological Resources Study (Richardson and McKenzie, 1992) similarly provides the most recent overview of these collections.

In 1994 a University Museums Review Committee was established, covering collections held in Australia's universities, including natural history items. The review was due for completion in the first half of 1995. The Commonwealth is funding a national inventory of secret/sacred indigenous objects held in museum collections, which is being implemented with advice from Museums Australia and in liaison with all relevant Aboriginal Land Councils.

Cultural collections: access and conservation

Museums have changed markedly in the two decades since the Pigott report. For example, all State and Federal museums now employ conservators for their collections (Anderson, in press). However, few specific undergraduate university and TAFE courses are available to train graduates for this specialist work (see Table 9.10), and the number of conservators is still small compared to the need (Anderson, in press). Regional and community museums face even greater difficulties in obtaining advice on conservation of collections.

Providing more resources to establish interlinked, readily accessible databases has helped to unlock the combined wealth of information contained in natural history collections throughout the country. The State and Commonwealth governments are funding these initiatives. In an attempt to increase access to collections without risk to their preservation, major museums are now investigating computer interactive programs. A working group established by the Heritage Collections Committee made substantial progress in this area during 1994 (Anderson, in press).

Redressing the imbalance of cultural collections

Museums increasingly have recognised the imbalances in their collections, and are changing the direction of their collecting programs to make them more representative of the whole Australian community (Galla, 1993). From the early 1980s, State museums progressively established sections specifically to collect historical material, which often focused on documenting the less-privileged in society (Anderson, in press). Museums have also become increasingly aware of their social roles and are more open to community participation. Initiatives include cooperative collecting ventures resulting in collections of specific relevance to the cultural traditions and contemporary practice of a range of ethnic and interest groups. Western Australia has established a Multicultural Heritage Task Force to encourage individuals and groups to care for their own cultural heritage, in their homes, in cultural centres, or in the State museum if desired.

Neither of the Commonwealth cultural statements (see page 9-43) provided secure funding for the proposed National Museum of Australia, the establishment of which was recommended in the Pigott report. With its proposed integrated approach to the environment, history and human settlement, such a museum would greatly assist in redressing many of the known gaps in the Distributed National Collection, and provide a holistic view that integrates the social values and meanings of objects.

Return of indigenous cultural property by museums

In 1993, the Council of Australian Museums Associations produced a policy document recognising the need for museums to broaden their roles. It urged increasing Aboriginal participation in collecting, keeping and researching indigenous artefacts and human remains. The policy advocates support for the establishment of indigenous community museums and keeping places, and for the loan of cultural material from their collections to such centres. The major museums are concerned to discuss with relevant indigenous groups the return of material in their collections. Many now have special areas for the care of restricted (secret/sacred) material, employ Aboriginal staff and liaison officers and provide special training for Aboriginal staff members.

Context of collections

The desirability of preserving objects *in situ*, if at all feasible, is now generally accepted so that the original physical context is retained with all that it has to tell us. Many museums now conduct extensive research, including oral history, in the course of their collecting and exhibiting activities. These activities have been particularly significant where there has been no prospect of preserving objects on site; at least their cultural context has been carefully recorded. Some institutions have specific oral history programs. Much information obtained through oral history programs about non-indigenous objects and places is likely to be as an adjunct to other activities rather than by specific design (McCarthy *et al.*, in press).

Future and synthesis

'Which are the places that our children's children will thank us for conserving? Conversely, which places will we be condemned for ignoring?'

(Dovey, in Blair, 1994)

Community attitude surveys show the present generation realises an obligation to protect things for future generations (Purdie, in press). Places and objects, and the meanings associated with them, are an integral part of Australia's natural and cultural heritage. Previously, heritage objects were not perceived as part of the environment despite their intrinsic heritage value and close links with place. Their inclusion in this report reflects their importance in that respect. In a society endeavouring to achieve the sustainable use of its resources, knowledge about the state of all our heritage resources is essential to help guide decisions about their sustainable use and to help determine those places and objects that should be saved for our children's children.

This chapter has reviewed the state of Australia's natural and cultural heritage, and where possible identified trends over the last one to two decades. State has been assessed in terms of knowledge, legal provisions, physical condition, conservation and community involvement, using a range of possible indicators (see Table 9.19). Major pressures currently affecting heritage and a range of responses to its state since the mid 1970s and early 1980s are also outlined. Table 9.20 summarises major findings.

A superficial glance at this Aboriginal stone quarry at Lake Moondarra, Queensland, might suggest only a scatter of naturally broken rock. Closer examination reveals an irreplaceable technological record of quarrying and the controlled shaping of stone through a series of steps to form an important tool — the hatchet head.

What is the state of Australia's natural and cultural heritage?

Thousands of places across the country, including 11 World Heritage properties, and millions of objects located in public and private collections have been identified as part of our heritage. The size of heritage registers and collections has grown considerably over the last one to two decades and continues to expand.

Current World Heritage listings do not reflect the mixture of natural and cultural heritage values in the Australian landscape. The Register of the National Estate, similarly is not fully comprehensive nor fully representative of our heritage places; it contains major geographic and thematic gaps. This situation is more serious in State and Territory registers of historic places, the majority of which post-date 1989. Records of Aboriginal and Torres Strait Islander places maintained under State and Territory indigenous

heritage legislation also do not fully cover some types of places. No quantifiable national data exist to provide a broad assessment of the physical state of Australia's heritage places.

Heritage collections are widely dispersed, with a large proportion housed in local museums. The number of objects in these has never been quantified exactly, and the most recent information indicates that the collections are not representative. For example, specimens relating to Australia's natural environment far exceed those representing human occupation, while fewer artefacts relate to non-indigenous history than to indigenous heritage. No one knows in detail the physical condition of the majority of material collections, nor the state of those collections at greatest risk. However, in view of limited conservation resources, the physical condition of many is likely to be deteriorating.

Legislative protection of indigenous and historic places has improved over the last decade through enactment of new State and Territory heritage laws. Many of these have also increased the legislative protection of objects. A range of indigenous places receives blanket legislative protection in all States and Territories, although many historic places have no formal protection under these laws and hence are potentially at risk. Some natural heritage places have received increased protection through Commonwealth, State and Territory laws. Commonwealth legislation now also protects classes of heritage objects from export.

The state of conservation practice has improved through the adoption of more targeted evaluation and management principles, improved standards of assessment and documentation, increasing numbers of heritage professionals and the availability of more heritage assistance programs. While some people oppose heritage listings, many individuals and community groups actively work to assist the conservation of Australia's heritage. However, the community does not have sufficient access to conservation and curation services, training and resources to implement the conservation measures necessary to ensure the protection of the heritage in its care.

Data availability

Reporting on the state of Australia's heritage has been hampered by major gaps in knowledge, with only poor data available for many of the indicators (see Table 9.19). The state of knowledge about material collections is particularly poor — it is not possible to compare collections accurately between many institutions because of poor documentation and the absence of national registers. This makes it difficult to assess the physical condition of objects, establish priority conservation needs against available resources nationally or address gaps in the collections. No national programs are currently in place to monitor the physical condition of Australia's heritage places or objects.

Similar data deficiencies hampered reporting on pressures and responses (see Table 9.20). Generally,

---

Table 9.19 Indicators considered in this chapter relevant to the state of heritage nationally and the adequacy of relevant data

	Indicators	Adequacy of data for indicator[1]	
		Places	Objects
State of knowledge	• number and type of heritage places listed in registers or heritage objects in collections	A	D
	• number and 'strength' of traditional indigenous languages	A	A
State of legislative protection	• number of international heritage conventions ratified by Australia	A	A
	• number and nature of relevant Commonwealth, State and Territory Acts	A	A
	• number of places/objects protected under relevant Acts	A	?
	• number of decisions under specific protective provisions of relevant Acts	?	?
	• level of statutory referrals from Commonwealth agencies for conservation advice about heritage places	A	Not applicable
State of physical condition	• condition of the 'fabric' of places/objects	D	D
State of conservation practice	• number managed under agreed management plans and with appropriate resources	D	D
	• level of financial assistance available	B (historic) D (indigenous)	?
	• size of the 'pool' of heritage professionals available	D	D
State of community participation	• level of involvement in relevant non government organisations	D	Not considered
	• level of involvement in heritage identification, evaluation and conservation	C–D	Not considered
	• objections to listings in heritage registers	C	Not applicable

Note:
1. A = Excellent; B = Good; C = Fair; D = Poor; ? = Uncertain (information not able to be obtained for this report). Indicators for which data were inadequate or uncertain will need to be addressed in future reports.

no national, quantifiable data were available on the magnitude, extent or relative importance of pressures on natural and cultural heritage places — even for major pressures. Where data were sought (Hyde, in press; Marshall and Pearson, in press; Ward, in press) relevant information was not readily available or was not of a uniform standard across the States and Territories. It was not possible to assess the level of resources available for identifying and protecting heritage places and objects in many instances, the distribution of resources between different areas of need, the participation and success rates of applicant groups or the degree of duplication or overlap in programs. The data gaps could not be attributed just to the short time available for collating information for this chapter. The inadequacy of existing data greatly increases the difficulty of assessing the effectiveness of responses in conserving heritage resources.

Aboriginal and Torres Strait Islander heritage

One of the major changes in public attitudes about heritage over the last one to two decades has been the increased recognition of, and importance placed on, the heritage of indigenous Australians. This has been manifest at the government level, for example, by support for the return of indigenous cultural material to custodians despite the complexities involved.

Although many changes have occurred since European settlement, the Australian landmass and its surrounding seas remain a significant cultural landscape for indigenous communities, whose visible expression of culture has become a strong focus for cultural tourism. Many non-indigenous Australians still lack an understanding of the importance of indigenous languages and cultures although education about them has increased at primary schools and tertiary institutions, and through popular publications.

Loss of traditional languages, low levels of involvement of indigenous peoples in the management of their heritage, and tourism are major pressures adversely affecting indigenous heritage. Programs designed to assist the maintenance of indigenous languages and cultures have increased, but these may not be sufficient to prevent the loss within a generation of many of the 90 traditional languages spoken today.

Over the last decade, more indigenous people have become involved in 'heritage areas' as administrative, technical and field staff in a range of organisations, and as members of the advisory bodies, boards of trustees, policy-making groups or executives of these institutions. However, the current levels and organisational infrastructures are still far from sufficient to ensure that the identification and management of indigenous places and objects are culturally appropriate. The potential positive impacts of tourism bypass many indigenous communities, which rarely have a strong role in its development.

▲

This floating gold dredge at Maldon, Victoria is an important element in the cultural landscapes of the Central Victorian Goldfields.

Linking heritage

The strong links between places, objects and associated meanings are not reflected in current institutional approaches to heritage identification and protection. Most oral history programs for example, conducted by major institutions, such as libraries, are poorly integrated with identification and documentation projects of heritage agencies or museums. Policies and programs for heritage places are rarely linked with those for material objects and hence do not encourage a balanced approach to heritage conservation. Similarly, while there are moves towards better integration in some areas, natural and cultural heritage are handled separately in much government legislation, administration and policy. This hinders rather than helps decision making intended to achieve the sustainable use of natural and cultural heritage resources. Developing and applying the concept of cultural landscapes to heritage management should help integration.

No national heritage conservation strategy exists to link places, objects and the values people attach to them, conceptually or in terms of policy. Such a strategy would facilitate the cooperation of governments, industry, business, voluntary groups and the community in developing agreed goals for heritage conservation and development. It would also expedite the electronic databasing and national linkages of heritage registers and collections essential to establish a national perspective. Such a broad view provides the necessary basis for management decisions on the sustainable use of heritage resources.

Linking governments

The Commonwealth, Territories and all States have enacted heritage laws covering indigenous places and objects. All except Tasmania also have laws for historic places. This highlights the importance of coordinated identification and conservation strategies within and between the different levels of government to eliminate duplication or neglect of key areas.

Our cultural resource is complex and rich and has many links with natural resources. A variety of agencies are responsible for cultural and natural

Table 9.20 Summary

Element of the Environment/ Pressure	Effect on state	Information availability	Response	Effectiveness of response
World Heritage places Focus on natural heritage values; opposition to listings; growth of tourism	7 properties listed for natural values, 4 for natural and cultural values	Data on impact of tourism poor	Lobbying to change cultural criteria; legislation to protect properties; IGAE used as framework for nominations and management	Uluru inscribed as cultural landscape; some nominations and management arrangements still being negotiated
Historic places – metropolitan areas Increased demolition, reuse, loss of context through development, rezoning etc	Large numbers of places recognised but still major imbalances in heritage registers; effects on physical condition can be positive or negative	No national data on magnitude and effect of pressures; no national data on condition of places	Heritage legislation enacted now in all except one state; review of Commonwealth heritage buildings; community protests to conserve places; targetted studies to address gaps; special assistance programs for conservation	Too early to assess effectiveness of legislative protection or review; condition of places receiving assistance improved; many places still require conservation
– rural areas Neglect through lack of identification or reduced rural economies	Fewer places recognised as heritage; many not receiving legislative protection and not eligible for heritage assistance	No national data on magnitude and effect of pressures; no national data on condition of places		
Indigenous places Insufficient management and administrative role for indigenous people; legislative focus on archaeological sites; cultural insensitivity; loss of traditional indigenous languages	Management of many places inappropriate; cultural values of places adversely affected; loss of traditional knowledge about places	Often poor but variable	Increased legislative protection; increased numbers of indigenous people in relevant government agencies and on boards; change in community attitudes; language maintenance programs	Not all types of places protected in some states; employment levels of indigenous people still low; too early to assess effectiveness of language programs
Natural places (see also previous chapters) Conflicting land uses; resource use in heritage areas; urban growth	Many places not managed appropriately for their values	National data lacking on magnitude of pressures, impact on heritage registers and condition of places	Targetted heritage studies; community protests for threatened places	Many places not conserved
Natural and cultural places generally Inadequate community involvement in heritage studies; lack of recognition of cultural values in natural areas; insufficient heritage assistance; insufficient heritage professionals; tourism	Imbalances in heritage registers; many places not managed appropriately for their values; inadequate conservation	National data lacking on magnitude of pressures, impact on heritage registers and condition of places	Community protests for threatened places; targetted heritage studies; development of new evaluation methods; heritage assistance programs; increase in tertiary training courses; development of government tourism strategies; tourism codes of practice	Improved documentation of places and integration of natural and cultural values; some places conserved but others still need assistance; too early to assess effectiveness of tourism responses
Objects *in situ* Inappropriate removal from original context; destruction of surroundings; tourism	Loss of context and knowledge; physical destruction of items	No national data on number, type or physical condition of heritage objects, nor on magnitude of pressures	Burra Charter promotes retention of objects *in situ*; legislation passed to protect indigenous objects in many states, to protect objects associated with shipwrecks, and to control export of objects; changed policies in some museums	Uncertain; no data available to assess effectiveness of many responses
Material collections (natural and cultural objects) Inadequate national co-ordination; biases in collecting policies of institutions; inadequate conservation facilities and expertise; inadequate documentation; lack of classification systems and standard nomenclature for cultural items	Major imbalances in collections; physical condition of many collections probably deteriorating	Poor national, quantifiable data on number, type and physical condition of objects, and on magnitude of pressures	Reviews of collections; legislation to control export of objects; national co-ordinating bodies established; government policies established but often lack specific funding for implementation; more conservation staff employed; changes in collecting policies	Some major pressures identified in 1974 still apply; too early to assess effectiveness of recent responses or no data available to assess adequately

Table 9.20 Summary (continued)

Element of the Environment/ Pressure	Effect on state	Information availability	Response	Effectiveness of response
Collections of Indigenous items Indigenous communities requesting relevant items to be returned to custodians; many indigenous communities lack conservation facilities, expertise and access to management advice	Returned objects located in appropriate cultural context but physical condition often at risk	Lack of information in many areas.	Government policies and funding programs to assist return of items and establishment of keeping places; changed museum policies; museums employing more indigenous people in relevant areas	Too early to assess effectiveness.
Living collections (biological) Inadequate national coordination	Imbalances in collections	National data on endangered species most readily available	Co-ordinating bodies established and implementing specific programs for endangered species; adoption of conservation strategies	Too early to assess effectiveness.

heritage in most States and Territories. Planning for integrated management needs to be undertaken at the regional level rather than at just the State or local government levels to achieve sustainable use of these resources. Although local governments have an increasing role in conserving historic places in most States, many currently appear to lack the necessary skills and resources to do this effectively.

Effective coordination between and within governments is essential to ensure that heritage values are considered in the early stages of government policies and program development. This is equally necessary in metropolitan, rural and remote regions. Documented heritage resources are concentrated in urban regions, and thus more places are affected by government programs. In rural areas, fewer opportunities may exist to retain places valued by the community. Remote regions may contain many natural and cultural heritage resources, but these are not generally recognised or valued as heritage and remain poorly documented.

Community involvement in heritage identification

Community involvement has increased over the past few years but remains limited among the general population. The heritage of groups of non-English-speaking background and places of social value to the wider community are not well represented in heritage registers. Oral history appears to be a neglected tool in heritage identification, although it is a significant means of documenting social value. Community-based heritage studies and provision of training and resources for local people in areas such as oral history research are not yet common. While identification of places significant to the community remains inadequate and management

Heritage week activities, such as the open day at Blundell's Cottage, Canberra, provide an opportunity for the community to enjoy and learn about their heritage.

decisions take poor account of social value, community opposition to the destruction of heritage places will continue to frustrate decision-makers and development bodies alike. Heritage registers and collections will continue to fail to reflect the historical and ongoing cultural development of the nation, and the natural and cultural heritage of all Australians.

The future

Despite numerous positive responses to assist the identification, documentation and conservation of Australia's heritage, it is too early to assess the effectiveness of many initiatives. National agreement is needed about which indicators are the most useful for measuring the state of our heritage resources before evaluating such initiatives. The technical reports compiled for this chapter will offer a firm basis for such discussion. The chapter itself should act as a catalyst for the recognition and sustainable use of natural and cultural heritage as part of Australia's environment.

References

Anderson, M. (1991). Heritage collections in Australia — report. *Report to the Heritage Collections Working Group.*

Anderson, M. (1992). Heritage collections in Australia — report, stage II (March 1992). *Report to the Heritage Collections Working Group.*

Anderson, M. (1993). Heritage collections in Australia — report, stage III. *Report to the Heritage Collections Working Group.*

Anderson, M. (in press). Material culture and the cultural environment: objects and places. *Report prepared for the Culture and Heritage Reference Group, Department of the Environment, Sport and Territories.*

Armstrong, H. (1994). Post-world war II migrant heritage places in Australia. *Unpublished Report to the Australian Heritage Commission.*

Australian Association of Environmental Education (in press). Culture and heritage: education programs. *Report prepared for the Culture and Heritage Reference Group, Department of the Environment, Sport and Territories.*

Australian Heritage Commission (in press). The Register of the National Estate: who, what, where? *Report prepared for the Culture and Heritage Reference Group, Department of the Environment, Sport and Territories.*

Australian Heritage Commission and Department of Conservation and Natural Resources, Victoria (1994). National Estate Values in the Central Highlands of Victoria. Draft Project Report. (Department of Conservation and Natural Resources: Melbourne).

Behr, M., Grant, M., and Haskovec, I.P. (1994). Bigger than Ben Hur. *Paper presented to Australian Archaeological Association Conference, Melbourne, November 1994.*

Blair, S. (ed.) (1994). People's places: identifying and assessing social value for communities. Report of the Social Value workshop held at the University of Melbourne on 20 October 1993. *Australian Heritage Commission, 1993 Technical Workshop Series* No. 6.

Blair, S., and Truscott, M. (1989). Cultural landscapes — their scope and their recognition. *Historic Environment,* 7(2), 3-8.

Boden, R. (in press). Culture and heritage: natural objects. *Report prepared for the Culture and Heritage Reference Group, Department of the Environment, Sport and Territories.*

Brown, V.A. (1994). 'Acting Globally. Supporting the Changing Role of Local Government in Integrating Environmental Management.' (Department of the Environment, Sport and Territories: Canberra.)

Council for Aboriginal Reconciliation (1993). 'Exploring Common Ground: Aboriginal Reconciliation and the Australian Mining Industry.' (AGPS: Canberra.)

CRASTE (1987). See Daniels, V.K. (1987)

Daniels V.K. (Chairperson) (1987). Windows onto worlds: studying Australia at tertiary level. *The Report of the Committee to Review Australian Studies in Tertiary Education,* 195–212.

Department of the Environment, Sport and Territories (1995). 'Submission to the House of Representatives Standing Committee on the Environment, Recreation and the Arts (HORSCERA) Inquiry into World Heritage Areas.' *Unpublished report, Department of the Environment Sport and Territories.*

Domicelj, J. (1992). Balancing the Register of the National Estate — the historic environment. Stage 3 report. *Unpublished report to the Australian Heritage Commission.*

Domicelj, J., Halliday, H., and James P. (1992). Australia's cultural estate — framework for the assessment of Australia's cultural properties against the World Heritage criteria. *Report to the Department of the Arts, Sport , the Environment and Territories.*

Driml, S., and Common, M. (1995). Economic and financial benefits of tourism in major protected areas. *Australian Journal of Environmental Management,* 2(1), 19-29.

Elliott and Shanahan Research (1993). An exploration of community attitudes and stakeholder attitudes towards the Australian Heritage Commission. *Unpublished Report to the Australian Heritage Commission.*

Fagg, M., and Wilson, J. (1994). 'Directory of Australian Botanic Gardens and Arboreta.' 2nd Edition. (Australian National Botanic Gardens: Canberra.)

Galla, A. (1993). Training as Access: Guidelines for the Development of Heritage Curricula and Cultural Diversity. (AGPS: Canberra.)

Griffiths, T.R. (1991). History and natural history: conservation movements in conflict? In 'The Humanities and the Australian Environment', ed. D.J. Mulvaney, pp. 87–110. (Australian Academy of Humanities: Canberra.)

Henderson, J., and Nash, D. (in press). Culture and heritage: indigenous languages. *Report prepared for the Culture and Heritage Reference Group, Department of the Environment, Sport and Territories.*

Hope, R.M. (Chairman) (1974). 'Report of the Committee of Inquiry into the National Estate.' (AGPS: Canberra.)

Hyde, G. (in press). Culture and heritage: tourism and the environment. *Report prepared for the Culture and Heritage Reference Group, Department of the Environment, Sport and Territories.*

Imber, D., Stevenson, G., and Wilks, L. (1991). A contingent valuation survey of the Kakadu Conservation Zone. *Resource Assessment Commission Research Paper* No. 3.

James, P. (1993). A guide to the legal protection of the National Estate in Australia. *Unpublished Report for the Australian Heritage Commission.*

Jones, P., and Sutton, P. (1986). 'Art and Land: Aboriginal Sculptures of the Lake Eyre Region.' (South Australian Museum: Adelaide.)

Kenderdine, S. (in press). Culture and heritage: shipwrecks and associated objects. *Report prepared for the Culture and Heritage Reference Group, Department of the Environment, Sport and Territories.*

Lennon, J. (in press). Case study of the cultural landscapes of the Central Victorian goldfields. *Report prepared for the Culture and Heritage Reference Group, Department of the Environment, Sport and Territories.*

Marshall, D. (1993). Review of the effectiveness of conservation advice under the Australian Heritage Commission Act. *Unpublished Report prepared for the Australian Heritage Commission.*

Marshall, D., and Pearson, M. (in press). Culture and heritage: historic environment. *Report prepared for the Culture and Heritage Reference Group, Department of the Environment, Sport and Territories.*

McCarthy, L., Ashton, P., and Graham, H. (in press). Culture and heritage: oral history. *Report prepared for the Culture and Heritage Reference Group, Department of the Environment, Sport and Territories.*

Olney, P.J.S., and Ellis, P. (eds) (1992). Zoos and Aquaria of the world. *International Zoo Yearbook* 32, 262–3.

Pigott, P.H. (Chairman) (1975). 'Museums in Australia. Report of the Committee of Inquiry on Museums and National Collections including the Report of the Planning Committee on the Gallery of Aboriginal Australia.' (AGPS: Canberra.)

Purdie, R. (in press). Natural and cultural heritage: community attitudes — annotated references. *Report prepared for the Culture and Heritage Reference Group, Department of the Environment, Sport and Territories.*

Reid, J. (1995). The Prime Minister's pre-election promise of World Heritage listing for the Lake Eyre basin: flight or flight of fancy? *Rangeland Journal,* **16,** 273–97.

Richardson, B.J., and McKenzie, A.M. (1992). Australia's biological collections and those who use them. *Australian Biologist,* 5, 19–30.

Robertson, M., Vang, K., and Brown, A.J. (1992). Wilderness in Australia — issues and options. *A Discussion Paper written for the Minister for the Arts, Sport, the Environment and Territories.*

Toyne, P. (1994). 'The Reluctant Nation. Environment, Law and Politics in Australia.' (Australian Broadcasting Corporation: Sydney.)

Yencken, D.G.D. (Chairman) (1985). Australia's national estate: the role of the Commonwealth. *Australian Heritage Commission Special Australian Heritage Publication Series* No. 1.

Ward, G. (in press). Culture and heritage: indigenous places and objects. *Report prepared for the Culture and Heritage Reference Group, Department of the Environment, Sport and Territories.*

Acknowledgments

We especially thank the following individuals who assisted in the preparation of this chapter:

Ms Margaret Anderson (Museum of Western Australia)

Mr Paul Ashton (University of New South Wales)

Mr Keith Baker (Photograph researcher)

Dr Robert Boden (Robert Boden and Associates)

Dr Sandy Blair (Australian Heritage Commission)

Dr Robert Bruce (Australian Heritage Commission)

Dr Kay Daniels (Department of Communication and the Arts)

Dr John Henderson (University of Western Australia)

Mr Geoff Hyde (Hyde King & Associates)

Ms Sarah Kenderdine (Western Australia Maritime Museum)

Ms Jane Lennon (Jane Lennon & Associates)

Dr Louella McCarthy (University of New South Wales)

Mr Duncan Marshall (Heritage consultant)

Dr David Nash (Australian Institute of Aboriginal and Torres Strait Islander Studies)

Dr Michael Pearson (Heritage Management Consultants)

Dr Graeme Ward (Australian Institute of Aboriginal and Torres Strait Islander Studies)

Mr Geoff Young (Australian Association of Environmental Education)

In addition, a large number of other people provided information, usually at very short notice. These include individuals in State and Territory government departments, private industry and voluntary organisations. Their assistance is also gratefully acknowledged. Commonwealth government departments and members of the Commonwealth/State ANZECC State of the Environment Reporting Taskforce also helped identify errors of fact or omission.

Referees

The following people reviewed the chapter in draft form and provided constructive comments.

Professor Sandra Bowdler (University of Western Australia)

Professor Graeme Davison (Australian National University)

Mr Phil Gordon (Australian Museum)

Mrs Barbara Hardy (Investigator Science and Technology Centre)

Professor Julie Marcus (Charles Sturt University)

Associate Professor Colin Mercer (Griffith University)

Emeritus Professor John Mulvaney (Australian Academy of the Humanities)

Ms Gaye Sculthorpe (Museum of Victoria)

Photo credits

(Note: Institution shown in brackets indicates the agency holding the material)

Page 9-1: *Artist* – Lin Onus (Australian Heritage Commission)

Page 9-4: Australian Heritage Commission

Page 9-5: *The Age*

Page 9-7: *(from left)* Kathie Atkinson; Mike Archer

Page 9-8: South Australian Museum

Page 9-9: National Trust of Victoria

Page 9-11: Murray Fagg (Australian Heritage Commission)

Page 9-15: Australian National Insect Collection, CSIRO

Page 9-16: Australian Heritage Commission

Page 9-18: (*from top*) Patrict Grant (courtesy of Godden
   Mackay P/L and the Sydney Cove Authority);
   Lauretta Zilles

Page 9-19: Australian Heritage Commission

Page 9-20: Joy McCann

Page 9-22: Murray Fagg

Page 9-23: Heritage Victoria

Page 9-24: Australian Heritage Commission

Page 9-26: Rosemary Purdie (Australian Heritage
   Commission)

Page 9-29: Celeste Thomson (Taronga Zoo)

Page 9-30: (*from top*)
   Patrick Baker (Western Australian Maritime Museum);
   Brian Richards (Western Australian Maritime
   Museum)

Page 9-32: Keith Baker

Page 9-33: Australian Heritage Commission

Page 9-34: Australian Heritage Commission

Page 9-35: (*from top*) *Bombala Times; The Age*

Page 9-38: City of Fremantle

Page 9-39: Pat Boldra

Page 9-41: (*from top*)
   Conservation Commission of the Northern Territory;
   D. Ziegeler (Australian Heritage Commission)

Page 9-42: Bruno David (Earthwatch)

Page 9-44: Kingsley Dixon

Page 9-45: Peter Hiscock

Page 9-47: Keith Baker

Page 9-49: Linda Young

Chapter 10

Towards ecological sustainability

Prepared by
The State of the Environment Advisory Council

Overall Message

- This is the first ever independent and comprehensive State of the Environment Report for Australia. It links land, water, air, plants and animals, human settlements and how we value them.

- An independent advisory council and seven expert groups prepared the report. It draws on the knowledge and skills of more than 200 eminent scientists and other experts.

- The report shows that Australia has a beautiful, diverse and often unique environment which is a priceless heritage and should be a source of pride to all Australians.

- Some aspects of the Australian environment are in relatively good condition by international standards. In some areas our approach to environmental management has won international recognition.

- In many other areas it is not possible to decide whether our environmental management is adequate. We urgently need better information and understanding, which will require data collection and research.

- The report also shows that Australia has some very serious environmental problems. If we are to achieve our goal of ecological sustainability, these problems need to be dealt with immediately. This will be no small task.

- The problems are the cumulative consequences of population growth and distribution, lifestyles, technologies and demands on natural resources over the last 200 years and more.

- No single government or sector is to blame for these problems. We are all responsible. Changes are needed in government policies and programs, corporate practices and personal behaviour.

- Australians are among the most environmentally aware people in the world. All sections of the community now recognise the need to do more to tackle environmental issues.

- Most of the problems identified in the report do have solutions. The report details many positive and successful initiatives.

- Our actions have been most effective where they have taken a comprehensive and systematic approach, integrating different aspects of the overall problem. By contrast, failures tend to be piecemeal efforts that treat symptoms rather than underlying causes.

- Australia has an international responsibility to protect its rich biological diversity and its unique environmental features such as the Great Barrier Reef and other World Heritage Areas. We also have a national responsibility towards future generations of Australians.

- Australia has a better opportunity than perhaps any other nation to protect its environment and use its natural and heritage resources sustainably. We need to do much more if we are not to lose this opportunity.

- Progress towards ecological sustainability requires recognition that human society is part of the ecological system and integration of ecological thinking into all social and economic planning.

Contents

Introduction

In this chapter the Advisory Council makes its assessment of the findings of the State of the Environment report and of our progress towards ecological sustainability. It has identified the key issues as: the need for an overall systems approach, biodiversity, water issues, land degradation, global climate change, coastal and urban environmental issues and finally social and cultural issues such as the well-being of indigenous Australians and protecting Australia's heritage.

The Council concludes that, despite positive achievements to date, some serious adverse trends need urgent attention.

Principles of ecological sustainability

Our traditional pattern of economic development has been at question since the publication of the reports of the Club of Rome, the Blueprint for Survival and the World Conservation Strategy, which coined the term 'sustainable development'.

Australian governments adopted the principle of Ecologically Sustainable Development, or ESD, as a major national strategy in 1992, following a national consultative process. This defined ESD as a pattern of development that improves the total quality of life, both now and in the future, in a way that maintains the ecological processes on which life depends (see the box below).

The National Strategy for Ecologically Sustainable Development

Core objectives

- to enhance individual and community well-being and welfare by following a path of economic development that safeguards the welfare of future generations

- to provide for equity within and between generations

- to protect biological diversity and maintain essential ecological processes and life-support systems

Guiding principles

- decision making processes should effectively integrate both long and short-term economic, environmental, social and equity considerations

- where there are threats of serious or irreversible environmental damage, lack of full scientific certainty should not be used as a reason for postponing measures to prevent environmental degradation

- the global dimensions of environmental impacts of actions and policies should be recognised and considered

- the need to develop a strong, growing and diversified economy, which can enhance the capacity for environmental protection should be recognised

- the need to maintain stand enhance international competitiveness in an environmentally sound manner should be recognised

- cost effective and flexible policy instruments should be adopted, such as improved valuation, pricing and incentive mechanisms

- decisions and actions should provide for broad community involvement on issues that affect them

Source: Council of Australian Governments, 1992.

While all the objectives of the National Strategy are relevant, the key one for this report is the third: to protect biodiversity and maintain essential ecological processes and life-support systems. Of the guiding principles in the National Strategy, the first two are central to this report, which has the function of analysing progress toward the goal of ecological sustainability.

While this is a national state of the environment report, it is also set in a context of our global obligations, recognising the third guiding principle.

What do we need to sustain?

Ecological sustainability requires the sustenance of a healthy and diverse ecosystem on behalf of existing and future generations of humans and other species. To achieve this, we need to sustain biodiversity, ecological integrity, 'natural capital' and social integrity.

Biodiversity

Biodiversity, as introduced in Chapter 4, is the variety of species, populations, habitats and ecosystems. There are sound practical and moral reasons for seeking to maintain biodiversity by measures such as the reservation of representative ecosystems and habitats, the protection of endangered species and populations, and maintaining biodiversity in areas where many human activities take place. Conserving biodiversity allows for further studies to improve understanding of ecosystems and their dynamics. Such protection provides a stockpile of genetic diversity for potential use in agriculture and medicine. This is of economic importance because many of the industries using genetic resources produce high-value commodities. In the USA some 25 per cent of prescriptions are filled with medicines whose active ingredients are extracted or derived from plants. In Europe, Japan, Australia, Canada and the USA, the market value for both prescriptions and over-the-counter medicines based on plants was estimated to be nearly $60 billion in 1995 (ANZECC Task Force on Biological Diversity, 1994). The stockpile of genetic diversity can be a crucial defence against predators or disease. When phylloxera threatened to wipe out the European and Australian wine industries in the nineteenth century, the grapes were restored using phylloxera-resistant root stock from California.

Conserving biodiversity maintains ecosystem productivity and function. It increases the chance of ecosystem stability in the face of climate change and assists the absorption of waste products. Equally importantly, it provides places for rest, recreation and (where appropriate) tourism; we should not underestimate the psychological value to humans of areas which are not exploited for economic gain.

By protecting biodiversity we recognise that we share this planet with countless millions of other species. Many people believe that other life-forms have intrinsic value, whether or not they are perceived to be useful to humans. Even if an endangered species had no economic value to us, there would still be a feeling in the community that it should be protected. There was, for example, a vigorous debate in the scientific community when it was proposed that the last remaining cultures of the smallpox virus should be destroyed.

The final argument for conserving biodiversity concerns the question of our state of knowledge. Do we have enough information to evaluate which natural resources need to be preserved for the future? In the absence of detailed information on which to base such decisions, it is prudent to err on the side of caution and attempt to retain our existing biodiversity.

Ecological integrity

Ecological integrity refers to the general health and resilience of natural life-support systems and encompasses their ability to assimilate wastes, such as pollution of air, water or soil, through basic natural cycles, such as the water, carbon and nitrogen cycles. It also includes the ability of ecosystems to withstand stresses, such as climate change or depletion of the stratospheric ozone layer. The effects of a growing human population, increasing resource use and consequent waste production have combined to impair the ability of natural ecosystems to provide these free services. The most dramatic examples in Australia in recent times have been large-scale nutrient overloads in waterways, culminating in an outbreak of blue-green algae that extended for 1000 kilometres in the Darling River — the longest outbreak ever, anywhere in the world. The underlying causes of these events are not yet fully understood, but they constitute a clear warning signal.

Maintaining the integrity of ecosystems requires a concerted local, national and international effort to increase energy and resource efficiency, encourage clean technologies and to implement appropriate pollution standards. It also requires, as discussed below, further research to obtain a better understanding of the impacts of human activities on natural systems.

Natural capital

A third consideration for sustainable development is the need to maintain our 'natural capital': the stock of productive soils, fresh water, the marine environment, forests, sub-soil assets and other resources needed for our survival and prosperity. In particular, we need to maintain our renewable resources. They are important indicators of the well-being of a community. They are part of our natural heritage and underpin much of our economic requirements and our cultural heritage. In simple terms, it could be argued that a society should live on the 'ecological interest' provided by the natural capital stock. By this standard, renewable resources would only be harvested at a rate no greater than the rate of natural replenishment.

Given our imperfect understanding of ecosystems, we are obliged to err on the side of caution. This

National environmental accounting

The Australian Bureau of Statistics is preparing a set of 'satellite accounts', which document changes in a range of natural assets such as land, water, forests and fish, as well as accounts associated with wastes and emissions. The satellite accounts, based on guidelines recently developed by the United Nations and other international organisations, will mesh with the conventional national income and production accounts. The complete system of accounts will provide a useful tool for assessing policy options for achieving ecologically sustainable development.

Environmental accounts can link economic information about various sectors and industries to environmental indicators, enabling us to assess the impacts of different patterns of economic growth and technological change on the nation's stock of environmental resources. Some economists believe that environmental accounts can be used to estimate 'green GDP'. The Gross Domestic Product, or GDP, is the sum of the total economic value of all transactions. 'Green GDP' adjusts the sum by including changes in natural resource stocks (depletion) and changes in services provided by the environment. These are dependent on environmental quality (degradation) as well as expenditures on environmental protection programs and projects. We clearly need better ecological, social and economic information if we are to manage our resources and environment in an integrated fashion.

means using renewable resources at a rate no greater than that needed to replenish stocks and using non-renewable resources with great care. It is possible to compensate future generations for losses, by producing other kinds of capital such as 'human capital' and physical capital (buildings, machinery, infrastructure). Such forms of capital are usually reproducible but may not be acceptable as substitutes for natural capital. In many cases, the destruction or degradation of natural capital is irreversible; it may not be possible to restore the environment, however much people are prepared to pay. A cautious approach is also necessary with 'cultural capital', which can be similarly destroyed or irreversibly degraded.

To be confident that our natural capital is wisely managed, we need a system of national accounts that informs us of changes in stocks of natural capital, as well as the flows of goods and services it provides (see the box above).

Social integrity

Sustainability refers not only to the natural environment; it also encompasses the resilience of social and cultural systems. Social and environmental issues are often intertwined. As a result of history and economic pressures, low-income households are often situated in degraded environments, near industrial plants and other hazards. Wealthy families have the economic freedom to move to more desirable environments.

Australia is a unique multicultural society with a distinctive and multi-layered cultural heritage. Preserving that heritage is important, as it helps us understand how we have reached our present position and gives us clues about our future choices. We need to examine our distinctive social structures to preserve important elements for sustainable development, such as our love of the bush and our traditions of participation and social justice. Where our social structures are inconsistent with ESD, we will need to consider changes.

The overall framework

To achieve ecological sustainability, we need to incorporate ecological principles into our decision-making processes. Some groups in society have benefited more than others from unsustainable practices. The transition to a sustainable society should be managed to redress the extreme imbalances in access to 'environmental goods' between different groups. The sustainable society must incorporate equity within this generation as well as equity between generations. This is an explicit provision of the National Strategy for Ecologically Sustainable Development.

Thus, any development strategy that aims to sustain biodiversity, ecological integrity and the stock of natural capital must recognise the primacy of ecological considerations.

Achieving environmental change

Alleviating the pressures on the environment will require a considered effort to modify human activities. Change in human behaviour can be achieved in a number of ways including: regulation or legal force, economic inducements, and changes in attitudes.

Legislation

All spheres of government have enacted various measures to protect the environment, ranging from planning restrictions at the local level to World Heritage legislation by the Commonwealth. Some aspects of environmental protection are so important that they need to have the force of law.

Economic incentives

Rewards to encourage certain behaviour or penalties to discourage it can induce change. Response measures to environmental concerns may include a lower level of taxation on unleaded petrol or significant fines for polluting waterways. Many people believe that economic incentives are the most efficient means of achieving desired outcomes.

Demonstration projects

Schemes that provide practical examples of cleaner production practices, are proving effective. These can illustrate the benefits to be gained from environmentally-sensitive cleaner production and from careful 'eco-design' of products.

Attitudinal change

Finally, change can occur as a result of alterations in social attitudes; for example, the increasing volume of material being offered for recycling by Australian householders, without any regulatory provisions or economic inducements. When community attitudes support change, strong legal measures are easier to enforce, and governments have more incentive to offer financial support. On the other hand, incentives or sanctions may be ineffective if there is community opposition to change.

Accurate information

Providing accurate information about the state of the environment to the community can alter attitudes, both by dispelling groundless fears and by alerting people to previously unsuspected problems. A central aim of this report is to provide the most accurate picture possible of the state of the environment in 1995 and to direct community attention toward the most serious problems affecting our natural and cultural environments.

Global environmental context

We now recognise that we are a part of a global ecosystem and an integrated global economy. This poses a new set of questions for the sustainability of development. Local environmental effects associated with productive activity do not necessarily stop at national boundaries. This became apparent with the issue of acid rain. Canadian emissions from the Sudbury smelter affected the north-eastern United States, while emissions from United States' mid-western cities drifted into Canada. The patchwork of small nations in western Europe has become criss-crossed by plumes of emissions, defying any possibility of national sovereignty over pollution. The depletion of the stratospheric ozone layer, changes to the atmosphere and depletion of the world's marine resources are now seen as problems of global scope, demanding the attention of all nations.

Reaction to the depletion of the ozone layer clearly demonstrates an international ability to take concerted action when the threat is proven. The science showing that release of chlorofluorocarbons (CFCs) could deplete the ozone layer was published in 1974. But political will to tackle the problem was blunted by the suggestion that the science was uncertain. Little then occurred until 1987, when the 'hole' in the ozone layer was definitively shown to have been caused by CFCs. After measurement of the dramatic drop in ozone levels near the South Pole, the response was rapid and we now have international agreement to phase out the release of CFCs and other chemicals that cause depletion of the ozone layer. Australia played a leading role in these negotiations.

Global climate change is a much more difficult problem. CFCs have substitutes, so there need not necessarily be a conflict between the goals of economic development and environmental protection. Global climate change is predominantly caused by the use of fossil hydrocarbon fuels. No cheap substitutes exist for oil, gas and coal, so maintaining the quality of the global environment seems likely to be at some short-term economic cost.

It is certain that a natural greenhouse effect keeps Earth warmer than it would otherwise be. It is also certain that human activities are increasing the amounts of greenhouse gases in the atmosphere. The Inter-governmental Panel on Climate Change (IPCC) has now concluded that these increases are enhancing the greenhouse effect, resulting in an additional warming of Earth's surface.

It is calculated with confidence that carbon dioxide emissions are responsible for about 60 per cent of the enhanced greenhouse effect, and that overall emissions would need to be reduced below the current level to stabilise the composition of the global atmosphere. The IPCC believes that stabilisation, even at double the present level of carbon dioxide, will be very difficult to achieve, as it will require an overall reduction in the emissions resulting from human activity. The IPCC has reached similar conclusions about other key greenhouse gases, methane and nitrous oxide — in each case, stabilisation of the atmospheric concentration at today's levels would require reductions in emissions from human activity. The trends are not hopeful at this time, with global emissions still increasing.

This raises the issue of the standard of proof required before taking action. Traditionally, environmental damage has been expected to meet the criterion used in criminal law of 'guilty beyond all reasonable doubt' before action is taken. In the case of depletion of atmospheric ozone, for example, concerted international action only began when the ozone 'hole' was detected. Many people are now arguing that we should apply the principle used in civil law of the 'balance of probabilities'. This would justify taking action much sooner. The United Nations Framework Convention on Climate Change and the National Greenhouse Response Strategy are based on the 'precautionary principle': the idea that lack of full scientific certainty should not delay our response if the consequences of delay might be serious or irreversible.

Various other signs indicate global environmental stress. The Worldwatch Institute argues that the health of bird species is an indicator of the state of the global atmosphere, in the same way that canaries once indicated unsatisfactory air quality in mines. Of the world's estimated 9600 bird species, only 3000 are stable or increasing in population; the other 6600 are in decline, vulnerable or threatened with extinction. One obvious cause is the loss of habitat such as wetlands. In the United States, for example, the area of wetlands has halved in recent years, leading to inevitable decline of species which rely on those areas. Frog populations appear to be declining in all parts of the world, though nobody can say why with any certainty (Tyler, 1994). Coral reefs appear to be in trouble around the world, threatened by run-off from urban areas as well as other pressures.

Serious concerns also exist at the global level about biodiversity. Estimates suggest that humans use, directly or indirectly, 35–40 per cent of terrestrial

The health of bird species and the state of coral reefs can provide an indication of the state of the global environment.

production of biomass (Vitousek *et al.*, 1986). The new estimate of the sustainability of fisheries argues that the current fish catch accounts for 24–35 per cent of the aquatic production in fresh water, upwelling and shelf systems (Pauly and Christensen, 1995). Thus, on land and on sea, we consume a large part of the total available photosynthetic product.

Key indicators of food production per head are falling at the global level (Worldwatch Institute, 1994 and 1995), leading to obvious concerns about the sustainability of the current level of population growth. Production of beef and mutton per head peaked in 1972 and has declined 13 per cent since, grain per head peaked in 1984 and has fallen 12 per cent, while seafood per head reached a peak in 1988 and has dropped nine per cent since then. Each year the human population increases by about 90 million. At a global level, farming and fishing are losing the battle to expand food production at the same rate as the population.

The Brundtland Commission concluded that two principal pressures are being exerted on the natural environment: population growth and associated poverty in the Third World, and unsustainable patterns of consumption in the industrialised world. Continuation of current trends will lead to inevitable difficulties, if not absolute limits.

The global economy

The recent liberalisation of Australian investment regulations and lowering of international trade barriers have stimulated areas of productive activity in which Australia has a comparative advantage.

The United States economist Michael Porter has refined the traditional theory of comparative advantage by arguing that those countries, and those companies, that show leadership in the move toward sustainable development will reap economic benefits as well as having reduced environmental impacts (Porter and van der Linde, 1995). Using the principle of sustainable development to guide the future search for comparative advantage is thus likely to benefit individual companies, and the economy as well as the environment.

In 1994, when the Australian government canvassed the possibility of discouraging use of fossil fuels by a modest carbon levy, opponents feared that it would reduce economic competitiveness. Little serious attempt was made to canvass the possible economic benefits that could be gained by improving our competitiveness in other areas of growing importance such as improved efficiency and renewable energy technologies. Several countries have now introduced carbon taxes. One argument used to support them is that they will encourage the development of greenhouse gas abatement technology.

We need to develop a framework that responds to both global and local agendas on the environment and the economy.

Australian context: constraints and driving forces

Australia's natural environment is a product of its geological history and the manifold impacts of human activity. Most of its physical features reflect the geological past of Australia as a very old, heavily weathered and extensively eroded landscape. As discussed below, variations in climate, especially rainfall, are part of the natural environment.

The long period of physical isolation from other land masses has resulted in a unique set of biota, with a large proportion of Australian flora and fauna found nowhere else on the planet. This rich endowment carries with it a heavy responsibility to ensure that our biological heritage is conserved, not just for future generations of Australians but for all people. As discussed earlier in this chapter, that biological heritage is an important economic resource, as well as an important source of resilience in the face of change.

Human population

The Australian cultural landscape is an important part of our heritage. For at least 50 000 years, people have lived here and modified the land by their actions. Since 1788, the impact of the human population has increased dramatically, both because of the rapid growth in the size of the population

and because of land use practices such as agriculture, mining and the establishment of urban areas. Australia is one of the most urbanised of all nations, with the vast majority of people living in a few cities. This concentration of population leads to heavy localised environmental pressures. As discussed in Chapter 3, however, population levels do not relate simply to environmental impact; some of the most serious environmental deterioration noted in this report is in lightly populated agricultural areas.

Australia has a relatively high rate of population growth by the standard of OECD countries, although it declined slightly from 1.2 per cent in 1991–92 to 1.1 per cent in 1993–94. This decline was due to a reduction in the level of immigration. The 'natural increase', the difference between births and deaths, has been consistent in recent years at 0.8 per cent per annum, while the net inward migration has declined from 0.5 per cent to 0.2 per cent per annum. Inward migration has had a significant effect on the population level because of the high level of intake maintained for almost 50 years. At the 1991 census, almost four million of the 17.3 million people in this country had been born overseas, mainly in Europe.

In 1995, the population passed 18 million, and demographic analysis by the Australian Bureau of Statistics revealed a rate of increase of about 200 000 per year. On that trend, Australia will have another million people by the year 2000. The growth is unevenly distributed, concentrating in Queensland, Western Australia and New South Wales. Although other States have lower rates of population increase, all are still growing. The ABS forecasts that the population will not stabilise before the year 2030, on current trends.

The cultural heritage of Australia is diverse and multi-layered, reflecting cumulative impacts over thousands of years, with increasing pressure on the natural world in recent times. The heritage of Aboriginal and Torres Strait Islander people is particularly distinctive and significant because of the close relationship between the people, their culture and the natural world. This relationship was severely disrupted by the arrival of European settlers, but there is now increasing recognition of its value and the respect it warrants. The Aboriginal and Torres Strait Islander people now own or control 15 per cent of the land mass of Australia. For this reason alone, it is crucially important that they are actively involved in land management decisions in Australia and in managing their heritage.

The 1994 Report of the House of Representatives Standing Committee for Long Term Strategies pointed to the need to evaluate the 'carrying capacity' of Australia for a range of assumptions about lifestyle.

Pattern of human activity

Human settlements

Two unusual features of Australia's population distribution affect its impact on the natural environment. The first is the concentration of population in the coastal zone, especially along the south-eastern and eastern coast-lines. The second is the concentration of population in a small number of urban centres. At the time of the 1991 census, more than 85 per cent of the population was located in urban areas; indeed, more than 60 per cent of the entire population is contained in the five large cities of Sydney, Melbourne, Brisbane, Adelaide and Perth.

Many inland communities are declining in population as settlement drifts toward the coast. This pattern of changing population distribution is straining the natural and cultural environments of the coastal zone as well as the cultural environment of inland Australia.

The more recent settlements are described as having very high 'metabolic flows'. In other words, they use large quantities of resources and produce large amounts of waste compared with the natural world and other advanced industrial nations. These metabolic flows per person have been increasing in recent years, compounding the heavy burdens imposed already on local environments.

Australian urban areas provide a comfortable living environment by international standards, and measure up well on indicators of 'livability'. However, one negative aspect of our towns and cities is that significant environmental problems are arising from the burden of waste products on natural systems. Another is that the loss of heritage places is reducing the 'livability' of our urban settlements by eroding our sense of place.

Lifestyles

A second major pressure on natural systems is the changing lifestyle expectations of the community. In Australian cities of 40 years ago most workers commuted relatively short distances by public transport. Emissions of fuel combustion products are much greater in our cities now, with people commuting long distances by car. Compared with an era of more food self-sufficiency when many households grew their own vegetables, modern centra-lised food production requires more fuel energy for production, processing and transport of food.

Changing patterns of domestic life impose new pressures. The size of the average household has contracted dramatically in recent years, so that half of all households now consist of one or two people, and two-thirds consist of three or fewer. That means that more housing is needed per head of population. It also means that people are more likely to travel to visit relatives, rather than living with them under the same roof.

Technological change

The third factor affecting human impact on the natural world is technology. Transport is one example. The move from public transport to the private car has increased fuel emissions significantly, as the fuel efficiency of our car fleet is about half that of public transport.

The fuel efficiency of vehicles has changed over time. Between 1960 and 1975, the average fuel

▲

efficiency of cars fell by about 25 per cent as they steadily became larger, more powerful and more extensively equipped with such accessories as air conditioning, automatic transmission and power steering. Between 1975 and 1987, technological improvement recovered the ground lost in the previous 15 years, returning the average fuel efficiency of the car fleet to the 1960 level. Further improvement has continued to result from changing patterns of consumer demand, though there are no specific regulations to encourage this process.

The average fuel efficiency of the Australian car fleet is poor by international standards. A significant factor is the average age of Australian cars. The improvements of recent years mean that new cars are generally more efficient than similar models which are ten years old. Australia has, by international standards, a very old fleet of private cars.

Economic activities and land use

Economic activities and changes in our approach to economic management create another set of pressures. Australia has a resource-intensive economy, with a strong export emphasis on minerals and agricultural produce. Expanding primary production will impose additional pressures on natural systems unless current practices are changed. Using fewer resources per unit of economic output could help make the Australian economy more sustainable.

The shape of our economy is determined partly by natural factors which give Australia a comparative advantage in various activities, and partly by a series of policy decisions that encourage those activities. For example, mineral royalties that are low by international standards encourage extraction — until recently, gold mining was exempted from taxation to stimulate that activity. In the area of research support, the traditional emphasis on primary industries has been of great assistance to their continuing competitiveness. Other examples are the pricing of irrigation water below the real cost of supply — the community provides a subsidy to water use for agriculture; and the provision of incentives to farmers to clear land.

Aboriginal and Torres Strait Islander people now own or control 15 per cent of the land mass of Australia. It is important that they are actively involved in land management decisions and in managing their heritage.

Human uses of forests affect the habitats and population sizes of some native species as well as increasing the risk of invasion by exotic species. A recent government report showed that more than 20 per cent of natural land cover in Australia has been cleared for agriculture and grazing in the period since European settlement, representing a significant loss of natural habitat (Graetz *et al.*, 1995). The clearing of land for agricultural purposes still continues in various parts of Australia, setting in train a complex set of changes to the land including erosion, loss of soil structure and alterations to soil chemistry. As discussed in Chapter 6, similar comments can be made about some cultivation practices.

Land clearance leads to loss of biodiversity and additional pressure on natural systems. It may alter the pattern of water flow by causing erosion, by raising the local water table or by increasing salinisation. Salinity problems can arise from other practices associated with agriculture, most obviously the diversion of the flow of natural watercourses, as discussed in Chapter 7.

Chapter 6 lists the factors currently affecting our land resources. Australia feeds the equivalent of about 50 million extra people through our cereal production, produces twice as much meat as we consume and our grazing lands yield a third of the entire world wool production. So the effective population pressure on our land resources is much greater than that associated with the number of inhabitants, which are relatively few by world standards.

The Australian landscape is itself an economic asset of great worth. The recent rapid growth of international tourism here is a direct reflection of the interest of other people in our natural and cultural heritage. The growth in tourist numbers puts considerable pressure on the natural environment. As the recent government report *Living on the Coast* put it, there is a real danger that this pressure will degrade the asset so that it will be less valuable, in economic as well as ecological terms. Tourism also has potential negative impacts on our cultural heritage, as discussed in Chapter 9.

The ecological impact of mining operations causes public concern. Modern mining practice is much improved, to the point where the Australian subsidiary of the Alcoa company was entered on the United Nations Environment Program's Global 500 Roll of Honour for its rehabilitation of land mined for bauxite. The most serious conflicts now occur where proposed mining ventures are located on sites of significant habitat, high biodiversity or special cultural significance.

Widespread agriculture and the nature of past practices have measurably changed the flow regimes of watercourses, while coastal development has altered estuaries and the marine environment. These examples show that analysis of variability and change must go beyond natural variations to include changes effected by human agents.

Many of these human activities lead to high average levels of consumption of fuel resources:

annual per head domestic use amounts to five tonnes of black coal, 2.5 tonnes of brown coal, 1000 litres of petrol and 600 litres of diesel fuel (ABS, 1995).

The nature of the changes to natural systems are in turn influenced by our social institutions. At the local level, the High Court of Australia in its 1992 'Mabo decision' rejected the doctrine of '*Terra Nullius*' (that Australia belonged to no one at the time of European settlement) and held that a form of native title rights and interests survived European settlement (but are subject to the sovereignity of the Crown). This is likely to alter the pattern of land holdings, and may in turn change land use and impact on the natural world. These changes may be beneficial, or may have mixed effects.

Thus, this report can analyse the impacts of current human activities, but those impacts will alter as the scale of population and the balance of activities change.

Problems of predicting environmental impacts

The Australian landscape is characterised by variation and change. The natural variability of the weather limits our capacity to predict the impacts of human actions. In South Australia in the late 19th century, for example, a run of unusually wet seasons led to the view that 'rain follows the plough'. We now know that is not true. Climate variability has been a constant problem for people in Australia, as it causes natural fluctuations in vegetation and fauna. Rural Australians recognise the poetic description of the country as one of 'droughts and flooding rains'.

It is not the poor soils of this continent or the variability of our climate that put pressure on our land resources, but human activities that fail to take sufficient account of those characteristics. Land use planning should recognise both the historic natural variability of rainfall and the anticipated effects of global climate change, which is likely to accentuate the variability of the climate. It must also take account of the possible reduction in biodiversity, as the loss of any species can affect natural cycles and so alter the pressure of existing activities on land resources.

Our understanding of the link between the El Niño Southern Oscillation and the Australian climate is improving. Detailed time-series data on grain production in Queensland, for example, show the influence of El Niño events on agricultural productivity. There is concern that the recent run of El Niño events may represent a trend rather than a random fluctuation in the climate pattern. Scientific knowledge does not yet allow us to say whether we are seeing a minor fluctuation or a significant change to the long-term trend.

Examples such as the decline in seagrass and the increase in algal blooms illustrate how difficult it is to assess human impact on the natural world. The time scale of responses to our actions is often long, measured in decades or centuries rather than years. The complex process of changing natural systems often defies simple analysis by cause and effect. It is rarely possible to refine our knowledge by controlled experiments. Environmental impact statements have a very poor record of predicting in advance the effects of human activity (Buckley, 1989).

This report contains many examples of known changes to complex ecosystems which cannot be attributed to a particular cause. In other instances, we know the cause, but have difficulty reversing an observed trend. Removal of vegetation in the Mallee has led to rising groundwater, and irrigation in the Murray–Darling Basin has led to increasing soil salinity. In both cases, it will take decades, possibly centuries, to restore the natural systems.

For the first time in our existence as a species, the size of the human population and its level of resource use has become a significant force for environmental change. At a global level, the scale of release of the chlorofluorocarbons and other compounds has measurably depleted the ozone layer which protects the natural systems from the sun's ultra-violet light. The widespread use of carbon fuels (coal, oil and gas) has changed the composition of the atmosphere so much that atmospheric scientists are confident we face inevitable changes to the climate.

Societal responses

The community undoubtedly believes that more attention should be paid to environmental issues. A 1994 survey by the Australian Bureau of Statistics found that 18 per cent rank environmental protection above economic growth, while 71 per cent believe the two are equally important — little different to the 70 per cent recorded in 1992. The ABS also found that community concern for the environment has increased 22 per cent over the last eight years (see Fig. 10.1). The 1992 survey was conducted at the time of the Earth Summit in Rio, which heightened community environmental awareness.

In recent years, all levels of government have begun to promote integrated approaches to environmental problems: cleaner production practices, recycling schemes, up-graded sewerage systems and improvements to water quality. Responses to environmental problems are by no means restricted to government. Community groups, for example, are involved in repairing degraded areas. Of added significance is the recent evidence that some corporations and employer representative bodies are accepting their environmental responsibilities. The private sector employs the majority of Australian workers, as well as using most of the resources and raw materials that flow through the economy. We are beginning to see the use of environmental auditing, more sustainable production methods and business plans that recognise environmental constraints (see Figs 10.2 and 10.3). The private sector still needs to do a lot more before it is seen to have accepted its full responsibility in the move toward sustainable development.

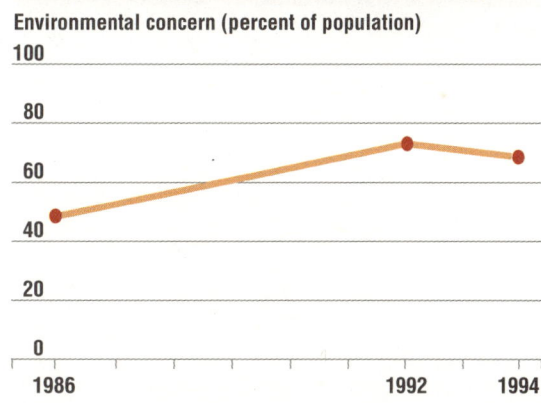

Figure 10.1 Community attitude to the environment

Environmental concern (percent of population)

Source: ABS, 1986 and 1995.

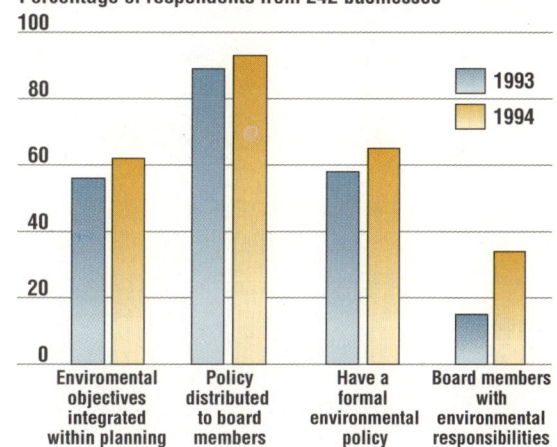

Figure 10.2 Indications that environmental management is a key business issue

Percentage of respondents from 242 businesses

Source: Coopers and Lybrand Consultants, 1994.

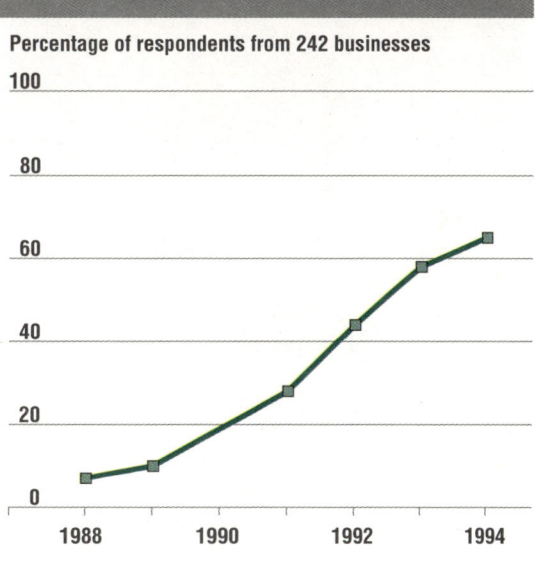

Figure 10.3 Businesses with environmental policies

Percentage of respondents from 242 businesses

Source: Coopers and Lybrand Consultants, 1994.

Figure 10.4 Decision making framework for ecological sustainability

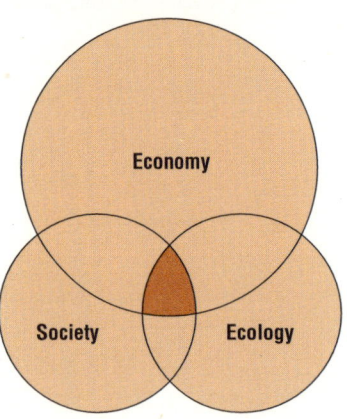

A. The predominant model of decision making in Australia until the 1980s. It gives primacy to economic decisions and assumes that environmental problems can always be solved if the economy is sound.

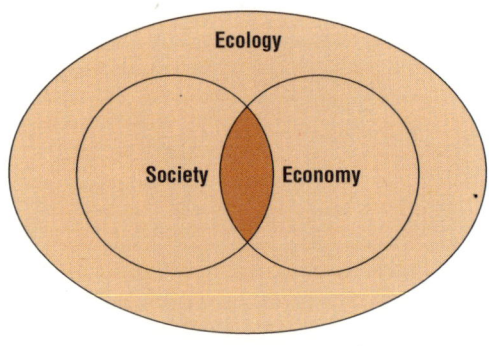

B. This model of decision making is a variant of the previous one. Use of this model began in the 1980s. It characterises the piecemeal approach towards ESD.

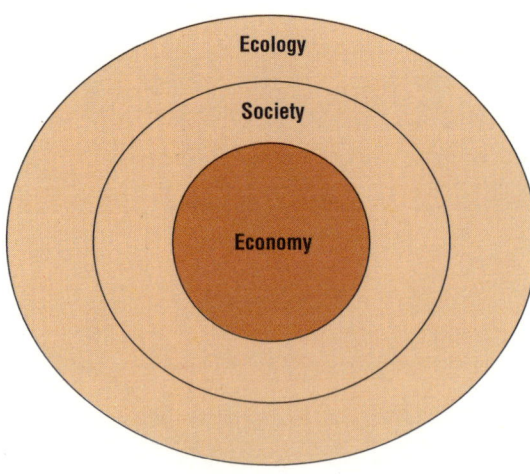

C. This is the decision making model needed for an ecologically sustainable future for Australia. It recognises that the economy is a sub-set of society, since many important aspects of society do not involve economic activity. Similarly, human society is totally constrained by the natural ecology of our planet. It requires integration of ecological thinking into all social and economic planning.

Source: Adapted from Lowe, 1994.

We base much of our decision making on the implicit premise that the prime consideration is economic and that environmental problems can always be solved if the economy is sound. Thus it is common to use a mental model like Fig. 10.4A, and the sort of model depicted by Fig. 10.4B is seen as a progressive variant. The only basis for sensible long-term decisions is the model shown in Fig. 10.4C. It recognises that the economy is a sub-set of society, since many important aspects of society do not involve economic activity. Similarly, the natural ecology of this planet totally encloses human society. Daly has made the same point, arguing that most poor planning arises from the palpably false premise that the ecology is part of the economy rather than vice versa (Daly, 1992).

Institutional arrangements

Historically, problems in our government systems, which have tended to use waterways as boundaries, have hindered rational management of catchments and marine ecosystems. This means that each complex ecosystem is managed as a series of isolated sectors, rather than as an integrated whole. A concern for integrated management of catchments has led to re-examination of the structural impediments to that approach. Some catchment management authorities have been formed, although so far without the power to prevent inappropriate patterns of development.

Concern is felt about the loss of biodiversity — caused mainly by the destruction of habitat — within our aquatic systems. The Murray–Darling Ministerial Council has just made a decision of great historic significance. It has agreed to place an upper limit on the extraction of water from our largest river system. This is a positive step in overcoming structural impediments to the management of complex ecological systems spanning jurisdictional boundaries.

Catchments are an extreme case of a general problem arising from the structure of government in Australia. As a federation of six States and two main Territories, Australia has so far shown little determination to take unified action on environmental issues. Many important decisions are the province of State governments, some of which have environmental protection low on their list of priorities. When one State takes a position of leadership on environmental issues, as South Australia has with container deposit legislation to reduce packaging waste, other States may see their own lower standards as offering a competitive advantage. Even common efficiency standards for electrical appliances have been delayed to allow for consensus between the Commonwealth and the States.

Impediments to concerted action also occur within each level of government. Resource development agencies and industry tend to press for economic development, leaving agencies charged with environmental protection to affect some compromise at the bureaucratic level. Thus the intentions of the National Strategy for Ecologically Sustainable Development are insufficiently translated into overall coordinated policies.

Key issues

The Advisory Council has identified the key issues arising from this report as: the need for an overall systems approach, biodiversity, land degradation, global climate change, water issues, coastal and urban environmental issues and finally social and cultural issues such as the well-being of indigenous Australians and protecting Australia's heritage.

Systems approach

The problems of catchment management provide a concrete example of the need for an integrated approach to management. Without that perspective, well-intentioned policies may inadvertently make the situation worse. Policy responses must consider the overall picture at the level of catchments and biophysical regions and have a cross-sectoral approach.

Similarly, agricultural advice needs to be based on a systems approach, rather than being directed to measures of short-term agricultural production. Some tillage practices have been associated with various forms of land degradation such as erosion, soil compaction and decline of soil structure. Growing awareness of these problems has led to improved management of agricultural land, but not universally. Integrated pest management appears to have better prospects of sustainable yields than a reliance on chemicals alone to defeat insect pests and weeds. In irrigated areas, unsustainable levels of water extraction threaten the health of riverine ecosystems, especially in seasons that are drier than average. A high priority for rational management is to improve understanding of both natural variability and the sustainability of different patterns of water allocation.

The problems of river systems literally flow on to estuaries and the ocean. If a river is heavily loaded with nutrients and sediments, it increases the burdens of sediments and nutrients in the sea. Estuaries tend to trap and accumulate pollutants. Thus the failure to address nutrient levels in run-off from agricultural areas or hydrocarbon waste in run-off from urban areas can reduce the capacity of an estuary to produce young fish and so reduce the productivity of a fishery.

These examples illustrate a general principle: the need for an approach that incorporates the complexity of natural systems as well as integrating natural and cultural environments. A coherent policy is unlikely to emerge from a compartmentalised approach where different government agencies and different levels of government all act separately.

Habitat loss and biodiversity

We have a rich and unique biota, with an unusually high proportion of local species found nowhere else on earth. Australia has accepted the responsibility of conserving its biological heritage for future generations. Through its ratification of the Convention on Biological Diversity and the development of the National Strategy for the Conservation of Australia's Biological Diversity, the

▲ Australia has a rich and unique biota, with an unusually high proportion of local species found nowhere else on earth.

Australian government is commited to preserving our biodiversity through *in situ* conservation measures and the sustainable management of natural systems.

This report identifies four key issues affecting biodiversity: clearing of vegetation, especially sub-tropical woodland; loss of native forests; threats to individual species (especially from feral animals and other introduced species); and the absence from the reserve system of representative ecosystems. The biodiversity of marine ecosystems is also threatened by the degradation and destruction of coastal habitat. Underlying all these issues is our limited knowledge of the diverse biota of Australia.

Clearing of vegetation

The most serious pressure on biodiversity is loss of habitat. Clearing of vegetation removes trees and other plant species directly, but also disrupts the web of life. It removes the food of herbivores, which are in turn the prey of carnivores. It removes sites for nesting and breeding. By exposing the soil to sun, wind and rain, it sets in train erosion and loss of topsoil, as well as changing the local micro-climate and thus influencing its acceptability as a habitat. It can also change the pattern of run-off after rain and so affect the water table; in one example quoted in Chapter 4, clearing of vegetation has caused saline groundwater to rise by more than 20 metres.

Clearance of native vegetation is the single greatest threat to terrestrial biodiversity. Clearing began with the establishment of agricultural enterprises such as cropping and grazing. Further losses have resulted from over-grazing, salinisation or changes in the availability of water. Chapter 2 refers to the 'development ethos' which led to the clearing of considerable areas of land. To some extent, this ethos is still influential.

In Queensland and Western Australia, leases often specifically require that land be cleared. Clearing is still going on in Queensland at the rate of about half a million hectares per year. In 1994 alone, permits were granted for the clearing of almost 1.1 million hectares, of which only about 0.4 million hectares was re-growth or woody weeds. These permits will remain valid for five years.

Experts advise that the scale of land clearing is causing serious loss of biodiversity as well as other effects such as increased emission of greenhouse gases. The issue is of such public concern that the Queensland government in 1995 announced its intention to introduce guidelines similar to those operating in South Australia, Western Australia, Victoria and New South Wales. Such guidelines effectively subject agriculture to the same sort of environmental restrictions as those routinely applied to other areas of economic activity.

Statistics from South Australia indicate that land clearing controls have protected significant areas of native vegetation that would otherwise have been lost.

Finally, bushland is being cleared on the fringe of major settlements as a result of urban spread. Clearing of sclerophyll forest in the south-east Queensland region is an inevitable consequence of a level of population growth greater than can be accommodated by the re-urbanisation discussed in Chapter 3. It represents the loss of the habitat of various plants and animals, including one which is almost a cultural icon in Australia: the koala.

The spread of suburbs, hobby farms and rural coastal development exerts related pressures on biodiversity. Scattered development into bushland clears habitat directly, but it also increases the problems of weeds and feral animals at the edges of the remaining uncleared bush. While there is a continuing debate about the degree of impact of feral cats on native animals, there is no dispute that cats are a significant problem. The pattern of development can also alter the fire regime, putting wildlife at greater risk (Flannery, 1995). The noise of human activity, especially the motor vehicles that make such scattered development possible, tends to drive away mammals.

Native forests

About 40 per cent of Australia's forests have been cleared in the 200 years since European settlers arrived, with another 35 per cent affected by logging. Political pressure is increasing to preserve as much as possible of the remaining 25 per cent of original forest as well as some areas affected by logging. Significant areas of forest are in reserves, national parks or World Heritage Areas, including a large fraction of our remaining tropical rainforest. However, even tropical rainforest is still being cleared for agriculture, tourism and housing.

The 1992 inquiry into the forest and timber industry by the Resource Assessment Commission (RAC) examined the options for 'old-growth' forests — forest that has never been logged. The RAC concluded that only about 14 per cent of all the unlogged forest is in conservation reserves, but this overall figure conceals a wide range of figures for different types of forests. Some 64 per cent of mangrove and swamp forests are in reserves, compared with only five per cent for south-eastern

dry forests and woodland (RAC, 1992). These disparities strongly suggest that the reserve system is mainly protecting the resources that are not commercially significant. The Australian government recommended as a target in 1995 that the secure reserves should contain at least 15 per cent of the pre-European area of each forest type, with the aim of preserving the biodiversity in forest ecosystems.

The Regional Forest Agreement process initiated by the government in 1995 will assess the extent of old-growth forests and seek to place in a national reserve system at least 60 per cent of all existing old-growth area, including at least 90 per cent of rare types of old-growth forest. In December 1995 the Commonwealth established Deferred Forest Areas with the stated intention of protecting potentially important forests and reserves while comprehensive regional assessments are completed.

The RAC concluded that clearing of old-growth forests was potentially a violation of the precautionary principle in that an irreplaceable resource would be destroyed. The National Biodiversity Council and other authoritative bodies can see no justification for any further clearing of such forest remnants.

Proper management of old-growth forests is difficult to achieve. It takes many decades or centuries to recreate an old-growth stand. A policy of total preservation is not sufficient to maintain old-growth forests into the future. They are but one part of a cycle of renewal and ageing in native forests. Old-growth stands will inevitably be lost through wildfires and thus we need to ensure that there is a pool of maturing forests ready to replace them.

As the issue of logging in native forests arouses great passion, the Advisory Council had hoped to be able to use the State of the Forests Report to reach some conclusions on the sustainability of current operations, but that report has been delayed by difficulties in obtaining data from some States. This is an example of the problem discussed earlier — institutional arrangements constituting an obstacle to sustainable management of natural systems. Some States continue to subsidise logging operations that are not economically viable (see for example NSW Public Accounts Comittee, 1990) and have been reluctant to provide data that would allow outside scrutiny of their management of the forest resource.

Loss of species

As already mentioned, the most serious pressures on biodiversity arise from the large-scale modification of habitat, particularly the removal of vegetation for urban development, agriculture and forestry. The rate of loss arising from these pressures is dangerously high. Chapter 4 reveals that five per cent of higher plants, seven per cent of reptiles, nine per cent of birds, nine per cent of fresh-water fish, 16 per cent of amphibians and 23 per cent of mammals are extinct, endangered or vulnerable.

It is not only the group of species listed as endangered or vulnerable that are affected. This is the visible peak of the pyramid, but equally serious problems result at lower levels. For example, Table 4.11 shows that 0.5 per cent of fresh-water fish species are extinct, 4.1 per cent endangered and 3.7 per cent vulnerable, making a total of about nine per cent in these categories of very serious threat. However, a further nine per cent of species are poorly known and 17 per cent rare, making a total of one-third of all known species to be a cause for concern. This report contains a comprehensive list of species that are extinct or under varying degrees of threat (see Appendix 2).

The process is rarely a simple one, although examples occur where targeted human intervention has endangered or removed a particular species, as with the cutting of red cedar or the hunting of large whales. More often, a sequence of changes results from an action such as land clearing.

A lack of diversity in small populations can lead to increased susceptibility to disease or other threats. Removal of nutrients by activities such as timber harvesting may threaten the overall health of the ecosystem, although not enough data on re-growth forests are available to say whether current logging operations are having that effect. Clearing of trees and other vegetation leads to changes in rainfall interception and evapotranspiration, with subsequent impacts on the soil and the water table.

Introduced species have significantly disrupted local ecosystems. As discussed in Chapter 6, the introduction of exotic species continues, even though we understand the consequent problems better. Between 1947 and 1985, more than 450 exotic grasses and legumes were introduced into northern Australia; only four of these species turned out to be useful without also causing weed problems. Feral cats are a major problem right across the continent. At least 18 exotic mammals have established wild populations in Australia, including rabbits, foxes, dogs, pigs, water buffalo, donkeys, goats and horses. They exert a pressure on biodiversity by eating native fauna and flora, as well as competing with native species for habitat. European carp degrade inland aquatic environments and some introduced marine species are adversely affecting marine habitats. Foxes have been blamed for the decline, and possibly the extinction, of several native animals. The recent, unplanned release of rabbit calicivirus has exposed a range of issues demonstrating both the complexity of ecological systems and the difficulty of understanding the many consequences of changing the abundance or health of one species within a system.

There are inspiring examples of success in reversing the decline in biodiversity. The recovery of the Lord Howe Island woodhen was achieved by a management plan developed after detailed ecological study. On the other hand, the recovery plan for the Western swamp tortoise is a classic example of treating the symptoms of decline rather than the cause. However much effort is put into

More than 20 per cent of Australia's natural land cover has been removed by clearing for agriculture since European settlement.

restoring the species, the plan is unlikely to succeed if the habitat continues to be depleted.

This example shows why protection of biodiversity must be part of the overall management of the ecological system. Chapter 4 discusses a variety of economic instruments that could be used to conserve biodiversity.

Lack of representative ecosystems in reserves

The reserve system has generally evolved through a series of ad hoc decisions to acquire parcels of land. Until recently there has been no strategic effort to ensure that all major biological communities are represented in protected areas. On the contrary, the tendency has been for land of marginal commercial value to be set aside as reserves. This issue was canvassed for the particular case of native forests on page 10-14.

Ecosystems most in danger of being displaced by crops or pastures often receive no protection; reserves occur where there are fewest competing claims on the land concerned rather than where the threats are greatest. Thus, they tend to be established on Crown land, which is often land that has not been released for other uses because it is too steep or too poorly drained.

The lack of a systematic approach means that the reserve system does not usually represent essential natural features, such as species or ecosystems. Inevitably some will be well represented and others overlooked. Priority in extending the reserve system should be the full representation of all major ecological communities.

We still have the opportunity to balance the conservation of biodiversity with social and economic objectives. The primary mechanism for ensuring that the loss of biodiversity is reduced and finally halted is the comprehensive implementation of the National Strategy for the Conservation of Australia's Biological Diversity and the National Strategy for the Conservation of Australian Species and Ecological Communities Threatened with Extinction.

Land degradation*

Land degradation is a complex issue and aspects of it are also discussed elsewhere in this report. Evaluation of the extent of degradation needs to take account of the intensity of land occupancy. The following criteria were used to assess land degradation:

- Roughly half the continental land cover retains essentially pre-European vegetation, but with numerous instances of genetic loss and physical erosion. These regions have been assessed relative to their pre-European ecological condition.

- Some 40 per cent of lands have been profoundly changed by complete clearing or thinning of native vegetation for agricultural and pastoral use. These lands have been evaluated relative to conditions required for optimal productivity under current systems of crop and livestock production.

- Five per cent of land is still forested, with multi-purpose values, including forest products, conservation and hydrology. This land was assessed against criteria framed to ensure retention and sustained future use.

Issue	Pressure	State	Response
Conservation of ecosystem integrity of desert, coastal, mountain reserves	Impact of numerous bore water points; invasions by vertebrate pests and weed; coastal development activities	30–40 per cent of floristic communities affected; 10 per cent severely (seagrass, woodlands, mangrove, rainforest, shrublands)	State legislation on clearing/pest control strategies; some increase in reserve size/number; very variable local government actions
Coastal and inland wetlands decreasing	Infrastructural impacts; off-site effects from rural–urban lands, drainage and water use schemes	Large reduction in area; increased exotic weeds and fish; increases in nutrient and heavy metal inflows	Increase in preservation orders — often ineffective, public concern (land care groups), more demand for recreational access
Increase in hobby farms with fragmented land management	Land-use intensified; land abandonment; inadequate local controls	All metropolitan regions affected; land and waterway pollution; reserve fragmentation	Varied —depending on local government control, planning and land values
Sustainability of rangelands	Overstocking by domestic animals in dry years; total grazing pressure greatly increased by vertebrate pests	Both positive and negative trends found in the 51 per cent of area surveyed; 2.3 per cent land irreversibly eroded; 15 per cent needs destocking to recover; weeds and small pest species increasing; large pests decreasing	Value of current pastoral systems being reassessed by governments and the private sector in light of pressures and developing alternative land uses (eg Aboriginal land uses and ecotourism); a National Strategy for Rangeland Management is being developed jointly by State, Territory and Commonwealth governments and non-government organisations, with extensive community consultation; it addresses the commercial viability of pastoralism, where appropriate; the draft strategy is expected to be released in mid 1996; new drought policies put onus on risk management
Sustainability of agricultural regions	Low profitability of 50 per cent of enterprises reduces resource care; major effects from loss of perennial vegetation, low use of lime and run-down in fertility; erosion from bare land and over-use of irrigation water	Secondary salinity increasing; saline ground-water rise increasing; soil acidification increasing; wind and water erosion stable; soil fertility declining in one third cropped land and increasing in 10 per cent; more lime and gypsum used	One-third of farmers in Landcare; catchment and regional plans for resource use in each State; changes to State resource legislation; controls on clearing; increases in water prices; controls on irrigation water use
Sustainability and multiple use of forest and woodland regions	Continued cutting and fragmentation of smaller remnants; altered fire regimes in large state-managed stands; low investment in plantation forestry	Historical overuse and genetic loss affect condition; smaller remnants are non-regenerative and still losing biodiversity; net area of clearing and mortality greater than area of tree-planting	Governments have agreed not to reduce forest estate further; poor reporting on biology acknowledged; comprehensive regional assessments initiated to manage forests; no tax concessions for plantations

*Synthesis by Ann Hamblin (CRC for Soil and Land Management, Adelaide), based on Chapters 4, 6 and 7.

Land degradation

Degradation of agricultural and pastoral land is a matter of considerable concern. The failure to take account of weather variability has led to overstocking of rangelands in dry years. Another serious pressure comes from introduced herbivores, such as rabbits and goats, and increases in populations of native herbivores such as kangaroo species. These factors have contributed to a situation in which more than two per cent of rangeland is irreversibly eroded and a further 15 per cent needs de-stocking to recover.

The condition of rangelands differs between regions. The Western Australia arid zone and particular areas of Queensland — Cape York and the Burdekin in the north and the Channel country and Maranoa/Warrego in the south-west of the State — are most severely affected. Local variability also occurs, with a tendency for most severe degradation to be along creek and river frontages. The problem may have a significant effect on the economic productivity of pastoral areas, on future alternatives for land use and on conservation values.

Much of Australia's agricultural land is also under pressure from erosion, loss of vegetation cover and overuse of irrigation water. Problems such as soil salinity, acidification and rising groundwater all appear to be increasing in severity. Soil fertility is declining in one-third of all cropped land, more than offsetting the improvement in the fertility of 10 per cent of land.

While remnants of native vegetation remain as narrow strips along road verges or isolated uncleared areas, the overall condition of land cover is a cause for concern. Some of this remnant vegetation appears to be in poor condition because seedling regeneration is limited by introduced herbivores or competition from introduced flora. In land not used for agriculture or grazing, invasion by weeds and vertebrate pests has had serious impacts.

Global climate change

Australia currently has the highest energy-related carbon dioxide emissions per unit of GDP of any OECD country. The various reasons for this include: our almost total reliance on the burning of fossil fuels as our primary energy source; the overall structure of the economy (taking advantage of our abundant natural resources) and subsidies to energy-intensive industries; land clearing for agriculture; the emphasis on the private car for personal transport and road transport for freight; and the 'tyranny of distance' between and within our cities.

The National Greenhouse Response Strategy provides a reasonable basis for addressing our obligations under the Framework Convention on Climate Change. However, there is little sign of a serious commitment to this Strategy. The most obvious problem is once again institutional. Energy and resource development agencies in some States, having charters that limit the possibility of considering broad issues such as global climate

Salt encrustation and soil erosion in the upper Kent River, south-west Western Australia.

change, are pressing ahead with plans to expand energy-intensive activities. At the recent Berlin conference of the parties to the Framework Convention it was clear that Australia and most other developed countries will fall short of returning their emissions to 1990 levels by the year 2000 as required under the Convention. International sympathy is unlikely for the view among Australian energy and resource development agencies that increasing our emissions is in the global interest.

Two recent initiatives are significant. The Commonwealth government released a package of response measures, *Greenhouse 21C,* and private sector peak bodies, in an endeavour to show that proposals such as a carbon levy will not be needed, began encouraging large companies to enter into cooperative agreements to limit carbon dioxide emissions. At the time of writing, it is too early to judge the effects of these initiatives.

Water issues: catchments and marine ecosystems

Chapters 6 to 8 of this Report detail the importance of managing Australia's natural assets on an integrated, large-scale basis. How we use land resources affects the state of inland waterways, which in turn affects the state of marine and estuarine environments. So, success will be more likely if future planning is based on biophysical regions, management of water systems on integrated catchments, and protection of the oceans on the basis of an integrated approach to the whole marine ecosystem. These conclusions are inescapable.

As Australia is the driest inhabited continent, it should not be surprising that a group of key issues relate to water and its management. The scarcity of water is an issue in many parts of the country, while water quality and nutrient loads are serious problems in many areas. Waste water disposal is a particular problem in urban areas, while some parts of rural Australia are seriously affected by rising groundwater. Finally, pollution of coastal waters is both a problem in itself and a source of concern because of its impact on fish stocks.

Water quantity

Future planning must accept the high natural variability of rainfall and run-off in Australia. The level of stress affecting the Murray–Darling Basin is partly due to its lack of water which in turn is in part the result of past failure to recognise the variability of run-off. It is also partly a consequence of the allocation for irrigation of much of the natural flow of these rivers and their tributaries.

Seventy-five years ago, about 15 per cent of the average natural flow of the Murray–Darling Basin was diverted for human use, leaving 85 per cent to flow naturally through the river system to the sea. Today, about 80 per cent of the average natural annual flow is diverted, leaving only 20 per cent, which is not adequate to sustain the natural systems of the river basin. But the situation is even worse than suggested by this average figure; the considerable variability of rainfall and run-off mean that the volume diverted in dry periods may equal the total available water, leaving little or none to sustain the natural systems of the river basin.

Environmental consequences have followed inevitably. With rivers and wetlands starved of water, riverine environments have been disrupted and wetlands destroyed or seriously disturbed, putting great pressure on the species reliant on those habitats. Irrigation is also associated with waterlogging and salinisation of agricultural land: about one-third of the land area irrigated in Victoria now suffers salinity problems.

The figures suggest that the problem lies with a model of decision making, based on short-term economic objectives of individuals rather than long-term State or national interests. Thus irrigation and town water supplies have taken precedence over riverine ecosystems for the water in our inland rivers. Natural ecosystems must retain as much water as they need to remain healthy — one reason is to ensure the quality of drinking water. Only if spare capacity remains, should extractive uses such as irrigation be permitted.

Water availability is likely to be an increasingly serious problem in southern Australia. The Great Artesian Basin supports much of inland eastern Australia. It lies underneath about one-fifth of the continent, typically at depths of more than one kilometre. Uncontrolled extraction from bores has depleted the resource significantly. About 20 per cent of bores in south-west Queensland have ceased to flow because of declining water pressure in the Basin. Average flow rates have steadily declined from 600 million litres per day early this century to less than half that level today. A program of capping bores to conserve groundwater is underway, but it will take several decades to complete at the planned rate of implementation.

Thus, the viability of those agricultural activities that rely on bore water from the Great Artesian Basin is in doubt. They have relied on the harvesting of water, which has taken hundreds of thousands of years to percolate through rock strata to inland Australia. That practice clearly cannot be sustained at the historic scale.

Water quality and nutrient loads

Several of the impacts on inland waters are a direct consequence of land management practices. Many waterways are affected by sediment loads borne by run-off. The scale of these loads is a direct consequence of management practice. Nutrients, especially phosphorus and nitrogen, as well as pesticide residues, are washed into waterways.

Dryland agriculture is associated with increasing levels of salt in waterways, causing problems for aquatic biota. Many of these native species are in serious decline, partly as a result of the changes discussed above and partly because of the pressure of introduced species. Exotic fish, plants and mammals all pose threats to native species. Some significant fish species are trout and European carp, while considerable damage has also been done along waterways by introduced plant species such as *Mimosa pigra*. In the face of these problems, several States have adopted an integrated approach to the management of whole catchments. Significant initiatives have followed, ranging from catchment management authorities to major engineering works like the Dawesville Channel, intended to revitalise the Peel–Harvey estuary in Western Australia.

As Australian soils have low levels of naturally-occurring nutrients, our agriculture depends on chemical fertilisers for the phosphorus needed by crops and pastures, and the nitrogen needs of many crops. There is no doubt that some of the phosphorus in inland waterways is due to the application of fertilisers, while some is due to removal of the element by erosion on cleared land. The relative contributions of these two mechanisms is disputed, but there is no disagreement that the dominant source of the phosphorus burden in inland rivers is agricultural land.

Land use practices have changed local hydrology. Land clearing has raised water tables and caused salinisation. Storing water in dams is a rational response to variable rainfall from the viewpoint of the individual landholder, but the cumulative effect of diverting large volumes of water from natural watercourses is considerable change to the riverine environment. The inland rivers of New South Wales have been seriously affected by the level of withdrawal of water, mainly for agricultural purposes.

Groundwater disturbance

Groundwater is an integral part of the water cycle. People living in about 60 per cent of the land area of Australia are totally dependent on groundwater, while those in another 20 per cent use it for the majority of their needs. The time scale for change in groundwater systems is very long. While surface water takes a few weeks to flow from the Great Dividing Range in Queensland to central Australia, water in the Great Artesian Basin takes millions of years to travel the same distance. Two serious problems are associated with groundwater: depletion of the resource in northern and central Australia and rising levels of groundwater in southern Australia.

Clearing has caused severe degradation of agricultural land. The replacement of deep-rooted vegetation, such as large trees, by shallow-rooted species such as crops, has caused groundwater levels to rise. Water-logging of the root zone can cause serious decline in economic yields in the short term. In some cases, the rising water levels are redistributing salt, leading to salinity problems. This can have devastating impacts on rural areas. The reclamation of agricultural lands affected by this condition may not be economically feasible, even over many decades. More than two million hectares of land in southern and western Australia are already affected. The levels of water tables are rising at rates as high as 50 centimetres per year. As the problem results from past indiscriminate clearing of land, the situation can be improved by planting salt-tolerant species of trees, but it requires an extensive and concerted approach.

Waste water disposal

A range of pollutants contaminate industrial and domestic waste streams and urban run-off. Few quantitative data are available, even at local levels, and there are no credible national figures for the resulting load on waterways and oceans. Chapter 8 quotes the impacts on Moreton Bay of suspended sediments, heavy metals, nutrients and pathogens in the run-off from the Brisbane River. A 1987 study estimated that 17 per cent of mangroves and 21 per cent of saltmarsh has been lost, with significant impacts also on seagrass meadows and algal communities. Despite growing concern about the state of Moreton Bay, crucial decisions remain fragmented among a host of different authorities.

Although the build-up of nutrients in waterways is a major problem, upgrading of waste water treatment is only proceeding slowly. Ocean outfalls continue to be the major form of disposal. In New South Wales, about 200 storm-water outlets discharge into the ocean adjacent to the Sydney metropolitan area.

Overall, it has been estimated that about 80 per cent of contaminants in the sea and estuaries have found their way into the water from the land. The main sources of contamination are storm-water systems, sewage effluent discharges and run-off from agricultural areas. In the absence of integrated systems of catchment management that include urban areas, there is little prospect of resolving this problem.

Coastal pollution and fish stocks

There are many adverse impacts on the coastal environment. An example is the increasing incidence and intensity of marine algal blooms, which have now been reported from most parts of

▲
Many areas of saltmarsh and mangroves have suffered from coastal development.

the coastline. Annual outbreaks in the Port River estuary, near Adelaide, have poisoned mussels and massive blooms of blue-green algae in the Peel–Harvey estuary in Western Australia are affecting fish catches and crab populations.

Shellfish beds in Tasmania and Victoria have been closed as a consequence of the less dramatic events now regularly occurring right around the coastline. Evidence suggests that non-toxic algal blooms are associated with the die-back of seagrass. While it is suggested that the increased frequency and size of algal blooms in the waters off east Queensland may be caused by nutrient run-off from the land, the data are not conclusive.

Coral reefs appear to be in decline around the world. A definitive picture of the state of the Great Barrier Reef cannot be established at this time, as the current monitoring programs have not been conducted for long enough to give clear results. Photographic comparisons cited in Chapter 8 show reductions in the level of hard coral cover on some reef flats in the inter-tidal zone. Crown-of-thorns starfish have damaged nearly 20 per cent of reefs to a greater or lesser extent in the past 30 years, although there is still debate about whether or not these outbreaks are related to human activity.

Overexploitation has put several species of fish at risk, reducing the breeding stock to dangerously low levels. Examples include the southern blue-fin tuna and eastern gemfish. The orange roughy and coral trout catches per unit of effort are also declining, although there is no evidence that breeding stock has reached dangerous levels; continued monitoring of the situation is needed. While the numbers of other creatures such as sea snakes and turtles are more difficult to estimate, the overall population of loggerhead turtles is declining, to the point at which regional populations are now threatened.

On the other hand, the marine area can provide examples of successful management of resources through scientific assessment and subsequent intervention. Management of the tropical rock lobster fishery involves Commonwealth–State cooperation as well as the involvement of another nation, Papua New Guinea. The goal of protecting the traditional way of life and livelihood of the Torres Strait Islander people has been achieved by restricting access by others. The fishery is still assessed as underexploited, or able to sustain higher rates of harvesting than the current activities.

Following assessments that the barramundi stocks of northern Australia were being seriously depleted, a management plan was introduced. While continuing to allow traditional fishing by local Aboriginal people, the plan reduced commercial exploitation to a level that appears to be sustainable. The development of such management plans for fisheries reflects increasing recognition of the need to exploit fish resources at rates below the maximum sustainable level.

The marine environment is also significantly affected by introduced species. Toxic marine algae threaten fisheries in Victoria and Tasmania, and concern is growing about the Northern Pacific Seastar, the spread of which is a hazard to scallop and abalone fisheries, aquaculture farms and marine life generally. Chapter 8 notes the invasion of saltmarshes by introduced species, capable of altering the natural ecology.

Coastal and urban environmental issues

Coastal settlement

Chapter 8 concludes that most estuaries well away from population centres experience little disturbance, but others are under heavy pressure. Regulations control emissions from point sources, such as major industrial operations, but significant problems arise from diffuse sources such as urban run-off from roofs, driveways and streets.

One factor likely to disturb the ecosystem of an estuary is land clearing. More than half the land area of the respective catchments have been cleared. Around 86 per cent of the estuaries in South Australia have more than 50 per cent of their land cleared, as do 60 per cent of those in Victoria and 37 per cent of those in New South Wales. Enclosed gulfs and bays near major cities, such as Port Phillip Bay in Victoria and Moreton Bay in Queensland, are particularly affected by run-off from urban areas.

These examples highlight a general problem: concentrated coastal settlement and its impacts on the coastal zone, especially in the east and south-east of the continent. Although Chapter 3 shows that environmental impacts per head are often higher in small settlements, urban environmental issues relate to the coastal zone and the major cities because these are the places where most people live. In recent decades the population of the non-metropolitan coastal zone has grown dramatically — it almost doubled in the 20-year period from 1971 to 1991. This urban growth involves clearing of land, damming of rivers for water supply or modification of flow regimes for flood control, as well as pollution of water from sewage and urban run-off. Other impacts include shore-line erosion associated with coastal engineering structures such

as breakwaters and sea walls. Above all else, the extensive development of the coastal zone is associated with the loss of bushland, degradation of coastal ecosystems, and impact on cultural heritage.

Developments ranging from dumping of rubbish to land 'reclamation' have damaged coastal saltmarshes in the inter-tidal zone. The loss of saltmarshes in the temperate zone of south-eastern Australia is linked to significant losses of endemic species. Extensive clearing of mangroves has occurred near urban areas, with further significant areas under threat from planned development.

A particularly serious problem appears to be the loss of seagrass beds, with more than 1600 square kilometres (about three per cent of the total known area of Australian seagrass) lost in recent decades around the coast. Much of this loss is attributed to floods and cyclones, with the remainder due to human changes such as elevated nutrient levels, run-off, dredging and reclamation.

Prospects for recovery are poor. There is no record of temperate seagrass recovering from damage or loss on reasonable time-scales, while the tropical and sub-tropical species take more than a decade to return under best conditions. Seagrass beds are critical habitats for turtles and dugong, as well as being important nursery areas for commercial and recreational fisheries. Their loss must cause concern, especially as the temperate seagrass areas may not respond successfully to changes in climate.

Thus a range of serious environmental problems affect our inland waters, estuaries and coastal regions. The biological and physical interactions mean that an integrated approach to these problems is needed, encompassing the chain of events from land management in catchments through to the coastal zone.

Towns and cities

Although the environmental problems of Australia's largest settlements are arguably less serious per head than those of many smaller settlements, urban problems in our cities are among the nation's most important environmental concerns. Four of these issues have been singled out for discussion in this chapter: the dispersed and low-density character of our settlements; waste minimisation and waste management; the retention of remnant vegetation and habitat; and changing patterns of population distribution.

The dispersed and low-density character of our settlements has created cities that depend heavily on motor vehicle transport. The rail networks in our cities, especially Sydney and Melbourne, are extensive but public transport patronage has declined significantly over the past 50 years. Research, funding and management for public transport have all been deficient. Commonwealth funding for transport has, until recently, concentrated on road building. Few attempts are made to integrate land use and transport planning across all possible modes, although it is rational to take a comprehensive approach to examining the best economic, environmental and social options for transport provision along particular corridors or in particular areas. Severe environmental and social impacts are the result. Australian cities use high levels of transport energy, generate high levels of pollutant emissions and create unequal conditions for those with little or no access to a car.

The long-term trend toward smaller, more numerous households means that growth is compounded by a tendency to spread further from the urban cores, and use more transport fuel. As transport fuel use is a major source of urban air pollution, this problem is directly linked to population growth and distribution. A recent poll showed that Australians see urban air quality as one of the most important environmental problems. While most indicators of air quality are acceptable on most days, the trends are worrying. Precursor conditions for photochemical smog formation apply in some of our cities, and in all our major cities the emission levels of the gases which cause the problem are likely to rise given current trends in demography and motor vehicle use. A recent Perth study concluded that the city is on the verge of a significant air quality problem, mainly due to vehicle emissions of gases which lead to the formation of photochemical smog in hot weather (Weir and Muriale, 1994). The report noted that Melbourne and Sydney experience more photo-chemical smog events than Perth, which exceeds the World Health Organisation guideline for one-hour ozone levels on about ten days each summer.

Important changes have occurred recently. Comparative research on Australian and Canadian cities is showing that high levels of public transport patronage are closely related to the adequacy and frequency of services (Mees, 1994). This is confirmed by initiatives that provide a good public transport service. As Chapter 3 notes, the new Northern Suburbs railway in Perth has experienced dramatic growth in patronage because it provides a journey that is faster and more comfortable than driving a car. The Better Cities program is helping to improve public transport, through such schemes as the reconstruction of the Brisbane–Gold Coast railway and the development of the Ultimo–Pyrmont light-rail link in Sydney. The same program is encouraging integrated land use and transport planning, as well as supporting public transport. Bicycle networks are being built and used. At the same time, however, road building proceeds apace. As discussed in Chapter 3, large freeway programs are being undertaken in many of our major cities, although this is contrary to transport planning in most developed countries and contrary to the National Strategy for Ecologically Sustainable Development.

A different strand of responses relates to reducing vehicle emissions by improving engine efficiency, reducing the size of cars or changing the fuel used. Government policy has encouraged the use of unleaded fuel, but little impact has been made so far on vehicle size or efficiency. With petrol less expensive than almost any other liquid bought in bulk, change is unlikely to be achieved by the current approach.

North coast, New South Wales; the population of the non-metropolitan coastal zone has grown dramatically over the past two decades.

A second urban environmental issue is waste minimisation and waste management. As cities grow, the larger population produces proportionally larger quantities of waste, putting greater pressure on natural and artificial waste management systems. Modern urban refuse systems are usually well engineered to minimise their impacts on natural systems, although some cases of groundwater contamination have occurred. More serious are the effects on the urban air-shed of combustion products and the impacts on waterways of run-off from residential, commercial and industrial premises. The oil industry has pointed out that more liquid hydrocarbons reach the ocean from urban run-off than from shipping accidents.

Urban areas also face problems of solid waste disposal. The average amount of domestic refuse is about one tonne per person per year, a major part of which is unnecessary packaging. Packaging makes up about one-third of the volume of domestic waste. Some cities have made significant steps toward kerbside recycling — a real success story in Australia. Although we have adopted a range of initiatives to achieve the national target of a 50 per cent reduction in waste going to landfill by the year 2000, we need to identify more materials for waste reduction, and place more emphasis on reducing the source, if we are to achieve this target. We also need better national information about the composition of the waste stream and waste generation rates, and effective waste minimisation methods. Once again, institutional arrangements inhibit a unified approach.

A third key issue is the retention of remnant vegetation and habitat. Some cities have retained their urban creeks and associated vegetation, providing corridors of wildlife habitat — Brisbane is a large-scale example, while Perth has a much higher proportion of remnant vegetation than many rural shires in the south-west of Western Australia. In Sydney and Melbourne, by contrast, many urban streams have been turned into concrete storm-water drains. The Better Cities program is demonstrating in a number of places that water-sensitive urban planning can rehabilitate creeks and turn storm-water into a positive element of the urban landscape. While our urban areas have extensive areas of parkland, much of this is planted with imported rather than native species. Remnant bushland in urban areas is under heavy pressure from development for suburbs, hobby farms and coastal settlement.

A fourth issue relates to changing patterns of population distribution. Vigorous debate continues about the relative impact of people in different types of human settlements. It is clear that there are pressures on natural systems associated with the drift to the coastal zone, especially in Queensland and northern New South Wales, and to particular urban areas such as Brisbane, Perth and the western outskirts of Sydney. The movement of population also affects natural and cultural heritage. Growing social problems are arising from the decline of rural communities.

Social and cultural issues

These issues are particularly relevant to the first core objective and the first guiding principle of the National ESD strategy. Four key issues have been identified under this broad heading: various forms of social breakdown, the physical and cultural well-being of indigenous Australians, the need to maintain continuity under conditions of rapid change, and protecting our heritage.

Signs of social stress

Australians enjoy a high living standard, with a wealth per head which is higher than the OECD average. However, this living standard is not shared equally, with a growing gap between rich and poor households. In the 15-year period from 1976 to 1991, average household incomes in wealthy neighbourhoods (the top five per cent by income) increased by 23 per cent ($12 500) while the average incomes in the poorest neighbourhoods declined by the same percentage, or $7600. The difference in annual income between median households and those in the top percentile doubled in the same period, from $23 000 to $46 000. Australia remains a more equitable society than others such as the United States, because of the various provisions known as the 'social wage'.

However, there is a worrying trend toward a geographical concentration of disadvantage and the appearance of 'pockets of poverty' in our cities. A contributing cause is the changing pattern of paid employment, with a tendency for unemployment to be concentrated in areas of disadvantage, while two-income households are common in more affluent areas. The trend away from work in agriculture and manufacturing toward professional jobs in the service sector places a high premium on formal qualifications. The overall unemployment rate at this time is about eight per cent, but the rate among young people is more than three times this level. There is increasing concern that Australia is developing an under-class of young people who have no real prospect of paid work, as a result of

which they are isolated from the mainstream community. Aboriginal youth is particularly disadvantaged in the labour market.

On the whole, Australians enjoy good health. Average life expectancies are significantly better than in New Zealand, the United Kingdom or the United States, and similar to those in Sweden. There are, however, marked differences for some groups. Aboriginal health is a serious concern. There are also general problems of physical and mental health. Youth suicide rates are among the highest in the world. Where once the archetypal Australian was bronzed and fit, today only a small minority of adult Australians get enough exercise to maintain basic cardio-vascular fitness. While mortality related to heart disease has declined dramatically in the last 25 years, associated with improvements in diet, cardio-vascular diseases still account for almost half of all deaths. Injuries are a major health problem. Each year about 2700 Australians die as a result of work-related health problems, while more than half a million suffer work-related injuries or illness. Motor vehicle accidents are a major cause of deaths and injuries, especially in rural Australia, despite dramatic reductions since the introduction of compulsory seat belts and random breath testing. Recognising the health effects and costs of tobacco smoking, restrictions on smoking are being introduced in many enclosed public places and most advertising of tobacco products has been banned.

Well-being and culture of indigenous Australians

Many indigenous Australians suffer much poorer health than the rest of the community. While the average life span of the Australian community as a whole has increased by about 20 years this century, the average life expectancy of Aboriginal people has not and is 16 to 18 years less than the average for the whole population. Aboriginal babies are, on average, 200 grams lighter than the overall mean for the population, and infant mortality rates are four times that of the national average. Mortality of young and middle-aged adult males is particularly high; men aged between 35 and 44 die at more than six times the rate of the non-Aboriginal population (Aboriginal and Torres Strait Islander Commission, 1994). There are various contributing factors and no simple solution, but the issue deserves urgent, concerted attention. Environmental factors are directly involved in some health problems; for example, nearly 200 Aboriginal or Torres Strait Islander communities do not have adequate sewerage, contributing to high rates of preventable communicable disease.

Traditional social structures are under pressure from the encroachment of the values of the broader community, undermining the standing of elders. Cultural diversity and national heritage are being eroded through the disappearance of traditional Aboriginal languages, most of which have been lost. Of the estimated 250 languages in use 200 years ago, only 90 are still spoken today and only 34 have 200 or more speakers. Only about 20 are still being taught to children and used as the primary means of communication. All of those are under threat, despite language maintenance programs, and could be lost within a generation. Like threatened species, they are unlikely to survive if the system which has supported them is under continuous attack.

Only about 800 places of significance to Aboriginal and Torres Strait Islander people are among the almost 11 000 places listed in the Register of the National Estate, although a much larger number of sites (some 85 000) are recognised under State or Territory lists or legislation. Existing listed conservation and heritage areas are skewed toward the south-east corner of the country, with relatively poor representation of the heritage of cultural groups living in the north and west, especially Aboriginal and Torres Strait Islander groups.

Cultural continuity in times of rapid change

Rapid technological change is causing associated social change. The cultural identity of the nation is being modified by a range of external influences, especially the mass media. At the same time, the changing balance of immigration to Australia is widening and enriching the cultural diversity of the country, leading to continuous re-evaluation of what it means to be an Australian. Australia is one of the most successful examples of multiculturalism anywhere in the world.

It is a challenge to maintain a resilient social structure and cultural identity. Since Australia itself can be seen as a social and cultural landscape, it can help people's sense of continuity in the face of change if they learn to acknowledge, value and protect their cultural heritage.

Protecting our heritage

Australia has a rich and complex heritage of places and objects and well-developed systems to identify, conserve and present that heritage on behalf of the public. Despite these advantages, much valued evidence of our cultural and natural history is being lost or mismanaged for two reasons. First, protective measures are incompletely and inconsistently applied across the country. Second, heritage administration is arbitrarily divided in several ways: between the natural and the cultural, between one type of place and another, between one type of object and another, or between places and objects with close associations.

The currently recognised set of heritage places is geographically unbalanced; it does not adequately reflect the heritage of the north and west of Australia. Chapter 9 lists several reasons for this imbalance. Often the heritage values are not recognised because of insufficient community involvement in the selection process. Sometimes there is inadequate recognition of social values in heritage studies. In other instances, there is insufficient assistance to the community to identify those aspects of heritage deserving protection.

The multicultural nature of Australian society and its values are inadequately reflected in heritage registers, museum collections and protective

▲
Results of vandalism on the Sun Theatre foyer, Yarraville, Vic. Cultural heritage needs to be acknowledged, valued and protected.

measures. This is despite recent efforts of heritage administrators. Minority groups have minimal access to heritage decision-making. In particular, the belief systems and relationships with place of indigenous communities, remain largely unrecognised.

More generally, the strong links between places, objects and the meanings they are given by Australians are rarely reflected in current policies and institutional approaches. No national heritage strategy exists to integrate natural and cultural values effectively. While our knowledge of Australia's heritage has improved significantly in the past decade, so that registers now list thousands of places of natural and cultural significance, there are still major thematic gaps.

Recent legislative measures have improved the protection of important heritage sites, but much remains to be done; places significant to indigenous peoples are still inadequately protected. Our collections of heritage objects have thematic gaps and some States have not yet enacted legislative protection for significant heritage objects.

Without a national overview of the physical condition of our known heritage places, or of the objects held in collections, we neither know what we are losing nor how fast it is going. No systematic processes are in place to monitor the loss of, or damage to, our national estate. Once again, an identifiable problem is the lack of coordination between the three levels of government. Heritage values need to be incorporated by all spheres of government into their policies and program development if they are to be maintained for future generations. It should not be overlooked that this approach includes integrating heritage values into the management of the natural environment.

In summary, it is probable that we will continue to lose significant aspects of our heritage. Losing our past makes it more difficult to evaluate our options for the future.

Information needs

It remains true, as reported by the Australian Science and Technology Council in 1990, that 'Australia lacks an integrated national system for measurement of environmental quality, a national database of sufficient calibre to assess and manage environmental quality, and appropriate national baseline data to evaluate the effectiveness of

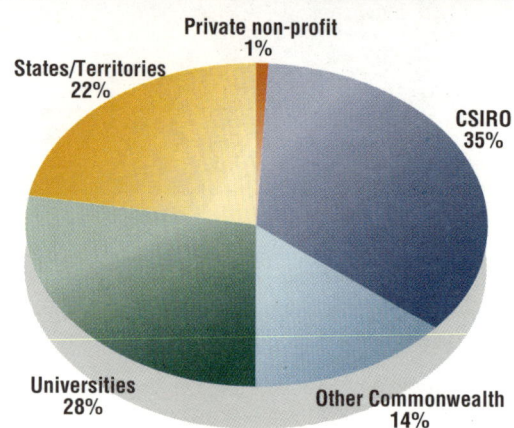

Figure 10.5 Research effort by non-business sector organisations 1990–91

Private non-profit 1%

States/Territories 22%

CSIRO 35%

Universities 28%

Other Commonwealth 14%

Source: Department of Industry, Technology and Commerce, 1994.

Table 10.1 Non-business sector spending on research and development, 1990–91

Research and development objective	$m	person years
Environmental research		
Natural ecosystems	76.1	1181
Climate	19.5	213
Water resources	19.4	298
Impacts, protection	18.6	274
Oceans	12.4	130
Atmosphere	10.7	129
Land, land use	9.5	137
Other	9.3	133
Sub total	**175.7**	**2 495**
Research into environmental aspects of economic development		
Agriculture, forestry and fishing	87.5	1231
Manufacturing	17.9	256
Mineral development	10.8	125
Energy	5.8	71
Construction	5.6	76
Commercial services	1.4	20
Transport	0.7	11
Other	8.2	105
Sub total	**137.9**	**1 895**
Environmental health		
	7.1	90
Total	**320.7**	**4 480**

Source: Department of Industry, Technology and Commerce, 1994.

strategies'. Without these requirements, we are unable to say whether our pattern of development is genuinely sustainable. Achieving sustainable development requires suitable information on which to assess whether activities are sustainable. Given the stated intention of Australian governments to develop our natural and cultural resources in an ecologically sustainable manner, we need to invest in understanding those resources and the consequences of human activities, existing and planned.

State of the Environment reporting is an important step towards improving our knowledge base for future deciisions. In many areas, the available data are inadequate to assess whether current activities are sustainable.

Scientific knowledge is most complete in describing the condition of the environment and the pressures on it. Evaluation of responses is much less clear and we need to refine the knowledge base for making future response decisions. Responses can be directed at the state of the environment, for example, cleaning up a polluted river. They can also be directed at the pressure(s) responsible; in this case, activities up-stream leading to the pollution. Responses to the pressures on the environment are generally more effective in the long run. Treating the state without taking account of the pressures is like treating a person's fever by applying cooling towels; it may reduce the observable symptoms, but does not address the underlying problem.

Considerable effort already goes into research and the collection of data relevant to environmental issues. For the latest year for which data are available, the non-business sector spent about 4500 person-years of effort and more than $320 million on environmental research and development. The distribution of the effort is analysed in Fig. 10.5 and Table 10.1. On top of this is the considerable research spending of the private sector and the prodigious efforts of voluntary groups.

Considerable research effort is being expended in two areas: natural ecosystems, mainly in the universities, and the environmental aspects of primary industry, mainly conducted by State governments and CSIRO. Some impressive achievements appear, such as the deserved reputation of world leadership in ecosystem monitoring by satellite imaging. Key indicators of environmental condition such as the rate of land clearing, siltation of rivers and the health of coral reefs are being monitored through this new technology.

However, some aspects of the distribution of research funds are difficult to reconcile with the goal of Ecologically Sustainable Development; for example, the relatively small effort on the environmental aspects of transport. Although it is an area with a major impact on urban air quality, it apparently attracts only about 10 per cent of the spending on environmental aspects of mineral development and about one per cent of the funds for study of agriculture's environmental impacts. The Australian government supports a transport research organisation, the Australian Road Research Board, which works mainly to improve road transport, but there is no body to fund parallel studies on public transport or related environmental issues such as air quality.

Despite the efforts of these thousands of full-time researchers and community volunteers, there are still large gaps of knowledge where we do not have enough information to judge whether the current pattern of development is sustainable. The earlier chapters of this report have detailed some priority areas for future effort in research and data collection. Four examples of information deficiencies illustrate this point.

Biodiversity

As Chapter 4 of this report shows, our current understanding of the biodiversity of Australia is limited. While we believe we have identified and described more than 90 per cent of vertebrate species and vascular plants, we have probably only identified about 50 per cent of the invertebrates and simpler plant species. Our knowledge of the fungi and bacteria is even more limited. With such a poor understanding of the existing biodiversity, it is not possible to say how much of it is threatened, nor what effects human activity may be having on the emergence of new species or natural selection between existing species.

The task of mapping the genetic diversity of Australia is a large one. Priority is being given to the species that are in decline or in danger of becoming extinct, but we do not have sufficient resources to preserve the current diversity. We also face the problem of inadequate knowledge about the nature of particular ecological systems to confidently establish operable recovery plans.

Air quality

Chapter 5 of this report concludes that it is impossible to assess whether all Australians are equally protected from air pollution. Data are not being collected for many important population centres, including Hobart and Darwin. As recent work on the Sydney air-shed has shown that transport of pollutants may occur between the city and the major industrial areas centred on Newcastle to the north and on Wollongong to the south, more extensive monitoring is clearly needed. Research on the formation of photochemical smog in the urban areas has also shown the need to collect different data, and Chapter 5 shows the need to monitor air toxics. As this report was being completed, the Environmental Protection Authority of Western Australia released the preliminary results of the Perth Photochemical Smog Study. The findings reveal that more extensive monitoring will be needed to identify long-term trends.

Much of the data that do exist are scattered around various industry bodies and government agencies, without an overall inventory of data holdings.

Community monitoring

In addition to formal research and data gathering networks, a huge volunteer sector exists. The Royal Australasian Ornithological Union coordinates a nation-wide network of amateur bird-watchers who provide constant streams of expert observations. About 6000 Bureau of Meteorology volunteers supplement the professional data-gathering capacity of the Bureau, 2000 of them collecting data daily. Through the Saltwatch program in the three eastern mainland States and the Ribbons of Blue program in Western Australia, adults and school-children test salt levels in local water supplies.

The Commonwealth government's Waterwatch scheme is providing nearly $3 million over three years to enable community groups to monitor physical, chemical and biological parameters of creeks and rivers. Some 3000 volunteers in south-east Queensland have collected data on local flora and fauna as part of a scheme organised by the State Department of Environment and Heritage. The Landcare program has funded hundreds of groups for monitoring activities.

Hopeful signs of change include such recent initiatives as the National Pollutant Inventory and the national approach to data collection and handling called for under the Intergovernmental Agreement on the Environment. A comprehensive national database on air quality indicators is an urgent priority.

Water quality

Chapter 7 shows that monitoring of water systems has been a low priority for most agencies, so that few long-term and nationally comparable water data are available and even basic water data do not exist in many areas. This is a cause for considerable concern, for it is impossible to say whether the water systems are being managed sustainably. Further data are needed to resolve questions about the relative contributions of various sources of nutrients affecting inland waterways.

Heritage

Chapter 9 identifies a lack of information about heritage objects and lack of monitoring of the condition of both heritage places and objects. No national monitoring scheme for either exists at present, nor for the state of traditional indigenous languages. Though the various heritage registers now list literally thousands of places of cultural importance, the picture has major geographic and subject gaps. More importantly, Chapter 9 concludes that heritage issues should figure more prominently in decision-making. There is little recognition of the need for information to allow heritage issues to be part of the political mainstream.

In conclusion, these information deficiencies together constitute a powerful case for an enhanced effort in monitoring, data collection and environmental research so that decision-makers can base their judgements on solid, credible information about the state of the environment.

Chapter 3 shows that it is not only the lack of information that matters. Information needs to be collected in particular ways to be effective. It is argued that the most appropriate way to assess the environmental performance of cities and other settlements is the metabolism model which measures the amounts of inputs (energy, water, food and other commodities) and outputs (wastes and emissions). If this approach is to be used in future, the information needs to be assembled in a suitable form.

For future decision-making, we need credible indicators of the state of the environment. An initial set of environmental indicators was developed during the preparation of this report. These will continue to be refined and further developed to provide a set of indicators that identify trends in environmental quality.

Summary

Current state of the environment

This overall summary assesses both what we are doing well and the problem areas. This report has identified positive trends and examples of good practice to be publicised and reinforced. Important questions which cannot be answered by the information now available suggest priorities for research and data collection. The Advisory Council has singled out for attention in this final summary the most important trends deserving urgent action. In particular, several aspects of current environmental management are not adequate to achieve the goal of ecological sustainability.

Most serious problems identified here seem to be due to mismanagement, usually through ignorance. We are beginning to develop the more sophisticated understanding needed to plan for a sustainable pattern of development. Integrated management is a crucial step toward that goal, recognising the need for environmental issues to be considered with economic planning and for all elements of the environment to be considered together.

Positive trends

This report identifies areas of the natural environment that are not a cause for major concern and, in some cases, are grounds for cheer.

Unlike most industrial nations, we have no widespread problem with air-borne sulfur dioxide or its serious consequence, acid rain. Our urban water quality is generally very good, as is the standard of our food, with low levels of chemical residues and metals. Away from major urban areas, our oceans and estuaries are in relatively good shape. There are some success stories of fisheries' management reducing the fishing pressure, either to restore overexploited stocks or to reduce the risk of overexploitation. Our urban housing is generally of good quality, and we have a relatively high level of social cohesion. While there are emerging problems of air quality in some cities, this report concludes that guidelines are probably satisfied most of the time. A system for protecting significant indigenous and historic places is now well established.

Adverse trends

A variety of problem areas are revealed in this report. One of the most serious is the loss of biodiversity as a result of the loss of habitat. This theme recurred through several chapters. In each of the main biological systems of land resources, inland waters, estuaries and the sea, the same problem is evident. The same pattern recurs in wetlands and saltmarshes, mangroves and bushland, inland creeks and estuaries. Our record of biodiversity loss is very poor by international standards. This is probably the most urgent issue in the whole field of environmental management. In many cases, the loss of habitat is continuing at an alarming rate, with associated inevitable loss of biodiversity.

Land degradation has emerged as an area of grave concern. In our rangelands, 2.3 per cent is irreversibly eroded and another 15 per cent needs destocking to recover. Much of our agricultural land is suffering from groundwater rise, salinity, acidification and declining fertility. Rates of soil erosion from agricultural land leave no grounds for complacency. Marginal cropping areas are characterised by shallow soils and variable climate; there is evidence of some degradation of these areas. The data show grounds for concern about the continuing productivity of much marginal agricultural land. These issues will have future economic as well as environmental impacts.

A series of problems relate to our urban areas. Overall planning is essential for the transition to sustainable development of our cities. Priorities include transport planning, urban waste management and curbing the impact of settlements on water systems. The waste products of urban areas and the spread of settlement along the coast threaten coastal systems. Cultural heritage is inevitably lost in this process. The issue of the impact of human activity on the coastal zone deserves urgent attention.

Our inland water systems are in poor shape, due to poor management in the past. Apart from recent initiatives in the Murray–Darling Basin, jurisdictional boundaries have precluded overall catchment management, while water use has not taken sufficient account of rainfall variability and the long-term needs of riverine systems. Levels of nutrients and sediment are a cause for concern, as are salt loads. The public is understandably anxious about algal blooms.

Despite strong global measures to phase out chlorofluorocarbons (CFCs), the stratospheric ozone 'hole' over the southern polar region has been at its largest and deepest during the past three or four years, thereby subjecting more Australian people and other species to increased ultra-violet light. In addition, the Intergovernmental Panel on Climate Change (IPCC) concluded in 1995 that the balance of evidence suggests a discernible human influence on climate change through an increase in anthropogenic greenhouse gas emissions (for example, carbon dioxide and methane) in combination with on-going impacts of sulfate aerosols from atmospheric pollution.

The report highlights the issue of health in Aboriginal communities. While in the population as a whole, such indicators as infant mortality and overall life expectancy reflect a high quality of life, they show that indigenous communities have not shared in the health benefits. This is a stark example of the pressures on our diverse cultural groups, needing urgent attention.

Where the judgement is complicated

Various problem areas are flagged by this report for investigation, as existing knowledge does not permit a clear assessment of the current situation. Others are shown as areas of concern rather than alarm. There is concern about the decline of some marine species, such as mammals, reptiles and certain types of fish. We cannot say whether some types of forest are adequately protected to ensure their survival.

Our reserve system is patchy, with a tendency to include areas of little economic potential while ecosystems on more productive land are rarely reserved. For a variety of situations discussed on pages 10-24 and 10-25, the available data do not permit a clear statement of the national picture; this applies to aspects of urban air quality, the natural and cultural heritage and many aspects of water systems. Better data and more credible models are needed for considered decisions to be made about complex environmental problems.

Responses and the path to sustainability

This report identifies areas in which our responses appear to be adequate, others in which they are questionable, and a third group in which they are either inadequate or counterproductive. It is important to consider this assessment as a whole.

There are many areas in which Australia is doing well. The listing of natural areas under the World Heritage Convention and their subsequent protection is a real success story, as is the increasing provision for other forms of reserve status. Some innovative structural solutions to complex management problems, such as the Great Barrier Reef Marine Park Authority and the Murray–Darling Basin Commission, are now internationally recognised as responses to these issues. In particular, the Great Barrier Reef Marine Park Authority is a good example of a multiple-use bioregion scale management regime, with a strong legislative base and tested inter-governmental arrangements. Prompt and purposeful action has been taken to phase out ozone-depleting substances, such as CFCs, earlier than required under the Montreal Protocol, and the Commonwealth government has instituted financial inducements to accelerate the transition to unleaded petrol. There have been successful initiatives to place management of significant elements of natural and cultural heritage — such as Uluṟu-Kata Tjuṯa, Kakadu and Jervis Bay — in the hands of the traditional owners.

The Landcare program involves people from both rural and urban communities.

The Landcare program is in many ways a commendable initiative, achieving desirable gains in the agricultural sector. However, the program strongly emphasises economic productivity, leading to concern about its ability to address wider issues such as water quality and loss of biodiversity. This is not a criticism of Landcare participants, but another example of the need for a systems approach to complex environmental issues.

While water extraction rates remain in many cases higher than can be sustained year-in, year-out without damage to natural systems, the emergence of catchment management bodies is a positive move forward, and the decision by the Murray–Darling Basin Ministerial Council in June 1995 to set a limit on extraction is a landmark for the sustainable management of waterways.

As part of a scheme to encourage a more responsible pattern of energy use, the Queensland government has introduced subsidies for solar hot water systems and compact fluorescent light bulbs, while the government of South Australia is publicly committed to an ambitious target of obtaining 25 per cent of its non-transport energy from renewable forms within ten years. The Rouse Hill development, on the western outskirts of Sydney, is being built around a dual reticulation system allowing 'grey water' from showers and sinks to be used for watering gardens and flushing toilets, in order to minimise impacts on the Hawkesbury River. Recent data show that more than $5 billion, or about 1.3 per cent of GDP, is now spent on environmental protection each year in Australia.

Kerb-side recycling schemes in urban areas have dramatically reduced the volume of waste and eased environmental problems associated with disposal of garbage. Better Cities demonstration programs, water efficiency appliance schemes, the National Pollutant Inventory and moves toward cleaner production are all positive initiatives that are reflected in improved environmental standards.

A second group of responses are questionable. Reform of the electricity industry is directed at achieving greater efficiency and industry competitiveness, more efficient use of resources and lower electricity prices. The reforms may lead to increased electricity use and associated pollution.

The move toward regulation of aircraft noise by taxation and legislation is positive, but it comes at a time of widespread concern about the noise impact of the new runway at Sydney airport.

Third, there are areas in which our responses seem either inadequate or counter-productive. Despite occasional successes such as the Murray–Darling Basin Commission, the national ability to manage the environment well is continually hamstrung by structural problems between different areas of government. Environmental standards vary from one State to another. State and Commonwealth governments frequently battle over these issues. The National Environment Protection Council was established in 1995 with the aim of developing national environmental protection measures.

Measures to combat the continuing depletion of biodiversity are inadequate. There are still some concerns about whether the fisheries management measures introduced are sufficient to reverse the decline in some stocks.

Where environmental protection appears to be in conflict with potential economic development, some government agencies still promote economic development with little regard for the environmental costs. While land clearing is restricted in some States, in others it is still tolerated or even encouraged. Old growth forests continue to be logged.

Urban planning in general and transport planning in particular remain problem areas, with few effective attempts to contain the outward urban sprawl or improve public transport. As this report was being finalised, three State governments were proposing massive new urban arterial roads, encouraging more cars to drive longer distances. There is little sign of any concerted attempt to redirect the pattern of consumption into a sustainable direction.

Although the need to moderate emissions of greenhouse gases such as carbon dioxide is now internationally recognised, our response at all levels of government remains half-hearted. The European Union is committed to increasing the contribution of renewable energy sources from the current five per cent to 15 per cent within ten years. Funding of a new Cooperative Research Centre for Renewable Energy Technologies has just been announced, but we need to move beyond research to implementing programs that will curb growth in emissions.

As a general point developed in two earlier chapters, we do not yet have an integrated ecosystem-based approach to the management of our resources. Until we develop that approach, environmental management will be characterised by ad hoc responses to urgent problems without a strategic vision to achieve the ultimate goal of ESD. Despite the adoption in 1992 of the

National Strategy for Ecologically Sustainable Development and the emergence of the National Strategy for the Conservation of Australia's Biological Diversity, there is little evidence that these strategies affect decision-making in any but the most perfunctory way. Similarly, cultural heritage considerations are not systematically integrated into the management of natural resources.

Finally, there is little sign that economic planning takes serious account of the ecological impact of the options available at any time; it is assumed that the first priority is a healthy economy, with the doubtful corollary that other problems will be solved by sensible deployment of the wealth created. The economy is a subset of human society which is in turn part of the ecological system. Progress toward sustainability requires recognition of that fundamental truth and a willingness to build ecological thinking into all social and economic planning.

References

Aboriginal and Torres Strait Islander Commission (1994). Indigenous Australia Today. (ATSIC: Canberra.)

Australian and New Zealand Environment and Conservation Council (1994). 'National Strategy for the Conservation of Australia's Biological Diversity.' (ANZECC: Canberra.)

Australian and New Zealand Environment and Conservation Council Task Force on Biological Diversity (1994). 'Access to Australia's Genetic Resources', (mimeo) (ANZECC: Canberra.)

Australian Bureau of Statistics (1986). Environmental Issues and Usage of National Parks, Australia, April 1986, Cat. No. 4115.0, (ABS: Canberra.)

Australian Bureau of Statistics (1995a). Cost of Environment Protection, Australia, Selected Industries, 1991–92, Table 1.1, Cat. No. 4603.0. (ABS: Canberra.)

Australian Bureau of Statistics (1995b). Environmental Issues, People's Views and Practices, June 1994, Cat. No. 4602.0. (ABS: Canberra.)

Australian Bureau of Statistics (1995c). Australian Demographic Statistics, March Quarter 1995, Cat. No. 3101.0. (ABS: Canberra.)

Australian Bureau of Statistics (1995). 'Pocket Year Book Australia.' (ABS: Canberra.)

Australian Science and Technology Council (1990). 'Environmental Research in Australia: the Issues.' (AGPS: Canberra.)

Buckley, R. (1989). What's wrong with EIA, *Search*, **20**, pp. 146–47.

Commission of Inquiry into the Conservation, Management and Use of Fraser Island and the Great Sandy Region (1991). *Report*, Queensland Government, Brisbane.

Commonwealth Environment Protection Agency (1994). National Pollutant Inventory: public discussion paper, CEPA, Canberra.

Commonwealth Environment Protection Agency (1992). National Waste Minimisation and Recycling Strategy, CEPA, Canberra.

Commonwealth of Australia (1992). 'National Greenhouse Response Strategy.' (AGPS: Canberra.)

Commonwealth of Australia High Court (1992). 'Eddie Mabo and Ors, plaintiffs and the State of Queensland, defendant: Order and Reasons for Judgement', High Court of Australia.

Commonwealth of Australia, House of Representatives Standing Committee for Long-Term Strategies (1994). 'Australia's Population 'Carrying Capacity': One Nation - Two Ecologies.' (AGPS: Canberra.)

Commonwealth of Australia (1994). 'Climate Change: Australia's national report under the United Nations Framework Convention on Climate Change.' (AGPS: Canberra.)

Coopers & Lybrand Consultants (1994). 'Environmental Management Practices - A survey of major Australian organisations.' (C&L Consultants: Sydney.)

Council of Australian Governments (1992). 'Intergovernmental Agreement on the Environment..'

Council of Australian Governments (1992). National Strategy for Ecologically Sustainable Development. (AGPS: Canberra.)

Daly H.E. (1992). 'Steady-state Economics.' (Earthscan Books: London.)

Daly, H.E., and Cobb, J.B. (1989). 'For the Common Good.' (Beacon Press: Boston.)

Department of the Environment, Sport and Territories (1995). 'Living on the Coast: the Commonwealth Coastal Policy.' (DEST: Canberra.)

Department of the Environment, Sport and Territories (1995). 'Greenhouse 21: a Plan of Action for a Sustainable Future.' (DEST: Canberra.)

Department of Industry, Science and Technology (1994). *Australian Science and Innovation Resources Brief.*

Endangered Species Advisory Council (1992). 'National Strategy for the Conservation of Australian Species and Ecological Communities Threatened with Extinction.' (ANCA: Canberra.)

Flannery, T. (1994). 'The Future Eaters.' (Reed Books: Sydney.)

Goldsmith, E. (1972). A blueprint for survival, *The Ecologist*.

Graetz, R.D., Wilson, M.A., and Campbell, S.K. (1995). Landcover Disturbance over the Australian Continent: a contemporary assessment, *Department of the Environment, Sport and Territories Biodiversity Series, Paper* No. 7.

Inter-governmental Panel on Climate Change (1990). Climate change: the scientific assessment, ed. J.T. Houghton, G.J. Jenkins, and J.J. Ephraums. (Cambridge University Press: Cambridge UK.)

Inter-governmental Panel on Climate Change (1992). Climate Change 1992: The Supplementary Report to the IPCC scientific assessment, ed. J.T. Houghton, B.A. Callander and S.K. Varney. (Cambridge University Press: Cambridge UK.)

Inter-governmental Panel on Climate Change (1995). 'Climate Change 1994: Radiative Forcing of Climate Change and An Evaluation of the IPCC IS92 Emission Scenarios', ed. J.T. Houghton, L.G. Meira Filho, J. Bruce, Hoesung Lee, B.A. Callander, E.Haites, N. Harris and K. Maskell. (Cambridge University Press: Cambridge UK.)

Inter-governmental Panel on Climate Change (in press). 'Climate Change: Second Assessment Report.' (Cambridge University Press: Cambridge UK.)

International Union for Nature and Natural Resources, United Nations Environment Programme and World Wildlife Fund (1980). 'World Conservation Strategy.' (IUCN: Geneva.)

Lowe, I. (1994). Performance Measurement. *Proceedings of the Fenner Conference on the Environment, November 1994.*

Meadows, D.H. *et al.*(1970). 'The Limits to Growth.' (Club of Rome/Pan Books: London.)

Mees, P. (1994). Toronto: paradigm re-examined. *Urban Policy and Research*, **12**(3), pp. 146–63.

Murray–Darling Basin Ministerial Council (1995). Council Considers Water Audit, *Murray–Darling Basin CommissionMedia Release* 30 June 1995.

New South Wales Parliament Public Accounts Committee (1990). Report on the Forestry Commission, *Report* No. 52.

Pauly, D. and Christensen, V. (1995). Primary production required to sustain global fisheries. *Nature*, **374**, pp. 255–7. (See also their correction in *Nature*, **376**, p. 279.)

Porter, M.E., and van der Linde, C. (1995). Green and competitive: ending the stalemate, *Harvard Business Review*, Sept–Oct. 1995, pp.120–34.

Resource Assessment Commission (1991). 'Kakadu Conservation Zone: Final Report.' (AGPS: Canberra.)

Resource Assessment Commission (1992). 'Forest and Timber Inquiry: Final Report. (AGPS: Canberra.)

Tyler, M. (1994). 'Australia's Frogs - A Natural History.' (Reed Books.)

United Nations Conference on Environment and Development (1992). 'Agenda 21.' (UNCED Secretariat: Geneva.)

World Commission on Environment and Development (1987). Our Common Future, (the Brundtland Report). (Oxford University Press: Oxford.)

Worldwatch Institute (1994). 'State of the World 1994.' (Earthscan Publications: London.)

Worldwatch Institute (1995). 'State of the World 1995.'(Earthscan Publications: London.)

Vitousek, P.M., Ehrlich, P.R., Ehrlich, A.H., and Matsom P.A. (1986). Human appropriation of the products of photosynthesis, *BioScience*, **36**, No. 6, pp. 368–73.

Weir, P., and Muriale, O. (1994). Development of an Inventory of Emissions for the Perth Airshed, *Proceedings of the 12th International Conference Clean Air Society of Australia and New Zealand, Perth*, **1**, pp. 457–68.

Acknowledgments

The Advisory Council thanks Ian Lowe for compiling this chapter which synthesises the key points emerging ·from the material prepared by the seven expert reference groups for Chapters 3 to 9 of the report.

The Advisory Council gratefully acknowledges the assistance of the State of the Environment Reporting Unit and Reference Group chairs in the preparation of this chapter.

Photo credits

Page 10-1: Australian Picture Library
Page 10-7: (*clockwise from left*) Nigel Holmes; Nick Alexander (Oryx Films); Nigel Holmes
Page 10-9: Uluṟu Kata Tjuṯa National Park
Page 10-13: Mike Prociv (Wetropics)
Page 10-15: Malcolm Paterson (CSIRO)
Page 10-17: Simon Neville
Page 10-19: Geoff Sainty
Page 10-21: Peter Newman
Page 10-23: Heritage Victoria
Page 10-27: Greening Australia, Victoria

Appendix 1

International treaties relating to the environment, conservation and heritage

Appendix 2

Australia's extinct, endangered and vulnerable plants and animals

Acronyms

Glossary

Subject Index

International treaties relating to the environment, conservation and heritage

Multilateral

Date and place of instrument	Head Treaty Title	Entry into force generally	Notes
25 January 1924, Paris	International Agreement for the Creation at Paris of an International office for dealing with Contagious Diseases of Animals	12 January 1925	The Agreement established the International Office of Epizootics. Acceded to by Australia 9 February 1924.
2 December 1946, Washington	International Convention for the Regulation of Whaling (International Whaling Convention)	10 November 1948	Established the International Whaling Commission, located in the UK. Signed for Australia 2 December 1946. Instrument of ratification deposited for Australia 1 December 1947. The Schedule to the Convention has been amended every year since entry into force. See also the 1956 Protocol to the Convention. Signed for Australia 19 November 1956. Instrument of ratification deposited for Australia 8 April 1957.
11 October 1947, Washington	Convention of the World Meteorological Organization (WMO)	23 March 1950	Signed for Australia 11 October 1947. Instrument of ratification deposited for Australia 14 March 1949. Extended to Norfolk Island from 1950 and Australian Antarctic Territory from 1955.
26 February 1948, Baguio	Agreement for the Establishment of the Indo-Pacific Fisheries Council (under the auspices of FAO)	9 November 1948	Instrument of acceptance deposited for Australia 10 March 1949. Entry into force for Australia 10 March 1949. Renamed the Indo-Pacific Fishery Commission in 1976.
6 March 1948, Geneva	Convention on the Inter-Governmental Maritime Consultative Organization (IMCO, later IMO)	17 March 1958	Signed for Australia 6 March 1948. Instrument of acceptance deposited for Australia 13 February 1952. Title of Convention amended to "Convention on the International Maritime Organization" in 1975. See also the 1964 Amendments to the Convention. Accepted for Australia 6 January 1965. See also the 1965 Amendments to the Convention. Instrument of acceptance deposited for Australia 23 June 1966. See also the 1975 and 1977 Amendments to the Convention. Instruments of acceptance deposited for Australia 29 May 1980. See also the 1979 Amendments to the Convention. Instrument of acceptance deposited for Australia 10 November 1980 See also the 1991 Amendments to the Convention. Instrument of acceptance deposited for Australia 1 July 1994. The Amendments are not yet in force.
6 December 1951, Rome	International Plant Protection Convention (under the auspices of FAO)	3 April 1952	Signed for Australia 30 April 1952. Instrument of ratification deposited for Australia 27 August 1952. Entry into force for Australia 27 August 1952. Extended to Norfolk Island by declaration in 1954. See also Supplementary Convention of 27 February 1956, below. See also 1979 Revised Text of Convention. Instrument of acceptance deposited for Australia 22 May 1981.
14 May 1954, The Hague	Convention for the Protection of Cultural Property in the Event of Armed Conflict (under the auspices of UNESCO)	7 August 1956	Signed for Australia 14 May 1954. Instrument of ratification deposited for Australia 19 September 1984. Entry into force for Australia 19 December 1984.
27 February 1956, Rome	Plant Protection Agreement for the South-East Asia and Pacific Region (under the auspices of FAO)	2 July 1956	Signed for Australia, without reservation as to ratification, 27 February 1956. Title of Agreement was amended to "Plant Protection Agreement for the Asia and Pacific Region" in 1979. See also the 1984 Amendments to the Agreement which extended the definition of the region to include China. Instrument of acceptance deposited for Australia 24 October 1989.

Date and place of instrument	Head Treaty Title	Entry into force generally	Notes
1 December 1959, Washington	The Antarctic Treaty	23 June 1961	Signed for Australia 1 December 1959. Instrument of ratification deposited for Australia 23 June 1961.
			See also the 1991 Protocol on Environmental Protection to the Treaty (Annexes I - IV). Signed for Australia 4 October 1991. Instrument of ratification deposited for Australia 6 April 1994. The Protocol is not yet in force.
			See also 1991 Annex V on Area Protection to the Protocol on Environment Protection to the Treaty. Instrument of approval deposited for Australia 6 April 1994. The Annex is not yet in force.
5 August 1963, London, Moscow, Washington	Treaty banning Nuclear Weapon Testing in the Atmosphere, in Outer Space and Under Water (Partial Test Ban Treaty)	10 October 1963	Signed for Australia 8 August 1963. Instruments of ratification deposited for Australia 12 November 1963 Entry into force for Australia 12 November 1963.
27 January 1967, London, Moscow, Washington	Treaty on Principles Governing the Activities of States in the Exploration and Use of Outer Space, including the Moon and other Celestial Bodies	10 October 1967	Signed for Australia 27 January 1967. Instruments of ratification deposited for Australia 10 October 1967.
			See also 1979 Agreement Governing the Activities of States on the Moon and other Celestial Bodies (under the auspices of the UN). Instrument of accession deposited for Australia 7 July 1986. Entry into force for Australia 6 August 1986.
29 November 1969, Brussels	International Convention on Civil Liability for Oil Pollution Damage (under the auspices of IMO)	19 June 1975	Signed for Australia 17 December 1970. Instrument of ratification, with declarations regarding claims on sovereign immunity, deposited for Australia 7 November 1983. Entry into force for Australia 5 February 1984.
			See also the 1973 Protocol to Convention. Instrument of accession deposited for Australia 7 November 1983. Entry into force for Australia 5 February 1984.
			Instrument of denunciation of Protocol, pursuant to Article VI, deposited for Australia 22 June 1988 with effect from the date on which Protocol enters into force. However, latter Protocol was denounced before it entered into force for Australia.
			See also 1992 Protocol to amend of the Convention. Instrument of accession deposited for Australia 9 October 1995.
29 November 1969, Brussels	International Convention relating to Intervention on the High Seas in Cases of Oil Pollution Casualties, 1969 (under the auspices of IMO)	6 May 1975	Signed for Australia 17 December 1970. Instrument of ratification, with declaration on coastal protection, deposited for Australia 7 November 1983. Entry into force for Australia 5 February 1984.
			See also 1973 Protocol relating to Intervention on the High Seas in Cases of Marine Pollution by Substances other than Oil. Instrument of accession, with declaration regarding intervention on the high seas, deposited for Australia 7 November 1983. Entry into force for Australia 5 February 1984.
17 November 1970, Paris	Convention on the Means of Prohibiting and Preventing the Illicit Import, Export and Transfer of Ownership of Cultural Property	24 April 1972	Adopted by UNESCO General Conference on 14 November 1970 by which date it is sometimes cited. Instrument of acceptance, with reservation to Article 10 regarding the maintenance of registers by antique dealers, deposited for Australia 30 October 1989. Entry into force for Australia 30 January 1990.
2 February 1971, Ramsar (Iran)	Convention on Wetlands of International Importance especially as Waterfowl Habitat	21 December 1975	The Convention, including the List of Wetlands of International Importance established under it, is administered by the International Union for Conservation of Nature and Natural Resources (IUCN). UNESCO is depositary for ratifications, accession, etc. Signed for Australia without reservation as to ratification 8 May 1974.
			See also the 1982 Protocol to Convention. Instrument of accession deposited for Australia 12 August 1983.
			See also the 1987 Amendments to the Convention. Instrument of acceptance deposited for Australia 25 July 1990.

Date and place of instrument	Head Treaty Title	Entry into force generally	Notes
11 February 1971, London, Moscow, Washington	Treaty on the Prohibition of the Emplacement of Nuclear Weapons and other Weapons of Mass Destruction on the Sea-bed and the Ocean Floor and in the Subsoil Thereof	18 May 1972	Signed for Australia 11 February 1971. Instruments of ratification deposited for Australia 23 January 1973. Entry into force for Australia 23 January 1973.
18 December 1971, Brussels	International Convention on the Establishment of an International Fund for Compensation for Oil Pollution Damage (under the auspices of IMO)	16 October 1978	Instrument of accession deposited for Australia 10 October 1994. Entry into force for Australia 8 January 1995. See also the 1976 Protocol to Convention. Instrument of accession deposited for Australia 10 October 1994. Entry into force for Australia 8 January 1995. See also 1992 Protocol to amend the Convention. Instrument of accession deposited for Australia 9 October 1995.
29 March 1972, London, Moscow, Washington	Convention on International Liability for Damage Caused by Space Objects	1 September 1972	Instruments of accession deposited for Australia 20 January 1975. Entry into force for Australia 20 January 1975.
10 April 1972, London, Moscow, Washington	Convention on the Prohibition of the Development, Production and Stockpiling of Bacteriological (Biological) and Toxin Weapons and on their Destruction	26 March 1975	Signed for Australia 10 April 1972. Instruments of ratification deposited for Australia 5 October 1977. Entry into force for Australia 5 October 1977.
1 June 1972, London	Convention for the Conservation of Antarctic Seals	11 March 1978	Signed for Australia 5 October 1972. Instrument of ratification deposited for Australia 1 July 1987. Entry into force for Australia 31 July 1987. Amendments to the Annex, done at London on 16 September 1988, entered into force 27 March 1990
23 November 1972, Paris	Convention for the Protection of the World Cultural and Natural Heritage (under the auspices of UNESCO)	17 December 1975	Adopted by UNESCO General Conference 16 November 1972 by which date it is sometimes cited. Instrument of ratification deposited for Australia 22 August 1974.
29 December 1972, London, Mexico City, Moscow, Washington	International Convention on the Prevention of Marine Pollution by Dumping of Wastes and Other Matter (London Convention)	30 August 1975	Known as London Dumping Convention until 1993. Signed for Australia, with declaration regarding Article VII(1)(c), 10 October 1973. Instrument of ratification deposited for Australia 21 August 1985. Entry into force for Australia 20 September 1985. Amendments to Annexes made in 1978, 1980 and 1989. These entered into force in 1979, 1981 and 1990 respectively. Amendments of 12 November 1993, concerning phasing out of sea disposal of industrial waste, entered into force generally 20 February 1994. Accepted for Australia by declaration deposited 15 February 1994, except in relation to jarosite waste where the option of dumping at sea will be retained for a short period after the January 1996 deadline but not beyond December 1997.
3 March 1973, Washington	Convention on International Trade in Endangered Species of Wild Fauna and Flora (CITES)	1 July 1975	Signed for Australia 21 September 1973. Instrument of ratification deposited for Australia 29 July 1976. Entry into force for Australia 27 October 1976. Appendices I, II and III are subject to continual revision. See also 1979 Amendment to Article XI.3(a) of the Convention. Instrument of acceptance deposited for Australia 1 July 1986. See also 1983 Amendment to Article XXI of the Convention. Instrument of acceptance deposited for Australia 13 November 1991. The Amendment is not yet in force.

Date and place of instrument	Head Treaty Title	Entry into force generally	Notes
12 April 1973, Rome	Statutes of the International Centre for the Study of the Preservation and Restoration of Cultural Property of 5 December 1956, as amended 24 April 1963 and 12 April 1973 (ICCROM - under the auspices of UNESCO)	10 May 1958	The Centre was established in Rome. Declaration of accession deposited for Australia 26 June 1975. Entry into force for Australia 26 June 1975.\n\nFurther amended by the General Assembly of the Centre 23 April 1979.
2 November 1973, London	International Convention for the Prevention of Pollution from Ships (MARPOL 1973) and Protocols I and II		See also Protocol of 1973 to the Convention. Signed for Australia, with declaration and subject to ratification, 24 December 1974. The Convention did not enter into force but see Protocol of 17 February 1978 (below).
18 November 1974, Paris	Agreement on an International Energy Program, as amended 5 February 1975 (under the auspices of OECD)	18 November 1974 (provisionally) 19 January 1976 (definitively)	The Agreement established the International Energy Agency. Instrument of accession, with explanatory statement, deposited for Australia 17 May 1979. Entry into force for Australia 27 May 1979.
12 June 1976, Apia (Western Samoa)	Convention on the Conservation of Nature in the South Pacific	26 June 1990	Instrument of accession deposited for Australia 28 March 1990. Federal statement deposited for Australia 15 November 1990.
18 May 1977, Geneva	Convention on the Prohibition of Military or any other Hostile Use of Environmental Modification Techniques (ENMOD – under the auspices of the UN)	5 October 1978	Adopted by UN General Assembly 10 December 1976 by which date it is sometimes cited. Signed for Australia 31 May 1978. Instrument of ratification deposited for Australia 7 September 1984. Entry into force for Australia 7 September 1984.
17 February 1978, London	Protocol of 1978 relating to the International Convention for the Prevention of Pollution from Ships of 2 November 1973, as amended (MARPOL Protocol)	2 October 1983 (Annex I: 2 October 1983; Annex II: 6 April 1987; Annex III: 1 July 1992; Annex V: 31 December 1988)	Signed for Australia 30 May 1979. Instrument of ratification, including Annexes I and II, deposited for Australia 14 October 1987. Entry into force for Australia 14 January 1988. Instrument of accession to Annex III (Regulations for the Prevention of Pollution by Harmful Substances Carried by Sea in Packaged Form) deposited for Australia 10 October 1994. Entry into force of Annex III for Australia 10 January 1995. Instrument of accession to Annex V (Regulations for the Prevention of Pollution by Garbage from Ships) deposited for Australia 14 August 1990. Entry into force of Annex V for Australia 14 November 1990.\n\nMARPOL Convention of 2 November 1973 (above) did not enter into force but pursuant to Article 1(b) of the MARPOL Protocol, the latter incorporated the terms of the Convention, including Protocols I and II, subject to modifications and additions made by MARPOL Protocol.\n\nThere have been various amendments to Annexures.
23 October 1978, Geneva	International Convention for the Protection of New Varieties of Plants of 2 December 1961, as amended 10 November 1972 and revised 23 October 1978	8 November 1981	Instrument of accession deposited for Australia 1 February 1989. Entry into force for Australia 1 March 1989.
23 June 1979, Bonn	Convention on the Conservation of Migratory Species of Wild Animals	1 November 1983	Instrument of accession, with federal statement, deposited for Australia 26 June 1991. Entry into force for Australia 1 September 1991.\n\nAmendments to Appendices I and II, done at Bonn in 1985, entered into force in January 1986; and done at Geneva in 1988 entered into force in January 1989.
10 July 1979, Honiara	South Pacific Forum Fisheries Agency Convention	9 August 1979	Accepted by signature for Australia 13 September 1979. Entry into force for Australia 13 October 1979.\n\nSee also 1983 Amendment to the Convention. Instrument of acceptance deposited for Australia, with explanatory note, 2 February 1987. The Amendment is not yet in force. The Amendment was revised in May 1989.

Date and place of instrument	Head Treaty Title	Entry into force generally	Notes
3 March 1980, Vienna	Convention on the Physical Protection of Nuclear Material (under the auspices of IAEA)	8 February 1987	Signed for Australia 22 February 1984. Instrument of ratification deposited for Australia 22 September 1987. Entry into force for Australia 22 October 1987.
20 May 1980, Canberra	Convention on the Conservation of Antarctic Marine Living Resources	7 April 1982	Signed for Australia 11 September 1980. Instrument of ratification deposited for Australia 6 May 1981. Australia is depositary for the Convention.
10 December 1982, Montego Bay	United Nations Convention on the Law of the Sea	16 November 1994	The Convention was adopted by the third UN Conference on the Law of the Sea (UNCLOS). Signed for Australia 10 December 1982. Instrument of ratification deposited for Australia 5 October 1994. Replaced Conventions (four) of 29 April 1958.

See also 1994 Agreement relating to the implementation of Part XI of the Convention. Signed for Australia 29 July 1994. Instrument of ratification deposited for Australia 5 October 1994. Provisional application of Convention will terminate 16 November 1998 if requirements of Article 6.1 are not fulfilled.

See also 1995 Agreement for the Implementation of the Convention relating to the Conservation and Management of Straddling Fish Stocks and Highly Migratory Fish Stocks. Signed for Australia, subject to ratification, 4 December 1995. The Agreement is not yet in force. |
| 18 November 1983, Geneva | International Tropical Timber Agreement, 1983 | 1 April 1985 (provisionally) | The Agreement established the International Tropical Timber Organization with headquarters at Yokohama. Instrument of accession deposited for Australia 16 February 1988. Entry into force for Australia 16 February 1988. Due to expire 1990 but extended to 1994. |
| 22 March 1985, Vienna | Vienna Convention for the Protection of the Ozone Layer | 22 September 1988 | Instrument of accession deposited for Australia 16 September 1987.

See also Montreal Protocol on Substances that that Deplete the Ozone Layer of 16 September 1987 (below). |
6 August 1985, Rarotonga	South Pacific Nuclear Free Zone Treaty	11 December 1986	Also known as Treaty of Rarotonga. Signed for Australia 6 August 1985. Instrument of ratification deposited for Australia 11 December 1986. Protocols 1, 2 and 3 to Treaty, done at Suva on 8 August 1986, provide for acceptance by certain nuclear weapon States of undertakings in relation to the Zone.
26 September 1986, Vienna	Convention on Assistance in the Case of a Nuclear Accident or Radiological Emergency (under the auspices of IAEA)	26 February 1987	Signed for Australia 26 September 1986. Instrument of ratification, with declaration pursuant to Article 8.9 that Australia would not be bound by Articles 8.2 and 8.3, deposited for Australia 22 September 1987. Entry into force for Australia 23 October 1987.
26 September 1986, Vienna	Convention on Early Notification of a Nuclear Accident (under the auspices of IAEA)	27 October 1986	Signed for Australia 26 September 1986. Instrument of ratification deposited for Australia 22 September 1987. Entry into force for Australia 23 October 1987.
24 November 1986, Noumea	Convention for the Protection of the Natural Resources and Environment of the South Pacific Region (SPREP)	22 August 1990	Signed for Australia 24 November 1987. Instrument of ratification deposited for Australia 19 July 1989.

See also 1986 Protocol for the Prevention of Pollution of the South Pacific Region by Dumping. Signed for Australia, subject to ratification, 24 November 1987.

See also 1986 Protocol concerning Co-operation in Combating Pollution Emergencies in the South Pacific Region. Signed for Australia 24 November 1987. Instrument of ratification deposited for Australia 19 July 1989.

See also 1993 Agreement establishing the South Pacific Regional Environment Programme (SPREP) [as an intergovernmental organisation]. Signed for Australia at Suva on 21 September 1993. Instrument of ratification deposited for Australia 18 October 1994. |

Date and place of instrument	Head Treaty Title	Entry into force generally	Notes
2 April 1987, Port Moresby	Treaty on Fisheries between the Governments of Certain Pacific Island States and the Government of the United States of America, and Agreed Statement on Observer Programme	15 June 1988	Signed for Australia 2 April 1987. Instrument of ratification deposited for Australia 18 May 1987. See also Agreement with the United States of America of 2 April 1987 (under Bilateral). See also 1992 Amendments to Annexures I and II. Notification of acceptance deposited for Australia 9 June 1993. The Amendments are not yet in force.
2 April 1987, Port Moresby	Agreement among Pacific Island States concerning the Implementation and Administration of the Treaty on Fisheries between the Governments of Certain Pacific Island States and the Government of the United States of America	15 June 1988	Signed for Australia 2 April 1987. Instrument of ratification deposited for Australia 18 May 1987. See also 1992 Amendments to Schedule 1 of Agreement. Instrument of acceptance deposited for Australia 14 July 1993. The Amendments are not yet in force.
16 September 1987 Montreal	Montreal Protocol on Substances that Deplete the Ozone Layer	1 January 1989	Signed for Australia 8 June 1988. Instrument of ratification deposited for Australia 19 May 1989. Entry into force for Australia 17 August 1989. See also 1992 Amendment to the Montreal Protocol. Instrument of acceptance deposited for Australia 11 August 1992. Entry into force for Australia 9 November 1992. See also 1994 Amendment to the Montreal Protocol. Instrument of acceptance deposited for Australia 30 June 1994. Entry into force for Australia 28 September 1994.
22 March 1989, Basel, Switzerland	Basel Convention on the Control of Transboundary Movements of Hazardous Wastes and their Disposal	5 May 1992	Instrument of accession deposited for Australia 5 February 1992.
24 November 1989, Wellington	Convention for the Prohibition of Fishing with Long Driftnets in the South Pacific	17 May 1991	Signed for Australia at Auckland 2 February 1990. Instrument of ratification deposited for Australia 6 July 1992. Entry into force for Australia 6 July 1992.
10 September 1990, Washington	Agreement concerning the Continuation of Marine Geoscientific Research and Mineral Resource Studies in the South Pacific Region (Tripartite Phase II Extended Agreement)	10 September 1990	The Agreement, concluded with New Zealand and the United States of America, was signed for Australia and entered into force on signature. Extended provisions of Phase II Agreement of 19 September 1984. Due to expire 9 September 1995.
30 November 1990, London	International Convention on Oil Pollution Preparedness, Response and Co-operation (under the auspices of IMO)	13 May 1995	Instrument of accession deposited for Australia 6 July 1992.
9 May 1992, New York	United Nations Framework Convention on Climate Change	21 March 1994	Signed for Australia at Rio de Janeiro, 4 June 1992. Instrument of ratification deposited for Australia 30 December 1992.
5 June 1992, Rio de Janeiro	Convention on Biological Diversity	29 December 1993	Signed for Australia 5 June 1992. Instrument of ratification deposited for Australia 18 June 1993.
9 July 1992, Honiara	Niue Treaty on Cooperation in Fisheries Surveillance and Law Enforcement in the South Pacific Region	20 May 1993	Signed for Australia 9 July 1992. Instrument of ratification deposited for Australia 3 September 1993. Entry into force for Australia 3 September 1993.
13 January 1993, Paris	Convention on the Prohibition of the Development, Production, Stockpiling and Use of Chemical Weapons and on their Destruction		Signed for Australia 13 January 1993. Instrument of ratification deposited for Australia 6 May 1994. The Convention is not yet in force.
10 May 1993, Canberra	Convention for the Conservation of Southern Bluefin Tuna	20 May 1994	Signed for Australia, Japan and New Zealand 10 May 1993. Instrument of ratification deposited for Australia 20 May 1994. Australia is depositary for Convention.
17 June 1994, Paris	United Nations Convention to Combat Desertification in those Countries experiencing Serious Drought and/or Desertification, particularly in Africa		Signed for Australia, subject to ratification, 14 October 1994. The Convention is not yet in force.

Date and place of instrument	Head Treaty Title	Entry into force generally	Notes
17 December 1994, Lisbon	Energy Charter Treaty	17 December 1994 (provisionally - for signatories accepting such application) Part VII: 17 December 1994 (provisionally for all signatories)	Signed for Australia, subject to ratification, 17 December 1994, with declaration pursuant to Article45(2)(a) not accepting provisional application of the Treaty, and declaration concerning trade-related investment measures. The Treaty is not yet in force definitively. See also 1994 Protocol to Treaty on Energy Efficiency and related Environmental Aspects. Signed for Australia, subject to ratification, 17 December 1994. The Protocol is not yet in force.
16 September 1995, Port Moresby	Regional Convention on Hazardous Wastes (Waigani Convention)		Also known as Convention to Ban the Importation into Forum Island Countries of Hazardous and Radioactive Wastes and to Control Transboundary Movement and and Management of Hazardous Wastes within the South Pacific Region. Signed for Australia, subject to ratification, 16 September 1995. The Convention is not yet in force.

Bilateral

Date and place of signature	Head Treaty Title	Entry into force	Notes
20 October 1986, Canberra	**CHINA** Agreement for the Protection of Migratory Birds and their Environment	1 September 1988	The Agreement entered into force when Notes were exchanged 26 December 1986 and 1 September 1988 pursuant to Article VI(1).
11 December 1989, Timor Sea	**INDONESIA** Treaty on the Zone of Cooperation in an Area between the Indonesian Province of East Timor and Northern Australia (Timor Gap Treaty)	9 February 1991	The Treaty entered into force 30 days following an exchange of Notes of 10 January 1991 pursuant to Article 32.
22 April 1992, Jakarta	Agreement relating to Cooperation in Fisheries	29 May 1993	The Agreement entered into force when Notes were exchanged 24 June 1992 and 29 May 1993 pursuant to Article 13.1.
8 September 1986, Hobart	**INTERNATIONAL ORGANISATIONS** Headquarters Agreement with the Commission for the Conservation of Antarctic Marine Living Resources	8 September 1986	The Agreement, concerning the Commission's headquarters in Hobart, entered into force on signature pursuant to Article 26.1.
6 February 1974, Tokyo	**JAPAN** Agreement for the Protection of Migratory Birds and Birds in Danger of Extinction and their Environment	30 April 1981	The Agreement entered into force when instruments of ratification were exchanged at Canberra 30 April 1981 pursuant to Article IX.2.
17 October 1979, Canberra	Agreement on Fisheries	1 November 1979	The Agreement entered into force on the date specified in Article XII. 1. Superseded Agreement of 27 November 1968. Subsidiary Agreements were done yearly between 1979 and 1993.
21 December 1994, Melbourne	Subsidiary Agreement (to Agreement of 27 October 1979) concerning Japanese Tuna Long-Line Fishing	21 December 1994	The Subsidiary Agreement entered into force on signature pursuant to Article IX. Due to expire 31 October 1995.
23 November 1983, Canberra	**KOREA, REPUBLIC OF** Agreement on Fisheries, and Exchange of Letters	24 November 1983	The Agreement entered into force on the date specified in an exchange of Notes of 24 November 1983 pursuant to Article XIII. The Exchange of Letters entered into force on the same date.
18 December 1978, Sydney	**PAPUA NEW GUINEA** Treaty concerning Sovereignty and Maritime Boundaries in the Area between the Two Countries, including the Area known as the Torres Strait, and Related Matters	15 February 1985	Also known as the Torres Strait Treaty. The Agreement entered into force when instruments of ratification were exchanged at Port Moresby 15 February 1985 pursuant to Article 32.
15 February 1990, Canberra	**RUSSIA** Agreement (with the Union of Soviet Socialist Republics) on Cooperation in the Field of Protection and Enhancement of the Environment	15 February 1990	The Agreement entered into force on signature pursuant to Article X.1.
2 April 1987, Port Moresby	**UNITED STATES OF AMERICA** Exchange of Notes constituting an Agreement on Access to the Australian Fishing Zone	2 April 1987	The Agreement entered into force on the date of the Note in reply. See also Agreements with the Pacific Islands of 2 April 1987 (under Multilateral).

Source: Condensed from list provided by the Department of Foreign Affairs and Trade (1995).

Australia's extinct, endangered and vulnerable plants and animals

This Appendix lists plant and animal species in Australia which are either recognised in the Commonwealth's *Endangered Species Protection Act 1992* (Schedule 1), or are currently being considered for listing. Species that are endangered are listed in Part 1 (see pages A11–12), those that are vulnerable are listed in Part 2 (see pages A13–16), and species that are presumed extinct are listed in Part 3 (see page A17). Conservation status recommended by the ANZECC Endangered Fauna Network, which are to be considered for changes to the Endangered Species Protection Act Schedule, are provided on pages A18–19.

Threatened species that may have been identified by action plans and overviews — being prepared under the National Endangered Species Program — for bats, cetaceans (whales and dolphins), seals, dugongs, amphibians, and non-marine invertebrates and non-vascular plants are not included in this Appendix. Also, this Appendix does not include the public nominations pending under the Commonwealth's Endangered Species Protection Act, such as the koala, the Tiger Quoll, Eastern Gemfish and others.

It should be remembered that these lists are compiled from the national perspective — species that are known to be threatened to some degree at the national level. They do not take into account species, subspecies and discrete populations that are either extinct or threatened on a regional, state, territory or local level. In addition, these lists do not reflect those plants and animals which may have already experienced significant declines in population and distribution, but are not recognised as threatened under existing legislation. Such a comprehensive analysis is possible, but was not undertaken for the purposes of this national report. Nor does this list include threatened freshwater and marine plants, marine inverterbrates or marine fish and lichens.

Commonwealth Endangered Species Protection Act, 1992 Schedule 1

Part 1 – Species that are endangered

Animals

Fish

Galaxias fontanus
  Swan Galaxias

Galaxias fuscus
  Brown Galaxias

Galaxias johnstoni
  Clarence Galaxias

Galaxias pedderensis
  Pedder Galaxias

Maccullochella ikei
  Eastern Freshwater Cod

Maccullochella macquariensis
  Trout Cod

Melanotaenia eachamensis
  Lake Eacham Rainbow Fish

Amphibians

Geocrinia vitellina
  Yellow-bellied Frog

Litoria spenceri
  Spotted Tree Frog

Rheobatrachus silus
  Gastric-brooding Frog

Rheobatrachus vitellinus
  Eungella Gastric-brooding Frog

Taudactylus acutirostris
  Sharp-snouted Day Frog

Taudactylus diurnus
  Southern Day Frog

Taudactylus eungellensis
  Eungella Day Frog

Reptiles

Aprasia aurita
  Legless Lizard

Aprasia parapulchella
  Pink-tailed Legless Lizard

Caretta caretta
  Loggerhead Turtle

Hoplocephalus bungaroides
  Broad-headed Snake

Pseudemydura umbrina
  Western Swamp Tortoise

Tiliqua adelaidensis
  Adelaide Blue-tongued Lizard

Birds

Cacatua pastinator pastinator
  Western Long-billed Corella

Calyptorhynchus banksii graptogyne
  South-eastern Red-tailed Black Cockatoo

Coracina tenuirostris melvillensis
  Melville Cicadabird

Cyanoramphus novaezelandiae cookii
  Norfolk Island Parrot

Diomedea exulans chionoptera
  Wandering Albatross (southern subspecies)

Erythrura gouldiae
  Gouldian Finch

Falcunculus frontatus whitei
  Kimberley Crested Shriketit

Geopsittacus occidentalis
  Night Parrot

Leipoa ocellata
  Mallee fowl

Lichenostomus melanops cassidix
  Helmeted Honeyeater

Manorina melanotis
  Black-eared Miner

Neophema chrysogaster
  Orange-bellied Parrot

Ninox novaeseelandiae undulata
  Norfolk Island Boobook Owl

Pardalotus quadragintus
  Forty-spotted Pardalote

Petrophassa smithii blaauwi
  Western Partridge Pigeon

Pezoporus wallicus flaviventris
  Western Ground Parrot

Poecilodryas superciliosa cerviniventris
  Derby White-browed Robin

Psephotus chrysopterygius
  Golden-shouldered Parrot

Psittaculirostris diophthalma coxeni
  Coxen's Fig-parrot

Pterodroma leucoptera leucoptera
  Gould's Petrel

Sterna albifrons
  Little Tern

Stipiturus malachurus intermedius
  Mount Lofty Southern Emu-wren

Sula abbotti
  Abbott's Booby

Turdus poliocephalus poliocephalus
  Norfolk Island Thrush

Xanthomyza phrygia
  Regent Honeyeater

Zosterops albogularis
  Norfolk Island Silvereye

Mammals

Balaenoptera musculus
  Blue Whale

Bettongia lesueur
  Burrowing Bettong

Bettongia penicillata
  Brush-tailed Bettong

Bettongia tropica
  Northern Bettong

Crocidura tenuata var. trichura
  Christmas Island Shrew

Dasyuroides byrnei
  Kowari

Dasyurus geoffroii
  Western Quoll

Eubalaena australis
  Southern Right Whale

Gymnobelideus leadbeateri
  Leadbeater's Possum

Isoodon auratus
  Golden Bandicoot

Lagorchestes hirsutus
  Rufous Hare-wallaby

Lagostrophus fasciatus
  Banded Hare-wallaby

Lasiorhinus krefftii
  Northern Hairy-nosed Wombat

Leporillus conditor
  Greater Stick-nest Rat

Megaptera novaeangliae
  Humpback Whale

Myrmecobius fasciatus
  Numbat

Notomys fuscus
  Dusky Hopping-mouse

Onychogalea fraenata
  Bridled Nailtail Wallaby

Parantechinus apicalis
  Dibbler

Perameles bougainville
  Western Barred Bandicoot

Perameles eremiana
  Desert Bandicoot

Petaurus gracilis
  Mahogany Glider

Phascogale calura
  Red-tailed Phascogale

Potorous longipes
  Long-footed Potoroo

Potorous tridactylus gilberti
  Gilbert's Potoroo

Pseudocheirus peregrinus occidentalis
  Western Ringtail Possum

Pseudomys oralis
  Hastings River Mouse

Pseudomys praeconis
  Shark Bay Mouse

Pseudomys shortridgei
  Heath Rat

Sminthopsis douglasi
  Julia Creek Dunnart

Zyzomys pedunculatus
  Central Rock-rat

Plants

Acacia cretacea

Acacia enterocarpa

Acacia leptalea Maslin ms.

Acacia pinguifolia

Acacia porcata

Acronychia littoralis

Adenanthos pungens

Adenanthos velutinus

Agrostis adamsonii

Agrostis limiteana

Alectryon ramiflorus

Allocasuarina defungens

Allocasuarina emuina

Allocasuarina portuensis

Allocasuarina thalassoscopica

Andersonia sp. Two Peoples Bay (G. Keighery 8229)

Anthocercis gracilis

Apatophyllum constablei

Aristida granitica

Arthraxon hispidus

Asterolasia elegans

Astrotricha roddii

Atalaya collina

Atriplex kochiana

Austromyrtus fragrantissima

Austromyrtus gonoclada

Ballantinia antipoda

Banksia brownii

Barbarea australis

Bentleya spinescens

Boronia granitica

Boronia repanda

Brachyscome muelleri

Cajanus mareebensis

Caladenia audasii

Caladenia busselliana Hopper and A.P. Brown ms.

Caladenia cristata

Caladenia elegans Hopper and A.P. Brown ms.

Caladenia formosa

Caladenia fragrantissima subsp. *orientalis*

Caladenia fulva

Caladenia hastata

Caladenia lowanensis

Caladenia robinsonii

Caladenia rosella

Caladenia tensa

Caladenia thysanochila

Caladenia viridescens Hopper and A.P. Brown ms.

Calochilus psednus

Calochilus richae

Calytrix breviseta subsp. *breviseta*

Centrolepis pedderensis

Chamaelaucium griffinii Marchant & Keighery ms.

Colobanthus curtisiae

Conostylis setigera subsp. *dasys*

Corchorus cunninghamii

Cyathea exilis

Cynanchum elegans

Cyperus ohwii

Danthonia popinensis

Darwinia apiculata

Darwinia carnea

Darwinia ferricola N.G. Marchant and Keighery ms.

Davidsonia sp. Mullumbimby-Currumbin Ck (A.G. Floyd 1595)

Daviesia bursarioides Crisp ms.

Daviesia microcarpa Crisp ms.

Daviesia oxylobium Crisp ms.

Daviesia purpurascens

Dendrobium antennatum

Dendrobium mirbelianum

Dendrobium nindii

Deyeuxia appressa

Digitaria porrecta

Diospyros mabacea

Diploglottis campbellii

Dipodium pictum

Diuris fragrantissima

Diuris micrantha

Diuris pallens

Dodonaea subglandulifera

Drakonorchis drakeoides Hopper and A.P. Brown ms.

Dryandra mimica

Dryandra sp. Kamballup (M. Pieroni s.n. 20/9/88)

Dryandra sp. Stirling Range (F. Lullfitz 3379)

Elaeocarpus williamsianus

Endiandra floydii

Epacris hamiltonii

Epacris stuartii

Epiblema grandiflorum var. *cyanea* Dixon ms.

Eremophila caerulea subsp. *merrallii* Chinnock ms.

Eremophila nivea

Eremophila veneta Chinnock ms.

Eremophila verticillata

Eriocaulon carsonii

Eucalyptus absita

Eucalyptus conglomerata

Eucalyptus crenulata

Eucalyptus crucis subsp *praecipua* Brooker and Hopper ms.

Eucalyptus cuprea Brooker and Hopper ms.

Eucalyptus dolorosa Brooker & Hopper ms.

Eucalyptus graniticola Brooker and Hopper ms.

Eucalyptus impensa Brooker and Hopper ms.

Eucalyptus morrisbyi

Eucalyptus phylacis

Eucalyptus recurva

Eucalyptus rhodantha

Euphrasia collina muelleri

Euphrasia collina osbornii

Euphrasia sp. Southport (W.M. Curtis 5/12/58)

Fimbristylis adjuncta

Fontainea oraria

Frankenia plicata

Gardenia actinocarpa

Gastrolobium callistachys

Gastrolobium glaucum

Gastrolobium graniticum

Gastrolobium hamulosum

Genoplesium rhyoliticum

Genoplesium tectum

Gentiana baeuerlenii

Gentianawingecarribiensis

Gonocarpus intricatus

Graptophyllum reticulatum

Grevillea batrachioides

Grevillea beadleana

Grevillea caleyi

Grevillea calliantha

Grevillea flexuosa

Grevillea iaspicula

Grevillea scapigera

Grevillea wilkinsonii

Habenaria macraithii

Hakea aculeata

Hakea pulvinifera

Haloragis eyreana

Haloragodendron lucasii

Homopholis belsonii

Huperzia carinata

Huperzia dalhousieana

Huperzia filiformis

Huperzia squarrosa

Hypocalymma longifolium

Isoglossa eranthemoides

Isopogon uncinatus

Kunzea rupestris

Lambertia echinata subsp. *echinata*

Lambertia fairallii

Lambertia orbifolia

Lepidium catapycnon

Lepidium hyssopifolium

Lepidium monoplocoides

Leucochrysum albicans subsp. *incanum*

Lomatia tasmanica

Macadamia sp.Woopen Creek (B. Hyland 3472)

Macrozamia lomandroides

Malaxis lawleri

Marsdenia coronata

Marsdenia longiloba

Meziella trifida

Muellerargia timorensis

Myoporum turbinatum

Neoroepera buxifolia

Ochrosia moorei

Olearia flocktoniae

Olearia hygrophila

Olearia microdisca

Persoonia nutans

Phaius bernaysii

Phalaenopsis rosenstromii

Phebalium daviesii

Phebalium equestre

Phebalium lachnaeoides

Phebalium obtusifolium

Pimelea rara

Pimelea spicata

Pityrodia scabra

Planchonella eerwah

Pleurophascum occidentale

Pomaderris cotoneaster

Prasophyllum chasmogamum

Prasophyllum concinnum

Prasophyllum diversiflorum

Prasophyllum petilum

Prasophyllum uroglossum

Prostanthera eurybioides

Psoralea parva

Pterostylis arenicola

Pterostylis despectans

Pterostylis gibbosa

Pterostylis sp. Dimboola (D.L. Jones 5333 ex P. Branwhite s.n.)

Pterostylis sp. Northampton (S.D. Hopper 3349)

Ptilotus beckerianus

Ptychosema pusillum

Ptychosperma bleeseri

Pultenaea pauciflora

Pultenaea trichophylla

Quassia sp. Mooney Creek (J. King s.n., 1949)

Randia moorei

Ranunculus prasinus

Restio abortivus

Romnalda strobilacea

Rutidosis leptorrhynchoides

Sclerolaena napiformis

Senecio behrianus

Sowerbaea multicaulis

Stemmacantha australis

Stylidium coroniforme

Stylidium scabridum

Swainsona recta

Tectaria devexa

Tetratheca gunnii

Tetratheca paynteri Alford ms.

Thelymitra dedmaniae

Thelymitra epipactoides

Thysanotus wangariensis

Triunia robusta

Tylophora linearis

Tylophora rupicola

Tylophora woollsii

Uromyrtus australis

Verticordia fimbrilepis

Verticordia harveyi

Verticordia hughanii

Verticordia plumosa var. *ananeotes*

Vrydagzynea paludosa

Westringia crassifolia

Xanthostemon formosus

Xerothamnella herbacea

Zieria adenophora

Zieria baeuerlenii J.A. Armstrong ms.

Zieria buxijugum J. Briggs and J.A. Armstrong ms.

Zieria formosa J. Briggs and J.A. Armstrong ms.

Zieria obcordata

Zieria parrisiae J. Briggs and J.A. Armstrong ms.

Zieria prostrata J.A. Armstrong ms.

Part 2 – Species that are vulnerable

Animals

Fish

Galaxias parvus
Swamp Galaxias

Galaxias tanycephalus
Saddled Galaxias

Nannoperca obscura
Yarra Pygmy Perch

Nannoperca variegata
Ewens Pygmy Perch

Prototroctes maraena
Australian Grayling

Pseudomugil mellis
Honey Blue-eye

Amphibians

Geocrinia alba
Creek Frog/White-bellied Frog

Philoria frosti
Mount Baw Baw Frog

Reptiles

Chelonia mydas
Green Turtle

Ctenophorus yinnietharra
Yinnietharra Rock-dragon

Ctenotus angusticeps
Airlie Island Ctenotus

Ctenotus lancelini
Lancelin Island Striped Skink

Delma impar
Striped Legless Lizard

Delma mitella
Legless Lizard

Delma torquata
Legless Lizard

Dermochelys coriacea
Leathery Turtle

Egernia stokesii aethiops
Baudin Island Spiny-tailed Skink

Eretmochelys imbricata
Hawksbill Turtle

Lepidochelys olivacea
Pacific Ridley

Morelia carinata
Rough-scaled Python

Ophidiocephalus taeniatus
Bronzebacked Legless Lizard

Pseudemoia palfreymani
Pedra Branka Skink

Rheodytes leukops
Fitzroy River Tortoise

Birds

Amytornis dorotheae
Carpentarian Grasswren

Amytornis textilis textilis
Western grasswren

Anous tenuirostris melanops
Lesser Noddy

Atrichornis clamosus
Noisy Scrub-bird

Casuarius casuarius
Southern Cassowary

Cereopsis novaehollandiae grisea
Recherche Cape Barren Goose

Charadrius rubricollis
Hooded Plover

Dasyornis brachypterus
Eastern Bristlebird

Dasyornis longirostris
Western Bristlebird

Eclectus roratus
Eclectus Parrot

Erythrotriorchis radiatus
Red Goshawk

Falcunculus frontatus leucogaster
South-west Crested Shriketit

Fregata andrewsi
Christmas Island Frigatebird

Lathamus discolor
Swift Parrot

Malurus leucopterus edouardi
Barrow Island Black-and-white Fairy-wren

Malurus leucopterus leucopterus
Dirk Hartog Black-and-white Fairy-wren

Ninox squamipila natalis
Christmas Island Hawk-owl

Pachycephala rufogularis
Red-lored Whistler

Pedionomus torquatus
Plains-wanderer

Polytelis alexandrae
Alexandra's Parrot

Stipiturus malachurus parimeda
Eyre Peninsula Southern Emu-wren

Strepera graculina crissalis
Lord Howe Island Currawong

Tricholimnas sylvestris
Lord Howe Island Woodhen

Turnix melanogaster
Black-breasted Button-quail

Turnix varia scintillans
Abrolhos Painted Button-quail

Mammals

Burramys parvus
Mountain Pygmy-possum

Dasycercus cristicauda
Mulgara

Dasyurus viverrinus
Eastern Quoll

Macroderma gigas
Ghost Bat

Macropus robustus isabellinus
Barrow Island Euro

Macrotis lagotis
Greater Bilby

Mesembriomys macrurus
Golden-backed Tree-rat

Notomys aquilo
Northern Hopping-Mouse

Perameles gunnii
Eastern Barred Bandicoot

Petrogale lateralis
Black-footed Rock-wallaby

Petrogale penicillata
Brush-tailed Rock-wallaby

Petrogale persephone
Proserpine Rock-wallaby

Pseudomys australis
Plains Rat

Pseudomys chapmani
Pebble-mound Mouse

Pseudomys occidentalis
Western Mouse

Pseudomys pillagaensis
Pilliga Mouse

Sminthopsis psammophila
Sandhill Dunnart

Xeromys myoides
False Water-rat

Plants

Acacia anomala

Acacia aphylla

Acacia araneosa

Acacia argutifolia

Acacia attenuata

Acacia axillaris

Acacia barattensis

Acacia bynoeana

Acacia calantha

Acacia carnei

Acacia chinchillensis

Acacia clunies-rossiae

Acacia constablei

Acacia courtii

Acacia crombiei

Acacia curranii

Acacia denticulosa

Acacia depressa

Acacia deuteroneura

Acacia eremophiloides

Acacia fleckeri

Acacia flocktoniae

Acacia forrestiana

Acacia georgensis

Acacia glandulicarpa

Acacia grandifolia

Acacia guymeri

Acacia handonis

Acacia imbricata

Acacia lanuginophylla R.S. Cowan & Maslin ms.

Acacia latzii

Acacia lauta

Acacia lobulata

Acacia menzelii

Acacia merrickiae

Acacia perangusta

Acacia peuce

Acacia pharangites

Acacia phasmoides

Acacia pickardii

Acacia pubescens

Acacia pubifolia

Acacia purpureapetala

Acacia pygmaea Maslin ms.

Acacia ramiflora

Acacia rhetinocarpa

Acacia ruppii

Acacia semicircinalis

Acacia simulans

Acacia sp. Dandaragan (S. van Leeuwen 269)

Acacia tenuinervis

Acacia undoolyana

Acacia vassalii

Acacia wardellii

Acronychia crassipetala

Acrophyllum australe

Actinotus schwarzii

Adenanthos cunninghamii

Adenanthos dobagii

Adenanthos ellipticus

Adenanthos eyrei

Adenanthos ileticos

Allocasuarina fibrosa

Allocasuarina glareicola

Allocasuarina tortiramula

Alloxylon flammeum

Ammobium craspedioides

Amorphospermum whitei

Amphibromus fluitans

Angophora robur

Anigozanthos bicolor subsp. minor

Anigozanthos humilis subsp. chrysanthus

Anigozanthos viridis subsp. terraspectans

Apatophyllum olsenii

Aphanes pentamera

Apium prostratum subsp. phillipii Keighery ms.

Aponogeton hexatepalus

Archidendron lovelliae

Aristida annua

Arytera dictyoneura

Asperula asthenes

Asplenium hookerianum

Asplenium pellucidum

Asplenium wildii

Astelia australiana

Asterolasia drummondii

Asterolasia grandiflora

Asterolasia nivea

Asterolasia phebalioides

Atriplex infrequens

Baeckea arbuscula

Baeckea crenatifolia

Baeckea sp. Mt Tozer (L.J. Brass 19348)

Baeckea sp. Pyramids (W.F.J. McDonald 357)

Baloghia marmorata

Banksia cuneata

Banksia goodii

Banksia oligantha

Banksia sphaerocarpa var. dolichostyla

Banksia tricuspis

Banksia verticillata

Basedowia tenerrima

Bertya ingramii

Bertya opponens

Bertya pinifolia

Bertya sharpeana

Beyeria subtecta

Billardiera mollis

Boronia adamsiana

Boronia deanei

Boronia keysii

Boronia revoluta

Borya mirabilis

Bosistoa monostylis

Bosistoa selwynii

Bosistoa transversa

Bossiaea oligosperma

Bothriochloa biloba

Bothriochloa bunyensis

Brachychiton vitifolius

Brachyscome ascendens

Brachyscome muelleroides

Brachyscome papillosa

Budawangia gnidioides

Bulbophyllum boonjee

Bulbophyllum globuliforme

Bulbophyllum gracillimum

Bulbophyllum longiflorum

Cadellia pentastylis

Caladenia bryceana

Caladenia caesarea subsp. maritima Hopper and A.P. Brown ms.

Caladenia caudata

Caladenia christineae Hopper and A.P. Brown ms.

Caladenia dilatata subsp. villosissima

Caladenia dorrienii

Caladenia excelsa Hopper and A.P. Brown ms.

Caladenia exstans Hopper and A.P. Brown ms.

Caladenia gladiolata

Caladenia harringtoniae Hopper and A.P. Brown ms.

Caladenia hoffmanii Hopper and A.P. Brown ms.

Caladenia huegelii

Caladenia insularis

Caladenia integra

Caladenia longii

Caladenia ovata

Caladenia rigida

Caladenia tesselata

Caladenia versicolor

Caladenia voigtii Hopper and A.P. Brown ms.

Caladenia wanosa

Calamus warburgii

Calectasia arnoldii Dixon ms.

Callistemon chisholmii

Callistemon formosus

Callitriche cyclocarpa

Callitris oblonga

Callitris sp. aff. *oblonga* Apsley River (A.M. Gray 22495)

Calophyllum bicolor

Calotis glandulosa

Calotis moorei

Calytrix gurulmundensis

Canarium acutifolium

Canthium costatum

Capparis thozetiana

Carronia pedicellata

Cassinia rugata

Centotheca philippensis

Centrolepis paludicola

Chamaelaucium griffinii

Chamaelaucium erythrochlora N.G. Marchant and Keighery ms.

Chamaelaucium roycei N.G. Marchant and Keighery ms.

Cheiranthera volubilis

Choricarpia subargentea

Clematis fawcettii

Codonocarpus pyramidalis

Comesperma oblongatum

Commersonia sp. Cardarga (G.P. Guymer 1642)

Conospermum toddii

Conostylis drummondii

Conostylis lepidospermoides

Conostylis micrantha

Conostylis misera

Conostylis rogeri

Conostylis seorsiflora subsp. trichophylla

Conostylis wonganensis

Coopernookia georgei

Corokia whiteana

Correa baeuerlenii

Correa calycina

Corybas limpidus

Cossinia australiana

Croton magneticus

Cryptocarya foetida

Cryptostylis hunteriana

Ctenopteris blechnoides

Ctenopteris walleri

Cupaniopsis shirleyana

Cupaniopsis tomentella

Cyperus semifertilis

Darwinia acerosa

Darwinia biflora

Darwinia collina

Darwinia macrostegia

Darwinia masonii

Darwinia meeboldii

Darwinia oxylepis

Darwinia sp. Stirling Range (G.J. Keighery 5732)

Darwinia squarrosa

Darwinia wittwerorum

Daviesia discolor

Daviesia euphorbioides

Daviesia megacalyx Crisp ms.

Daviesia pseudaphylla Crisp ms.

Daviesia speciosa Crisp ms.

Daviesia spiralis

Dendrobium bigibbum

Dendrobium callitrophyllum

Dendrobium carronii

Dendrobium johannis

Dendrobium tozerensis

Denhamia parvifolia

Desmodium acanthocladum

Dichanthium queenslandicum

Dichanthium setosum

Dillwynia tenuifolia

Dioclea reflexa

Diplazium cordifolium

Diuris aequalis

Diuris drummondii

Diuris praecox

Diuris purdiei

Diuris recurva

Diuris sheaffiana

Diuris venosa

Dodonaea rupicola

Drakaea concolor Hopper and A.P. Brown ms.

Drakaea confluens Hopper and A.P. Brown ms.

Drakaea elastica

Drakaea micrantha Hopper and A.P. Brown ms.

Drakonorchis barbarella Hopper and A.P. Brown ms.

Drosera fimbriata

Drosera schizandra

Drummondita ericoides

Drummondita hassellii var. longifolia

Dryandra serratuloides

Ectrosia blakei

Ehretia microphylla

Eleocharis blakeana

Eleocharis obicis
Eleocharis retroflexa
Endiandra cooperiana
Endiandra hayesii
Epacris apsleyensis
Epacris barbata
Epacris glabella
Epacris grandis
Epacris limbata
Epacris sp. Dans Hill (S. J. Jarmen HO 32456)
Epilobium brunnescens
Eremophila barbata
Eremophila denticulata
Eremophila inflata
Eremophila microtheca
Eremophila prostrata Chinnock ms.
Eremophila racemosa
Eremophila resinosa
Eremophila subteretifolia Chinnock ms.
Eremophila ternifolia
Eremophila tetraptera
Eremophila virens
Eremophila viscida
Eriocaulon australasicum
Eriocaulon pusillum
Eriostemon ericifolius
Eriostemon sp. Mt Tozer (L.J. Brass 19483)
Eriostemon wonganensis
Erythranthera pumila
Eucalyptus aquatica
Eucalyptus articulata Brooker and Hopper ms.
Eucalyptus argophloia
Eucalyptus argutifolia
Eucalyptus balanites
Eucalyptus beardiana
Eucalyptus bennettiae
Eucalyptus benthamii
Eucalyptus blaxellii
Eucalyptus brevipes
Eucalyptus burdettiana
Eucalyptus cadens
Eucalyptus camfieldii
Eucalyptus ceracea
Eucalyptus cerasiformis
Eucalyptus coronata
Eucalyptus crispata
Eucalyptus crucis subsp. *crucis*
Eucalyptus erectifolia
Eucalyptus glaucina
Eucalyptus goniantha subsp. *goniantha*

Eucalyptus hallii
Eucalyptus imlayensis
Eucalyptus infera
Eucalyptus insularis
Eucalyptus johnsoniana
Eucalyptus kabiana
Eucalyptus kartzoffiana
Eucalyptus langleyi
Eucalyptus lateritica
Eucalyptus leprophloia Brooker and Hopper ms.
Eucalyptus leptoloma
Eucalyptus macrorrhyncha subsp. *cannonii*
Eucalyptus mckieana
Eucalyptus merrickiae
Eucalyptus mooreana
Eucalyptus nicholii
Eucalyptus olivacea Brooker and Hopper ms.
Eucalyptus paedoglauca
Eucalyptus parramattensis subsp. *decadens*
Eucalyptus parvifolia
Eucalyptus pruiniramis
Eucalyptus pulverulenta
Eucalyptus pumila
Eucalyptus rameliana
Eucalyptus raveretiana
Eucalyptus rhodops
Eucalyptus robertsonii subsp. *hamaespherica*
Eucalyptus rubida barbigororum
Eucalyptus rubida canobolensis
Eucalyptus scoparia
Eucalyptus sp. Norseman (S.D. Hopper 2736)
Eucalyptus steedmanii
Eucalyptus sturgissiana
Eucalyptus suberea
Eucalyptus synandra
Eucalyptus tetrapleura
Eucalyptus virens
Eucalyptus xanthope
Eucryphia sp. Mt Bartle Frere (M. Godwin C1158)
Euphrasia amphisysepala
Euphrasia bella
Euphrasia eichleri
Euphrasia phragmostoma
Euphrasia semipicta
Floydia praealta
Fontainea australis
Fontainea rostrata
Fontainea venosa

Gastrolobium appressum
Gastrolobium tomentosum
Gentiana bredboensis
Genus nov. sp. Boonjee (B. Hyland 2519; Family: MYRTACEAE)
Germainia capitata
Glycine latrobeana
Gnaphalium nitidulum
Goodenia macbarronii
Goodenia megasepala
Goodenia quadrifida
Grammitis reinwardtii
Graptophyllum ilicifolium
Grevillea cirsiifolia
Grevillea dryandroides
Grevillea evansiana
Grevillea glossadenia
Grevillea inconspicua
Grevillea infecunda
Grevillea infundibularis
Grevillea involucrata
Grevillea kennedyana
Grevillea prostrata
Grevillea rivularis
Grevillea saccata
Grevillea scortechinii
Grevillea shiressii
Grevillea treueriana
Grevillea venusta
Gymnema brevifolium
Habenaria xanthantha
Hakea megalosperma
Hakea pulvinifera
Hakea sp. Kowmung River (M. Doherty 17-24)
Hakea sp. Mariala Scientific Reserve (C. Sandercoe 507)
Hakea trineura
Haloragis exalata
Halosarcia bulbosa
Halosarcia flabelliformis
Helicteres sp. Glenluckie Creek (N. Byrnes 1280)
Hemiandra gardneri
Hemiandra rutilans
Hemigenia viscida
Hensmania chapmanii
Hexaspora pubescens
Hibbertia crispula
Hibbertia sp. Porongurups (R.D. Hoogland 12186)
Hicksbeachia pinnatifolia
Hodgkinsonia frutescens
Homoranthus darwinioides
Homoranthus decumbens

Homoranthus montanus
Homoranthus porteri
Huperzia lockyeri
Huperzia marsupiiformis
Huperzia phlegmarioides
Huperzia prolifera
Hydrocharis dubia
Hydrocotyle lemnoides
Idiospermum australiense
Indigofera efoliata
Ipomoea sp. Stirling (P. Latz 10408)
Isopogon fletcheri
Ixodia achillaeoides subsp. *arenicola*
Jagera javanica
Jedda multicaulis
Kelleria laxa
Kennedia beckxiana
Kennedia glabrata
Kennedia macrophylla
Kennedia retrorsa
Kunzea cambagei
Lasiopetalum longistamineum
Lasiopetalum micranthum
Lastreopsis walleri
Lawrencia buchananensis
Laxmannia jamesii
Lechenaultia chlorantha
Lechenaultia laricina
Lechenaultia pulvinaris
Lechenaultia superba
Lepidium pseudopapillosum
Leptorhynchos gatesii
Leptospermum deanei
Leptospermum thompsonii
Leucopogon cuspidatus
Leucopogon exolasius
Leucopogon obtectus
Limosella granitica
Lindsaea pulchella
Livistona drudei
Livistona mariae
Logania diffusa
Logania insularis
Macadamia integrifolia
Macadamia ternifolia
Macadamia tetraphylla
Macadamia sp. Iron Range (B. Hyland 3102)
Macropteranthes montana
Macrozamia fearnsidae
Macrozamia macdonnellii
Maireana cheelii
Maireana melanocarpa

Malacocera gracilis
Medicosma elliptica
Medicosma obovata
Melaleuca groveana
Melaleuca kunzeoides
Melaleuca sciotostyla
Mesua sp. Boonjee (A.K. Irvine 1218)
Microcorys eremophiloides
Microlepidium alatum
Micromyrtus blakelyi
Micromyrtus minutiflora
Microstrobos fitzgeraldii
Microtis globula
Minuria tridens
Myoporum cordifolium
Myoporum latisepalum
Myriophyllum coronatum
Myriophyllum implicatum
Myriophyllum petraeum
Myriophyllum porcatum
Neisosperma kilneri
Newcastelia velutina
Nicotiana burbidgeae
Notelaea lloydii
Olearia astroloba
Olearia cordata
Olearia macdonnellensis
Olearia pannosa subsp. *pannosa*
Oreodendron biflorum
Oreoporanthera petalifera
Owenia cepiodora
Ozothamnus eriocephalus
Ozothamnus tesselatus
Pandanus spiralis var. *flammeus*
Paspalidium grandispiculatum
Persicaria elatior
Persoonia acerosa
Persoonia amaliae
Persoonia glaucescens
Persoonia marginata
Phaius australis
Phaius tancarvilliae
Phebalium lowanense
Phebalium ralstonii
Phebalium rhytidophyllum
Phebalium sympetalum
Phebalium whitei
Phyllota humifusa
Picris evae
Pimelea leptospermoides
Pimelea spinescens subsp. *spinescens*
Pimelea venosa
Pityrodia augustensis

Plectranthus gratus
Pleuropappus phyllocalymmeus
Plinthanthesis rodwayi
Poa sallacustris
Polyscias bellendenkerensis
Pomaderris brunnea
Pomaderris clivicola
Pomaderris halmaturina subsp. *halmaturina*
Pomaderris pallida
Pomaderris parrisiae
Pomaderris sericea
Pomaderris subplicata
Pomatocalpa marsupiale
Prasophyllum frenchii
Prasophyllum morganii
Prasophyllum pallidum
Prasophyllum truncatum
Prasophyllum validum
Prasophyllum wallum
Prostanthera calycina
Prostanthera carrickiana
Prostanthera cineolifera
Prostanthera cryptandroides
Prostanthera densa
Prostanthera discolor
Prostanthera magnifica
Prostanthera sp. Dunmore (D.M. Gordon 84)
Prostanthera sp. Mt Tinbeerwah (C. Sandercoe C1256)
Prostanthera stricta
Prostanthera teretifolia
Pterostylis bicornis
Pterostylis cucullata
Pterostylis pulchella
Pterostylis cobarensis
Pterostylis tenuissima
Ptilotus maconochie
Pultenaea aristata
Pultenaea baeuerlenii
Pultenaea campbellii
Pultenaea glabra
Pultenaea parrisiae J. Briggs and Crisp ms.
Pultenaea parviflora
Pultenaea selaginoides
Pultenaea setulosa
Pultenaea stuartiana
Pultenaea villifera var. *glabrescens*
Pultenaea williamsoniana
Quassia bidwillii
Ranunculus anemoneus
Restio longipes
Rhagodia acicularis

Rhamphicarpa australiensis
Rhaphidospora bonneyana
Rhizanthella gardneri
Ricinocarpos gloria-medii
Ricinocarpos trichophorus
Ristantia gouldii
Roycea pycnophylloides
Rulingia procumbens
Rulingia prostrata
Rutidosis heterogama
Rutidosis leiolepis
Sagina sp. Mt Anne (A.M. Buchanan 5115)
Sarcochilus fitzgeraldii
Sarcochilus hartmannii
Sarcochilus hirticalcar
Sarcochilus weinthalii
Sauropus macranthus
Sclerolaena blakei
Sclerolaena walkeri
Senecio garlandii
Senecio laticostatus
Senecio macrocarpus
Senecio megaglossus
Solanum carduiforme
Solanum dunalianum
Sophora fraseri
Sowerbaea subtilis
Spathoglottis plicata
Spirogardnera rubescens
Spyridium coactilifolium
Spyridium eriocephalum var. *glabrisepalum*
Spyridium leucopogon
Spyridium microphyllum
Spyridium obcordatum
Stackhousia annua
Stawellia dimorphantha
Stemona angusta
Stipa metatoris
Stylidium galioides
Stylidium plantagineum
Swainsona minutiflora
Swainsona murrayana
Swainsona plagiotropis
Swainsona pyrophila
Symplocos baeuerlenii
Symplocos sp. Imbil-Beenleigh (W.J. MacDonald 3832)
Syzygium hodgkinsoniae
Syzygium moorei
Syzygium paniculatum
Syzygium velarum
Taraxacum cygnorum
Tasmannia glaucifolia
Tasmannia purpurascens

Tephrosia leveillei
Tetratheca aphylla
Tetratheca glandulosa
Tetratheca harperi
Tetratheca juncea
Thelymitra matthewsii
Thelymitra psammophila
Thelymitra stellata
Thesium australe
Thomasia montana
Thomasia sp. York (A.S. George 8075)
Thryptomene wittweri
Tinospora tinosporoides
Toechima pterocarpum
Tribonanthes purpurea
Trichanthodium baracchianum
Trichoglottis australiensis
Trigonostemon inopinatus
Trymalium minutiflorum
Tylophora williamsii
Vanda hindsii
Velleia perfoliata
Verreauxia verreauxii
Verticordia creba
Verticordia helichrysantha
Verticordia staminosa
Villarsia calthifolia
Westringia davidii
Westringia parvifolia
Wrixonia schultzii
Wurmbea tubulosa
Wurmbea sp. Cape Naturaliste (S.D. Hopper 5871)
Xanthostemon oppositifolius
Xanthostemon youngii
Xerothamnella parvifolia
Xyris sp. Stirling Range (G.J. Keighery 7951)
Zeuxine polygonoides
Zieria citriodora J.A. Armstrong ms.
Zieria collina
Zieria covenyi J.A. Armstrong ms.
Zieria granulata
Zieria ingramii J.A. Armstrong ms.
Zieria involucrata
Zieria murphyi
Zieria obovata (C.T. White) J.A. Armstrong ms.
Zieria rimulosa
Zieria tuberculata J.A. Armstrong ms.
Zieria verrucosa J.A. Armstrong ms.

Part 3 – Species that are presumed extinct

Animals

Fish
nil

Amphibians
nil

Reptiles
nil

Birds

Aplonis fusca
  Norfolk Island Starling

Columba vitiensis godmanae
  Lord Howe Pigeon

Cyanoramphus novaezelandiae erythrotis
  Macquarie Island Parakeet

Cyanoramphus novaezelandiae subflavescens
  Lord Howe Parakeet

Dasyornis broadbenti littoralis
  South-western Rufous Bristlebird

Dromaius baudinianus
  Kangaroo Island Emu

Dromaius minor
  Dwarf Emu/King Island Emu

Drymodes superciliaris colcloughi
  Roper River Scrub-robin

Gerygone insularis
  Lord Howe Warbler

Hemiphaga novaeseelandiae spadicea
  New Zealand Pigeon (Norfolk Island Race)

Lalage leucopyga leucopyga
  Norfolk Island Long-tailed Triller

Nestor productus
  Norfolk Island Kaka

Ninox novaeseelandiae albaria
  Lord Howe Boobook Owl

Notornis alba
  White Gallinule

Psephotus pulcherrimus
  Paradise Parrot

Rallus pectoralis clelandi
  Lewin's Water Rail (western race)

Rallus philippensis maquariensis
  Macquarie Island Rail

Rhipidura cervina
  Lord Howe Fantail

Turdus xanthopus vinitinctus
  Lord Howe Island Vinous-tinted Thrush

Zosterops strenua
  Robust White-eye

Mammals

Caloprymnus campestris
  Desert Rat-kangaroo

Chaeropus ecaudatus
  Pig-footed Bandicoot

Conilurus albipes
  White-footed Rabbit-rat

Lagorchestes asomatus
  Central Hare-wallaby

Lagorchestes leporides
  Eastern Hare-wallaby

Leporillus apicalis
  Lesser Stick-nest Rat

Macropus greyi
  Toolache Wallaby

Macrotis leucura
  Lesser Bilby

Notomys amplus
  Short-tailed Hopping-mouse

Notomys longicaudatus
  Long-tailed Hopping-mouse

Notomys macrotis
  Big-eared Hopping-mouse

Notomys mordax
  Darling Downs Hopping-mouse

Onychogalea lunata
  Crescent Nailtail Wallaby

Potorous platyops
  Broad-faced Potoroo

Pseudomys fieldi
  Alice Springs Mouse

Pseudomys gouldii
  Gould's Mouse

Pteropus brunneus
  Percy Island Flying Fox

Rattus macleari
  Christmas Island Rat

Rattus nativitatus
  Christmas Island Rat

Thylacinus cynocephalus
  Thylacine

Plants

Acacia murruboensis
Acacia prismifolia
Acacia volubilis
Acanthocladium dockeri
Acianthus ledwardii
Amphibromus whitei
Argentipallium spiceri
Beyeria lepidopetala
Caladenia atkinsonii
Caladenia pumila
Calothamnus accedens
Centrolepis caespitosa
Choristemon humilis
Coleanthera virgata
Deyeuxia drummondii
Deyeuxia lawrencei
Dicrastylis morrisonii
Eriostemon falcatus
Euphorbia carissoides
Euphrasia arguta
Euphrasia sp. Tamworth (Rupp s.n. -/9/1904)
Frankenia conferta
Frankenia decurrens
Frankenia parvula
Glyceria drummondii
Grevillea divaricata
Gyrostemon reticulatus
Haloragis platycarpa
Hemigenia exilis
Hemigenia obtusa
Hutchinsia tasmanica
Hydatella leptogyne
Hypsela sessiliflora
Lasiopetalum rotundifolium
Lepidium drummondii
Lepidium peregrinum
Leptomeria dielsiana
Leptomeria laxa
Leucopogon cryptanthus
Leucopogon marginatus
Menkea draboides
Nemcia lehmannii
Oberonia attenuata
Olearia oliganthema
Opercularia acolytantha
Ozothamnus selaginoides
Persoonia prostrata
Phlegmatospermum drummondii
Pimelea spinescens subsp. *pubiflora*
Platysace dissecta
Plectrachne bromoides
Prasophyllum subbisectum
Pseudanthus nematophorus
Ptilotus fasciculatus
Ptilotus pyramidatus
Rapanea sp. Richmond River (J.H. Maiden and J.L. Boorman NSW 26751)
Scaevola attenuata
Scaevola macrophylla
Schoenus natans
Senecio georgianus
Stylidium merrallii
Stylidium neglectum
Tetraria australiensis
Tetratheca elliptica
Tetratheca fasciculata
Thomasia gardneri
Trachymene scapigera
Trianthema cypseloides

Australian and New Zealand Environment and Conservation Council Endangered Fauna Network conservation status recommendations which are to be considered for changes to the Endangered Species Protection Act Schedules

Taxon	*Current*	*New*
Birds		
Emu (Tasmanian subspecies) *Dromaius novaehollandiae diemenensis*	–	X
Grey-headed Albatross *Diomedea chrysostoma*	–	V
Kermadec Petrel (western subspecies) *Pterodroma neglecta neglecta*	–	V
Soft-plumaged Petrel (northern subspecies) *Pterodroma mollis deceptionis*	–	V
Blue Petrel *Halobaena caerulea*	–	V
Fairy Prion (southern subspecies) *Pachyptila turtur subantarctica*	–	V
White-bellied Storm-Petrel (Australasian subspecies) *Fregetta grallaria grallaria*	–	V
Heard Island Shag *Phalacrocorax nivalis*	–	V
Macquarie Island Shag *Phalacrocorax purpurascens*	–	V
Brown Goshawk (Christmas Island subspecies) *Accipiter fasciatus natalis*	–	V
Wedge-tailed Eagle (Tasmanian subspecies) *Aquila audax fleayi*	–	E
Mallee fowl *Leipoa ocellata*	E	V
Buff-banded Rail (Cocos (Keeling) Island subspecies) *Gallirallus philippensis andrewsi*	–	E
Buff-breasted Button-quail *Turnix olivei*	–	E
Antarctic Tern (Indian and Atlantic Ocean subspecies) *Sterna vittata vittata*	–	V
Antarctic Tern (New Zealand subspecies) *Sterna vittata bethunei*	–	E
Partridge Pigeon (eastern subspecies) *Geophaps smithii smithii*	–	V
Squatter Pigeon (southern subspecies) *Geophaps scripta scripta*	–	V
Eclectus Parrot (Australian subspecies) *Eclectus roratus macgillivrayi*	V	delete
Superb Parrot *Polytelis swainsonii*	–	V
Regent Parrot (eastern subspecies) *Polytelis anthopeplus anthopeplus*	–	V
Glossy Black-Cockatoo (Kangaroo Island subspecies) *Calyptorhynchus lathami halmaturinus*	–	E
Carnaby's Black-Cockatoo *Calyptorhynchus latirostris*	–	V
Western Long-billed Corella (southern subspecies) *Cacatua pastinator pastinator*	E	V
Masked Owl (northern subspecies) *Tyto novaehollandiae kimberli*	–	V
Masked Owl (Melville Island subspecies) *Tyto novaehollandiae melvillensis*	–	V
Noisy Scrub-bird *Atrichornis clamosus*	V	E
Purple-crowned Fairy-wren (western subspecies) *Malurus coronatus coronatus*	–	V
Mallee Emu-wren *Stipiturus mallee*	–	V
Carpentarian Grasswren *Amytornis dorotheae*	V	delete
Thick-billed Grasswren (eastern subspecies) *Amytornis textilis modestus*	–	V
Thick-billed Grasswren (Gawler Ranges subspecies) *Amytornis textilis myall*	–	V
Western Bristlebird *Dasyornis longirostris*	V	E
Brown Thornbill (King Island subspecies) *Acanthiza pusilla archibaldi*	–	E
Slender-billed Thornbill (western subspecies) *Acanthiza iredalei iredalei*	–	V
White-browed Robin (western subspecies) *Poecilodryas superciliosa cerviniventris*	E	delete
Scarlet Robin (Norfolk Island subspecies) *Petroica multicolor multicolor*	–	V
Western Whipbird (western heath subspecies) *Psophodes nigrogularis nigrogularis*	–	E
Western Whipbird (eastern subspecies) *Psophodes nigrogularis leucogaster*	–	V
Western Whipbird (western mallee subspecies) *Psophodes nigrogularis oberon*	–	E
Crested Shrike-tit (northern subspecies) *Falcunculus frontatus whitei*	E	V
Crested Shrike-tit (western subspecies) *Falcunculus frontatus leucogaster*	V	delete
Golden Whistler (Norfolk Island subspecies) *Pachycephala pectoralis xanthoprocta*	–	V
Cicadabird (north-western subspecies) *Coracina tenuirostris melvillensis*	E	delete
Star Finch (eastern subspecies) *Neochmia ruficauda ruficauda*	–	E
Crimson Finch (white-bellied subspecies) *Neochmia phaeton evangelinae*	–	V
Black-throated Finch (southern subspecies) *Poephila cincta cincta*	–	V
Marsupials		
Eastern Quoll *Dasyurus viverrinus*	V	delete
Proserpine Rock-wallaby *Petrogale persephone*	V	E
Western Ringtail *Pseudocheirus occidentalis* (currently listed as a subspecies now to be listed as a full species)	E	V
Sandhill Dunnart *Sminthopsis psammophila*	V	E
Rodents		
Thevenard Island Mouse *Leggadina* affin. *lakedownensis*	–	E

Taxon	Current	New
Frogs		
Orange-bellied Frog *Geocrinia vitellina*	E	V
Creek Frog or White-bellied Frog *Geocrinia alba*	V	E
Armoured Mistfrog *Litoria lorica*	–	E
Waterfall Frog *Litoria nannotis*	–	E
Mountain Mistfrog *Litoria nyakalensis*	–	E
Lace-eyed Tree Frog *Nyctimystes dayi*	–	E
Tinkling Frog *Taudactylus reophylis*	–	E
Reptiles		
Mary River Tortoise *Elusior macruros*	–	E
Lancelin Island Skink *Ctenotus lancelini*	V	E
Mallee Worm-lizard *Aprasia aurita* (common name change)	E	E
Blue Mountains Water Skink *Eulamprus leuraensis*	–	E
Dreeite Water Skink *Eulamprus tympanum ssp. nov.* (basalt plains, VIC)	–	E
Allan's Lerista *Lerista allanae*	–	E
Woma *Aspidites ramsayi* (south-western WA)	–	E
Western Spiny-tailed Skink *Egernia stokesii badia*	–	E
Black-striped Snake *Simoselaps calonotus*	–	E
Hermite Island Worm-lizard *Aprasia rostrata rostrata*	–	V
Christmas Island Blind Snake *Ramphotyphlops exocoeti*	–	V
Lord Howe Island Skink *Pseudemoia lichenigera*	–	V
Christmas Island Gecko *Lepidodactylus listeri*	–	V
Long-legged Worm-skink *Anomalopus mackayi*	–	V
Krefft's Tiger Snake *Notechis ater ater* (Flinders Ranges, SA)	–	V
Broad-headed Snake *Hoplocephalus bungaroides*	E	V
Namoi River Elseya *Elseya sp. nov.* (Namoi River, NSW)	–	V
Bellinger River Emydura *Emydura signata* (Bellinger River, NSW)	–	V
Pygmy Copperhead *Austrelaps labialis* (Adelaide, SA)	–	V
Collared Delma *Delma torquata* (common name change)	V	V
Mount Cooper Striped Lerista *Lerista vittata*	–	V
Three-toed Snake-tooth Skink *Coeranoscincus reticulatus*	–	V
Ornamental Snake *Denisonia maculata*	–	V
Dunmall's Snake *Furina dunmalli*	–	V
Lord Howe Island Gecko *Christinus guentheri*	–	V
Pedra Branca Skink *Niveoscincus palfreymani* (scientific name change)	V	V
Striped-tailed Delma *Delma labialis*	–	V
Flinders Ranges Worm-lizard *Aprasia pseudopulchella*	–	V
Western Australian Carpet Python *Morelia spilota imbricata*	–	V
Rough-scaled Python *Morelia carinata*	V	delete
Lake Cronin Snake *Echiopsis atriceps*	–	V
Bardick *Echiopsis curta* (population east of Adelaide)	–	V
Fitzroy Tortoise *Rheodytes leukops* (common name change)	V	V
Pernatty Knob-tail *Nephrurus deleani*	–	V
Border Thick-tailed Gecko *Underwoodisaurus sphyrurus*	–	V
Short-nosed Snake *Elapognathus minor*	–	V
Hamelin Ctenotus *Ctenotus zastictus*	–	V
Great Desert Skink *Egernia kintorei*	–	V
Jurien Bay Rock-skink *Egernia pulchra longicauda*	–	V
Houtman Abrolhos Spiny-tailed Skink *Egernia stokesii stokesii*	–	V
Bronzeback Snake-lizard *Ophidiocephalus taeniatus* (common name change)	V	V
Brigalow Scaly-foot *Paradelma orientalis*	–	V
Fish (fresh water)		
Barred Galaxias *Galaxias fuscus* (common name change)	E	E
Swamp Galaxias *Galaxias parvus*	V	delete
Dwarf Galaxias *Galaxiella pusilla*	–	V
Red-Finned Blue-Eye *Scaturiginichthys vermeilipinnis*	–	E
Clarence River Cod *Maccullochella ikei* (common name change)	E	E
Mary River Cod *Maccullochella peeli mariensis*	–	E
Elizabeth Springs Goby *Chlamydogobius sp. A*	–	E
Murray Hardyhead *Craterocephalus fluviatilis*	–	V
Oxleyan Pygmy Perch *Nannoperca oxleyana*	–	V

Note: E – Endangered;   V – Vulnerable;   X – Presumed Extinct

Acronyms

AAS
Australian Academy of Science

ABARE
Australian Bureau of Agricultural and Resource Economics

ABS
Australian Bureau of Statistics

ACT
Australian Capital Territory

AGPS
Australian Government Publishing Service

AHC
Australian Heritage Commission

AIATSIS
Australian Institute of Aboriginal and Torres Strait Islander Studies

AIHW
Australian Institute of Health and Welfare

ANCA
Australian Nature Conservation Agency

ANOP
Australian National Opinion Polls

ANZECC
Australian and New Zealand Environment and Conservation Council

AQIS
Australian Quarantine Inspection Service

ASTEC
Australian Science and Technology Council

ATSIC
Aboriginal and Torres Strait Islander Commission (of Commonwealth)

AUSLIG
Australian Surveying and Land Information Group

AWQG
Australian Water Quality Guidelines (of ANZECC)

BoM
Bureau of Meteorology (of Commonwealth)

BRS
Bureau of Resource Sciences (of Commonwealth)

CaLM
Conservation and Land Management (present or former name of government departments in several States)

CAMBA
China/Australia Migratory Birds Agreement

CAT
Centre for Appropriate Technology (in Alice Springs, Northern Territory)

CITES
Convention on International Trade in Endangered Species of Wild Fauna and Flora

COAG
Council of Australian Governments (Council of Commonwealth and State heads of government and the head of the Australian Local Government Association)

CRC
Cooperative Research Centre

CSD
Commission on Sustainable Development (of UN)

CSIRO
Commonwealth Scientific and Industrial Research Organisation

DEET
Department of Employment, Education and Training (of Commonwealth)

DEST
Department of the Environment, Sport and Territories (of Commonwealth)

DHRD
Department of Housing and Regional Development (of Commonwealth)

DPIE
Department of Primary Industries and Energy (of Commonwealth)

EEZ
Exclusive Economic Zone

EPA
Environment Protection Authority or Agency (NSW, SA, VIC, and WA have an Environment Protection Authority, while the Commonwealth has an Environment Protection Agency)

ERIN
Environmental Resources Information Network (of DEST)

ERP
Estimated Resident Population

ESCAP
Economic and Social Commission for Asia and the Pacific (of United Nations)

ESD
Ecologically Sustainable Development

FCCC
Framework Convention on Climate Change (of United Nations)

GBRMPA
Great Barrier Reef Marine Park Authority (of Commonwealth)

GIS
Geographic Information System

IBRA
Interim Biogeographic Regionalisation for Australia

ICICPA
Independent Committee of Inquiry into Competition Policy in Australia

ICOMOS
International Council for Monuments and Sites

IGAE
Intergovernmental Agreement on the Environment
(Commonwealth/State/Local government
agreement)

IPCC
Intergovernmental Panel on Climate Change
(international panel under auspices of WMO and
UNEP)

ISTP
Institute for Science and Technology Policy
(of Murdoch University, Western Australia)

IUCN
World Conservation Union (formerly the
International Union for the Conservation of
Nature and Natural Resources)

JAMBA
Japan/Australia Migratory Birds Agreement

JANIS
Joint Australia and New Zealand Environment and
Conservation Council, the Ministerial Council of
Forests, Fisheries and Aquaculture, and the
National Forest Policy Statement Implementation
Subcommittee

MARPOL
International Convention for the Prevention of
Pollution from Ships

MDBC
Murray–Darling Basin Commission
(Commonwealth/State commission)

NEPC
National Environment Protection Council

NGGI
National Greenhouse Gas Inventory

NGGIC
National Greenhouse Gas Inventory Committee

NGRS
National Greenhouse Response Strategy

NHMRC, NH & MRC
National Health and Medical Research Council

NSW
New South Wales

NT
Northern Territory

OCS
Offshore Constitutional Settlement

OECD
Organisation for Economic Cooperation and
Development

QDPI
Queensland Department of Primary Industries
(since March 1996, Queensland Department of
Natural Resources)

QLD
Queensland

RAC
Resource Assessment Commission
(former Commonwealth commission)

RFA
Regional Forest Agreement
(Commonwealth/State agreement)

SA
South Australia

SLA
Statistical Local Area

SoE
State of the Environment

SoMER
State of the Marine Environment Report
(of DEST)

TAS
Tasmania

UNCLOS
United Nations Convention on the Law of the Sea

UNEP
United Nations Environment Program

UNESCO
United Nations Educational Scientific and Cultural
Organization

VIC
Victoria

WA
Western Australia

WHO
World Health Organization
(of United Nations)

WMO
World Meteorological Organization
(of United Nations)

WWF
World Wide Fund for Nature

Glossary

Aboriginal and Torres Strait Islander places
(as used in chapter 9) sites, areas or regions of significance to indigenous peoples including places with archaeological traces, ceremonial, story and other places with particular traditional or contemporary associations; places reflecting the historic interaction of indigenous peoples with non-indigenous peoples; may be single sites, site complexes or landscapes *see* **historic places, natural places**

Aboriginal English
a term covering many types of English as spoken by indigenous people, especially in northern Australia; the first language of thousands of Australians

acid deposition
the deposition on the earth's surface, either in dry or wet form, of substances derived from natural and human-induced emissions of various compounds, especially those of sulfur and nitrogen which have been transformed by chemical processes in the atmosphere

acid gases and acid precursors
gases such as sulfur dioxide and oxides of nitrogen which undergo chemical reactions in the atmosphere and are transformed to sulfate and nitrate particles (acid gas precursors) as well as gaseous sulfuric and nitric acids (acid gases)

adaptation
a particular part of the anatomy, a physiological process, or a behaviour pattern that improves an organism's chances to survive and reproduce

adaptation measures
management measures or options in response to, or anticipation of climate change (as a result of increasing emissions of greenhouse gases) to reduce or avoid adverse consequences or to take advantage of beneficial changes *see* **mitigation measures**

aerosol
a suspension of particles, other than water or ice, in the atmosphere and ranging in size from approximately $10^{-3}\mu$m to larger than 10 μm in radius; may be either natural or caused by human activity and most of the latter are usually considered to be pollutants

after housing poverty
households falling below standards developed by the 1975 Henderson Poverty Inquiry after meeting housing costs are described as being in 'after housing poverty' or financial housing stress (Jones, R. (1994); The Housing Needs of Indigenous Australians (1991), *Centre for Aboriginal Economic Policy Research, Australian National University Research Monograph* No. 8)

age-standardised death rates
summary measures that allow comparison of populations with different age distributions, either different populations at the same time or the same population at different times; they are calculated by applying the observed age-specific death rates for the population of interest to a given reference population, and represent the total death rate that would be observed in the population of interest, if it had the same age distribution as the reference population

agricultural land
any land on which crops or pastures are cultivated or domestic stock are grazed

air toxics
pollutants present at very low concentrations, known to cause or suspected of causing long-term health effects in humans

airshed
a body of air bounded by topographical and/or meteorological features in which a contaminant, once emitted, is contained

Airtrak
a smog monitoring instrument developed by CSIRO in the late 1980s which simultaneously measures all major components of photochemical smog and combines the measurements with information on weather conditions; it then predicts how the smog will develop as the day progresses

algal blooms
a sudden proliferation of microscopic algae in water bodies, stimulated by the input of nutrients such as phosphates

allele
a form of a gene, where multiple such forms occur

allergen
a substance inducing an allergic reaction

alluvial
arising from sediments deposited from flowing water

ambient air
surrounding outdoor air

ambient air quality guideline
a level to which atmospheric concentrations of a substance should be reduced to avoid undesirable effects on human health, well-being and/or vegetation; guidelines are generally regarded as advisory rather than mandatory requirements

anthropogenic
of human origin or human induced; can be used in the context of emissions that are produced as a result of human activities

aquaculture
the commercial growing of marine (mariculture) or freshwater animals and plants in water

arable land
land that is, or has the potential to be, cultivated for crop production

arid zone
often arbitrarily defined in Australia as those areas receiving less than 250 mm of annual rainfall in the south and 350 mm (or sometimes higher) in the north

assimilative capacity
the ability of the natural environment to absorb or deal with external pressures such as emissions

produced as a result of human activities without leading to permanent and/or significant change

atmosphere
composite layer of colourless, odourless gases, known as air, surrounding the Earth; it shows distinct vertical zonation *see* **troposphere, stratosphere**

atmospheric stability
a parcel of air is termed stable, neutral or unstable according to its motion when displaced upward adiabatically (with no exchange of heat with its environment) from its equilibrium position — if stable the parcel will tend to return to its equilibrium position, if neutral it remains constant, and if unstable it will accelerate upwards; the major meteorological factor controlling vertical mixing of air pollutants

Australian Fishing Zone (AFZ)
proclaimed 200 nautical mile wide zone around the coast, within which Australia controls domestic and foreign access to fish resources

ballast water
water carried in tanks to maintain stability when a ship is lightly loaded, it is normally discharged to the sea when the ship is loaded with cargo

bare fallow
a fallow period in which the soil has no vegetative or crop residue cover

basalt
dark coloured rocks of volcanic origin

baseline information
information relating to a specific time or defined area of land or water, from which trends or changes can be assessed

benthic
associated with (attached to or buried in) the sea floor

Better Cities
a series of demonstration programs in each Australian State with seed funds from the Commonwealth Government, which is designed to show how to make better cities through integrated planning with sustainability and social justice goals

biodiversity
the variety of all life-forms: the different plants, animals and micro-organisms, the genes they contain and the ecosystems they form; often considered at three levels: genetic diversity, species diversity and ecosystem diversity

biogeochemical cycle
the movement of chemical elements between organisms and non-living compartments of atmosphere, aquatic systems and soils

biogeographic region
an extensive region distinguished from adjacent regions by its broad physical and biological characteristics

biological control
controlling a pest by the use of its natural enemies

biomass
the quantity of organic matter within an ecosystem (usually expressed as dry weight for unit area or volume)

biomass burning
the combustion of organic waste matter, burning in slash-and-burn cultivation, fuel-wood use and land clearing through forest burning

bloom
a proliferation of plants (for example, macroalgae or phytoplankton) during favourable growing conditions generated by availability of nutrients or sunlight

bore capping
permanent or temporary closure of discharging artesian bore

Bos indicus
the species of humped cattle native to India and Africa, including modern breeds such as the Brahman

brackish water
water that is saline, but less so than sea water; it may be suitable for selective irrigation and watering of livestock

broadacre farms
commercial farms producing relatively low value crops such as wool, sheep meat, beef, cereals, on large areas

Burra Charter
a document prepared by the Australian Committee for the International Council for Monuments and Sites (Australia ICOMOS) to guide conservation philosophy and practice for cultural heritage places in Australia; its full name is The Australia ICOMOS Charter for the Conservation of Places of Cultural Significance

bushfire
a term used to describe almost any form of fire burning out of control *see* **prescribed fire**

by-catch
species taken incidentally in a fishery where other species are the target; may be of lesser value than the target species and are often discarded

calcareous
composed of, or containing lime or limestone

carrying capacity
the maximum population size that can be supported indefinitely by a given environment

catchment
the area determined by topographic features within which rainfall will contribute to run-off at a particular point under consideration

CFCs (chlorofluorocarbons)
synthetic products, which do not occur naturally and contain chlorine and fluorine; commonly used in various industrial processes and as refrigerants and, prior to 1990, as a propellant gas for sprays; deplete ozone in the stratosphere and are powerful greenhouse gases

chenopod shrubland
areas dominated by shrubs of the Chenopodiaceae family (commonly known as saltbushes and bluebushes)

chlorophyll
the green pigment in plants that functions in photosynthesis by absorbing light from the sun

cleaner production
a worldwide industry trend towards reduced resource inputs and waste outputs in the production process; in Australia, the Commonwealth EPA has set up a program to demonstrate cleaner production processes

clearfelling
the removal of all trees on a specified cutting area *see* **coupe**; in many cases some trees are retained for environmental protection or conservation reasons

clearing
removing vegetation, particularly trees and shrubs, from a landscape, often with the intention of replacing it with plants regarded to be more directly useful to humans

climate
the synthesis of the day-to-day weather conditions in a given area; the actual climate is characterised by long-term statistics of the state of the atmosphere in an area

climate change
under the terms of the United Nations Framework Convention on Climate Change, the term means a change of climate which is attributed directly or indirectly to human activity that alters the composition of the global atmosphere and which is in addition to natural climate variability observed over comparable time periods

climate variability
the natural year to year and season to season variation of the climate system

codes of practice (relating to forestry)
sets of guidelines adopted by forest management agencies concerned with minimising impacts of forestry operations on the environment (for example, on soil erosion) and with worker safety

coliform
a group of bacteria originating from animal (including human) intestines and used as an indicator of the sanitary quality of water

compaction
the reduction in bulk volume of sediments owing to the increased weight of overlying materials or by impact on the surface layers such as by machinery or livestock

comprehensiveness (as used in chapter 9)
the extent to which heritage registers or collections include all significant places or objects of a particular type

condition indicator
something that describes the quality of the environment and the quality and quantity of natural resources; highlights changes in environmental conditions over time

conservation (as used in chapter 9)
all the processes of looking after places or objects so as to retain their heritage significance

conservation farming (tillage)
farming systems designed to reduce run off so that water storage in the soil is maximised and soil erosion is reduced

contour banks
small banks cut in the soil that are aligned close to the contour and convey water across the slope to a waterway or drain designed to resist erosion

coupe
an area of forest harvested in a single operation; usually 10 to 300 hectares harvested over a single season

creole
a language developed when a new generation takes a pidgin for its first language, and extends and develops it so that it is capable of a full range of expression *see* **pidgin**

cultural mapping
the identification and recording of the cultural resources and activities of a community or region

cultural tourism
travel for essentially cultural motivations, which may include travel for specific purposes, for example, to attend festivals or to visit sites or monuments, or may be more broadly motivated by the desire to experience cultural diversity or to immerse oneself in the culture of a region

db(A)
decibels of noise levels measured using the electronic 'A-weighting' filter incorporated in sound level measuring devices; the frequency response of this filter is similar to that of human hearing; the level of sound in db(A) is an accurate measure of the loudness of that sound

denitrification
the process by which nitrogen, which would otherwise be available to plants, is converted to a gaseous form and lost from the soil

desertification
the degradation of land in arid, semi-arid, and other areas with a dry season; caused primarily by over-exploitation and inappropriate land use interacting with climatic variations

diatom
a single-celled microscopic alga with two ornate interfitting outer shells containing silica

diffuse-source pollution
pollution from sources such as an eroding paddock, urban or suburban lands and forests; spread out, and often not easily identified or managed

dinoflagellate
a microscopic, single-celled organism that moves using two whip-like hairs called flagella. Many are photosynthetic, considered to be algae, and form part of the phytoplankton

discharge
(as used in Chapter 7)
the volume of water which flows through a cross-section of a stream

dispersion
(as used in Chapter 5)
the spread of pollutants caused by atmospheric mixing, transportation or turbulence (random fluctuations in wind velocity)

Distributed National Collection
the aggregate of collections of objects located in major Commonwealth, State and Territory collecting institutions, as well as those held in community, regional and specialist museums, libraries, schools and private collections

divertible resources
the volume of water that can be diverted on a sustained basis into conventional water supply systems or to substantial private users, using existing storage and potential dam sites

domestic animals
animals directly managed by humans *see* **feral**

doughnut effect
the loss of population and urban services from inner city areas to outer areas

drainage
the interception and/or removal of surface and/or groundwater from a given area by natural or artificial means

drainage division
the primary subdivision of groups of catchments *see* Fig. 7.1

dredge spoil
sediments and materials removed from the seabed during dredging

drip irrigation
a method of irrigation by a pipe system which provides water to individual plants by means of a drip emitter

dry sclerophyll
a type of eucalypt forest found in moderate rainfall (less than 1000 mm per year) areas. Sometimes called 'open forests' *see* **wet sclerophyll**

dryland cropping
cropping without irrigation, usually in areas of relatively low rainfall

dryland salinity
soil salinity levels high enough to affect plant growth; occurs as a result of natural soil forming process (primary salinity) or in disturbed landscapes through clearing or other activities that interfere with the water and salinity balance and lead to shallow watertables; hydrological response to the replacement of deep-rooted perennial native vegetation with shallow rooted annuals which use less water; as a consequence more rainfall enters the groundwater, causing watertables to rise; where these rise to within 1–2 m of the soil surface, salinisation occurs as a result of evapotranspiration and direct evaporation; can result in both stream and soil salinity

eco-tourism
nature-based tourism which involves education and interpretation of the natural environment and is managed to be ecologically sustainable

ecological footprint
the ecological impact of cities, including the direct local effects and the indirect regional and global effects due to the resources they use and the wastes they produce

ecological sustainability
the capacity of ecosystems to maintain their essential processes and functions and to retain their biological diversity without impoverishment

ecologically mature forests
stands of trees approaching the limit of their life span which show little increase in biomass and usually support a high biodiversity *see* **old-growth forests**

Ecologically Sustainable Development (ESD)
development that improves the total quality of life, both now and in the future, in a way that maintains the ecological processes on which life depends *(for the ESD core objectives and guiding principles, see Council of Australian Governments (1992) 'National Strategy for Ecologically Sustainable Development', AGPS, Canberra)*

ecology
the scientific study of living organisms and their relationships to one another and their environment

economic efficiency
the extent to which managers are able to make optimum use of resources in production by accounting for the relative prices of resources and products; prices are used to select from a number of technically efficient combinations of resources *see* **technical efficiency**; where market prices do not coincide with community values the result is the selection of products and practices that are socially inefficient

ecosystem
a dynamic complex of plant, animal and micro-organism communities and their non-living environment interacting as a functional unit

ecosystem services
the role played by organisms in creating a healthy environment for human beings, from production of oxygen to soil formation and maintenance of water quality

effluent
(i) a discharge or emission of liquid or gas or other waste product
(ii) description of a stream network which draws water out of or away from a river or water body

El Niño
a warm water current which periodically flows southwards along the coast of Ecuador and Peru in South America, replacing the usually cold northwards flowing current; occurs once every five to seven years usually during the Christmas season (the name refers to the Christ child). Occasionally (eg 1925, 1972–73, 1982–83 and 1990–94) the

occurrence is major and prolonged; the opposite phase of an El Niño event is called a La Niña *see* **ENSO**

emissions
(as used in Chapter 5)
substances such as gases, or particles discharged into the atmosphere as a result of natural processes or human activities, including those from chimneys, elevated point sources and tailpipes of motor vehicles

endemic
native to a particular area and found nowhere else

enhanced greenhouse effect
the addition to the natural greenhouse effect resulting from human activities such as the burning of fossil fuels and land clearing, which increase the atmospheric levels of greenhouse gases such as carbon dioxide, methane, nitrous oxide, ozone and CFCs *see* **greenhouse effect.**

ENSO (El Niño–Southern Oscillation)
a suite of events that occur at the time of an El Niño; at one extreme of the cycle, when the central Pacific Ocean is warm and the atmospheric pressure over Australia is relatively high, the ENSO causes drought conditions over eastern Australia *see* **El Niño, Southern Oscillation**

environmental indicator
physical, chemical, biological or socio-economic measures that can be used to assess natural resources and environmental quality

environmental stress
the damaging influence of human activities on the environment (for example, through pollution or consumption of natural resources) or that generated by natural events such as storms or droughts

ephemeral
organisms that have a short life span or a watercourse that does not flow all the time

epiphyte
a plant or other organism that lives on the surface of plants without deriving nutrition from them

equivalence scales (Henderson) and equivalent final income
household income data does not provide a very realistic picture of relative living standards because it fails to take the numbers of people in households or their different needs into account; the **Henderson equivalence scales** provide a way of factoring these differences into the assessment of final household income and of providing figures which more accurately reflect household living standards; **equivalent final income** is derived by applying equivalence scales to more conventional measures of disposable household income

erodible
susceptible to erosion

estuary
area of an inlet or river mouth that is influenced by the tides and also by fresh water from the land; area where fresh and salt waters mix

eutrophication
process by which waters become enriched with nutrients, primarily nitrogen and phosphorus, which stimulate the growth of aquatic flora and/or fauna

evapotranspiration
water withdrawn from soil by evaporation and/or plant transpiration

Exclusive Economic Zone (EEZ)
a concept recognised under the United Nations Law of the Sea, whereby coastal States assume jurisdiction over the exploration and exploitation of marine resources extending 200 nautical miles from the shore or baseline

exotic species
an animal or plant that has been introduced to a region (compare with native or indigenous species)

externalities (external costs)
costs (or benefits) arising from the decisions of an individual which impact on people other than that individual; for example, the costs of salinity that may arise downstream as a result of the agricultural practices used by a farmer upstream *see* **off-site impacts**

exurbanisation
the development of rural areas, which although normally considered to be outside the reach of the metropolitan area, contain more than 20 per cent of households with people commuting to the city for work

fallow
a phase when land is not being actively cropped

family
in the hierarchical classification of organisms, a group of species of common descent higher than the genus and lower than the order; hence a group of genera

fauna
the entire animal life of a region *see* **flora**

feral animals
animals that have reverted to a wild state from domestication (for example, feral cats, pigs, donkeys etc)

fire regime
the pattern of fires at a location; includes the frequency, intensity and seasonality of the fires

flood levee
a constructed embankment designed to prevent flooding of selected areas

flora
the entire plant life of a region *see* **fauna**

flow modification
to change natural sequence of flows in or into a waterbody

flux
the rate at which heat (energy, radiation, carbon dioxide, water vapour etc) flows across unit area eg heat flux is the flow of heat in a heat exchange process

forest
a tree covered area in which at least 60 per cent of the land surface is covered by tree crowns

fossil fuel
any hydrocarbon deposit that can be burned for heat or power, such as coal, oil and natural gas; produces carbon dioxide when burnt

freehold tenure
land owned privately *see* **leasehold land**

fresh water
water containing no significant amounts of salts, potable water suitable for all normal uses
see **potable water**

fugitive emissions
in the context of the National Inventory of Greenhouse Gas Emissions, these are greenhouse gases emitted from fuel production, processing, transmission, storage and distribution processes and include emissions from oil and natural gas exploration, venting, flaring as well as the mining of black coal

gene
the basic unit of heredity

genome
all the genes of a particular organism or species

gentrification
colonisation of an urban area by a higher socio-economic group

Geographic information system (GIS)
a package of computer programs specifically designed to deal with data that are spatially related; a set of tools for collecting, storing, retrieving, manipulating, analysing and displaying mapped data from the real world

GJ (gigajoule)
one thousand million joules

GL (gigalitre)
one thousand million litres

globalisation
the economic and social process whereby local markets and cultures are increasingly dominated by global markets and culture

grassland
areas dominated by grasses and with few or no trees

Great Artesian Basin
an enormous store of underground water underlying much of the drier regions of eastern Australia

greenhouse effect
a popular term used to describe the role of atmospheric trace gases — water vapour, carbon dioxide, methane, nitrous oxide, ozone, in keeping the Earth's surface warmer than it would be otherwise *see* **enhanced greenhouse effect**

grey water
waste water useable for a limited range of purposes, such as playing field irrigation or industrial cooling

groundwater
water occurring below the ground surface

gully erosion
a form of erosion involving the formation of deep, steep-sided channels or gullies which cannot be removed by cultivation *see* **rill erosion, sheet erosion**

habitat
the place where an animal or a plant normally lives and reproduces

hardwood
timber from sources other than pines and cypresses; includes timbers from eucalypts, wattles and most rainforest species *see* **softwood**

HCFCs (Hydrochlorofluorocarbons)
transitional replacements for CFCs; they are also greenhouse gases

headwaters
the upper parts of a river drainage system

heath
a vegetation dominated by small shrubs with small hard leaves

heavy metal
metallic element with relatively high atomic mass (over 5.0 specific gravity), such as lead, cadmium, arsenic and mercury; generally toxic in relatively low concentrations to plant and animal life

hectare (ha)
10 000 square metres

herbarium
a systematically arranged collection of dried plants

herbivore
an animal that consumes plants

heritage (as used in Chapter 9)
those places, objects and indigenous languages that have aesthetic, historic, scientific or social significance or other special value for future generations as well as for the community today

heritage registers
registers of places maintained by State and Territory heritage agencies administering laws designed to protect Australia's natural and cultural heritage

high pressure system
an atmospheric pressure distribution in which there is a high central pressure relative to the surroundings; the circulation about the centre of the high pressure is anticlockwise in the southern hemisphere (clockwise in the northern hemisphere) and the weather is usually fine and calm

historic places
(as used in chapter 9) those sites, areas or regions of heritage significance demonstrating physical characteristics or other associations with important events, developments or cultural phases in Australia's history since the arrival of non-indigenous people; individual structures such as buildings, archaeological sites, and cultural landscapes *see* **Aboriginal and Torres Strait Islander places, natural places**

hummock grass
spinifex grasses usually growing together as large rounded 'hummocks' which can be several metres across, often forming rings with a central dead or decaying patch; hummock grasslands are largely confined to the arid interior and to infertile soils

hybrid
the offspring of two animals or plants of different varieties, species or genera

hydrocarbon
an organic molecule containing hydrogen and carbon; the major components of petroleum

hypersaline
more salty than seawater

impoundment
a pond, lake, tank, basin, or other space, either natural or created in whole or part by the building of engineering structures, which is used for storage, regulation and control of water

in situ
(as used in Chapter 9)
the location of biological, physical or material culture objects in their original physical and cultural context

index of economic resources, and index for education and occupation
belonging to a family of four indices — the Socio-Economic Indexes (*sic*) for Areas (SEIFA) — designed to provide a range of new summary measures of social and economic well being across regions of Australia

indicator species
a species used to assess the health of an ecosystem.

indigenous species
species that are native to (that is, occur naturally) in a region *see* **exotic species**

indoor air quality
the totality of attributes of indoor air that affect a person's health and well-being

infant mortality rate
the number of deaths of infants in the first 12 months of life per 1000 live births

infiltration
the passage of water through the soil surface and into the soil matrix

inshore waters
waters of the shallower part of the continental shelf

intertidal
between the levels of low and high tide; (the intertidal zone is often called the littoral zone in Australia)

inversion
usually refers to a thermal inversion, in which air temperature increases rather than decreases with height *see* **atmospheric stability**

invertebrate
an animal without a backbone composed of vertebrae; examples include insects, worms, snails, mussels, prawns and cuttlefish *see* **vertebrate**

joule
a unit of energy *see* **GJ** (gigajoule), **PJ** (petajoule), **MJ** (megajoule), **KJ** (kilojoule)

keeping place
a special place or structure used by Aboriginal and Torres Strait Islander communities to house important cultural items, for example ceremonial objects; it may also house significant materials returned to traditional owners from museum and other collections, including in some cases human remains

KJ (kilojoule)
one thousand joules

KL (kilolitre)
one thousand litres, or one cubic metre.

L (litre)
a unit of volume equal to 1/1000 of one cubic metre

land cover
the physical state of the land surface, including vegetation, soil, rock and human made structures

Landcare
a voluntary and cooperative movement that brings together rural people, government agencies and others with an interest in the long term health of the land; the term was first used in Victoria in 1986 but spread nationally after 1988 when the Australian Conservation Foundation and the National Farmers' Federation persuaded the Commonwealth Government to provide significant financial support

laser levelling
the use of a rotating laser beam to control land levelling to achieve a land surface of uniform slope

leasehold land
land owned by governments on behalf of the people they represent but leased to specified people or organisations for a specific purpose; about 50% of Australia, mostly in the drier regions, comes under some form of leasehold; governments retain a variety of controls over how leasehold land is used

littoral
of or pertaining to a shore, especially a seashore; littoral zone — the specific zone of the sea floor lying between high and low tide levels (intertidal)

livability
those qualities of urban life and social amenity that are represented by income (including the social wage), employment, education, housing, accessibility, community and health

mallee
small multi-stemmed eucalypts that often dominate semi-arid and arid areas

mangrove
(i) a plant (belonging to any of a wide range of species, mainly trees and shrubs) that grows in sediment regularly inundated by seawater
(ii) a community (forest, woodland, shrubland) of such plants

marginal water
water that is suitable for watering of livestock, irrigation, and other general uses

material culture collections
collections of objects of cultural significance housed in museums and other collecting institutions

materials conservation
the processes involved in the conservation and preservation of the physical material of objects and the physical fabric of structures or places

medic
species of the genus *Medicago*, which includes lucerne; many help add (fix) nitrogen to the soil

microgram (μg)
1×10^{-6} grams

minimum tillage
see **reduced tillage, zero tillage**

mitigation measures
(with respect to climate change)
management measures or options for responding to climate change (as a result of increasing emissions of greenhouse gases) in which the growth of greenhouse gas concentrations in the atmosphere is slowed or reversed by limiting emissions of greenhouse gases or enhancement of greenhouse sinks; examples include reafforestation, fuel switching from high to low carbon dioxide-emitting-fuels and increased energy efficiency *see* **adaptation measures**

mixing
(as used in Chapter 5)
a process associated with turbulent flow where air parcels are subject to random fluctuations in the wind velocity

MJ (megajoule)
one million joules

ML (megalitre)
one million litres, or 1000 cubic metres, (1 acre foot = 1234 cubic metres)

monitoring
routine counting, testing or measuring of environmental factors or biota to determine their status or condition

monoculture
the cultivation of a single species, usually a single crop on land

multiple use
managing an area to achieve multiple goals or multiple outputs; for example timber production, water and recreational opportunities

National Estate
'those places, being components of the natural environment of Australia, or the cultural environment of Australia, that have aesthetic, historic, scientific or social significance or other special value for future generations as well as for the present community' (*Australian Heritage Commission Act 1975*, Section 4)

National Estate Grants Program
the Commonwealth Government's major program to assist the identification, conservation and presentation of national estate (heritage) places across Australia

natural places
(as used in chapter 9) those sites, areas or regions for which the heritage significance is based on their natural biological and physical features; may also have cultural heritage values *see* **Aboriginal and Torres Strait Islander places, historic places**

neritic
located in the surface or middle depths of the water, over the continental shelf

neutron moisture meter
an instrument used for measuring the volumetric water content of materials by counting the rate of absorption of thermal neutrons; widely used to measure field soils for irrigation scheduling purposes

nitrogen fixation
the conversion of gaseous nitrogen into more complex molecules that can be used by plants and other organisms; often carried out by micro-organisms in the soil or closely associated with some plant species (for example the legumes or pea relatives)

nominal prices
nominal prices, incomes, costs or interest rates are not adjusted for inflation (purchasing power); everyday market prices *see* **real prices**

nutrient enrichment
the increase of nourishing substances

off-site impacts
consequences of an action or decision that occur beyond the area (for example the farm or catchment) under consideration *see* **externalities**

old-growth forests
forests dominated by mature trees and with little or no evidence of any disturbance such as logging, road building or clearing *see* box on page 6.20 for more detail

open woodlands
an area with scattered trees in which the tree crowns cover less than about 30% of land surface *see* **woodland**

oral history
a record of information gathered in oral form, usually on tape, as the result of a planned interview

organochlorine
a hydrocarbon compound containing chlorine. Includes many pesticides and industrial chemicals

ozone
a gas with molecules comprising three atoms of oxygen; in the **stratosphere** it occurs naturally and provides a protective layer shielding the Earth from ultraviolet radiation; in the **troposphere**, it is usually formed from **anthropogenic** emissions and is a major component of **photochemical smog**; ozone is also a greenhouse gas

particles
very small pieces of solid or liquid matter, such as soot, dust, smoke, or mist etc

particulate matter
in this report, the term particulate matter has been used interchangeably with particles

pastoral areas
those areas used predominantly for grazing livestock with little or no cultivation or improved pastures

pathogen
agent causing disease

PCBs (polychlorinated biphenyls)
a group of chlorinated organic compounds that are non-corroding and resistant to heat and biological degradation; used as insulation in electrical equipment; can accumulate in some species and disrupt reproduction

pelagic
associated with the surface or middle depths of a body of water

per capita consumption
the average amount of a commodity used per person

perennial plants
plants that live for more than one year

pest
an animal, or sometimes a plant, occurring where it is not wanted by humans *see* **weed**

photochemical smog
air pollution caused by chemical reactions among various substances and pollutants in the atmosphere in the presence of sunlight; **ozone** is a major constituent

photosynthesis
the biochemical process in plants and certain other organisms by which energy from the sun, captured by chlorophyll, powers the production of organic matter from carbon dioxide and water, releasing oxygen

pidgin
a restricted form of language which has relatively limited vocabulary and grammatical devices and which is not anyone's first language; generally developed as a means of communication between peoples of different language backgrounds *see* **creole**

PJ (petajoule)
one thousand million million (10^{15} joules)

plankton
free-drifting, suspended organisms; usually small plants (phytoplankton), or animals (zooplankton)

plume
(as used in Chapter 5)
a sometimes visible or measurable discharge of a pollutant from a given source; sources can range from car exhausts to whole urban areas

PM10/PM2.5
see **TSP**

point-source pollution
pollution from an easily discernible, single source such as a factory *see* **diffuse-source pollution**

pollution
the direct or indirect alteration of the physical, thermal, biological or radioactive properties of any part of the environment in such a way as to create a hazard or potential hazard to the health, safety or welfare of any living species

population
a group of individuals of the same species, forming a breeding unit and sharing a habitat

potable water
water suitable for drinking

precautionary principle
where there are threats of serious or irreversible environmental damage, lack of full scientific certainty should not be used as a reason for postponing measures to prevent environmental degradation

precipitation
any form or all forms of liquid or solid water particles that fall from the atmosphere and reach the Earth's surface; includes drizzle, rain, snow, snow pellets, ice crystals, ice pellets and hail

prescribed fire
a fire deliberately lit and controlled by humans, usually as part of a land management program; for example to reduce the chance of uncontrollable bushfires, or to control weeds

preservation
(as used in chapter 9) maintaining the physical material of places or objects in their existing state and retarding deterioration

pressure indicators
measures that can be used to describe both positive and negative pressures on the environment, including the quality and quantity of natural resources; such pressures can be caused by human inaction as well as action

primary productivity
the rate at which plants produce organic matter through photosynthesis

primary salinity
salinity mobilised in the landscape from a change in the hydrologic cycle either from climate change or clearing

primary treatment of waste water
the first step in sewage treatment to remove large solid objects by screens and sediment and organic matter by settlement *see* **secondary treatment**, **tertiary treatment**

productivity (biological)
the rate of accumulation of organic material in an ecosystem

productivity growth (economic)
the sum of technological change and change in

economic efficiency; 'total factor productivity' is a measure of productivity growth for an industry which takes account of all the resources used (or factors of production) and all products produced

protocol
a formal arrangement defining procedures

protozoans
a group of single-celled animals

public realm
those parts of rural and urban areas that are not owned privately, such as streets, parks, public transport, squares and public buildings; privately owned landscapes and streetscapes which have a public dimension can also be included; it can also incorporate resources such as air, water and biodiversity.

pulp logs
logs that are used to produce woodchips or wood-based products such as chipboard *see* sawlogs

rangelands
areas of native grasslands, shrublands and woodlands that cover a large proportion of the arid and semi-arid regions, and also include tropical savanna woodlands; regular cropping is not practised and the predominant agricultural use, if any, is grazing of sheep and cattle on native vegetation

re-urbanisation
the redevelopment taking place in the existing city and suburbs rather than on the fringe of the city

real prices
real prices, incomes, costs or interest rates are adjusted for inflation; for example, if the nominal interest rate is 10% and inflation is 6% as judged by the CPI, the real interest rate is 4% *see* **nominal prices**

recharge
the action by which water is added to a rock layer either naturally or artificially

recirculation
the recycling of pollutants over an area or within an airshed in response to reversal in the wind regime usually due to diurnal thermal changes such as the land-breeze/sea-breeze regime; in Australian urban airsheds it is a major cause of air pollution episodes

recruitment
(to a fishery) the addition of new individuals to the fished component of a stock

reduced (or minimum) tillage
a soil management system in which tillage is avoided as much as possible

refugia
areas that have been protected from environmental changes such as increasing aridity or fire

Register of the National Estate
a national heritage register that covers significant natural, historic and Aboriginal and Torres Strait Islander places across Australia *see* **National Estate**

representativeness
(as used in chapter 9) the extent to which every significant type of place or object is represented in heritage registers or collections

residence time
the period for which a substance remains in its active form in the air

response indicator
indicator that shows the extent to which society is responding to environmental changes and concerns; includes changes in attitude and individual and collective actions aimed at mitigating, adapting to or reversing negative impacts on the environment and reversing environmental damage already caused; also includes actions to improve the preservation and conservation of the environment

rill erosion
a form of erosion involving formation of shallow gutters which may be removed by cultivation *see* **gully erosion, sheet erosion**

ring barking
killing of trees by cutting a narrow strip deep beneath the bark; sometimes called girdling

river regulation
the formulation and execution of a specific operating plan for flow modification of water in a river system; may involve the creation of impoundments and diversions, and the control and flow to and from such storages

run-off
that portion of precipitation not immediately absorbed into or detained upon the soil and which thus becomes a surface flow

rural (tree) dieback
the gradual decline in vigour and eventual death of many trees in rural settings; there are several different causes; some of them related to current agricultural practices such as fertilisation

salinisation
the process by which soluble salt levels in the soil increase to the point where plant growth is affected *see* **secondary salinity**

salinity
the concentration of salts in water

saltbush
see **chenopod shrubland**

saltmarsh
saltwater wetland occupied mainly by herbs and dwarf shrubs, characteristically able to tolerate extremes of environmental conditions, notably waterlogging and salinity

savanna
a vegetation type with scattered trees over a grassland, usually found in subtropical areas

sawlogs
logs that can been sawn to produce sawn timber, sleepers, poles etc *see* also **pulp logs**

seagrass
flowering plant adapted to living wholly submerged in seawater; not true grasses, but many have a grass-like form

seaweed
macro-algae (not flowering plants) occurring in the sea; typical examples are kelps, Neptune's necklace and sea-lettuce

secondary salinity
human induced, largely believed to be related to irrigation, results from rise in naturally saline watertable to less than one metre from the root zone, causing salinisation *see* **salinisation**

secondary treatment of waste-water
after primary treatment, removal of biodegradable organic matter from sewage by bacteria and other micro-organisms, activated sludge or trickle filters; also removes about 30 per cent of phosphorus and 50 per cent of nitrate *see* **primary treatment, tertiary treatment**

sediment
(i) solid material settled from suspension in the water, (ii) solid material, both mineral and organic, that is in suspension, is being transported, or has been moved from its site of origin by water, air or ice and has come to rest on the land or sea floor

seed banks
the seed naturally available at a site; most of it is stored in the soil, but some may be in protective fruits such as banksia 'cones'

semi-arid zone
lands where rainfall is so low and unreliable that crops cannot be grown with any reliability *see* **arid zone**

sheet erosion
the removal of a fairly uniform layer of soil from the land surface by raindrop splash and/or run-off *see* **rill erosion, gully erosion**

shrubland
an area dominated by short, multi-stemmed plants; a typical example is the chenopod shrublands but sometimes the 'mallee' is classified as a shrubland

siltation
deposition of sediments from water in channels, harbours, etc.

sinks
processes or places that remove pollutants or greenhouse gases from the atmosphere

sodic soils
soils with a high proportion of sodium that cause poor physical conditions; about 30% of Australian soils

softwood
timbers from trees such as pines and cypresses; in Australia most softwood comes from pine plantations *see* **hardwood**

SOI (Southern Oscillation Index)
an indicator based on the pressure gradient between the quasi-stationary low pressure region over Indonesia and the centre of the subtropical high pressure cell over the eastern Pacific Ocean; traditionally, Darwin and Tahiti are used as the sites for determining the magnitude of the Southern Oscillation; a negative SOI is associated with higher than normal pressures over Darwin and drought conditions over much of eastern Australia *see* **Southern Oscillation**

soil acidification
a gradual increase in the acidity of a soil as a consequence of a variety of natural processes and management actions

Southern Oscillation
a fluctuation in the atmospheric circulation, in particular over the tropical areas of the Pacific and Indian Oceans; in general, when atmospheric pressures are high over the eastern Pacific Ocean they tend to be low in the eastern Indian Ocean and vice versa; the fluctuation between the two produces a marked variation in parameters such as the sea-surface temperature and rainfall over a wide area of the Pacific and has a cycle of 2–7 years; the phenomenon is strongly linked to the **El Niño** *see* **SOI**

species
a group of plants, animals or micro-organisms that have a high degree of similarity and generally can interbreed only amongst themselves to produce fertile offspring, so that they maintain their 'separateness' from other such groups

standardised mortality ratios
an estimate of the number of deaths expected for each human settlement type if that population were to experience the same age-specific death rates as the Australian-born population; the ratio of the number of deaths observed to the number expected is known as the standardised mortality ratio (SMR); an SMR greater than 1.0 indicates a level of mortality higher than that in the Australian-born population, while an SMR less than 1.0 indicates a lower level of mortality

State of the Environment reporting
a process that provides a scientific assessment of environmental conditions, focusing on the impacts of human activities, their significance for the environment and societal responses to the identified trends

stock
(in fisheries) a group of individuals of a **species** that can be regarded as an entity for management or assessment purposes; commonly a distinct local **population**; some species form a single stock, others several distinct stocks

stocking rate
the number of animals carried per unit area of land on a year-long basis

stratosphere
the region of the atmosphere roughly 15 to 50 km above the Earth's surface where typically the temperature changes little or increases with height; its thermal structure is determined by radiation balance and is generally very stable

Glossary

. . . .

streamflow
a smaller watercourse than a river, usually forms the link between a drainage line and a river in a natural catchment flow path

strip cropping
strips of various crops and fallow land aligned at right angles to the direction of water flow; this slows the flow of surface water and reduces soil loss

subsurface
pertaining to, formed, or occurring, underneath the surface of the earth

subtidal
below low-water level

subtropical ridge
the belt of high pressure systems (highs or anticyclones) which pass from west to east between latitudes 25 and 45 degrees in both hemispheres

suburbanisation
the development of new suburbs in undeveloped sites usually on the fringe of the city

surface water
water that remains at or close to the land surface

surfactant
a material that facilitates and accentuates the emulsifying, wetting and other surface-modifying properties of substances

suspended solids
any solid substance present in water in an undissolved state, usually contributing directly to turbidity *see* **sediment**

sustainable infrastructure plan
a way of ensuring that settlement infrastructure is helping to shift settlements towards sustainability

sustainable yield
a management goal in which the rate of harvesting does not exceed the rate of renewal of the resource over a prescribed time

symbiotic
a close association between the individuals of pairs of species often leading to mutual gains

taxonomy
the categorisation and naming of animals and plants, animal and plant groups and the relationships between them; a group of organisms so named (for example, a **species**, a **family**, etc.) is called a taxon (plural taxa)

technical efficiency
the physical relationship between resources and products; achieved when a maximum level of output is obtained from a given level of inputs; the level of maximum output achievable may be increased by technological progress *see* **eonomic efficiency**

tertiary treatment of waste water
the removal of nitrates, phosphates, chlorinated compounds, salts, acids, metals and toxic organics after secondary treatment of sewage *see* **primary treatment, secondary treatment**

tillage
mechanical disturbance of the soil by using various implements to alter the soil structure; usually done to create a seedbed, kill weeds or increase water entry

Top End
the northern section of the Northern Territory.

total factor productivity
see **productivity growth**

trajectory
a line drawn to represent the path of an air parcel over an interval of time

translocation
the movement of organisms from their native range to another bioregion

trend
a general direction or tendency; an indication of change (or its absence) in a property or condition

troposphere
the lower layer of the atmosphere extending to roughly 15 km above the Earth's surface where typically the temperature decreases with height; nearly all clouds form and weather processes are found in this region

TSP (total suspended particles)
includes all particles from the smallest up to 50 μm in diameter; subcategories within the TSP range include those particles less than 10 μm in diameter known as PM10, and those smaller than 2.5 μm and known as PM2.5

turbidity
(i) a measure of the extent to which passage of light through water is reduced by suspended matter (ii) the cloudy conditions caused by suspended solids in liquids

tussock grass
grasses in which individuals occur in discrete, compact clumps or bunches

Type Specimen
a specimen of a plant or animal species which is the designated representative of a taxon *see* **taxonomy**

ultraviolet (UV) radiation
electromagnetic radiation of higher frequencies and shorter wavelengths than visible light; ultraviolet radiation is divided into three ranges: UV-A (320–400 nm), UV-B (280–320 nm) and UV-C (40–290 nm)

upwelling
divergence of water currents or the movement of surface water away from land can lead to a 'welling-up' of deeper water which commonly is richer in nutrients than is surface water

urban agglomeration
the metropolitan area and its surrounding ex-urban areas of development where more than 20 per cent of households commute to the city for work

urbanisation
the shift of population from rural to urban areas

vascular plants
a grouping of plants that includes ferns, the gymnosperms (for example, pines) and flowering plants

vector
a disease carrier

vegetable fault
plant material (for example, seeds, thorns) that becomes trapped in wool and reduces the value of a fleece

vertebrate
an animal with a backbone composed of vertebrae, examples include mammals, fishes, frogs, amphibians, reptiles and birds *see* **invertebrate**

visibility
the greatest distance at which an object can be seen and identified with the naked eye in any particular circumstances, or, in the case of night observations, can be seen and identified if the general illumination is raised to the normal daylight level

VOCs (volatile organic compounds)
organic compounds with boiling points between 50°C and 260°C; in this report VOCs also include formaldehyde and pesticides

Walker Cell
an east-west air circulation confined to equatorial regions of the Pacific Ocean and driven principally by the oceanic temperature gradient

water abstraction
the removal of water from a natural water body for human use

water repellence
a condition of porous materials where the surface repels water lying on it and so inhibits infiltration; drought may induce water repellency in soils

water resources
water in various forms, such as groundwater, surface water, snow, and ice, at present in the land phase of the hydrological cycle — some parts may be renewable seasonally but others may be effectively being mined

waterlogging
the saturation of soils with water; often associated with insufficient oxygen for good plant growth

watertable
a surface defined by the level to which water rises in an open well or piezometer

weather
the day-to-day changing atmospheric conditions, which in synthesis constitute the climate of a region

weed
a plant species growing where it is not wanted by humans

wet sclerophyll
a type of eucalypt forest found in high rainfall (more than 1000 mm per year) areas; sometimes called 'tall-open forests' *see* **dry sclerophyll**

wetland
the land area alongside fresh and salt waters, that is flooded all or part of the time; marine and estuarine wetlands include tidal basins, saltmarshes and mangroves

wilderness
remote areas that are substantially undisturbed by colonial and modern technological society and that are large enough to enable the long-term protection and integrity of their natural systems and biological diversity

woodland
an area with scattered trees where the portion of the land surface covered by the crowns is more than 30 per cent (open woodland) but less than 60 per cent (forest)

woody weeds
shrubby plants (both native and exotic) that have increased in numbers to be a problem for pastoralists in parts of the arid and semi-arid zones

World Heritage
a term applied to sites of outstanding universal natural or cultural significance which are included on the World Heritage List

zero tillage
a production system in which there is no tillage at all; many Australians use the term incorrectly to describe what should be referred to as reduced tillage *see* **reduced tillage**

zooxanthellae
unicellular dinoflagellates that live within the tissues of certain corals, giant clams and other invertebrates

Subject index

. . . .